Nonparametric Finance

WILEY SERIES IN PROBABILITY AND STATISTICS

Established by *Walter A. Shewhart and Samuel S. Wilks*

Editors: *David J. Balding, Noel A. C. Cressie, Garrett M. Fitzmaurice, Geof H. Givens, Harvey Goldstein, Geert Molenberghs, David W. Scott, Adrian F. M. Smith, Ruey S. Tsay*

Editors Emeriti: *J. Stuart Hunter, Iain M. Johnstone, Joseph B. Kadane, Jozef L. Teugels*

The *Wiley Series in Probability and Statistics* is well established and authoritative. It covers many topics of current research interest in both pure and applied statistics and probability theory. Written by leading statisticians and institutions, the titles span both state-of-the-art developments in the field and classical methods.

Reflecting the wide range of current research in statistics, the series encompasses applied, methodological and theoretical statistics, ranging from applications and new techniques made possible by advances in computerized practice to rigorous treatment of theoretical approaches.

This series provides essential and invaluable reading for all statisticians, whether in academia, industry, government, or research.

A complete list of titles in this series can be found at
http://www.wiley.com/go/wsps

Nonparametric Finance

Jussi Klemelä

This edition first published 2018
© 2018 John Wiley & Sons, Inc.

The right of Jussi Klemelä to be identified as the author of this work has been asserted in accordance with law.

Registered Offices
John Wiley & Sons, Inc., 111 River Street, Hoboken, NJ 07030, USA

Editorial Office
111 River Street, Hoboken, NJ 07030, USA

For details of our global editorial offices, customer services, and more information about Wiley products visit us at www.wiley.com.

Wiley also publishes its books in a variety of electronic formats and by print-on-demand. Some content that appears in standard print versions of this book may not be available in other formats.

Library of Congress Cataloging-in-Publication Data

Names: Klemelä, Jussi, 1965- author.
Title: Nonparametric finance / by Jussi Sakari Klemelä.
Description: Hoboken, NJ : John Wiley & Sons, Inc., [2018] | Series: Wiley
   series in probability and statistics | Includes bibliographical references
   and index. |
Identifiers: LCCN 2017038242 (print) | LCCN 2017046064 (ebook) | ISBN
   9781119409113 (pdf) | ISBN 9781119409120 (epub) | ISBN 9781119409106
   (cloth)
Subjects: LCSH: Finance—Statistical methods. | Finance—Mathematical models.
Classification: LCC HG176.5 (ebook) | LCC HG176.5 .K553 2018 (print) | DDC
   332.01/51954–dc23
LC record available at https://lccn.loc.gov/2017038242

Cover design by Wiley
Cover image: (Background) © Michel Leynaud/Gettyimages;
(Image grid) Courtesy of Jussi Klemelä

Set in 10/12pt WarnockPro by SPi Global, Chennai, India

Printed in the United States of America

10 9 8 7 6 5 4 3 2 1

Contents

Preface

We study applications of nonparametric function estimation into risk management, portfolio management, and option pricing.

The methods of nonparametric function estimation have not been commonly used in risk management. The scarcity of data in the tails of a distribution makes it difficult to utilize the methods of nonparametric function estimation. However, it has turned out that some semiparametric methods are able to improve purely parametric methods.

Academic research has paid less attention to portfolio selection, as compared to the attention that has been paid to risk management and option pricing. We study applications of nonparametric prediction methods to portfolio selection. The use of nonparametric function estimation to reach practical financial decisions is an important part of machine learning.

Option pricing might be the most widely studied part of quantitative finance in academic research. In fact, the birth of modern quantitative finance is often dated to the 1973 publication of the Black–Scholes option pricing formula. Option pricing has been dominated by parametric methods, and it is especially interesting to provide some insights of nonparametric function estimation into option pricing.

The book is suitable for mathematicians and statisticians who would like to know about applications of mathematics and statistics into finance. In addition, the book is suitable for graduate students, researchers, and practitioners of quantitative finance who would like to study some underlying mathematics of finance, and would like to learn new methods. Some parts of the book require fluency in mathematics.

Klemelä (2014) is a book that contains risk management (volatility prediction and quantile estimation) and it describes methods of nonparametric regression, which can be applied in portfolio selection. In this book, we cover those topics and also include a part about option pricing.

The chapters are rather independent studies of well-defined topics. It is possible to read the individual chapters without a detailed study of the previous material.

The research in the book is reproducible, because we provide R-code of the computations. It is my hope, that this makes it easier for students to utilize the book, and makes it easier for instructors to adapt the material into their teaching.

The web page of the book is available in http://jussiklemela.com/statfina/.

Helsinki, Finland   *Jussi Klemelä*
June 2017

1

Introduction

Nonparametric function estimation has many useful applications in quantitative finance. We study four areas of quantitative finance: statistical finance, risk management, portfolio management, and pricing of securities.[1]

A main theme of the book is to study quantitative finance starting only with few modeling assumptions. For example, we study the performance of nonparametric prediction in portfolio selection, and we study the performance of nonparametric quadratic hedging in option pricing, without constructing detailed models for the markets. We use some classical parametric methods, such as Black–Scholes pricing, as benchmarks to provide comparisons with nonparametric methods.

A second theme of the book is to put emphasis on the study of economic significance instead of statistical significance. For example, studying economic significance in portfolio selection could mean that we study whether prediction methods are able to produce portfolios with large Sharpe ratios. In contrast, studying statistical significance in portfolio selection could mean that we study whether asset returns are predictable in the sense of the mean squared prediction error. Studying economic significance in option pricing could mean that we study whether hedging methods are able to well approximate the payoff of the option. In contrast, studying statistical significance in option pricing could mean that we study the goodness-of-fit of our underlying model for asset prices. Studying statistical significance can be important for understanding the underlying reasons for economic significance. However, the study of economic significance is of primary importance, and the study of statistical significance is of secondary importance.

---

1 The quantitative finance section of preprint archive "arxiv.org" contains four additional sections: computational finance, general finance, mathematical finance, and trading and market microstructure. We cover some topics of computational finance that are useful in derivative pricing, such as lattice methods and Monte Carlo methods. In addition, we cover some topics of mathematical finance, such as the fundamental theorems of asset pricing.

A third theme of the book is the connections between the various parts of quantitative finance.

1) There are connections between risk management and portfolio selection: In portfolio selection, it is important to consider not only the expected returns but also the riskiness of the assets. In fact, the distinction between risk management and portfolio selection is not clear-cut.
2) There are connections between risk management and option pricing: The prices of options are largely influenced by the riskiness of the underlying assets.
3) There are connections between portfolio management and option pricing: Options are important assets to be included in a portfolio. In addition, multiperiod portfolio selection and option hedging can both be casted in the same mathematical framework.

Volatility prediction is useful in risk management, option pricing, and portfolio selection. Thus, volatility prediction is a constant topic throughout the book.

1.1 Statistical Finance

Statistical finance makes statistical analysis of financial and economic data.

Chapter 2 contains a description of the basic financial instruments, and it contains a description of the data sets that are analyzed in the book.

Chapter 3 studies univariate data analysis. We study univariate financial time series, but ignore the time series properties of data. A decomposition of a univariate distribution into the central part and into the tail parts is an important theme of the chapter.

1) We use different estimators for the central part and for the tails. Nonparametric density estimation is efficient at the center of a univariate distribution, but in the tails of the distribution the scarcity of data makes nonparametric estimation difficult. When we combine a nonparametric estimator for the central part and a parametric estimator for the tails then we obtain a semiparametric estimator for the distribution.
2) We use different visualization methods for the central part and for the tails. We apply two basic visualization tools: (1) kernel density estimates and (2) tail plots. Kernel density estimates can be used to visualize and to estimate the central part of the distribution. Tail plots are an empirical distribution based tool, and they can be used to visualize the tails of the distribution.

Chapter 4 studies multivariate data analysis. Multivariate data analysis considers simultaneously several time series, but the time series properties are ignored, and thus the analysis can be called cross-sectional. A basic concept is the copula, which makes it possible to compose a multivariate distribution into the part that describes the dependence and into the parts that describe the marginal distributions. We can estimate the marginal distributions using nonparametric methods, but to estimate dependence for a high-dimensional distribution it can be useful to apply parametric models. Combining nonparametric estimators of marginals and a parametric estimator of the copula leads to a semiparametric estimator of the distribution. Note that there is an analogy between the decomposition of a multivariate distribution into the copula and the marginals, and between the decomposition of a univariate distribution into the tails and the central area.

Chapter 5 studies time series analysis. Time series analysis adds the elements of dependence and time variation into the univariate and multivariate data analysis. Completely nonparametric time series modeling tends to become quite multidimensional, because dependence over k consecutive time points leads to the estimation of a k-dimensional distribution. However, a rather convenient method for time series analysis is obtained by taking as a starting point a univariate or a multivariate parametric model, and estimating the parameter using time localized smoothing. For example, we can apply time localized least squares or time localized maximum likelihood.

Chapter 6 studies prediction. Prediction is a central topic in time series analysis. The previous observations are used to predict the future observations. A distinction is made between moving average type of predictors and state space type of predictors. Both types of predictors can arise from parametric time series modeling: moving average and GARCH $(1, 1)$ models lead to moving average predictors, and autoregressive models lead to state space predictors. It is easy to construct nonparametric moving average predictors, and nonparametric regression analysis leads to nonparametric state space predictors.

1.2 Risk Management

Risk management studies measurement and management of financial risks. We concentrate on the market risk, which means the risk of unfavorable moves of asset prices.[2]

Chapter 7 studies volatility prediction. Prediction of volatility means in our terminology that the square of the return of a financial asset is predicted. The volatility prediction is extremely useful in almost every part of quantitative

---

2 Other relevant types of risk are credit risk, liquidity risk, and operational risk. Credit risk means the risk of the default of a debtor and the risks resulting from downgrading the rating of a debtor. Liquidity risk means the risk from additional cost of liquidating a position when buyers are rare. Operational risk means the risk caused by natural disasters, failures of the physical plant and equipment of a firm, failures in electronic trading, clearing or wire transfers, trading and legal liability losses, internal and external theft and fraud, inappropriate contractual negotiations, criminal mismanagement, lawsuits, bad advice, and safety issues.

finance: we can apply volatility prediction in quantile estimation, and volatility prediction is an essential tool in option pricing and in portfolio selection. In addition, volatility prediction is needed when trading with variance products. We concentrate on the following three methods:

1) GARCH models are a classical and successful method to produce volatility predictions.
2) Exponentially weighted moving averages of squared returns lead to volatility predictions that are as good as GARCH $(1, 1)$ predictions.
3) Nonparametric state space smoothing leads to improvements of GARCH $(1, 1)$ predictions. We apply kernel regression with two explanatory variables: a moving average of squared returns and a moving average of returns. The response variable is a future squared return. A moving average of squared returns is in itself a good volatility predictor, but including a kernel regression on top of moving averages improves the predictions. In particular, we can take the leverage effect into account. The leverage effect means that when past returns have been low, then the future volatility tends to be higher, as compared to the future volatility when the past returns have been high.

Chapter 8 studies estimation of quantiles. The term *value-at-risk* is used to denote upper quantiles of a loss distribution of a financial asset. Value-at-risk at level $0.5 < p < 1$ has a direct interpretation in risk management: it is such value that the probability of losing more has a smaller probability than $1 - p$. We concentrate on the following three main classes of quantile estimators:

1) The empirical quantile estimator is a quantile of the empirical distribution. The empirical quantile estimator has many variants, since it can be used in conditional quantile estimation and it can be modified by kernel smoothing. In addition, empirical quantiles can be combined with volatility based and excess distribution based methods, since empirical quantiles can be used to estimate the quantiles of the residuals.
2) Volatility based quantile estimators apply a location-scale model. A volatility estimator leads directly to a quantile estimator, since estimation of the location is less important. The performance of volatility based quantile estimators depends on the choice of the base distribution, whose location and scale is estimated. However, in a time series setting the use of the empirical quantiles of the residuals provides a method that bypasses the problem of the choice of the base distribution.
3) Excess distribution based quantile estimators model the tail parametrically. These estimators ignore the central part of the distribution and model only the tail part parametrically. The tail part of the distribution is called the excess distribution. Extreme value theory can be used to justify the choice of the generalized Pareto distribution as the model for the excess distribution. Empirical work has confirmed that the generalized Pareto distribution

provides a good fit in many cases. In a time series setting the estimation can be improved if the parameters of the excess distribution are taken to be time changing. In addition, in a time series setting we can make the estimation more robust to the choice of the parametric model by applying the empirical quantiles of the residuals. In this case, the definition of a residual is more involved than in the case of volatility based quantile estimators.

1.3   Portfolio Management

Portfolio management studies optimal security selection and capital allocation. In addition, portfolio management studies performance measurement.

Chapter 9 discusses some basic concepts of portfolio theory.

1) A major issue is to introduce concepts for the comparison of wealth distributions and return distributions. The comparison can be made by the Markowitz mean–variance criterion or by the expected utility. We need to define what it means that a return distribution is better than another return distribution. This is needed both in portfolio selection and in performance measurement.

2) A second major issue is the distinction between the one period portfolio selection and multiperiod portfolio selection. We concentrate on the one period portfolio selection, but it is instructive to discuss the differences between the approaches.

Chapter 10 studies performance measurement.

1) The basic performance measures that we discuss are the Sharpe ratio, certainty equivalent, and the alpha of an asset.

2) Graphical tools are extremely helpful in performance measurement. The performance measures are sensitive to the time period over which the performance is measured. The graphical tools address the issue of the sensitivity of the time period to the performance measures. The graphical tools help to detect periods of good performance and the periods of bad performance, and thus they give clues for searching explanations for good and bad performance.

Chapter 11 studies Markowitz portfolio theory. Markowitz portfolios are such portfolios that minimize the variance of the portfolio return, under a minimal requirement for the expected return of the portfolio. Markowitz portfolios can be utilized in dynamic portfolio selection by predicting the future returns, future squared returns, and future products of returns of two assets, as will be done in Chapter 12.

Chapter 12 studies dynamic portfolio selection. Dynamic portfolio selection means in our terminology such trading where the weights of the portfolio are rebalanced at the beginning of each period using the available information. Dynamic portfolio selection utilizes the fact that the expected returns, the expected squared returns (variances), and the expected products of returns (covariances) change in time. The classical insight of efficient markets has to be modified to take into account the predictability of future returns and squared returns.

1) First, we discuss how prediction can be used in portfolio selection. Time series regression can be applied in portfolio selection both when we use the maximization of the expected utility and when we use mean–variance preferences. In the case of the maximization of the expected utility, we predict the future utility transformed returns with time series regression. In the case of mean–variance preferences we predict, the future returns, squared returns, and products of returns.

2) The Markowitz criterion can be seen as decomposing the expected utility into the first two moments. The decomposition has the advantage that different methods can be used to predict the returns, squared returns, and products of returns. The main issue is to study the different types of predictability of the mean and the variance. In fact, most of the predictability comes from the variance part, whereas the expectation part has a much weaker predictability.

   a) We need to use different prediction horizons for the prediction of the returns and for the prediction of the squared returns. For the prediction of the returns we need to use a prediction horizon of 1 year or more. For the prediction of squared returns we can use a prediction horizon of 1 month or less.

   b) We need to use different prediction methods for the prediction of the returns and for the prediction of the squared returns. For the prediction of the returns, it is useful to apply such explanatory variables as dividend yield and term spread. For the prediction of the squared returns we can apply GARCH predictors or exponentially weighted moving averages.

1.4   Pricing of Securities

Pricing of securities considers valuation and hedging of financial securities and their derivatives.

Chapter 13 studies principles of asset pricing. We start the chapter by a heuristic introduction to pricing of securities, and discuss such concepts as absolute pricing, relative pricing using arbitrage, and relative pricing using "statistical arbitrage."[3]

---

3 The term *statistical arbitrage* refers often to pairs trading and to the application of mean reversion. We use term *statistical arbitrage* more generally, to refer to cases where two payoffs are close to each other with high probability. Thus, also term *probabilistic arbitrage* could be used.

1) The first main topic is to state and prove the first fundamental theorem of asset pricing in discrete time models, and to state the second fundamental theorem of asset pricing. These theorems provide the foundations on which we build the development of statistical methods of asset pricing. We give a constructive proof of the first fundamental theorem of asset pricing, instead of using tools of abstract functional analysis. The constructive proof of the first fundamental theorem of asset pricing turns out to be useful, because the method can be applied in practise to price options in incomplete models. The construction uses the Esscher martingale measure, and it is a special case of using utility functions to price derivatives.

2) The second main topic is to discuss evaluation of pricing and hedging methods. The basic evaluation method will be to measure the hedging error. The hedging error is the difference between the payoff of the derivative and the terminal value of the hedging portfolio. By measuring the hedging error, we simultaneously measure the modeling error and the estimation error. Minimizing the hedging error has economic significance, whereas modeling error and estimation error are underlying statistical concepts. Thus, emphasizing the hedging error is an example of emphasizing economic significance instead of statistical significance.

Chapter 14 studies pricing by arbitrage. The principle of arbitrage-free pricing combines two different topics: pricing of futures and pricing of options in complete models, like binary models and the Black–Scholes model.

1) A main topic is pricing in multiperiod binary models. First, these models introduce the idea of backward induction, which is an important numerical tool to value options in the Black–Scholes model, and which is an important tool in quadratic hedging. Second, these models lead asymptotically to the Black–Scholes prices.

2) A second main topic is to study the properties of Black–Scholes hedging. We illustrate how hedging frequency, strike price, expected return, and volatility influence the hedging error. These illustrations give insight into hedging methods in general, and not only into Black–Scholes hedging.

3) A third main topic is to study how Black–Scholes pricing and hedging performs with various volatility predictors. Black–Scholes pricing and hedging provides a benchmark, against which we can measure the performance of other pricing methods. Black–Scholes pricing and hedging assumes that the stock prices have a log-normal distribution with a constant volatility. However, when we combine Black–Scholes pricing and hedging with a time changing GARCH $(1, 1)$ volatility, then we obtain a method that is hard to beat.

Chapter 15 gives an overview of several pricing methods in incomplete models. Binary models and the Black–Scholes model are complete models, but we are interested in option pricing when the model makes only few restrictions on the underlying distribution of the stock prices. Chapter 16 is devoted to quadratic hedging, and in Chapter 15 we discuss pricing by utility maximization, pricing by absolutely continuous changes of measures, pricing in GARCH models, pricing by a nonparametric method, pricing by estimation of the risk neutral density, and pricing by quantile hedging.

1) A main topic is to introduce two general approaches for pricing derivatives in incomplete models: the method of utility functions and the method of an absolutely continuous change of measure (Girsanov's theorem). For some Gaussian processes and for some utility functions these methods coincide. The method of utility functions can be applied to construct a nonparametric method of pricing options, whereas Girsanov's theorem can be applied in the case of some processes with Gaussian innovations, such as some GARCH processes.

2) A second main topic is to discuss pricing in GARCH models. GARCH $(1, 1)$ model gives a reasonable fit to the distribution of stock prices. Girsanov's theorem can be used to find a natural pricing function when it is assumed that the stock returns follow a GARCH $(1, 1)$ process. Heston–Nandi modification of the standard GARCH $(1, 1)$ model leads to a computationally attractive pricing method. Heston–Nandi model has been rather popular, and it can be considered as a discrete time version of continuous time stochastic volatility models.

Chapter 16 studies quadratic hedging. In quadratic hedging the price and the hedging coefficients are determined so that the mean squared hedging error is minimized. The hedging error means the difference between the terminal value of the hedging portfolio and the value of the option at the expiration.

1) A main aim of the chapter is to derive recursive formulas for quadratically optimal prices and hedging coefficients. It is important to cover both the global and the local quadratic hedging. Local quadratic hedging leads to formulas that are easier to implement than the formulas of global quadratic hedging. Quadratic hedging has some analogies with linear least squares regression, but quadratic hedging is a version of sequential regression, which is done in a time series setting. In addition, quadratic hedging does not assume a linear model, but we are searching the best linear approximation in the sense of the mean squared error.

2) A second main aim of the chapter is to implement quadratic hedging. This will be done only for local quadratic hedging. We implement local quadratic hedging nonparametrically, without assuming any model for the underlying distribution of the stock prices. Although quadratic hedging finds an optimal linear approximation for the payoff of the option, the quadratically optimal price and hedging coefficients have a nonlinear dependence on volatility, and thus nonparametric approach may lead to a better fit for these nonlinear functions than a parametric modeling.

Chapter 17 studies option strategies. Option strategies provide a large number of return distributions to choose from, so that it is possible to create a portfolio that is tailored to the expectations and the risk profile of each investor. We discuss such option strategies as vertical spreads, strangles, straddles, butterflies, condors, and calendar spreads. Options can be combined with stocks to create covered calls and protective put. Options can be combined with bonds to create capital guarantee products. We give insight into these option strategies by estimating the return distributions of the strategies.

Chapter 18 describes interest rate derivatives. The market of interest rate derivatives is even larger than the market of equity derivatives. Interest rate forwards include forward zero-coupon bonds, forward rate agreements, and swaps. Interest rate options include caps and floors.

Part I

Statistical Finance

2

Financial Instruments

The basic assets which are traded in financial markets include stocks and bonds. A large part of financial markets consists of trading with derivative assets, like futures and options, whose prices are derived from the prices of the basic assets. Stock indexes can be considered as derivative assets, since the price of a stock index is a linear combination of the prices of the underlying stocks. A stock index is a more simple derivative asset than an option, whose terminal price is a nonlinear function of the price of the underlying stock.

In addition, we describe in this section the data sets which are used throughout the book to illustrate the methods.

2.1 Stocks

Stocks are securities representing an ownership in a corporation. The owner of a stock has a limited liability. The limited liability implies that the price of a stock is always nonnegative, so that the price S_t of a stock at time t satisfies

$$0 \le S_t < \infty.$$

Stock issuing companies have a variety of legal forms depending on the country of domicile of the company.[1] Common stock typically gives voting rights in company decisions, whereas preferred stock does not typically give voting rights, but the owners of preferred stocks are entitled to receive a certain amount of dividend payments before the owners of common stock can receive any dividends.

---

1 Statistical data of stock prices is usually available only for the stocks that are publicly traded in a stock exchange. In UK the companies whose stocks are publicly traded are called public limited companies (PLC), and in Germany they are called Aktiengesellschaften (AG). The companies whose owners have a limited liability but whose stocks are not publicly traded are called private companies limited by shares (Ltd), and Gesellschaft mit beschränkter Haftung (GmbH).

Nonparametric Finance, First Edition. Jussi Klemelä.
© 2018 John Wiley & Sons, Inc. Published 2018 by John Wiley & Sons, Inc.

2.1.1   Stock Indexes

We define a stock index, give examples of the uses of stock indexes, and give examples of popular stock indexes.

2.1.1.1   Definition of a Stock Index

The price of a stock index is a weighted sum of stock prices. The value I_t of a stock index at time t is calculated by formula

$$I_t = C \sum_{i=1}^{d} n_i S_t^i, \tag{2.1}$$

where C is a constant, d is the number of stocks in the index, n_i is the number of shares of stock i, and S_t^i is a suitably adjusted price of stock i at time t, where $i = 1, \dots, d$. Note that $n_i S_t^i$ is the market capitalization of stock i. The definition of a stock index involves three parameters: constant C, numbers n_i, and values S_t^i:

1) The constant C can be chosen, for example, to make the value of the index equal to 100 at a given past day. When the constitution of the index is changed, then the constant C is changed, to keep the index equal to 100 at the chosen day.
2) The numbers n_i can equal the total number of shares of stock i, but they can also be equal to the number of freely floating stocks. Float market capitalization excludes stocks which are not freely floating (cannot be bought in the open market).
3) The values S_t^i are calculated differently depending on whether the index is a price return index or a total return index. Price return indexes are calculated without regard to cash dividends but total return indexes are calculated by reinvesting cash dividends. The adjusted closing price of a stock is the closing price of a stock which is adjusted to cash dividends, stock dividends, stock splits, and also to more complex corporate actions, such as rights offerings. The calculation of the adjusted closing price is often made by data providers.

2.1.1.2   Uses of Stock Indexes

Stock indexes can be used to summarize information about stock markets. Stock indexes can also be used as a proxy for the market index when testing and applying finance theories. The market index is the stock index which sums the values of all companies worldwide. Stock indexes are traded in futures markets and in exchanges as exchange traded funds (ETF). Furthermore, investment banks provide financial instruments whose values depend on stock indexes.

2.1.1.3 Examples of Stock Indexes

Dow Jones Industrial Average Dow Jones Industrial Average is an index where the prices are not weighted by the number of shares, and thus Dow Jones Industrial Average is an exception of the rule (2.1). Dow Jones Industrial Average is just a sum of the prices of the components, multiplied by a constant.

S&P 500 S&P 500 was created at March 4, 1957. It was calculated back until 1928 and the basis value was taken to be 10 from 1941 until 1943. The S&P 500 index is a price return index, but there exists also total return versions (dividends are invested back) and net total return versions (dividends minus taxes are invested back) of the S&P 500 index. The S&P 500 is a market value weighted index: prices of stocks are weighted according to the market capitalizations of the companies. Since 2005 the index is float weighted, so that the market capitalization is calculated using only stocks that are available for public trading.

Nasdaq-100 Nasdaq-100 is calculated since January 31, 1985. The basis value was at that day 250. Nasdaq-100 is a price index, so that the dividends are not included in the value of the index. Nasdaq-100 is a different index than Nasdaq Composite, which is based on 3000 companies. Nasdaq-100 is calculated using the 100 largest companies in Nasdaq Composite. Nasdaq-100 is a market value weighted index, but the influence of the largest companies is capped (the weight of any single company is not allowed to be larger than 24%).

DAX 30 DAX 30 (Deutscher AktienindeX) was created at July 1, 1988. The basis value is 1000 at December 31, 1987. DAX 30 is a performance index (dividends are reinvested in calculating the value of the index). DAX 30 stock index is a market value weighted index of 30 largest German companies. Market value is calculated using only free floating stocks (stocks that are not owned by an owner which has more than 5% of stocks). The largeness of a company is measured by taking into account both the free floating market value and the transaction volume (total value of the stocks that are exchanged in a given time period). The weight of any single company is not allowed to be larger than 10%.

2.1.2 Stock Prices and Returns

Statistical analysis of stock markets is usually done from time series of returns. Before defining a return time series we describe the initial price data in its raw form, as it is evolving in a stock exchange, and we describe some methods of sampling of prices.

2.1.2.1 Initial Price Data

During the opening hours of an exchange the stocks are changing hands at irregular time points. The stock exchange receives bid prices with volumes (numbers of stocks one is willing to buy with the given bid price) from buyers, and ask prices with volumes from the sellers. The exchange has an algorithm which allocates the stocks from the sellers to the buyers. The allocation happens when there are bid prices and ask prices that meet each other (ask prices that are smaller or equal to bid prices). The algorithms of stock allocation take into account the arrival times of the orders, the volumes of the orders, and the types of the orders.

The most common order types are the market order and the limit order. A market order expresses the intention to buy the stock at the lowest ask price, or the intention to sell the stock at the highest bid price. A limit order expresses the intention to buy the stock at the lowest ask price, under the condition that the ask price is lower than the given limit price, or the intention to sell the stock at the highest bid price, under the condition that the bid price is higher than the given limit price.

2.1.2.2 Sampling of Prices

The price changes at irregular time intervals in a stock exchange, but for the purpose of a statistical analysis we typically sample price at equispaced intervals.

To obtain a time series of daily prices, we can pick the closing price of each trading day. The closing price can be considered as the consensus reached between the sellers and the buyers about the fair price, taking into account all information gathered during the day. An alternative method would choose the opening price.

However, depending on the purpose of the analysis, we can sample data once in a second, once in 10 days, or once in a month, for example. Note that when the sampling interval is longer (monthly, quarterly, or yearly), the number of observations in a return time series will be smaller, and thus the statistical conclusions may be more vague. Note also, that the distribution of the returns may vary depending on the sampling frequency.

It is not obvious how to define equispaced sampling, since we can measure the time as the physical time, trading time, or effective trading time:

1) The physical time is the usual time in calendar days. Assume that we want to sample data once in 20 days. If we use the physical time, then we calculate all calendar days.

2) The trading time or market time takes into account only the time when markets are open. For example, when we want to sample data once in 20 days and we use trading time, then we calculate only the trading days (not all calendar days). However, information is accumulating also during the weekends (and during the night), which would be an argument in favor of physical time.

3) The effective trading time takes into account that the market activity is not uniform during market hours. To define the sampling interval, we could take into account the number of transactions, or the volume of the transactions. The effective trading time is interesting especially when we gather intraday data, but it can be used also in the case of longer sampling intervals, to correct for diminishing market activity during summer or at the end of year.[2]

Sampling daily closing prices can be interpreted as using the trading time, because weekends and holidays are ignored in the daily sampling. Since there is roughly the same number of trading days in every week and every month, we can interpret sampling the weekly and monthly closing prices both as using the physical time and using the trading time. Discussion about scales in finance is provided by Mantegna and Stanley (2000).

2.1.2.3 Stock Returns
Let us consider a time series $S_0, \dots, S_T$ of stock prices, sampled at equispaced time points. We can calculate gross returns, net returns, or logarithmic returns.

1) Gross returns (price relatives) are defined by

$$\frac{S_{t+1}}{S_t},$$

2) net returns (relative price differences) are defined by

$$\frac{S_{t+1} - S_t}{S_t},$$

3) logarithmic returns (continuously compounded returns) are defined by

$$\log\left(\frac{S_{t+1}}{S_t}\right),$$

where $t = 0, \dots, T-1$.

Gross returns are positive numbers like 1.02 (when the stock rose 2%) or 0.98 (when the stock fell 2%). Value zero for a gross return means bankruptcy. The gross returns have a concrete interpretation: starting with wealth W_t and buying a stock with price S_t leads to the wealth $W_{t+1} = W_t \times S_{t+1}/S_t$.

Net returns are obtained from gross returns by subtracting one, and thus net returns are numbers larger than -1. Net returns are numbers like 0.02 (when the stock rose 2%) or -0.02 (when the stock fell 2%). Value -1 for a net return means bankruptcy.

---

2 Let V_u be the number or the volume of the transactions at time u. After sampling time t_i is chosen, we can determine the next sampling time t_{i+1} by

$$t_{i+1} = \min\left\{t : \sum\{V_u : t_i \le u \le t\} \ge C\right\},$$

where $C > 0$ is a constant.

Logarithmic returns are obtained from gross returns by taking the logarithm.[3] A logarithmic return can take any real value, but typically logarithmic returns are close to net returns, because $\log(x) \approx x - 1$ when $x \approx 1$. Value $-\infty$ for a logarithmic return means bankruptcy. The logarithmic function is an example of a utility function, as discussed in Section 9.2.2. We will consider taking the logarithm as an application of a utility function, and apply mainly gross returns. However, there are some reasons for the use of logarithmic returns. First, we can derive approximate distributions for the stock price by applying limit theorems for the sum of the logarithmic returns, which makes the study of logarithmic returns interesting. Indeed, we can write

$$S_T = S_0 \exp\left\{ \sum_{t=0}^{T-1} \log\left(\frac{S_{t+1}}{S_t}\right) \right\}. \tag{2.2}$$

See (3.49) for a more detailed derivation of the log-normal model for stock prices. Second, taking logarithms of returns transforms the original time series of prices to a stationary time series, as explained in the connection of Figure 5.1.

For a statistical modeling we need typically a stationary time series. Stationarity is defined in Section 5.1. For example, autoregressive moving average processes (ARMA) and generalized autoregressive conditional heteroskedasticity (GARCH) models, defined in Section 5.3, are stationary time series models. The original time series of stock prices is not a stationary time series, but it can be argued that a return time series is close to stationarity.[4]

Note that we can write, analogously to (2.2),

$$S_T = S_0 + \sum_{t=0}^{T-1}(S_{t+1} - S_t).$$

Thus, we can derive approximate distributions for the stock price by applying limit theorems for the sum of the price differences. See (3.46) for a more detailed derivation of the normal model for stock prices. The time series of price differences is not a stationary time series, as discussed in the connection of Figure 5.2. However, for short time periods a time series of price differences can be approximately stationary. Thus, modeling price differences instead of returns can be reasonable.

———

3 We take the logarithm to be the natural logarithm, with e (Euler's number or Napier's constant) as the basis. The logarithmic functions with other bases could be used as well.

4 Time series $\{Y_t\}$ is called strictly stationary, if $(Y_1, \dots, Y_t)$ and $(Y_{1+k}, \dots, Y_{t+k})$ are identically distributed for all $t, k \in \{0, \pm1, \pm2, \dots\}$. Stationarity means, roughly speaking, that every subperiod of the time series has similar statistical characteristics. For example, consider a stock whose price is 1\$, which then rises to have a price of 100\$. The change of 1\$ is very large at the beginning of the period but moderate at the end of the period. Thus, the time series of prices is not stationary.

2.2   Fixed Income Instruments

One unit of currency today is better than one unit of currency tomorrow. Fixed income research studies how much one should pay today, in order to receive a cash payment at a future day.

Fixed income instruments are described in more detail in Chapter 18. Here we give an overview of zero-coupon bonds, coupon paying bonds, interest rates, and of calculation of bond returns.

2.2.1   Bonds

Bonds include zero-coupon bonds and coupon bearing bonds.

1) A zero-coupon bond, or a pure discount bond, is a certificate which gives the owner a nominal amount P (principal) at the future maturity time T. Typically we take $P = 1$.
2) Coupon bearing bonds make regular payments (coupons) before the final payment at the maturity. A coupon bond can be defined as a series of payments $P_1, \ldots, P_n$ at times $T_1, \ldots, T_n$. The terminal payment contains the principal and the final coupon payment.[5]

A zero-coupon bond is a more basic instrument than a coupon bond, because a coupon bond can be defined as a portfolio of zero-coupon bonds. Let $C(t_0, T_n)$ be the price of a coupon bond which starts at t_0 and makes payments $P_1, \ldots, P_n$ at times $T_1 < \cdots < T_n$, where $T_1 > t_0$. It holds that

$$C(t_0, T_n) = \sum_{i=1}^{n} P_i Z(t_0, T_i),$$

where $Z(t_0, T_i)$ are the prices of zero-coupon bonds starting at t_0 with maturity T_i, and with principal $P = 1$.

The cash flow generated by a bond is determined when the bond is issued. The bond can be traded before its maturity and its price can fluctuate before the maturity. For example, the price of a zero-coupon bond with the nominal amount P is equal to P at the maturity, but its price fluctuates until the maturity is reached. The price fluctuates as a function of interest rate fluctuation. Thus, bonds bear interest rate risk if they are not kept until maturity. If the bonds are kept until maturity they bear the inflation risk and the risk of the default of the issuer.

Bonds can be divided by the issuer. The main classes are government bonds, municipal bonds, and corporate bonds. Credit rating services give credit ratings

---

5 For example, a 5 year 4% semi-annual coupon bond with 1000$ face value makes ten 20$ payments every 6 months and the final payment of 1000$. Thus $P_i = 20\$$ for $i = 1, \ldots, n-1$ and the last payment is $P_n = 1020\$$, where $n = 10$.

to the bond issuers. Credit ratings help the investors to evaluate the probability of the payment default. Credit rating services include Standard & Poor's and Moody's.

US Treasury securities are backed by the US government. US Treasury securities include Treasury bills, Treasury notes, and Treasury bonds.

1) Treasury bills are zero-coupon bonds with original time to maturity of 1 year or less.[6]
2) Treasury notes are coupon bonds with original time to maturity between 2 and 10 years.
3) Treasury bonds are coupon bonds with original time to maturity of more than 10 years.

Widely traded German government bonds include Bundesschatzanweisungen (Schätze), which are 2 year notes, Bundesobligationen (Bobls), which are 5 year notes, and Bundesanleihen (Bunds and Buxl), which are 10 and 30 year bonds.

There are many types of fixed income securities. Callable bonds are such bonds that allow the bond issuer to purchase the bond back from the bondholders. The callable bonds make it possible for the issuer to retire old high-rate bonds and issue new low-rate bonds. Floating rate bonds (floaters) are such bonds whose rates are adjusted periodically to match inflation rates. Treasury STRIPS are such fixed income securities where the principal and the interest component of US Treasury securities are traded as separate zero coupon securities. The acronym STRIPS means separate trading of registered interest and principal securities.

2.2.2 Interest Rates

Interest rates are the basis for many financial contracts. We can separate between the government rates and the interbank rates. The government rates are deduced from the bonds issued by the governments and the interbank rates are obtained from the rates at which deposits are exchanged between banks.

Libor (London interbank offered rate) and Euribor (Euro interbank offered rate) are important interbank rates. Eonia (Euro overnight index average) is an overnight interest rate within the eurozone, but unlike the Euribor and Libor does not include term loans. Eonia is similar to the federal funds rate in the US. Sonia (Sterling overnight index average) is the reference rate for overnight unsecured transactions in the Sterling market.

Euribor and Libor are comparable base rates. Euribor rates are trimmed averages of interbank interest rates at which a collection of European banks are

---

6 The Treasury issues bills with times to maturity of 13 weeks, 26 weeks, and 52 weeks (3-month bills, 6-month bills, and 1-year bills). 13-week bills and 26-week bills are auctioned once a week and 52-week bills are auctioned once a month.

prepared to lend to one another. Libor rates are trimmed averages of interbank interest rates at which a collection of banks on the London money market are prepared to lend to one another. Euribor and Libor rates come in different maturities. In contrast to Euribor rates, the Libor rates come in different currencies. Euribor and Libor rates are not based on actual transactions, whereas Eonia is based on actual transactions. A study published in May 2008 in The Wall Street Journal suggested that the banks may have understated the borrowing costs. This led to reform proposals concerning the calculation of the Libor rates.

The Eonia rate is the rate at which banks provide unsecured loans to each other with a duration of 1 day within the Euro area. The Eonia rate is a volume weighted average of transactions on a given day and it is computed by the European Central Bank by the close of the real-time gross settlement on each business day. Eonia can be considered as the 1 day Euribor rate or as the Euro version of overnight index swaps (OIS). The Eonia panel consists of over 50 mostly European banks. The banks are chosen to the panel based on their premium credit rating and the high volume of their money market transactions conducted within the Eurozone. Banks on the Eonia panel are the same banks included in the Euribor panel.

Euribor rates are used as a reference rate for euro-denominated forward rate agreements, short term interest rate futures contracts, and interest rate swaps. Libor rates are used for Sterling and US dollar-denominated instruments.

2.2.2.1 Definitions of Interest Rates

The different definitions of interest rate are discussed in detail in Chapter 18. As an example we can consider a loan where the interest is paid at the end of a given period, and the interest is quoted in annual rate. Rate conventions determine how the quoted annual rate relates to the actual payment. Maybe the most common convention is to pay $P \times rT/360$, where P is the principal, r is the annual rate, and T is the number of calendar days of the deposit or loan. Note that loan rates are either rates that apply to a loan starting now until a given expiry, or forward rates, that are rates applying to a loan starting in the future for a given period of time.

Rates are quoted in percents but they are compared in basis points, where a basis point is 0.01%, that is, 1% is 100 basis points.

2.2.2.2 The Risk Free Rate

The risk free rate is different depending on the investment horizon. For one day horizon the risk free rate could be the Eonia rate or the rate of a bank account, and for 1 month horizon the risk free rate could be the rate of 1 month government bond.

2.2.3 Bond Prices and Returns

A 10 year zero-coupon bond has the time to maturity of 10 years at the emission, after 1 year the time to maturity is 9 years, after 2 years the time to maturity is 8 years, and so on. The price of the zero-coupon bond is fluctuating according to the fluctuation of the interest rates, until the price equals the nominal value at the maturity. Thus, the price of the 10 year zero-coupon bond gives information about the 10 year interest rate at the emission, after 1 year the price of the bond gives information about the 9 year interest rate, after 2 years the price of the bond gives information about the 8 year interest rate, and so on.

Information of the bond markets is given by data providers in terms of the yields. The yield of a zero-coupon bond is defined as

$$Y(t, T) = -\frac{1}{T-t} \log Z(t, T), \tag{2.3}$$

where $T - t$ is the time to maturity in fractions of a year, and $Z(t, T)$ is the bond price with $Z(T, T) = 1$. The price of a bond can be written in terms the yield as

$$Z(t, T) = \exp\{-(T - t)Y(t, T)\}.$$

See Section 18.1.2 for a discussion of the yield of a zero-coupon bond.

Let $s < t \le T$, where T is the expiration day of the zero-coupon bond. The prices are $Z(s, T)$ and $Z(t, T)$. The return of a bond trader is equal to

$$\frac{Z(t, T)}{Z(s, T)} = \frac{\exp\{-(T - t)Y(t, T)\}}{\exp\{-(T - s)Y(s, T)\}}$$
$$= \exp\{(T - s)[Y(s, T) - Y(t, T)] + (t - s)Y(t, T)\}, \tag{2.4}$$

where we used the fact $T - t = T - s - (t - s)$.

Data providers give a time series $Y_0, \dots, Y_n$ of yields of a τ year bond, where

$$Y_i = -\frac{1}{\tau} \log Z(t_i, t_i + \tau),$$

where $t_0 < \cdots < t_n$ are the time points of sampling. How to obtain a time series $R_0, \dots, R_n$ of the returns of a bond investor? Let us denote $t_i = s$, $t_{i+1} = t$, and $T - s = \tau$. Then $Y(s, T) = Y_i$. Let us make approximation

$$Y(t, T) = Y(t_{i+1}, t_i + \tau) \approx Y(t_{i+1}, t_{i+1} + \tau) = Y_{i+1}.$$

Then (2.4) implies

$$R_i \approx \exp\{\tau(Y_i - Y_{i+1}) + (t_{i+1} - t_i)Y_{i+1}\}, \tag{2.5}$$

where $t_{i+1} - t_i$ is the length of the sampling interval in fractions of a year. For example, with monthly sampling $t_{i+1} - t_i = 1/12$.

2.3 Derivatives

Derivatives are financial assets whose payoff is defined in terms of more basic assets. We describe first forwards and futures, and after that we describe options. For many assets trading with derivatives is more active than trading with the basic assets. For example, exchange rates and commodities are traded more actively in the future markets than in the spot markets.

Over-the-counter (OTC) derivatives are traded directly between two counterparties. Exchange traded derivatives are traded in an exchange, which acts as an intermediary party between the traders.

2.3.1 Forwards and Futures

First we define forwards and futures. After that we give examples of some actively traded futures. Forwards are derivatives traded over the counter whereas futures contracts are traded on exchanges. The underlyings of a forward or a futures contract can be stocks (single-stock futures), commodities, currencies, interest rates, or stock indexes, for example.

2.3.1.1 Forwards

A forward is a contract written at time t_0, with a commitment to accept delivery of (or to deliver) the specified number of units of the underlying asset at a future date T, at forward price F_{t_0}, which is determined at t_0.

At time t_0 nothing changes hands, all exchanges will take place at time T. A long position is a commitment to accept the delivery at time T. A short position is a commitment to deliver the contracted amount. The current price of the underlying is called the spot price.

2.3.1.2 Futures

A futures contract can be considered as a special case of a forward contract. An instrument is called a futures contract if the trading is done in a futures exchange, where the forward commitment is made through a homogenized contract so that the size of the underlying asset, the quality of the underlying asset, and the expiration date are preset. In addition, futures exchanges require a daily mark-to-market of the positions.

A futures exchange acts as an intermediary between the participants of a futures contract. The existence of the intermediary minimizes the risk of the default of the participants of the contract. When a participant enters a futures contract the exchange requires to put up an initial amount of liquid assets into the margin account. Marking to market means that the daily futures price is settled daily so that the exchange will draw money out of one party's margin account and put it into the others so that the daily loss or profit is taken into account. If the margin account goes below a certain value, then a margin call is made and the account owner must add money to the margin account. In contrast to futures contracts, forward contracts may not require any marking to market until the expiration day.

A futures contract can be settled with cash or with the delivery of the underlying. For example, if the underlying of the futures contract is a stock index, then the futures contract is usually settled with cash. A futures contract can be closed before the expiration day by entering the opposite direction futures contract.

On the delivery date, the amount exchanged is not the specified price on the contract but the spot value (i.e., the original value agreed upon, since any gain or loss has already been previously settled by marking to market).

The situation where the price of a commodity for future delivery is higher than the spot price, or where a far future delivery price is higher than a nearer future delivery, is known as contango. The reverse, where the price of a commodity for future delivery is lower than the spot price, or where a far future delivery price is lower than a nearer future delivery, is known as backwardation.

2.3.2 Options

We describe calls and puts, applications of options, and some exotic options.

2.3.2.1 Calls and Puts

The buyer of a call option receives the right to buy the underlying instrument and the buyer of a put option receives the right to sell the underlying instrument.

An European call option gives the right to buy an asset at the given expiration time T at the given strike price K. An European put option gives the right to sell an asset at the given expiration time T at the given strike price K. Let us denote with C_t the price of an European call option at time t and with S_t the price of the asset. The value C_T of the European call option at the expiration time T is equal to

$$C_T = \max\{S_T - K, 0\}.$$

Let us denote with P_t the price of a put option at time t. The value of the European put option at the expiration time T is equal to

$$P_T = \max\{K - S_T, 0\}.$$

American options have a different mode concerning the right to exercise the option than the European options. American call and put options can be exercised at any time before the expiration date, whereas European options can be exercised only at the expiration day. Thus an American option is more expensive than the corresponding European option. When we use the term "option" without a further qualification, then we refer to an European option.

The following terminology is used to describe options.

- A call option is out of the money if $S_t < K$. A call option is at the money if $S_t = K$. A call option is in the money if $S_t > K$. A call option is deep out of the money (deep in the money) if $S_t \ll K$ ($S_t \gg K$).
  The moneyness of a call option is defined as S_t/K. The moneyness of a put option is defined as K/S_t.[7]
- Before the expiration time T the price of a call option satisfies

$$C_t > (S_t - K)_+;$$

see (14.10). The difference $C_t - (S_t - K)_+$ is called the time value of the option. The value $(S_t - K)_+$ is called the intrinsic value. Thus,

$$C_t = \text{time value} + \text{intrinsic value}. \tag{2.6}$$

2.3.2.2 Applications of Options
Options can serve at least the following purposes:

1) Options can be used to create a large number of different payoffs. Some payoffs applied in option trading are described in Chapter 17. For example, buying a call and a put with the same strike price and the same expiration creates a straddle position which profits from large positive or negative movements of the underlying.
2) Options can provide insurance. With options it is possible to create a payoff which cuts the losses that could occur without using of the options. Buying a put option gives an insurance in the case one has to sell in a future time an asset one possesses. Buying a call option gives an insurance in the case when one has to buy in a future time an asset one does not possess. Examples of providing insurance with options include the following:
   - Buying a put option on a stock gives an insurance policy for an investor. If an investor owns a stock, buying a put option will cut the future possible losses.
   - Buying a put option on an exchange rate gives an insurance policy for a company receiving payments on a foreign currency in future.
3) Call options can be used to give a compensation to managers, since the payoff of a call option is positive only when the stock price is larger than the strike price.
4) Options make leveraging possible, since option trading requires a small initial capital as compared to stock trading.[8]

---

7 Sometimes moneyness is defined by $S_t/(Ke^{-r(T-t)})$ and $Ke^{-r(T-t)}/S_t$, where $T - t$ is the time to expiration in fractions of year and r is the annualized short term interest rate.
8 Suppose that the stock price is $S_t = 100$, the strike price is $K = 105$, and the call price is $C_t = 5$. If the stock price rises to $S_T = 110$ at the expiration time of the call option, then the owner of the stock has the return of 10% but the owner of the call option has the return of $(110 - 105)/5 = 100\%$.

2.3.2.3 Exotic Options

We say that an option is exotic if it is not an European or an American call or put option.

Bermudan Options There exists three basic modes concerning the right to exercise the option: European, American, and Bermudan. A Bermudan option can be exercised at some times or time periods before the expiration. whereas European options can be exercised only at the expiration date, and American options can be exercised at any time before the expiration.

Asian Options The value of an Asian call option at the expiration is

$$C_T = \max\{0, M_T - K\},$$

where

$$M_T = \frac{1}{n}\sum_{i=1}^{n} S_{t_i}$$

with $t_1 < \cdots < t_n \leq T$ being a collection of predetermined time points. Asian options are more resistant to manipulation than European options: The value of an European option at the expiration depends on the value of the underlying asset at one time point (the expiration date), whereas the value of an Asian option depends on the values of the underlying asset at several time points.

Barrier Options Barrier options disappear if the underlying either exceeds, or goes under the barrier. Alternatively, a barrier option could have value only if it has exceeded, or went under the barrier. Knock-in options come into existence if some barrier is hit and knock-out options cease to exist if some barrier is hit. One speaks of up-and-out, down-and-out, up-and-in, down-and-in options. For example, a knock-out option on stock S_t, written at time 0, with expiration time T, has the payoff

$$B_T = \begin{cases} 0, & \text{when } \max_{0 \leq t \leq T} S_t \geq H, \\ \max\{0, S_T - K\}, & \text{otherwise}, \end{cases}$$

where $K > 0$ is the strike price, and $H > K$ is the barrier. Barrier options are cheaper than the corresponding European options, which makes them useful.

Multiasset Options Multiasset options involve many underlying assets and many strike prices. We give some examples of multiasset options.

1) A call can be generalized to a multiasset option with payoff

$$\max\left\{S_T^1 - K_1, S_T^2 - K_2, 0\right\},$$

where S^1 and S^2 are the underlying assets and $K_1, K_2 > 0$ are strike prices. A payoff can have elements of a call and a put:

$$\max\{K_1 - S_T^1, S_T^2 - K_2, 0\}.$$

2) The payoff of an option on a linear combination can be written as

$$f\left(\sum_{i=1}^{d} w_i S^i\right),$$

where $f : \mathbf{R} \to \mathbf{R}$ is a payoff function, $S^1, \ldots, S^d$ are assets, and $w_1, \ldots, w_d$ are weights. For example, an option on a linear combination can be an option on an index or an option on a spread.

3) Outperformance options are calls on the maximum and puts on the minimum. We have that

$$\max\{S^1, S^2\} = S^1 + \max\{S^2 - S^1, 0\}.$$

Thus, the payoff of an outperformance option can be written as a payoff of a linear combination of the underlying and an option on the spread between the underlyings.

4) The payoff of a univariate digital option is $I_{[K,\infty)}(S_T)$, where $K > 0$ is the strike price. The option pays one unit at the maturity time if the value of the underlying exceeds the strike price. The bivariate digital option pays one unit if both of the underlyings exceed the respective strike prices. The payoff is

$$I_{[K_1,\infty)\times[K_2,\infty)}\left(S_T^1, S_T^2\right).$$

5) The payoff of an option written on a basket can be written as

$$G\left(\psi\left(S_T^1, \ldots, S_T^N\right)\right),$$

where G is a univariate function and ψ is a multivariate function. For example, $G(x) = (x - K)_+$ and $\psi(S^1, \ldots, S^d) = \min(S^1, \ldots, S^d)$, or $\psi(S^1, \ldots, S^d) = \sum_{i=1}^{d} w_i S^i$.

2.4 Data Sets

We describe the data sets which are used to illustrate the methods throughout the book. Some additional data are described in Section 6.3.

2.4.1 Daily S&P 500 Data

The daily S&P 500 data consists of the daily closing prices starting at January 4, 1950 and ending at April 2, 2014, which gives 16,046 daily observations.[9]

Figure 2.1 shows (a) the daily closing prices S_t and (b) the returns $R_t = S_t/S_{t-1}$ of S&P 500.

---

9 The data is obtained from Yahoo (http://finance.yahoo.com/) with ticker ^GSPC.

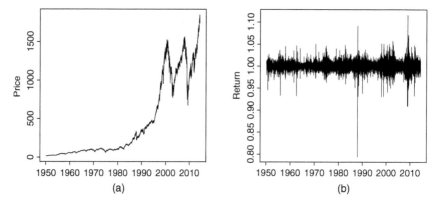

Figure 2.1 *S&P 500 index.* (a) Daily closing prices of S&P 500 and (b) daily returns.

2.4.2 Daily S&P 500 and Nasdaq-100 Data

The S&P 500 and Nasdaq-100 data consists of the daily closing prices starting at October 1, 1985 and ending at May 21, 2014, which gives 7221 daily observations.[10]

Figure 2.2 shows (a) the normalized prices and (b) a scatter plot of the returns of S&P 500 and Nasdaq-100. S&P 500 prices is shown with black and the Nasdaq-100 prices is shown with red. The prices are normalized so that they start with value one for both indexes. (Note that the normalized price is the cumulative wealth when the initial wealth is one.)

2.4.3 Monthly S&P 500, Bond, and Bill Data

The data consists of the monthly returns of S&P 500 index, monthly returns of US Treasury 10 year bond, and monthly rates of US Treasury 1 month bill. The data starts at May 1953 and ends at December 2013, which gives 728 monthly observations.[11] The 10 year bond returns are calculated from the yields as in (2.5).

Figure 2.3 shows (a) cumulative wealth and (b) a scatter plot of returns of S&P 500 and 10 year bond. The cumulative wealth is $W_t = \prod_{i=1}^{t} R_i$, where R_i are the gross returns. The cumulative wealth of S&P 500 is shown with black, 10 year bond with red, and 1 month bill with blue. Figure 2.4 shows (a) the treasury bill rates (blue) and (b) the yields of 10 year Treasury bond (red).

---

10 The data is obtained from Yahoo (http://finance.yahoo.com/) with tickers ^GSPC and ^NDX.
11 The data is obtained from http://www.hec.unil.ch/agoyal/ (Amit Goyal).

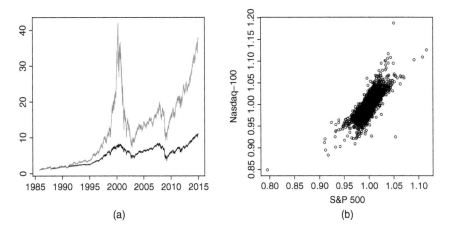

(a)

(b)

Figure 2.2 *S&P 500 and Nasdaq-100 indexes.* (a) The prices of S&P 500 (black) and Nasdaq-100 (red). The prices are normalized to start at value one. (b) A scatter plot of the daily returns of S&P 500 and Nasdaq-100.

2.4.4 Daily US Treasury 10 Year Bond Data

The US Treasury 10 year bond data consists of the daily yields starting at January 2, 1962 and ending at March 3, 2014, which gives 13,006 daily observations.[12] We have described the US 10 year Treasury bonds in Section 2.2.1.

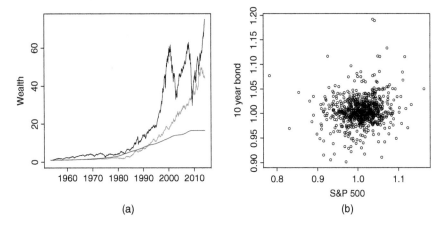

(a)

(b)

Figure 2.3 *S&P 500, US Treasury 10 year bond, and 1 month bill.* (a) The cumulative wealth of S&P 500 (black), 10 year bond (red), and 1 month bill (blue). The cumulative wealths are normalized to start at value one. (b) A scatter plot of monthly returns of S&P 500 and 10 year bond.

---

12 The data is obtained from Federal Reserve Bank of St. Louis with ticker DGS10, see the web site http://research.stlouisfed.org/. There were 13,590 days when the market is open but the data was missing in 584 days.

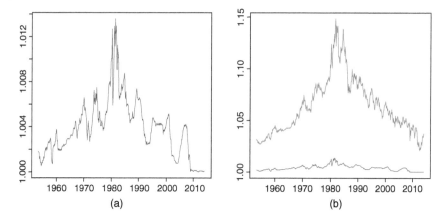

Figure 2.4 *US Treasury bill rates and 10 year bond yields.* (a) Treasury bill rates (blue). (b) Yields of 10 year Treasury bond (red).

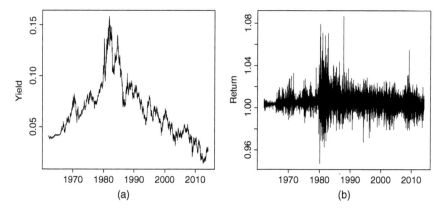

Figure 2.5 *10 year US Treasury bond.* (a) Daily yields of the 10 year US Treasury bond and (b) daily returns of the bond.

Figure 2.5 shows (a) the daily yields and (b) the daily returns of the US 10 year Treasury bond. The 10 year bond returns are calculated from the yields as in (2.5).

2.4.5 Daily S&P 500 Components Data

The S&P 500 components data consists of daily closing prices of 312 stocks, which were components of S&P 500 at May 23, 2014. The data starts September 30, 1991 and ends at May 23, 2014. There are 5707 daily observations.

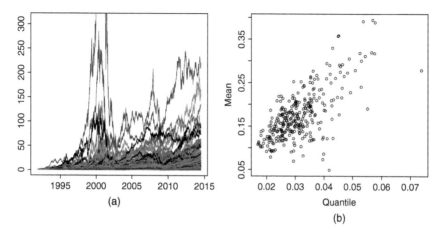

Figure 2.6 *S&P 500 components.* (a) Time series of the normalized prices of the components. (b) A scatter plot of (q_i, μ_i), where q_i are the 95% empirical quantiles of the negative returns, and μ_i are the annualized sample means of the returns.

Figure 2.6(a) shows the normalized prices of the stocks. The prices are normalized to have value one at the beginning. Panel (b) shows a scatter plot of points (q_i, μ_i), where q_i are the 95% empirical quantiles of the negative returns of the ith stock, and μ_i are the annualized sample means of the returns of the ith stock.[13]

---

13 That is, q_i satisfies approximately $P(R_t^i \le -q_i) = 0.05$, where $R_t^i = S_t/S_{t-1} - 1$ is the net return of the ith stock, and μ_i is approximately $250 \times ER_t^i$.

3

Univariate Data Analysis

Univariate data analysis studies univariate financial time series, but ignoring the time series properties of data. Univariate data analysis studies also cross-sectional data. For example, returns at a fixed time point of a collection of stocks is a cross-sectional univariate data set.

A univariate series of observations can be described using such statistics as sample mean, median, variance, quantiles, and expected shortfalls. These are covered in Section 3.1.

The graphical methods are explained in Section 3.2. Univariate graphical tools include tail plots, regression plots of the tails, histograms, and kernel density estimators. We use often tail plots to visualize the tail parts of the distribution, and kernel density estimates to visualize the central part of the distribution. The kernel density estimator is not only a visualization tool but also a tool for estimation.

We define univariate parametric models like normal, log-normal, and Student models in Section 3.3. These are parametric models, which are alternatives to the use of the kernel density estimator.

For a univariate financial time series it is of interest to study the tail properties of the distribution. This is done in Section 3.4. Typically the distribution of a financial time series has heavier tails than the normal distributions. The estimation of the tails is done using the concept of the excess distribution. The excess distribution is modeled with exponential, Pareto, gamma, generalized Pareto, and Weibull distributions. The fitting of distributions can be done with a version of maximum likelihood. These results prepare us to quantile estimation, which is considered in Chapter 8.

Central limit theorems provide tools to construct confidence intervals and confidence regions. The limit theorems for maxima provide insight into the estimation of the tails of a distribution. Limit theorems are covered in Section 3.5.

Section 3.6 summarizes the univariate stylized facts.

Nonparametric Finance, First Edition. Jussi Klemelä.

3.1 Univariate Statistics

We define mean, median, and mode to characterize the center of a distribution. The spread of a distribution can be measured by variance, other centered moments, lower and upper partial moments, lower and upper conditional moments, quantiles (value-at-risk), expected shortfall, shortfall, and absolute shortfall.

We define both population and sample versions of the statistics. In addition, we define both unconditional and conditional versions of the statistics.

3.1.1 The Center of a Distribution

The center of a distribution can be defined using the mean, the median, or the mode. The center of a distribution is an unknown quantity that has to be estimated using the sample mean, the sample median, or the sample mode. The conditional versions of theses quantities take into account the available information. For example, if we know that it is winter, then the expected temperature is lower than the expected temperature when we know that it is summer.

3.1.1.1 The Mean and the Conditional Mean

The population mean is called the expectation. The population mean can be estimated by the arithmetic mean. The conditional mean is estimated using regression analysis.

The Population Mean The population mean (expectation) of random variable $Y \in \mathbf{R}$, whose distribution is continuous, is defined as

$$EY = \int_{-\infty}^{\infty} y f_Y(y) \, dy, \tag{3.1}$$

where $f_Y : \mathbf{R} \to \mathbf{R}$ is the density function of Y.[1] Let $X \in \mathbf{R}^d$ be an explanatory random variable (random vector). The conditional expectation of Y given $X = x$ can be defined by

$$E(Y \mid X = x) = \int_{-\infty}^{\infty} y f_{Y \mid X=x}(y) \, dy,$$

where $f_{Y \mid X=x}(y) : \mathbf{R} \to \mathbf{R}$ is the conditional density.[2]

---

1 The density function $f_Y : \mathbf{R} \to \mathbf{R}$ is a function which satisfies (1) $f_Y(y) \geq 0$ for almost all $y \in \mathbf{R}$, and (2) $P(Y \in A) = \int_A f_Y(y) \, dy$ for measurable $A \subset \mathbf{R}$. Thus, we can express all probabilities as integrals of f_Y.

2 The conditional density is defined as

$$f_{Y \mid X=x}(y) = \frac{f_{X,Y}(x,y)}{f_X(x)}, \quad y \in \mathbf{R}, \quad x \in \mathbf{R}^d, \tag{3.2}$$

The population mean of random variable $Y \in \mathbf{R}$, whose distribution is discrete with the possible values $y_1, \dots, y_N$, is defined as

$$EY = \sum_{i=1}^{N} y_i P(Y = y_i). \tag{3.3}$$

The conditional expectation can be defined as

$$E(Y \mid X = x) = \sum_{i=1}^{N} y_i P(Y = y_i \mid X = x).$$

The Sample Mean Given a sample $Y_1, \dots, Y_T$ from the distribution of Y, the mean EY can be estimated with the sample mean (the arithmetic mean):

$$\bar{Y} = \frac{1}{T} \sum_{t=1}^{T} Y_t. \tag{3.4}$$

Regression analysis studies the estimation of the conditional expectation. In regression analysis, we observe values $X_1, \dots, X_T$ of the explanatory random variable (random vector), in addition to observing values $Y_1, \dots, Y_T$ of the response variable. Besides linear regression there exist various nonparametric methods for the estimation of the conditional expectation. For example, in kernel regression the arithmetic mean in (3.4) is replaced by a weighted mean

$$\hat{f}(x) = \sum_{t=1}^{T} p_t(x) Y_t,$$

where $p_t(x)$ is a weight that is large when X_t is close to x and small when X_t is far away from x. Now $\hat{f}(x)$ is an estimate of the conditional mean $f(x) = E(Y \mid X = x)$, for $x \in \mathbf{R}^d$. Kernel regression and other regression methods are described in Section 6.1.2.

The Annualized Mean The return of a portfolio is typically estimated using the arithmetic mean and it is expressed as the annualized mean return. Let $S_{t_0}, \dots, S_{t_n}$ be observed stock prices, sampled at equidistant time points. Let $R_{t_i} = (S_{t_i} - S_{t_{i-1}})/S_{t_{i-1}}$, $i = 1, \dots, n$, be the net returns. Let the sampling interval be $\Delta t = t_i - t_{i-1}$. The annualized mean return is

$$\frac{1}{\Delta t} \frac{1}{n} \sum_{i=1}^{n} R_{t_i}. \tag{3.5}$$

---

when $f_X(x) > 0$, where $f_{X,Y} : \mathbf{R}^{d+1} \to \mathbf{R}$ is the joint density of (X, Y), and $f_X : \mathbf{R}^d \to \mathbf{R}$ is the density of X:

$$f_X(x) = \int_{\mathbf{R}} f_{X,Y}(x, y)\, dy, \quad x \in \mathbf{R}^d.$$

If $f_X(x) = 0$, then $f_{Y \mid X=x}(y) = 0$.

For the monthly returns $\Delta t = 1/12$. For the daily returns $\Delta t = 1/250$, because there are about 250 trading days in a year. Sampling of prices and several definitions of returns are discussed in Section 2.1.2.

The Geometric Mean Let $S_0, \ldots, S_T$ be the observed stock prices and let $R_t = S_t/S_{t-1}, t = 1, \ldots, T$, be the gross returns. The geometric mean is defined as

$$\left(\prod_{t=1}^{T} R_t \right)^{1/T}.$$

The logarithm of the geometric mean is equal to the arithmetic mean of the logarithmic returns:

$$\frac{1}{T} \sum_{t=1}^{T} \log R_t.$$

Note that $W_t = \prod_{i=1}^{t} R_i$ is the cumulative wealth at time t when we start with wealth 1. Thus,

$$\frac{1}{T} \log W_T = \frac{1}{T} \sum_{t=1}^{T} \log R_t.$$

3.1.1.2 The Median and the Conditional Median

The median can be defined in the case of a continuous distribution function of a random variable $Y \in \mathbf{R}$ as the number $\mathrm{median}(Y) \in \mathbf{R}$ satisfying

$$P(Y \le \mathrm{median}(Y)) = 0.5.$$

Thus, the median is the point that divides the probability mass into two equal parts. Let us define the distribution function $F : \mathbf{R} \to \mathbf{R}$ by

$$F(y) = P(Y \le y).$$

When F is continuous, then

$$\mathrm{median}(Y) = F^{-1}(0.5).$$

In general, covering also the case of discrete distributions, we can define the median uniquely as the generalized inverse of the distribution function:

$$\mathrm{median}(Y) = \inf\{y : F(y) \ge 0.5\}. \tag{3.6}$$

The conditional median is defined using the conditional distribution function

$$F_{Y|X=x}(y) = P(Y \le y \,|\, X = x),$$

where X is a random vector taking values in $\mathbf{R}^d$. Now we can define

$$\mathrm{median}(Y \,|\, X = x) = \inf\{y : F_{Y|X=x}(y) \ge 0.5\}, \tag{3.7}$$

where $x \in \mathbf{R}^d$.

The sample median of observations $Y_1, \ldots, Y_T \in \mathbf{R}$ can be defined as the observation that has as many smaller observations as larger observations:

$$\text{median}(Y_1, \ldots, Y_T) = Y_{([T/2]+1)}, \tag{3.8}$$

where $Y_{(1)} \leq \cdots \leq Y_{(T)}$ is the ordered sample and $[x]$ is the largest integer smaller or equal to x. The sample median is a special case of an empirical quantile. Empirical quantiles are defined in (8.21)–(8.23).

3.1.1.3 The Mode and the Conditional Mode

The mode is defined as an argument maximizing the density function of the distribution of a random variable:

$$\text{mode}(Y) = \underset{y \in \mathbf{R}}{\text{argmax}}\, f_Y(y), \tag{3.9}$$

where $f_Y : \mathbf{R} \to \mathbf{R}$ is the density function of the distribution of Y. The density f_Y can have several local maxima, and the use of the mode seems to be interesting only in cases where the density function is unimodal (has one local maximum). The conditional mode is defined as an argument maximizing the conditional density:

$$\text{mode}(Y \mid X = x) = \underset{y \in \mathbf{R}}{\text{argmax}}\, f_{Y \mid X=x}(y).$$

A mode can be estimated by finding a maximizer of a density estimate:

$$\widehat{\text{mode}}(Y) = \underset{y \in \mathbf{R}}{\text{argmax}}\, \hat{f}_Y(y),$$

where $\hat{f}_Y : \mathbf{R} \to \mathbf{R}$ is an estimator of the density function f_Y. Histograms and kernel density estimators are defined in Section 3.2.2.

3.1.2 The Variance and Moments

Variance and higher order moments characterize the dispersion of a univariate distribution. To take into account only the left or the right tail we define upper and lower partial moments and upper and lower conditional moments.

3.1.2.1 The Variance and the Conditional Variance

The variance of random variable Y is defined by

$$\text{Var}(Y) = E(Y - EY)^2 = EY^2 - (EY)^2. \tag{3.10}$$

The standard deviation of Y is the square root of the variance of Y. The conditional variance of random variable Y is equal to

$$\text{Var}(Y \mid X = x) = E\{[Y - E(Y \mid X = x)]^2 \mid X = x\} \tag{3.11}$$
$$= E(Y^2 \mid X = x) - [E(Y \mid X = x)]^2. \tag{3.12}$$

The conditional standard deviation of Y is the square root of the conditional variance.

The Sample Variance The sample variance is defined by

$$\widehat{\text{Var}}(Y) = \frac{1}{T} \sum_{i=1}^{T} (Y_i - \bar{Y})^2 = \frac{1}{T} \sum_{i=1}^{T} Y_i^2 - \bar{Y}^2, \tag{3.13}$$

where $Y_1, \ldots, Y_T$ is a sample of random variables having identical distribution with Y, and $\bar{Y}$ is the sample mean.[3]

The Annualized Variance The sample variance and the standard deviation of portfolio returns are typically annualized, analogously to the annualized sample mean in (3.5). Let $S_{t_0}, \ldots, S_{t_n}$ be the observed stock prices, sampled at equidistant time points. Let $R_{t_i} = (S_{t_i} - S_{t_{i-1}})/S_{t_{i-1}}$, $i = 1, \ldots, n$, be the net returns. Let the sampling interval be $\Delta t = t_i - t_{i-1}$. The annualized sample variance of the returns is

$$\frac{1}{\Delta t} \frac{1}{n} \sum_{i=1}^{n} (R_{t_i} - \bar{R})^2,$$

where $\bar{R} = n^{-1} \sum_{i=1}^{n} R_{t_i}$. For the monthly returns $\Delta t = 1/12$. For the daily returns $\Delta t = 1/250$, because there are about 250 trading days in a year. Sampling of prices and several definitions of returns are discussed in Section 2.1.2.

3.1.2.2 The Upper and Lower Partial Moments

The definition of the variance of random variable $Y \in \mathbf{R}$ can be generalized to other centered moments

$$E|Y - EY|^k,$$

for $k = 1, 2, \ldots$. The variance is obtained when $k = 2$. The centered moments take a contribution both from the left and the right tail of the distribution. The lower partial moments take a contribution only from the left tail and the upper partial moments take a contribution only from the right tail. For example, if we are interested only in the distribution of the losses, then we use the lower partial moments of the return distribution, and if we are interested only in the distribution of the gains, then we use the upper partial moments. The upper partial moment is defined as

$$\text{UPM}_{\tau,k}(Y) = E(Y - \tau)_+^k = E[(Y - \tau)^k I_{[\tau,\infty)}(Y)], \tag{3.14}$$

where $k = 0, 1, 2, \ldots$, $(x)_+ = \max\{x, 0\}$, and $\tau \in \mathbf{R}$. The lower partial moment is defined as

$$\text{LPM}_{\tau,k}(Y) = E(\tau - Y)_+^k = E[(\tau - Y)^k I_{(-\infty,\tau]}(Y)]. \tag{3.15}$$

---

3 The sample variance is often defined as $(T - 1)^{-1} \sum_{i=1}^{T} (Y_i - \bar{Y})^2$, because this is an unbiased estimator of the population variance. For large and moderate T it does not matter whether the divisor is T or $T - 1$.

When Y has density f_Y, we can write

$$\text{UPM}_{\tau,k}(Y) = \int_{\tau}^{\infty} (y - \tau)^k f_Y(y) \, dy, \quad \text{LPM}_{\tau,k}(Y) = \int_{-\infty}^{\tau} (\tau - y)^k f_Y(y) \, dy.$$

For example, when $k = 0$, then

$$\text{UPM}_{\tau,0}(Y) = P(Y \geq \tau), \quad \text{LPM}_{\tau,0}(Y) = P(Y \leq \tau),$$

so that the upper partial moment is equal to the probability that Y is greater or equal to τ, and the lower partial moment is equal to the probability that Y is smaller or equal to τ. For $k = 2$ and $\tau = EY$ the partial moments are called the upper and lower semivariance of Y. For example, the lower semivariance is defined as

$$E[(Y - EY)^2 I_{(-\infty,EY]}(Y)]. \tag{3.16}$$

The square root of the lower semivariance can be used to replace the standard deviation in the definition of the Sharpe ratio, or in the Markowitz criterion.

The sample centered moments are

$$\frac{1}{T} \sum_{t=1}^{T} |Y_i - \bar{Y}|^k,$$

where $\bar{Y}$ is the sample mean. The sample upper and the sample lower partial moments are

$$\widehat{\text{UPM}}_{\tau,k}(Y) = \frac{1}{T} \sum_{i=1}^{T} (Y_i - \tau)_{+}^k, \quad \widehat{\text{LPM}}_{\tau,k}(Y) = \frac{1}{T} \sum_{i=1}^{T} (\tau - Y_i)_{+}^k. \tag{3.17}$$

For example, when $k = 0$ we have

$$\widehat{\text{LPM}}_{\tau,0}(Y) = \frac{N(\tau)}{T},$$

where

$$N(\tau) = \#\{Y_i : i = 1, \dots, T, \quad Y_i \leq \tau\}. \tag{3.18}$$

3.1.2.3 The Upper and Lower Conditional Moments

The upper conditional moments are the moments conditioned on the right tail of the distribution and the lower conditional moments are the moments conditioned on the left tail of the distribution. The upper conditional moment is defined as

$$\text{UCM}_{\tau,k}(Y) = E[(Y - \tau)^k \mid Y - \tau \geq 0]$$

and the lower conditional moment is defined as

$$\text{LCM}_{\tau,k}(Y) = E[(\tau - Y)^k \mid \tau - Y \geq 0], \tag{3.19}$$

where $k = 0, 1, 2, \dots$ and $\tau \in \mathbf{R}$ is a target rate.

The sample lower conditional moment is

$$\widehat{\mathrm{LCM}}_{\tau,k}(Y) = \frac{1}{N(\tau)} \sum_{i=1}^{T} (\tau - Y_i)_+^k,$$ (3.20)

where $N(\tau)$ is defined in (3.18). Note that in (3.17) the sample size is the denominator but in (3.20) we have divided with the number of observations in the left tail.

We can condition also on an external variable X and define conditional on X versions of both upper and lower moments, and upper and lower conditional moments.

3.1.3 The Quantiles and the Expected Shortfalls

The quantiles are applied under the name value-at-risk in risk management to characterize the probability of a tail event. The expected shortfall is a related measure for a tail risk.

3.1.3.1 The Quantiles and the Conditional Quantiles
The pth quantile is defined as

$$Q_p(Y) = \inf\{y : F(y) \geq p\},$$ (3.21)

where $0 < p < 1$ and $F(y) = P(Y \leq y)$ is the distribution function of Y. The value-at-risk is defined in (8.3) as a quantile of a loss distribution. For $p = 1/2$, $Q_p(Y)$ is equal to median(Y), defined in (3.6). In the case of a continuous distribution function, we have

$$F(Q_p(Y)) = p$$

and thus it holds that

$$Q_p(Y) = F^{-1}(p),$$

where F^{-1} is the inverse of F. The pth conditional quantile is defined replacing the distribution function of Y with the conditional distribution function of Y:

$$Q_p(Y|X = x) = \inf\{y : F_{Y|X=x}(y) \geq p\}, \quad x \in \mathbf{R}^d,$$ (3.22)

where $0 < p < 1$ and $F_{Y|X=x}(y) = P(Y \leq y | X = x)$ is the conditional distribution function of Y.

The empirical quantile is defined as

$$\hat{Q}_p = Y_{(\lceil pT \rceil)},$$ (3.23)

where $Y_{(1)} \leq Y_{(2)} \leq \cdots \leq Y_{(T)}$ is the ordered sample and $\lceil x \rceil$ is the smallest integer $\geq x$. We give equivalent definitions of the empirical quantile in Section 8.4.1. Chapter 8 discusses various estimators of quantiles and conditional quantiles.

3.1.3.2 The Expected Shortfalls
The expected shortfall is a measure of risk that aggregates all quantiles in the right tail (or in the left tail). When Y has a continuous distribution function, then the expected shortfall for the right tail is

$$\text{ES}_p(Y) = E(Y \mid Y \geq Q_p(Y)) = \frac{1}{1-p} E(Y I_{[Q_p(Y),\infty)}(Y)), \tag{3.24}$$

where $0 < p < 1$. Thus, the pth expected shortfall is the conditional expectation under the condition that the random variable is larger than the pth quantile. The term "tail conditional value-at-risk" is sometimes used to denote the expected shortfall. In the general case, when the distribution of Y is not necessarily continuous, the expected shortfall for the right tail is defined as

$$\text{ES}_p(Y) = \frac{1}{1-p} \int_p^1 Q_u(Y)\, du, \quad 0 < p < 1. \tag{3.25}$$

The equality of (3.24) and (3.25) for the continuous distributions is proved in McNeil *et al.* (2005, lemma 2.16). In fact, denoting $q_p = Q_p(Y)$,

$$E[Y I_{(q_p,\infty)}(Y)] = E[F^{-1}(U) I_{(q_p,\infty)}(F^{-1}(U))]$$
$$= E[F^{-1}(U) I_{[p,1)}(U)]$$
$$= \int_p^1 F^{-1}(u)\, du,$$

where $U \sim \text{Uniform}([0,1])$ and we use the fact that $F^{-1}(U) \sim Y$.[4] Finally, note that $P(Y \geq Q_p(Y)) = 1 - p$ for continuous distributions.

The expected shortfall for the left tail is

$$\text{ES}_p(Y) = \frac{1}{p} \int_0^p Q_u(Y)\, du, \quad 0 < p < 1.$$

When Y has a continuous distribution function, then the expected shortfall for the left tail is

$$\text{ES}_p(Y) = E(Y \mid Y \leq Q_p(Y)) = \frac{1}{p} E(Y I_{(-\infty,Q_p(Y)]}(Y)). \tag{3.26}$$

This expression shows that in the case of a continuous distribution function, $p\text{ES}_p(Y)$ is equal to the expectation that is taken only over the left tail, when the left tail is defined as the region that is on the left side of the pth quantile of

---

4 We have that $P(F^{-1}(U) \leq x) = P(U \leq F(x)) = F(x)$.

the distribution. Note that the expected shortfall for the left tail is related to the lower conditional moment of order $k = 1$ and target rate $\tau = Q_p(Y)$:

$$ES_p(Y) = Q_p(Y) - E(Q_p(Y) - Y \mid Y \leq Q_p(Y))$$
$$= Q_p(Y) - LCM_{Q_p(Y),1},$$

where the lower conditional moment $LCM_{Q_p(Y),1}$ is defined in (3.19).[5]

The expected shortfall for the right tail, as defined in (3.24), can be estimated from the data $Y_1, \ldots, Y_T$ by

$$\widehat{ES}_p(Y) = \frac{1}{T - m + 1} \sum_{i=m}^{T} Y_{(i)}, \tag{3.27}$$

where $Y_{(1)} \leq \cdots \leq Y_{(T)}$ and $m = \lceil pT \rceil$, with, for example, $p = 0.95$ or $p = 0.99$. When the expected shortfall is for the left tail, as defined by (3.26), then we define the estimator as

$$\widehat{ES}_p(Y) = \frac{1}{m} \sum_{i=1}^{m} Y_{(i)}, \tag{3.28}$$

where $m = \lceil pT \rceil$ with, for example, $p = 0.05$ or $p = 0.01$.

3.2 Univariate Graphical Tools

We consider sequence $Y_1, \ldots, Y_T \in \mathbf{R}$ of real numbers, and assume that the sequence is a sample from a probability distribution. We want to visualize the sequence in order to discover properties of the underlying distribution. We divide the graphical tools to those that are based on the empirical distribution function and the empirical quantiles, and to those that are based on

---

5 Analogously to the definition of lower partial moments in (3.15), we can define the absolute shortfall as

$$AS_p(Y) = E(YI_{(-\infty, Q_p(Y)]}(Y)).$$

The absolute shortfall for the left tail is related to the lower partial moment of order $k = 1$ and target rate $\tau = Q_p(Y)$:

$$AS_p(Y) = pQ_p(Y) - E((Q_p(Y) - Y)I_{(-\infty, Q_p(Y)]}(Y))$$
$$= pQ_p(Y) - LPM_{Q_p(Y),1}.$$

The absolute shortfall is estimated from observations $Y_1, \ldots, Y_T$ by

$$\widehat{AS}_p(Y) = \frac{1}{T} \sum_{i=1}^{m} Y_{(i)},$$

where $Y_{(1)} \leq \cdots \leq Y_{(T)}$ is the ordered sample and $m = \lceil pT \rceil$. Here, we divide by T, but in the estimator (3.28) of the expected shortfall we divide by m.

the estimation of the underlying density function. The distribution function and quantiles based tools give more insight about the tails of the distribution, and the density based tools give more information about the center of the distribution.

A two-variate data can be visualized using a scatter plot. For a univariate data there is no such obvious method available. Thus, visualizing two-variate data may seem easier than visualizing univariate data. However, we can consider many of the tools to visualize univariate data to be scatter plots of points

$$(Y_i, \text{level}(Y_i)), \quad i = 1, \dots, T, \tag{3.29}$$

where level : $\{Y_1, \dots, Y_T\} \to \mathbf{R}$ is a mapping that attaches a real value to each data point $Y_i \in \mathbf{R}$. Thus, in a sense we visualize univariate data by transforming it into a two-dimensional data.

3.2.1 Empirical Distribution Function Based Tools

The distribution function of the distribution of random variable $Y \in \mathbf{R}$ is

$$F(x) = P(Y \le x), \quad x \in \mathbf{R}.$$

The empirical distribution function can be considered as a starting point for several visualizations: tail plots, regression plots of tails, and empirical quantile functions. We use often tail plots. Regression plots of tails have two types: (1) plots that look linear for an exponential tail and (2) plots that look linear for a Pareto tail.

3.2.1.1 The Empirical Distribution Function
The empirical distribution function $\hat{F}$, based on data $Y_1, \dots, Y_T$, is defined as

$$\hat{F}(x) = \frac{1}{T} \#\{Y_i : Y_i \le x, \quad i = 1, \dots, T\}, \tag{3.30}$$

where $x \in \mathbf{R}$, and #A means the cardinality of set A. Note that the empirical distribution function is defined in (8.20) using the indicator function. An empirical distribution function is a piecewise constant function. Plotting a graph of an empirical distribution function is for large samples practically the same as plotting the points

$$(Y_{(i)}, i/T), \quad i = 1, \dots, T, \tag{3.31}$$

where $Y_{(1)} \le \cdots \le Y_{(T)}$ are the ordered observations. Thus, the empirical distribution function fits the scheme of transforming univariate data to two-dimensional data as in (3.29).

Figure 3.1 shows empirical distribution functions of S&P 500 net returns (red) and 10-year bond net returns (blue). The monthly data of S&P 500 and US Treasury 10-year bond returns is described in Section 2.4.3. Panel (a) plots the points (3.31) and panel (b) zooms to the lower left corner, showing the

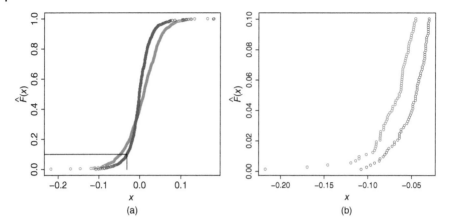

Figure 3.1 *Empirical distribution functions.* (a) Empirical distribution functions of S&P 500 returns (red) and 10-year bond returns (blue); (b) zooming at the lower left corner.

empirical distribution function for the $[T/10]$ smallest observations; the empirical distribution function is shown on the range $x \in (-\infty, \hat{q}_p)$, where $\hat{q}_p$ is the pth empirical quantile for $p = 0.1$. Neither of the estimated return distributions dominates the other: The S&P 500 distribution function is higher at the left tail but lower at the right tail. That is, S&P 500 is more risky than the 10-year bond. Note that Section 9.2.3 discusses stochastic dominance: a first return distribution dominates stochastically a second return distribution when the first distribution function takes smaller values everywhere than the second distribution function.

3.2.1.2 The Tail Plots
The left and right tail plots can be used to visualize the heaviness of the tails of the underlying distribution. A smooth tail plot can be used to visualize simultaneously a large number of samples. The tail plots are almost the same as the empirical distribution function, but there are couple of differences:

1) In tail plots we divide the data into the left tail and the right tail, and we visualize separately the two tails.
2) In tail plots the y-axis shows the number of observations and a logarithmic scale is used for the y-axis.

Tail plots have been applied in Mandelbrot (1963), Bouchaud and Potters (2003), and Sornette (2003).

The Left and the Right Tail Plots The observations in the left tail are

$$\mathcal{L} = \{Y_i : Y_i < u, \; i = 1, \dots, T\},$$

where $u = \hat{q}_p$ is the pth empirical quantile for $0 < p < 1/2$. For the left tail plot we choose the level

$$\text{level}(Y_i) = \#\{Y_j : Y_j \le Y_i, Y_j \in \mathcal{L}\}, \quad Y_i \in \mathcal{L}. \tag{3.32}$$

Thus, the smallest observation has level one, the second smallest observation has level two, and so on. Note that level(Y_i) is often called the rank of Y_i. The left tail plot is the two-dimensional scatter plot of the points $(Y_i, \text{level}(Y_i))$, $Y_i \in \mathcal{L}$, when the logarithmic scale is used for the y-axis.

The observations in the right tail are

$$\mathcal{R} = \{Y_i : Y_i > u, \ i = 1, \dots, T\},$$

where $u = \hat{q}_p$ is the pth empirical quantile for $1/2 < p < 1$. We choose the level of Y_i as the number of observations larger or equal to Y_i:

$$\text{level}(Y_i) = \#\{Y_j : Y_j \ge Y_i, Y_j \in \mathcal{R}\}, \quad Y_i \in \mathcal{R}. \tag{3.33}$$

Thus, the largest observation has level one, the second largest observation has level two, and so on. The right tail plot is the two-dimensional scatter plot of the points $(Y_i, \text{level}(Y_i))$, $Y_i \in \mathcal{R}$, when the logarithmic scale is used for the y-axis.

The left tail plot can be considered as an estimator of the function

$$L(x) = TF(x), \tag{3.34}$$

where F is the underlying distribution function and $x \in (-\infty, u]$. Indeed, for the level in (3.32) we have that level(Y_i) $= T\hat{F}(Y_i)$. The right tail plot can be considered as an estimator of the function

$$R(x) = T(1 - F(x)), \tag{3.35}$$

where $x \in [u, \infty)$. For the level in (3.33) we have that level(Y_i) $\approx T(1 - \hat{F}(Y_i))$.

Figure 3.2 shows the left and right tail plots for the daily S&P 500 data, described in Section 2.4.1. Panel (a) shows the left tail plot and panel (b) shows the right tail plot. The black circles show the data points. The y-axis is logarithmic. The colored curves show the population versions (3.34) and (3.35) for the Gaussian distribution (red) and for the Student distributions with degrees of freedom $v = 3, 4, 5, 6$ (blue).[6] We can see that for the left tail Student's distribution with degrees of freedom $v = 3$ gives the best fit, but for the right tail degrees of freedom $v = 4$ gives the best fit.

---

6 The Gaussian curve in the left tail plot shows the function $x \mapsto T\Phi((x - \hat{\mu})/\hat{\sigma})$, where Φ is the distribution function of the standard Gaussian distribution, $\hat{\mu}$ is the sample mean, and $\hat{\sigma}$ is the sample standard deviation. In the right tail plot the function is $x \mapsto T(1 - \Phi((x - \hat{\mu})/\hat{\sigma}))$. The Student curves in the left tail plot are $x \mapsto TF_v((x - \hat{\mu})/\hat{\sigma})$, where F_v is the distribution function of the Student distribution with degrees of freedom v, and $\hat{\sigma} = \hat{s}/\sqrt{v/(v-2)}$, where $\hat{s}$ is the sample standard deviation. The Student distributions are defined in (3.53). Note that when $Y \sim N(\mu, \sigma^2)$, then the distribution function of Y is $\Phi((x - \mu)/\sigma)$. When $Y \sim t(v, \mu, \sigma^2)$, then the distribution function of Y is $F_v((x - \mu)/\sigma)$.

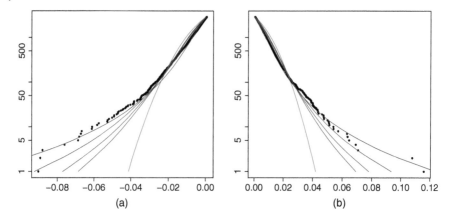

Figure 3.2 *Left and right tail plots.* (a) The left tail plot for S&P 500 returns; (b) the right tail plot. The red curve shows the theoretical Gaussian curve and the blue curves show the Student curves for the degrees of freedom $v = 3$–6.

A left tail plot and a right tail plot can be combined into one figure, at least when both the left and the right tails are defined by taking the threshold to be the sample median $u = \hat{q}_{0.5}$ (see Figures 14.24(a) and 14.25(a)).

Smooth Tail Plots Figure 3.3 shows smooth tail plots for the S&P 500 components data, described in Section 2.4.5. Panel (a) shows left tail plots and panel (b) shows right tail plots. The gray scale image visualizes with one picture all tail plots of the stocks in the S&P 500 components data. The red points show the tail plots of S&P 500 index, which is also shown in Figure 3.2. Note that the x-axes have the ranges $[-0.1, 0]$ and $[0, 0.1]$, so that the extreme observations are not shown. Note that instead of the logarithmic scale of y-values $1, \dots, [T/2]$, we have used values $\log(1), \log(2), \dots, \log([T/2])$ on the y-axis. We can see that the index has lighter tails than most of the individual stocks.

In a smooth tail plot we make an image that simultaneously shows several tail plots. Let us have m stocks and T returns for each stock. We draw a separate left or right tail plot for each stock. Plotting these tail plots in the same figure would cause overlapping, and we would see only a black image. That is why we use smoothing. We divide the x-axis to 300 grid points, say. The y-axis has $[T/2]$ grid points. Thus, we have $300 \times [T/2]$ pixels. For each x-value we compute the value of a univariate kernel density estimator at that x-value. Each kernel estimator is constructed using m observations. This is done for each $[T/2]$ rows, so that we evaluate $[T/2]$ estimates at 300 points. See Section 3.2.2 about kernel density estimation. We choose the smoothing parameter using the normal reference rule and use the standard Gaussian kernel. The values of the density estimate are raised to the power of 21 before applying the gray scale.

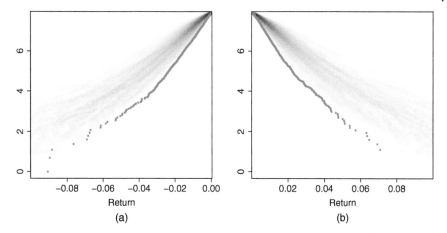

Figure 3.3 *Smooth tail plots.* The gray scale images show smooth tail plots of a collection of stocks in the S&P 500 index. The red points show the tail plots of the S&P 500 index. (a) A smooth left tail plot; (b) a smooth right tail plot.

3.2.1.3 Regression Plots of Tails

Regression plots are related to the empirical distribution function, just like tail plots, but now the data is transformed so that it lies on $[0, \infty)$, both in the case of the left tail and in the case of the right tail. We use the term "regression plot" because these plots suggest fitting linear regression curves to the data. We distinguish the plot for which exponential tails looks linear and the plot for which Pareto tails look linear.

Plots which Look Linear for an Exponential Tail Let the original observations be $Y_1, \ldots, Y_T$. Let $u \in \mathbf{R}$ be a threshold. We choose u to be an empirical quantile $\hat{q}_p$ for some $p \in (0, 1)$: $\hat{q}_p = Y_{(m)}$ for $m = [pT]$, where $Y_{(1)} < \cdots < Y_{(T)}$ are the ordered observations. Let $\mathcal{T}_l$ be the left tail and $\mathcal{T}_r$ be the right tail, transformed so that the observations lie on $[0, \infty)$:

$$\mathcal{T}_l = \{u - Y_i : Y_i \leq u\}, \quad \mathcal{T}_r = \{Y_i - u : Y_i \geq u\}.$$

For the left tail $u = \hat{q}_p$ for $p \in (0, 1/2)$ and for the right tail $u = \hat{q}_p$ for $p \in (1/2, 1)$. Let us denote by $\mathcal{T}$ either the left tail or the right tail. Denote

$$n = \#\mathcal{T}.$$

Let

$$\hat{F}(z) = \frac{1}{n+1} \#\{Z_i \in \mathcal{T} : Z_i \leq z\}$$

be the empirical distribution function, based on data $\mathcal{T}$. Note that in the usual definition of the empirical distribution function we divide by n, but now we divide by $n+1$ because we need that $\hat{F}(Z_i) < 1$, in order to take the logarithm of $1 - \hat{F}(Z_i)$. Denote

$$\mathcal{T} = \{Z_1, \dots, Z_n\}.$$

Assume that the data is ordered:

$$Z_1 < \cdots < Z_n.$$

We have that

$$\hat{F}(Z_i) = \frac{i}{n+1}.$$

The regression plot that is linear for exponential tails is a scatter plot of the points[7]

$$\{(Z_i, \log(1 - \hat{F}(Z_i))) : Z_i \in \mathcal{T}\}. \tag{3.36}$$

Figure 3.4 shows scatter plots of points in (3.36). We use the S&P 500 daily data, described in Section 2.4.1. Panel (a) plots data in the left tail with $p = 10\%$ (black), $p = 5\%$ (red), and $p = 1\%$ (blue). Panel (b) plots data in the right tail with $p = 90\%$ (black), $p = 95\%$ (red), and $p = 99\%$ (blue).

The data looks linear for exponential tails and convex for Pareto tails. The exponential distribution function is $F(x) = 1 - \exp\{-x/\beta\}$ for $x \geq 0$, where $\beta > 0$. The exponential distribution function satisfies

$$\log(1 - F(x)) = -\frac{x}{\beta} I_{(0,\infty)}(x).$$

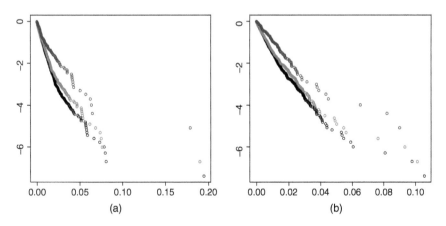

(a) (b)

Figure 3.4 *Regression plots which are linear for exponential tails: S&P 500 daily returns.* (a) Left tail with $p = 10\%$ (black), $p = 5\%$ (red), and $p = 1\%$ (blue); (b) right tail with $p = 90\%$ (black), $p = 95\%$ (red), and $p = 99\%$ (blue).

---

7 Denote $p_i = \hat{F}(Z_i)$ and $q_{p_i} = Z_i$. Then we can write (3.36) as a plot of points $(q_{p_i}, \log(1 - p_i))$. The plot of points $(-\log(1 - p_i), q_{p_i})$ is called the return level plot; see Coles (2004, pp. 49, 81).

Plotting the curve

$$x \mapsto -\frac{x}{\beta} \qquad (3.37)$$

for $x \geq 0$ and for various values of $\beta > 0$ shows how well the exponential distributions fit the tail. The Pareto distribution function for the support $[0, \infty)$ is $F(x) = 1 - (u/(x+u))^{\alpha}$ for $x \geq 0$, where $\alpha > 0$; see (3.74). The Pareto distribution function satisfies

$$\log(1 - F(x)) = -\alpha \log \left(\frac{x+u}{u}\right) I_{(0,\infty)}(x).$$

Plotting the curve

$$x \mapsto -\alpha \log \left(\frac{x+u}{u}\right) \qquad (3.38)$$

for $x \geq 0$ and for various values of $\alpha > 0$ shows how well the Pareto distributions fit the tail.[8]

Figure 3.5 shows how parametric models are fitted to the left tail, defined by the pth empirical quantile with $p = 5\%$. We use the S&P 500 daily data, as described in Section 2.4.1. Panel (a) shows fitting of exponential tails: we show functions (3.37) for three values of parameter β. Panel (a) shows fitting of Pareto tails: we show functions (3.38) for three values of parameter α.

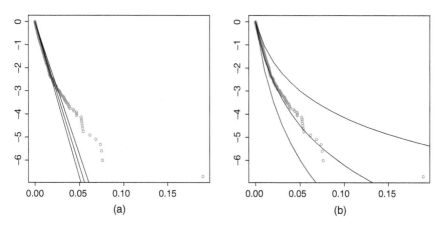

Figure 3.5 *Fitting of parametric families for data that is linear for exponential tails.* The data points are from left tail of S&P 500 daily returns, defined by the pth empirical quantile with $p = 0.05$. (a) Fitting of exponential distributions; (b) fitting of Pareto distributions.

---

8 The generalized Pareto distribution is defined in (3.83). The distribution function F satisfies

$$\log(1 - F(x)) = \begin{cases} -\frac{1}{\xi} \log \left(1 + \frac{\xi x}{\beta}\right) I_{(0,\infty)}(x), & \xi > 0, \\ -\frac{x}{\beta} I_{[0,\infty)}(x), & \xi = 0, \end{cases}$$

where $\beta > 0$.

The middle values of the parameters are the maximum likelihood estimates, defined in Section 3.4.2.

Plots which Look Linear for a Pareto Tail Let

$$T_l = \{Y_i/u : Y_i \le u\}, \quad T_r = \{Y_i/u : Y_i \ge u\}.$$

For the right tail we assume that $u > 0$ and for the left tail we assume that $u < 0$. Let us denote by T either the left tail or the right tail. Denote

$$T = \{Z_1, \dots, Z_n\}.$$

Assume that the data is ordered: $Z_1 < \cdots < Z_n$. The regression plot that is linear for Pareto tails is a scatter plots of the points

$$\left\{ (\log Z_i, \log(1 - \hat{F}(Z_i))) : Z_i \in T \right\}. \tag{3.39}$$

Figure 3.6 shows scatter plots of points in (3.39). We use the S&P 500 daily data, described in Section 2.4.1. Panel (a) plots data in the left tail with $p = 10\%$ (black), $p = 5\%$ (red), and $p = 1\%$ (blue). Panel (b) plots data in the right tail with $p = 90\%$ (black), $p = 95\%$ (red), and $p = 99\%$ (blue).

The data looks linear for Pareto tails and concave for exponential tails. The exponential distribution function for the support $[u, \infty)$ is $F(x) = 1 - \exp\{-(x - u)/\beta\}$ for $x \ge u$, where $\beta > 0$. The exponential distribution function satisfies

$$\log(1 - F(x)) = -\frac{x - u}{\beta} I_{(u,\infty)}(x).$$

Plotting the curve

$$x \mapsto -\frac{|u|}{\beta} (e^x - 1)$$

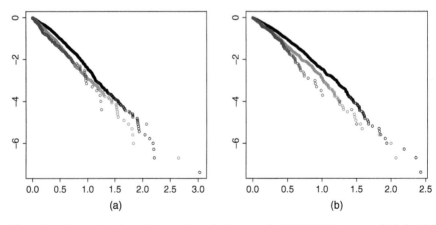

Figure 3.6 *Regression plots which are linear for Pareto tails: S&P 500 daily returns.* (a) Left tail with $p = 10\%$ (black), $p = 5\%$ (red), and $p = 1\%$ (blue); (b) right tail with $p = 90\%$ (black), $p = 95\%$ (red), and $p = 99\%$ (blue).

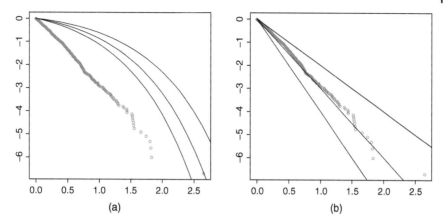

Figure 3.7 *Fitting of parametric families for data that is linear for Pareto tails.* The data points are from left tail of S&P 500 daily returns, defined by the *p*th empirical quantile with $p = 0.05$. (a) Fitting of exponential distributions; (b) fitting of Pareto distributions.

for $x \geq 0$ and for various values of $\beta > 0$ shows how well the exponential distributions fit the tail. The Pareto distribution function for the support $[u, \infty)$ is $F(x) = 1 - (u/x)^\alpha$ for $x \geq u$, where $\alpha > 0$. The Pareto distribution function satisfies

$$\log(1 - F(x)) = -\alpha \log\left(\frac{x}{u}\right) I_{(u,\infty)}(x).$$

Plotting the curve

$$x \mapsto -\alpha x$$

for $x \geq 0$ and for various values of $\alpha > 0$ shows how well the Pareto distributions fit the tail.

Figure 3.7 shows how parametric models are fitted to the left tail, defined by the *p*th empirical quantile with $p = 5\%$. We use the S&P 500 daily data, described in Section 2.4.1. Panel (a) shows fitting of exponential tails: we show functions (3.37) for three values of parameter β. Panel (a) shows fitting of Pareto tails: we show functions (3.38) for three values of parameter α. The middle values of the parameters are the maximum likelihood estimates, defined in Section 3.4.2.

3.2.1.4 The Empirical Quantile Function

The *p*th quantile of the distribution of the random variable $Y \in \mathbf{R}$ is defined in (3.21) as

$$Q_p = \inf\{y : F(y) \geq p\},$$

where $0 < p < 1$ and $F(y) = P(Y \leq y)$ is the distribution function of Y. The empirical quantile can be defined as

$$\hat{Q}_p = \inf\{y : \hat{F}(y) \geq p\},$$

where $\hat{F}$ is the empirical distribution function, as defined in (3.30); see (8.21). Section 8.4.1 contains equivalent definitions of the empirical quantile.

The quantile function is

$$p \mapsto Q_p, \quad p \in (0, 1).$$

For continuous distributions the quantile function is the same as the inverse of the distribution function. The empirical quantile function is

$$p \mapsto \hat{Q}_p, \quad p \in (0, 1), \tag{3.40}$$

where $\hat{Q}_p$ is the empirical quantile. A quantile function can be used to compare return distributions. A first return distribution dominates a second return distribution when the first quantile function takes higher values everywhere than the second quantile function. See Section 9.2.3 about stochastic dominance.

Plotting a graph of the empirical quantile function is close to plotting the points

$$\left(i/T, Y_{(i)}\right), \quad i = 1, \dots, T, \tag{3.41}$$

where $Y_{(1)} < \cdots < Y_{(T)}$ are the ordered observations.

Figure 3.8 shows empirical quantile functions of S&P 500 returns (red) and 10-year bond returns (blue). The monthly data of S&P 500 and US Treasury 10-year bond returns is described in Section 2.4.3. Panel (a) plots the points (3.41) and panel (b) zooms at the lower left corner, showing the empirical quantile on the range $p \in (0, 0.1)$. Neither of the estimated return distributions dominates the other: The S&P 500 returns have a higher median and higher upper quantiles, but they have smaller lower quantiles. That is, S&P 500 is more risky than 10-year bond.

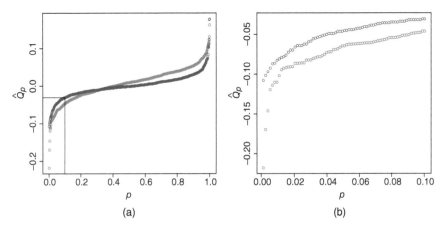

Figure 3.8 *Empirical quantile functions.* (a) Empirical quantile functions of S&P 500 returns (red) and 10-year bond returns (blue); (b) zooming to the lower left corner.

3.2.2 Density Estimation Based Tools

We describe both histograms and kernel density estimators.

3.2.2.1 The Histogram

A histogram estimator of the density of $X \in \mathbf{R}^d$, based on identically distributed observations $X_1, \ldots, X_T$, is defined as

$$\hat{f}(x) = \sum_{i=1}^{M} \frac{n_i/T}{\text{volume}(R_i)} I_{R_i}(x), \quad x \in \mathbf{R}^d, \tag{3.42}$$

where $\{R_1, \ldots, R_M\}$ is a partition on $\mathbf{R}^d$ and

$$n_i = \#\{i : X_i \in R, \quad i = 1, \ldots, T\}$$

is the number of observations in R_i. The partition is a collection of sets $R_1, \ldots, R_M$ that are (almost surely) disjoint and they cover the space of the observed values $X_1, \ldots, X_T$.[9]

Figure 3.9(a) shows a histogram estimate using S&P 500 returns. We use the S&P 500 monthly data, described in Section 2.4.3. The histogram is

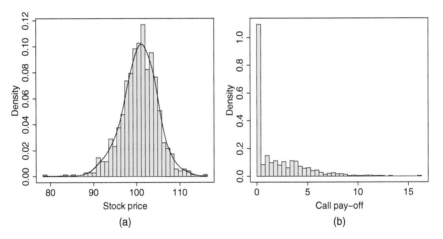

Figure 3.9 *Histogram estimates.* (a) A histogram of historically simulated S&P 500 prices. A graph of kernel density estimate is included. (b) A histogram of historically simulated call option pay-offs.

---

9 In the univariate case the partition to the intervals of equal length can be defined by

$$R_i = [a_i, b_i], \quad a_i = X_{(1)} + \delta(i-1), \quad b_i = a_i + \delta, \quad \delta = (X_{(T)} - X_{(1)})/M,$$

where $X_{(1)} = \min\{X_i\}$ and $X_{(T)} = \max\{X_i\}$. Then the histogram can be written as

$$\hat{f}(x) = \frac{1}{T\delta} \sum_{i=1}^{M} n_i \, I_{R_i}(x), \quad x \in \mathbf{R}.$$

constructed from the data $100 \times R_t$, $t = 1, \ldots, T$, where R_t are the monthly gross returns. Panel (b) shows a histogram constructed from the historically simulated pay-offs of the call option with the strike price 100. The histogram is constructed from the data $\max\{100R_t - 100, 0\}$, $t = 1, \ldots, T$. Panel (a) includes a graph of a kernel density estimate, defined in (3.43). The histogram in panel (b) illustrates that a histogram is convenient to visualize the density of data that is not from a continuous distribution; for this data the value 0 has a probability about 0.5.

3.2.2.2 The Kernel Density Estimator

The kernel density estimator $\hat{f}(x)$ of the density function $f : \mathbf{R}^d \to \mathbf{R}$ of random vector $X \in \mathbf{R}^d$, based on identically distributed data $X_1, \ldots, X_T \in \mathbf{R}^d$, is defined by

$$\hat{f}(x) = \frac{1}{T} \sum_{i=1}^{T} K_h(x - X_i), \quad x \in \mathbf{R}^d, \tag{3.43}$$

where $K : \mathbf{R}^d \to \mathbf{R}$ is the kernel function, $K_h(x) = K(x/h)/h^d$, and $h > 0$ is the smoothing parameter.[10]

We can also take the vector smoothing parameter $h = (h_1, \ldots, h_d)$ and $K_h(x) = K(x_1/h_1, \ldots, x_d/h_d)/\prod_{i=1}^{d} h_i$. The smoothing parameter of the kernel density estimator can be chosen using the normal reference rule:

$$h_i = \left(\frac{4}{d+2} \right)^{1/(d+4)} T^{-1/(d+4)} \hat{\sigma}_i, \tag{3.44}$$

for $i = 1, \ldots, d$, where $\hat{\sigma}_i$ is the sample standard deviation for the ith variable; see Silverman (1986, p. 45). Alternatively, the sample variances of the marginal distributions can be normalized to one, so that $\hat{\sigma}_1 = \cdots = \hat{\sigma}_d = 1$.

Figure 3.10(a) shows kernel estimates of the distribution of S&P 500 monthly net returns (blue) and of the distribution of US 10-year bond monthly net returns (red). The data set of monthly returns of S&P 500 and US 10-year bond is described in Section 2.4.3. Panel (b) shows kernel density estimates of S&P

---

10 The definition of the kernel density estimator can be motivated in the following way. The density a point $x \in \mathbf{R}^d$ can be approximated by

$$f(x) \approx \frac{P(B_h(x))}{\lambda(B_h(x))},$$

where $B_h(x) = \{y \in \mathbf{R}^d : \|x - y\| \le h\}$, $h > 0$ is small, and $\lambda(B_h(x))$ is the Lebesgue measure of $B_h(x)$. We have that

$$\frac{P(B_h(x))}{\lambda(B_h(x))} \approx \frac{1}{\lambda(B_h(x))} \frac{1}{T} \sum_{i=1}^{T} I_{B_h(x)}(X_i) = \frac{1}{T} \sum_{i=1}^{T} K_h(x - X_i),$$

when $K(x) = I_{B_1(0)}(x)$. We arrive into (3.43) by allowing other kernel functions than only the indicator function.

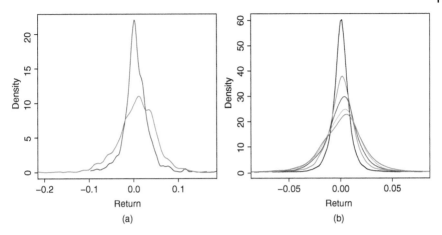

Figure 3.10 *Kernel density estimates of distributions of asset returns.* (a) Estimates of the distribution of S&P 500 monthly returns (blue) and of US 10-year bond monthly returns (red); (b) estimates of S&P 500 net returns with periods of 1–5 trading days (colors black–green).

500 net returns with periods of 1–5 trading days (colors black–green). We use S&P 500 daily data of Section 2.4.1 to construct returns for the different horizons.

3.3 Univariate Parametric Models

We describe normal and log-normal distributions, Student distributions, infinitely divisible distributions, Pareto distributions, and models that interpolate between exponential and polynomial tails. We consider also the estimation of the parameters, in particular, the estimation of the tail index.

3.3.1 The Normal and Log-normal Models

After defining the normal and log-normal distributions, we discuss how the central limit theorem can be used to justify that these distributions can be used to model stock prices.

3.3.1.1 The Normal and Log-normal Distributions

A univariate normal distribution can be parameterized with the expectation $\mu \in \mathbf{R}$ and the standard deviation $\sigma > 0$. When X is a random variable with a normal distribution we write

$$X \sim N(\mu, \sigma^2).$$

The density of the normal distribution $N(\mu, \sigma^2)$ is

$$f(x) = \frac{1}{\sigma\sqrt{2\pi}} \exp\left\{ \frac{(x - \mu)^2}{2\sigma^2} \right\},$$

where $x \in \mathbf{R}$. The parameters μ and σ can be estimated by the sample mean and sample standard deviation.

When $\log X \sim N(\mu, \sigma^2)$, then it is said that X has a log-normal distribution, and we write

$$X \sim \text{lognorm}(\mu, \sigma^2).$$

The density function of a log-normal distribution is

$$f(x) = \frac{1}{x\sigma\sqrt{2\pi}} \exp\left\{-\frac{(\log x - \mu)^2}{2\sigma^2}\right\}, \qquad (3.45)$$

where $x > 0$. Thus, log-normally distributed random variables are positive (almost surely). The expectation of a log-normally distributed random variable X is

$$EX = e^{\mu + \sigma^2/2}.$$

For $k \geq 1$, $EX^k = e^{k\mu + k^2\sigma^2/2}$. Given observations $X_1, \ldots, X_n$ from a log-normal distribution, the parameters μ and σ can be estimated using the sample mean and sample standard deviation computed from the observations $\log X_1, \ldots, \log X_n$.

Note that a linear combination of log-normal variables is not log-normally distributed, but a product of log-normally distributed random variables is log-normally distributed, because a linear combination of normal variables is normally distributed.

3.3.1.2 Modeling Stock Prices

We can justify heuristically the normal distribution for the differences of stock prices using the central limit theorem. The central limit theorem can also be used to justify the log-normal model for the gross returns (which amounts to a normal model for the logarithmic returns). Let us consider time interval $[0, T]$ and let $t_i = iT/n$ for $i = 0, \ldots, n$, so that $S_{t_0}, \ldots, S_{t_n}$ is an equally spaced sample of stock prices, where $t_0 = 0$ and $t_n = T$. The time interval between the sampled prices is $\Delta t = t_{i+1} - t_i = T/n$.

1) *Normal model.* We may write the price at time t_i, $1, \ldots, n$, as

$$S_{t_i} = S_0 + \sum_{j=0}^{i-1}(S_{t_{j+1}} - S_{t_j}). \qquad (3.46)$$

If the price increments $S_{t_{j+1}} - S_{t_j}$ are i.i.d. with expectation m and variance s^2, then an application of the central limit theorem gives the approximation[11]

$$S_{t_i} - S_0 \sim N(t_i\mu, t_i\sigma^2), \qquad (3.47)$$

---

11 We have approximately that $i^{-1/2}(S_{t_i} - S_0 - im)/s \sim N(0, 1)$. Thus, approximately $S_{t_i} - S_0 \sim N(im, is^2)$. We can write $im = t_i m/\Delta t$ and $is^2 = t_i s^2/\Delta t$.

where $\mu = m/\Delta t$, and $\sigma^2 = s^2/\Delta t$. Equation (3.47) defines the Gaussian model for the asset prices. Under the normal model we have

$$S_{t_i} = S_0 + t_i \mu + \sqrt{t_i}\, \sigma Z, \tag{3.48}$$

where $Z \sim N(0,1)$ is a random variable that has the standard normal distribution.

2) *Log-normal model.* We may write the asset price at time t_i, $i = 1, \ldots, n$, as

$$S_{t_i} = S_0 \cdot \prod_{j=0}^{i-1} \frac{S_{t_{j+1}}}{S_{t_j}} = S_0 \cdot \exp \left\{ \sum_{j=0}^{i-1} \log \left(\frac{S_{t_{j+1}}}{S_{t_j}} \right) \right\}. \tag{3.49}$$

If $\log(S_{t_{j+1}}/S_{t_j})$ are i.i.d. with expectation m and variance s^2, then an application of the central limit theorem gives the approximation[12]

$$\log \frac{S_{t_i}}{S_0} \sim N(t_i \mu, t_i \sigma^2), \tag{3.50}$$

where

$$\mu = m/\Delta t, \quad \sigma^2 = s^2/\Delta t. \tag{3.51}$$

This is equivalent to saying that S_{t_i} is log-normally distributed with parameters $t_i \mu + \log S_0$ and $\sqrt{t_i}\sigma$:

$$\frac{S_{t_i}}{S_0} \sim \text{lognorm} \left(t_i \mu, t_i \sigma^2 \right).$$

Equation (3.50) defines the log-normal model for the asset prices. Under the log-normal model we have

$$S_{t_i} = S_0 \exp \left\{ t_i \mu + \sqrt{t_i}\, \sigma Z \right\}, \tag{3.52}$$

where $Z \sim N(0,1)$ is a random variably that has the standard normal distribution.

Parameter μ in (3.51) is called the annualized mean of the logarithmic returns and parameter σ is called the annualized volatility. For the daily data $1/\Delta t = 250$ and for the monthly data $1/\Delta t = 12$, when we take $T = 1$.

Figure 3.11 shows estimates of the densities of stock price S_T using the data of S&P 500 daily prices, described in Section 2.4.1. In panel (a) $T = 20/250$, which equals 20 trading days, and in panel (b) $T = 2$ years. The normal density is shown with black and the log-normal density is shown with red. We take $S_0 = 100$, and for the purpose of fitting a normal distribution for the price increments we change the price data to $\tilde{S}_{t_i} = 100 \times S_{t_i}/S_{t_{i-1}}$. For the normal model the

---

12 We have approximately that $i^{-1/2}(\log S_{t_i} - \log S_0 - im)/s \sim N(0,1)$. Thus, approximately $\log S_{t_i} - \log S_0 \sim N(im, is^2)$. We can write $im = t_i m/\Delta t$ and $is^2 = t_i s^2/\Delta t$.

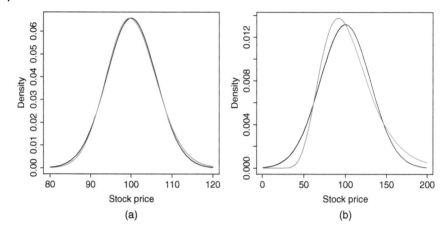

Figure 3.11 *Normal and log-normal densities.* Shown are a normal density (black) and a log-normal density (red) of the distribution of the stock price S_T, when $S_0 = 100$. In panel (a) $T = 20/250$, which equals 20 trading days, and in panel (b) $T = 2$ years.

estimate $\hat{m}_1$ is the sample mean and $\hat{s}_1$ is the sample standard deviation of the daily increments. Then we arrive at the distribution

$$S_T \sim N \left(S_0 + \frac{T\hat{m}_1}{\Delta t}, \frac{T\hat{s}_1^2}{\Delta t} \right),$$

where $\Delta t = 1/250$. For the log-normal model the estimate $\hat{m}_2$ is the sample mean and $\hat{s}_2$ is the sample standard deviation of the logarithmic daily returns. Then we arrive at the distribution

$$S_T \sim \text{lognorm} \left(\log S_0 + \frac{T\hat{m}_2}{\Delta t}, \frac{T\hat{s}_2^2}{\Delta t} \right).$$

The log-normal density is skewed to the left and the right tail is heavier than the left tail. The normal density is symmetric with respect to the mean.

Log-normally distributed random variables take only positive values, but normal random variables can take negative values. Note, however, that the tail of the normal distribution is so thin that the probability of negative values can be very small. Thus, the positivity of log-normal distributions is not a strong argument in favor of their use to model prices.

The Gaussian model for the increments of the stock prices was used by Bachelier (1900). The continuous time limit of the log-normal model is the Black–Scholes model, that is used in option pricing. The log-normal model is applied in (14.49) to derive a price for options. A log-normal distribution allows for greater upside price movements than downside price movements. This leads to the fact that in the Black–Scholes model 105 call has more value than 95 put when the stock is at 100. See Figure 14.4 for the illustration of the asymmetry.

3.3.2 The Student Distributions

The density of the standard Student distribution with degrees of freedom $v > 0$ is given by

$$f(x) = c \left(1 + \frac{x^2}{v}\right)^{-(v+1)/2}, \tag{3.53}$$

for $x \in \mathbf{R}$, where the normalization constant is equal to

$$c = \frac{\Gamma((v+1)/2)}{(v\pi)^{1/2}\Gamma(v/2)},$$

and the gamma function is defined by $\Gamma(u) = \int_0^\infty x^{u-1}e^{-x}\, dx$ for $u > 0$. When X follows the Student distribution with degrees of freedom v, then we write

$$X \sim t(v).$$

3.3.2.1 Properties of Student Distributions

Let $X \sim t(v)$. If $v > 1$ then $E|X| < \infty$ and $EX = 0$. If $v > 2$, then

$$\mathrm{Var}(X) = \frac{v}{v-2}. \tag{3.54}$$

We have that $E|X|^k < \infty$ only when $0 < k < v$. In fact, a Student density has tails

$$f(x) \asymp |x|^{-1-v}, \tag{3.55}$$

as $|x| \to \infty$.[13] Thus, Student densities have Pareto tails, as defined in Section 3.4.

We can consider three-parameter location-scale Student families. When $X \sim t(v)$, then $Y = \mu + \sigma X$ follows a location-scale Student distribution, and we write[14]

$$Y \sim t(v, \mu, \sigma^2).$$

Note that for $v > 1$, $EY = \mu$ but σ^2 is not the variance of Y. Instead,

$$\mathrm{Var}(Y) = \frac{v}{v-2}\, \sigma^2,$$

due to (3.54).[15]

---

13 Notation $a_x \asymp b_x$ means that $0 < \liminf_{x\to\infty}(a_x/b_x) \le \limsup_{x\to\infty}(a_x/b_x) < \infty$.
14 Random variable Y has the density $f((x - \mu)/\sigma)/\sigma$, where f is the density of $t(v)$ distribution. The distribution function is $F((x - \mu)/\sigma)$, where F is the distribution function of $t(v)$ distribution.
15 Thus, an estimate of σ^2 is

$$\hat{\sigma}^2 = \frac{v-2}{v}\, s^2,$$

where s^2 is the sample variance, when we assume that v is known. Analogously, in simulations we have to note that when $Y \sim t(v)$, then

$$\mu + \frac{\sigma Y}{\sqrt{v/(v-2)}}$$

has mean μ and variance σ^2.

When $v \to \infty$, then the Student density approaches the Gaussian density. Indeed, $(1 + t^2/v)^{-(v+1)/2} \to \exp\{-t^2/2\}$, as $v \to \infty$, since $(1 + a/v)^v \to e^a$, when $v \to \infty$.

A student distributed random variable $X \sim t(v)$ can be written as

$$X = \frac{Z}{\sqrt{W/v}},$$

where $Z \sim N(0, 1)$, and W has χ^2-distribution with degrees of freedom v. Thus, Student distributions belong to the family of normal variance mixture distributions (scale-mixtures of normal distribution), as defined in Section 4.3.3.

3.3.2.2 Estimation of the Parameters of a Student Distribution

Let us observe $Y_1, \dots, Y_T$ from a Student distribution $t(v, \mu, \sigma^2)$ with the density function $f(x; v, \mu, \sigma)$. The maximum likelihood estimates are maximizers of the likelihood over $v > 0$, $\mu \in \mathbb{R}$, and $\sigma > 0$. Equivalently, we can minimize the negative log-likelihood. Assuming the independence of the observations, the negative log-likelihood is equal to

$$l(v, \mu, \sigma) = - \sum_{i=1}^{T} \log f(Y_i; v, \mu, \sigma).$$

We apply the restricted maximum likelihood estimator that minimizes

$$l(v, \hat{\mu}, \sigma) \tag{3.56}$$

over $v > 0$ and $\sigma > 0$, where $\hat{\mu}$ is the sample mean.

Figure 3.12 studies how the return horizon affects the maximum likelihood estimates for the Student family. We consider the data of daily S&P 500 returns, described in Section 2.4.1. The data is used to consider return horizons up to 40 days. Panel (a) shows the estimates of parameter v as a function of return horizon in trading days. Panel (b) shows the estimates of σ as a function of the return horizon. We see that the estimates are larger for the longer return horizons but there is fluctuation in the estimates.

Figure 3.13 shows the estimates of the degrees of freedom and the scale parameter for each series of daily returns in the S&P 500 components data, described in Section 2.4.5. We get an individual estimate of v and σ for each stock. Panel (a) shows a kernel density estimate and a histogram estimate of the distribution of $\hat{v}$. Panel (b) shows the estimates of the distribution of $\hat{\sigma}$.[16] The maximizers of the kernel estimates (modes) are indicated by the blue lines. The most stocks has $\hat{v} \approx 3.5$, but the estimates vary as $\hat{v} \in [1.5, 5]$.

---

16 The smoothing parameter is chosen using the normal reference rule, and the kernel function is the standard normal density. In fact, we show the values $\sqrt{250} \times 100 \times \hat{\sigma}$.

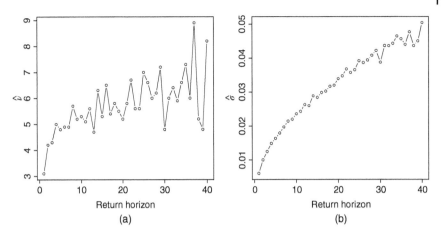

Figure 3.12 *Parameter estimates for various return horizons.* The maximum likelihood estimates of (a) v and (b) σ as a function of the return horizon in trading days.

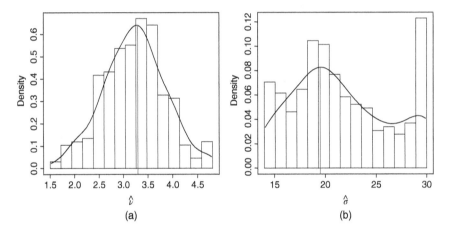

Figure 3.13 *Distribution of estimates $\hat{v}$ and $\hat{\sigma}$.* (a) A kernel density estimate and a histogram of the distribution of $\hat{v}$; (b) the estimates of the distribution of $\hat{\sigma}$. The maximizers of the kernel estimates are indicated by the blue lines.

3.4 Tail Modeling

The normal, log-normal, and Student distributions provide models for the complete return distribution. These models assume that the return distribution is approximately symmetric. We consider an approach where the left tail, the right tail, and the central area are modeled and estimated separately. There are at least two advantages with this approach:

1) We may better estimate distributions whose left tail is different from the right tail. For example, it is possible that the distribution of losses is different from the distribution of gains.
2) We may apply different estimation methods for different parts of the distribution. For example, we may apply nonparametric methods for the estimation of the central part of the distribution and parametric methods for the estimation of the tails.

In risk management, we are mainly interested in the estimation of the left tail (the probability of losses). In portfolio selection, we might be interested in the complete distribution.

A semiparametric approach for the estimation of the complete return distribution estimates the left and the right tails of the distribution using a parametric model, but the central region of the distribution is estimated using a kernel estimator, or some other nonparametric density estimator. It is a nontrivial problem to make a good division of the support of the distribution into the area of the left tail, into the area of the right tail, and into the central area.

3.4.1 Modeling and Estimating Excess Distributions

We model the left and the right tails of a return distribution parametrically. The estimation of the parameters can be done using maximum likelihood, or by a regression method, for example.

3.4.1.1 Modeling Excess Distributions

Let $f(x; \theta)$ be a parameterized family of density functions whose support is $[0, \infty)$. This family will be used to model the tails of the density $g : \mathbf{R} \to \mathbf{R}$ of the returns.

To estimate the right tail, we assume that the density function $g : \mathbf{R} \to \mathbf{R}$ of the returns satisfies

$$g(x)I_{[u,\infty)}(x) = (1 - p)f(x - u; \theta) \tag{3.57}$$

for some θ, where u is the pth quantile of the return density: $1 - p = \int_u^\infty g$, and the probability p satisfies $0.5 < p < 1$.[17] To estimate the left tail we assume that the density function $g : \mathbf{R} \to \mathbf{R}$ of the returns satisfies

$$g(x)I_{(-\infty,u]}(x) = pf(u - x; \theta) \tag{3.58}$$

for some θ, where u is the pth quantile of the return density: $p = \int_{-\infty}^u g$, and $0 < p < 0.5$.

The assumptions can be expressed using the concept of the excess distribution with threshold $u > 0$. Let $G : \mathbf{R} \to \mathbf{R}$ be the distribution function of the

---

17 The indicator function $I_A : \mathbf{R} \to \mathbf{R}$ is defined for any $A \subset \mathbf{R}$ by $I_A(x) = 1$, when $x \in A$, and $I_A(x) = 0$, when $x \notin A$.

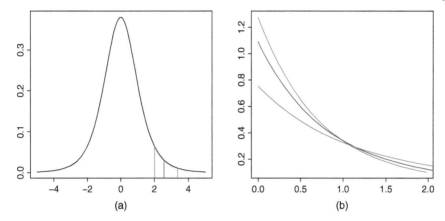

Figure 3.14 *Excess distributions.* (a) The density function of *t*-distribution with degrees of freedom five. The green, blue, and red vectors indicate the location of quantiles q_p for $p = 0.95, p = 0.99$, and $p = 0.999$. (b) The right excess distributions for $u = q_p$.

returns and let $g : \mathbf{R} \to \mathbf{R}$ be the density function of the returns. Let X be the return. Now $P(X \leq x) = G(x)$. The distribution function of the excess distribution with threshold u is

$$G_u(x) = P(X - u \leq x \,|\, X > u) = \frac{G(x + u) - G(u)}{1 - G(u)}. \tag{3.59}$$

The density function of the excess distribution with threshold u is

$$g_u(x) = \frac{g(x + u)}{1 - G(u)} \, I_{[0,\infty)}(x). \tag{3.60}$$

Thus, the assumption in (3.57) says that

$$g_u(x) = f(x; \theta)$$

for some θ. Limit theorems for threshold exceedances are discussed in Section 3.5.2.

Figure 3.14 illustrates the definition of an excess distribution. Panel (a) shows the density function of *t*-distribution with degrees of freedom five. The green, blue, and red vectors indicate the location of quantiles q_p for $p = 0.95$, $p = 0.99$, and $p = 0.999$. Panel (b) shows the right excess distributions for $u = q_p$. The choice of the threshold u affects the goodness-of-fit, and this issue will be addressed in the following sections.

3.4.1.2 Estimation
Estimation is done by first identifying the data coming from the left tail, and the data coming from the right tail. Second, the data is transformed onto $[0, \infty)$. Third, we can apply any method of fitting parametric models.

Identifying the Data in the Tails We choose threshold u of the excess distribution to be an estimate of the pth quantile. For the estimation of the left tail we need to estimate the pth quantile for $0 < p < 0.5$, and for the estimation of the right tail we need to estimate the pth quantile for $0.5 < p < 1$. The data in the left tail and the right tail are

$$\mathcal{L} = \{Y_i : Y_i \leq u\}, \quad \mathcal{R} = \{Y_i : Y_i \geq u\}, \tag{3.61}$$

where u are estimates of a lower and an upper quantile, respectively. We use the empirical quantile to estimate the population quantile. Let $Y_1, \dots, Y_T$ be the sample from the distribution of the returns, and let $Y_{(1)} \leq \cdots \leq Y_{(T)}$ be the ordered sample. The empirical quantile is

$$\hat{q}_p = Y_{([pT])},$$

where $[x]$ is the integer part of $x \in \mathbf{R}$. See Section 3.1.3 and Chapter 8 for more information about quantile estimation. Now the data in the left tail and the right tail can be written as

$$\mathcal{L} = \{Y_{(1)}, \dots, Y_{([pT])}\}, \quad \mathcal{R} = \{Y_{([pT])}, \dots, Y_{(T)}\}. \tag{3.62}$$

The Basic Principle of Fitting Tail Models Assume that we have an estimation procedure for the estimation of the parameter θ of the family $f(\cdot; \theta)$, $\theta \in \Theta$. The family consists of densities whose support is $[0, \infty)$, and it is used to model the left or the right part of the density, as written in assumptions (3.58) and (3.57). We need a procedure for the estimation of the parameter $\theta = \theta_{left}$ in model (3.58), or the parameter $\theta = \theta_{right}$ in model (3.57). We apply the estimation procedure for estimating θ using data

$$\{u - Y_i : Y_i \leq u\}, \quad \{Y_i - u : Y_i \geq u\}.$$

Maximum Likelihood in Tail Estimation We use the method of maximum likelihood for the estimation of the tails under the assumptions (3.57) and (3.58). We write the likelihood function under the assumption of independent and identically distributed observations, but we apply the maximum likelihood estimator for time series data. Thus, the method may be called pseudo maximum likelihood. Time series properties will be taken into account in Chapter 8, where quantile estimation is studied using tail modeling. The likelihood is maximized separately using the data in the left tail and in the right tail.

The family $f(\cdot, \theta)$, $\theta \in \Theta$, models the excess distribution. The maximum likelihood estimator for the parameter of the left tail is

$$\hat{\theta}_{left} = \underset{\theta}{\text{argmax}} \prod_{Y_i \in \mathcal{L}} f(u - Y_i; \theta), \tag{3.63}$$

where $u = \hat{q}_p$ for $0 < p < 0.5$ and $f(\cdot, \theta)$ has support $[0, \infty)$. The maximum likelihood estimator for the parameter of the right tail is

$$\hat{\theta}_{right} = \underset{\theta}{\text{argmax}} \prod_{Y_i \in \mathcal{R}} f(Y_i - u; \theta), \tag{3.64}$$

where $u = \hat{q}_p$ for $0.5 < p < 1$.

3.4.2 Parametric Families for Excess Distributions

We describe the following one- and two-parameter families:

1) *One-parameter families.* The exponential and Pareto distributions.
2) *Two-parameter families.* The gamma, generalized Pareto, and Weibull distributions.

Furthermore, we describe a three parameter family which contains many one- and two-parameter families as special cases.

The exponential distributions have a heavier tail than the normal distributions. The Pareto distributions have a heavier tail than the exponential distributions, but an equally heavy tail as the Student distributions. The Pareto densities have polynomial tails, the exponential densities have exponential tails, and the gamma densities have densities whose heaviness is between the Pareto and the exponential densities.

3.4.2.1 The Exponential Distributions

The exponential densities are defined as

$$f(x) = \frac{1}{\beta} \exp\left\{-\frac{x}{\beta}\right\} I_{[0,\infty)}(x), \tag{3.65}$$

where $\beta > 0$ is the scale parameter. The parameter $\lambda = 1/\beta > 0$ is called the rate parameter. The distribution function and the quantile function are

$$F(x) = \left(1 - \exp\left\{-\frac{x}{\beta}\right\}\right) I_{[0,\infty)}(x), \quad F^{-1}(p) = -\beta \log(1-p) I_{[0,1)}(p).$$

The expectation and the variance are

$$EX = \beta, \quad \text{Var}(X) = \beta^2, \tag{3.66}$$

where X is a random variable following the exponential distribution.

Maximum Likelihood Estimation: Exponential Distribution When we observe $Y_1, \ldots, Y_T$, which are i.i.d. with exponential distribution, then the maximum likelihood estimator is[18]

$$\hat{\beta} = \frac{1}{T} \sum_{i=1}^{T} Y_i. \tag{3.67}$$

---

18 The likelihood function is

$$L(\beta) = \beta^{-T} \exp\left\{-\frac{1}{\beta} \sum_{i=1}^{T} Y_i\right\}.$$

The logarithmic likelihood is

$$\log L(\beta) = -T \log \beta - \frac{1}{\beta} \sum_{i=1}^{T} Y_i.$$

Putting the derivative equal to zero and solving the equation gives the maximum likelihood estimator.

Regression Method: Exponential Distribution Regression plots were shown in Figures 3.4 and 3.5. We study further the regression method for fitting an exponential distribution.

For exponential distributions the logarithm of the survival function $1 - F$ is a linear function, which can be used to visualize data and to estimate the parameter of the exponential distribution (see Section 3.2.1). Let $Y_1, \ldots, Y_T$ be a sample from an exponential distribution and assume $Y_1 < \cdots < Y_T$. Let $\hat{F}$ be the empirical distribution function, based on the observations $Y_1, \ldots, Y_T$, defined as $\hat{F}(x) = \#\{Y_i \leq x\}/(T + 1)$. The empirical distribution function is defined in (3.30), but we modify the definition so that the divisor is $T + 1$ instead of T. We use the facts that (for the ordered data)

$$1 - \hat{F}(Y_i) = 1 - i/(T + 1), \quad 1 - F(Y_i) = e^{-Y_i/\beta}.$$

Thus,

$$-Y_i \approx \beta \log(1 - i/(T + 1)).$$

The least squares estimator of β is[19]

$$\hat{\beta} = -\frac{\sum_{i=1}^T Y_i \log(1 - i/(T+1))}{\sum_{i=1}^T (\log(1 - i/(T+1)))^2}. \tag{3.69}$$

Now we can write

$$\hat{\beta} = \sum_{i=1}^T w_i Y_i,$$

where

$$w_i = \frac{\log((T+1)/(T+1-i))}{\sum_{i=1}^T (\log((T+1)/(T+1-i)))^2}. \tag{3.70}$$

Thus, more weight is given to the observations in the extreme tails.[20]

Figure 3.15 shows the fitting of regression estimates for the S&P 500 daily returns, described in Section 2.4.1. Panel (a) considers the left tail and panel (b) the right tail. The tails are defined by the pth empirical quantiles for

---

[19] In the regression model $Y_i = \beta X_i + \epsilon_i$, $i = 1, \ldots, T$, the least squares estimator of β is

$$\hat{\beta} = \frac{\sum_{i=1}^T Y_i X_i}{\sum_{i=1}^T X_i^2}. \tag{3.68}$$

[20] Note that, unlike in the case of the maximum likelihood estimator, we do not obtain an estimator for the rate parameter $\lambda = 1/\beta$ by $1/\hat{\beta}$. Instead, an estimator for $\lambda = 1/\beta$ follows from $\log(1 - i/(T + 1)) \approx -\lambda Y_i$. Thus, the least squares estimator of λ is

$$\hat{\lambda} = -\frac{\sum_{i=1}^n Y_i \log(1 - i/(T+1))}{\sum_{i=1}^T Y_i^2}. \tag{3.71}$$

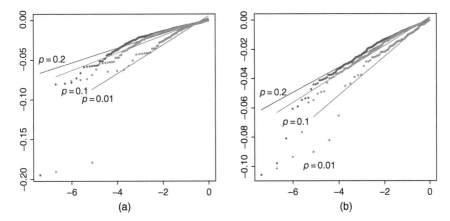

Figure 3.15 *Exponential model for S&P 500 daily returns: Regression fits.* Panel (a) considers the left tail and panel (b) the right tail. We show the regression data and the fitted regression lines for $p = 10\%/90\%$ (blue), $5\%/95\%$ (green), and $1\%/99\%$ (red).

$p = 10\%/90\%$ (blue), $5\%/95\%$ (green), and $1\%/99\%$ (red). We also show the fitted linear regression lines.

3.4.2.2 The Pareto Distributions

We define first the class of Pareto distributions with the support $[u, \infty)$, where $u > 0$. The class of Pareto distributions with support $[0, \infty)$ is obtained by translation.

The Pareto distributions are parameterized by the tail index $\alpha > 0$. Parameter $u > 0$ is taken to be known, but in the practice of tail estimation u is used to define the tail area and u chosen by a quantile estimator. The density function is

$$f(x) = \frac{\alpha}{u}\left(\frac{x}{u}\right)^{-1-\alpha} I_{[u,\infty)}(x), \tag{3.72}$$

where $\alpha > 0$ is the tail index. The distribution function and the quantile function are

$$F(x) = \left[1 - \left(\frac{u}{x}\right)^{\alpha}\right] I_{[u,\infty)}(x), \quad F^{-1}(p) = u(1-p)^{-1/\alpha} I_{[0,1)}(p). \tag{3.73}$$

Pareto Distributions as Excess Distributions Assumption (3.57) says that the excess distribution is modeled with a parametric distribution whose support is $[0, \infty)$. The density function of a Pareto distribution can be moved by the translation $f(x) \mapsto f(x + u)$ to have the support $[0, \infty)$, which gives the density function[21]

$$f(x) = \frac{\alpha}{u}\left(\frac{x+u}{u}\right)^{-1-\alpha} I_{[0,\infty)}(x). \tag{3.74}$$

---

21 The distribution function and the quantile function are

$$F(x) = \left[1 - \left(\frac{u}{x+u}\right)^{\alpha}\right] I_{[0,\infty)}(x), \quad F^{-1}(p) = [u(1-p)^{-1/\alpha} - 1] I_{[0,1)}(p).$$

Now we could consider $u > 0$ as the scaling parameter, which leads to the two-parameter Pareto distributions, which are called the generalized Pareto distributions, and defined in (3.82) and (3.84).

Maximum Likelihood Estimation: Pareto Distribution When Y follows the Pareto distribution with parameters $\alpha > 0$ and $u > 0$, then $X = \log(Y/u)$ follows the exponential distribution with scale parameter $\beta = 1/\alpha$. Indeed, $P(Y > x) = (x/u)^{-\alpha}$ and thus $P(X > x) = P(Y > ue^x) = e^{-\alpha x}$. We observed in (3.67) that scale parameter β of the exponential distribution can be estimated with $\hat{\beta} = T^{-1} \sum_{i=1}^{T} X_i$. Thus, the maximum likelihood estimator of $1/\alpha$ is

$$\widehat{1/\alpha} = 1/\hat{\alpha} = \frac{1}{T} \sum_{i=1}^{T} \log(Y_i/u).$$

The maximum likelihood estimator of the shape parameter α of the Pareto distribution is[22]

$$\hat{\alpha} = \left(\frac{1}{T} \sum_{i=1}^{T} \log(Y_i/u) \right)^{-1}. \tag{3.75}$$

We are more interested in estimating $1/\alpha$, since it appears in the quantile function.

Regression Method: Pareto Distribution Regression plots were shown in Figures 3.6 and 3.7. We study further the regression method for fitting a Pareto distribution.

Let us consider the estimation of the tail index $\alpha > 0$ and the inverse $1/\alpha$. The basic idea is that the logarithm of the distribution function F or the logarithm of the survival function $1 - F$ are linear in α: From (3.78) we get that $\log(1 - F(x)) = \log L(x) - \alpha \log x$, and from (3.79) we get that $\log F(-x) = \log L(x) - \alpha \log x$.

Let $Y_1, \dots, Y_T$ be a sample from a Pareto distribution and assume

$$Y_1 < \cdots < Y_T.$$

---

22 The likelihood function is

$$L(\alpha) = \prod_{i=1}^{T} \frac{\alpha u^{\alpha}}{Y_i^{\alpha+1}},$$

where it is assumed that $Y_1, \dots, Y_T$ are i.i.d. Pareto distributed random variables. Taking logarithms leads to

$$\log L(\alpha) = T \log(\alpha) + T\alpha \log u - (\alpha + 1) \sum_{i=1}^{T} \log Y_i.$$

Differentiating with respect to α and setting the derivative equal to zero gives the maximum likelihood estimator.

Let $\hat{F}$ be the empirical distribution function, based on the observations $Y_1, \ldots, Y_T$, defined as $\hat{F}(x) = \#\{Y_i \leq x\}/(T+1)$. The empirical distribution function is defined in (3.30), but we modify the definition so that the divisor is $T + 1$ instead of T. We use the facts that

$$1 - \hat{F}(Y_i) = 1 - i/(T+1), \quad 1 - F(Y_i) = (Y_i/u)^{-\alpha}.$$

Thus,

$$-\log(Y_i/u) \approx (1/\alpha)\log(1 - i/(T+1)).$$

The least squares estimator of $1/\alpha$ is

$$\widehat{1/\alpha} = -\frac{\sum_{i=1}^{T}[\log(Y_i/u) \cdot \log(1 - i/(T+1))]}{\sum_{i=1}^{T}(\log(1 - i/(T+1)))^2}; \tag{3.76}$$

see (3.68) for the least squared formula. The estimator of $1/\alpha$ can be written as

$$\widehat{1/\alpha} = \sum_{i=1}^{T} w_i \log(Y_i/u),$$

where w_i is defined in (3.70). More weight is given to the observations in the extreme tails.

To estimate α, instead of $1/\alpha$, we use

$$\log(1 - i/(T+1)) \approx -\alpha \log(Y_i/u).$$

The least squares estimator of α is

$$\hat{\alpha} = -\frac{\sum_{i=1}^{T}[\log(Y_i/u) \cdot \log(1 - i/(T+1))]}{\sum_{i=1}^{T}(\log(Y_i/u))^2}. \tag{3.77}$$

Figure 3.16 shows the fitting of regression estimates for the S&P 500 daily returns, described in Section 2.4.1. Panel (a) considers the left tail and panel (b) the right tail. The tails are defined by the pth empirical quantiles for $p = 10\%/90\%$ (blue), $5\%/95\%$ (green), and $1\%/99\%$ (red). We also show the fitted linear regression lines. If the tails are Pareto tails, then the points should be on a straight line whose slope is equal to α. We can see that the slopes increase when we move to the more extreme parts of the tail (p decreases).

Pareto Tails The Student distributions have Pareto tails, as written in (3.55). The Lévy distributions with $0 < \alpha < 2$ have Pareto tails, as written in (3.94).

A distribution of random variable $X \in \mathbf{R}$ with distribution function $F : \mathbf{R} \to [0, 1]$ is said to have a Pareto right tail when

$$P(X \geq x) = 1 - F(x) = L(x) \, x^{-\alpha}, \tag{3.78}$$

for $x > 0$, for some $\alpha > 0$, where $L : (0, \infty) \to (0, \infty)$ is a slowly varying function at $+\infty$:

$$\lim_{x \to \infty} \frac{L(\lambda x)}{L(x)} = 1,$$

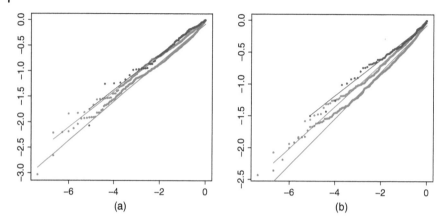

Figure 3.16 *Pareto model for S&P 500 daily returns: Regression fits.* Panel (a) considers the left tail and panel (b) the right tail. We show the regression data and the fitted regression lines for $p = 10\%/90\%$ (blue), $5\%/95\%$ (green), and $1\%/99\%$ (red).

for all $\lambda > 0$.[23] A distribution is said to have a Pareto left tail when

$$P(X \leq -x) = F(-x) = L(x)\, x^{-\alpha}, \tag{3.79}$$

for $x > 0$, for some $\alpha > 0$, where $L : (0, \infty) \to (0, \infty)$ is a slowly varying function.

For example, if density function $f : \mathbf{R} \to \mathbf{R}$ satisfies

$$f(x) = Cx^{-1-\alpha}$$

for $x \geq u$, where $u \geq 0$, $\alpha > 0$, and $C > 0$, then the distribution has a Pareto right tail. If

$$f(x) = C(-x)^{-1-\alpha}$$

for $x \leq -u$, where $u \geq 0$, $\alpha > 0$, and $C > 0$, then the distribution has a Pareto left tail.

3.4.2.3 The Gamma Distributions

For the gamma distributions the density functions have a closed form expression but the distribution functions and the maximum likelihood estimator cannot be written in a closed form.

The gamma densities are defined as

$$f(x) = C(\kappa, \beta) \cdot x^{\kappa-1} \exp\left\{-\frac{x}{\beta}\right\} I_{[0,\infty)}(x), \tag{3.80}$$

---

23 If $0 < \lim_{x \to \infty} L(\lambda x)/L(x) < \infty$ for all λ, then $L : (0, \infty) \to (0, \infty)$ is called regularly varying.

where $\kappa > 0$, $\beta > 0$ and the normalization constant is

$$C(\kappa, \beta) = \frac{1}{\beta^\kappa \Gamma(\kappa)},$$

where $\Gamma(x) = \int_0^\infty t^{x-1} e^{-t} \, dt$ is the gamma function. The distribution function is

$$F(x) = \frac{\gamma(\kappa, x/\beta)}{\Gamma(\kappa)} I_{[0,\infty)}(x),$$

where the lower incomplete gamma function is defined as

$$\gamma(\kappa, x) = \int_0^x t^{\kappa-1} e^{-t} \, dt$$

for $x \geq 0$ and $\gamma > 0$.

When $\kappa = 1$, then we obtain the family of exponential distributions. When $\kappa > 1$, then the gamma densities have a tail that is heavier than the exponential densities but lighter than the Pareto densities. When $0 < \kappa < 1$, then the gamma densities have a tail that is lighter than the exponential densities.

Assuming independent and identically distributed observations $Y_1, \dots, Y_T$ the logarithmic likelihood is

$$\log L(\kappa, \beta) = -T\kappa \log \beta - T \log \Gamma(\kappa)$$
$$+ (\kappa - 1) \sum_{i=1}^T \log Y_i - \frac{1}{\beta} \sum_{i=1}^T Y_i. \tag{3.81}$$

The maximum likelihood estimator of parameter β, given κ, is

$$\hat{\beta}(\kappa) = \frac{1}{\kappa T} \sum_{i=1}^T Y_i.$$

The maximum likelihood estimator of κ is the maximizer of $\log L(\kappa, \hat{\beta}(\kappa))$ over $\kappa > 0$. The maximum likelihood estimator of β is $\hat{\beta} = \hat{\beta}(\hat{\kappa})$.

3.4.2.4 The Generalized Pareto Distributions

The one-parameter Pareto distributions were defined in (3.73) and (3.72). We define the two-parameter generalized Pareto distributions, which contain the exponential distributions as a limiting case.

The density functions, distribution functions, and quantile functions have a closed form expression but the maximum likelihood estimator does not have a closed form expression.

The density functions of the generalized Pareto distributions are

$$f(x) = \begin{cases} \frac{1}{\beta}\left(1 + \frac{\xi x}{\beta}\right)^{-1/\xi - 1} I_{[0,\infty)}(x), & \xi > 0, \\ \frac{1}{\beta} \exp\left\{-\frac{x}{\beta}\right\} I_{[0,\infty)}(x), & \xi = 0, \end{cases} \tag{3.82}$$

where $\beta > 0$ and $\xi \geq 0$. The distribution functions are

$$F(x) = \begin{cases} 1 - \left(1 + \frac{\xi x}{\beta}\right)^{-1/\xi} I_{[0,\infty)}(x), & \xi > 0, \\ 1 - \exp\left\{-\frac{x}{\beta}\right\} I_{[0,\infty)}(x), & \xi = 0. \end{cases} \tag{3.83}$$

The quantile functions are

$$F^{-1}(p) = \begin{cases} \frac{\beta}{\xi}[(1-p)^{-\xi} - 1] \, I_{[0,1)}(x), & \xi > 0, \\ -\beta \log(1-p) I_{[0,1)}(x), & \xi = 0. \end{cases}$$

When $\xi = 0$, then the distributions are exponential distributions, defined in (3.65).

The generalized Pareto distribution can be defined for the cases $\xi < 0$. In this case the support is $[0, -\beta/\xi]$. See (3.101) for the distribution function and (8.65) for the density function. The generalized Pareto distributions are obtained as limit distributions for threshold exceedances (see Section 3.5.2).

For the calculation of the maximum likelihood estimation it is convenient to use the following parameterization. We define the class of generalized Pareto distributions using the tail index $\alpha > 0$ (shape parameter) and the scaling parameter $\sigma > 0$ by defining the density function as

$$f(x) = \frac{\alpha}{\sigma}\left(1 + \frac{x}{\sigma}\right)^{-1-\alpha} I_{[0,\infty)}(x). \tag{3.84}$$

The parameters of the generalized Pareto distribution (3.84) are related to the parameterization in (3.83) by $\alpha = 1/\xi$ and $\sigma = \beta/\xi$. Note that the densities (3.84) can be obtained heuristically from a translation of the one-parameter Pareto distributions, as written in (3.74).

The maximum likelihood estimator cannot be expressed in a closed form but we can reduce the numerical maximization of the two-variate likelihood function to the numerical maximization of a univariate function. For the computation of the maximum likelihood estimator, we use the parameterization of the density as in (3.84).

The logarithmic likelihood function for i.i.d. observations $Y_1, \ldots, Y_T$ is

$$\log L(\alpha, \sigma) = T \log\left(\frac{\alpha}{\sigma}\right) - (1 + \alpha) \sum_{i=1}^{T} \log\left(1 + \frac{Y_i}{\sigma}\right). \tag{3.85}$$

Setting the partial derivative equal to zero and solving for α gives[24]

$$\hat{\alpha}(\sigma) = \left[\frac{1}{T} \sum_{i=1}^{T} \log\left(1 + \frac{Y_i}{\sigma}\right)\right]^{-1}.$$

---

24 The partial derivative with respect to α is

$$\frac{\partial}{\partial \alpha} \log L(\alpha, \sigma) = \frac{T}{\alpha} - \sum_{i=1}^{T} \log\left(1 + \frac{Y_i}{\sigma}\right).$$

The maximum likelihood estimator $\hat{\sigma}$ for σ is the maximizer of the univariate function $\log L(\hat{a}(\sigma), \sigma)$ over $\sigma > 0$. The maximum likelihood estimator for α is $\hat{\alpha} = \hat{a}(\hat{\sigma})$. The maximum likelihood estimators for ξ and β are

$$\hat{\xi} = 1/\hat{\alpha}, \quad \hat{\beta} = \hat{\xi}\hat{\sigma}.$$

3.4.2.5 The Weibull Distributions

For the Weibull distributions the density functions, distribution functions, and quantile functions have a closed form expression but the maximum likelihood estimator cannot be written in a closed form.

The Weibull densities are defined as

$$f(x) = \frac{\kappa}{\beta} \cdot \left(\frac{x}{\beta}\right)^{\kappa-1} \exp\left\{-\left(\frac{x}{\beta}\right)^{\kappa}\right\} I_{[0,\infty)}(x), \tag{3.86}$$

where $\kappa > 0$ is the shape parameter and $\beta > 0$ is the scale parameter. The distribution function is

$$F(x) = \left(1 - \exp\left\{-\left(\frac{x}{\beta}\right)^{\kappa}\right\}\right) I_{[0,\infty)}(x).$$

The quantile function is

$$F^{-1}(p) = \beta(-\log(1-p))^{1/\kappa} I_{[0,1)}(p).$$

For $\kappa = 1$ we obtain the exponential distribution. The Weibull distributions are also called stretched exponential distributions because $1 - F(x)$ is a stretched exponential function.

The maximum likelihood estimator cannot be expressed in a closed form but we can reduce the numerical maximization of the two-variate likelihood function to the numerical maximization of a univariate function. The logarithmic likelihood function for i.i.d. observations $Y_1, \ldots, Y_T$ is

$$\log L(\kappa, \beta) = T\log(\kappa/\beta) + (\kappa - 1)\sum_{i=1}^{T}\log(Y_i/\beta) - \sum_{i=1}^{T}(Y_i/\beta)^{\kappa}. \tag{3.87}$$

Setting the partial derivative equal to zero and solving for β gives[25]

$$\hat{\beta}(\kappa) = \left(\frac{1}{T}\sum_{i=1}^{T}Y_i^{\kappa}\right)^{1/\kappa}.$$

The maximum likelihood estimator $\hat{\kappa}$ for κ is the maximizer of the univariate function $\log L(\hat{\beta}(\kappa), \kappa)$ over $\kappa > 0$. The maximum likelihood estimator for β is $\hat{\beta} = \hat{\beta}(\hat{\kappa})$.

---

25 The partial derivative with respect to β is

$$\frac{\partial}{\partial\beta}\log L(\kappa, \beta) = \kappa\beta^{-1}\left(\beta^{-\kappa}\sum_{i=1}^{T}Y_i^{\kappa} - T\right).$$

3.4.2.6 A Three Parameter Family

A flexible family for the modeling of the right tail is defined in Malevergne and Sornette (2005, p. 57) by density functions

$$f(x) = C(a, b, c, u) \cdot x^{-(a+1)} \exp\{-(x/c)^b\}\, I_{[u,\infty)}(x), \qquad (3.88)$$

where $u > 0$ is the starting point of the distribution, $a \in \mathbb{R}$, and $b, c \in [0, \infty)$. When $b = 0$, then $a > 0$. The normalization constant $C(a, b, c, u)$ has the expression

$$C(a, b, c, u) = \frac{c^a b}{\Gamma(-a/b, (u/c)^b)},$$

where $\Gamma(a, x) = \int_x^\infty t^{a-1} e^{-t}\, dt$ is the nonnormalized incomplete Gamma function.

The family contains several sub-families:

1) The exponential density is obtained when $u = 0$, $a = -1$, $b = 1$, and $c > 0$. The exponential densities are $f(x) = c^{-1} \cdot \exp\{-x/c\} I_{[0,\infty)}(x)$, where $c > 0$. We defined exponential densities in (3.65).
2) The Pareto density is obtained when $a > 0$ and $b = 0$. The Pareto densities are $f(x) = au^a \cdot x^{-(a+1)}\, I_{[u,\infty)}(x)$, where $a > 0$ and $u > 0$. We defined Pareto densities in (3.72).
3) The gamma density is obtained by choosing $u = 0$ and $b = 1$. The gamma densities are $f(x) = [c^a/\Gamma(-a)]\, x^{-(a+1)}\, e^{-x/c} I_{[0,\infty)}(x)$, where $a < 0$ and $c > 0$. The gamma densities were defined in (3.80).
4) The Weibull density is obtained when $a = -b$, $b > 0$, and $c > 0$:

   $$f(x) = C(b, c, u) \cdot x^{b-1} \exp\{-(x/c)^b\} I_{[u,\infty)}(x),$$

   where $C(b, c, u) = b/c^b \exp\{(u/c)^b\}$. The Weibull densities were defined in (3.86).
5) The incomplete gamma density is obtained when $a = b = 1$ and $c > 0$:

   $$f(x) = C(c, u) \cdot x^{-2} \exp\{-(x/c)\} I_{[u,\infty)}(x),$$

   where $C(c, u) = c/\Gamma(-1, u/c)$.

The Pareto density and the stretched exponential density can be interpolated smoothly by the log-Weibull density

$$f(x) = C(a, b) \cdot x^{-1} [\log_e(x/u)]^{b-1} \exp\{-a[\log_e(x/u)]^b\}\, I_{[u,\infty)}(x),$$

where $C(a, b, u) = ab$.

3.4.3 Fitting the Models to Return Data

We fit models first to S&P 500 returns, and then to a collection of individual stocks in S&P 500. Fitting of the distributions gives background for the quantile estimation of Chapter 8.

3.4.3.1 S&P 500 Daily Returns: Maximum Likelihood

We fit one-parameter models (exponential and Pareto) and two-parameter models (gamma, generalized Pareto, and Weibull) to the tails of S&P 500 daily returns. The S&P 500 daily data is described in Section 2.4.1.

We study maximum likelihood estimators (3.63) and (3.64). The estimates are constructed using data

$$\{\hat{q}_p - Y_i : Y_i \le \hat{q}_p\}, \quad \{Y_i - \hat{q}_{1-p} : Y_i \ge \hat{q}_{1-p}\},$$

for the left and the right tails, respectively. Threshold $\hat{q}_p$ is the pth empirical quantile, and $\hat{q}_{1-p}$ is the $(1-p)$th empirical quantile, where $0 < p < 0.5$. The estimators $\hat{\theta}_{left}$ and $\hat{\theta}_{right}$ depend on the parameter p.

To show the sensitiveness of the estimates with respect the parameter p we plot the values of the estimates as a function of p. These plots are related to the Hill's plot, which name is used in the case of estimating parameter α of the Pareto distribution.

To characterize the goodness of fit we show tail plots, as defined in Section 3.2.1. The tail plots include both the observations and the fitted curves, for several values of p.

The one-parameter models indicate that the left tail is heavier than the right tail. However, the two-parameter families seem to give much better fits than the one-parameter families.

The Exponential Model The maximum likelihood estimator of the parameter of the exponential distribution is given in (3.67). The estimators for the parameters of the left tail and the right tail are obtained from (3.63) and (3.64) as

$$\hat{\beta}_{left} = \frac{1}{\#\mathcal{L}} \sum_{Y_i \in \mathcal{L}} (\hat{q}_p - Y_i), \quad \hat{\beta}_{right} = \frac{1}{\#\mathcal{R}} \sum_{Y_i \in \mathcal{R}} (Y_i - \hat{q}_{1-p}),$$

where $\mathcal{L}$ and $\mathcal{R}$ are defined in (3.61) and (3.62). The estimates $\hat{\beta}_{left}$ and $\hat{\beta}_{right}$ are related to the estimates of the expected shortfall in (3.28) and (3.27).

Figure 3.17 shows estimates of the parameter β and $1/\beta$ of the exponential distribution. Panel (a) shows estimates of $100 \times \beta$ and panel (b) shows estimates of $1/\beta$, as a function of p. Parameter β occurs in the quantile function, and is more important in quantile estimation, but for the convenience of the reader we also show the estimates of the rate parameter $1/\beta$. The red curves show the maximum likelihood estimates for the left tail, and the blue curves show the maximum likelihood estimates for the right tail. In addition, we show the values of the regression estimates (3.69) and (3.71). The pink curves show the regression estimates for the left tail, and the green curves show the regression estimates for the right tail. We see that the estimates for β are larger for the left tail than for the right tail. This indicates that the left tail is heavier than the right tail. The estimates become smaller when p increases.

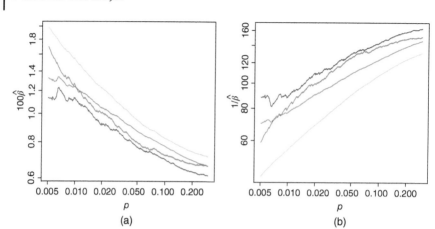

Figure 3.17 *Exponential model for S&P 500 daily returns: Parameter estimates.* Panel (a) shows estimates of $100 \times \beta$ and panel (b) shows estimates of $1/\beta$, as a function of p. Red and blue: the maximum likelihood estimates; pink and green: the regression estimates; red and pink: the left tail; blue and green: the right tail.

The regression estimates are larger than the maximum likelihood estimates. For the estimates of $1/\beta$ the behavior is opposite.

Figure 3.18 shows tail plots, defined in Section 3.2.1. Panel (a) shows the left tail plots and panel (b) shows the right tail plots. The red and green points show the observed data and the black lines show the exponential distribution functions when parameter β is estimated with maximum likelihood. The four black

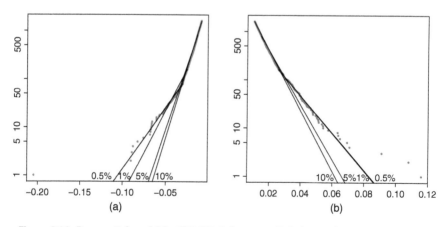

Figure 3.18 *Exponential model for S&P 500 daily returns: Tail plots with maximum likelihood.* Panel (a) shows the left tail plots and panel (b) shows the right tail plots. The red and green points show the observed data and the black lines show the exponential fits with $p = 0.5\%$, 1%, 5%, and 10%.

curves show the cases $p = 0.5\%, 1\%, 5\%$, and 10%. The tails are fitted better with small values of p.

The Pareto Model The maximum likelihood estimator of the parameter of the Pareto distribution is given in (3.75).[26] The estimators for the parameters of the left and the right tails are obtained from (3.63) and (3.64) as

$$\hat{\alpha}_{left} = \left(\frac{1}{\#\mathcal{L}} \sum_{Y_i \in \mathcal{L}} \log(Y_i/u) \right)^{-1}, \quad \hat{\alpha}_{right} = \left(\frac{1}{\#\mathcal{R}} \sum_{Y_i \in \mathcal{R}} \log(Y_i/u) \right)^{-1}, \quad (3.89)$$

where $\mathcal{L} = \{Y_i : Y_i \le u\}$ with $u = \hat{q}_p$ for the left tail, and $\mathcal{R} = \{Y_i : Y_i \ge u\}$ with $u = \hat{q}_{1-p}$ for the right tail. Now $0 < p < 0.5$. The maximum likelihood estimators are called Hill's estimators.[27]

Figure 3.19 shows estimates of the parameter $1/\alpha$ and α of the Pareto distribution. Panel (a) shows estimates of $100/\alpha$ and panel (b) shows estimates of α, as a function of p. The plot in panel (b) is known as Hill's plot. Parameter $1/\alpha$ occurs in the quantile function, and is more important in quantile estimation, but for the convenience of the reader we also show the estimates of parameter α. The red curves show the maximum likelihood estimates for the left tail and the blue curves show the maximum likelihood estimates for the right tail. In addition, we show the values of regression estimates of $100/\alpha$, defined in (3.76), and the values of regression estimates of α, defined in (3.77). The pink curves show the regression estimates for the left tail and the green curves show the regression estimates for the right tail. We see that the estimates of $1/\alpha$ are larger for the left tail than for the right tail, which means that the left tail is estimated to be heavier than the right tail. The estimates of $1/\alpha$ become larger when p increases. The regression estimates of $1/\alpha$ are smaller than the maximum likelihood estimates. For the estimates of α the behavior is opposite.

Figure 3.20 shows tail plots. Panel (a) shows the left tail plots and panel (b) shows the right tail plots. The red and green points show the observed data and the black curves show the Pareto distribution functions when parameter α is estimated with maximum likelihood. The four black curves show the cases $p = 0.5\%, 1\%, 5\%$, and 10%.

---

26 For the Pareto distribution translated to have support $[0, \infty)$ the maximum likelihood estimator is $\hat{\alpha} = (T^{-1} \sum_{i=1}^{T} \log((Y_i + u)/u))^{-1}$.

27 The computation of the estimates can be done in the following way. Let $0 < p < 0.5$ and $m = [pT]$. Let $Y_{(1)} < \cdots < Y_{(T)}$ be the ordered sample. The Hill's estimators are

$$\hat{\alpha}_{right} = \frac{m}{\sum_{i=T-m+1}^{T} \log(Y_{(i)}/Y_{(T-m+1)})} \quad (3.90)$$

and

$$\hat{\alpha}_{left} = \frac{m}{\sum_{i=1}^{m} \log(Y_{(i)}/Y_{(m)})}. \quad (3.91)$$

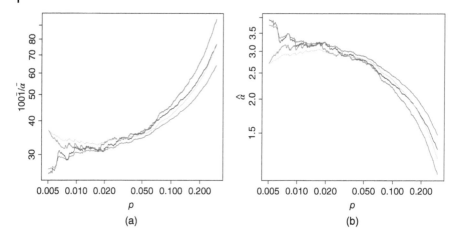

Figure 3.19 *Pareto model for S&P 500 daily returns: Parameter estimates.* Panel (a) shows estimates of $100/\alpha$ and panel (b) shows estimates of α as a function of p. Red and blue: the maximum likelihood estimates; pink and green: the regression estimates; red and pink: the left tail; blue and green: the right tail.

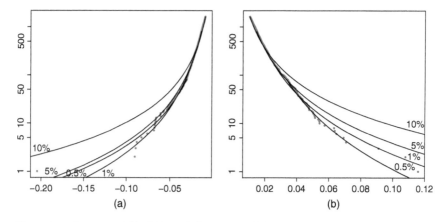

Figure 3.20 *Pareto model for S&P 500 daily returns: Tail plots with maximum likelihood.* Panel (a) shows the left tail plots and panel (b) shows the right tail plots. The red and green points show the observed data and the black curves show the fits with $p = 0.5\%$, 1%, 5%, and 10%.

The Gamma Model The gamma densities are defined in (3.80). The maximum likelihood estimators for the scale parameter $\beta > 0$ and for the shape parameter $\kappa > 0$ of a gamma distribution do not have a closed form expression, but the computation can be done by minimizing a univariate function. We get the maximum likelihood estimates for the parameters of the left tail and the right

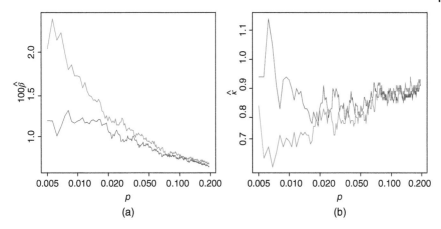

Figure 3.21 *Gamma model for S&P 500 daily returns: Parameter estimates.* Panel (a) shows estimates of $100 \times \beta$ and panel (b) shows estimates of κ, as a function of p. Red: the left tail; blue: the right tail.

tail by applying the numerical procedure for the observations

$$\{\hat{q}_p - Y_i : Y_i \le \hat{q}_p\}, \quad \{Y_i - \hat{q}_{1-p} : Y_i \ge \hat{q}_{1-p}\}, \tag{3.92}$$

for the left and the right tails, respectively, where $0 < p < 0.5$.

Figure 3.21(a) shows estimates of 100β and panel (b) shows estimates of κ. The red curves show the estimates for the left tail, and the blue curves show the estimates for the right tail. We see that the estimates for β are larger for the left tail than for the right tail. The estimates become smaller when p increases.

Figure 3.22 shows tail plots. Panel (a) shows the left tail plots and panel (b) shows the right tail plots. The red and green points show the observed data and the black curves show the gamma distribution functions when parameters are estimated with maximum likelihood. The four black curves show the cases $p = 0.5\%$, 1%, 5%, and 10%.

The Generalized Pareto Model The density of a generalized Pareto distribution is given in (3.82). The maximum likelihood estimators for the scale parameter $\beta > 0$ and for the shape parameter $\xi \ge 0$ of a generalized Pareto distribution do not have a closed form expression, but the computation can be done by minimizing a univariate function. We get the maximum likelihood estimates for the parameters of the left tail and the right tail by applying the numerical procedure for the observations in (3.92).

Figure 3.23(a) shows estimates of 100β, and panel (b) shows estimates of ξ. The red curves show the estimates for the left tail, and the blue curves show the estimates for the right tail. The estimates of β become smaller when p increases.

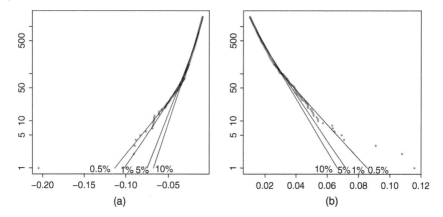

Figure 3.22 *Gamma model for S&P 500 daily returns: Tail plots with maximum likelihood.* Panel (a) shows the left tail plots and panel (b) shows the right tail plots. The red and green points show the observed data and the black lines show the fits with $p = 0.5\%, 1\%, 5\%$, and 10%.

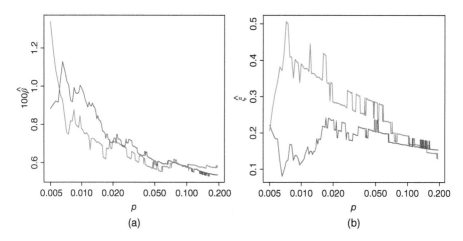

Figure 3.23 *Generalized Pareto model for S&P 500 daily returns: Parameter estimates.* Panel (a) shows estimates of $100 \times \beta$ and panel (b) shows estimates of ξ, as a function of p. Red shows the estimates for the left tail, and blue shows them for the right tail.

Figure 3.24 shows tail plots. Panel (a) shows the left tail plots and panel (b) shows the right tail plots. The red and green points show the observed data and the black curves show the distribution functions when parameters are estimated using maximum likelihood. The four black curves show the cases $p = 0.5\%, 1\%, 5\%$, and 10%. The fitted curves do not change in a monotonic order when p is decreased.

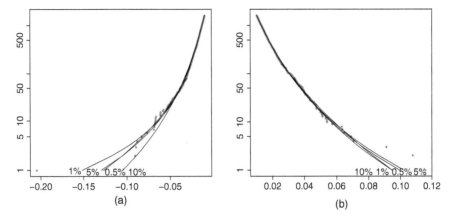

Figure 3.24 *Generalized Pareto model for S&P 500 daily returns: Tail plots with maximum likelihood.* Panel (a) shows the left tail plots and panel (b) shows the right tail plots. The red and green points show the observed data and the black curves show the fits with $p = 0.5\%$, 1%, 5%, and 10%.

The Weibull Model The Weibull densities are given in (3.86). The maximum likelihood estimators for the scale parameter $\beta > 0$ and for the shape parameter $\kappa >$ of a Weibull distribution do not have a closed form expression, but the computation can be done by minimizing a univariate function. We get the maximum likelihood estimates for the parameters of the left tail and the right tail by applying the numerical procedure for the observations in (3.92).

Figure 3.25(a) shows estimates of 100β, and panel (b) shows estimates of κ. The red curves show the estimates for the left tail, and the blue curves show the estimates for the right tail. The estimates of β become smaller when p increases.

Figure 3.26 shows tail plots. Panel (a) shows the left tail plots and panel (b) shows the right tail plots. The red and green points show the observed data and the black curves show the distribution functions when parameters are estimated using maximum likelihood. The four black curves show the cases $p = 0.5\%$, 1%, 5%, and 10%.

3.4.3.2 Tail Index Estimation for S&P 500 Components
We study fitting of the Pareto model for the daily returns of stocks in S&P 500 index. S&P 500 components data is described in Section 2.4.5.

Figure 3.27 shows how $\hat{\alpha}_{left}$ and $\hat{\alpha}_{right}$ are distributed. The estimators are defined in (3.89); these are Hill's estimators for the left and right Pareto indexes. Panel (a) shows the distribution of the estimates of the left tail index and panel (b) shows the distribution of the estimates of the right tail index. We have computed the estimates for each 312 stocks in the S&P 500 components data set, and the kernel density estimator is applied for this data set of 312 observations. This is done for $p = 0.05, 0.06, \ldots, 0.2$. The smoothing parameter

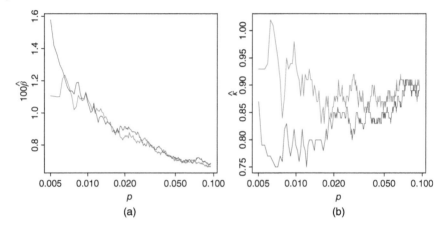

Figure 3.25 *Weibull model for S&P 500 daily returns: Parameter estimates.* Panel (a) shows estimates of $100 \times \beta$ and panel (b) shows estimates of ξ, as a function of p. Red shows the estimates for the left tail, and blue shows them for the right tail.

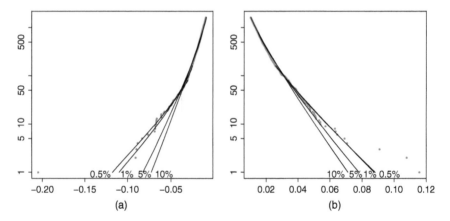

Figure 3.26 *Weibull model for S&P 500 daily returns: Tail plots with maximum likelihood.* Panel (a) shows the left tail plots and panel (b) shows the right tail plots. The red and green points show the observed data and the black curves show the fits with $p = 0.5\%, 1\%, 5\%$, and 10%.

is chosen by the normal reference rule, and the standard Gaussian kernel function is used. A smaller p gives a smaller estimate of α.

Figure 3.28 shows a scatter plot of points $(\hat{\alpha}_{left}, \hat{\alpha}_{right})$, when the estimates are computed for each stock in the S&P 500 components data. We have used $p = 0.1$. There are about the same number of stocks for which the left tail index is smaller than the right tail index

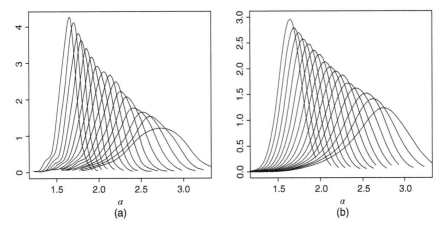

Figure 3.27 *Density estimates of the distribution of Hill's estimates.* (a) Distribution of the left tail index; (b) the right tail index. Hill's estimates are calculated for the 312 stocks and kernel estimates are calculated from 312 estimated values of α. There is an kernel estimate for each $p = 0.05, 0.06, \ldots, 0.2$.

Figure 3.28 *A scatter plot of estimates of α.* We show a scatter plot of points $(\hat{\alpha}_{left}, \hat{\alpha}_{right})$ for the stocks in the S&P 500 components data. The red line shows the points with $y = x$.

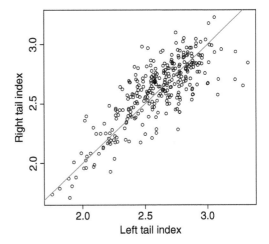

3.5 Asymptotic Distributions

First we describe central limit theorems and second we describe limit theorems for the excess distribution. The limit distributions of the central limit theorems can be used to model the complete return distribution of a financial asset and the limit distributions for the excess distribution can be used to model the tail areas of the return distribution of a financial asset.

3.5.1 The Central Limit Theorems

We applied a central limit theorem for sums in (3.46) and (3.49) to justify the normal and the log-normal model for the stock prices. In a similar way we can apply the central limit theorems to justify alternative models for the stock prices. When the variance of the summands is finite the limit is a normal distribution, but if the variance is not finite, the limit distributions can have heavier tails than the normal distributions.

We describe first a central limit theorem for sums of independent but not necessarily identically distributed random variables. The limit distributions belong to the class of infinitely divisible distributions. Second we describe central limit theorems for sums of independent and identically distributed random variables. Now the limit distributions belong to the class of stable distributions. The class of stable distributions is a subset of the class of infinitely divisible distributions. The stable distributions include the normal distributions but they include also heavy tailed distributions, which can be used to describe phenomena where both very large and very small values can be observed, like the stock returns.

Third we consider the case of sums of dependent random variables. When the dependence is weak, then a convergence towards a normal distribution occurs, but the asymptotic variance is affected by the dependence.

We do not apply stable distributions or infinitely divisible distributions to model return distributions, but it is useful to note that heavy tailed distributions arise already from central limit theorems, and not only from limit distributions for the excess distribution.

3.5.1.1 Sums of Independent Random Variables

The Khintchine theorem states that for a distribution to be a limit distribution of a sum of independent (but not necessarily identically distributed) random variables it is necessary and sufficient that the distribution is infinitely divisible; see Billingsley (2005, pp. 373–374) and Breiman (1993, p. 191).

The infinitely divisible distributions are such that a random variable following an infinitely divisible distribution can be represented as a sum of n i.i.d. random variables for each natural number n. In other words, a distribution function F is infinitely divisible if for each n there is a distribution function F_n such that F is the n-fold convolution $F_n * \cdots * F_n$.[28] For example, the normal, Poisson, and gamma distributions are infinitely divisible but the uniform distributions

---

28 The convolution of functions $f : \mathbf{R} \to \mathbf{R}$ and $g : \mathbf{R} \to \mathbf{R}$ is defined as $f * g(x) = \int_{-\infty}^{\infty} f(y)g(x - y)\, dy$.

The characteristic function of an infinitely divisible distribution is the nth power of some characteristic function. The characteristic function ϕ of a probability distribution P on $\mathbf{R}$ is defined by $\phi(t) = \int_{-\infty}^{\infty} e^{itx}\, dP(x)$, where $t \in \mathbf{R}$. The characteristic function of an infinitely divisible distribution can be found in Breiman (1993, p. 194). See also Billingsley (2005, p. 372).

are not. See Billingsley (2005, Chapter 5) and Breiman (1993, Section 9.5) about infinitely divisible distributions.

Let $(Y_{nk})_{k=1,\ldots,n}$, $n = 1, 2, \ldots$, be a triangular array of row-wise independent random variables which satisfy

$$\max_{k=1,\ldots,n} P(|Y_{nk}| > \epsilon) \longrightarrow 0,$$

as $n \to \infty$, for every $\epsilon > 0$. Then $\sum_{k=1}^{n} Y_{nk}$ can be normalized to converge to an infinitely divisible distribution.

3.5.1.2 Sums of Independent and Identically Distributed Random Variables

For a distribution to be a limit distribution of a sum of independent and identically distributed random variables it is necessary and sufficient that the distribution is stable.

Stable Distributions A random variable is said to have a stable distribution, if for every natural number n and for $X_1, \ldots, X_n$ independent and with the same distribution as X, there are constants $a_n > 0$ and b_n such that

$$X = a_n(X_1 + X_2 + \cdots + X_n) + b_n$$

holds in distribution; see Breiman (1993, p. 199). Stable distributions are infinitely divisible distributions, because the distribution function of X is the n-fold convolution of F_n, where F_n is the distribution function of $a_n X_1 + b_n/n$. In particular, the sum of two independent and identically distributed stable random variables has also a stable distribution.

Density functions of stable distributions cannot be written in a closed form in general. The characteristic function of a stable distribution is

$$\psi(t) = \exp\left\{ i\mu t - |\sigma t|^{\alpha} \left(1 + i\beta \, \frac{t}{|t|} \, \Psi_{\alpha}(t) \right) \right\}, \quad t \in \mathbf{R},$$

where

$$\Psi_{\alpha}(t) = \begin{cases} \tan(\pi\alpha/2), & \alpha \neq 1, \\ \frac{2}{\pi} \log_e |t|, & \alpha = 1. \end{cases}$$

Note that $t/|t|$ is the sign of t, and we can define $0/|0| = 0$. Parameter $0 < \alpha \leq 2$ is the exponent of the distribution, which is related to the heaviness of the tails, $\mu \in \mathbf{R}$ is the location term, $\sigma > 0$ is the scale factor, and $-1 \leq \beta \leq 1$ is the asymmetry parameter (skewness parameter). When $\beta = 0$, then distribution is symmetric, when $\beta > 0$, then distribution is skewed to the right, and when $\beta < 0$, the distribution is skewed to the left. See Breiman (1993, p. 204).

The analytical form of the density is known for $\alpha = 2$ (Gaussian), $\alpha = 1$, $\beta = 0$ (Cauchy), and $\alpha = 1/2$, $\beta = 1$ (Lévy–Smirnov or Lévy). The density of the Cauchy distribution is given by

$$f(x) = \frac{1}{\pi\sigma} \frac{1}{1 + (x - \mu)^2/\sigma^2}, \quad x \in \mathbf{R}.$$

The Cauchy distribution is the Student distribution for the degrees of freedom $v = 1$. The density of the Lévy–Smirnov distribution is given by

$$f(x) = \left(\frac{\sigma}{2\pi}\right)^{1/2} \frac{1}{(x-\mu)^{3/2}} \exp\left\{-\frac{\sigma}{2(x-\mu)}\right\}, \quad x > \mu.$$

Symmetric stable distributions are stable distributions with location parameter $\mu = 0$ and skewness parameter $\beta = 0$. The characteristic function of a symmetric stable distribution is

$$\psi(t) = \exp\{-|\sigma t|^{\alpha}\}, \quad t \in \mathbf{R},$$

where $0 < \alpha < 2$ and $\sigma > 0$. The density of a symmetric stable distribution can be written as a series expansion

$$f(x) = \sum_{k=1}^{\infty} \frac{(-1)^{k+1}}{\pi k!} \frac{a_{\alpha}^{k}}{x^{1+k\alpha}} \Gamma(1+k\alpha) \sin(\pi\alpha k/2), \quad x \in \mathbf{R}, \tag{3.93}$$

where a_{α} is defined through

$$A_{\alpha} = \begin{cases} \alpha\Gamma(\alpha-1)\frac{\sin(\pi\alpha/2)}{\pi} a_{\alpha}, & 1 < \alpha < 2, \\ (1-\alpha)\Gamma(\alpha)\frac{\sin(\pi\alpha/2)}{\pi\alpha} a_{\alpha}, & 0 < \alpha < 1. \end{cases}$$

Symmetric stable distributions have the power-law behavior of the tails:

$$f(x) \sim \frac{C_{\alpha}}{|x|^{1+\alpha}}, \quad x \to \pm\infty. \tag{3.94}$$

Equation (3.94) gives the leading asymptotic term in (3.93). For the distributions with Pareto tails the kth moment does not exist if $k \geq \alpha$. This implies that the variance of a symmetric stable distribution is always infinite, and the mean is infinite when $\alpha \leq 1$. The mode is used as the location parameter of the symmetric stable distributions (symmetric stable distributions are unimodal).

Convergence to a Stable Distribution The central limit theorems were presented in Gnedenko and Kolmogorov (1954), Feller (1957), and Feller (1966). We follow the exposition of Embrechts *et al.* (1997, Theorem 2.2.15). Assume that $Y_1, \ldots, Y_n$ are independent and identically distributed with the same distribution as Y.

1) Assume that $EY^2 < \infty$. Then,

$$(\sigma n^{1/2})^{-1}\left(\sum_{i=1}^{n} Y_i - n\mu\right) \xrightarrow{d} N(0,1),$$

where $\mu = EY$ and $\sigma^2 = \text{Var}(Y)$.
2) Assume that

$$L(x) = \int_{|y|\leq x} y^2 dF(y)$$

is slowly varying.[29] Let b_n be the solution of the equation

$$Q(b_n) = n^{-1}, \tag{3.95}$$

where

$$Q(x) = P(|Y| > x) + x^{-2} \int_{|y| \le x} y^2 \, dF(y).$$

Then,

$$b_n^{-1} \left(\sum_{i=1}^n Y_i - n\mu \right) \xrightarrow{d} N(0, 1),$$

where $\mu = EY$. It holds that $b_n = n^{1/2} L(n)$, for a slowly varying function L.

3) Assume that the distribution function F of Y satisfies

$$F(-x) = x^{-\alpha}(c_1 + o(1))L(x), \quad 1 - F(x) = x^{-\alpha}(c_2 + o(1))L(x),$$

as $x \to \infty$, where L is slowly varying, and $c_1, c_2 \ge 0$, $c_1 + c_2 > 0$. Let b_n be the solution of (3.95). Then,

$$b_n^{-1} \left(\sum_{i=1}^n Y_i - a_n \right) \xrightarrow{d} G_\alpha,$$

where $a_n = n \int_{|y| \le b_n} y \, dF(y)$ and G_α is a stable distribution with $0 < \alpha < 2$.

3.5.1.3 Sums of Dependent Random Variables

We apply a limit theorem for dependent random variables in Sections 6.2.2 and 10.1.2.

Let $(Y_t)_{t \in \mathbb{Z}}$ be a strictly stationary time series. We define the weak dependence in terms of a condition on the α-mixing coefficients. Let $\mathcal{F}_i^j$ denote the sigma algebra generated by random variables $Y_i, \dots, Y_j$. The α-mixing coefficient is defined as

$$\alpha_n = \sup_{A \in \mathcal{F}_{-\infty}^0, B \in \mathcal{F}_n^\infty} |P(A \cap B) - P(A)P(B)|,$$

where $n = 1, 2, \dots$. Now we can state the central limit theorem. Let $E|Y_t|^\delta < \infty$ and $\sum_{j=1}^\infty \alpha_j^{1 - 2/\delta} < \infty$ for some constant $\delta > 2$. Then,

$$n^{-1/2} \sum_{i=1}^n (Y_i - EY_i) \xrightarrow{d} N(0, \sigma^2), \tag{3.96}$$

where

$$\sigma^2 = \sum_{j=-\infty}^\infty \gamma(j) = \gamma(0) + 2 \sum_{j=1}^\infty \gamma(j),$$

---

29 We call function $L : (0, \infty) \to (0, \infty)$ slowly varying function at $+\infty$ if $\lim_{x \to \infty} L(\lambda x)/L(x) = 1$ for all $\lambda > 0$.

$\gamma(j) = \text{Cov}(Y_t, Y_{t+j})$, and we assume that $\sigma^2 > 0$. Ibragimov and Linnik (1971, Theorem 18.4.1) gave necessary and sufficient conditions for a central limit theorem under α-mixing conditions. A proof for our statement of the central limit theorem in (3.96) can be found in Peligrad (1986); see also Fan and Yao (2005, Theorem 2.21) and Billingsley (2005, Theorem 27.4).

3.5.2 The Limit Theorems for Maxima

Since we have modeled the excess distribution parametrically, it is of special interest that the limit distribution of the excess distribution is a generalized Pareto distribution; this limit theorem is stated in (3.102). The weak convergence of maxima is related to the convergence of the excess distribution.

3.5.2.1 Weak Convergence of Maxima

Let the real valued random variables $Y_1, \ldots, Y_n$ be independent and identically distributed, and denote the maximum

$$M_n = \max\{Y_1, \ldots, Y_n\}.$$

Sometimes convergence in distribution holds in the sense that there exists sequences (c_n) and (d_n) where $c_n > 0$ and $d_n \in \mathbf{R}$ so that

$$P\left(\frac{M_n - d_n}{c_n} \leq x\right) \xrightarrow{d} F\left(\frac{x - b}{a}\right), \tag{3.97}$$

for all $x \in \mathbf{R}$, as $n \to \infty$, where F is a distribution function, $b \in \mathbf{R}$, and $a > 0$. The Fisher–Tippett–Gnedenko theorem states that if the convergence in (3.97) holds, then F can only be a Fréchet, Weibull, or Gumbel distribution function. See Fisher and Tippett (1928), Gnedenko (1943), and Embrechts *et al.* (1997, p. 121).

To derive the result for the minimum we use the fact that for

$$m_n = \min\{Y_1, \ldots, Y_n\}$$

we have $m_n = -\max\{-Y_1, \ldots, -Y_n\}$. Let us denote

$$L_n = \max\{-Y_1, \ldots, -Y_n\}$$

so that $m_n = -L_n$. Now,

$$P(m_n \leq x) = P(-L_n \leq x) = P(L_n \geq -x) = 1 - P(L_n < -x). \tag{3.98}$$

3.5.2.2 Extreme Value Distributions

The Fréchet distribution functions are

$$\Phi_\alpha(x) = \begin{cases} 0, & x \leq 0, \\ \exp\{-x^{-\alpha}\}, & x > 0, \end{cases} \tag{3.99}$$

where $\alpha > 0$. The Weibull distribution functions are

$$\Psi_\alpha(x) = \begin{cases} \exp\{-(-x)^\alpha\}, & x \le 0, \\ 1, & x > 0, \end{cases}$$

where $\alpha > 0$. The Gumbel distribution function is

$$\Lambda(x) = \exp\{-e^{-x}\}, \quad x \in \mathbf{R}.$$

These distributions are called the extreme value distributions.
Define

$$H_\xi = \begin{cases} \Phi_{1/\xi}, & \text{if } \xi > 0, \\ \Lambda, & \text{if } \xi = 0, \\ \Psi_{-1/\xi}, & \text{if } \xi < 0. \end{cases}$$

Then,

$$H_\xi(x) = \begin{cases} \exp\{-(1+\xi x)^{-1/\xi}\}, & \xi \ne 0, \\ \exp\{-e^{-x}\}, & \xi = 0, \end{cases} \tag{3.100}$$

where H_ξ is defined on set $\{x : 1 + \xi x > 0\}$. This is known as the Jenkinson–von Mises representation of the extreme value distributions, or the generalized extreme value distribution; see Embrechts *et al.* (1997, p. 152). We obtain the parametric class of possible limit distributions

$$H_\xi\left(\frac{x - \mu}{\sigma}\right),$$

where $\xi \in \mathbf{R}$ is the shape parameter, $\mu \in \mathbf{R}$, and $\sigma > 0$. The support of the distribution is $\{x : 1 + \xi(x - \mu)/\sigma > 0\}$.

Using (3.98), we obtain the class of limit distribution functions for the minima. The limit distribution functions are

$$\tilde{H}_\xi\left(\frac{x - \mu}{\sigma}\right),$$

where $\xi \in \mathbf{R}$, $\mu \in \mathbf{R}$, $\sigma > 0$, and $\tilde{H}_\xi(x) = 1 - H_\xi(-x)$. Distribution function $\tilde{H}_\xi$ is defined on set $\{x : 1 - \xi x > 0\}$.

3.5.2.3 Convergence to an Extreme Value Distribution

If the distribution that generated the observations $Y_1, \ldots, Y_n$ has polynomial tails, then (3.97) holds and the limit distribution of the maximum belongs to the Fréchet class. More precisely, if

$$1 - F(x) = x^{-\alpha} L(x)$$

for some slowly varying function L, then a normalized maximum converges to a Fréchet distribution Ψ_α; see Embrechts *et al.* (1997, p. 131).

Let $x_F = \sup\{x : F(x) < 1\}$ be the endpoint of the distribution of Y. If $x_F < \infty$ and

$$1 - F(x_F - x^{-1}) = x^{-\alpha} L(x)$$

for some slowly varying function L, then a normalized maximum converges to a Weibull distribution Ψ_α; see Embrechts *et al.* (1997, p. 135). The equation

$$\Psi_\alpha(-x^{-1}) = \Phi_\alpha(x), \quad x > 0$$

explains the relation between the convergence to a Fréchet distribution and to a Weibull distribution.

If the distribution which generated the observations is exponential, normal, or log-normal, then (3.97) holds and the limit distribution of the maximum is the Gumbel distribution. See Embrechts *et al.* (1997, p. 145).

3.5.2.4 Generalized Pareto Distributions

The distribution function of the generalized Pareto distribution is

$$G_{\xi,\beta}(x) = \begin{cases} 1 - (1 + \xi x/\beta)^{-1/\xi}, & \xi \neq 0, \\ 1 - \exp\{-x/\beta\}, & \xi = 0, \end{cases} \tag{3.101}$$

where $\beta > 0$. When $\xi \geq 0$, then $0 \leq x < \infty$. When $\xi < 0$, then $0 \leq x \leq -\beta/\xi$. When $\xi = 0$, then the distributions are exponential distributions. Note that

$$G_{\xi,\beta}(x) = 1 - \log H_\xi(x/\beta),$$

where H_ξ is the distribution function of a generalized extreme value distribution, as defined in (3.100). Parameter ξ is a shape parameter and parameter β is a scale parameter. The Pareto distributions were defined in (3.73) and (3.83).

3.5.2.5 Convergence to a Generalized Pareto Distribution

Let $Y \in \mathbf{R}$ be a random variable and let F be the distribution function of Y. We define the excess distribution with threshold u as the distribution with the distribution function

$$F_u(x) = P(Y - u \leq x \mid Y > u) = \frac{F(x + u) - F(u)}{1 - F(u)}.$$

We can typically approximate the distribution function F_u with the distribution function of a generalized Pareto distribution. This follows from the Gnedenko–Pickands–Balkema–de Haan theorem; see Embrechts *et al.* (1997, p. 158). Let $X_F = \sup\{x : F(x) < 1\}$. The Gnedenko–Pickands–Balkema–de Haan theorem states that

$$\lim_{u \to x_F} \sup_{0 \leq x < x_F - u} |F_u(x) - G_{\xi,\beta(u)}(x)| = 0 \tag{3.102}$$

for some positive function $\beta(u)$ if and only if F belongs to the maximum domain of attraction of H_ξ, where $\xi \in \mathbf{R}$. To say that F belongs to the maximum domain of attraction of H_ξ means that (3.97) holds for some sequences $\{c_n\}$ and $\{d_n\}$.

The basic idea of deriving the limit distribution of the excess distribution from the limit distribution of the maximum comes from the Poisson approximation. The Poisson approximation states that

$$\lim_{n\to\infty} n(1 - F(u_n)) = \tau$$

and

$$\lim_{n\to\infty} P(M_n \leq u_n) = \exp\{-\tau\}$$

are equivalent, where $0 \leq \tau \leq \infty$, (u_n) is a sequence of real numbers, and $M_n = \max\{Y_1, \ldots, Y_n\}$ is the maximum of i.i.d. random variables; see Embrechts *et al.* (1997, p. 116).[30]

When the distribution function of the maximum M_n can be approximated by

$$G(x) = \exp\left\{-\left[1 + \xi\left(\frac{x - \mu}{\sigma}\right)\right]^{-1/\xi}\right\}$$

for some $\mu, \sigma > 0$ and ξ, then F_u can be approximated by the distribution function

$$H(x) = 1 - \left(1 + \frac{\xi y}{\tilde{\sigma}}\right)^{-1/\xi}$$

defined on set $\{x : x > 0 \text{ and } 1 + \xi x / \tilde{\sigma} > 0\}$, where

$$\tilde{\sigma} = \sigma + \xi(u - \mu).$$

3.6 Univariate Stylized Facts

The heaviness of the tails is one of the main univariate stylized facts. There are several questions related to the heaviness of the tails. We give a list of the observations that can be obtained from the figures of this chapter, and give some references to the literature.

1) *How heavy are the tails of S&P 500 returns?*
   Figure 2.1(b) shows a time series of S&P 500 daily returns. To highlight the heaviness of the tails we can compare the real time series with the simulated time series in Figure 3.29. Panel (a) shows uncorrelated observations

---

30 Indeed,

$$P(M_n \leq u_n) = F^n(u_n) = (1 - (1 - F(u_n)))^n \to \exp\{-\tau\},$$

if $\lim_{n\to\infty} n(1 - F(u_n)) = \tau$. Also, because $-\log(1 - x) \sim x$,

$$n(1 - F(u_n)) \sim -n\log(1 - (1 - F(u_n))) = \log P(M_n \leq u_n) \to \tau,$$

if $\lim_{n\to\infty} P(M_n \leq u_n) = \exp\{-\tau\}$. (We can argue that now $1 - F(u_n) \to 0$.) We have assumed that $0 < \tau < \infty$.

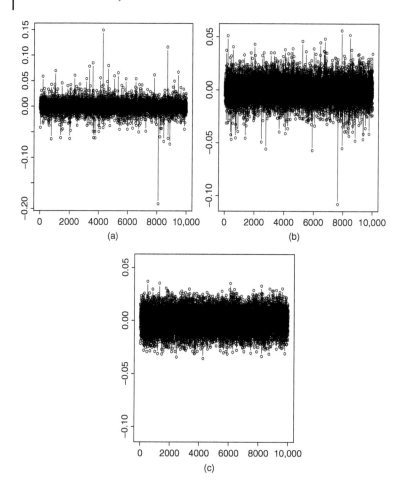

Figure 3.29 *Simulated i.i.d. time series.* We have simulated 10,000 observations. (a) Student's t-distribution with degrees of freedom $v = 3$; (b) Student's t-distribution with degrees of freedom $v = 6$; (c) Gaussian distribution. The mean of the observations is zero and the standard deviation is equal to the standard deviation of the S&P 500 returns.

whose distribution is the t-distribution with three degrees of freedom, in panel (b) the t-distribution has six degrees of freedom, and in panel (c) the distribution of the observations is Gaussian.[31]

_____

31 The simulated data has mean zero and standard deviation equal to the sample standard deviation of the S&P 500 returns. The variance of the t-distribution is $v/(v-2)$, where v is the degrees of freedom, and thus we have multiplied the simulated data from the t-distribution with $\sqrt{(v-2)/v}\hat{\sigma}$, where $\hat{\sigma}$ is the sample standard deviation of the S&P 500 returns.

Figure 3.2 shows tail plots of S&P 500 daily returns: t-distribution with degrees of freedom three and four gives reasonable fits both for the left tail and the right tails.

Figure 3.4 shows exponential regression plots of S&P 500 daily returns: The tails seem to be heavier than the exponential tails.

Figure 3.5 shows exponential regression plots of S&P 500 daily returns, and fits both exponential and Pareto distributions: Pareto fits seem to be better.

Figure 3.6 shows Pareto regression plots of S&P 500 daily returns: The tails seem to fit reasonably well for the Pareto model.

Figure 3.7 shows Pareto regression plots of S&P 500 daily returns, and fits both exponential and Pareto distributions: Pareto fits seem to be better.

Figure 3.13 shows how estimates of parameters v and σ of Student distribution for S&P 500 components are distributed: The mode of $\hat{v}$ is about 3.5 and the range of values of the estimates is about $\hat{v} \in [1.5, 4.5]$.

Figure 3.27 shows kernel density estimates of the distribution of the estimates of Pareto left tail index and Pareto right tail index for S&P 500 components: The choice of parameter p has a significant influence on the value of the estimate, but we are in the range $\hat{\alpha} \in [1.5, 2.5]$.

2) *How the heaviness of the tails varies across asset classes (stocks, bonds, indexes)?*

Figure 2.5(b) shows a times series of US 10-year bond monthly returns. The time series can be compared to the times series of S&P 500 daily returns in Figure 2.1(b), or to the simulated time series in Figure 3.29.

Figure 3.1 shows empirical distribution functions of S&P 500 and US 10-year bond monthly returns: S&P 500 seems to have heavier tails than 10-year bond.

Figure 3.3 shows smooth tail plots of the daily returns of S&P 500 components and of S&P 500 index: The individual components seem to have heavier tails than the index.

Figure 3.8 shows empirical quantile functions of S&P 500 and US 10-year bond monthly returns: S&P 500 seems to have heavier tails than 10-year bond.

Figure 3.10(a) shows kernel density estimates of S&P 500 and US 10-year bond monthly returns: These estimates do not reveal information about the tails, but in the central area 10-year bond seems to be more concentrated around zero than S&P 500. Cont (2001) reports that returns of US Treasury bonds are positively skewed, whereas the returns of stock indices are negatively skewed.

Bouchaud (2002) reports that the tails of the stock returns have Pareto (power-law) tails $x^{-1-\alpha}$, where α is approximately 3, but emerging markets can have α smaller than 2. Cont (2001) notes that the tail index varies between 2 and 5 that excludes the Gaussian and the stable laws with infinite variance. The standard deviation of daily returns is 3% for stocks, 1% for stock indices, and 0.03% for short term interest rates; see Bouchaud (2002).

3) *What is the best model for the tails?*

Figures 3.17–3.26 study fitting of parametric models to the tails of S&P 500 returns. In particular, tail plots are shown for the exponential distribution in Figure 3.18, for the Pareto distribution in Figure 3.20, for the gamma distribution in Figure 3.22, for the generalized Pareto distribution in Figure 3.24, and for the Weibull distribution in Figure 3.26. Two-parameter families give reasonable fits, in particular, the generalized Pareto distribution gives a good fit.

Malevergne and Sornette (2005) give a review of fitting Pareto distributions, stretched exponentials and log-Weibull distributions.

4) *Are the left tail and the right equally heavy?*

The parameter estimates for fitting models to the daily returns of S&P 500 indicate that the left tail is heavier than the right tail (see Figures 3.17, 3.19, 3.21, 3.23, and 3.25).

Figure 3.28 shows values of estimates of Pareto tail index α for S&P 500 components, both for the left and right tail: There seems to be equal amount of stocks with a larger left tail index as there are stocks with a larger right tail index.

Cont (2001) reports that gains and loss are asymmetric; large drawdowns are observed but not equally large upward movements.

5) *How the heaviness of the tails is affected by the return horizon?*

Figure 3.12 shows values of estimates of parameters of t-distribution (degrees of freedom ν and scaling parameter σ) for various return horizons of S&P 500 returns: the estimates increase from $\hat{\nu} = 3$ for daily returns to $\hat{\nu} = 5, \ldots, 9$ for 2-month returns. Also $\hat{\sigma}$ increases with the return horizon. Figure 3.10(b) shows kernel density estimates of the S&P 500 return distribution when the return horizon varies between one and five days.

Cont (2001) observes that the distribution of returns looks more and more like a Gaussian distribution when the time scale is increased.

4

Multivariate Data Analysis

Multivariate data analysis studies simultaneously several time series, but the time series properties are ignored, and thus the analysis can be called cross-sectional.

The copula is an important concept of multivariate data analysis. Copula models are a convenient way to separate multivariate analysis to the purely univariate and to the purely multivariate components. We compose a multivariate distribution into the part that describes the dependence and into the parts that describe the marginal distributions. The marginal distributions can be estimated efficiently using nonparametric methods, but it can be useful to apply parametric models to estimate dependence, for a high-dimensional distribution. Combining nonparametric estimators of marginals and a parametric estimator of the copula leads to a semiparametric estimator of the distribution.

Multivariate data can be described using such statistics as linear correlation, Spearman's rank correlation, and Kendall's rank correlation. Linear correlation is used in the Markowitz portfolio selection. Rank correlations are more natural concepts to describe dependence, because they are determined by the copula, whereas linear correlation is affected by marginal distributions. Coefficients of tail dependence can capture whether the dependence of asset returns is larger during the periods of high volatility.

Multivariate graphical tools include scatter plots, which can be combined with multidimensional scaling and other dimension reduction methods.

Section 4.1 studies measures of dependence. Section 4.2 considers multivariate graphical tools. Section 4.3 defines multivariate parametric distributions such as multivariate normal, multivariate Student, and elliptical distributions. Section 4.4 defines copulas and models for copulas.

4.1 Measures of Dependence

Random vectors $X, Y \in \mathbf{R}^d$ are said to be independent if

$$P(X \in A, \ Y \in B) = P(X \in A) \cdot P(Y \in B),$$

Nonparametric Finance, First Edition. Jussi Klemelä.
© 2018 John Wiley & Sons, Inc. Published 2018 by John Wiley & Sons, Inc.

for all measurable $A, B \subset \mathbf{R}^d$. This is equivalent to

$$P(X \in A \mid Y \in B) = P(X \in A),$$

for all measurable $A, B \subset \mathbf{R}^d$, so knowledge of Y does not affect the probability evaluations of X. The complete dependence between random vectors X and Y occurs when there is a bijection $G : \mathbf{R}^d \to \mathbf{R}^d$ so that

$$Y = G(X) \qquad (4.1)$$

holds almost everywhere. When the random vectors are not independent and not completely dependent we may try to quantify the dependency between two random vectors. We may say that two random vectors have the same dependency when they have the same copula, and the copula is defined in Section 4.4.

Correlation coefficients are defined between two real valued random variables. We define three correlation coefficients: linear correlation ρ_L, Spearman's rank correlation ρ_S, and Kendall's rank correlation ρ_τ. All of these correlation coefficients satisfy

$$\rho(X_1, X_2) \in [-1, 1],$$

where X_1 and X_2 are real valued random variables. Furthermore, if X_1 and X_2 are independent, then $\rho(X_1, X_2) = 0$ for any of the correlation coefficients. Converse does not hold, so that correlation zero does not imply independence.

Complete dependence was defined by (4.1). Both for the Spearman's rank correlation and for the Kendall's rank correlation we have that

$$|\rho(X_1, X_2)| = 1 \text{ if and only if } X_1 \text{ and } X_2 \text{ are completely dependent,} \quad (4.2)$$

where $\rho = \rho_S$ or $\rho = \rho_\tau$. In the case of real valued random variables the complete dependency can be divided into comonotonicity and countermonotonicity. Real-valued random variables X_1 and X_2 are said to be comonotonic if there is a strictly increasing function $g : \mathbf{R} \to \mathbf{R}$ so that $X_2 = g(X_1)$ almost everywhere. Real-valued random variables X_1 and X_2 are said to be countermonotonic if there is a strictly decreasing function $g : \mathbf{R} \to \mathbf{R}$ so that $X_2 = g(X_1)$ almost everywhere. Both for the Spearman's rank correlation and for the Kendall's rank correlation we have that $\rho(X_1, X_2) = 1$ if and only if X_1 and X_2 are comonotonic, and $\rho(X_1, X_2) = -1$ if and only if X_1 and X_2 are countermonotonic, where $\rho = \rho_S$ or $\rho = \rho_\tau$.

The linear correlation coefficient ρ_L does not satisfy (4.2). However, we have that

$$|\rho_L(X_1, X_2)| = 1 \text{ if and only if } X_2 = a + bX_1 \text{ for some } a, b \in \mathbf{R}. \quad (4.3)$$

If $\rho_L(X_1, X_2) = 1$, then $b > 0$. If $\rho_L(X_1, X_2) = -1$, then $b < 0$.

4.1.1 Correlation Coefficients

We define linear correlation ρ_L, Spearman's rank correlation ρ_S, and Kendall's rank correlation ρ_τ.

4.1.1.1 Linear Correlation

The linear correlation coefficient between real valued random variables X_1 and X_2 is defined as

$$\rho_L(X_1, X_2) = \frac{\text{Cov}(X_1, X_2)}{\text{sd}(X_1)\,\text{sd}(X_2)}, \tag{4.4}$$

where the covariance is

$$\text{Cov}(X_1, X_2) = E[(X_1 - EX_1)(X_2 - EX_2)],$$

and the standard deviation is $\text{sd}(X_k) = \sqrt{\text{Var}(X_k)}$.

We noted in (4.3) that the linear correlation coefficient characterizes linear dependency. However, (4.2) does not hold for the linear correlation coefficient. Even when X_1 and X_2 are completely dependent, it can happen that $|\rho(X_1, X_2)| < 1$. For example, let $Z \sim N(0,1)$, $X_1 = e^Z$, and $X_2 = e^{\sigma Z}$, where $\sigma > 0$. Then,

$$\rho_L(X_1, X_2) = \frac{e^\sigma - 1}{\sqrt{(e-1)(e^{\sigma^2}-1)}},$$

and $\rho_L(X_1, X_2) = 1$ only for $\sigma = 1$, otherwise $0 < \rho_L(X_1, X_2) < 1$; the example is from McNeil *et al.* (2005, p. 205).

Let us assume that X_1 and X_2 have continuous distributions and let us denote with F the distribution function of (X_1, X_2) and with F_1 and F_2 the marginal distribution functions. Then,

$$\text{Cov}(X_1, X_2) = \int_{\mathbf{R}^2} (F(u,v) - F_1(u)F_2(v))\,du\,dv \tag{4.5}$$

$$= \int_{[0,1]^2} [C(u,v) - uv]\,dF_1^{-1}(u)\,dF_2^{-1}(v),$$

where $C(u,v) = F(F_1^{-1}(u), F_2^{-1}(v))$, $u, v \in [0,1]$, is the copula of the distribution of (X_1, X_2), as defined in (4.29). Equation (4.5) is called Höffding's formula, and its proof can be found in McNeil *et al.* (2005, p. 203). Thus, the linear correlation is not solely a function of the copula, it depends also on the marginal distributions F_1 and F_2.

The linear correlation coefficient can be estimated with the sample correlation. Let $X_{1,1}, \ldots, X_{1,T}$ be a sample from the distribution of X_1 and $X_{2,1}, \ldots, X_{2,T}$ be a sample from the distribution of X_2. The sample correlation coefficient is defined as

$$\hat{\rho}_L = \frac{1}{s_1 s_2} \cdot \frac{1}{T} \sum_{i=1}^{T} (X_{1,i} - \overline{X}_1)(X_{2,i} - \overline{X}_2), \tag{4.6}$$

where $\overline{X}_k = T^{-1} \sum_{i=1}^{T} X_{k,i}$ and $s_k^2 = T^{-1} \sum_{i=1}^{T} (X_{k,i} - \overline{X}_k)^2$. An alternative estimator is defined in (4.10).

4.1.1.2 Spearman's Rank Correlation

Spearman's rank correlation (Spearman's rho) is defined by

$$\rho_S(X_1, X_2) = \rho(F_1(X_1), F_2(X_2)),$$

where F_k is the distribution function of X_k, $k = 1, 2$. If X_1 and X_2 have continuous distributions, then

$$\rho_S(X_1, X_2) = 12 \int_{\mathbb{R}^2} [F(u, v) - F_1(u)F_2(v)] \, dF_1(u) \, dF_2(v)$$

$$= 12 \int_{[0,1]^2} [C(u, v) - uv] \, du \, dv,$$

where $C(u, v) = F(F_1^{-1}(u), F_2^{-1}(v))$, $u, v \in [0, 1]$, is the copula as defined in Section 4.4 (see McNeil *et al.*, 2005, p. 207).[1] Thus, Spearman's correlation coefficient is defined solely in terms of the copula.

We have still another way of writing Spearman's rank correlation. Let $X = (X_1, X_2)$, $Y = (Y_1, Y_2)$, and $Z = (Z_1, Z_2)$, let X, Y, Z have the same distribution, and let X, Y, Z be independent. Then,

$$\rho_S(X_1, X_2) = 2(P[(X_1 - Y_1)(X_2 - Z_2) > 0] - P[(X_1 - Y_1)(X_2 - Z_2) < 0]).$$

The sample Spearman's rank correlation can be defined as the sample linear correlation coefficient between the ranks. Let $X_{1,1}, \dots, X_{1,T}$ be a sample from the distribution of X_1 and $X_{2,1}, \dots, X_{2,T}$ be a sample from the distribution of X_2. The rank of observation $X_{i,k}$, $i = 1, \dots, T$, $k = 1, 2$, is

$$\mathrm{rank}(X_{k,i}) = \#\{X_{k,j} : X_{k,j} \le X_{k,i}, j = 1, \dots, T\}.$$

That is, $\mathrm{rank}(X_{i,k})$ is the number of observations of the kth variable smaller or equal to $X_{k,i}$.[2] Let us use the shorthand notation

$$r_k(i) = \mathrm{rank}(X_{k,i}),$$

so that $r_k : \{1, \dots, T\} \to \{1, \dots, T\}$, $k = 1, 2$. Then the sample Spearman's rank correlation can be written as

$$\hat{\rho}_S = \hat{\rho}_L(\{r_1(1), \dots, r_1(T)\}, \{r_2(1), \dots, r_2(T)\}),$$

---

1 We have also $\rho_S(X_1, X_2) = 12 \int_{[0,1]^2} C(u, v) \, du \, dv - 3 = 12 \int_{[0,1]^2} u \cdot v \, dC(u, v) - 3$.
2 Let $X_{k,(1)} < \cdots < X_{k,(T)}$ be the ordered observations, $k = 1, 2$. The ranks can be defined as

$$\mathrm{rank}(X_{k,i}) = \begin{cases} 1, & \text{when } X_{k,i} = X_{k,(1)}, \\ 2, & \text{when } X_{k,i} = X_{k,(2)}, \\ \vdots \end{cases}$$

where $\hat{\rho}_L$ is the sample linear correlation coefficient, defined in (4.6). Since $\sum_{i=1}^T i = (T+1)T/2$ and $\sum_{i=1}^T i^2 = T(T+1)(2T+1)/6$, we can write

$$\hat{\rho}_S = \frac{12}{T(T^2-1)} \sum_{i=1}^T \left(r_1(i) - \frac{1}{2}(T+1)\right)\left(r_2(i) - \frac{1}{2}(T+1)\right).$$

4.1.1.3 Kendall's Rank Correlation

Let $X = (X_1, X_2)$ and $Y = (Y_1, Y_2)$, let X and Y have the same distribution, and let X and Y be independent. Kendall's rank correlation (Kenadall's tau) is defined by

$$\rho_\tau(X_1, X_2)$$
$$= P[(X_1 - Y_1)(X_2 - Y_2) > 0] - P[(X_1 - Y_1)(X_2 - Y_2) < 0]. \qquad (4.7)$$

When X_1 and X_2 have continuous distributions, we have

$$\rho_\tau(X_1, X_2) = 2P[(X_1 - Y_1)(X_2 - Y_2) > 0] - 1,$$

and we can write

$$\rho_\tau(X_1, X_2) = 4 \int_{\mathbb{R}^2} F(u, v)\, dF(u, v) - 1$$
$$= 4 \int_{[0,1]^2} C(u, v)\, dC(u, v) - 1,$$

where $C(u, v) = F(F_1^{-1}(u), F_2^{-1}(v))$, $u, v \in [0, 1]$, is the copula as defined in Section 4.4 (see McNeil *et al.*, 2005, p. 207).

Let us define an estimator for $\rho_\tau(X_1, X_2)$. Let $X_{1,1}, \dots, X_{1,T}$ be a sample from the distribution of X_1 and $X_{2,1}, \dots, X_{2,T}$ be a sample from the distribution of X_2. Kendall's rank correlation can be written as

$$E \operatorname{sign}[(X_1 - Y_1)(X_2 - Y_2)],$$

where $\operatorname{sign}(t) = 1$, if $t \geq 0$ and $\operatorname{sign}(t) = -1$, if $t < 0$. This leads to the sample version

$$\hat{\rho}_\tau = \frac{2}{T(T-1)} \sum_{1 \leq i < j \leq T} \operatorname{sign}((X_{1,i} - X_{1,j})(X_{2,i} - X_{2,j})). \qquad (4.8)$$

The computation takes longer than for the sample linear correlation and for the sample Spearman's correlation.

4.1.1.4 Relations between the Correlation Coefficients

We have a relation between the linear correlation and the Kendall's rank correlation for the elliptical distributions. Let $X = (X_1, X_2)$ be a bivariate random vector. For all elliptical distributions with $P(X = 0) = 0$,

$$\rho_\tau(X_1, X_2) = \frac{2}{\pi} \arcsin \rho_L(X_1, X_2), \qquad (4.9)$$

where ρ_τ is the Kendall's rank correlation, as defined in (4.7), and ρ_L is the linear correlation, as defined in (4.4) (see McNeil *et al.*, 2005, p. 217). This relationship can be applied to get an alternative and a more robust estimator for the estimator (4.6) of linear correlation. Define the estimator as

$$\hat{\rho}_L = \sin\left(\frac{\pi}{2}\,\hat{\rho}_\tau\right), \tag{4.10}$$

where $\hat{\rho}_\tau$ is the estimator (4.8).

For the distributions with a Gaussian copula, we also have a relation between the Spearman's rank correlation and the linear correlation. Let $X = (X_1, X_2)$ be a distribution with a Gaussian copula and continuous margins. Then,

$$\rho_S(X_1, X_2) = \frac{6}{\pi}\arcsin\left(\frac{1}{2}\rho_L(X_1, X_2)\right),$$

and (4.9) holds also (see McNeil *et al.*, 2005, p. 215).

Figure 4.1 studies linear correlation and Spearman's rank correlation for S&P 500 and Nasdaq-100 daily data, described in Section 2.4.2. Panel (a) shows a moving average estimate of linear correlation (blue) and Spearman's rank correlation (yellow). We use the one-sided moving average defined as

$$\hat{\rho}_{L,t} = \frac{\sum_{i=1}^{t} p_i(t) R_{1,i} R_{2,i}}{\sqrt{\sum_{i=1}^{t} p_i(t) R_{1,i}^2 \sum_{i=1}^{t} p_i(t) R_{2,i}^2}},$$

where $R_{1,i}$ are the S&P 500 centered returns and $R_{2,i}$ are the Nasdaq-100 centered returns. The weights $p_i(t)$ are one for the last 500 observations, and zero for the other observations. See (6.5) for a more general moving average. The moving average estimator $\hat{\rho}_{S,t}$ is the Spearman's rho computed from the 500

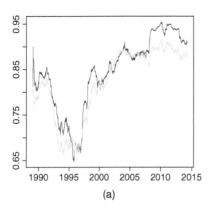 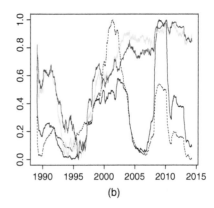

(a) (b)

Figure 4.1 *Linear and Spearman's correlation, together with volatility.* (a) Time series of moving average estimates of correlation between S&P 500 and Nasdaq-100 returns, with linear correlation (blue) and Spearman's rho (yellow); (b) we have added moving average estimates of the standard deviation of S&P 500 (black solid) and Nasdaq-100 (black dashed).

previous observations. Panel (b) shows the correlation coefficients together with the moving average estimates of the standard deviation of S&P 500 returns (solid black line) and Nasdaq-100 returns (dashed black line). All time series are scaled to take values in the interval $[0, 1]$. We see that there is some tendency that the inter-stock correlations increase in volatile periods.

4.1.2 Coefficients of Tail Dependence

The coefficient of upper tail dependence is defined for random variables X_1 and X_2 with distribution functions F_1 and F_2 as

$$\lambda_u = \lim_{q \uparrow 1} P\left(X_2 > F_2^{-1}(q) \mid X_1 > F_1^{-1}(q)\right),$$

where F_1^{-1} and F_2^{-1} are the generalized inverses. Similarly, the coefficient of lower tail dependence is

$$\lambda_l = \lim_{q \downarrow 0} P\left(X_2 \leq F_2^{-1}(q) \mid X_1 \leq F_1^{-1}(q)\right).$$

See McNeil *et al.* (2005, p. 209).

4.1.2.1 Tail Coefficients in Terms of the Copula

The coefficients of upper and lower tail dependence can be defined in terms of the copula. Let F_1 and F_2 be continuous. We have that

$$
\begin{aligned}
P\left(X_2 > F_2^{-1}(q), X_1 > F_1^{-1}(q)\right) \\
= P(q < F_1(X_2) \leq 1, q < F_2(X_2) \leq 1) \\
= C(1, 1) - C(1, q) - C(q, 1) + C(q, q) \\
= 1 - 2q + C(q, q).
\end{aligned}
$$

Also,

$$P\left(X_1 > F_1^{-1}(q)\right) = 1 - P(F_1(X_1) \leq q) = 1 - q.$$

Thus, the coefficient of upper tail dependence is

$$\lambda_u = \lim_{q \uparrow 1} \frac{1 - 2q + C(q, q)}{1 - q}. \tag{4.11}$$

We have that

$$P\left(X_2 \leq F_2^{-1}(q), X_1 \leq F_1^{-1}(q)\right) = P(F_1(X_2) \leq q, F_2(X_2) \leq q) = C(q, q).$$

Also,

$$P\left(X_1 \leq F_1^{-1}(q)\right) = P(F_1(X_1) \leq q) = q.$$

Thus, the coefficient of lower tail dependence for continuous F_1 and F_2 is equal to

$$\lambda_l = \lim_{q \downarrow 0} \frac{C(q, q)}{q}. \tag{4.12}$$

4.1.2.2 Estimation of Tail Coefficients

Equations (4.11) and (4.12) suggest estimators for the coefficients of tail dependence. We can estimate the upper tail coefficient nonparametrically, using

$$\hat{\lambda}_u = \frac{1 - 2q + \hat{C}(q, q)}{1 - q},$$

where $\hat{C}$ is the empirical copula, defined in (4.38), and $q < 1$ is close to 1. We can take, for example, $q = 1 - k/T$, where $k = \sqrt{T}$. The coefficient of lower tail dependence can be estimated by

$$\hat{\lambda}_l = \frac{\hat{C}(q, q)}{q},$$

where $q > 0$ is close to zero. We can take, for example, $q = k/T$, where $k = \sqrt{T}$. These estimators have been studied in Dobric and Schmid (2005), Frahm *et al.* (2005), and Schmidt and Stadtmüller (2006).

Figure 4.2 studies tail coefficients for S&P 500 and Nasdaq-100 daily data, described in Section 2.4.2. Panel (a) shows the tail coefficients as a function of q for lower tail coefficients (red) and as a function of $1 - q$ for upper tail coefficients (blue). Panel (b) shows a moving average estimate of the lower tail coefficients. The tail coefficient is estimated using the window of the latest 1000 observations, for $q = 0.01$.

4.1.2.3 Tail Coefficients for Parametric Families

The coefficients of lower and upper tail dependence for the Gaussian distributions are zero. The coefficients of lower and upper tail dependence for the

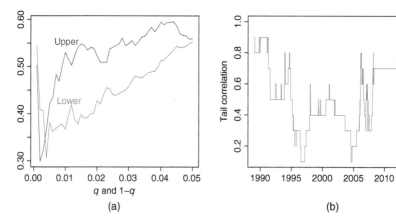

Figure 4.2 *Tail coefficients for S&P 500 and Nasdaq-100 returns.* (a) Tail coefficients as a function of q for lower tail coefficients (red) and as a function of $1 - q$ for upper tail coefficients (blue); (b) time series of moving average estimates of lower tail coefficients.

Student distributions with degrees of freedom v and correlation coefficient ρ are

$$\lambda_u = \lambda_l = 2t_{v+1}\left(-\sqrt{\frac{(v+1)(1-\rho)}{1+\rho}}\right),$$

where t_{v+1} is the distribution function of the univariate t-distribution with $v + 1$ degrees of freedom, and we assume that $\rho > -1$; see McNeil *et al.* (2005, p. 211).

4.2 Multivariate Graphical Tools

First, we describe scatter plots and smooth scatter plots. Second, we describe visualization of correlation matrices with multidimensional scaling.

4.2.1 Scatter Plots

A two-dimensional scatter plot is a plot of points $\{X_1, \dots, X_T\} \subset \mathbf{R}^2$.

Figure 4.3 shows scatter plots of daily net returns of S&P 500 and Nasdaq-100. The data is described in Section 2.4.2. Panel (a) shows the original data and panel (b) shows the corresponding scatter plot after copula preserving transform with standard normal marginals, as defined in (4.36).

When the sample size is large, then the scatter plot is mostly black, so the visuality of density of the points in different regions is obscured. In this case it is possible to use histograms to obtain a smooth scatter plot. A multivariate histogram is defined in (3.42). First we take square roots $f_i = \sqrt{n_i}$ of the bin

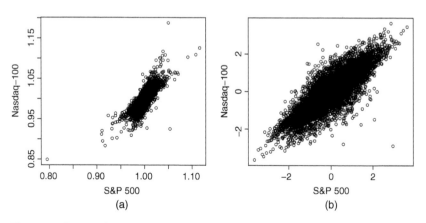

(a)   (b)

Figure 4.3 *Scatter plots.* Scatter plots of the net returns of S&P 500 and Nasdaq-100. (a) Original data; (b) copula transformed data with marginals being standard normal.

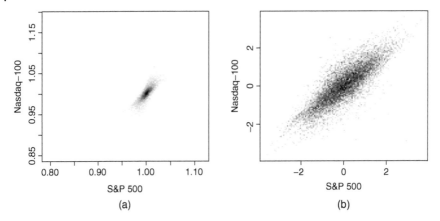

Figure 4.4 *Smooth scatter plots.* Scatter plots of the net returns of S&P 500 and Nasdaq-100. (a) Original data; (b) copula transformed data with marginals being standard normal.

counts n_i and then we define $g_i = 1 - (f_i - \min_i(f_i) + 0.5)/(\max_i(f_i) - \min_i(f_i) + 0.5)$. Now $g_i \in [0, 1]$. Values g_i close to one are shown in light gray, and values g_i close to zero are shown in dark gray. See Carr *et al.* (1987) for a study of histogram plotting.

Figure 4.4 shows smooth scatter plots of daily net returns of S&P 500 and Nasdaq-100. The data is described in Section 2.4.2. Panel (a) shows a smooth scatter plot of the original data and panel (b) shows the corresponding scatter plot after copula preserving transform when the marginals are standard Gaussian.

4.2.2 Correlation Matrix: Multidimensional Scaling

First, we define the correlation matrix. Second, we show how the correlation matrix may be visualized using multidimensional scaling.

4.2.2.1 Correlation Matrix

The correlation matrix is the $d \times d$ matrix whose elements are the linear correlation coefficients $\rho_L(X_i, X_j)$ for $i, j = 1, \dots, d$. The sample correlation matrix is the matrix whose elements are the sample linear correlation coefficients.

The correlation matrix can be defined using matrix notation. The covariance matrix of random vector $X = (X_1, \dots, X_d)'$ is defined by

$$\text{Cov}(X) = E[(X - EX)(X - EX)']. \tag{4.13}$$

The covariance matrix is the $d \times d$ matrix whose elements are $\text{Cov}(X_i, X_j)$ for $i, j = 1, \dots, d$, where we denote $\text{Cov}(X_i, X_i) = \text{Var}(X_i)$. Let

$$D = \text{diag}(1/\text{sd}(X_1), \dots, 1/\text{sd}(X_d))$$

be the diagonal matrix whose diagonal is the vector of the inverses of the standard deviations. Then the correlation matrix is

$$\mathrm{Cor}(X) = D\ \mathrm{Cov}(X)D.$$

The covariance matrix can be estimated by the sample covariance matrix

$$\widehat{\mathrm{Cov}}(X) = \frac{1}{T} \sum_{i=1}^{T} (X_i - \overline{X})(X_i - \overline{X})', \tag{4.14}$$

where $X_1, \dots, X_T \in \mathbf{R}^d$ are identically distributed observations whose distribution is the same as the distribution of X, and $\overline{X} = T^{-1} \sum_{i=1}^{T} X_i$ is the arithmetic mean.

4.2.2.2 Multidimensional Scaling

Multidimensional scaling makes a nonlinear mapping of data $X_1, \dots, X_T \in \mathbf{R}^d$ to $\mathbf{R}^2$, or to any space $\mathbf{R}^k$ with $2 \le k < d$. We can define the mapping $Q : \{X_1, \dots, X_T\} \to \mathbf{R}^2$ of multidimensional scaling in two steps:

1) Compute the pairwise distances $\|X_i - X_j\|$, $i \ne j$.
2) Find points $Q(X_1), \dots, Q(X_T) \in \mathbf{R}^2$ so that $\|Q(X_i) - Q(X_j)\| = \|X_i - X_j\|$ for $i \ne j$.

In practice, we may not be able to find a mapping that preserves the distances exactly, but we find a mapping $Q : \{X_1, \dots, X_T\} \to \mathbf{R}^2$ so that the stress functional

$$\sum_{1 \le i < j \le T} (\|X_i - X_j\| - \|Q(X_i) - Q(X_j)\|)^2$$

is minimized. Sammon's mapping uses the stress functional

$$\sum_{1 \le i < j \le T} \frac{(\|X_i - X_j\| - \|Q(X_i) - Q(X_j)\|)^2}{\|X_i - X_j\|}.$$

This stress functional emphasizes small distances. Numerical minimization is needed to solve the minimization problems.

Multidimensional scaling can be used to visualize correlations between time series. Let $X_i = (R_1^i, \dots, R_T^i)$ be the time series of returns of company i, where $i = 1, \dots, d$. When we normalize the time series of returns so that the vector of returns has sample mean zero and sample variance one, then the Euclidean distance is equivalent to using the correlation distance. Indeed, let

$$Y_i = \frac{X_i - \overline{X}_i}{s(X_i)},$$

where $\overline{X}_i = T^{-1} \sum_{t=1}^{T} X_t^i$ and $s^2(X_i) = T^{-1} \sum_{t=1}^{T} (X_t^i)^2 - \overline{X}_i^2$. Now

$$\frac{1}{T} \|Y_i - Y_j\|^2 = 2[1 - \hat{\rho}_L(X_i, X_j)],$$

where $\hat{\rho}_L(X_i, X_j)$ is the sample linear correlation. Thus, we apply the multidimensional scaling for the norm

$$\|Y_i - Y_j\|_{2,T}^2 = \frac{1}{T} \|Y_i - Y_j\|^2 = \frac{1}{T} \sum_{t=1}^{T} \left(Y_t^i - Y_t^j\right)^2,$$

which is obtained by dividing the Euclidean norm by $\sqrt{T}$. Since

$$-1 \le \hat{\rho}_L(X_i, X_j) \le 1,$$

we have that

$$0 \le \|Y_i - Y_j\|_{2,T} \le 2.$$

Zero correlation gives $\|Y_i - Y_j\|_{2,T} = \sqrt{2}$, positive correlations give $0 \le \|Y_i - Y_j\|_{2,T} < \sqrt{2}$, and negative correlations give $\sqrt{2} < \|Y_i - Y_j\|_{2,T} \le 2$.

Figure 4.5 studies correlations of the returns of the components of DAX 30. We have daily observations of the components of DAX 30 starting at January 02, 2003 and ending at May 20, 2014, which makes 2892 observations. Panel (a) shows the correlation matrix as an image. We have used R-function "image." Panel (b) shows the correlations with multidimensional scaling. We have used R-function "cmdscale." The image of the correlation matrix is not as helpful as the multidimensional scaling. For example, we see that the return time series of Volkswagen with the ticker symbol "VOW" is an outlier. The returns of Fresenius and Fresenius Medical Care ("FRE" and "FME") are highly correlated.

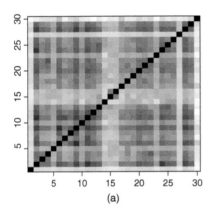 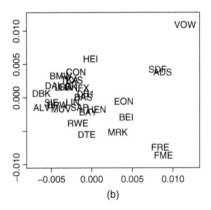

(a)   (b)

Figure 4.5 *Correlations of DAX 30.* (a) An image of the correlation matrix for DAX 30; (b) correlations for DAX 30 with multidimensional scaling.

4.3 Multivariate Parametric Models

We give examples of multivariate parametric models. The examples include Gaussian and Student distributions (t-distributions). More general families are normal variance mixture distributions and elliptical distributions.

4.3.1 Multivariate Gaussian Distributions

A d-dimensional Gaussian distribution can be parametrized with the expectation vector $\mu \in \mathbf{R}^d$ and the $d \times d$ covariance matrix Σ. When random vector X follows the Gaussian distribution with parameters μ and Σ, then we write $X \sim N(\mu, \Sigma)$ or $X \sim N_d(\mu, \Sigma)$. We say that a Gaussian distribution is the standard Gaussian distribution when $\mu = 0$ and $\Sigma = I_d$. The density function of the Gaussian distribution is

$$f(x) = (2\pi)^{-d/2} |\Sigma|^{-1/2} \exp\left\{ -\frac{1}{2}(x - \mu)'\Sigma^{-1}(x - \mu) \right\}, \tag{4.15}$$

where $x \in \mathbf{R}^d$ and $|\Sigma|$ is the determinant of Σ. The characteristic function of the Gaussian distribution is

$$\psi(t) = E \exp(it'X) = \exp\left\{ it'\mu - \frac{1}{2}t'\Sigma t \right\}, \tag{4.16}$$

where $t \in \mathbf{R}^d$.

A linear transformation of a Gaussian random vector follows a Gaussian distribution: When $X \sim N_d(\mu, \Sigma)$, A is $k \times d$ matrix, and a is a k vector, then

$$AX + a \sim N_k(A\mu + a, A\Sigma A'). \tag{4.17}$$

Also, when $X \sim N(\mu_1, \Sigma_1)$ and $Y \sim N(\mu_2, \Sigma_2)$ are independent, then

$$X + Y \sim N(\mu_1 + \mu_2, \Sigma_1 + \Sigma_2).$$

Both of these facts can be proved using the characteristic function.[3]

4.3.2 Multivariate Student Distributions

A d-dimensional Student distribution (t-distribution) is parametrized with degrees of freedom $\nu > 0$, the expectation vector $\mu \in \mathbf{R}^d$, and the $d \times d$ positive definite symmetric matrix Σ. When random vector X follows the t-distribution

---

3 The characteristic function of $AX + a$ is

$$\psi_{AX+a}(t) = E \exp(it'(AX + a)) = e^{it'a} E \exp(i(At)'X) = e^{it'a}\psi_X(At)$$

$$= \exp\left\{ it'(A\mu + a) - \frac{1}{2}t'A\Sigma A't \right\},$$

where ψ_X is the characteristic function of X.

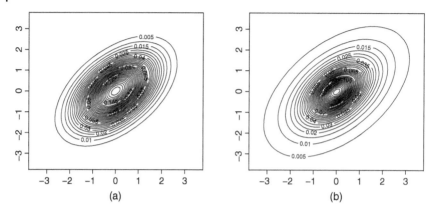

Figure 4.6 *Gaussian and Student densities.* (a) Contour plot of the Gaussian density with marginal standard deviations equal to one and correlation 0.5; (b) Student density with degrees of freedom 2 and correlation 0.5.

with parameters v, μ, and Σ, then we write $X \sim t(v, \mu, \Sigma)$ or $X \sim t_d(v, \mu, \Sigma)$. The density function of the multivariate t-distribution is

$$f(x) = c|\Sigma|^{-1/2}\left(1 + v^{-1}(x - \mu)'\Sigma^{-1}(x - \mu)\right)^{-(v+d)/2}, \tag{4.18}$$

where

$$c = \frac{\Gamma((v + d)/2)}{(\pi v)^{d/2}\Gamma(v/2)}. \tag{4.19}$$

The multivariate Student distributed random vector has the covariance matrix

$$\mathrm{Cov}(X) = \frac{v}{v - 2}\,\Sigma,$$

when $v > 2$.

When $v \to \infty$, then the Student density approaches a Gaussian density. Indeed, $(1 + t/v)^{-(d+v)/2} \to \exp\{-t/2\}$, as $v \to \infty$, since $(1 + a/v)^v \to e^a$, when $v \to \infty$. The Student density has tails $f(x) \asymp \|x\|^{-(d+v)}$, as $\|x\| \to \infty$.

Figure 4.6 compares multivariate Gaussian and Student densities. Panel (a) shows the Gaussian density with marginal standard deviations equal to one and correlation 0.5. Panel (b) shows the density of t-distribution with degrees of freedom 2 and correlation 0.5. The density contours are in both cases ellipses but the Student density has heavier tails.

4.3.3 Normal Variance Mixture Distributions

Random vector $X \in \mathbf{R}^d$ follows a Gaussian distribution with parameters μ and Σ when $\Sigma = AA'$ for a $d \times d$ matrix A and

$$X \sim \mu + AZ,$$

where $Z \sim N_d(0, I_d)$ follows the standard Gaussian distribution. This leads to the definition of a normal variance mixture distribution. We say that $X \in \mathbf{R}^d$ follows a normal variance mixture distribution when

$$X \sim \mu + \sqrt{W}AZ,$$

where $Z \sim N_d(0, I_d)$ follows the standard Gaussian distribution, and $W \geq 0$ is a random variable independent of Z. It holds that

$$EX = \mu$$

and

$$\mathrm{Cov}(X) = EW \cdot \Sigma,$$

where $\Sigma = AA'$. When random vector X follows the normal variance mixture distribution with parameters μ, Σ, and F_W, where F_W is the distribution function on W, then we write $X \sim M(\mu, \Sigma, F_W)$.

The density function can be calculated as

$$f_X(x) = \int_0^\infty f_{X,W}(x, w) \, dw = \int_0^\infty f_{X|W=w}(x) f_W(w) \, dw$$

$$= |\Sigma|^{-1/2} g((x - \mu)' \Sigma^{-1}(x - \mu)), \tag{4.20}$$

where $f_{X,W}$ is the density of (X, W), f_W is the density of W, $f_{X|W=w}$ is the density of X conditional on $W = w$, and $g : \mathbf{R} \to \mathbf{R}$ is defined by

$$g(x) = (2\pi)^{-d/2} \int_0^\infty w^{-d/2} \exp\left\{-\frac{x}{2w}\right\} f_W(w) \, dw. \tag{4.21}$$

The characteristic function is obtained, using (4.16), as

$$\psi_X(t) = E \exp(it'X) = EE(\exp(it'X) \mid W)$$

$$= E \exp\left\{it'\mu - \frac{1}{2}Wt'\Sigma t\right\} = e^{it'\mu} \phi_W\left(\frac{1}{2}t'\Sigma t\right),$$

where $\phi_W(t) = E \exp(-tW)$.

The family of normal variance mixtures $M(\mu, \Sigma, F_W)$ is closed under linear transformations: When $X \sim M_d(\mu, \Sigma, F_W)$, A is $k \times d$ matrix, and a is a k vector, then

$$AX + a \sim M_k(A\mu + a, A\Sigma A', F_W). \tag{4.22}$$

This can be seen using the characteristic function, similarly as in (4.17).

Let W be such random variable that vW^{-1} follows the χ^2-distribution with degrees of freedom $v > 0$. Then the normal variance mixture distribution is the multivariate t-distribution $t_d(v, \mu, \Sigma)$, where $\Sigma = AA'$, as defined in Section 4.3.2.

4.3.4 Elliptical Distributions

The density function of an elliptical distribution has the form

$$f(x) = |\det(\Sigma)|^{-1/2} g\{(x - \mu)'\Sigma^{-1}(x - \mu)\}, \quad x \in \mathbf{R}^d, \tag{4.23}$$

where $g : [0, \infty) \to [0, \infty)$ is called the density generator, Σ is a symmetric positive definite $d \times d$ matrix, and $\mu \in \mathbf{R}^d$. Since Σ is positive definite, it has inverse Σ^{-1} that is positive definite, which means that for all $z \in \mathbf{R}^d$, $z'\Sigma^{-1}z > 0$. Thus, g needs to be defined only on the nonnegative real axis. Let $g_1 : [0, \infty) \to [0, \infty)$ be such that $\int_0^\infty t^{d/2-1} g_1(t)\, dt < \infty$. Then $g = c \cdot g_1$ is a density generator when c is chosen by

$$c^{-1} = \int_{\mathbf{R}^d} g_1(\|x\|^2)\, dx = \text{volume}(S_d)\, 2^{-1} \int_0^\infty t^{d/2-1} g_1(t)\, dt, \tag{4.24}$$

where $S_d = \{x \in \mathbf{R}^d : \|x\| = 1\}$. We give examples of density generators.

1) From (4.15) we see that the Gaussian distributions are elliptical and the Gaussian density generator is

$$g(t) = c \cdot \exp\{-t/2\}, \quad t \in \mathbf{R}, \tag{4.25}$$

where $c = (2\pi)^{-d/2}$.

2) From (4.20) we see that the normal variance mixture distributions are elliptical and the normal variance mixture density generator is given in (4.21).

3) From (4.18) we see that the t-distributions are elliptical and the Student density generator is

$$g(t) = c \cdot (1 + t/v)^{-(d+v)/2}, \quad t \in \mathbf{R}, \tag{4.26}$$

where $v > 0$ is the degrees of freedom, and c is defined in (4.19). The Student density generator has tails $g(t) \asymp t^{-(d+v)/2}$, as $t \to \infty$, and thus the density function is integrable when $v > 0$, according to (4.24).

Let $\Sigma = AA'$, where A is a $d \times d$ matrix and let

$$X = \mu + AY,$$

where Y follows a spherical distribution with density $f(x) = g(\|x\|^2)$. Then X follows an elliptical distribution with density (4.23). When random vector X follows the elliptical distribution with parameters μ, Σ, and F_Y, where F_Y is the distribution function on Y, then we write $X \sim E_d(\mu, \Sigma, F_Y)$. The family of elliptical distributions is closed under linear transformations: When $X \sim E_d(\mu, \Sigma, F_Y)$, A is a $k \times d$ matrix, and a is a k vector, then

$$AX + a \sim E_k(A\mu + a, A\Sigma A', F_Y). \tag{4.27}$$

This can be seen using the characteristic function, similarly as in (4.17).

4.4   Copulas

We can decompose a multivariate distribution into a part that describes the dependence and into parts that describe the marginal distributions. This decomposition helps to estimate and analyze multivariate distributions, and it helps to construct new parametric and semiparametric models for multivariate distributions.

The distribution function $F : \mathbf{R}^d \to \mathbf{R}$ of random vector $(X_1, \dots, X_d)$ is defined by

$$F(x_1, \dots, x_d) = P(X_1 \le x_1, \dots, X_d \le x_d),$$

where $(x_1, \dots, x_d) \in \mathbf{R}^d$. The distribution functions $F_1 : \mathbf{R} \to \mathbf{R}$, ..., $F_d : \mathbf{R} \to \mathbf{R}$ of the marginal distributions are defined by

$$F_1(x_1) = P(X_1 \le x_1), \dots, F_d(x_d) = P(X_d \le x_d),$$

where $x_1, \dots, x_d \in \mathbf{R}$.

A copula is a distribution function $C : [0, 1]^d \to [0, 1]$ whose marginal distributions are the uniform distributions on $[0, 1]$. Often it is convenient to define a copula as a distribution function $C : \mathbf{R}^d \to [0, 1]$ whose marginal distributions are the standard normal distributions. Any distribution function $F : \mathbf{R}^d \to \mathbf{R}$ may be written as

$$F(x_1, \dots, x_d) = C(F_1(x_1), \dots, F_d(x_d)),$$

where F_i, $i = 1, \dots, d$, are the marginal distribution functions and C is a copula. In this sense we can decompose a distribution into a part that describes only the dependence and into parts that describe the marginal distributions.

We show in (4.29) how to construct a copula of a multivariate distribution and in (4.31) how to construct a multivariate distribution function from a copula and marginal distribution functions. We restrict ourselves to the case of continuous marginal distribution functions. These constructions were given in Sklar (1959), who considered also the case of noncontinuous margins. For notational convenience we give the formulas for the case $d = 2$. The generalization to the cases $d > 2$ is straightforward.

4.4.1   Standard Copulas

We use the term "standard copula," when the marginals of the copula have the uniform distributions on $[0, 1]$. Otherwise, we use the term "nonstandard copula."

4.4.1.1   Finding the Copula of a Multivariate Distribution

Let X_1 and X_2 be real valued random variables with distribution functions $F_1 : \mathbf{R} \to [0, 1]$ and $F_2 : \mathbf{R} \to [0, 1]$. Let $F : \mathbf{R}^2 \to [0, 1]$ be the distribution function of (X_1, X_2), and assume that F_1 and F_2 are continuous. Then,

$$
\begin{aligned}
F(x_1, x_2) &= P(X_1 \le x_1, X_2 \le x_2) \\
&= P(F_1(X_1) \le F_1(x_1), F_2(X_2) \le F_2(x_2)) \\
&= C(F_1(x_1), F_2(x_2)),
\end{aligned}
\tag{4.28}
$$

where

$$
\begin{aligned}
C(u, v) &= P(F_1(X_1) \le u, F_2(X_2) \le v) \\
&= F\left(F_1^{-1}(u), F_2^{-1}(v)\right),
\end{aligned}
\tag{4.29}
$$

and $u, v \in [0, 1]$. We call $C : [0, 1]^2 \to [0, 1]$ in (4.29) the copula of the joint distribution of X_1 and X_2. Copula C is the distribution function of the vector $(F_1(X_1), F_2(X_2))$, and $F_1(X_1)$ and $F_2(X_2)$ are uniformly distributed random variables.[4]

The copula density is

$$
c(u, v) = \frac{f\left(F_1^{-1}(u), F_2^{-1}(v)\right)}{f_1\left(F_1^{-1}(u)\right) \cdot f_2\left(F_2^{-1}(v)\right)}
\tag{4.30}
$$

because $(\partial/\partial u)F_i^{-1}(u) = 1/f_i(F_i^{-1}(u))$, where f is the density of F and f_1, and f_2 are the densities of F_1 and F_2, respectively.

4.4.1.2 Constructing a Multivariate Distribution from a Copula

Let $C : [0, 1]^2 \to [0, 1]$ be a copula, that is, it is a distribution function whose marginal distributions are uniform on $[0, 1]$. Let $F_1 : \mathbf{R} \to [0, 1]$ and $F_2 : \mathbf{R} \to [0, 1]$ be univariate distribution functions of continuous distributions. Define $F : \mathbf{R}^2 \to [0, 1]$ by

$$
F(x_1, x_2) = C(F_1(x_1), F_2(x_2)).
\tag{4.31}
$$

Then F is a distribution function whose marginal distributions are given by distribution functions F_1 and F_2. Indeed, Let (U_1, U_2) be a random vector with distribution function C. Then,

$$
\begin{aligned}
C(F_1(x_1), F_2(x_2)) &= P(U_1 \le F_1(x_1), U_2 \le F_2(x_2)) \\
&= P\left(F_1^{-1}(U_1) \le x_1, F_2^{-1}(U_2) \le x_2\right)
\end{aligned}
$$

and $F_i^{-1}(U_i) \sim F_i$ for $i = 1, 2$, because $U_i \sim \text{Uniform}([0, 1])$.[5]

4.4.2 Nonstandard Copulas

Typically a copula is defined as a distribution function with uniform marginals. However, we can define a copula so that the marginal distributions of the copula

---

4 We have that $P(F_1(X_1) \le t) = P(X_1 \le F_1^{-1}(t)) = F_1(F_1^{-1}(t)) = t$, for $t \in [0, 1]$, since F_1 is assumed to be strictly increasing.
5 We have $P(F^{-1}(U_i) \le t) = P(U_i \le F_i(t)) = F_i(t)$, for $t \in \mathbf{R}$.

is some other continuous distribution than the uniform distribution on $[0, 1]$. It turns out that we get simpler copulas by choosing the marginal distributions of a copula to be the standard Gaussian distribution.

As in (4.28) we can write distribution function $F : \mathbf{R}^2 \to [0, 1]$ as

$$F(x_1, x_2) = C(\Phi^{-1}(F_1(x_1)), \Phi^{-1}(F_2(x_2))),$$

where $\Phi : \mathbf{R} \to \mathbf{R}$ is the distribution function of the standard Gaussian distribution and

$$\begin{aligned} C(u, v) &= P(\Phi^{-1}(F_1(X_1)) \leq u, \Phi^{-1}(F_2(X_2)) \leq v) \\ &= F\left(F_1^{-1}(\Phi(u)), F_2^{-1}(\Phi(v))\right), \end{aligned} \tag{4.32}$$

$u, v \in \mathbf{R}$. Now $C : \mathbf{R}^2 \to [0, 1]$ is a distribution function whose marginals are standard Gaussians, because $F_i(X_i)$ follow the uniform distribution on $[0, 1]$ and thus $\Phi^{-1}(F_i(X_i))$ follow the standard Gaussian distribution.

Conversely, given a distribution function $C : \mathbf{R}^2 \to [0, 1]$ with the standard Gaussian marginals, and univariate distribution functions F_1 and F_2, we can define a distribution function $F : \mathbf{R}^2 \to [0, 1]$ with marginals F_1 and F_2 by the formula

$$F(x_1, x_2) = C(\Phi^{-1}(F_1(x_1)), \Phi^{-1}(F_2(x_2))).$$

The copula density is

$$\begin{aligned} c(u, v) = &f\left(F_1^{-1}(\Phi(u)), F_2^{-1}(\Phi(v))\right) \\ &\times \frac{\phi(u)\phi(v)}{f_1\left(F_1^{-1}(\Phi(u))\right) \cdot f_2\left(F_2^{-1}(\Phi(v))\right)}, \end{aligned} \tag{4.33}$$

where f is the density of F, f_1 and f_2 are the densities of F_1 and F_2, respectively, and ϕ is the density of the standard Gaussian distribution.

4.4.3 Sampling from a Copula

We do not have observations directly from the distribution of the copula but we show how to transform the sample so that we get a pseudo sample from the copula. Scatter plots of the pseudo sample can be used to visualize the copula. The pseudo sample can also be used in the maximum likelihood estimation of the copula. Before defining the pseudo sample, we show how to generate random variables from a copula.

4.4.3.1 Simulation from a Copula

Let random vector $X = (X_1, X_2)$ have a continuous distribution. Let $F_k(t) = P(X_k \leq t)$, $k = 1, 2$, be the distribution functions of the margins of X. Now

$$Z = (F_1(X_1), F_2(X_2)) \tag{4.34}$$

is a random vector whose marginal distributions are uniform on $[0, 1]$. The distribution function of this random vector is the copula of the distribution of $X = (X_1, X_2)$. Thus, if we can generate a random vector X with distribution F, we can use the rule (4.34) to generate a random vector Z whose distribution is the copula of F. Often the copula with uniform marginals is inconvenient due to boundary effects. We may get statistically more tractable distribution by defining

$$Z = (\Phi^{-1}(F_1(X_1)), \Phi^{-1}(F_2(X_2))),$$

where Φ is the distribution function of the standard Gaussian distribution. The components of Z have the standard Gaussian distribution.

4.4.3.2 Transforming the Sample

Let us have data $X_1, \dots, X_T \in \mathbf{R}^2$ and denote $X_i = (X_{i1}, X_{i2})$. Let the rank of observation X_{ik}, $i = 1, \dots, T$, $k = 1, 2$, be

$$\text{rank}(X_{ik}) = \#\{X_{jk} : X_{jk} \le X_{ik}, j = 1, \dots, T\}.$$

That is, $\text{rank}(X_{ik})$ is the number of observations of the kth variable smaller or equal to X_{ik}. We normalize the ranks to get observations on $[0, 1]^2$:

$$Z_i = \left(\frac{\text{rank}(X_{i1})}{T+1}, \frac{\text{rank}(X_{i2})}{T+1} \right), \tag{4.35}$$

for $i = 1, \dots, T$. Now $\{Z_{1k}, \dots, Z_{Tk}\} = \{1/(T+1), \dots, T/(T+1)\}$ for $k = 1, 2$. In this sense we can consider the observations as a sample from a distribution whose margins are uniform distributions on $[0, 1]$. Often the standard Gaussian distribution is more convenient and we define

$$Z_i = \left(\Phi^{-1}\left(\frac{\text{rank}(X_{i1})}{T+1} \right), \Phi^{-1}\left(\frac{\text{rank}(X_{i2})}{T+1} \right) \right), \tag{4.36}$$

for $i = 1, \dots, T$.

4.4.3.3 Transforming the Sample by Estimating the Margins

We can transform the data $X_1, \dots, X_T \in \mathbf{R}^2$ using estimates of the marginal distributions. Let $\hat{F}_1$ and $\hat{F}_2$ be estimates of the marginal distribution functions F_1 and F_2, respectively. We define the pseudo sample as

$$Z_i = \left(\hat{F}_1(X_{i1}), \hat{F}_2(X_{i2}) \right), \tag{4.37}$$

where $i = 1, \dots, T$. The estimates $\hat{F}_1$ and $\hat{F}_2$ can be parametric estimates. For example, assuming that the kth marginal distribution is a normal distribution, we would take $\hat{F}_k(t) = \Phi((x - \hat{\mu}_k)/\hat{\sigma}_k)$, where Φ is the distribution function of the standard normal distribution, $\hat{\mu}_k$ is the sample mean of $X_{1k}, \ldots, X_{Tk}$, and $\hat{\sigma}_k$ is the sample standard deviation. If $\hat{F}_k$ are the empirical distribution functions

$$\hat{F}_k(t) = \frac{1}{T} \sum_{i=1}^{T} I_{(-\infty,t]}(X_{ik}),$$

then we get almost the same transformation as (4.35), but $T + 1$ is now replaced by T:

$$Z_i = \left(\frac{\text{rank}(X_{i1})}{T}, \frac{\text{rank}(X_{i2})}{T} \right).$$

4.4.3.4 Empirical Copula
The empirical distribution function $\hat{F} : \mathbf{R}^2 \to [0, 1]$ is calculated using a sample $X_1, \ldots, X_T \in \mathbf{R}^2$ of identically distributed observations, and we define

$$\hat{F}(x_1, x_2) = \frac{1}{T} \sum_{i=1}^{T} I_{(-\infty,x_1]}(X_{i1}) \cdot I_{(-\infty,x_2]}(X_{i2}),$$

where we denote $X_i = (X_{i1}, X_{i2})$.

The empirical copula is defined similarly as the empirical distribution function. Now,

$$\hat{C}(u_1, u_2) = \frac{1}{T} \sum_{i=1}^{T} I_{[0,u_1]}(Z_{i1}) \cdot I_{[0,u_2]}(Z_{i2}), \tag{4.38}$$

where Z_i are defined in (4.37).

4.4.3.5 Maximum Likelihood Estimation
Pseudo samples are needed in maximum likelihood estimation. In maximum likelihood estimation we assume that the copula has a parametric form. For example, the copula of the normal distribution, given in (4.39), is parametrized with the correlation matrix, which contains $d(d-1)/2$ parameters. Let $C(u_1, \ldots, u_d; \theta)$ be the copula with parameter $\theta \in \Theta$. The corresponding copula density is $c(u_1, \ldots, u_d; \theta)$, as given in (4.30). Let us have independent and identically distributed observations $X_1, \ldots, X_T$ from the distribution of F. We calculate the pseudo sample $Z_1, \ldots, Z_T$ using (4.35) or (4.37). A maximum likelihood estimate is a value $\hat{\theta}$ maximizing

$$\prod_{i=1}^{T} c_\theta(Z_i; \theta)$$

over $\theta \in \Theta$.

4.4.4 Examples of Copulas

We give examples of parametric families of copulas. The examples include the Gaussian copulas and the Student copulas.

4.4.4.1 The Gaussian Copulas

Let $X \sim N(\mu, \Sigma)$ be a d-dimensional Gaussian random vector, as defined in Section 4.3.1. The copula of X is

$$C(u_1, \dots, u_d) = \Phi_P(\Phi^{-1}(u_1), \dots, \Phi^{-1}(u_d)), \tag{4.39}$$

where Φ_P is the distribution function of $N(0, P)$ distribution, P is the correlation matrix of X, and Φ is the distribution function of $N(0, 1)$ distribution.

Indeed, let us denote $\Delta = \mathrm{diag}(\sigma_1, \dots, \sigma_d)$, where σ_i is the standard deviation of X_i. Then $\Delta^{-1}(X - \mu)$ follows the distribution $N(0, P)$.[6] Let F be the distribution function of X. Then, using the notation $(X \leq x) = (X_1 \leq x_1, \dots, X_d \leq x_d)$,

$$F(x) = P(X \leq x) = P(\Delta^{-1}(X - \mu) \leq \Delta^{-1}(x - \mu))$$
$$= \Phi_P(\Delta^{-1}(x - \mu))$$
$$= \Phi_P\left(\frac{x_1 - \mu_1}{\sigma_1}, \dots, \frac{x_d - \mu_d}{\sigma_d}\right),$$

where $\mu = (\mu_1, \dots, \mu_d)$. Also,[7]

$$F_i^{-1}(u_i) = \mu_i + \sigma_i \Phi^{-1}(u_i),$$

for $i = 1, \dots, d$. Thus,

$$F\left(F_1^{-1}(u_1), \dots, F_d^{-1}(u_d)\right) = \Phi_P\left(\Phi^{-1}(u_1), \dots, \Phi^{-1}(u_d)\right).$$

Thus, (4.29) leads to (4.39).

Figure 4.7 shows perspective plots of the densities of the Gaussian copula. The margins are uniform on $[0, 1]$. The correlation parameter is in panel (a) $\rho = 0.1$ and in panel (b) $\rho = 0.8$. Figure 4.7 shows that the perspective plots of the copula densities are not intuitive, because the probability mass is concentrated near the corners of the square $[0, 1]^2$, especially when the correlation is high. From now on we will show only pictures of copulas with standard Gaussian margins, as defined in (4.32), because these give more intuitive representation of the copula.

4.4.4.2 The Student Copulas

Let $X \sim t(\nu, \mu, \Sigma)$ be a d-dimensional t-distributed random vector, as defined in Section 4.3.2. The copula of X is

$$C(u_1, \dots, u_d) = T_{\nu, P}\left(t_\nu^{-1}(u_1), \dots, t_\nu^{-1}(u_d)\right),$$

---

6 Random vector $\Delta^{-1}(X - \mu)$ has expectation zero and covariance matrix $P = \Delta^{-1}\mathrm{Cov}(X)\Delta^{-1}$.
7 If $y_1 = F_1^{-1}(u_1)$, then $u_1 = F_1(y_1) = \Phi((y_1 - \mu_1)/\sigma_1)$, which leads to $y_1 = \mu_1 + \sigma_1 \Phi^{-1}(u_1)$.

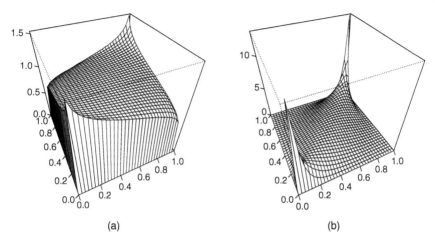

Figure 4.7 *Gaussian copulas*. Perspective plots of the densities of the Gaussian copula with the correlation (a) 0.1 and (b) 0.8. The margins are uniform on $[0, 1]$.

where $T_{\nu,P}$ is the distribution function of $t(\nu, 0, P)$ distribution, P is the correlation matrix of X, and t_ν is the distribution function of the univariate t-distribution with degrees of freedom ν.

Indeed, the claim follows similarly as in the Gaussian case for

$$P = \Delta^{-1} \Sigma \Delta^{-1},$$

where $\Delta = \operatorname{diag}(\sigma_1, \dots, \sigma_d)$ and σ_i is the square root of the ith element in the diagonal of Σ. The matrix P is indeed the correlation matrix, since

$$\operatorname{Cor}(X) = \Gamma^{-1} \operatorname{Cov}(X) \Gamma^{-1} = \Delta^{-1} \Sigma \Delta^{-1},$$

where $\Gamma = \operatorname{diag}(\operatorname{sd}(X_1), \dots, \operatorname{sd}(X_d))$, $\operatorname{sd}(X_k) = \sqrt{\operatorname{Var}(X_k)}$.

Figure 4.8 shows contour plots of the densities of the Student copula when the margins are standard Gaussian. The correlation is $\rho = 0.5$. The degrees of freedom are in panel (a) two and in panel (b) four. The Gaussian and Student copulas are similar in the main part of the distribution but they differ in the tails (in the corners of the unit square). The Gaussian copula has independent extremes (asymptotic tail independence) but the Student copula generates concomitant extremes with a nonzero probability. The probability of concomitant extremes is larger when the degrees of freedom is smaller and the correlation coefficient is larger.

4.4.4.3 Other Copulas

We define Gumbel and Clayton copulas. These are examples of Archimedean copulas. Gaussian and Student copulas are examples of elliptical copulas.

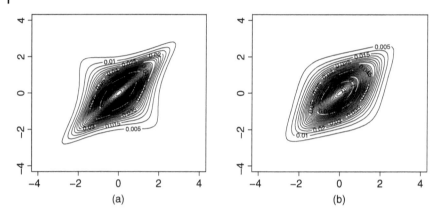

Figure 4.8 *Student copula with standard Gaussian margins.* Contour plots of the densities of the Student copula with degrees of freedom (a) 2 and (b) 4. The correlation is $\rho = 0.5$.

The Gumbel–Hougaard Copulas The Gumbel–Hougaard or the Gumbel family of copulas is defined by

$$C_{gh}(u; \theta) = \exp\left\{ -\left[\sum_{i=1}^{d} (-\log_e u_i)^\theta \right]^{1/\theta} \right\}, \quad u \in [0, 1]^d,$$

where $\theta \in [1, \infty)$ is the parameter. When $\theta = 1$, then $C_{gh}(u; \theta) = u_1 \cdots u_d$ and when $\theta \to \infty$, then $C_{gh}(u; \theta) \to \min\{u_1, \dots, u_d\}$.

Figure 4.9 shows contour plots of the densities with the Gumbel copula when $\theta = 1.5$, $\theta = 2$, and $\theta = 4$. The marginals are standard Gaussian.

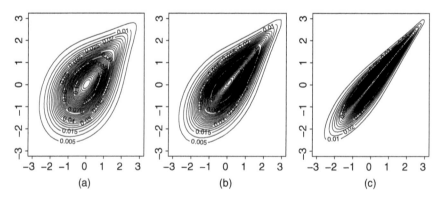

Figure 4.9 *Gumbel copula.* Contour plots of the densities of the Gumbel copula with $\theta = 1.5$, $\theta = 2$, and $\theta = 4$. The marginals are standard Gaussian.

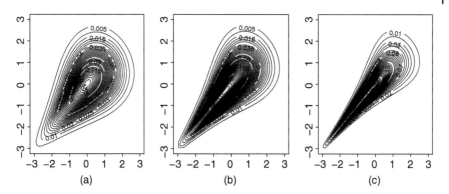

Figure 4.10 *Clayton copula.* Contour plots of the densities of the Clayton copula with $\theta = 1$, $\theta = 2$, and $\theta = 4$. The marginals are standard Gaussian.

The Clayton Copulas Clayton's family of copulas is defined by

$$C_{cl}(u; \theta) = \left(1 - d + \sum_{i=1}^{d} u_i^{-\theta} \right)^{-1/\theta}, \quad u \in [0, 1]^d, \tag{4.40}$$

where $\theta > 0$. When $\theta = 0$, we define $C_{cl}(u; \theta) = \prod_{i=1}^{d} u_i$. When the parameter θ increases, then the dependence between coordinate variables increases. The dependence is larger in the negative orthant. The Clayton family was discussed in Clayton (1978).

Figure 4.10 shows contour plots of the densities with the Clayton copula when $\theta = 1$, $\theta = 2$, and $\theta = 4$. The marginals are standard Gaussian.

Elliptical Copulas Elliptical distributions are defined in Section 4.3.4. An elliptical copula is obtained from an elliptical distribution F by the construction (4.29). The Gaussian copula and the Student copula are elliptical copulas.

Archimedean Copulas Archimedean copulas have the form

$$C(u) = \phi^{-1}(\phi(u_1) + \cdots + \phi(u_d)), \quad u \in [0, 1]^d,$$

where $\phi : [0, 1] \to [0, \infty)$ is strictly decreasing, continuous, convex, and $\phi(1) = 0$. For C to be a copula, we need that $(-1)^i \partial^i \phi^{-1}(t)/\partial t^i \geq 0$, $i = 1, \ldots, d$. The function ϕ is called the generator. The product copula, Gumbel copula, Clayton copula, and Frank copula are all Archimedean copulas and we have:

- product copula: $\phi(t) = \log_e t$,
- Gumbel copula: $\phi(t) = (-\log_e t)^\theta$,
- Clayton copula: $\phi(t) = t^{-\theta} - 1$,
- Frank copula: $\phi(t) = -\log_e[(e^{-\theta t} - 1)/(e^{-\theta} - 1)]$.

The density of an Archimedean copula is

$$c(u) = \psi(\phi(u_1) + \cdots + \phi(u_d))\phi'(u_1) \cdots \phi'(u_d),$$

where ψ is the second derivative of ϕ^{-1}:

$$\psi(y) = -\frac{\phi''(\phi^{-1}(y))}{(\phi'(\phi^{-1}(y)))^3},$$

because $(\partial/\partial x)\phi^{-1}(x) = 1/\phi'(\phi^{-1}(x))$. We have:

- Gumbel copula: $\phi'(t) = -\theta t^{-1}(-\log_e t)^{\theta-1}$, $\phi''(t) = \theta t^{-2}(-\log_e t)^{\theta-2}(\theta - 1 - \log_e t)$, $\phi^{-1}(t) = \exp(-t^{1/\theta})$,
- Clayton copula: $\phi'(t) = -\theta t^{-\theta-1}$, $\phi''(t) = \theta(\theta + 1)t^{-\theta-2}$, $\phi^{-1}(t) = (t + 1)^{-1/\theta}$,
- Frank copula: $\phi'(t) = \theta e^{-\theta t}/(e^{-\theta t} - 1)$, $\phi''(t) = \theta^2 e^{-\theta t}/(e^{-\theta t} - 1)^2$, $\phi^{-1}(t) = -\theta^{-1}\log_e[(e^{-\theta} - 1)e^{-t} + 1]$.

4.4.4.4 Empirical Results

Research on testing the hypothesis of Gaussian copula and other copulas on financial data has been done in Malevergne and Sornette (2003), and summarized in Malevergne and Sornette (2005). They found that the Student copula is a good model for foreign exchange rates but for the stock returns the situation is not clear.

Patton (2005) takes into account the volatility clustering phenomenon. He filters the marginal data by a GARCH process and shows that the conditional dependence structure between Japanese Yen and Euro is better described by Clayton's copula than by the Gaussian copula. Note, however, that the copula of the residuals is not the same as the copula of the raw returns and many filters can be used (ARCH, GARCH, and multifractal random walk). Using the multivariate multifractal filter of Muzy *et al.* (2001) leads to a nearly Gaussian copula.

Breymann *et al.* (2003) show that the daily returns of German Mark/Japanese Yen are best described by a Student copula with about six degrees of freedom, when the alternatives are the Gaussian, Clayton's, Gumbel's, and Frank's copulas. The Student copula seems to provide an even better description for returns at smaller time scales, when the time scale is larger than 2 h. The best degrees of freedom is four for the 2-h scale.

Mashal and Zeevi (2002) claim that the dependence between stocks is better described by a Student copula with 11–12 degrees of freedom than by a Gaussian copula.

5

Time Series Analysis

Time series analysis can be used to analyze both a univariate time series and a vector time series. We are interested in estimating the dependence between consecutive observations. In a vector time series there is both cross-sectional dependence and time series dependence, which means that the components of the vector depend on each other at any given point of time, and the future values of the vectors depend on the past values of the vectors.

Model free time series analysis can estimate the joint distribution of

$$(Y_t, Y_{t-1}, \dots, Y_{t-k}), \tag{5.1}$$

for some $k \geq 1$. The estimation could be done using nonparametric multivariate density estimation. A different model free approach models at the first step the distribution of $(Y_t, Y_{t-1}, \dots, Y_{t-k})$ parametrically, using density $f(\cdot, \theta) : \mathbf{R}^{k+1} \to \mathbf{R}$. At the second step, parameter θ is taken to be time dependent. This leads to a semiparametric time series analysis, because we combine a cross-sectional parametric model with a time varying estimation of the parameter. Time localized maximum likelihood or time localized least squares can be used to estimate the parameter. Of particular interest is to estimate a univariate excess distribution $f(\cdot, \theta)$ with a time varying θ, because this leads to time varying quantile estimation.

Prediction is one of the most important applications of time series analysis. In prediction it is useful to use regression models

$$Y_t = f(Y_{t-1}, \dots, Y_{t-k}) + \epsilon_t, \tag{5.2}$$

where $k \geq 1$ and ϵ_t is noise. For the estimation of f we can use nonparametric regression. We study prediction with models (5.2) in Chapter 6.

Autoregressive moving average processes (ARMA) models are classical parametric models for time series analysis. It is of interest to find formulas of conditional expectation in ARMA models, because these formulas for conditional expectation can be used to construct predictors. The formulas for conditional expectation in ARMA models give insight into different types of predictors:

Nonparametric Finance, First Edition. Jussi Klemelä.
© 2018 John Wiley & Sons, Inc. Published 2018 by John Wiley & Sons, Inc.

AR models lead to state space prediction, and MA models lead to time space prediction.

Prediction of future returns of a financial asset is difficult, but prediction of future absolute returns and future squared returns is feasible. Generalized autoregressive conditional heteroskedasticity (GARCH) models are applied in the prediction of squared returns. Prediction of future squared returns is called volatility prediction. Prediction of volatility is applied in Chapter 7.

We concentrate on time series analysis in discrete time, but we define also some continuous time stochastic processes, like the geometric Brownian motion, because it is a standard model in option pricing.

Section 5.1 discusses strict stationarity, covariance stationarity, and auto-covariance function. Section 5.2 studies model free time series analysis. Section 5.3 studies parametric time series models, in particular, ARMA and GARCH processes. Section 5.4 considers models for vector time series. Section 5.5 summarizes stylized facts of financial time series.

5.1 Stationarity and Autocorrelation

A time series (stochastic process) is a sequence of random variables, indexed by time. We define time series models for double infinite sequences

$$\{Y_t\}, \quad t \in \mathcal{Z} = \{0, \pm 1, \pm 2, \dots\}.$$

A time series model can also be defined for a one-sided infinite sequence $\{Y_t\}$, where $t \in \mathcal{N}_0 = \{0, 1, \dots\}$ or $t \in \mathcal{N} = \{1, 2, \dots\}$. A realization of a time series is a finite sequence $Y_1, \dots, Y_T$ of observed values. We use the term "time series" both to denote the underlying stochastic process and a realization of the stochastic process. Besides a sequence of real valued random variables, we can consider a vector time series, which is a sequence $\{X_t\}$ of random vectors $X_t \in \mathbf{R}^d$.[1]

5.1.1 Strict Stationarity

Time series $\{Y_t\}$ is called strictly stationary, if $(Y_1, \dots, Y_t)$ and $(Y_{1+k}, \dots, Y_{t+k})$ are identically distributed for all $t, k \in \{0, \pm 1, \pm 2, \dots\}$. This means that for a strictly stationary time series all finite dimensional marginal distributions are equal.

Figure 5.1(a) shows a time series of S&P 500 daily prices, using data described in Section 2.4.1. The time series has an exponential trend and is not station-ary. The exponential trend can be removed by taking logarithms, as shown in

---

1 Spatial statistics considers a collection of random variables Y_t, indexed by a spatial location: $t \in \mathbf{R}^2$ or $t \in \mathbf{R}^3$. We can also consider a collection of random elements indexed by a space of functions, or indexed by an abstract space.

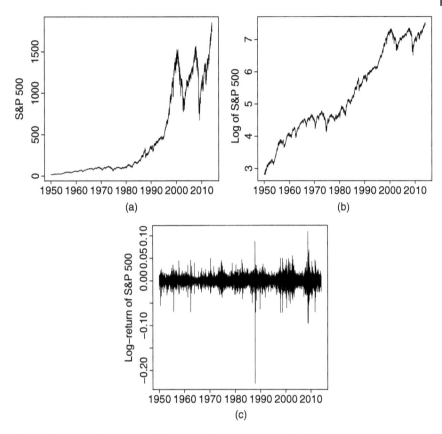

Figure 5.1 *Removing a trend: Differences of logarithms.* (a) S&P 500 prices; (b) logarithms of S&P 500 prices; (c) differences of the logarithmic prices.

panel (b), but after that we have a time series with a linear trend. The linear trend can be removed by taking differences, as shown in panel (c), which leads to the time series of logarithmic returns, which already seems to be a stationary time series. Figure 2.1(b) shows that the gross returns seem to be stationary.

Figure 5.2(a) shows a time series of differences of S&P 500 prices, which is not a stationary time series. Panels (b) and (c) show short time series of price differences, which seem to be approximately stationary. Thus, we could also define the concept of approximate stationarity.

Figure 5.3 studies a time series of squares of logarithmic returns, computed from the daily S&P 500 data, which is described in Section 2.4.1. The squared logarithmic returns are often modeled as a stationary GARCH(1, 1) time series. However, we can also model the squared logarithmic returns with a signal plus noise model

$$Y_t = \mu_t + \epsilon_t,$$

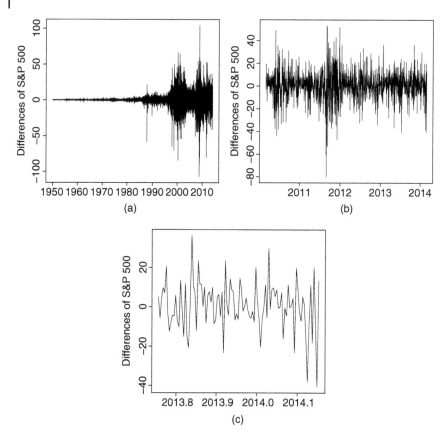

Figure 5.2 *Removing a trend: Differencing.* (a) Differences of S&P 500 prices over 65 years; (b) differences over 4 years; (c) differences over 100 days.

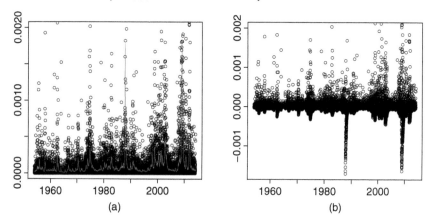

Figure 5.3 *Removing a trend: Subtracting a moving average.* (a) A times series of squared returns and a moving average of squared returns (red); (b) squared returns minus the moving average of squared returns.

where $Y_t = [\log(S_t/S_{t-1})]^2$, μ_t is a deterministic trend, and ϵ_t is stationary white noise. We can estimate the trend μ_t with a moving average $\hat{\mu}_t$. Moving averages are defined in Section 6.1.1. Panel (a) shows time series Y_t (black circles) and $\hat{\mu}_t$ (red line). Panel (b) shows $Y_t - \hat{\mu}_t$. Panel (b) suggests that subtracting the moving average could lead to stationarity. We use the one-sided exponential moving average in (6.3) with smoothing parameter $h = 40$.

5.1.1.1 Random Walk

Random walk is a discrete time stochastic process $\{Y_t\}$ defined by

$$Y_t = Y_{t-1} + \epsilon_t, \quad t = 1, 2, \dots,$$

where Y_0 is a random variable or a fixed value, and $\{\epsilon_t\}$ is distributed as $\text{IID}(\mu, \sigma^2)$. We have that

$$Y_t = Y_0 + \sum_{k=1}^{t} \epsilon_k, \quad t = 1, 2, \dots.$$

If Y_0 is a constant, then $EY_t = Y_0 + \mu t$ and $\text{Var}(Y_t) = t\sigma^2$. Thus, random walk is not strictly stationary (and not covariance stationary). We obtain a Gaussian random walk if $\{\epsilon_t\}$ is Gaussian white noise. If $Y_0 = 0$, then a Gaussian random walk satisfies $Y_t \sim N(t\mu, t\sigma^2)$.

Figure 5.4(a) shows the time series of S&P 500 prices over a period of 100 days. Panel (b) shows a simulated Gaussian random walk of length 100, when the initial value is 0. A random walk leads to a time series that has a stochastic trend. A stochastic trend is difficult to distinguish from a deterministic trend. A time series of stock prices resembles a random walk. Also a time series of a dividend price ratio in Figure 6.7(a) resembles a random walk.

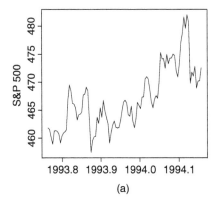 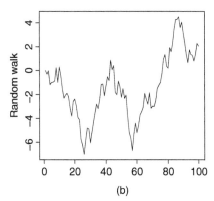

(a) (b)

Figure 5.4 *Stochastic trend.* (a) Prices of S&P 500 over 100 days; (b) simulated random walk of length 100, when the initial value is 0.

Geometric random walk is a discrete time stochastic process defined by

$$Y_t = Y_0 \prod_{k=1}^{t} \epsilon_k, \quad t = 1, 2, \ldots,$$

where $\epsilon_1, \epsilon_2, \ldots$ are i.i.d and Y_0 is independent of $\epsilon_1, \epsilon_2, \ldots$.

5.1.2 Covariance Stationarity and Autocorrelation

We define autocovariance and autocorrelation first for scalar time series and then for vector time series.

5.1.2.1 Autocovariance and Autocorrelation for Scalar Time Series

We say that a time series $\{Y_t\}$ is covariance stationary, if EY_t is a constant, not depending on t, and $\mathrm{Cov}(Y_t, Y_{t+k})$ depends only on k but not on t. A covariance stationary time series is called also second-order stationary.

If $EY_t^2 < \infty$ for all t, then strict stationary implies covariance stationarity. There exists time series that are strictly stationary but for which covariance is not defined.[2] Covariance stationarity does not imply strict stationarity. For a Gaussian time series, strict stationarity and covariance stationarity are equivalent. By a Gaussian time series, we mean a time series whose all finite dimensional marginal distributions have a Gaussian distribution.

For a covariance stationary time series the autocovariance function is defined by

$$\gamma(k) = \mathrm{Cov}(Y_t, Y_{t-k}),$$

where $k = 1, 2, \ldots$. The covariance stationarity implies that $\gamma(k)$ depends only on k and not on t. The autocorrelation function is defined as

$$\rho(k) = \mathrm{Cor}(Y_t, Y_{t-k}) = \frac{\gamma(k)}{\gamma(0)},$$

where $k = 1, 2, \ldots$.

The sample autocovariance with lag k, based on the observations $Y_1, \ldots, Y_T$, is defined as

$$\hat{\gamma}(k) = \frac{1}{T} \sum_{t=1}^{T-k} (Y_t - \bar{Y})(Y_{t+k} - \bar{Y}),$$

where $\bar{Y} = T^{-1} \sum_{i=1}^{T} Y_t$.[3] The sample autocorrelation with lag k is defined as

$$\hat{\rho}(k) = \frac{\hat{\gamma}(k)}{\hat{\gamma}(0)}. \tag{5.3}$$

---

2 We have $E(Y_t Y_{t-k}) \le [EY_t^2 EY_{t-k}^2]^{1/2}$ and $E|Y_t| \le [EY_t^2]^{1/2}$ by Cauchy's inequality, so that if $EY_t^2 < \infty$ for all t, then $\mathrm{Cov}(Y_t, Y_{t-k})$ is defined for all t and k.
3 We have divided intentionally by T and not by $T - k$, although divisor $T - k$ would lead to an unbiased estimator.

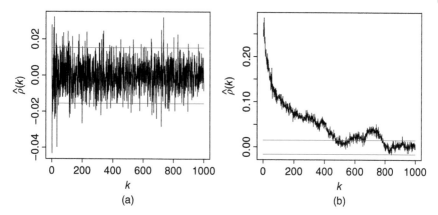

Figure 5.5 *S&P 500 autocorrelation.* (a) The sample autocorrelation function $k \mapsto \hat{\rho}(k)$ of S&P 500 returns for $k = 1, \ldots, 1000$; (b) the sample autocorrelation function for absolute returns. The red lines indicate the 95% confidence band for the null hypothesis of i.i.d process.

Figure 5.5 shows sample autocorrelation functions for the daily S&P 500 index data, described in Section 2.4.1. Panel (a) shows the sample autocorrelation function $k \mapsto \hat{\rho}(k)$ for the return time series $Y_t = R_t = (S_t - S_{t-1})/S_{t-1}$ and panel (b) shows the sample autocorrelation function for the time series of the absolute returns $Y_t = |R_t|$. The lags are on the range $k = 1, \ldots, 1000$.

If $Y_1, Y_2, \ldots$ are i.i.d. with mean zero, then

$$\sqrt{T}\,(\hat{\rho}(1), \ldots, \hat{\rho}(k)) \xrightarrow{d} N\left(0, I_k\right),$$

as $T \to \infty$; see Brockwell and Davis (1991). Thus, if $Y_1, Y_2, \ldots$ are i.i.d. with mean zero, then about $1 - \alpha$ of the observed values $Y_1, \ldots, Y_T$ should be inside the band

$$\pm z_{1-\alpha/2} T^{-1/2},$$

where z_α is the α-quantile for the standard normal distribution. Figure 5.5 has the red lines at the heights $\pm z_{1-\alpha/2} T^{-1/2}$, where we have chosen $\alpha = 0.05$, so that $z_{1-\alpha/2} \approx 1.96$.

The Box–Ljung test can be used to test whether the autocorrelations are zero for a stationary time series $Y_1, Y_2, \ldots$. The null hypothesis is that $\rho(k) = 0$ for $k = 1, \ldots, h$, where $h \geq 1$. Let us have observed time series $Y_1, \ldots, Y_T$. The test statistics is

$$Q(h) = T(T+2) \sum_{k=1}^{h} \frac{\hat{\rho}^2(k)}{T-k},$$

where $\hat{\rho}(k)$ is defined in (5.3). The test rejects the null hypothesis of zero autocorrelations if

$$Q(h) > \chi^2_{h,1-\alpha},$$

where $\chi^2_{h,1-\alpha}$ is the $1-\alpha$-quantile of the χ^2-distribution with degrees of freedom h. We can compute the observed p-values

$$p_h = 1 - F_h(Q(h)),$$

for $h = 1, 2, \ldots$, where F_h is the distribution function of the χ^2-distribution with degrees of freedom h. Small observed p-values indicate that the observations are not compatible with the null hypothesis.

5.1.2.2 Autocovariance for Vector Time Series

Let $X_t = (X_{t,1}, X_{t,2})'$ be a vector time series with two components. Vector time series $\{X_t\}$ is covariance stationary when the components $\{X_{t,1}\}$ and $\{X_{t,2}\}$ are covariance stationary and

$$\text{Cov}(X_{t,1}, X_{s,2}) = \text{Cov}(X_{t+h,1}, X_{s+h,2}) \tag{5.4}$$

for all $t, s, h \in \mathbf{Z}$. Thus, vector time series $\{X_t\}$ is covariance stationary when EX_t is a vector of constants, not depending on t, and the covariance

$$\text{Cov}(X_t, X_{t+h}) = \begin{bmatrix} \text{Cov}(X_{t,1}, X_{t+h,1}) & \text{Cov}(X_{t,1}, X_{t+h,2}) \\ \text{Cov}(X_{t,2}, X_{t+h,1}) & \text{Cov}(X_{t,2}, X_{t+h,2}) \end{bmatrix},$$

depends only on h but not on t for $t, h \in \mathbf{Z}$.

For a covariance stationary time series the autocovariance function is defined by

$$\Gamma(h) = \text{Cov}(X_t, X_{t+h}). \tag{5.5}$$

For a scalar covariance stationary time series $\{Y_t\}$ we have

$$\gamma(h) = \text{Cov}(Y_t, Y_{t+h}) = \text{Cov}(Y_{t-h}, Y_t) = \gamma(-h).$$

However, the autocovariance function of a vector time series satisfies[4]

$$\Gamma(h) = \Gamma(-h)'. \tag{5.6}$$

5.2 Model Free Estimation

Univariate and multivariate descriptive statistics and graphical tools can be applied to get insight into a distribution of a time series. We can apply k-variate

---

4 Combining

$$\Gamma(-h)' = \begin{bmatrix} \text{Cov}(X_{t,1}, X_{t-h,1}) & \text{Cov}(X_{t,2}, X_{t-h,1}) \\ \text{Cov}(X_{t,1}, X_{t-h,2}) & \text{Cov}(X_{t,2}, X_{t-h,2}) \end{bmatrix}$$

and (5.4) implies (5.6).

descriptive statistics and graphical tools to the k-dimensional marginal distributions of a time series. This is discussed in Section 5.2.1.

Univariate and multivariate density estimators and regression estimators can be applied to time series data. We can apply k-variate estimators to the k-dimensional marginal distributions of a time series. This is discussed in Section 5.2.2, by assuming that the time series is a Markov process of order $k \geq 1$.

Section 5.2.3 considers modeling time series with a combination of parametric and nonparametric methods. First a static parametric model is posed on the observations and then the time dynamics is introduced with time space or state space smoothing. The approach includes both local likelihood, covered in Section 5.2.3.1, and local least squares method, covered in Section 5.2.3.2. We apply local likelihood and local least squares to estimate time varying tail index in Section 5.2.3.3.

5.2.1 Descriptive Statistics for Time Series

Univariate statistics, as defined in Section 3.1, can be used to describe time series data $Y_1, \dots, Y_T \in \mathbf{R}$. Using univariate statistics, like sample mean and sample variance, is reasonable if Y_t are identically distributed.

Multivariate statistics, as defined in Section 4.1, can be used to describe vector time series data $X_1, \dots, X_T \in \mathbf{R}^d$. Again, the use of multivariate statistics like sample correlation is reasonable if X_t are identically distributed.

Multivariate statistics can be used also for univariate time series data $Y_1, \dots, Y_T \in \mathbf{R}$ if we create a vector time series from the initial univariate time series. We can create a two-dimensional vector time series by defining

$$X_t = (Y_t, Y_{t-k}), \tag{5.7}$$

for some $k \geq 1$. Now we can compute a sample correlation coefficient, for example, from data $X_{k+1}, \dots, X_T$. This is reasonable if X_t are identically distributed. The requirement that X_t in (5.7) are identically distributed follows from strict stationarity of $\{Y_t\}$.

5.2.2 Markov Models

We have defined strict stationarity in Section 5.1.1. A strictly stationary time series $\{Y_t\}_{t \in \mathbf{Z}}$ can be defined by giving all finite dimensional marginal distributions. That is, to define the distribution of a strictly stationary time series we need to define the distributions

$$(Y_1, \dots, Y_k)$$

for all $k \geq 1$. If the time series is IID$(0, \sigma^2)$, then we need only to define the distribution of Y_1. We say that the time series is a Markov process, if

$$P(Y_t \in A \mid Y_{t-1}, Y_{t-2}, \dots) = P(Y_t \in A \mid Y_{t-1}).$$

To define a Markov process we need to define the distribution of Y_t and (Y_t, Y_{t+1}). More generally, we say that the time series is a Markov process of order $k \geq 1$, if

$$P(Y_t \in A \mid Y_{t-1}, Y_{t-2}, \ldots) = P(Y_t \in A \mid Y_{t-1}, \ldots, Y_{t-k}).$$

To define a Markov process of order k we need to define the distributions of Y_t, (Y_t, Y_{t+1}), $\ldots$, $(Y_t, \ldots, Y_{t+k})$.

To estimate nonparametrically the distribution of a Markov process of order $k \geq 1$, we can estimate the distributions of Y_t, (Y_t, Y_{t+1}), $\ldots$, $(Y_t, \ldots, Y_{t+k})$ nonparametrically.

5.2.3 Time Varying Parameter

Let $Y_1, \ldots, Y_T \in \mathbf{R}$ be a time series. Let $f_\theta : \mathbf{R} \to \mathbf{R}$ be a density function, where θ is a parameter, and $\theta \in \Theta \subset \mathbf{R}^p$. We could ignore the time series properties and assume that $Y_1, \ldots, Y_T$ are independent and identically distributed with density f_θ.

However, we can assume that parameter $\theta = \theta_t$ changes in time. Then the observations are not identically distributed, but Y_t has density f_{θ_t}. In practice, we do not specify any dynamics for θ_t, but construct estimates $\hat{\theta}_t$ using non-parametric smoothing.

Note that even when we would assume independent and identically distributed observations, with time series data the parameter estimate is changing in time, because at time t the estimate $\hat{\theta}_t$ is constructed using data $Y_1, \ldots, Y_t$. This is called sequential estimation.

5.2.3.1 Local Likelihood

If $Y_1, \ldots, Y_T$ are independent with density f_θ, then the density of $(Y_1, \ldots, Y_T)$ is

$$f(y_1, \ldots, y_T) = \prod_{i=1}^{T} f_\theta(y_i).$$

The maximum likelihood estimator of θ is the value $\hat{\theta}$ maximizing

$$\sum_{i=1}^{T} \log f_\theta(Y_i)$$

over $\theta \in \Theta$. We can find a time varying estimator $\hat{\theta}_t$ using either time space or state space localization. The local likelihood approach has been studied in Spokoiny (2010). The localization is discussed more in Sections 6.1.1 and 6.1.2.

Time Space Localization Let

$$p_i(t) = \frac{K((t-i)/h)}{\sum_{j=1}^{t} K((t-j)/h)}, \tag{5.8}$$

where $h > 0$ is the smoothing parameter and $K : [0, \infty) \to \mathbf{R}$ is a kernel function. For example, we can take $K(x) = \exp(-x)\, I_{[0,\infty)}(x)$. Let $\hat{\theta}_t$ be the value maximizing

$$\sum_{i=1}^{t} p_i(t) \log f_\theta(Y_i) \tag{5.9}$$

over $\theta \in \Theta$.

For example, let us consider the model

$$Y_t = \mu_t + \sigma_t \epsilon_t,$$

where $\epsilon_t \sim N(0, 1)$ are i.i.d. Denote $\theta = (\mu, \sigma^2)$ and $f_\theta(x) = \phi((x - \mu)/\sigma)/\sigma$, where $\phi(x) = (2\pi)^{-1/2} \exp\{-x^2/2\}$ is the density of the standard normal distribution. Let $\Theta = \mathbf{R} \times (0, \infty)$. Now

$$\sum_{i=1}^{t} p_i(t) \log f_\theta(Y_i) = \log (2\pi)^{-1/2} + \log \sigma^{-1} - \frac{1}{2\sigma^2} \sum_{i=1}^{t} p_i(t)(Y_i - \mu)^2.$$

Then $\hat{\theta}_t = (\hat{\mu}_t, \hat{\sigma}_t^2)$, where

$$\hat{\mu}_t = \sum_{i=1}^{t} p_i(t) Y_i, \qquad \hat{\sigma}_t^2 = \sum_{i=1}^{t} p_i(t) Y_i^2 - \hat{\mu}_t^2. \tag{5.10}$$

State Space Localization Let us observe the state variables $X_1, \dots, X_T$ in addition to observing time series $Y_1, \dots, Y_T$. Let

$$p_i(t) = \frac{K((X_t - X_i)/h)}{\sum_{j=1}^{t} K((X_t - X_j)/h)}, \tag{5.11}$$

where $h > 0$ is the smoothing parameter and $K : \mathbf{R}^d \to \mathbf{R}$ is a kernel function. We can take $K = \phi$ the density of the standard normal distribution. Let $\hat{\theta}_t$ be the value maximizing (5.9) over $\theta \in \Theta$.

For example, let us consider the model

$$Y_t = \mu(X_t) + \sigma(X_t)\, \epsilon_t,$$

where $\epsilon_t \sim N(0, 1)$ are i.i.d. The model can be written as

$$Y \mid X = x \sim N(\mu(x), \sigma(x)).$$

Denote $\theta = (\mu, \sigma^2)$ and $f_\theta(y) = \phi((y - \mu)/\sigma)/\sigma$, where ϕ is the density of the standard normal distribution. Then $\hat{\theta}_t = (\hat{\mu}_t, \hat{\sigma}_t^2)$, as defined in (5.10).

5.2.3.2   Local Least Squares
Let us consider a linear model with time changing parameters. Let us observe the explanatory variables $Z_1, \dots, Z_T$ in addition to observing time series $Y_1, \dots, Y_T$. Consider the model

$$Y_t = \alpha_t + \beta_t' Z_t + \epsilon_t,$$

where $\alpha_t \in \mathbf{R}$, $\beta_t \in \mathbf{R}^p$ are time dependent constants, $Z_t \in \mathbf{R}^p$ is the vector of explanatory variables, and ϵ_t is an error term.

We define the estimates of the time varying regression coefficients as the values $\hat{\alpha}_t$ and $\hat{\beta}_t$ minimizing

$$\sum_{i=1}^{t} (Y_i - \alpha - \beta'Z_i)^2 \, p_i(t),$$

where $p_i(t)$ is the time space localized weight defined in (5.8). When we observe in addition the state variables $X_1, \dots, X_T$, then we can use the state space localized weight $p_i(t)$, defined in (5.11).

5.2.3.3   Time Varying Estimators for the Excess Distribution

We discussed tail modeling in Section 3.4. The idea in tail modeling is to fit a parametric model only to the data in the left tail or to the data in the right tail. We can add time space or state space localization to the tail modeling. As before, $Y_1, \dots, Y_T$ is a time series.

Local Likelihood in Tail Estimation   Let family f_θ, $\theta \in \Theta$, model the excess distribution of a return distribution, where $f_\theta : [0, \infty) \to \mathbf{R}$. This means that if $g : \mathbf{R} \to \mathbf{R}$ is the density of the return distribution then we assume for the left tail that

$$g(x)I_{(-\infty, u]}(x) = p f_\theta(u - x)$$

for some θ, where $0 < p < 0.5$, and u is the pth quantile of the return density: $p = \int_{-\infty}^{u} g$. For the right tail the corresponding assumption is

$$g(x)I_{[u, \infty)}(x) = (1 - p)f_\theta(x - u)$$

for some θ, where $0.5 < p < 1$, and u is the pth quantile of the return density: $1 - p = \int_{u}^{\infty} g$.

The local maximum likelihood estimator for the parameter of the left tail is obtained from (3.63) as

$$\hat{\theta}_{left,t} = \operatorname*{argmax}_{\theta \in \Theta} \sum_{i : Y_i \in \mathcal{L}_t} p_i(t) \log f_\theta (u - Y_i), \tag{5.12}$$

where $u = \hat{q}_p$ is the empirical quantile computed from $Y_1, \dots, Y_t$, $0 < p < 0.5$, and

$$\mathcal{L}_t = \{Y_i : Y_i \leq u, \ i = 1, \dots, t\}.$$

The time space localized weights $p_i(t)$ are modified from (5.8) as

$$p_i(t) = \frac{K((t - i)/h)}{\sum_{j : j \in \mathcal{L}_t} K((t - j)/h)}, \tag{5.13}$$

where $h > 0$ is the smoothing parameter and $K : [0, \infty) \to \mathbf{R}$ is a kernel function. If there is available the state variables $X_1, \ldots, X_T$, then we can use the state space localized weights, modified from (5.11) as

$$p_i(t) = \frac{K((X_t - X_i)/h)}{\sum_{j:j \in \mathcal{L}_t} K((X_t - X_j)/h)}, \tag{5.14}$$

where $h > 0$ is the smoothing parameter and $K : \mathbf{R}^d \to \mathbf{R}$ is a kernel function.

The local maximum likelihood estimator for the parameter of the right tail is obtained from (3.64) as

$$\hat{\theta}_{right,t} = \operatorname*{argmax}_{\theta \in \Theta} \sum_{i:Y_i \in \mathcal{R}_t} p_i(t) \log f_\theta(Y_i - u),$$

where $u = \hat{q}_p$ is the empirical quantile, $0.5 < p < 1$, and

$$\mathcal{R}_t = \{Y_i : Y_i \geq u, \quad i = 1, \ldots, t\}.$$

The weights are obtained from (5.13) and (5.14) by replacing $\mathcal{L}_t$ with $\mathcal{R}_t$.

For example, let us assume that the excess distribution is the Pareto distribution, as defined in (3.74) as

$$f_\alpha(x) = \frac{\alpha}{u} \left(\frac{x + u}{u} \right)^{-1-\alpha} I_{[0,\infty)}(x),$$

where $\alpha > 0$ is the shape parameter. The maximum likelihood estimator has the closed form expression (3.75). The local maximum likelihood estimators are

$$\hat{\alpha}_{left,t} = \left(\sum_{Y_i \in \mathcal{L}_t} p_i(t) \log(Y_i/u) \right)^{-1}, \tag{5.15}$$

and

$$\hat{\alpha}_{right,t} = \left(\sum_{Y_i \in \mathcal{R}_t} p_i(t) \log(Y_i/u) \right)^{-1}, \tag{5.16}$$

where $0 < p < 0.5$, and $u = \hat{q}_p$. For the left tail we assume that $u < 0$, and for the right tail we assume that $u > 0$. These are the time varying Hill's estimators.

Figure 5.6 studies time varying Hill's estimates for the S&P 500 daily data, described in Section 2.4.1. Panel (a) shows the estimates for the left tail index and panel (b) shows the estimates for the right tail index. Sequentially calculated Hill's estimates are shown in black, time localized Hill's estimates with $h = 500$ are shown in blue, and the case with $h = 100$ is shown in yellow. The exponential kernel function is used. The estimation is started after there are 4 years of data. The tails are defined by the empirical quantile $u = \hat{q}_p$ with $p = 10\%$ and $p = 90\%$.

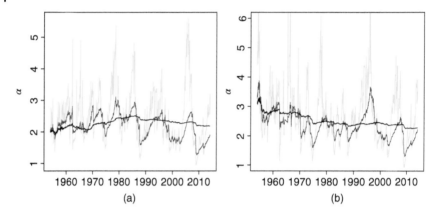

Figure 5.6 *Time varying Hill's estimator.* (a) Left tail index; (b) right tail index. The black curves show sequentially calculated Hill's estimates, the blue curves show the time localized estimates with $h = 500$ and the yellow curves have $h = 100$.

Time Varying Regression Estimator for Tail Index Let $Y_1, \dots, Y_t$ be the observed time series at time t. The regression estimator for the parameter $\alpha > 0$ of the Pareto distribution is given in (3.77). Let

$$\mathcal{L}_t = \{Y_i : Y_i \leq u, \quad i = 1, \dots, t\}.$$

The local regression estimator of the parameter of the left tail is

$$\hat{\alpha}_{left} = -\frac{\sum_{i:Y_i \in \mathcal{L}_t} p_{(i)}(t)[\log(Y_{(i)}/u) \cdot \log(i/(t+1))]}{\sum_{i:Y_i \in \mathcal{L}_t} p_{(i)}(t)(\log(Y_{(i)}/u))^2},$$

where $u = \hat{q}_p$ is the empirical quantile and we assume $u < 0$. The weights $p_{(i)}(t)$ are obtained from (5.13) and (5.14) by replacing index i with the index (i), so that the weights correspond to the ordering $Y_{(1)} \leq \cdots \leq Y_{(t)}$.

The local regression estimator of the parameter of the right tail is

$$\hat{\alpha}_{right} = -\frac{\sum_{i:Y_i \in \mathcal{R}_t} p_{(i)}(t)[\log(Y_{(i)}/u) \cdot \log(i/(t+1))]}{\sum_{i:Y_i \in \mathcal{R}_t} p_{(i)}(t)(\log(Y_{(i)}/u))^2},$$

where $Y_{(1)} \geq \cdots \geq Y_{(t)}$ are the observations in reverse order,

$$\mathcal{R}_t = \{Y_i : Y_i \geq u, \quad i = 1, \dots, t\},$$

$u = \hat{q}_p$ is the empirical quantile, $0.5 < p < 1$, and we assume $u > 0$.

Figure 5.7 studies time varying regression estimates for the tail index using the S&P 500 daily data, described in Section 2.4.1. Panel (a) shows the estimates for the left tail index and panel (b) shows the estimates for the right tail index. Sequentially calculated regression estimates are shown in black, time localized estimates with $h = 500$ are shown in blue, and the case with $h = 100$ is shown in

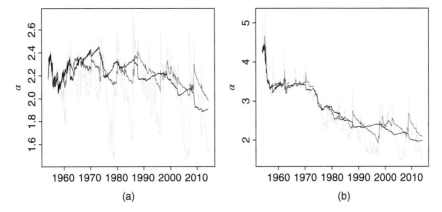

Figure 5.7 *Time varying regression estimator*. Time series of estimates of the tail index are shown. (a) Left tail index; (b) right tail index. The black curves show sequentially calculated regression estimates, the blue curves show the time localized estimates with $h = 500$ and the yellow curves have $h = 100$.

yellow. The standard Gaussian kernel function is used. The estimation is started after there are 4 years of data. The tails are defined by the empirical quantile $u = \hat{q}_p$ with $p = 10\%$ and $p = 90\%$.

5.3 Univariate Time Series Models

We discuss first ARMA (autoregressive moving average) processes and after that we discuss conditional heteroskedasticity models. Conditional heteroskedasticity models include ARCH (autoregressive conditional heteroskedasticity) and GARCH (generalized autoregressive conditional heteroskedasticity) models. The ARMA, ARCH, and GARCH processes are discrete time stochastic processes. We discuss also continuous time stochastic processes, because geometric Brownian motion and related continuous time stochastic processes are widely used in option pricing.

Brockwell and Davis (1991) give a detailed presentation of linear time series analysis, Fan and Yao (2005) give a short introduction to ARMA models and a more detailed discussion of nonlinear models. Shiryaev (1999) presents results of time series analysis that are useful for finance.

5.3.1 Prediction and Conditional Expectation

Our presentation of discrete time series analysis is directed towards giving prediction formulas: these prediction formulas are used in Chapter 7 to provide benchmarks for the evaluation of the methods of volatility prediction. Chapter 6 studies nonparametric prediction.

Let $\{Y_t\}$ be a time series with $t = 0 \pm 1, \pm 2, \ldots$. We take the conditional expectation

$$E(Y_{t+\eta} | Y_t, Y_{t-1}, \ldots) \tag{5.17}$$

to be the best prediction of $Y_{t+\eta}$, given the observations $Y_t, Y_{t-1}, \ldots$, where $\eta \geq 1$ is the prediction step. Using the conditional expectation as the best predictor can be justified by the fact that the conditional expectation minimizes the mean squared error. In fact, the function g minimizing

$$E(Y_{t+\eta} - g(Y_t, Y_{t-1}, \ldots))^2 \tag{5.18}$$

is the conditional expectation: $g(Y_t, Y_{t-1}, \ldots) = E(Y_{t+\eta} | Y_t, Y_{t-1}, \ldots)$.[5]

Besides predicting the value $Y_{t+\eta}$, we consider also predicting the squared value $Y_{t+\eta}^2$.

In the following text, we give expressions for $E(Y_{t+\eta} | Y_t, Y_{t-1}, \ldots)$ in the ARMA models and for $E(Y_{t+\eta}^2 | Y_t, Y_{t-1}, \ldots)$ in the ARCH and GARCH models. These expressions depend on the unknown parameters of the models. In order to apply the expressions we need to estimate the unknown parameters and insert the estimates into the expressions.

The conditional expectation whose condition is the infinite past is a function $g(Y_t, Y_{t-1}, \ldots)$ of the infinite past. Since we have available only a finite number of observations, we have to truncate these functions to obtain a function $\tilde{g}(Y_t, \ldots, Y_1)$. It would be more useful to obtain formulas for

$$E(Y_{t+\eta} | Y_t, Y_{t-1}, \ldots, Y_1)$$

and

$$E\left(Y_{t+\eta}^2 | Y_t, Y_{t-1}, \ldots, Y_1\right).$$

However, these formulas are more difficult to derive than the formulas where the condition of the conditional expectation is the infinite past.

5.3.2 ARMA Processes

ARMA processes are defined in terms of an innovation process. After defining innovation processes, we define MA (moving average) processes and AR (autoregressive) processes. ARMA processes are obtained by combining autoregressive and moving average processes.

---

5 Function $g(x) = E(Y | X = x)$ minimizes $E(Y - g(X))^2$ over measurable g. Indeed,

$$E(g(X) - Y)^2 = E(g(X) - E(Y | X))^2 + E(E(Y | X) - Y)^2,$$

because $E[(g(X) - E(Y | X))(E(Y | X) - Y)] = 0$. Thus, $E(g(X) - Y)^2$ is minimized with respect to g by choosing $g(x) = E(Y | X = x)$. Note that the conditional expectation defined as $g(x) = E(Y | X = x)$ is a real-valued function of x, but $E(Y | X)$ is a real-valued random variable, which can be defined as $E(Y | X) = g(X)$.

5.3.2.1 Innovation Processes

Innovation processes are used to build more complex processes, like ARMA and GARCH processes. We define two innovation processes: a white noise process and an i.i.d. process.

We say that $\{\epsilon_t\}_{t\in\mathbf{Z}}$ is a white noise process and write $\{\epsilon_t\}_{t\in\mathbf{Z}} \sim \mathrm{WN}(0, \sigma^2)$ if

1) $E\epsilon_t = 0$,
2) $E\epsilon_t^2 = \sigma^2$,
3) $E\epsilon_t\epsilon_{t+k} = 0$ for $k \neq 0$,

where $0 < \sigma^2 < \infty$ is a constant. A white noise is a Gaussian white noise if $\epsilon_t \sim N(0, \sigma^2)$.

We say that $\{\epsilon_t\}_{t\in\mathbf{Z}}$ is an i.i.d. process and write $\{\epsilon_t\}_{t\in\mathbf{Z}} \sim \mathrm{IID}(0, \sigma^2)$ if

1) $E\epsilon_t = 0$,
2) $E\epsilon_t^2 = \sigma^2$,
3) ϵ_t and ϵ_{t+k} are independent for $k \neq 0$.

An i.i.d. process is also a white noise process. A Gaussian white noise is also an i.i.d. process.

5.3.2.2 Moving Average Processes

We define first a moving average process of a finite order, then we give prediction formulas, and finally define a moving average process of infinite order.

MA(q) Process We use MA(q) as a shorthand notation for a moving average process of order q. A moving average process $\{Y_t\}$ of order $q \geq 0$ is a process satisfying

$$Y_t = \epsilon_t + b_1\epsilon_{t-1} + \cdots + b_q\epsilon_{t-q},$$

where $b_1, \ldots, b_q \in \mathbf{R}$, $\{\epsilon_t\} \sim \mathrm{WN}(0, \sigma^2)$ is a white noise process, and $t = 0, \pm 1, \pm 2, \ldots$.

Figure 5.8 illustrates the definition of MA(q) processes. In panel (a) $q = 1$ and in panel (b) $q = 2$. When $q = 1$, then Y_t and Y_{t+1} have one common white noise-term, but Y_t and Y_{t+2} do not have common white noise-terms. When $q = 2$, then Y_t and Y_{t+1} have two common white noise-terms, Y_t and Y_{t+2} have one common white noise-term, and Y_t and Y_{t+3} do not have common white noise-terms.

We have that

$$EY_t = 0, \quad \mathrm{Var}(Y_t) = \sigma^2 \left(1 + b_1^2 + \cdots + b_q^2\right), \tag{5.19}$$

and

$$EY_tY_{t+k} = \begin{cases} \sigma^2 \sum_{j=0}^{q-k} b_j b_{k+j}, & k = 1, \ldots, q, \\ 0, & k > q, \end{cases} \tag{5.20}$$

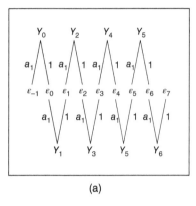 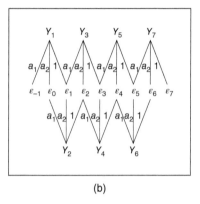

(a)   (b)

Figure 5.8 *The definition of a MA(q) process.* (a) MA(1) process; (b) MA(2) process.

where $b_0 = 1$. Thus, MA(q) process is such that a correlation exists between Y_t and Y_{t+k} only if $|k| \leq q$. Equations (5.19) and (5.20) show that MA(q) process is covariance stationary.

If we are given a covariance function $\gamma : \{0, 1, \dots\} \to \mathbf{R}$, which is such that $\gamma(k) = 0$ for $k > q$, we can construct a MA(q) process with this covariance function by solving σ^2 and $b_1, \dots, b_q$ from the $q + 1$ equations

$$\begin{cases} \gamma(0) &=& \sigma^2 \left(1 + b_1^2 + \cdots + b_q^2\right), \\ \gamma(1) &=& \sigma^2 (b_1 + b_1 b_2 + \cdots + b_{q-1} b_q), \\ &\vdots& \\ \gamma(q) &=& \sigma^2 b_q. \end{cases}$$

Prediction of MA Processes The conditional expectation $E(Y_{t+\eta} | \mathcal{F}_t)$ is the best prediction of $Y_{t+\eta}$ for $\eta \geq 1$, given the infinite past $Y_t, Y_{t-1}, \dots$, in the sense of the mean squared prediction error, as we mentioned in Section 5.3.1. We denote $E(Y_{t+\eta} | \mathcal{F}_t) = E(Y_{t+\eta} | Y_t, Y_{t-1}, \dots)$. The best linear prediction in the sense of the mean squared error is given in (6.19). We can use that formula when the covariance function of the MA(q) process is first estimated.

A recursive prediction formula for the MA(q) process can be derived as follows. We have that

$$E(Y_{t+\eta} | \mathcal{F}_t) = \begin{cases} b_\eta \epsilon_t + b_{\eta+1} \epsilon_{t-1} + \cdots + b_q \epsilon_{t-q-\eta}, & 1 \leq \eta \leq q, \\ 0, & \eta > q, \end{cases}$$

because $E(\epsilon_{t+k} | \mathcal{F}_t) = 0$ for $k = 1, \dots, \eta$. The noise terms $\epsilon_t, \dots, \epsilon_{t-q-\eta}$ are not observed, but we can write

$$\begin{aligned} \epsilon_t &=& Y_t - b_1 \epsilon_{t-1} - \cdots - b_q \epsilon_{t-q}, \\ \epsilon_{t-1} &=& Y_{t-1} - b_1 \epsilon_{t-2} - \cdots b_q \epsilon_{t-q-1}, \\ &\vdots& \end{aligned}$$

This leads to a formula for $E(Y_{t+\eta} | \mathcal{F}_t)$ in terms of the infinite past $Y_t, Y_{t-1}, \ldots$. For example, for the MA(1) process $Y_t = \epsilon_t + b\epsilon_{t-1}$ we have

$$E(Y_{t+\eta} | \mathcal{F}_t) = \begin{cases} b \sum_{k=0}^{\infty} (-1)^k b^k Y_{t-k}, & \eta = 1, \\ 0, & \eta \geq 2. \end{cases} \quad (5.21)$$

The prediction formula for prediction step $\eta = 1$ is a version of exponential moving average, which is defined in (6.7).

We can obtain a recursive prediction for practical use in the following way. Define $\hat{e}_i = 0$, when $i \leq 0$ and

$$\hat{e}_i = Y_i - b_1 \hat{e}_{i-1} - \cdots - b_q \hat{e}_{i-q},$$

when $i = 1, \ldots, t$. Finally we define the η-step prediction as

$$\hat{Y}_{t+\eta} = \begin{cases} b_\eta \hat{e}_t + b_{\eta+1} \hat{e}_{t-1} + \cdots + b_q \hat{e}_{t-q-\eta}, & 1 \leq \eta \leq q, \\ 0, & \eta > q. \end{cases}$$

For example, for the MA(1) process $Y_t = \epsilon_t + b\epsilon_{t-1}$ we get the truncated formulas

$$\hat{Y}_{t+\eta} = \begin{cases} b \sum_{k=0}^{t-1} (-1)^k b^k Y_{t-k}, & \eta = 1, \\ 0, & \eta \geq 2. \end{cases}$$

In the implementation the parameters $b_1, \ldots, b_q$ have to be replaced by their estimates.

MA(∞) Process A moving average process of infinite order is defined as

$$Y_t = \mu + \sum_{j=0}^{\infty} b_j \epsilon_{t-j}.$$

The series converges in mean square if [6]

$$\sum_{j=0}^{\infty} b_j^2 < \infty.$$

We have that

$$EY_t = \mu, \quad \text{Var}(Y_t) = \sigma^2 \sum_{j=0}^{\infty} b_j^2, \quad (5.22)$$

and

$$E(Y_t Y_{t+k}) = \sigma^2 \sum_{j=0}^{\infty} b_j b_{k+j}, \quad k \geq 0. \quad (5.23)$$

---

6 The convergence in mean square means that there is a random variable Y with $EY^2 < \infty$ such that $E(S_n - Y)^2 \to 0$, as $n \to \infty$, where $S_n = \sum_{j=0}^{n} b_j \epsilon_{t-j}$. We can check the convergence using the Cauchy criterion: For all $\delta > 0$ there is n_δ so that $E\left(\sum_{j=n+1}^{n+p} b_j \epsilon_{t-j}\right)^2 \leq \delta$, when $n \geq n_\delta$, for all $p \geq 1$.

Equations (5.22) and (5.23) imply that MA(∞) process is covariance stationary. MA(∞) process can be used to study the properties of AR processes. For example, if we can write an AR process as a MA(∞) process, this shows that the AR process is covariance stationary.

5.3.2.3 Autoregressive Processes

An autoregressive process $\{Y_t\}$ of order $p \geq 1$ is a process satisfying

$$Y_t = a_1 Y_{t-1} + a_2 Y_{t-2} + \cdots + a_p Y_{t-p} + \epsilon_t, \tag{5.24}$$

where $a_1, \ldots, a_p \in \mathbf{R}$, $\{\epsilon_t\} \sim \mathrm{WN}(0, \sigma^2)$ is a white noise process, and $t = 0, \pm 1, \pm 2, \ldots$ We assume that ϵ_t is uncorrelated with $Y_{t-1}, Y_{t-2}, \ldots$ We use AR(p) as a shorthand notation for an autoregressive process of order p.

The autocovariance function of an AR(p) process can be computed recursively. Multiply (5.24) by Y_{t-k} from both sides and take expectations to get

$$\gamma(k) = a_1 \gamma(k-1) + \cdots + a_p \gamma(k-p), \tag{5.25}$$

where $k \geq 0$. The first values $\gamma(0), \ldots, \gamma(p)$ can be solved from the $p+1$ equations. After that, the values $\gamma(k)$ for $k \geq p+1$ can be computed recursively from (5.25).

Prediction of AR Processes Let us consider the prediction of $Y_{t+\eta}$ for $\eta \geq 1$ when the process is an AR(p) process. The best prediction of $Y_{t+\eta}$, given the observations $Y_t, Y_{t-1}, \ldots$, is denoted by

$$\mathrm{pred}_t(\eta) = E(Y_{t+\eta} \,|\, \mathcal{F}_t),$$

where we denote $E(Y_{t+\eta} \,|\, \mathcal{F}_t) = E(Y_{t+\eta} \,|\, Y_t, Y_{t-1}, \ldots)$. We start with the one-step prediction. The best prediction of Y_{t+1}, given the observations $Y_t, Y_{t-1}, \ldots$, is

$$\mathrm{pred}_t(1) = E(Y_{t+1} \,|\, \mathcal{F}_t) = a_1 Y_t + a_2 Y_{t-1} + \cdots + a_p Y_{t-p+1}, \tag{5.26}$$

because $E(\epsilon_{t+1} \,|\, \mathcal{F}_t) = 0$. For the two-step prediction the best predictor is

$$\begin{aligned}
\mathrm{pred}_t(2) &= E(Y_{t+2} \,|\, \mathcal{F}_t) \\
&= E[E(Y_{t+2} \,|\, \mathcal{F}_{t+1}) \,|\, \mathcal{F}_t] \\
&= E[a_1 Y_{t+1} + a_2 Y_t + \cdots + a_p Y_{t-p+2} \,|\, \mathcal{F}_t] \\
&= a_1 \mathrm{pred}_t(1) + a_2 Y_t + \cdots + a_p Y_{t-p+2}.
\end{aligned}$$

The general prediction formula is

$$\mathrm{pred}_t(\eta) = a_1 \mathrm{pred}_t(\eta - 1) + a_2 \mathrm{pred}_t(\eta - 2) + \cdots + a_p \mathrm{pred}_t(\eta - p).$$

The best prediction is calculated recursively, using the value of $\mathrm{pred}_t(1)$ in (5.26), and the fact that $\mathrm{pred}_t(\eta) = Y_{t+\eta}$ for $\eta \leq 0$.

For example, for the MA(1) process $Y_t = aY_{t-1} + \epsilon_t$ we have

$$E\left(Y_{t+\eta} \,|\, Y_t, Y_{t-1}, \ldots\right) = E\left(Y_{t+\eta} \,|\, Y_t, \ldots, Y_1\right) = a^\eta Y_t. \tag{5.27}$$

5.3.2.4 ARMA Processes

We define an autoregressive moving average process $\{Y_t, t = 0, \pm1, \pm2, \dots\}$, of order (p,q), $p, q \geq 0$, as a process satisfying

$$Y_t = a_1 Y_{t-1} + a_2 Y_{t-2} + \cdots + a_p Y_{t-p} + u_t,$$

where $a_1, \dots, a_p \in \mathbf{R}$ and $\{u_t\}_{t \in \mathbb{Z}}$ is a MA(q) process. We use ARMA(p,q) as a shorthand notation for an autoregressive moving average process of order (p,q).

Stationarity, Causality, and Invertability of ARMA Processes Let $\{Y_t\}$ be an ARMA(p,q) process with

$$Y_t = a_1 Y_{t-1} + a_2 Y_{t-2} + \cdots + a_p Y_{t-p} + u_t,$$
$$u_t = b_0 \epsilon_t + b_1 \epsilon_{t-1} + \cdots + b_q \epsilon_{t-q}.$$

Denote

$$a(z) = 1 - a_1 z - a_2 z^2 - \cdots - a_p z^p,$$
$$b(z) = b_0 + b_1 z + b_2 z^2 + \cdots + b_q z^q,$$

where $z \in C$, and C is the set of complex numbers. If $a(z) \neq 0$ for all $z \in C$ such that $|z| = 1$, then there exists the unique stationary solution

$$Y_t = \sum_{j=-\infty}^{\infty} \psi_j \epsilon_{t-j},$$

where the coefficients ψ_j are obtained from the equation

$$\frac{b(z)}{a(z)} = \sum_{j=-\infty}^{\infty} \psi_j z^j,$$

where $r^{-1} < |z| < r$ for some $r > 1$; see Brockwell and Davis (1991, Theorem 3.1.3).

The condition for the covariance stationarity does not guarantee that the ARMA(p,q) process would be suitable for modeling. Let us consider the AR(1) model

$$Y_t = a Y_{t-1} + \epsilon_t,$$

where $\{\epsilon_t\} \sim \text{WN}(0, \sigma^2)$. The AR(1) model is covariance stationary if and only if $|a| \neq 1$. This can be seen in the following way. Let us consider first the case $|a| < 1$. We can write recursively

$$Y_t = a Y_{t-1} + \epsilon_t$$
$$= a^2 Y_{t-2} + a\epsilon_{t-1} + \epsilon_t$$
$$\vdots$$
$$= a^{k+1} Y_{t-k-1} + a^k \epsilon_{t-k} + \cdots + a\epsilon_{t-1} + \epsilon_t,$$

where $k \geq 0$. Since $|a| < 1$, we get the MA(∞) representation[7]

$$Y_t = \sum_{j=0}^{\infty} a^j \epsilon_{t-j},$$

which implies that $\{Y_t\}$ is covariance stationary. Let us then consider the case $|a| > 1$. Since $Y_{t+1} = aY_t + \epsilon_{t+1}$, we can write recursively

$$Y_t = a^{-1}Y_{t+1} - a^{-1}\epsilon_{t+1}$$
$$= a^{-2}Y_{t+2} - a^{-2}\epsilon_{t+2} - a^{-1}\epsilon_{t+1}$$
$$\vdots$$
$$= a^{-k-1}Y_{t+k+1} - a^{-k-1}\epsilon_{t+k+1} - \cdots - a^{-1}\epsilon_{t+1},$$

where $k \geq 0$. Since $|a| > 1$, we get the MA(∞) representation[8]

$$Y_t = -\sum_{j=1}^{\infty} a^{-j} \epsilon_{t+j},$$

which implies that $\{Y_t\}$ is covariance stationary. The latter case $|a| > 1$ is not suitable for modeling because Y_t is a function of future innovations ϵ_{t+j} with $j \geq 1$.

We define causality of the process to exclude examples like the AR(1) model with $|a| > 1$. An ARMA(p, q) process is called causal if there exists constants $\{\psi_j\}$ such that $\sum_{j=0}^{\infty} |\psi_j| < \infty$ and

$$Y_t = \sum_{j=0}^{\infty} \psi_j \epsilon_{t-j}, \quad t = 0, \pm 1, \pm 2, \ldots$$

Let the polynomials $a(z)$ and $b(z)$ have no common zeroes. Then $\{Y_t\}$ is causal if and only if $a(z) \neq 0$ for all $z \in C$ such that $|z| \leq 1$.[9] This has been proved in Brockwell and Davis (1991, Theorem 3.1.1). The coefficients $\{\psi_j\}$ are determined by

$$\sum_{j=0}^{\infty} \psi_j z^j = \frac{b(z)}{a(z)}.$$

Thus, under the conditions that $a(z)$ and $b(z)$ have no common zeroes and $a(z) \neq 0$ for all $z \in C$ such that $|z| \leq 1$, we have expressed an ARMA(p, q) process as an infinite order moving average process. Thus, an ARMA(p, q) process is covariance stationary under these conditions.

---

7 For each $k \geq 0$ we have $Y_t - \sum_{j=0}^{\infty} a^j \epsilon_{t-j} = a^{k+1}Y_{t-k-1} - \sum_{j=k+1}^{\infty} a^j \epsilon_{t-j}$, where $a^{k+1}Y_{t-k-1} \to 0$ and $\sum_{j=k+1}^{\infty} a^j \epsilon_{t-j} \to 0$ in the mean square, as $k \to \infty$, because $|a| < 1$.

8 For each $k \geq 0$ we have $Y_t + \sum_{j=1}^{\infty} a^{-j} \epsilon_{t+j} = a^{-k-1}Y_{t+k+1} + \sum_{j=k+2}^{\infty} a^{-j} \epsilon_{t+j}$, where $a^{-k-1}Y_{t+k+1} \to 0$ and $\sum_{j=k+2}^{\infty} a^{-j} \epsilon_{t+j} \to 0$ in the mean square, as $k \to \infty$, because $|a| > 1$.

9 The condition $\sum_{j=0}^{\infty} |\psi_j| < \infty$ implies that $\sum_{j=0}^{\infty} |\psi_j|^2 < \infty$, which implies, in turn, that $\sum_{j=0}^{\infty} \psi_j \epsilon_{t-j}$ converges in mean square.

An ARMA(p, q) process is called invertible if there exists constants $\{\pi_j\}$ such that $\sum_{j=0}^{\infty} |\pi_j| < \infty$ and

$$\epsilon_t = \sum_{j=0}^{\infty} \pi_j Y_{t-j}, \quad t = 0, \pm1, \pm2, \ldots$$

Let the polynomials $a(z)$ and $b(z)$ have no common zeroes. Then $\{Y_t\}$ is invertible if and only if $b(z) \neq 0$ for all $z \in C$ such that $|z| \leq 1$. This has been proved in Brockwell and Davis (1991, Theorem 3.1.2). The coefficients $\{\pi_j\}$ are determined by

$$\sum_{j=0}^{\infty} \pi_j z^j = \frac{a(z)}{b(z)}.$$

Prediction of ARMA Processes The prediction formulas for ARMA processes given the infinite past can be found in Hamilton (1994, p. 77). For the ARMA(1,1) process $Y_t = aY_{t-1} + \epsilon_t + b\epsilon_{t-1}$ we have

$$E(Y_{t+\eta} | Y_t, Y_{t-1}, \ldots) = a^{\eta-1}(a+b) \sum_{k=0}^{\infty} (-1)^k b^k Y_{t-k}, \tag{5.28}$$

where $\eta \geq 1$; see Shiryaev (1999, p. 151). Note that the prediction formula (5.21) of the MA(1) process and the prediction formula (5.27) of the AR(1) process follow from (5.28).

5.3.3 Conditional Heteroskedasticity Models

Time series $\{Y_t\}$ satisfies the conditional heteroskedasticity assumption if

$$Y_t = \sigma_t \epsilon_t, \quad t = 0, \pm1, \pm2, \ldots, \tag{5.29}$$

where $\{\epsilon_t\}$ is an IID(0, 1) process and $\{\sigma_t\}$ is the volatility process. The volatility process is a predictable random process, that is, σ_t is measurable with respect to the sigma-field $\mathcal{F}_{t-1}$ generated by the variables $Y_{t-1}, Y_{t-2}, \ldots$. We also assume that ϵ_t is independent of $Y_{t-1}, Y_{t-2}, \ldots$. Then,

$$E\left(Y_t^2 | \mathcal{F}_{t-1}\right) = E\left(\sigma_t^2 \epsilon_t^2 | \mathcal{F}_{t-1}\right) = \sigma_t^2 E\left(\epsilon_t^2 | \mathcal{F}_{t-1}\right)$$
$$= \sigma_t^2 E\left(\epsilon_t^2\right) = \sigma_t^2. \tag{5.30}$$

Thus, σ_t^2 is the best prediction of Y_t^2 in the mean squared error sense. Also, for $\eta \geq 1$,

$$E\left(Y_{t+\eta}^2 | \mathcal{F}_t\right) = E\left(\sigma_{t+\eta}^2 \epsilon_{t+\eta}^2 | \mathcal{F}_t\right)$$
$$= E\left[E\left(\sigma_{t+\eta}^2 \epsilon_{t+\eta}^2 | \mathcal{F}_{t+\eta-1}\right) | \mathcal{F}_t\right]$$
$$= E\left[\sigma_{t+\eta}^2 E\left(\epsilon_{t+\eta}^2 | \mathcal{F}_{t+\eta-1}\right) | \mathcal{F}_t\right]$$
$$= E\left[\sigma_{t+\eta}^2 | \mathcal{F}_t\right]. \tag{5.31}$$

Thus, the best prediction of $\sigma_{t+\eta}^2$ gives the best prediction of $Y_{t+\eta}^2$, in the mean squared error sense.

ARCH and GARCH processes are examples of conditional heteroskedasticity models.

5.3.3.1 ARCH Processes

Process $\{Y_t\}$ is an ARCH(p) process (autoregressive conditional heteroskedasticity process of order $p \geq 0$), if $Y_t = \epsilon_t \sigma_t$, where $\{\epsilon_t\}$ is an IID$(0, 1)$ process and

$$\sigma_t^2 = \alpha_0 + \sum_{i=1}^{p} \alpha_i Y_{t-i}^2, \tag{5.32}$$

where $\alpha_0 > 0$ and $\alpha_1, \ldots, \alpha_p \geq 0$. As a special case, the ARCH(1) process is defined as

$$Y_t = \epsilon_t \sqrt{\alpha_0 + \alpha_1 Y_{t-1}^2} \ .$$

The ARCH model was introduced in Engle (1982) for modeling UK inflation rates. The ARCH(p) process is strictly stationary if $\sum_{i=1}^{p} \alpha_i < 1$; see Fan and Yao (2005, Theorem 4.3) and Giraitis *et al.* (2000).

Let us consider the prediction of $Y_{t+\eta}^2$ for $\eta \geq 1$ when the process is an ARCH(p) process. The best prediction of $Y_{t+\eta}^2$, given the observations $Y_t, Y_{t-1}, \ldots,$ is denoted by

$$\mathrm{pred}_t(\eta) = E\left(Y_{t+\eta}^2 \,|\, \mathcal{F}_t\right).$$

We start with the one-step prediction. The best prediction of Y_{t+1}^2, given the observations $Y_t, Y_{t-1}, \ldots,$ using the inference in (5.30), is

$$\mathrm{pred}_t(1) = E\left(Y_{t+1}^2 \,|\, \mathcal{F}_t\right) = E\left(\sigma_{t+1}^2 \epsilon_{t+1}^2 \,|\, \mathcal{F}_t\right) = \sigma_{t+1}^2 E\left(\epsilon_{t+1}^2 \,|\, \mathcal{F}_t\right)$$
$$= \sigma_{t+1}^2 = \alpha_0 + \alpha_1 Y_t^2 + \alpha_2 Y_{t-1}^2 + \cdots + \alpha_p Y_{t-p+1}^2, \tag{5.33}$$

because $E(\epsilon_{t+1}^2 \,|\, \mathcal{F}_t) = 1$. For the two-step prediction we use (5.30) to obtain the best predictor

$$\mathrm{pred}_t(2) = E\left(Y_{t+2}^2 \,|\, \mathcal{F}_t\right) = E\left(\sigma_{t+2}^2 \,|\, \mathcal{F}_t\right)$$
$$= E\left[\alpha_0 + \alpha_1 Y_{t+1}^2 + \alpha_2 Y_t^2 + \cdots + \alpha_p Y_{t-p+2}^2 \,|\, \mathcal{F}_t\right]$$
$$= \alpha_0 + \alpha_1 \mathrm{pred}_t(1) + \alpha_2 Y_t^2 + \cdots + \alpha_p Y_{t-p+2}^2.$$

The general prediction formula is

$$\mathrm{pred}_t(\eta) = E\left(Y_{t+\eta}^2 \,|\, \mathcal{F}_t\right) = E\left(\sigma_{t+\eta}^2 \,|\, \mathcal{F}_t\right)$$
$$= \alpha_0 + \alpha_1 E_t Y_{t+\eta-1}^2 + \cdots + \alpha_p E_t Y_{t+\eta-p}^2,$$
$$= \alpha_0 + \alpha_1 \mathrm{pred}_t(\eta - 1) + \cdots + \alpha_p \mathrm{pred}_t(\eta - p), \tag{5.34}$$

where we denote $E_t = E(\,\cdot\,|\mathcal{F}_t)$. The best prediction is calculated recursively, using the value of $\mathrm{pred}_t(1)$ in (5.33), and the fact that $\mathrm{pred}_t(\eta) = Y_{t+\eta}^2$ for $\eta \leq 0$.

The best η-step prediction in the ARCH(1) model is

$$E\left(Y_{t+\eta}^2 \mid \mathcal{F}_t\right) = \bar{\sigma}^2 + \alpha_1^{\eta-1}\left(\sigma_{t+1}^2 - \bar{\sigma}^2\right) = \alpha_0 \frac{1-\alpha_1^\eta}{1-\alpha_1} + \alpha_1^\eta Y_t^2, \qquad (5.35)$$

where we assumed condition $\alpha_1 < 1$, which guarantees stationarity, and we denote $\bar{\sigma}^2 = EY_t^2 = \alpha_0/(1-\alpha_1)$.[10]

5.3.3.2 GARCH Processes

Process $\{Y_t\}$ is a GARCH(p, q) process (generalized autoregressive conditional heteroskedasticity process of order $p \geq 0$ and $q \geq 0$), if

$$Y_t = \epsilon_t \sigma_t, \qquad (5.37)$$

where $\{\epsilon_t\}$ is an IID$(0, 1)$ process and

$$\sigma_t^2 = \alpha_0 + \sum_{i=1}^p \alpha_i Y_{t-i}^2 + \sum_{i=1}^q \beta_i \sigma_{t-i}^2,$$

where $\alpha_0 > 0$, $\alpha_1, \dots, \alpha_p \geq 0$, and $\beta_1, \dots, \beta_q \geq 0$. As a special case we get the GARCH(1, 1) model, where

$$\sigma_t^2 = \alpha_0 + \alpha_1 Y_{t-1}^2 + \beta \sigma_{t-1}^2. \qquad (5.38)$$

The GARCH model was introduced in Bollerslev (1986). The GARCH(p, q) process is strictly stationary if

$$\sum_{i=1}^q \alpha_i + \sum_{j=1}^p \beta_j < 1; \qquad (5.39)$$

see Fan and Yao (2005, Theorem 4.4) and Bougerol and Picard (1992).

---

10 The prediction formula of the ARCH(1) model follows from the prediction formula of the GARCH(1, 1) model, which is given in (5.40). We can also use the calculation in Shiryaev (1999, p. 59), which gives that

$$\sigma_{t+\eta}^2 = \alpha_0 + \alpha_1 \sigma_{t+\eta-1}^2 \epsilon_{t+\eta-1}^2$$

$$= \alpha_0 + \alpha_1 \left(\alpha_0 + \alpha_1 \sigma_{t+\eta-2}^2 \epsilon_{t+\eta-2}^2\right) \epsilon_{t+\eta-1}^2$$

$$\vdots$$

$$= \alpha_0 + \alpha_0 \sum_{j=1}^{\eta-1} \prod_{i=1}^j \alpha_1 \epsilon_{t+j-i+1}^2 + \sigma_t^2 \prod_{i=1}^\eta \alpha_1 \epsilon_{t+\eta-i}^2.$$

Thus,

$$E\left(\sigma_{t+\eta}^2 \mid \mathcal{F}_t\right) = \alpha_0 + \alpha_0 \sum_{j=1}^{\eta-1} \alpha_1^j + \alpha_1^\eta Y_t^2$$

$$= \alpha_0 \frac{1-\alpha_1^\eta}{1-\alpha_1} + \alpha_1^\eta Y_t^2. \qquad (5.36)$$

Thus, the best η-step prediction of Y_t^2 in ARCH(1) model is given in (5.34), where $\eta \geq 1$ and we used (5.31) and (5.36).

The best one-step prediction of the squared value is obtained from (5.30) as

$$E\left(Y_{t+1}^2 \mid \mathcal{F}_t\right) = \sigma_{t+1}^2.$$

In the GARCH(1, 1) model the best η-step prediction of the squared value, in the mean squared error sense, is

$$E\left(Y_{t+\eta}^2 \mid \mathcal{F}_t\right) = \bar{\sigma}^2 + (\alpha_1 + \beta)^{\eta-1}\left(\sigma_{t+1}^2 - \bar{\sigma}^2\right), \quad \eta \geq 1, \tag{5.40}$$

where we assumed condition $\alpha_1 + \beta < 1$, which guarantees strict stationarity, and we denote the unconditional variance by

$$\bar{\sigma}^2 = EY_t^2 = \frac{\alpha_0}{1 - \alpha_1 - \beta}. \tag{5.41}$$

Let us show (5.40) for $\eta \geq 2$. Let us denote $E(\cdot \mid \mathcal{F}_t) = E_t$. We have

$$\sigma_{t+\eta}^2 = \alpha_0 + \alpha_1 Y_{t+\eta-1}^2 + \beta \sigma_{t+\eta-1}^2$$

and $\alpha_0 = (1 - \alpha_1 - \beta)\bar{\sigma}^2$. Thus,

$$\sigma_{t+\eta}^2 - \bar{\sigma}^2 = \alpha_1 \left(Y_{t+\eta-1}^2 - \bar{\sigma}^2\right) + \beta \left(\sigma_{t+\eta-1}^2 - \bar{\sigma}^2\right).$$

Thus, using (5.31),

$$E_t\left(\sigma_{t+\eta}^2 - \bar{\sigma}^2\right) = (\alpha_1 + \beta)E_t\left(\sigma_{t+\eta-1}^2 - \bar{\sigma}^2\right)$$

$$\vdots$$

$$= (\alpha_1 + \beta)^{\eta-1}E_t\left(\sigma_{t+1}^2 - \bar{\sigma}^2\right)$$

$$= (\alpha_1 + \beta)^{\eta-1}\left(\sigma_{t+1}^2 - \bar{\sigma}^2\right).$$

We have shown (5.40), since $E_t(\sigma_{t+\eta}^2) = E_t(Y_{t+\eta}^2)$, by (5.31). We can also write the best prediction of $Y_{t+\eta}^2$ in the GARCH(1, 1) model as

$$E\left(Y_{t+\eta}^2 \mid \mathcal{F}_t\right) = \alpha_0 \frac{1 - (\alpha_1 + \beta_1)^\eta}{1 - \alpha_1 - \beta_1} + (\alpha_1 + \beta_1)^{\eta-1}\left(\alpha_1 Y_t^2 + \beta_1 \sigma_t^2\right), \tag{5.42}$$

where $\eta \geq 1$.

The prediction formulas (5.40) and (5.42) are written in terms of σ_{t+1}^2. We have the following formula for σ_{t+1}^2 in a strictly stationary GARCH(1, 1) model:

$$\sigma_{t+1}^2 = \frac{\alpha_0}{1 - \beta} + \alpha_1 \sum_{k=0}^{\infty} \beta^k Y_{t-k}^2, \tag{5.43}$$

where we assume $\alpha_1 + \beta < 1$ to ensure strict stationarity. More generally, for the GARCH(p, q) model we have

$$\sigma_t^2 = \frac{\alpha_0}{1 - \sum_{j=1}^{p} \beta_j} + \sum_{k=1}^{\infty} d_k Y_{t-k}^2, \tag{5.44}$$

where d_k are obtained from the equation

$$\sum_{k=1}^{\infty} d_k z^i = \frac{\sum_{i=1}^{q} \alpha_i z^i}{1 - \sum_{j=1}^{p} \beta_j z^j},$$

for $|z| \le 1$; see Fan and Yao (2005, Theorem 4.4).

5.3.3.3 ARCH(∞) Model

GARCH$(1, 1)$ can be considered a special case of the ARCH(∞) model, since (5.43) can be written as

$$\sigma_t^2 = \alpha + \sum_{k=1}^{\infty} \beta_k Y_{t-k}^2,$$

where $\beta_k = \alpha_1 \beta^{k-1}$ and $\alpha = \alpha_0/(1 - \beta)$. We can obtain a more general ARCH(∞) model by defining

$$\sigma_t^2 = \alpha + \sum_{k=1}^{\infty} \psi_k(\theta) m(Y_{t-k}), \tag{5.45}$$

where $\alpha \in \mathbf{R}$, $\theta \in \mathbf{R}^p$, and $m : \mathbf{R} \to \mathbf{R}$ is called a news impact curve. More generally, following Linton (2009), the news impact curve can be defined as the relationship between σ_t^2 and $y_{t-1} = y$ holding past values σ_{t-1}^2 constant at some level σ^2. In the GARCH$(1, 1)$ model the news impact curve is

$$m(y, \sigma^2) = \alpha_0 + \alpha_1 y^2 + \beta \sigma^2.$$

The ARCH(∞) model in (5.45) has been studied in Linton and Mammen (2005), where it was noted that the estimated news impact curve is asymmetric for S&P 500 return data. The asymmetric news impact curve can be addressed by asymmetric GARCH processes.

5.3.3.4 Asymmetric GARCH Processes

Time series of asset returns show a leverage effect. Markets become more active after a price drop: large negative returns are followed by a larger increase in volatility than in the case of large positive returns. In fact, past price changes and future volatilities are negatively correlated. This implies a negative skew to the distribution of the price changes.

The leverage effect is taken into account in the model

$$\sigma_t^2 = \alpha_0 + \alpha_1 (\epsilon_{t-1} - \gamma \sigma_{t-1})^2 + \beta \sigma_{t-1}^2 \tag{5.46}$$

$$= \alpha_0 + \alpha_1 \frac{\left(Y_{t-1} - \gamma \sigma_{t-1}^2\right)^2}{\sigma_{t-1}^2} + \beta \sigma_{t-1}^2,$$

where $\gamma \in \mathbf{R}$ is the skewness parameter. The model was applied in Heston and Nandi (2000) to price options.[11] When $\log(S_t/S_{t-1}) = \sigma_t \epsilon_t$, then under (5.46)

$$\text{Cov}_{t-1}(\sigma_{t+1}, \log S_t) = -2\alpha_1 \gamma \sigma_t.$$

When $\gamma > 0$, then negative values of ϵ_{t-1} lead to larger increase in volatility than positive values of the same size of ϵ_{t-1}. Now the unconditional variance is

$$\bar{\sigma}^2 = EY_t^2 = \frac{\alpha_0 + \alpha_1}{1 - \alpha_1 \gamma^2 - \beta}. \qquad (5.49)$$

5.3.3.5 The Moment Generating function

We need the moment generating function in order to compute the option prices when the stock follows an asymmetric GARCH(1, 1) process. We follow Heston and Nandi (2000). Let

$$Y_t - Y_{t-1} = r + \lambda \sigma_t^2 + \sigma_t \epsilon_t,$$

where $r \in \mathbf{R}$, $\lambda \in \mathbf{R}$, (ϵ_t) are i.i.d. $N(0, 1)$, and

$$\sigma_t^2 = \alpha_0 + \alpha_1(\epsilon_{t-1} - \gamma \sigma_{t-1})^2 + \beta \sigma_{t-1}^2. \qquad (5.50)$$

For example, when the logarithmic returns follow the asymmetric GARCH(1, 1) process, then

$$\log S_t - \log S_{t-1} = r + \lambda \sigma_t^2 + \sigma_t \epsilon_t,$$

so that $Y_t = \log S_t$. We want to find the moment generating function

$$f(t, T, \phi) = E_t \exp\{\phi Y_T\},$$

where $t \leq T$ and $E_t = E(\,\cdot\,|F_t)$ is the conditional expectation at time t.

We have that

$$f(T, T, \phi) = \exp\{\phi Y_T\}. \qquad (5.51)$$

Also,

$$f(T-1, T, \phi) = \exp\left\{ \phi\left[Y_{T-1} + r + \lambda \sigma_T^2\right] + \frac{1}{2}\phi^2 \sigma_T^2 \right\} \qquad (5.52)$$

because the moment generating function of $z \sim N(0, 1)$ is $E\exp\{\phi z\} = \exp\{\phi^2/2\}$, and $E\exp\{\phi(\mu + \sigma z)\} = \exp\{\phi\mu + \sigma^2\phi^2/2\}$.

---

11 Engle and Ng (1993) define the nonlinear asymmetric GARCH model

$$\sigma_t^2 = \alpha_0 + \alpha_1 \sigma_{t-1}^2(\epsilon_{t-1} - \gamma)^2 + \beta \sigma_{t-1}^2, \qquad (5.47)$$

which is for $\gamma = 0$ equal to the GARCH(1, 1) model. Engle and Ng (1993) have defined the VGARCH model

$$\sigma_t^2 = \alpha_0 + \alpha_1(\epsilon_{t-1} - \gamma)^2 + \beta \sigma_{t-1}^2. \qquad (5.48)$$

Menn and Rachev (2009) propose the GARMAX model that can also cope with the leverage effect.

For $t \leq T$ we have

$$f(t, T, \phi) = \exp\left\{\phi Y_t + A(t, T, \phi) + B(t, T, \phi)\sigma_{t+1}^2\right\}, \tag{5.53}$$

where A and B are defined by the recursive formulas

$$A(T, T, \phi) = B(T, T, \phi) = 0,$$
$$A(t, T, \phi) = \phi r + A(t+1, T, \phi) + B(t+1, T, \phi)\alpha_0$$
$$-\frac{1}{2}\log[1 - 2\alpha_1 B(t+1, T, \phi)],$$
$$B(t, T, \phi) = \phi(\lambda + \gamma) - \frac{1}{2}\gamma^2 + \beta B(t+1, T, \phi)$$
$$+\frac{(\phi - \gamma)^2/2}{1 - 2\alpha_1 B(t+1, T, \phi)}.$$

The cases $t = T$ and $t = T - 1$ were proved in (5.51) and (5.52). Let $t \leq T - 2$. Let us make the induction assumption that the formulas hold at time $t + 1$. Now,

$$f(t, T, \phi) = E_t \exp\{\phi Y_T\}$$
$$= E_t E_{t+1} \exp\{\phi Y_T\}$$
$$= E_t f(t+1, T, \phi) \tag{5.54}$$
$$= E_t \exp\left\{\phi Y_{t+1} + A(t+1, T, \phi) + B(t+1, T, \phi)\sigma_{t+2}^2\right\}.$$

Insert values

$$Y_{t+1} = Y_t + r + \lambda\sigma_{t+1}^2 + \sigma_{t+1}\epsilon_{t+1},$$
$$\sigma_{t+2}^2 = \alpha_0 + \alpha_1(\epsilon_{t+1} - \gamma\sigma_{t+1})^2 + \beta\sigma_{t+1}^2$$

to get

$$f(t, T, \phi)$$

$$= E_t \exp\left\{ \phi Y_t + \phi r + A(t+1, T, \phi) + B(t+1, T, \phi)\alpha_0 \right.$$

$$+ B(t+1, T, \phi)\alpha_1\left(\epsilon_{t+1} - \gamma\sigma_{t+1} + \frac{\phi\sigma_{t+1}}{2B(t+1, T, \phi)\alpha_1}\right)^2 \tag{5.55}$$

$$\left. + \left(\phi(\lambda + \gamma) + B(t+1, T, \phi)\beta - \frac{\phi^2}{4B(t+1, T, \phi)\alpha_1}\right)\sigma_{t+1}^2\right\}.$$

When $\epsilon \sim N(0, 1)$, then

$$E\exp\{a(\epsilon + b)^2\} = \exp\left\{-\frac{1}{2}\log(1 - 2a) + \frac{ab^2}{1 - 2a}\right\}.$$

Equating terms in (5.54) and (5.55) gives the result.[12]

---

12 Denote for shortness $B(t+1) = B(t+1, T, \phi)$. We have

$$ab^2 = \left(\frac{\phi^2}{4B(t+1)\alpha_1} - \phi\gamma + \gamma^2 B(t+1)\alpha_1\right)\sigma_{t+1}^2.$$

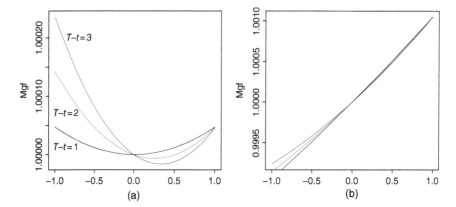

Figure 5.9 *Moment generating functions under GARCH.* We show functions $E_t \exp\{\phi \log S_T\}$, where (a) $S_t = 1$ and (b) $S_t = 1.001$. The case $T - t = 1$ is with black, $T - t = 2$ is with red, and $T - t = 3$ is with blue.

Figure 5.9 shows moment generating functions $\phi \to E_t \exp\{\phi \log S_T\}$. In panel (a) the current stock price is $S_t = 1$, and in panel (b) $S_t = 1.001$. The one period moment generating function ($T - t = 1$) is with black, two period ($T - t = 2$) is with red, and three period ($T - t = 3$) is with blue. The parameters α_0, α_1, β, and γ are estimated from the daily S&P 500 daily data of Section 2.4.1, using model (5.46).

Note that under the usual GARCH(1, 1) model

$$\sigma_t^2 = \alpha_0 + \alpha_1 Y_{t-1}^2 + \beta \sigma_{t-1}^2$$

functions A and B are defined by the recursive formulas

$$A(T, T, \phi) = B(T, T, \phi) = 0,$$
$$A(t, T, \phi) = A(t + 1, T, \phi) + B(t + 1, T, \phi)\alpha_0$$
$$- \frac{1}{2} \log \left[1 - 2\alpha_1 B(t + 1, T, \phi)\sigma_{t+1}^2 \right],$$

Thus,

$$\frac{ab^2}{1 - 2a} - \frac{\phi^2 \sigma_{t+1}^2}{4B(t + 1)\alpha_1} = \frac{\sigma_{t+1}^2}{1 - 2B(t + 1)\alpha_1} (\phi^2/2 - \phi\gamma + \gamma^2 B(t + 1)\alpha_1),$$

because

$$\frac{1}{1 - 2B(t + 1)\alpha_1} \frac{\phi^2}{4B(t + 1)\alpha_1} - \frac{\phi^2}{4B(t + 1)\alpha_1} = \frac{\phi^2/2}{1 - 2B(t + 1)\alpha_1}.$$

Finally,

$$\frac{\gamma^2 B(t + 1)\alpha_1}{1 - 2B(t + 1)\alpha_1} = \frac{\gamma^2/2}{1 - 2B(t + 1)\alpha_1} - \frac{1}{2}\gamma^2.$$

$$B(t, T, \phi) = \lambda\phi + \beta B(t+1, T, \phi) + \frac{\phi^2/2}{1 - 2\alpha_1 B(t+1, T, \phi)\sigma_{t+1}^2}.$$

This means that $A(t, T, \phi)$ and $B(t, T, \phi)$ depend on the unobserved sequence $\sigma_{t+1}, \dots, \sigma_T$, unlike in the case of model (5.50).

5.3.3.6 Parameter Estimation

We discuss first estimation of the ARCH processes, and then extend the discussion to the GARCH processes.

Parameter Estimation for ARCH Processes Estimation of the parameters of ARCH(p) model can be done using the method of maximum likelihood, if we make an assumption about the distribution of innovation ϵ_t. When we have observed $Y_1 = y_1, \dots, Y_T = y_T$, then the likelihood function is

$$L(\alpha_0, \dots, \alpha_p) = f_{Y_1, \dots, Y_p}(y_1, \dots, y_p) \prod_{t=p+1}^{T} f_{Y_t \mid Y_{t-1} = y_{t-1}, \dots, Y_1 = y_1}(y_t).$$

Let us ignore the term $f_{Y_1, \dots, Y_p}(y_1, \dots, y_p)$ and define the conditional likelihood

$$L_p(\alpha_0, \dots, \alpha_p) = \prod_{t=p+1}^{T} f_{Y_t \mid Y_{t-1} = y_{t-1}, \dots, Y_1 = y_1}(y_t).$$

Let us denote the density of ϵ_t by $f_\epsilon : \mathbf{R} \to \mathbf{R}$. Then the conditional density of $Y_t = \sigma_t \epsilon_t$, given $Y_{t-1}, \dots, Y_1$, is

$$f_{Y_t \mid Y_{t-1} = y_{t-1}, \dots, Y_1 = y_1}(y_t) = f_{Y_t \mid Y_{t-1}, \dots, Y_{t-p}}(y) = \frac{1}{\sigma_t} f_\epsilon\left(\frac{y}{\sigma_t}\right),$$

where

$$\sigma_t^2 = \alpha_0 + \alpha_1 Y_{t-1}^2 + \cdots + \alpha_p Y_{t-p}^2.$$

The parameters are estimated by maximizing the conditional likelihood, and we get

$$(\hat{\alpha}_0, \dots, \hat{\alpha}_p) = \underset{\alpha_0, \dots, \alpha_p}{\operatorname{argmax}} \log L_p(\alpha_0, \dots, \alpha_p),$$

where the logarithm of the conditional likelihood is

$$\log L_p(\alpha_0, \dots, \alpha_p) = -\frac{1}{2} \sum_{t=p+1}^{T} \log \sigma_t^2 + \sum_{t=p+1}^{T} \log f_\epsilon\left(\frac{y_t}{\sigma_t}\right). \tag{5.56}$$

If we assume that ϵ_t has the standard normal distribution $\epsilon_t \sim N(0, 1)$, then $f_\epsilon(x) = \exp\{-x^2/2\}/\sqrt{2\pi}$ and

$$(\hat{\alpha}_0, \dots, \hat{\alpha}_p) = \operatorname{argmin}_{\alpha_0, \dots, \alpha_p} \sum_{t=p+1}^{T} \left(\log \sigma_t^2 + \frac{y_t^2}{\sigma_t^2}\right). \tag{5.57}$$

Parameter Estimation for GARCH Processes In the GARCH(p, q) model we can use, similarly to (5.57),

$$(\hat{\alpha}_0, \ldots, \hat{\alpha}_p, \hat{\beta}_1, \ldots, \hat{\beta}_q)$$

$$= \operatorname{argmin}_{\alpha_0, \ldots, \alpha_p, \beta_1, \ldots, \beta_q} \sum_{t=r+1}^{T} \left(\log \tilde{\sigma}_t^2 + \frac{y_t^2}{\tilde{\sigma}_t^2} \right), \qquad (5.58)$$

where $r \geq \max\{p, q\}$. Unlike in the ARCH(p) model, σ_t^2 is a sum of infinitely many terms, and we need to truncate the infinite sum in order to be able to calculate the conditional likelihood. The value $\tilde{\sigma}_{r+1}^2$ can be chosen as the sample variance using $Y_1, \ldots, Y_r$, and $\tilde{\sigma}_t^2$ for $t \geq r + 2$ can be computed using the recursive formula. Then $\tilde{\sigma}_t^2$ is a function of $Y_1^2, \ldots, Y_{t-1}^2$ and of the parameters.

5.3.3.7 Fitting the GARCH(1, 1) Model

We fit the GARCH$(1, 1)$ model for S&P 500 index and for individual stocks of S&P 500.

S&P 500 Daily Data Figure 5.10 shows tail plots of the residuals Y_t/σ_t, where σ_t is the estimated volatility in the GARCH$(1, 1)$ model. Panel (a) shows the left tail plot and panel (b) the right tail plot. The black points show the residuals, the red curves show the standard normal distribution function, and the blue curves show the Student distributions with degrees of freedom 3, 6, and 12. Figure 3.2 shows the corresponding plots for the S&P 500 returns. We see that the standard normal distribution fits well the central area of the distribution of the residuals, but the tails may be better fitted with a Student distribution.

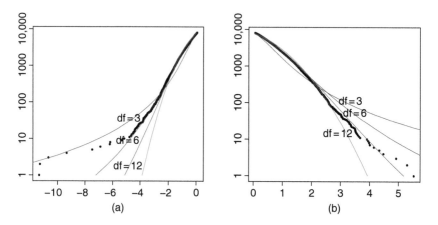

Figure 5.10 *GARCH(1,1) residuals: Tail plots.* (a) Left tail plot; (b) right tail plot. The red curves show the standard normal distribution function, and the blue curves show the Student distributions with degrees of freedom 3, 6, and 12.

S&P 500 Components Data We compute GARCH estimates for daily S&P 500 components data, described in Section 2.4.5. Estimates are computed both for the GARCH$(1,1)$ model and for the Heston–Nandi modification of the GARCH$(1,1)$ model, defined in (5.46).[13] Both models have parameters α_0, α_1, and β. The Heston–Nandi model has the additional skewness parameter γ.

Figure 5.11(a) shows a scatter plot of $(\log \hat{\alpha}_0, \log \hat{\alpha}_0^{hn})$, where $\hat{\alpha}_0$ are estimates of α_0 in the GARCH$(1,1)$ model and $\hat{\alpha}_0^{hn}$ are estimates of α_0 in the Heston–Nandi model. The red points show the estimates for daily S&P 500 data, described in Section 2.4.1. Panel (b) shows a scatter plot of $(\hat{\alpha}_1, \hat{\alpha}_1^{hn})$. We see that the estimates of α_1^{hn} are of the order $\sigma^2 \hat{\alpha}_1$.

Figure 5.12(a) shows a scatter plot of $(\hat{\beta}, \hat{\beta}^{hn})$, where $\hat{\beta}$ are estimates of β in the GARCH$(1,1)$ model, and $\hat{\beta}^{hn}$ are estimates of β in the Heston–Nandi model. We leave out outliers with small estimates for β. Panel (b) shows a histogram of estimates $\hat{\gamma}$ of γ in the Heston–Nandi model. The red points and the lines show the estimates for daily S&P 500 data, described in Section 2.4.1. We see that estimates of β are close to 1, and they are more linearly related in the two models than the estimates of α_1 and α_0. Also, we see that the estimates of the skewness parameter γ are positive for almost all S&P 500 components, with the median value about 2.5. This indicates that high negative returns increase volatility more than the positive returns.

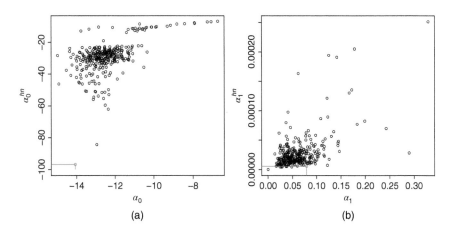

(a)   (b)

Figure 5.11 *GARCH(1,1) estimates versus Heston–Nandi estimates: α_0 and α_1. (a) A scatter plot of $(\log \hat{\alpha}_0, \log \hat{\alpha}_0^{hn})$; (b) a scatter plot of $(\hat{\alpha}_1, \hat{\alpha}_1^{hn})$, where $\hat{\alpha}_0$ and $\hat{\alpha}_1$ are estimates in the GARCH$(1,1)$ model, and $\hat{\alpha}_0^{hn}$ and $\hat{\alpha}_1^{hn}$ are estimates in the Heston–Nandi model.*

---

13 Maximum likelihood estimates for GARCH$(1,1)$ model are computed using R-package "tseries," and the estimates for Heston–Nandi model are computed using R-package "fOptions."

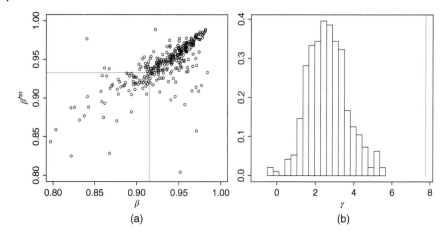

Figure 5.12 *GARCH(1,1) estimates versus Heston–Nandi estimates: β and γ.* (a) A scatter plot of $(\hat{\beta}, \hat{\beta}^{hn})$, where $\hat{\beta}$ are estimates in the GARCH(1, 1) model, and $\hat{\beta}^{hn}$ are estimates in the Heston–Nandi model. Panel (b) shows a histogram of estimates $\hat{\gamma}$ of γ in the Heston–Nandi model.

5.3.4 Continuous Time Processes

The geometric Brownian motion is used to model stock prices in the Black–Scholes model. We do not go into details about continuous time models, but we think that it is useful to review some basic facts about continuous time models. In particular, the geometric Brownian motion appears as the limit of a discrete time binomial model.

5.3.4.1 The Brownian Motion

Stochastic process W_t, $0 \le t \le T$, is called the standard Brownian motion, or the standard Wiener process, if it has the following properties:

1) $W_0 = 0$ with probability one,
2) $W_t \sim N(0, t)$,
3) $W_t - W_s$ is independent of W_s for $0 \le s < t \le T$.

The Brownian motion leads to the process

$$X_t = \mu t + \sigma W_t,$$

where $\mu \in \mathbf{R}$ is drift and $\sigma > 0$ is volatility. We can use the notation of stochastic differential equations:

$$dX_t = \mu dt + \sigma dW_t.$$

5.3.4.2 Diffusion Processes and Itô's Lemma

The diffusion Markov process is defined as

$$X_t = X_0 + \int_0^t a(u, X_u)du + \int_0^t b(u, X_u)dW_u, \tag{5.59}$$

where $0 \le t \le T$, X_0 is a random variable, and

$$\int_0^T |a(t, X_t)| \, dt < \infty, \quad \int_0^T b^2(t, X_t) \, dt < \infty$$

with probability one; see Shiryaev (1999, p. 237). A definition of the stochastic integrals with respect to the Brownian motion can be found in Shiryaev (1999, p. 252).[14] The definition of the process can be written with the shorthand notation of the stochastic differential equations:

$$dX_t = a(t, X_t) \, dt + b(t, X_t) \, dW_t, \quad 0 \le t \le T. \tag{5.60}$$

For example, a mean reverting model is defined as

$$dX_t = \lambda(\mu - X_t) \, dt + \sigma X_t \, dW_t, \quad 0 \le t \le T.$$

Let X_t be a diffusion process as in (5.60), and let $Y_t = F(t, X_t)$, where F is continuously differentiable with respect to the first argument and two times continuously differentiable with respect to the second argument. Furthermore, we assume that $\partial F/\partial x > 0$. Then Y_t is a diffusion Markov process with

$$dY_t = \alpha(t, Y_t)dt + \beta(t, Y_t)dW_t, \tag{5.61}$$

where

$$\alpha(t, y) = \frac{\partial F(t, x)}{\partial t} + a(t, x) \frac{\partial F(t, x)}{\partial x} + \frac{1}{2} b^2(t, x) \frac{\partial^2 F(t, x)}{dx^2},$$

$$\beta(t, y) = b(t, x) \frac{\partial F(t, x)}{\partial x},$$

---

14 For a simple function

$$f(t, \omega) = Y_0(\omega)I_0(t) + \sum_{i=1}^m Y_i(\omega)I_{(r_i, s_i]}(t)$$

the stochastic integral is defined as

$$I_t(f) = \int_0^t f(s, \omega)dB_s = \sum_{i=1}^m Y_i(\omega)(W_{s_i \wedge t} - W_{r_i \wedge t}),$$

where $Y_i(\omega)$ are random variables, $0 \le s_i < r_i$, and we denote $x \wedge t = \min\{x, t\}$. The stochastic integral can be defined for "square integrable" random functions $f(t, \omega)$ as the "limit" of integrals $I_t(f_n)$ of simple functions f_n, "approximating" function f.

and t, x, and y are related by $y = F(t, x)$. The expression for Y_t follows from Itô's lemma; see Shiryaev (1999, p. 263).[15]

5.3.4.3 The Geometric Brownian Motion

The geometric Brownian motion is the stochastic process

$$S_t = S_0 \exp\left\{ \left(\mu - \frac{1}{2} \sigma^2 \right) t + \sigma W_t \right\}, \quad 0 \le t \le T, \tag{5.62}$$

where W_t is the standard Brownian motion, $\mu \in \mathbf{R}$, and $\sigma > 0$. The stochastic differential equation of the geometric Brownian motion is

$$dS_t = \mu S_t \, dt + \sigma S_t \, dW_t, \quad 0 \le t \le T. \tag{5.63}$$

The fact that the solution of the stochastic differential equation in (5.63) is given in (5.62) follows from Itô's formula. Indeed, we consider diffusion process $X_t = \log(S_t)$, $X_0 = \log S_0$, $a(t, X_t) = \mu - \sigma^2/2$, and $b(t, X_t) = \sigma$. Then Itô's formula implies that $S_t = e^{X_t}$ is a diffusion process with $\alpha(t, S_t) = \mu S_t$ and $\beta(t, S_t) = \sigma S_t$.

5.3.4.4 Girsanov's Theorem

Let $(\Omega, \mathcal{F}, (\mathcal{F}_t)_{t \ge 0}, P)$ be a filtered probability space and let $(W_t, \mathcal{F}_t)_{t \ge 0}$ be a Brownian motion. Let $(a_t, \mathcal{F}_t)_{t \ge 0}$ be a stochastic process with $P(\int_0^t a_s^2 ds < \infty) = 1$, for $0 \le t \le T < \infty$. We construct a process $(Z_t, \mathcal{F}_t)_{t \ge 0}$ by setting

$$Z_t = \exp\left\{ \int_0^t a_s dW_s - \frac{1}{2} \int_0^t a_s^2 ds \right\}.$$

If $E \exp\{\frac{1}{2} \int_0^t a_s^2 ds\} < \infty$, then $E Z_T = 1$. We can define a probability measure $\tilde{P}_T$ on $(\Omega, \mathcal{F}_T)$ by

$$\tilde{P}_T(A) = E(Z_T I_A),$$

---

15 Let us consider the case $Y_t = F(X_t)$, so that we can write Itô's lemma as

$$dY_t = F_x \, dX_t + \frac{1}{2} F_{xx} \, b^2(t, X_t) \, dt,$$

where F_x and F_{xx} are the first and the second derivatives. Taylor expansion gives

$$F(t, X_0 + \Delta X_t) - F(t, X_0) \approx F_x \Delta X_t + \frac{1}{2} F_{xx} (\Delta X_t)^2,$$

where $\Delta X_t = X_t - X_0$. If the changes have zero mean, $E(\Delta X_t)^2 \asymp b(t, X_t)^2 \Delta$. Thus, in the stochastic case the second-order term is not of a smaller order than the first-order term, whereas in the deterministic case the second-order term is of a smaller order than the first-order term. The Itô's lemma holds for the class of Itô processes. An Itô process is defined as

$$X_t = X_0 + \int_0^t a(u, \omega) du + \int_0^t b(u, \omega) dW_u.$$

Itô processes are more general than diffusion processes, because in diffusion processes dependence on ω is through $X_u(\omega)$; see Shiryaev (1999, p. 257).

where $A \in \mathcal{F}_T$. Let $P_T = P|\mathcal{F}_T$ be the restriction of P to $\mathcal{F}_T$. Measure $\tilde{P}_T$ is equivalent to P_T. Girsanov's theorem states that

$$\tilde{W}_t = W_t - \int_0^t a_s ds, \qquad (5.64)$$

defines a Brownian motion $(\tilde{W}_t, \mathcal{F}_t, \tilde{P}_T)_{t \leq T}$; see Shiryaev (1999, p. 269). A proof can be found in Shiryaev (1999, Chapter VII, Section 3b).

5.4 Multivariate Time Series Models

The multivariate GARCH model is defined for vector time series $\{Y_t\}$ that has d components. It is assumed that $\{Y_t\}$ is strictly stationary and

$$Y_t = \Sigma_t^{1/2} \epsilon_t, \quad t = 0, \pm1, \pm2, \dots, \qquad (5.65)$$

where $\Sigma_t^{1/2}$ is the square root of a positive definite covariance matrix Σ_t, Σ_t is measurable with respect to the sigma-algebra generated by $Y_{t-1}, Y_{t-2}, \dots$, and ϵ_t is a d-dimensional i.i.d. process with $E\epsilon_t = 0$ and $\text{Var}(\epsilon_t) = I_d$, where I_d is the $d \times d$ identity matrix.

The square root of Σ_t can be defined by writing the eigenvalue decomposition $\Sigma_t = Q_t \Lambda_t Q_t'$, where Λ_t is the diagonal matrix of the eigenvalues of Σ_t and Q_t is the orthogonal matrix whose columns are the eigenvectors of Σ_t. Then we define $\Sigma_t^{1/2} = Q_t \Lambda_t^{1/2} Q_t'$, where $\Lambda_t^{1/2}$ is the diagonal matrix obtained from Λ_t by taking square root of each element. We can define $\Sigma_t^{1/2}$ also as a Cholesky factor of Σ_t.

Multivariate GARCH (MGARCH) processes are reviewed in McNeil *et al.* (2005, Section 4.6), Bauwens *et al.* (2006), and Silvennoinen and Teräsvirta (2009). Below we write the models only for the case $d = 2$, so that $Y_t = (Y_{t,1}, Y_{t,2})$. The multivariate GARCH models are denoted with MGARCH(p, q). We restrict ourselves to the first-order models with $p = q = 1$. The multivariate GARCH models are based on (5.65) but differ in the definition of the recursive formula for Σ_t.

5.4.1 MGARCH Models

First we define the VEC model and two restrictions of it: the diagonal VEC model and the Baba–Engle–Kraft–Kroner (BEKK) model. Then we define the constant correlation model and the dynamic conditional correlation model.

Let us denote $\sigma_{t,1}^2 = \text{Var}(Y_{t,1})$, $\sigma_{t,2}^2 = \text{Var}(Y_{t,2})$, and $\sigma_{t,12} = \text{Cov}(Y_{t,1}, Y_{t,2})$. The VEC model and the diagonal VEC model were introduced in Bollerslev *et al.* (1988). The VEC model assumes that

$$\sigma_{t,1}^2 = a_0 + a_1 Y_{t-1,1}^2 + a_2 Y_{t-1,2}^2 + a_3 Y_{t-1,1} Y_{t-1,2}$$
$$+ b_1 \sigma_{t-1,1}^2 + b_2 \sigma_{t-1,2}^2 + b_3 \sigma_{t-1,12},$$

$$\sigma_{t,2}^2 = c_0 + c_1 Y_{t-1,1}^2 + c_2 Y_{t-1,2}^2 + c_3 Y_{t-1,1} Y_{t-1,2}$$
$$+ d_1 \sigma_{t-1,1}^2 + d_2 \sigma_{t-1,2}^2 + d_3 \sigma_{t-1,12},$$
$$\sigma_{t,12} = e_0 + e_1 Y_{t-1,1}^2 + e_2 Y_{t-1,2}^2 + e_3 Y_{t-1,1} Y_{t-1,2}$$
$$+ f_1 \sigma_{t-1,1}^2 + f_2 \sigma_{t-1,2}^2 + f_3 \sigma_{t-1,12}.$$

This model has 21 parameters $a_0, \ldots, f_3$. Since the model has a large number of parameters, it is useful to consider models with less parameters. The diagonal VEC model has only nine parameters and assumes that

$$\sigma_{t,1}^2 = a_0 + a_1 Y_{t-1,1}^2 + b\sigma_{t-1,1}^2, \tag{5.66}$$
$$\sigma_{t,2}^2 = c_0 + c_1 Y_{t-1,2}^2 + d\sigma_{t-1,2}^2, \tag{5.67}$$
$$\sigma_{t,12} = e_0 + e_1 Y_{t-1,1} Y_{t-1,2} + f\sigma_{t-1,12}. \tag{5.68}$$

Thus, in the diagonal VEC model the components of Y_t follow univariate GARCH models. The BEKK model was introduced in Engle and Kroner (1995). The model has 11 parameters and it can be written more easily with the matrix notation as

$$\Sigma_t = G_0 + G' Y_{t-1} Y_{t-1}' G + H' \Sigma_{t-1} H,$$

where G_0 is a symmetric 2×2 matrix and G and H are 2×2 matrices. The BEKK model is obtained from the VEC model by restricting the parameters. We can express the parameters $a_1, \ldots, f_3$ of the VEC model in terms of the parameters of the BEKK model as follows:

$$a_1 = G_{11}^2, a_2 = G_{12}^2, a_3 = 2G_{11}G_{12}, b_1 = H_{11}^2, b_2 = H_{12}^2, b_3 = 2H_{11}H_{12},$$
$$c_1 = G_{22}^2, c_2 = G_{21}^2, c_3 = 2G_{22}G_{21}, d_1 = H_{22}^2, d_2 = H_{21}^2, d_3 = 2H_{22}H_{21},$$
$$e_1 = G_{11}G_{21}, e_2 = G_{22}G_{12}, e_3 = G_{11}G_{22} + G_{12}G_{21},$$
$$f_1 = H_{11}H_{21}, f_2 = H_{22}H_{12}, f_3 = H_{11}H_{22} + H_{12}H_{21},$$

where we denote the elements of G by G_{ij} and the elements of H by H_{ij}.

The recursive formula for Σ_t can be written by using the correlation matrix P_t. Let Δ_t be the diagonal matrix of the standard deviations of Σ_t. The correlation matrix P_t, corresponding to Σ_t, is such that $\Sigma_t = \Delta_t P_t \Delta_t$.

The constant correlation MGARCH model, introduced in Bollerslev (1990), is such that the components of Y_t follow univariate GARCH models, and the correlation matrix is constant. That is, $\Sigma_t = \Delta_t P \Delta_t$ and $\Delta_t = \mathrm{diag}(\sigma_{t,1}, \sigma_{t,2})$, where P is the constant correlation matrix. The constant correlation GARCH model assumes the univariate GARCH models for the components, as in (5.66) and (5.67), and

$$P_t = \rho.$$

The dynamic conditional correlation MGARCH model, introduced in Engle (2002), is such that the components of Y_t follow univariate GARCH models

and

$$\rho_t = e_0 + e_1 \tilde{Y}_{t-1,1} \tilde{Y}_{t-1,2} + f \rho_{t-1}, \tag{5.69}$$

where $\tilde{Y}_t = \Delta_t^{-1} Y_t, e_0, e_1, f \geq 0, e_1 + f < 1$. Engle (2002) suggests to estimate

$$\hat{e}_0 = (1 - \hat{e}_1 - \hat{f}) \bar{\rho}_t,$$

where $\bar{\rho}_t = t^{-1} \sum_{i=1}^{t} \tilde{Y}_{i,1} \tilde{Y}_{i,2}$ is the sample covariance with $\tilde{Y}_{i,1} = Y_{i,1}/\hat{\sigma}_{i,1}$ and $\tilde{Y}_{i,2} = Y_{i,2}/\hat{\sigma}_{i,2}$. We do not typically have $-1 \leq \hat{\rho}_t \leq 1$, and thus the conditional correlation is estimated from

$$\varrho_t = \frac{\rho_t}{\sigma_{t,1} \sigma_{t,2}},$$

where $\sigma_{t,1}^2 = e_0 + e_1 \tilde{Y}_{t-1,1}^2 + f \sigma_{t-1,1}^2$ and $\sigma_{t,2}^2 = e_0 + e_1 \tilde{Y}_{t-1,2}^2 + f \sigma_{t-1,2}^2$.

5.4.2 Covariance in MGARCH Models

The recursive equation (5.68) in the stationary diagonal VEC model implies that

$$\sigma_{t,12} = \frac{e_0}{1-f} + e_1 \sum_{k=1}^{\infty} f^{k-1} Y_{t-k,1} Y_{t-k,2}.$$

This follows similarly as in the case of GARCH(1, 1) model (see (5.43) and (5.44)). The recursive equation (5.69) in the stationary dynamic conditional correlation GARCH model implies similarly that

$$\rho_t = \frac{e_0}{1-f} + e_1 \sum_{k=1}^{\infty} f^{k-1} \tilde{Y}_{t-k,1} \tilde{Y}_{t-k,2},$$

where $\tilde{Y}_t = (Y_{t,1}/\sigma_{t,1}, Y_{t,2}/\sigma_{t,2})$.

Given the observations $Y_1 = (Y_{1,1}, Y_{1,2}), \ldots, Y_T = (Y_{T,1}, Y_{T,2})$, we estimate the parameters, similarly to GARCH(p, q) estimation in (5.58), by maximizing the conditional modified likelihood,

$$\log_e \tilde{L}_r(a_0, a_1, \ldots, e_1, f) = -\frac{1}{2} \sum_{t=r+1}^{T} \log_e |\tilde{\Sigma}_t| + \sum_{t=r+1}^{T} \log f_\epsilon \left(\tilde{\Sigma}_t^{-1/2} Y_t \right),$$

where $r \geq 1$, f_ϵ is the density of the standard normal bivariate distribution $N(0, I_2)$, and $\tilde{\Sigma}_t$ is the truncated covariance, with elements $\tilde{\sigma}_{t,1}^2, \tilde{\sigma}_{t,2}^2, \tilde{\sigma}_{t,12}$, where

$$\tilde{\sigma}_{t,12} = \frac{e_0}{1-f} + e_1 \sum_{k=1}^{t} f^{k-1} Y_{t-k,1} Y_{t-k,2},$$

and $\tilde{\sigma}_{t,1}^2, \tilde{\sigma}_{t,2}^2$ are defined similarly.

Given the data $Y_0, \ldots, Y_{t-1}$, the MGARCH(1, 1) estimator for the conditional covariance is

$$\hat{\sigma}_{t,12} = \frac{\hat{e}_0}{1-\hat{f}} + \hat{e}_1 \sum_{k=0}^{t-1} \hat{f}^k Y_{t-k-1,1} Y_{t-k-1,2}, \tag{5.70}$$

where the parameter estimators $\hat{e}_0$, $\hat{e}_1$, and $\hat{f}$ are are calculated with the maximum likelihood method.

5.5 Time Series Stylized Facts

Time series models of financial time series should be such that they are able to capture stylized facts. We describe the stylized facts mainly using the daily S&P 500 index data, described in Section 2.4.1. Stylized facts of financial time series are studied by Cont (2001) and Bouchaud (2002).

1) *Returns are uncorrelated.*
   Figure 5.5(a) shows the sample autocorrelation function for the S&P 500 returns. Sample autocorrelations are small, although they are not completely inside the 95% confidence band.
   When the time scale is shorter than tens of minutes, there can be considerable correlation; see Cont (2001) and Bouchaud (2002).
2) *Absolute returns are correlated.*
   Figure 5.5(b) shows the sample autocorrelation function for the absolute S&P 500 returns. The sample autocorrelation goes inside the 95% confidence band after the lag of 500 days, but does not stay inside the band.
   The decay of the autocorrelation of absolute returns has roughly a power law with an exponent in range [0.2, 0.4]; see Cont (2001).
   Since absolute returns are correlated, we can claim that the time series of returns does not consist of independent observations, although they are uncorrelated. The autocorrelation can also be seen in scatter plots. Figure 5.13 shows scatter plots of absolute returns. Panel (a) shows the

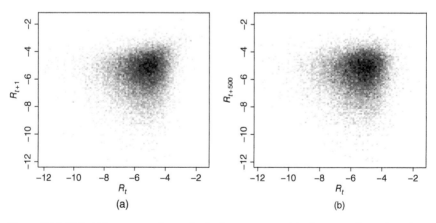

Figure 5.13 *S&P 500 scatter plots of absolute returns.* (a) Scatter plot of points $\left(\log |R_t|, \log |R_{t+1}|\right)$; (b) scatter plot of points $\left(\log |R_t|, \log |R_{t+400}|\right)$.

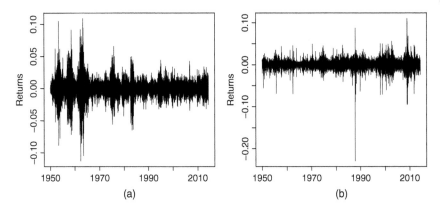

Figure 5.14 *Simulated GARCH*(1, 1) *returns and S&P 500 returns.* (a) A time series of simulated returns from a GARCH(1, 1) model; (b) the time series of S&P 500 returns.

scatter plot of points $(\log |R_t|, \log |R_{t+1}|)$, $t = 1, \dots, T - 1$. Panel (b) shows the scatter plot of points $(\log |R_t|, \log |R_{t+400}|)$, $t = 1, \dots, T - 400$.

3) *Volatility is clustered.*

There are localized outbursts of volatility. The bursts of high volatility last for some time, and then the volatility returns to more normal levels.

Figure 5.14 shows simulated GARCH(1, 1) returns and real S&P 500 returns. Panel (a) shows a time series of returns that are simulated from the GARCH(1, 1) model with parameters being equal to the estimates from S&P 500 daily data. The first return is simulated from the distribution $N(0, \hat{\alpha}_0/(1 - \hat{\alpha}_1 - \hat{\beta}))$. Panel (b) shows the time series of logarithmic S&P 500 returns. S&P 500 data is described in Section 2.4.1. Figure 3.29 shows the corresponding simulated i.i.d. Gaussian returns.

The decay of volatility correlation is slow. The volatility correlation can be defined as the autocorrelation of squared returns, and the autocorrelation of the squared returns shows similar behavior as the autocorrelation of the absolute returns. Volatility displays a positive autocorrelation over several days; see Cont (2001) and Bouchaud (2002).

4) *Extreme returns appear in clusters.*

Figure 5.15 shows the 10 largest and the 10 smallest returns of S&P 500. The largest returns are shown in blue and the smallest returns are shown in red. We can see that the biggest losses and the biggest gains occur at the same dates.

5) *Leverage effect.*

Markets become more active after a price drop; past price changes and future volatilities are negatively correlated. This implies a negative skew to the distribution of the price changes. The leverage effect has been taken into account in the VGARCH model in Engle and Ng (1993) and in the

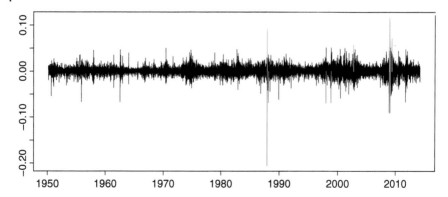

Figure 5.15 *S&P 500 returns.* The 10 smallest returns are shown in red and the 10 largest returns are shown in green.

VGARCH related option pricing in Heston and Nandi (2000). We study asymmetric GARCH models in Section 5.3.3.

Figures 5.11 and 5.12 study parameter fitting in the basic GARCH(1, 1) model and in an asymmetric GARCH(1, 1). Figure 5.12(b) shows that the skewness parameter tends to be positive for S&P 500 components.

6) *Conditional heavy tails.*

Even after correcting the returns for volatility clustering, the residual time series still has heavy tails. The residuals may be calculated, for example, via GARCH-type models.

Figure 5.10 shows the tails of the residuals when GARCH(1, 1) is fitted to S&P 500 daily data.

7) *The kurtosis has slow decay.*

This means that the autocorrelation of the fourth power of the returns has slow decay; see Bouchaud (2002).

8) *Volatility and volume are correlated.*

Volatility and the volume of the activity have long-ranged correlations; see Cont (2001) and Bouchaud (2002).

6

Prediction

We concentrate on prediction with nonparametric smoothing. Nonparametric smoothing can be divided into time and state space smoothing. Time space smoothing means that we use moving averages and state space smoothing means that we use kernel regression with state variables (external explanatory variables). The prediction formulas of ARMA and GARCH processes are related to time and state space smoothing. These prediction formulas are given in Section 5.3, where time series models are considered. Section 5.2 considers the combination of time and state space smoothing with parametric models.

Our emphasis will be more on the economic significance than on the statistical significance. We say that a prediction method is economically significant if it can produce portfolios with significantly higher Sharpe ratios than the Sharpe ratios of the portfolios that are constructed without prediction methods. However, looking at the sum of squared prediction errors can give insights into the underlying reasons for the economic significance.

The classical theory of efficient markets says that the asset returns are unpredictable. If the asset returns were predictable, investors would buy the assets whose predicted returns are high, and eventually this buying would increase the prices and distort the predicted returns. However, it is possible that risk aversion of investors makes the asset returns predictable. For example, in a recession the expected returns could be high but investors are not able to fully utilize the high expected returns, because they have to worry about bankruptcy or unemployment. This could keep the expected returns high by preventing the extensive buying of risky assets. Also, some investment methods require a level of sophistication that is not available to many investors. For example, utilizing momentum effect or volatility trading are not available to many investors, which can keep the expected returns high for some well-known investment strategies.

Section 6.1 studies methods of prediction. Section 6.2 considers forecast evaluation. Section 6.3 reviews some typical predictive variables to be used in asset return prediction. Section 6.4 studies the prediction of S&P 500 and 10-year bond returns.

Nonparametric Finance, First Edition. Jussi Klemelä.
© 2018 John Wiley & Sons, Inc. Published 2018 by John Wiley & Sons, Inc.

6.1 Methods of Prediction

Time space smoothing is covered in Section 6.1.1 and state space smoothing is covered in Section 6.1.2, where we also explain linear prediction under covariance stationarity.

6.1.1 Moving Average Predictors

We give the prediction formula of MA(1) process (moving average process of order one) in (5.21). This prediction formula is an exponentially weighted moving average of the previous values of the time series. Now we give a general definition of a moving average.

6.1.1.1 One-Sided Moving Average

Let us observe the values $Y_1, \dots, Y_t$ of a time series. We define the moving average prediction of $Y_{t+\eta}$ for $\eta \geq 1$ as

$$\hat{f}(t) = \frac{1}{h+1} \sum_{i=t-h}^{t} Y_i,$$

where $h = 0, 1, 2, \dots$. More generally, a one-sided moving average can be defined as

$$\hat{f}(t) = \sum_{i=1}^{t} p_i Y_i, \tag{6.1}$$

where the weights satisfy

$$p_1 \leq \cdots \leq p_t, \quad \sum_{i=1}^{t} p_i = 1. \tag{6.2}$$

To get a flexible class of moving averages we use a kernel function $K : [0, \infty) \to \mathbf{R}$ and smoothing parameter $h > 0$. We can take, for example, $K(x) = \exp(-x)I_{[0,\infty)}(x)$.[1] The one-sided moving average is

$$\hat{f}(t) = \sum_{i=1}^{t} p_i(t) Y_i, \tag{6.3}$$

where

$$p_i(t) = \frac{K((t-i)/h)}{\sum_{j=1}^{t} K((t-j)/h)}. \tag{6.4}$$

Note that the prediction step $\eta \geq 1$ does not show up in the definition of the one-sided moving average. The prediction step η can affect the

---

[1] Note that Gijbels *et al.* (1999) use half-kernels, which are kernel functions that are zero in their positive arguments, like $K(x) = \exp(x)I_{(-\infty,0]}(x)$.

choice of the smoothing parameter h. It would be natural to choose a large smoothing parameter h when the prediction step η is large. Then the predictor with a long prediction horizon would be close to the arithmetic mean $t^{-1} \sum_{i=1}^{t} Y_i$.

6.1.1.2 Exponential Moving Average

The exponential moving average is a one-sided moving average obtained by taking $K(x) = \exp(-x) I_{[0,\infty)}(x)$ and $h = -1/\log \gamma$, where $0 < \gamma < 1$. That is, the exponential moving average is defined as

$$\hat{f}(t) = \sum_{i=1}^{t} p_i(t) Y_i, \tag{6.5}$$

$$p_i(t) = \frac{\exp\{(i-t)/h\}}{\sum_{j=1}^{t} \exp\{(j-t)/h\}}, \tag{6.6}$$

where $h > 0$ is the smoothing parameter. Now the estimator (6.3) is equal to[2]

$$\hat{f}(t) = \frac{1-\gamma}{1-\gamma^t} \sum_{i=1}^{t} \gamma^{t-i} Y_i = \frac{1-\gamma}{1-\gamma^t} \sum_{i=0}^{t-1} \gamma^i Y_{t-i}. \tag{6.7}$$

We get a slightly different exponential moving average by making the recursive definition

$$\mathrm{ma}(t) = (1-\gamma)Y_t + \gamma\,\mathrm{ma}(t-1), \tag{6.8}$$

where $0 \le \gamma \le 1$. This leads to

$$\mathrm{ma}(t) = (1-\gamma) \sum_{i=1}^{t} \gamma^{t-i} Y_i,$$

when the moving average is calculated from $Y_t, \ldots, Y_1$, and we choose the initial value $\mathrm{ma}(1) = (1-\gamma)Y_1$. With infinite past the moving average is

$$\mathrm{ma}(t) = (1-\gamma) \sum_{i=0}^{\infty} \gamma^i Y_{t-i}.$$

Figure 6.1 compares the risk-free rate to the exponentially weighted moving averages of monthly returns of (a) S&P 500 monthly gross returns and (b) US Treasury 10-year bond monthly gross returns. The data is described in Section 2.4.3. The black time series shows the 1-month T-bill rate as a gross return. The moving averages of S&P 500 returns have $h = 30$ (red) and $h = 1000$ (blue). The moving averages of 10-year bond returns have $h = 30$ (red) and $h = 60$ (blue). In the case of 10-year bond, we can note that the

---

2 We have that $\gamma = \exp(-1/h)$ and $\exp\left(-(T-i)/h\right) = \gamma^{T-i}$. Using the geometric series summation formula $\sum_{j=0}^{T-1} r^j = (1-r^T)/(1-r)$, $0 < r < 1$, we have $\sum_{j=1}^{T} \gamma^{T-i} = (1-\gamma^T)(1-\gamma)$.

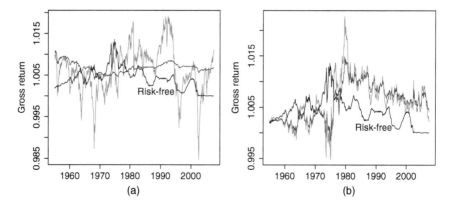

Figure 6.1 *Risk-free rate and moving averages.* (a) Moving averages of S&P 500 monthly gross returns with small h (red) and large h (blue); (b) moving averages of 10-year bond monthly gross returns with small h (red) and large h (blue). The black time series shows the 1-month T-bill rate.

moving averages are almost always lower than the risk-free rate until about 1980, and after that the moving averages are almost always higher than the risk-free rate. The moving averages of S&P 500 returns show a somewhat similar pattern but less clearly. Note that the moving average with $h = 1000$ is almost equal to the sequentially calculated sample mean.

6.1.2 State Space Predictors

A state space predictor is a predictor that is obtained from a regression function estimator. A regression function $f : \mathbf{R}^d \to \mathbf{R}$ is defined as the conditional expectation

$$f(x) = E(Z|X = x),$$

where $Z \in \mathbf{R}$ is the response variable and $X \in \mathbf{R}^d$ is the explanatory variable. A regression function estimator is a function $\hat{f} : \mathbf{R}^d \to \mathbf{R}$, which is computed from regression data $(X_1, Z_1), \ldots, (X_n, Z_n)$, consisting of identically distributed observations.

Let us observe vector time series

$$(X_1, Y_1), \ldots, (X_t, Y_t),$$

where $Y_i \in \mathbf{R}$, and $X_i \in \mathbf{R}^d$ contains information that is available at time i. For example, the state variable X_i can be a sequence of the previous observations of Y_i, so that

$$X_i = (Y_i, \ldots, Y_{i-d+1}), \tag{6.9}$$

where $d \geq 1$ and for $d \geq 2$ we assume to have $Y_0, \ldots, Y_{2-d}$ available.

We use this data to make a prediction of $Y_{t+\eta}$, where $\eta \geq 1$ is the prediction horizon. We construct regression data

$$(X_1, Y_{1+\eta}), \dots, (X_{t-\eta}, Y_t). \tag{6.10}$$

Let us denote $n = T - \eta$ and $Z_t = Y_{t+\eta}$, so that the regression data in (6.10) can be written as

$$(X_1, Z_1), \dots, (X_n, Z_n), \tag{6.11}$$

where $X_t \in \mathbf{R}^d$ are observations from d explanatory variables and $Z_t \in \mathbf{R}$ are observations from the response variable. If $\hat{f}$ is a regression function estimator, we predict the value $Y_{t+\eta}$ by

$$\hat{f}(X_t), \tag{6.12}$$

using the value X_t of the state variable observed at time t.

We define the linear least squares regression function estimator and the kernel regression estimator.

6.1.2.1 Linear Regression

Linear least squares regression function estimator is

$$\hat{f}(x) = \hat{\alpha} + \hat{\beta}'x, \quad x \in \mathbf{R}^d,$$

where $\hat{\alpha} \in \mathbf{R}$ and $\hat{\beta} \in \mathbf{R}^d$ are obtained as the minimizers of

$$\sum_{i=1}^{n} (Z_i - \alpha - \beta'X_i)^2.$$

Least Squares Solution The solution can be written as

$$\hat{\alpha} = \bar{Z} - \hat{\beta}'\bar{X}, \tag{6.13}$$

and

$$\hat{\beta} = \left[\sum_{i=1}^{n} (X_i - \bar{X})(X_i - \bar{X})' \right]^{-1} \sum_{i=1}^{n} (X_i - \bar{X})(Z_i - \bar{Z})', \tag{6.14}$$

where

$$\bar{X} = \frac{1}{n} \sum_{i=1}^{n} X_i, \quad \bar{Z} = \frac{1}{n} \sum_{i=1}^{n} Z_i.$$

In the case $d = 1$, we have

$$\hat{\alpha} = \bar{Z} - \hat{\beta}\bar{X}, \quad \hat{\beta} = \frac{\sum_{i=1}^{n} (X_i - \bar{X})(Z_i - \bar{Z})}{\sum_{i=1}^{n} (X_i - \bar{X})^2}. \tag{6.15}$$

There are more formulas for linear regression in Section 10.4.1.

Linear Prediction Under Covariance Stationarity Let us discuss the idea of autore-
gression in (6.9) more carefully. Let $\{Y_i\}$ be a covariance stationary time series
with $EY_i = 0$ and covariance function $\gamma(k) = EY_iY_{i+k}$, where $k \in \mathbf{Z}$. We want to
find the best linear prediction of $Y_{t+\eta}$ when data $Y_1, \ldots, Y_t$ is available, where
$\eta \geq 1$ is the prediction horizon. We define the best linear prediction of $Y_{t+\eta}$ to be

$$\beta_1 Y_t + \cdots + \beta_d Y_{t-d+1}, \tag{6.16}$$

where $1 \leq d \leq t$, and $\beta_1, \ldots, \beta_d \in \mathbf{R}$ minimize

$$E(Y_{t+\eta} - \beta_1 Y_t - \cdots - \beta_d Y_{t+d+1})^2. \tag{6.17}$$

Let us denote $X = (Y_t, \ldots, Y_{t-d+1})'$, $Y = Y_{t+\eta}$, and $\beta = (\beta_1, \ldots, \beta_d)'$. Then we can
write (6.17) in matrix notation as

$$E(Y - \beta'X)^2.$$

A minimizer satisfies $E(XX')\beta = E(XY)$. If $E(XX')$ is invertible, then a minimizer satisfies

$$\beta = [E(XX')]^{-1}E(XY). \tag{6.18}$$

We have that

$$E(XY) = (\gamma(\eta), \ldots, \gamma(\eta + d - 1))'$$

and matrix $E(XX')$ is the $d \times d$ matrix whose elements are

$$[\gamma(i - j)]_{i,j=1,\ldots,d}.$$

First, we can implement the predictor using the usual least squares estimator,
which replaces the expectations in (6.18) with the sample means, similarly as in
(6.14). Second, we can implement the predictor by estimating the autocovari-
ance function γ. Several parametric models for γ can be used. Let $\hat{\beta}_1, \ldots, \hat{\beta}_d$ be
the estimates of the coefficients $\beta_1, \ldots, \beta_d$. The predictor can be written as

$$\hat{Y}_{t+\eta} = \hat{\beta}_1 Y_t + \cdots + \hat{\beta}_t Y_{t-d+1}. \tag{6.19}$$

Formula (6.18) involves the inverse of $d \times d$ matrix. This matrix is large when
d is big. We could take $d = t$. Brockwell and Davis (1991, Section 5.2) define
recursive algorithms for computing $\hat{Y}_{t+\eta}$ in (6.16) which avoid the computation
of the inverse of $E(XX')$. These recursive algorithms are the Durbin–Levinson
algorithm and the innovations algorithm. Brockwell and Davis (1991, Proposition 5.1.1) states that matrix EXX' is nonsingular for every d, if $\gamma(0) > 0$ and
$\gamma(k) \to 0$ as $k \to \infty$.

6.1.2.2 Kernel Regression
Let us define kernel regression estimator when regression data is given in (6.11)
as $(X_1, Z_1), \ldots, (X_n, Z_n)$. Kernel regression estimator is

$$\hat{f}(x) = \sum_{i=1}^{n} p_i(x) Z_i, \tag{6.20}$$

where the weights are

$$p_i(x) = \frac{K((x - Z_i)/h)}{\sum_{j=1}^{n} K((x - Z_j)/h)}, \tag{6.21}$$

$K : \mathbf{R}^d \to \mathbf{R}$ is a kernel function and $h > 0$ is the smoothing parameter.

The weights $p_i(x)$ are normalized to sum to one, and when $K \geq 0$, then the weights satisfy $p_i(x) \geq 0$. The smoothing parameter may be chosen using the normal reference rule in (3.44):

$$h_i = \left(\frac{4}{d+2} \right)^{1/(d+4)} n^{-1/(d+4)} \hat{\sigma}_i,$$

for $i = 1, \ldots, d$, where $\hat{\sigma}_i$ is the sample standard deviation for the ith variable.

The idea behind the kernel regression estimator is that the weight $p_i(x)$ is large for those Z_i for which X_i is close to x. Remember that the predictor was defined in (6.12) as $\hat{f}(X_t)$, where X_t is the current value of the predictive variable. This means that we search those time points i where the state X_i is similar to the current state X_t, and give a large weight to the corresponding Z_i.

Figure 6.2 illustrates the idea of searching for time points whose state is similar to the current state. The predictive variable X_t is the dividend price ratio, defined in (6.31). The green bands indicate the time periods where the dividend yield has been close to the last value of the dividend price ratio. That is, those times where the dividend price ratio X_t is in the range $[X_t - h, X_t + h]$ are indicated, where X_t is the last value of the dividend price ratio. The black horizontal line shows the value X_t, and the red horizontal lines show the interval $[X_t - h, X_t + h]$. Panel (a) shows the case $h = 0.003$ and panel (b) shows the case $h = 0.001$.

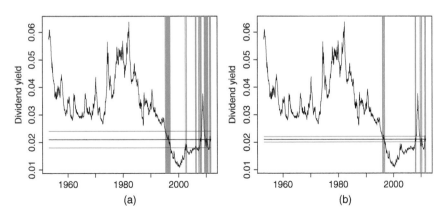

Figure 6.2 *Looking for times with similar states.* The time periods similar to the current state in terms of the dividend price ratio are shown with green. (a) $h = 0.003$; (b) $h = 0.001$.

6.2 Forecast Evaluation

The sum of squared prediction errors can be used to evaluate the performance of a predictor and to compare the performance of two predictors. Often the question arises whether there exist predictability. For example, is it possible to predict stock returns? To answer the question positively, we need to construct a predictor that performs better than a simple benchmark, constructed under the assumption of no predictability. The benchmark is usually the sample average. We would like to test whether an observed smaller prediction error is statistically significant.

6.2.1 The Sum of Squared Prediction Errors

First we define the out-of-sample sum of squared prediction errors and the in-sample sum of squared prediction errors. Then we discuss visual tools to study sums of squared prediction errors.

6.2.1.1 Out-of-Sample Sum of Squares
An out-of-sample sum of squared prediction errors can be defined as recursive, fixed, or rolling.

The Recursive Out-of-Sample Sum of Squares The sequential (or recursive) out-of-sample sum of squares of prediction errors is defined as

$$\mathrm{SSPE}(\hat{f}) = \sum_{t=t_0}^{T-\eta} (Y_{t+\eta} - \hat{f}_t(X_t))^2, \tag{6.22}$$

where $1 < t_0 \le T - \eta$, $\eta \ge 1$ is the prediction horizon, and $\hat{f}_t$ is estimated using the data $(X_i, Y_{i+\eta})$, $i = 1, \dots, t - \eta$. We can normalize the sum of squared prediction errors to get a coefficient of determination. A coefficient of determination compares the performance of a predictor to the performance of a sample mean. The sequential out-of-sample coefficient of determination is defined as

$$R^2 = 1 - \frac{\mathrm{SSPE}(\hat{f})}{\mathrm{SSPE}(\bar{Y})}, \quad \mathrm{SSPE}(\bar{Y}) = \sum_{t=t_0}^{T-\eta} (Y_{t+\eta} - \bar{Y}_t)^2, \tag{6.23}$$

where $\bar{Y}_t = t^{-1} \sum_{i=1}^{t} Y_i$ is the arithmetic mean using the t first observations. When $R^2 > 0$, then the regression forecast is diagnosed to be more accurate than the historical average. The inequality

$$R^2 \le 0$$

is equivalent to the inequality

$$\mathrm{SSPE}(\bar{Y}) \le \mathrm{SSPE}(\hat{f}).$$

The Fixed Out-of-Sample Sum of Squares   In the definition (6.22) of the sum of squared prediction errors the estimate $\hat{f}_t(X_t)$ is updated constantly. A computationally less expensive sum of squared prediction errors can be defined by dividing the sample into an estimation set and into a test set. The predictor is constructed using the estimation set and the sum of squared prediction errors is computed using the test set:

$$\text{SSPE}_{test}(\hat{f}) = \sum_{t=t_0}^{T-\eta} (Y_{t+\eta} - \hat{f}_{t_0}(X_t))^2, \tag{6.24}$$

where $\hat{f}_{t_0}$ is computed using the estimation data $(X_t, Y_{t+\eta})$, $t = 1, \ldots, t_0 - \eta$. The test data is $(X_t, Y_{t+\eta})$, $t = t_0, \ldots, T - \eta$.

The Rolling Out-of-Sample Sum of Squares   The third version of the out-of-sample sum of squared prediction errors is obtained when the predictor is updated at every time point, but the predictor uses always the same number of past observations. The predictor uses windows of observations that are rolled over the available data. Let predictor $\hat{f}_{s,t}$ be constructed using the data $(X_i, Y_{i+\eta})$, $i = s, \ldots, t$. Define

$$\text{SSPE}_{roll}(\hat{f}) = \sum_{t=t_0}^{T-\eta} (Y_{t+\eta} - \hat{f}_{t-t_0+1,t}(X_t))^2. \tag{6.25}$$

Now the sum of squared prediction errors is computed for the estimator that is constructed using exactly t_0 observations at every time point.

The rolling out-of-sample sum of squared prediction errors can be used to study whether a prediction method is better than another prediction method uniformly over all sample sizes t_0. It is possible that a prediction method is better than another method for small sample sizes, and worse for large sample sizes.

6.2.1.2   In-Sample Sum of Squares

We can distinguish between the in-sample and the out-of-sample sum of squared prediction errors. The in-sample sum of squares of prediction errors is defined as

$$\text{SSPE}_{in}(\hat{f}) = \sum_{t=1}^{T-\eta} (Y_{t+\eta} - \hat{f}_T(X_t))^2,$$

where $\hat{f}_T$ is computed using the complete data $(X_t, Y_{t+\eta})$, $t = 1, \ldots, T - \eta$. Thus, the predictor $\hat{f}_T(X_t)$ is constructed using the same data as is used to measure the accuracy of the predictor. The in-sample sum of squared prediction errors is sometimes used, although it could give a too optimistic view of the performance of a predictor.[3]

6.2.1.3 Visual Diagnostics

We define time series

$$D_t = \text{SSPE}(\hat{f})_t - \text{SSPE}(\bar{Y})_t, \quad t = t_0, \dots T - \eta, \tag{6.26}$$

where

$$\text{SSPE}(\hat{f})_t = \sum_{i=t_0}^{t} (Y_{i+\eta} - \hat{f}_i(X_i))^2, \quad \text{SSPE}(\bar{Y})_t = \sum_{i=t_0}^{t} (Y_{i+\eta} - \bar{Y}_i)^2.$$

Time series $\{D_t\}$ reveals useful information about the time periods where the prediction is accurate and about the time periods where it is inaccurate. This graphical diagnostics has been applied in Goyal and Welch (2003, 2008).

If $D_t - D_u < 0$, then predictor $\hat{f}_i$ performs better than the sequential sample average $\bar{Y}_i$ over time period $[u, t]$, where $t > u$. If $D_t - D_u > 0$, then the sequential average is better over time period $[u, t]$. Indeed,

$$D_t - D_u = \sum_{i=u+1}^{t} (Y_{i+\eta} - \hat{f}_i(X_i))^2 - \sum_{i=u+1}^{t} (Y_{i+\eta} - \bar{Y}_i)^2,$$

where $t > u$. Thus, we search for time periods $[u, t]$ which are such that $D_t < D_u$, to find periods of good prediction performance.

6.2.2 Testing the Prediction Accuracy

We are interested in testing the null hypothesis that the sample average is a better predictor than a more sophisticated predictor $\hat{f}$. Thus, the null hypothesis is that the expected sum of squared prediction errors for the sample average is less than the expected sum of squared prediction errors for predictor $\hat{f}$:

$$H_0 : E(\text{SSPE}(\bar{Y})) \leq E(\text{SSPE}(\hat{f})), \tag{6.27}$$
$$H_1 : E(\text{SSPE}(\bar{Y})) > E(\text{SSPE}(\hat{f})).$$

---

3 In regression analysis the in-sample sum of squares of prediction errors is sometimes called the sum of the squared residuals. The number $\text{SS}_{tot} = \sum_{t=1}^{T-\eta} (Y_{t+\eta} - \bar{Y})^2$ is called the total sum of squares. In linear regression we can write $R^2 = \text{SS}_{reg}/\text{SS}_{tot}$, where $\text{SS}_{reg} = \sum_{t=1}^{T-\eta} (\hat{f}(X_t) - \bar{Y})^2$ is the explained sum of squares. Number R^2 takes values in $[0, 1]$ when the intercept is included in the model, and it measures how well the linear model fits the data. In linear regression with a single explanatory variable R^2 is equal to the square of the correlation coefficient between the observations $Y_{t+\eta}$ and the fitted values $\hat{f}(X_t)$.

6.2.2.1 Diebold–Mariano Test

The test statistic of Diebold and Mariano (1995) can be used in testing when $\text{SSPE}_{test}(\hat{f})$ is defined in (6.24). Let us denote the regression forecast of $Y_{t+\eta}$ by $\hat{Y}_{t+\eta}$ and the forecast based on historical average by $\bar{Y}$. Let us denote

$$e_t = \hat{Y}_{t+h} - Y_{t+h}, \quad \epsilon_t = \bar{Y} - Y_{t+h}. \tag{6.28}$$

We get the time series of loss differentials

$$d_t = \epsilon_t^2 - e_t^2.$$

The null hypothesis and the alternative hypothesis are

$$H_0 : Ed_t \le 0, \quad H_1 : Ed_t > 0. \tag{6.29}$$

Since we are using $\text{SSPE}_{test}(\hat{f})$, defined in (6.24), then we can assume that d_t are identically distributed, and then the null and the alternative hypothesis are equivalent to the hypotheses in (6.27).

We apply the central limit theorem for dependent random variables, as stated in (3.96). Under the null hypothesis and under the assumptions of the central limit theorem, we have

$$(T - \eta - t_0 + 1)^{-1/2} \sum_{t=t_0}^{T-\eta} d_t \xrightarrow{d} N(0, \sigma^2),$$

as $T \to \infty$, where

$$\sigma^2 = \sum_{k=-\infty}^{\infty} \gamma(k), \quad \gamma(k) = Ed_0 d_k.$$

We can use the estimate

$$\hat{\sigma}^2 = \hat{\gamma}(0) = (T - \eta - t_0 + 1)^{-1} \sum_{t=t_0}^{T-\eta} (d_t - \bar{d})^2,$$

where $\bar{d} = (T - \eta - t_0 + 1)^{-1} \sum_{t=t_0}^{T-\eta} d_t$. We can also use an estimate that takes the serial correlation into account:

$$\hat{\sigma}^2 = \sum_{k=-(T-1)}^{T-1} w(k)\hat{\gamma}(k),$$

where $w(k) = \max\{0, 1 - k/h\}$ and $h > 0$ is a suitable smoothing parameter. Let us choose the test statistics

$$D = \hat{\sigma}^{-1}(T - t_0 + 1)^{-1/2} \sum_{t=t_0}^{T} d_t.$$

When we observe $D = d_{obs}$, then the p-value is calculated by $P(D > d_{obs}) \approx 1 - \Phi(d_{obs})$, where Φ is the distribution function of the standard normal distribution.[4]

The asymptotics of the Diebold–Mariano test statistic is not as straightforward when $\text{SSPE}(\hat{f})$ is the recursive sum of squared prediction errors as in (6.22), because then the assumptions of the central limit theorem do not hold; see West (1996). In fact, in the sequential case d_t are not identically distributed because the predictor is constructed using at each step one more observation than in the previous step.

West (2006) reviews the alternative asymptotics. We have considered the case where t_0 is fixed, and $T \to \infty$. We can consider the case where both $t_0 \to \infty$ and $T \to \infty$. Then we have to consider separately the cases where $(T - t_0)/t_0 \to 0$ and $(T - t_0)/t_0 \to \infty$. When $(T - t_0)/t_0 \to 0$, then the estimation error involved in the construction of the predictors is negligible, and we can typically replace the predictor $\hat{f}_{t_0}$ with its limit. The asymptotics of the recursive sum of squared prediction errors in (6.22) can be derived by separating the estimation error involved in the construction of the predictors and the estimation error involved in estimating the performance of the limit of the predictor.

6.2.2.2 Tests Using Sample Correlation and Covariance

Let $X_t = \epsilon_t + e_t$ and $Z_t = \epsilon_t - e_t$, where ϵ_t and e_t are defined in (6.28). Then, $\text{Cov}(X_t, Z_t) = E\epsilon_t^2 - Ee_t^2$, when $E\epsilon_t = Ee_t = 0$. Thus, the hypothesis in (6.29) are equivalent to the hypotheses

$$H_0 : \text{Cov}(X_t, Z_t) \le 0, \quad H_1 : \text{Cov}(X_t, Z_t) > 0,$$

where the covariance can also be replaced by the correlation. This was noted by Granger and Newbold (1977). We can derive the distribution, or the

---

4 Diebold and Mariano (1995) note that we could also test the hypothesis where the median replaces the expectation:

$$H_0 : \text{median}(d_t) \le 0, \quad H_1 : \text{median}(d_t) > 0.$$

In this case, we can use the sign test or the Wilcoxon's signed-rank test; see Lehmann (1975). Diebold and Mariano (1995) note also that when ϵ_t and e_t are zero mean, Gaussian, serially uncorrelated, and contemporaneously uncorrelated, then

$$Y = \frac{\sum_{t=t_0}^{T-\eta} \epsilon_t^2}{\sum_{t=t_0}^{T=\eta} e_t^2} \sim F(n, n),$$

where $F(n, n)$ is the F-distribution with degrees of freedom $n = T - \eta - t_0 + 1$. We can use test statistics Y to test the null hypothesis $H_0 : E\epsilon_t^2/Ee_t^2 \le 1$. Large values of Y lead to the rejection of the null hypothesis.

asymptotic distribution, of the sample covariance or the sample correlation coefficient under various assumption. For example, when ϵ_t and e_t are zero mean, Gaussian, and serially uncorrelated, then

$$\frac{\hat{\rho}_{xz}}{\sqrt{\left(1 - \hat{\rho}_{xz}^2\right)/(n-1)}} \sim t_{n-1},$$

where $\hat{\rho}_{xz}$ is the sample correlation coefficient, n is the sample size used in the calculation of the sample correlation coefficient, and t_{n-1} is the t-distribution with $n-1$ degrees of freedom; see Hogg and Craig (1978, p. 300). Meese and Rogoff (1988) use the sample covariance as a test statistics and apply the asymptotic distribution of the sample covariance, as given in Priestley (1981, p. 692). The asymptotic distribution of the sample correlation is given in Brockwell and Davis (1991).

6.3 Predictive Variables

We describe macroeconomic indicators that can be used to predict asset returns. The indicators are risk indicators (default spreads, credit spreads, and volatility indexes), interest rate variables (term spreads and real yield), stock market indicators (dividend price ratio, dividend yield, earnings, and valuation metrics), and sentiment indicators.[5]

6.3.1 Risk Indicators

Risk indicators include default spreads, credit spreads, and volatility indexes. The global financial stress index (GFSI), introduced by BofA Merrill Lynch, is an example of a further risk indicator.

6.3.1.1 Default Spread

The default spread is defined as the difference

$$BAA_t - AAA_t,$$

where BAA_t is the yield of the BAA rated companies and AAA_t is the yield of the AAA rated companies.

Figure 6.3 shows (a) the monthly time series of default spread and (b) the differenced time series.[6]

---

5 I wish to thank Kari Vatanen for helpful discussions concerning predictive variables.
6 We use the data provided by Amit Goyal in the web page http://www.hec.unil.ch/agoyal/.

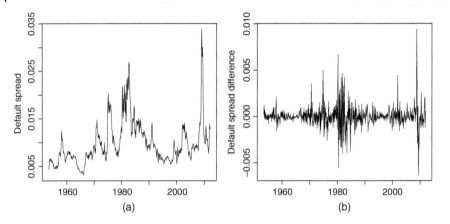

Figure 6.3 *Default spread.* (a) Time series of the default spread; (b) time series of the differences.

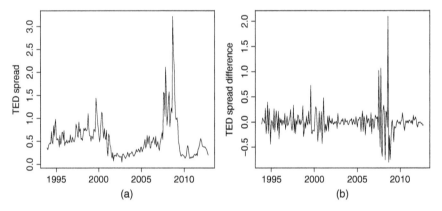

Figure 6.4 *TED spread.* (a) Time series of TED spread; (b) time series of the differences.

6.3.1.2 Credit Spreads

The short term financing expenses of banks are captured by the Treasury bill Eurodollar difference (TED) spread. TED spread is defined as the difference between the 3-month US Libor rate and the 3-month US T-bill rate.

Figure 6.4 shows (a) the time series of TED spread and (b) the differenced time series of TED spread.[7]

The Libor-OIS spread is the difference between the 3-month Libor rate and the OIS rate. OIS is an acronym for the overnight index swap rate, which shows market expectations of future interest rates set by central banks.

---

7 The data is obtained from the St. Louis Fred site. The ticker symbol for the 3-month US Libor rate is BBUSD3M and the ticker symbol for the 3-month US T-bill rate is FRTBW3M.

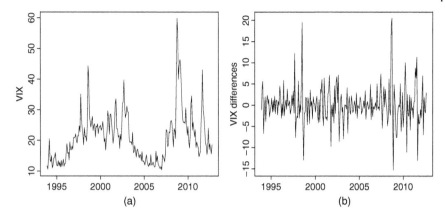

Figure 6.5 *VIX index*. (a) Time series of the VIX index; (b) time series of the differences.

6.3.1.3 Volatility Indexes

Chicago board of options exchange (CBOE) created volatility index (VIX) in 1993. The original VIX was constructed using the implied volatilities of eight different S&P 100 (OEX) option series so that, at any given time, it represented the implied volatility of a hypothetical at-the-money S&P 100 option with 30 days to expiration. The historical prices exist from 1986. Since only at-the-money options were used, no information about the volatility skew was incorporated.

The new VIX is based on S&P 500 index option prices and it incorporates information from the volatility skew by using a wider range of strike prices rather than just at-the-money series. See (14.66) for the formula of the new CBOE VIX index. Other volatility indexes include the JPMorgan Forex (FX) volatility index.

Figure 6.5 shows (a) the time series of VIX index from CBOE and (b) the differenced time series of VIX index.

6.3.2 Interest Rate Variables

In this section, we describe the term spread and the real yield.

6.3.2.1 Term Spread

A term spread is the difference between the yield of a longer maturity bond and a shorter maturity bond. The yield of a zero-coupon bond is defined in (2.3) and in Section 18.1.2. The term spread can be defined as

$$\text{Tbond}_t - \text{Tbill}_t, \tag{6.30}$$

where Tbond_t is the yield of the US Treasury 10-year and Tbill_t is the yield of the US Treasury 1-month bill.

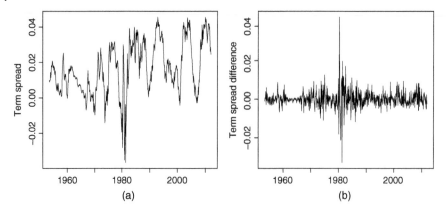

Figure 6.6 *Term spread*. (a) The time series of term spread; (b) the time series of the differences of term spread.

Figure 6.6 shows in panel (a) the monthly time series of the term spread and in panel (b) the differences of the term spread.[8]

The term spread could also be the difference between the yield of the 10-year bond and the yield of the 2-year bond, or the difference between the yield of the 10-year bond and the 3-month bond.

6.3.2.2 Real Yield
The real yield is the yield corrected with the expected inflation. The real yield can be obtained from the US Government 10-year TIPS yield data, available since about 1997.

6.3.3 Stock Market Indicators

Stock market indicators include the dividend price ratio, the dividend yield, earnings, valuation, and relative valuation.

6.3.3.1 Dividend Price Ratio and Dividend Yield
The dividend price ratio of the S&P 500 index is defined as

$$\mathrm{DP}_t = \frac{D_t}{S_t},$$

(6.31)

where D_t is the dollar value of the dividends paid by the S&P 500 companies during the last 12 months and S_t is the value of the S&P 500 index.[9]

---

8 We use the data provided by Amit Goyal in the web page http://www.hec.unil.ch/agoyal/. The bond yield data can be obtained from the St. Louis Fred site with the ticker GS10. The St. Louis yields have range about 1.53–15.32, so they have to be divided by 100 to get the yields in percentages.

9 This terminology is used by Goyal and Welch (2003), who define the yearly dividend price ratio as $\mathrm{DP}_t = D_t/S_t$, where D_t and S_t are values at the end of the year. Note that sometimes the logarithmic ratio $\log \mathrm{DP}_t$ is used.

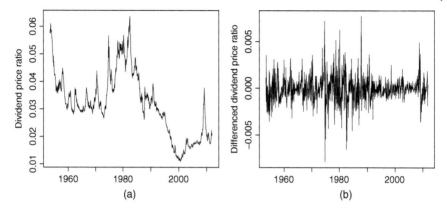

Figure 6.7 *Dividend price ratio.* (a) Time series of the dividend price ratio; (b) time series of the differences.

Figure 6.7 shows (a) the time series of the dividend price ratio and (b) the time series of differences.[10]

The dividend yield is defined as

$$DY_t = \frac{D_t}{S_{t-1}},$$

where D_t is the total amount of dividends paid by the companies of the index during the year and S_{t-1} is the value of the index at the beginning of the year.

The dividend price ratio and the dividend yield have lost importance as stock market indicators, because many companies use stock buy backs instead of paying dividends.

6.3.3.2 Valuation in Stock Markets

Earnings yield is earnings per share divided by the share price. Usually earnings is taken to be the net income for the most recent 12-month period.

Earnings yield is reciprocal to the price earnings ratio (P/E ratio). The trailing P/E is the price divided by the trailing 12 month earnings per share.

The price book ratio (P/B ratio) is the stock price divided by the book value per share. The book value is taken to be the value of total assets minus the value of intangible assets and liabilities.

6.3.3.3 Relative Valuation

We can compare prices of the large capital stocks versus small capital stocks. This can be done by comparing S&P 500 and Russell 2000, because S&P 500 contains large capital stocks and Russell 2000 contains small capital stocks.

---

10 We use the data provided by Amit Goyal in the web page http://www.hec.unil.ch/agoyal/.

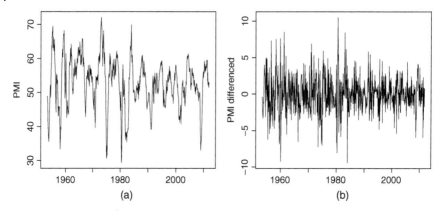

Figure 6.8 *Purchasing managers index.* (a) Time series of the PMI; (b) time series of the differences.

6.3.4 Sentiment Indicators

Sentiment indicators include purchasing managers indexes (PMI), and investor and consumer sentiment indexes.

6.3.4.1 Purchasing Managers Index

PMI are obtained by surveying purchasing managers of private sector companies. There are several regional versions of PMI.

Figure 6.8 shows the time series of the US purchasing managers index and the time series of differences.[11]

6.3.4.2 Investor and Consumer Sentiment

The investor sentiment can be measured with the Sentix index, which is available from http://www.sentix.de. Consumer sentiment indexes include Conference Board Consumer Confidence and University of Michigan Survey of Consumer Confidence.

6.3.5 Technical Indicators

We define technical indicators for a time series $S_1, \ldots, S_t$ of (monthly) closing prices.

---

11 The data is obtained from the Institute for Supply Management (ISM) web page http://www
.ism.ws/ISMReport/content.cfm?ItemNumber=10752, but the page is not available anymore. The
purchasing managers index is a weighted average of five sub-indexes: production level, new orders
from customers, whether supplier deliveries are coming faster or slower, inventories, and employment level. The weights of the sub-indexes are 0.25, 0.3, 0.15, 0.1, and 0.2. The managers respond to
the survey either with "better," "same," or "worse." The index values can be from 0 to 100. The index
values are obtained by taking the percentage of responses that reported better conditions than the
previous month and adding the half of the percentage of responses that reported no change in
conditions.

Many of the technical indicators are based on moving averages, and the moving averages themselves can be used as technical indicators. Moving averages are defined in Section 6.1.1. In statistical finance moving averages are typically computed from returns or squared returns, but in technical analysis moving averages of stock prices are used. The basic moving average is

$$M_t(k) = \frac{1}{k} \sum_{i=0}^{k-i} S_{t-i}.$$

Often $k = 3$ or $k = 6$ months. The exponentially weighted moving average is defined as

$$E_t(k) = \lambda \sum_{i=0}^{\infty} (1 - \lambda)^i S_{t-i} = \lambda S_t + (1 - \lambda)E_{t-1}(k),$$

where $\lambda = 2/(k + 1)$.

The trend is defined as

$$\log \frac{S_t}{M_t(12)}.$$

Figure 6.9 shows the time series of S&P 500 trend and the time series of differences.

Moving average convergence divergence (MACD) is defined as

$$C_t(k_1, k_2) = M_t(k_1) - M_t(k_2),$$

where typically $k_1 = 26$ and $k_2 = 12$. This is a difference of two moving averages, where we subtract from a slow period a fast period.

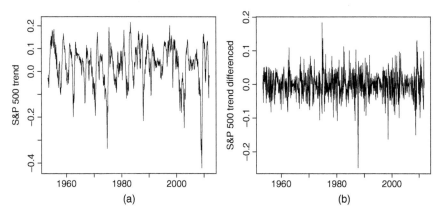

Figure 6.9 *S&P 500 trend.* (a) Time series of the S&P 500 trend; (b) time series of the differences.

MACD signal line is defined as

$$D_t(k, k_1, k_2) = \frac{1}{k} \sum_{i=0}^{k-i} C_{t-i}(k_1, k_2).$$

This is a moving average of MACD. Typically $k = 9$. A buy signal is generated when MACD crosses above the MACD signal line.

MACD histogram is defined as

$$C_t(k_1, k_2) - D_t(k, k_1, k_2).$$

This is the difference between the MACD line and the MACD signal line. A buy signal is generated when the MACD histogram crosses the zero line.

Stochastic oscillator is defined as

$$\text{FAST}_t(k) = \frac{S_t - \min_{i=0,\dots,k-1} S_{t-i}}{\max_{i=0,\dots,k-1} S_{t-i} - \min_{i=0,\dots,k-1} S_{t-i}}.$$

Typically $k = 5$. We can calculate the moving average of the stochastic oscillator. A buy signal is generated when any oscillator crosses below a threshold (say 20) and then crosses above the same threshold. In the minimum we can use the daily lows, and in the maximum we can use the daily highs, instead of the closing prices,

Relative strength index is defined as

$$\text{RSI}_t(k) = 100 \times \left(1 - \left(1 + \frac{k^{-1} \sum_{i=0}^{k-1} S_{t-i}^{up}}{k^{-1} \sum_{i=0}^{k-1} S_{t-i}^{down}} \right)^{-1} \right),$$

where

$$S_t^{up} = \begin{cases} S_t, & \text{if } S_t > S_{t-1}, \\ 0, & \text{otherwise,} \end{cases} \qquad S_t^{down} = \begin{cases} S_t, & \text{if } S_t \leq S_{t-1}, \\ 0, & \text{otherwise.} \end{cases}$$

Typically, $k = 9$, $k = 14$, or $k = 25$. A buy signal is generated when the relative strength index crosses below a lower band of 30, for example, since this situation is characterized as oversold.

Money flow index $\text{MFI}_t(k)$ is similar to the relative strength index $\text{RSI}_t(k)$, but now volume weighted price is used, instead of the price. Define $S_t^{typ} = (S_t^h + S_t^l + S_t)/3$ to be the typical price, where S_t^l is the daily low and S_t^h is the daily high. Let $\text{MF}_t = \text{Vol}_t \cdot S_t^{typ}$ be the strength of the money flow. Money flow index $\text{MFI}_t(k)$ is the moving average of MF_t. Typically $k = 15$. When $\text{MFI}_t(k)$ crosses below threshold 30, for example, then the market is oversold, and one should buy.

6.4 Asset Return Prediction

We study the prediction of S&P 500 returns and US Treasury 10-year bond returns using monthly data, described in Section 2.4.3. We use the predictive variables, which are described in Section 6.3. The gross return of an asset is defined as

$$R_{t+1} = \frac{S_{t+1}}{S_t},$$

where S_t is the price of the asset. The s period return is defined as

$$R_{t+s}^{(s)} = \frac{S_{t+s}}{S_t} = \prod_{u=t+1}^{t+s} R_u, \tag{6.32}$$

where $s \geq 1$. Note that $\{R_t^{(s)}\}$ is a monthly time series, even when the return horizon is longer than 1 month. The monthly time series $\{R_t\}$ has typically only a small autocorrelation, but using a prediction horizon s longer than 1 month creates additional autocorrelation to the time series $\{R_t^{(s)}\}$ due to the overlap: $R_t^{(s)}$ and $R_{t+1}^{(s)}$ are products which have $s - 1$ common terms $R_{t-s+2}, \dots, R_t$.[12]

We consider the prediction of $R_{t+s}^{(s)}$ at time t, where $R_{t+s}^{(s)}$ is the s period return, defined in (6.32). To make the prediction, we use regression

$$R_{t+s}^{(s)} = f(X_t) + \epsilon_{t+s}, \tag{6.33}$$

where $X_t \in \mathbf{R}^d$ is a vector of predicting variables, $f : \mathbf{R}^d \to \mathbf{R}$ is the unknown regression function, and ϵ_{t+s} is random noise. For example, in linear regression $f(x) = \alpha + \beta' x$, where $\alpha \in \mathbf{R}$ and $\beta \in \mathbf{R}^d$ are unknown regression coefficients.

In portfolio selection the rebalancing of the portfolio weights is often done at least monthly. Thus, we need a prediction of 1-month returns. The prediction of 1-month returns can be obtained from the regression in (6.33) by setting $s = 1$, but it turns out that there are better alternatives to make the prediction.

The first alternative to obtain a prediction for 1-month returns is to use regression model

$$(R_{t+s}^{(s)})^{1/s} = f(X_t) + \epsilon_{t+s}. \tag{6.34}$$

The fitted value $\hat{f}(X_t)$ gives the prediction

$$\hat{R}_{t+1} = \hat{f}(X_t). \tag{6.35}$$

The second alternative is to obtain the s period prediction $\hat{R}_{t+s}^{(s)} = \hat{f}(X_t)$ from (6.33), and define the one period prediction as[13]

$$\hat{R}_{t+1} = \left(\hat{R}_{t+s}^{(s)} \right)^{1/s}. \tag{6.36}$$

The two alternative prediction methods seem to give similar results in portfolio selection, as studied in Chapter 12.

---

12 Note that it is a different thing to predict $R_{t+\eta} = S_{t+\eta}/S_t$ at time t, and to predict $R_{t+s}^{(s)}$ at time t. Thus, we use notation s instead of notation η, to differentiate between these two concepts of prediction horizon.

13 Sometimes the prediction $\hat{R}_{t+s}^{(s)}$ is negative, and in this case we take $\hat{R}_{t+1} = (\max\{0, \hat{R}_{t+s}^{(s)}\})^{1/s}$.

6.4.1 Prediction of S&P 500 Returns

We consider the prediction of S&P 500 returns for various prediction horizons, using dividend price ratio and term spread as predictors. Dividend price ratio is defined in (6.31) and term spread is defined in (6.30). Predictor X_t in (6.33) is either dividend price ratio, term spread, or the pair of dividend price ratio and term spread.

6.4.1.1 S&P 500 Returns

Figure 6.10 shows S&P 500 1-month returns (black time series), 1-year returns (red time series), and 5-year returns (green time series). All time series have 1-month frequency. Panel (a) shows the returns $R_{t+s}^{(s)}$, defined in (6.32). Panel (b) shows the times series $(R_{t+s}^{(s)})^{1/s}$. We see that the times series with longer return horizons are smoother and have higher autocorrelation than the time series with shorter return horizons.

Figure 6.11 shows autocorrelations for several lags and horizons. Panel (a) shows autocorrelations for lags $l = 1, 2, 3$. The black curve has $l = 1$, the red curve has $l = 2$, and the blue curve has $l = 3$. The x-axis shows the prediction horizon $s = 1, 2, \ldots, 60$, and the y-axis shows the autocorrelation $\mathrm{Cor}(R_t^{(s)}, R_{t+l}^{(s)})$. For horizons $s = 1, 2$ the autocorrelations are close to zero, but when horizon increases, the autocorrelation starts to increase rapidly towards one, for all lags $l = 1, 2, 3$.

6.4.1.2 Linear Regression for Predicting S&P 500 Returns

Figure 6.12 shows R^2 for predicting S&P 500 returns using dividend price ratio and term spread as predictors in linear regression. The out-of-sample

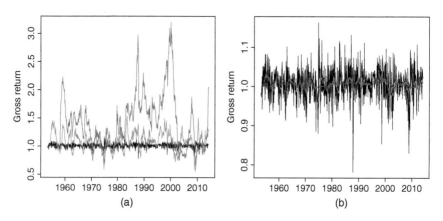

Figure 6.10 *S&P 500 returns for various horizons.* Panel (a) shows the returns $R_{t+s}^{(s)}$, defined in (6.32). Panel (b) shows the times series $(R_{t+s}^{(s)})^{1/s}$. The black time series show 1-month returns, the red time series show 1-year returns, and the green time series show 5-year returns.

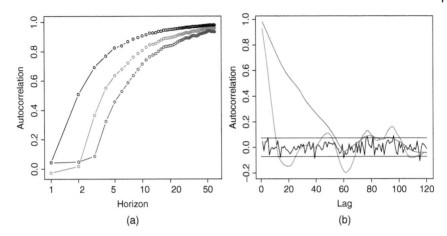

Figure 6.11 *S&P 500 return autocorrelations for various horizons and lags.* (a) We show autocorrelations $\text{Cor}(R_t^{(s)}, R_{t+l}^{(s)})$, where lag is $l = 1$ (black), $l = 2$ (red), and $l = 3$ (blue). The x-axis shows return horizon $s = 1, 2, \ldots, 60$. (b) We show autocorrelations for lags $l = 1, 2, \ldots, 120$, for horizons $s = 1$ (black), $s = 12$ (red), and $s = 60$ (green).

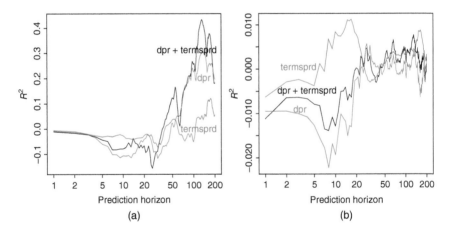

Figure 6.12 R^2 *of linear regression when predicting S&P 500 returns.* Predictors are the dividend price ratio (red), the term spread (green), and both the dividend price ratio and the term spread (black). Panel (a) shows R^2 from regression (6.33) when predicting s month returns, and Panel (b) shows R^2 from regression (6.34) when predicting 1-month returns.

coefficient of determination R^2 is defined in (6.23). Panel (a) shows R^2 from regression (6.33), where $R_{t+s}^{(s)}$ is predicted, with x-axis showing the prediction horizon s. Panel (b) computes R^2 from regression (6.34), where $(R_{t+s}^{(s)})^{1/s}$ is the response variable, 1-month returns are predicted, and x-axis shows the horizon s that is used in fitting the regression coefficients. The red curve shows

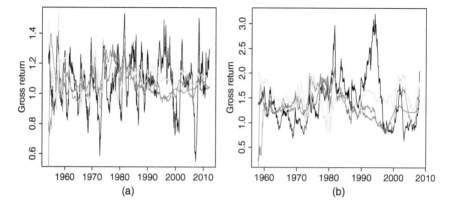

Figure 6.13 *Time series of predictions and realized values.* (a) Prediction horizon of 1 year; (b) prediction horizon of 5 years. The black time series show the realized values $R^{(s)}_{t+s}$, the green time series show the predictions when the predictor is dividend price ratio, the yellow time series show the predictions when the predictor is term spread, and red time series show the predictions when the predictors are both dividend price ratio and term spread.

R^2 when dividend price ratio is predictor, The green curve shows R^2 when term spread is predictor, and the black curve shows R^2 when both dividend price ratio and term spread are predictors. Panel (a) shows that R^2 is large when returns with large horizon s are predicted. Panel (b) shows that when 1-month returns are predicted, then R^2 increases above zero when parameter s is about 10–20, and after that increasing s does not improve R^2.

Figure 6.13 shows both the time series of realized values and the time series of several predictions. In panel (a) the prediction horizon is 1 year and in panel (b) the prediction horizon is 5 years. The time series of realized values $R^{(s)}_{t+s}$ is shown as black curves. The time series of predicted values $\hat{\alpha} + \hat{\beta}X_t$ is shown as a green curve when the predictor is dividend price ratio and as a yellow curve when the predictor is term spread. The time series of predicted values is red when the predictors are both dividend price ratio and term spread.

Figure 6.14 shows the regression data and the fitted regression functions as pink lines, when dividend price ratio is the predictor. Panel (a) shows the case of the prediction horizon of 1 year and panel (b) shows the case of the prediction horizon of 5 years. Since we are doing out-of-sample prediction, there are many fitted regression functions, and we show them all. A new regression function is fitted always when a new data point is added. The blue time series show the regression function that is fitted using all the data.

Figure 6.15 shows the regression data and the fitted regression functions as pink lines, when term spread is the predictor. Panel (a) shows the case of the prediction horizon of 1 year and panel (b) shows the case of the prediction

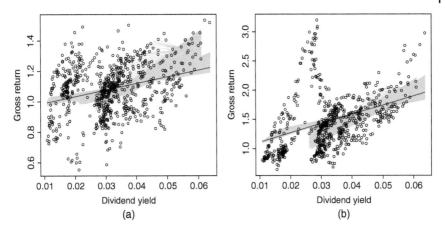

Figure 6.14 *Dividend price ratio as a predictor: Scatter plots and regression functions.* (a) Prediction horizon of 1 year; (b) prediction horizon of 5 years. Scatter plots show the points $(X_t, R_{t+s}^{(s)})$. The pink lines show the fitted regression functions.

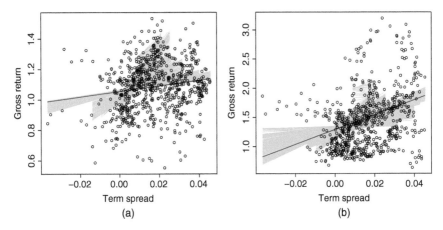

Figure 6.15 *Term spread as a predictor: Scatter plots and regression functions.* (a) Prediction horizon of 1 year; (b) prediction horizon of 5 years. Scatter plots show the points $(X_t, R_{t+s}^{(s)})$. The pink lines show the fitted regression functions.

horizon of 5 years. The blue time series show the regression function which is fitted using all the data.

6.4.2 Prediction of 10-Year Bond Returns

We study linear regression when the explanatory variables are the dividend price ratio and term spread (difference of the yield of the US Treasury 10-year bond and the US Treasury 1-month bill). The response variables are the gross returns of the US Treasury 10-year bond for a given horizon, defined as

$$Y_{t+s} = R_{t+s}^{(s)} = \prod_{i=t+1}^{t+\eta} \exp(R_{bond,i}),$$

where $R_{bond,i}$ is defined in (2.5). We fit the linear regression $Y_{t+s} = \alpha + \beta X_t + \epsilon_{t+s}$ using monthly data.

6.4.2.1 10-Year Bond Returns

Figure 6.16 shows US Treasury 10-year bond 1-month returns (black time series), 1-year returns (red time series), and 5-year returns (green time series). All time series have 1-month frequency. Panel (a) shows the returns $R_{t+s}^{(s)}$, defined in (6.32). Panel (b) shows the times series $(R_{t+s}^{(s)})^{1/s}$. We see that the times series with longer return horizons are smoother and have higher autocorrelation than the time series with shorter return horizons.

Figure 6.17 shows autocorrelations for several lags and horizons. Panel (a) shows autocorrelations for lags $l = 1, 2, 3$. The black curve has $l = 1$, the red curve has $l = 2$, and the blue curve has $l = 3$. The x-axis shows the prediction horizon $s = 1, 2, \ldots, 60$, and the y-axis shows the autocorrelation $\text{Cor}(R_t^{(s)}, R_{t+l}^{(s)})$. For horizons $s = 1, 2$ the autocorrelations are close to zero, but when horizon increases, the autocorrelation starts to increase rapidly towards one, for all lags $l = 1, 2, 3$.

6.4.2.2 Linear Regression for Predicting 10-Year Bond Returns

Figure 6.18 shows R^2 for predicting the US Treasury 10-year bond returns using dividend price ratio and term spread as predictors in linear regression.

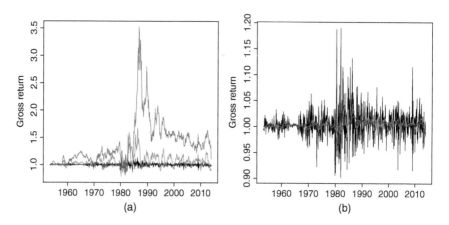

Figure 6.16 *Ten-year bond returns for various horizons.* Panel (a) shows the returns $R_{t+s}^{(s)}$, defined in (6.32). Panel (b) shows the times series $(R_{t+s}^{(s)})^{1/s}$. The black time series show 1-month returns, the red time series show 1-year returns, and the green time series show 5-year returns.

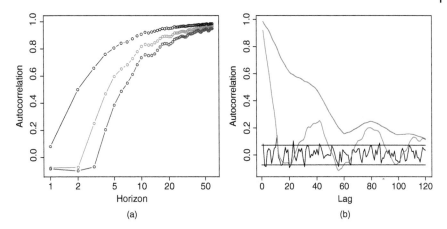

Figure 6.17 *Ten-year bond autocorrelations for various horizons and lags.* (a) We show autocorrelations $\text{Cor}(R_t^{(s)}, R_{t+l}^{(s)})$, where lag is $l = 1$ (black), $l = 2$ (red), and $l = 3$ (blue). The x-axis shows return horizon $s = 1, 2, \ldots, 60$. (b) We show autocorrelations for lags $l = 1, 2, \ldots, 120$, for horizons $s = 1$ (black), $s = 12$ (red), and $s = 60$ (green).

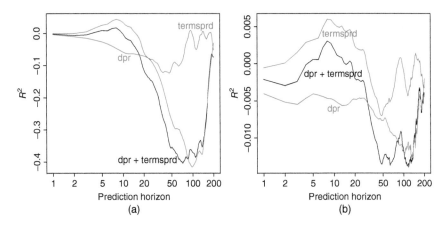

Figure 6.18 R^2 *of linear regression when predicting 10-year bond returns.* Predictors are the dividend price ratio (red), the term spread (green), and both the dividend price ratio and the term spread (black). Panel (a) shows R^2 from regression (6.33) when predicting s-month returns and Panel (b) shows R^2 from regression (6.34) when predicting 1-month returns.

The coefficient of determination R^2 is defined in (6.23). Panel (a) shows R^2 from regression (6.33) and x-axis is the prediction horizon. panel (b) calculates R^2 from regression (6.34), when predicting 1-month returns, and x-axis shows the horizon that is used in fitting the regression coefficients. The red curve shows R^2 when dividend price ratio is the predictor, The green curve shows R^2 when term spread is the predictor, and the black curve shows R^2 when both dividend price ratio and term spread are the predictors. Panel (a) shows that increasing

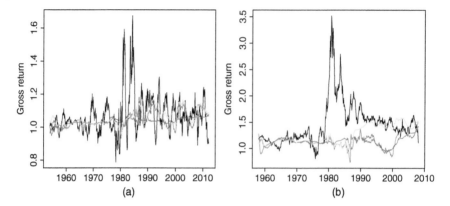

Figure 6.19 *Time series of predictions and realized values.* (a) Prediction horizon of 1 year; (b) prediction horizon of 5 years. The black time series show the realized values $R_{t+s}^{(s)}$, the green time series show the predictions when the predictor is dividend price ratio, the yellow time series show the predictions when the predictor is term spread, and red time series show the predictions when the predictors are both dividend price ratio and term spread.

prediction horizon s does not increase R^2. Panel (b) shows that when predicting 1-month returns, then R^2 is larger than zero for values about $s = 10$.

Figure 6.19 shows both the time series of realized values and the time series of several predictions. In panel (a) the prediction horizon is 1 year and in panel (b) the prediction horizon is 5 years. The time series of realized values $R_{t+s}^{(s)}$ is shown as black curves. The time series of predicted values $\hat{\alpha} + \hat{\beta}'X_t$ is shown as a green

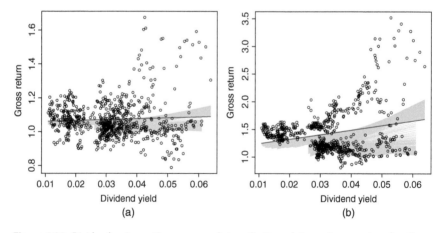

Figure 6.20 *Dividend price ratio as a predictor: Scatter plots and regression functions.* (a) Prediction horizon of 1 year; (b) prediction horizon of 5 years. Scatter plots show the points $(X_t, R_{t+s}^{(s)})$. The pink lines show the fitted regression functions.

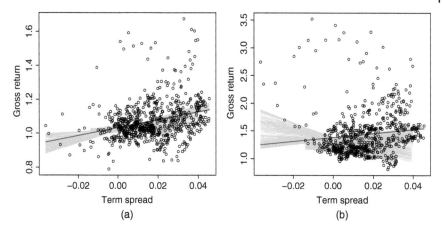

Figure 6.21 *Term spread as a predictor: Scatter plots and regression functions.* (a) Prediction horizon of 1 year; (b) prediction horizon of 5 years. Scatter plots show the points $(X_t, R^{(s)}_{t+s})$. The pink lines show the fitted regression functions.

curve when the predictor is dividend price ratio and as a yellow curve when the predictor is term spread. The time series of predicted values is red when the predictors are both dividend price ratio and term spread.

Figure 6.20 shows the regression data and the fitted regression functions as pink lines, when dividend price ratio is the predictor. Panel (a) shows the case of the prediction horizon of 1 year and panel (b) shows the case of the prediction horizon of 5 years. Since we are doing out-of-sample prediction, there are many fitted regression functions, and we show them all. A new regression function is fitted always when a new data point is added. The blue time series show the regression function that is fitted using all the data.

Figure 6.21 shows the regression data and the fitted regression functions as pink lines, when term spread is the predictor. Panel (a) shows the case of the prediction horizon of 1 year and panel (b) shows the case of the prediction horizon of 5 years. The blue time series show the regression function that is fitted using all the data.

Part II

Risk Management

7

Volatility Prediction

We use the term volatility prediction to mean the prediction of the squared return

$$R^2_{t+\eta},\tag{7.1}$$

where t is the current time, $\eta \geq 1$ is the prediction horizon, and

$$R_i = \frac{S_i}{S_{i-1}} - 1 \quad \text{or} \quad R_i = \log \frac{S_i}{S_{i-1}},$$

are either the net return or the log return of an asset with prices S_i. A closely related concept is the estimation of the conditional variance

$$\text{Var}_t(R_{t+\eta}) = E_t\left(R^2_{t+\eta}\right) - (E_t R_{t+\eta})^2,$$

where $E_t(\cdot) = E(\cdot|F_t)$ is the conditional expectation. Since the squared conditional expectation $(E_t R_{t+\eta})^2$ is often negligible as compared to $E_t(R^2_{t+\eta})$, the estimation of the conditional variance is close to the estimation of the conditional expectation of the squared return

$$E_t\left(R^2_{t+\eta}\right).$$

The conditional expectation of $R^2_{t+\eta}$ is the best prediction of $R^2_{t+\eta}$ in the mean squared error sense, as explained in (5.17) and (5.18), and thus the estimation of the conditional expectation of the squared return leads to a predictor of (7.1).

GARCH(1, 1) predictors and moving average predictors lead often to good predictions of volatility, for short prediction horizons. This is in contrast to the prediction of the returns, which are more difficult to predict, even for short prediction horizons; see Section 6.4 about prediction of returns.[1]

1 Note that it is a different thing to predict $R_{t+\eta} = S_{t+\eta}/S_{t+\eta-1}$ at time t, and to predict $R^{(\eta)}_{t+\eta} = S_{t+\eta}/S_t$ at time t. However, for the logarithmic returns we have

$$\log R^{(\eta)}_{t+\eta} = \log \frac{S_{t+\eta}}{S_t} = \sum_{i=t}^{t+\eta-1} \log \frac{S_{i+1}}{S_i} = \sum_{i=t}^{t+\eta-1} \log R_{i+1}.$$

GARCH(1, 1) predictors and moving average predictors use only historical returns to predict future volatility. However, if a company or a central bank is known to make an announcement in a near future, then this knowledge can be made to predict higher volatility. This kind of additional information can be utilized in state space predictors. We study state space predictors of volatility, but use as states only statistics computed from the previous asset prices.

VIX index can be a useful predictor for future volatility, because it expresses the expectations of the markets for the future volatility. VIX index is discussed in Section 6.3.1 and in (14.66).

A main result of the chapter is to show that GARCH(1, 1) predictor can be improved by a kernel regression predictor, which uses a moving average of the squared returns and a moving average of the returns as predictive variables. The moving average of the squared returns is in itself a predictor which performs as well as the GARCH(1, 1) predictor, but adding information about the past returns improves the prediction. We show that ARCH(p) and moving average predictors perform better than GARCH(1, 1) predictor for moderate prediction horizons (10 days), but the improved performance comes to a large extent from a single event (autumn 1987). In contrast, the performance of the kernel regression predictor is better than GARCH(1, 1) predictor over all time periods. The study is made using the daily S&P 500 data, described in Section 2.4.1.

When we study the performance of volatility predictors it will be seen that the crash of October 1987 has a large influence on the performance. Many studies remove one or more observations around October 1987 and consider these to be outliers. We have not removed any observations for two reasons. (1) The financial crisis of 2008 has almost as large influence on the performance as October 1987, and we would need to remove other observations as outliers, which would lead to a methodological discussion about the definition of an outlier. (2) When we use the differences of cumulative sums of squared prediction errors, then we can identify the influence of each time point, and obtain a description of the performance which is not contaminated by any single financial crash.

Andersen *et al.* (2006) contains a review of volatility prediction, which includes a comprehensive list of references.

Section 7.1 reviews applications of volatility prediction. Section 7.2 discusses the measurement of performance of volatility predictors. Section 7.3 studies

---

Thus, if the logarithmic returns are conditionally uncorrelated,

$$E_t\left(\log R_{t+\eta}^{(\eta)}\right)^2 = \sum_{i=t}^{t+\eta-1} E_t(\log R_{i+1})^2.$$

Thus, a prediction of the squared long horizon logarithmic return can be constructed from predictions of squared one-step logarithmic returns. However, note that for longer horizons the net returns and the logarithmic returns are not close to each other, because approximation $\log x \approx x - 1$ holds only for x close to one.

generalized autoregressive conditional heteroscedasticity (GARCH) and autoregressive conditional heteroscedasticity (ARCH) predictors of volatility. Section 7.4 considers the use of moving averages in volatility prediction. Section 7.5 considers the application of linear and kernel regression in volatility prediction.

7.1 Applications of Volatility Prediction

Volatility prediction can be applied in variance and volatility trading, in covariance trading, in portfolio selection, in quantile estimation, and in option pricing. In addition, prediction of volatility can be applied by credit institutes to measure risk and to set the risk premium.

7.1.1 Variance and Volatility Trading

Volatility can be traded with variance and volatility swaps. A variance swap is a forward contract that pays

$$V_T - K$$

at the expiration date T, where K is the delivery price, and V_T is the realized variance, defined by

$$V_T = \sum_{t=t_0+1}^{T} [\log(S_t/S_{t-1})]^2,$$

where t_0 is the starting day of the contract, and S_t are the prices of the underlying financial asset. A volatility swap pays at the expiration

$$\sqrt{V_T} - L,$$

where L is the delivery price.

Variance and volatility swaps are traded over the counter (OTC), but Chicago Board Options Exchange (CBOE) offers variance futures for the realized variance of the S&P 500 index, calculated with the daily returns of the index.

To make an investment decision for a variance or a volatility swap, we have to estimate the distribution, or the conditional distribution, of the random variables $V_T - K$ or $\sqrt{V_T} - L$. More simply, we can estimate the expectation, or the conditional expectation, of the random variables $V_T - K$ or $\sqrt{V_T} - L$.

7.1.2 Covariance Trading

Variance swaps open an opportunity to covariance trading if we have an access to a variance swap of an index and to variance swaps of its constituents. Let us consider an index whose net returns are

$$R_t = pR_t^1 + qR_t^2,$$

where R_t^i are the net returns of the index constituents and p and q are the weights of the constituents. Let us define the realized covariance as

$$C_T = \sum_{t=t_0+1}^{T} R_t^1 R_t^2.$$

Then,

$$C_T = \frac{1}{2pq} \left(V_T - p^2 V_T^1 - q^2 V_T^2 \right),$$

where $V_T = \sum_{t=t_0}^{T} R_t^2$ is the realized variance of the index and $V_T^i = \sum_{t=t_0}^{T} (R_t^i)^2$ the realized variances of the index constituents, $i = 1, 2$.

If we have three variance swaps which pay $V_T - K$, $V_T^1 - K_1$, and $V_T^2 - K_2$ at the expiration, then we can compose a contract whose components are the three contracts with the weights $1/(2pq)$, $-p/(2q)$, and $-q/(2p)$. This contract pays

$$C_T - M$$

at the expiration, where $M = (K - p^2 K_1 - q^2 K_2)/(2pq)$. The portfolio can be called a covariance swap. To make an investment decision for the portfolio, we have to estimate the distribution, or the conditional distribution, of the random variable $C_T - M$. More simply, we can estimate the expectation, or the conditional expectation, of the random variable $C_T - M$.

7.1.3 Quantile Estimation

Volatility-based quantile estimation is considered in Section 8.5. Volatility estimation can be applied in quantile estimation, because a standard deviation estimate can be used to construct a quantile estimate. Namely, consider the location-scale model

$$Y = \mu + \sigma \, \epsilon,$$

where $\mu \in \mathbf{R}$, $\sigma > 0$, and ϵ is a random variable with a continuous distribution. Now

$$P(Y \leq x) = P\left(\epsilon \leq \frac{x - \mu}{\sigma} \right) = F_\epsilon \left(\frac{x - \mu}{\sigma} \right),$$

where F_ϵ is the distribution function of ϵ. If ϵ has a continuous distribution, then F_ϵ is strictly increasing and the inverse function F_ϵ^{-1} exists. The pth quantile $Q_p(Y)$ of Y satisfies $P(Y \leq Q_p(Y)) = p$, and we can solve this equation to get

$$Q_p(Y) = \mu + \sigma \, F_\epsilon^{-1}(p). \tag{7.2}$$

Thus, for a known F_ϵ, we get from the estimates $\hat{\mu}$ and $\hat{\sigma}$ the estimate

$$\hat{Q}_p(Y) = \hat{\mu} + \hat{\sigma} \, F_\epsilon^{-1}(p).$$

7.1.4 Portfolio Selection

Mean–variance preferences are considered in Section 9.2.1. Let R_{t+1}^p be the return of a portfolio for the time period $[t, t+1]$. The portfolio weights can be chosen to optimize the Markowitz criterion

$$E_t R_{t+1}^p - \frac{\gamma}{2} \operatorname{Var}_t \left(R_{t+1}^p \right),$$

where $\gamma \geq 0$ is the risk aversion parameter, and E_t and Var_t mean the conditional expectation and conditional variance. To apply the Markowitz criterion, we have to estimate both the expected mean $E_t R_{t+1}^p$ and the variance $\operatorname{Var}_t(R_{t+1}^p)$.

7.1.5 Option Pricing

The prediction of volatility can be applied in option pricing. For example, the Black–Scholes price given in (14.58) depends on the distribution of the stock only through its volatility. Although the Black–Scholes price is derived under the assumption of constant volatility, we can insert a predicted volatility to the Black–Scholes pricing function (see Section 14.5).

7.2 Performance Measures for Volatility Predictors

Let $\hat{\sigma}_{t+\eta}^2$ be the predictor of the squared return $R_{t+\eta}^2$, estimated using the data available at time t, where $\eta \geq 1$ is the prediction horizon. We use sometimes notation

$$\hat{f}(t, \eta) = \hat{\sigma}_{t+\eta}^2$$

for the predictor. When $\eta = 1$, then we use sometimes the notation $\hat{f}(t) = \hat{\sigma}_{t+1}^2$. When we have observed returns $R_1, \ldots, R_T$, then the mean of squared prediction errors is defined as

$$\text{MSPE} = \frac{1}{T - t_0 - \eta} \sum_{t=t_0+1}^{T-\eta} \left| \hat{\sigma}_{t+\eta}^2 - R_{t+\eta}^2 \right|^2, \tag{7.3}$$

where $1 \leq t_0 \leq T - 1$, and $\hat{\sigma}_{t+\eta}^2$ is computed using data $R_1, \ldots, R_t$.

In order to compare two predictors, it is useful to plot the time series of differences of cumulative sums of squared prediction errors of the two predictors. The time series is defined as

$$Y_t = \sum_{i=t_0+1}^{t} \left| \hat{f}^1(i, \eta) - R_{i+\eta}^2 \right|^2 - \sum_{i=t_0+1}^{t} \left| \hat{f}^2(i, \eta) - R_{i+\eta}^2 \right|^2, \tag{7.4}$$

where $t_0 \leq t \leq T - \eta$, and $\hat{f}^1(i, \eta)$ and $\hat{f}^2(i, \eta)$ are the two predictors. This allows us to find whether a predictor is uniformly better than the other, or whether the first predictor is better in some periods and worse in others, as explained in connection of (6.26).

Andersen *et al.* (2006, p. 830) writes that "realized volatility provides *the* natural benchmark for forecast evaluation purposes." The realized volatility is the sum $\sum_{\eta=1}^{e} R_{t+\eta}^2$ of squared returns, for some $e \geq 1$. However, we evaluate the forecasts of individual squared returns $R_{t+\eta}^2$, for several horizons $\eta \geq 1$. This is done for two reasons. (1) The predictors of realized volatility are sums of predictors of individual squared returns $R_{t+\eta}^2$. Thus, we obtain a good predictor for the realized volatility from good predictors of $R_{t+\eta}^2$. (2) We need to choose the predictor differently for each different horizon η. When only the realized volatility is used for evaluation, then we are not able as easily to analyze how the performance of different predictors differs when the horizon is changed.

7.3 Conditional Heteroskedasticity Models

In Section 7.3.1, we recall the GARCH(1, 1) predictors of the squared returns and realized volatility, and apply them for the S&P 500 daily data, described in Section 2.4.1. In Section 7.3.2, we study ARCH predictors.

7.3.1 GARCH Predictor

We apply GARCH(1, 1) model for the logarithmic returns $R_t = \log(S_t/S_{t-1})$, where S_t is an asset price. The GARCH(p, q) model is defined in (5.37), and the GARCH(1, 1) model is defined as

$$R_t = \sigma_t \epsilon_t, \quad \sigma_t^2 = \alpha_0 + \alpha_1 R_{t-1}^2 + \beta \sigma_{t-1}^2,$$

where $\{\epsilon_t\}$ is an IID(0, 1) process. The parameters α_0, α_1, and β of the GARCH (1, 1) model are estimated using the maximum likelihood estimator, as defined in (5.58). We denote the estimators by $\hat{\alpha}_0$, $\hat{\alpha}_1$, and $\hat{\beta}$.

7.3.1.1 Predicting the Squared Returns
Let $R_1, \ldots, R_t$ be the observed logarithmic returns. Let us consider the prediction of $R_{t+\eta}^2$, where t is the current time and $\eta \geq 1$ is the prediction horizon. We use as the predictor an estimator of the conditional expectation

$$E_t R_{t+\eta}^2,$$

where E_t means the conditional expectation, conditional on the information available at time t. The conditional expectation is the best prediction in the sense of the mean squared error, as explained in (5.17) and (5.18).

In the GARCH(1, 1) model with $\alpha_1 + \beta < 1$, the conditional expectation $E_t R_{t+\eta}^2$ is obtained from (5.40). We define the GARCH(1, 1) predictor of $R_{t+\eta}^2$ by replacing the unknown population parameters with their estimates and obtain the predictor

$$\hat{f}(t, \eta) = \hat{\sigma}^2 + (\hat{\alpha}_1 + \hat{\beta})^{\eta-1} \left(\hat{\sigma}^2_{t+1} - \hat{\sigma}^2 \right),$$ (7.5)

where

$$\hat{\sigma}^2 = \frac{\hat{\alpha}_0}{1 - \hat{\alpha}_1 - \hat{\beta}}$$

and the formula for $\hat{\sigma}^2_{t+1}$ is obtained from (5.43): formula for $\hat{\sigma}^2_{t+1}$ is obtained by truncating the infinite sum and replacing the unknown population parameters with their estimates, which gives

$$\hat{\sigma}^2_{t+1} = \frac{\hat{\alpha}_0}{1 - \hat{\beta}} + \hat{\alpha}_1 \sum_{k=0}^{t-1} \hat{\beta}^k R^2_{t-k}.$$ (7.6)

In particular, the one-step predictor is

$$\hat{f}(t, 1) = \hat{\sigma}^2_{t+1}.$$ (7.7)

7.3.1.2 Predicting the Realized Volatility

Let us consider the prediction of the η-step realized volatility

$$V_{t,\eta} \overset{\text{def}}{=} R^2_{t+1} + \cdots + R^2_{t+\eta},$$

where t is the current time and $\eta \geq 1$. An estimator of the conditional expectation of the realized volatility is used to predict the realized volatility. Let us denote[2]

$$\sigma^2_{t,\eta} \overset{\text{def}}{=} E_t V_{t,\eta} = \sum_{k=1}^{\eta} E_t \left(R^2_{t+k} \right).$$

Using (5.40), we have the expression

$$\sigma^2_{t,\eta} = \eta \bar{\sigma}^2 + \left(\sigma^2_{t+1} - \bar{\sigma}^2 \right) \sum_{k=1}^{\eta} (\alpha_1 + \beta)^{k-1}.$$

We can write $\sum_{k=1}^{\eta} (\alpha_1 + \beta)^{k-1} = (1 - (\alpha_1 + \beta)^{\eta})/(1 - \alpha_1 - \beta)$. The estimator for the expected realized variance $E_t V_{t,\eta}$ is obtained by replacing the unknown population parameters with estimators:

$$\hat{\sigma}^2_{t,\eta} = \eta \hat{\sigma}^2 + \frac{1 - (\hat{\alpha}_1 + \hat{\beta})^{\eta}}{1 - \hat{\alpha}_1 - \hat{\beta}} \left(\hat{\sigma}^2_{t+1} - \hat{\sigma}^2 \right).$$

---

2 In the GARCH(1, 1) model, the R_t are conditionally uncorrelated and thus it holds also that $\sigma^2_{t,\eta} = E_t \left(\sum_{k=1}^{\eta} R_{t+k} \right)^2$.

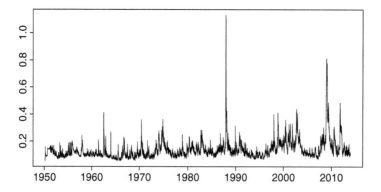

Figure 7.1 *S&P 500 volatility process.* The time series of estimated volatility $\sqrt{250}\hat{\sigma}_t$ in the GARCH(1, 1) model.

7.3.1.3 S&P 500 Volatility Prediction with GARCH(1, 1)

We consider the daily S&P 500 data, described in Section 2.4.1.

Figure 7.1 shows the time series $\sqrt{250}\hat{\sigma}_t$ of the estimated annualized volatility, where $\hat{\sigma}_t$ is calculated sequentially with the formula (7.6).

Figure 7.2 studies the distribution of $\sqrt{250}\hat{\sigma}_t$, where $\hat{\sigma}_t$ is calculated sequentially with the formula (7.6). Panel (a) shows a tail plot, as defined in Section 3.2.1, and panel (b) shows a kernel density estimate, as defined in Section 3.2.2. The blue lines show the annualized sample standard deviation,

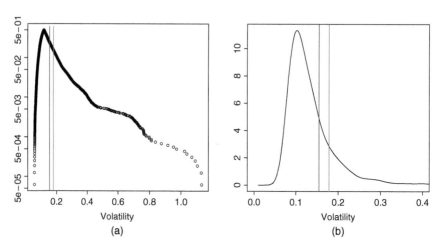

Figure 7.2 *Distribution of S&P 500 volatility predictions.* (a) A tail plot and (b) a kernel density estimate computed from $\sqrt{250}\hat{\sigma}_t$. The blue lines show the annualized sample standard deviation, and the red lines show the annualized unconditional standard deviation.

Figure 7.3 *Error criterion.* Function $q \mapsto$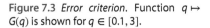
$G(q)$ is shown for $q \in [0.1, 3]$.

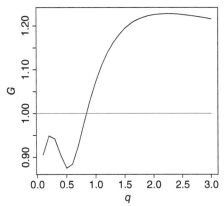

and the red lines show the annualized unconditional standard deviation, as defined by (5.41).

Figure 7.3 shows the function $q \mapsto G(q)$, where

$$G(q) = \frac{\sum_{t=1}^{T} \left| \hat{\sigma}_t^2 - R_t^2 \right|^{1/q}}{\sum_{t=1}^{T} \left| R_t \right|^{2/q}},$$

where $q > 0$. The values of q where $G(q)$ is larger than one are such that the zero estimator $\hat{\sigma}_t \equiv 0$ has smaller error than the GARCH(1, 1) estimator. Thus, it is reasonable to choose q so that $G(q)$ is smaller than one. We will choose the value $q = 0.5$ and measure the performance of volatility prediction by calculating the mean of the squared prediction errors, as defined in (7.3).

7.3.2 ARCH Predictor

We apply ARCH(p) model for the logarithmic returns $R_t = \log(S_t/S_{t-1})$, where S_t is an asset price. The ARCH(p) model is defined in (5.32) as

$$R_t = \sigma_t \epsilon_t, \quad \sigma_t^2 = \alpha_0 + \alpha_1 R_{t-1}^2 + \cdots + \alpha_p R_{t-p}^2,$$

where $\{\epsilon_t\}$ is an IID(0, 1) process. The parameters $\alpha_0, \ldots \alpha_p$ are estimated with the maximum likelihood estimator, as defined in (5.57). We denote the estimators by $\hat{\alpha}_0, \ldots, \hat{\alpha}_p$.

7.3.2.1 Predicting the Squared Returns

The best one-step prediction for the ARCH(p) model was given in (5.33). We use the predictor

$$\hat{f}_p(t) = \hat{\alpha}_0 + \hat{\alpha}_1 R_t^2 + \cdots + \hat{\alpha}_p R_{t-p+1}^2. \tag{7.8}$$

The parameter estimates $\hat{\alpha}_0, \ldots, \hat{\alpha}_p$ are computed from observations $R_t, R_{t-1}, \ldots$ using maximum likelihood.

A recursive formula for the best η-step prediction for ARCH(p) model was given in (5.34). We use the predictor

$$\hat{f}_p(t,\eta) = \hat{\alpha}_0 + \hat{\alpha}_1 \text{pred}_t(\eta-1) + \cdots + \hat{\alpha}_p \text{pred}_t(\eta-p),$$

where $\text{pred}_t(1)$ is given in (7.8) and $\text{pred}_t(\eta) = Y^2_{t+\eta}$ for $\eta \le 0$. Now $\hat{f}_p(t,\eta)$ predicts $R^2_{t+\eta}$.

7.3.2.2 S&P 500 Volatility Prediction with ARCH(p)

We consider the daily S&P 500 data, described in Section 2.4.1.

Figure 7.4[3] compares ARCH(p) predictions to GARCH(1,1) predictions. Panel (a) shows functions

$$p \mapsto \frac{\text{MSPE}^{arch}(\eta,p)}{\text{MSPE}^{garch}(\eta)},$$

where

$$\text{MSPE}^{arch}(\eta,p) = \frac{1}{T-t_0-\eta} \sum_{t=t_0+1}^{T-\eta} \left| \hat{f}_p(t,\eta) - R^2_{t+\eta} \right|^2,$$

and $\text{MSPE}^{garch}(\eta)$ is the mean of the squared prediction errors of the GARCH(1,1) estimate, when GARCH(1,1) estimate is defined in (7.5). Panel (b) shows the time series of differences

$$\text{CSPE}^{arch}(\eta)_t - \text{CSPE}^{garch}(\eta)_t$$

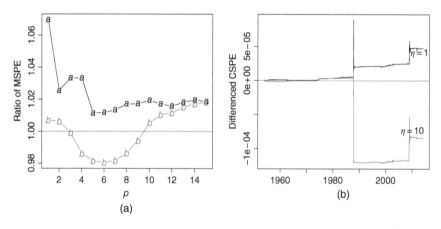

Figure 7.4 Comparison of ARCH(p) and GARCH(1,1). (a) Function $p \mapsto \text{MSPE}^{arch}(\eta,p)/\text{MSPE}^{garch}(\eta)$. (b) Time series of CSPE$^{arch}(\eta)_t - \text{CSPE}^{garch}(\eta)_t$. The prediction horizons are $\eta = 1$ (black with "a") and $\eta = 10$ (red with "b").

---

3 The green horizontal line is in the following figures a reference line to help the orientation, drawn either at height zero or at height one.

between the cumulative sums of the squared prediction errors of ARCH(p) and GARCH($1, 1$) predictor (see (7.4) for the formula). The ARCH predictor has $p = 6$. The prediction horizons are $\eta = 1$ (black with "a") and $\eta = 10$ (red with "b"). Panel (a) shows that for prediction horizon $\eta = 1$ the GARCH($1, 1$) predictor has a better mean squared prediction error, but for prediction horizon $\eta = 10$ ARCH(p) predictor seems better, for p about six. However, panel (b) shows that the better performance of ARCH for $\eta = 10$ comes from a single event: the market crush of October 1987.

7.4 Moving Average Methods

Let $R_1, \dots, R_t$ be the observed time series of financial returns. We consider the prediction of $R^2_{t+\eta}$, where $\eta \geq 1$ is the prediction horizon. A moving average prediction of $R^2_{t+\eta}$ is a weighted arithmetic mean of the squares of the previous returns. We have defined the moving average predictors in Section 6.1.1. In the data analyzes, we use the daily S&P 500 data, described in Section 2.4.1.

7.4.1 Sequential Sample Variance

Before giving results for the exponential moving average, we study the prediction of the η-step ahead squared return by using the sequentially computed sample variance. At time t the prediction of $R^2_{t+\eta}$ is the sample variance, computed from the observations $R_1, \dots, R_t$. The sample variance can be considered as the limit of a moving average when the window width $h \to \infty$ (or $\gamma \uparrow 1$). The sequentially computed standard deviation is defined as

$$\left(\frac{1}{t} \sum_{i=1}^{t} R_t^2 - \bar{R}_t^2 \right)^{1/2}, \quad \bar{R}_t = t^{-1} \sum_{i=1}^{t} R_i.$$

Figure 7.5 compares the time series of the sequentially computed standard deviation and the GARCH($1, 1$) stationary standard deviation. The black curve shows the time series of annualized sequentially computed standard deviations. The blue curve shows the GARCH($1, 1$) annualized stationary volatility. The stationary volatility is defined in (5.41) as

$$\left(\frac{\hat{\alpha}_{0,t}}{1 - \hat{\alpha}_{t,1} - \hat{\beta}_t} \right)^{1/2},$$

where $\hat{\alpha}_{0,t}$, $\hat{\alpha}_{t,1}$, and $\hat{\beta}_t$ are the GARCH($1, 1$) estimates, computed from the returns $R_1, \dots, R_t$.

Figure 7.6 compares prediction errors between the sequentially computed standard deviation and the GARCH($1, 1$) predictor. Panel (a) shows the function

$$\eta \mapsto \frac{\mathrm{MSPE}^{seq}(\eta)}{\mathrm{MSPE}^{garch}(\eta)},$$

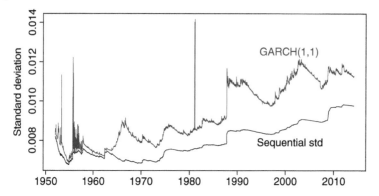

Figure 7.5 *Sequential and GARCH(1,1) standard deviations.* The black curve shows the sequentially computed sample standard deviations and the blue curve shows the sequentially computed GARCH(1, 1) stationary standard deviations.

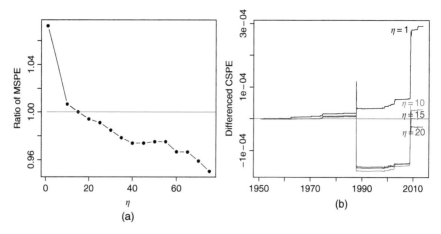

Figure 7.6 *Comparison of the sequential sample variance and GARCH(1,1) predictors.* (a) Shown is the function $\eta \mapsto \text{MSPE}^{seq}(\eta)/\text{MSPE}^{garch}(\eta)$. (b) Shown are the time series $\text{CSPE}_t^{seq}(\eta) - \text{CSPE}_t^{garch}(\eta)$ of the differences of the cumulative sums of prediction errors. We show the cases $\eta = 1$ (black), $\eta = 10$ (red), $\eta = 15$ (green), and $\eta = 20$ (blue).

where $\text{MSPE}^{seq}(\eta)$ and $\text{MSPE}^{garch}(\eta)$ are the means of squared prediction errors when prediction horizon is η (see (7.3) for the formula). The GARCH(1, 1) predictor is defined in (7.5). The parameters of GARCH(1, 1) model are estimated sequentially. Panel (b) shows the difference between the sum of the cumulative squared prediction errors of the sequential standard deviation and the sum of the cumulative squared prediction errors of GARCH(1, 1) (see (7.4) for the formula). We show the cases $\eta = 1$ (black), $\eta = 10$ (red), $\eta = 15$ (green), and $\eta = 20$ (blue). We see from panel (a) that the GARCH(1, 1) prediction is better when $\eta \leq 15$. The sample variance is better when $\eta \geq 15$. In fact, GARCH(1, 1) becomes exponentially worse when η increases, but sequential sample variance becomes linearly worse when η increases. We see from panel (b) that the longer horizons have a singular performance loss at end of 1980s.

7.4.2 Exponentially Weighted Moving Average

The one-sided moving average predictor was defined in (6.3). In volatility prediction, the one-sided moving average is equal to

$$\hat{f}(t) = \sum_{i=1}^{t} p_i(t) R_i^2, \tag{7.9}$$

where the weights are

$$p_i(t) = \frac{K((t-i)/h)}{\sum_{j=1}^{t} K((t-j)/h)}, \tag{7.10}$$

where $K : [0, \infty) \to \mathbf{R}$ is the kernel function and $h > 0$ is the smoothing parameter.

The exponential moving average was defined in (6.7). The exponential moving average is a one-sided moving average obtained by taking $K(x) = \exp(-x) I_{[0,\infty)}(x)$ and $h = -1/\log \gamma$, where $0 < \gamma < 1$. In volatility prediction the exponential moving average is equal to

$$\hat{f}(t) = \frac{1-\gamma}{1-\gamma^t} \sum_{i=1}^{t} \gamma^{t-i} R_i^2. \tag{7.11}$$

The prediction step η influences the choice of the smoothing parameter h. It is natural to choose a large smoothing parameter h when the prediction step η is large. Then a long horizon predictor is almost equal to the arithmetic mean $t^{-1} \sum_{i=1}^{t} R_i^2$.

Figure 7.7 compares volatility prediction with exponentially weighted moving averages to the GARCH(1, 1) prediction. We use the daily S&P 500 data, described in Section 2.4.1. Panel (a) shows functions

$$h \mapsto \frac{\text{MSPE}^{ewma}(\eta, h)}{\text{MSPE}^{garch}(\eta)},$$

where $\text{MSPE}^{ewma}(\eta, h)$ and $\text{MSPE}^{garch}(\eta, h)$ are the means of squared prediction errors when prediction horizon is η (see (7.3) for the formula). The h-axis is logarithmic. The exponentially weighted moving average is defined in (7.11), using smoothing parameter $h = -1/\log \gamma$. The symbols "a," "b," and "c" correspond to prediction horizons $\eta = 1$, $\eta = 10$, and $\eta = 20$. We can note that for the horizon $\eta = 1$, the GARCH(1, 1) predictions are better, whereas for the horizons $\eta = 10$ and $\eta = 20$, the exponentially weighted moving average is

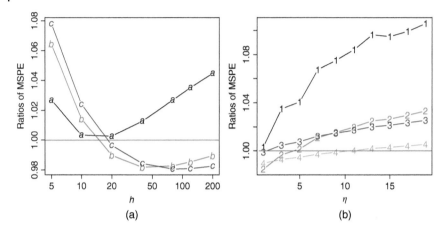

Figure 7.7 *Comparison of EWMA and GARCH(1,1) using MSPE.* Shown are the ratios $\text{MSPE}^{ewma}(\eta, h)/\text{MSPE}^{garch}(\eta)$. (a) The x-axis shows values of smoothing parameter h. The symbols "a," "b," and "c" correspond to $\eta = 1$, $\eta = 10$, and $\eta = 20$. (b) The x-axis shows values of prediction horizon η. The symbols "1," "2," "3," and "4" correspond to the smoothing parameters $h = 10$, $h = 40$, $h = 100$, and $h = 200$.

better, when the smoothing parameter is sufficiently large. When the horizon increases, then the optimal smoothing parameter of the EWMA estimator becomes larger. Using the Gaussian kernel gives almost similar results, but the results are slightly worse. Panel (b) shows functions

$$\eta \mapsto \frac{\text{MSPE}^{ewma}(\eta, h)}{\text{MSPE}^{garch}(\eta)}.$$

The symbols "1,",...,"4" correspond to the smoothing parameters $h = 10$, $h = 40$, $h = 100$, and $h = 200$. The smoothing parameter $h = 200$ gives the best results, except for $\eta = 1$.

Figure 7.8 shows the time series of differences

$$\text{CSPE}^{ewma}_t(\eta) - \text{CSPE}^{garch}_t(\eta)$$

between the sum of the cumulative squared prediction errors of a moving average predictor and GARCH(1, 1) predictor (see (7.4) for the formula). Panel (a) shows the complete time series and panel (b) shows the beginning of the time series. We show the cases $\eta = 1$ (black), $\eta = 10$ (red), and $\eta = 20$ (blue). The corresponding smoothing parameters are $h = 20$, $h = 80$, and $h = 80$. We see that the crash of October 1987 makes GARCH(1, 1) worse than the moving average, but in more typical periods GARCH(1, 1) tends to perform better.

Figure 7.9 shows the time series of differences

$$\text{CSPE}^{ewma}_t - \text{CSPE}^{seq}_t$$

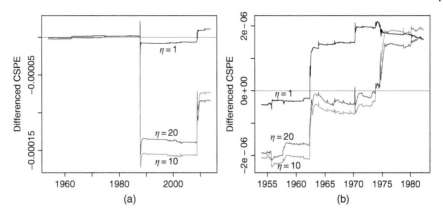

Figure 7.8 *Comparison of EWMA and GARCH(1,1) using CSPE.* Shown are the time series $CSPE_t^{ewma}(\eta) - CSPE_t^{garch}(\eta)$ of the differences of the cumulative sums of prediction errors. (a) The complete time series; (b) the beginning of the time series. We show the cases $\eta = 1$ (black), $\eta = 10$ (red), and $\eta = 20$ (blue).

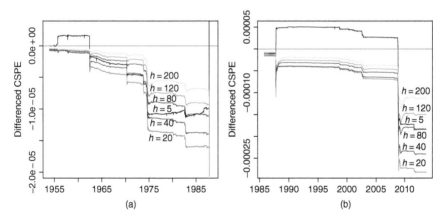

Figure 7.9 *Comparison of smoothing parameters of EWMA.* Shown are the time series $CSPE_t^{ewma} - CSPE_t^{seq}$ of the differences of the cumulative sums of prediction errors. (a) The first part of the time series; (b) the second part of the time series. The smoothing parameter takes values $h = 5$ (black), $h = 20$ (red), $h = 40$ (blue), $h = 80$ (dark green), $h = 120$ (turquoise), and $h = 200$ (pink).

between the sum of the cumulative squared prediction errors of a moving average predictor and the predictor based on sequentially computed sample variances (see (7.4) for the formula). Panel (a) shows the first part of the time series and panel (b) shows the second part of the time series. The smoothing parameter takes values $h = 5$ (black), $h = 20$ (red), $h = 40$ (blue), $h = 80$ (dark green), $h = 120$ (turquoise),and $h = 200$ (pink). The prediction horizon is one day. The smoothing parameter $h = 20$ gives overall best results. The cumulative prediction error and smoothing parameter are moving in tandem.

7.4.2.1 Asymmetric Exponentially Weighted Moving Average

We get a slightly different moving average than (7.11) by making a similar to (6.8) recursive definition

$$\hat{f}(t) = (1 - \gamma)R_t^2 + \gamma\hat{f}(t - 1)$$

where $0 \le \gamma \le 1$. The recursive definition can be modified to take the leverage effect into account:

$$\hat{f}(t) = (1 - \gamma)\hat{f}(t - 1)\left(\frac{R_t}{\sqrt{\hat{f}(t - 1)}} - \lambda\right)^2 + \gamma\hat{f}(t - 1), \tag{7.12}$$

where $\lambda \in \mathbf{R}$ is the skewness parameter. This is analogous to the GARCH-type model in (5.47). The smoothing parameters h and γ are related by

$$h = -1/\log\gamma.$$

Figure 7.10 compares volatility prediction with asymmetric exponentially weighted moving averages to the GARCH(1, 1) prediction. We use the daily S&P 500 data, described in Section 2.4.1. We show functions

$$h \mapsto \frac{\text{MSPE}^{ewma}(\eta, h, \lambda)}{\text{MSPE}^{garch}(\eta)},$$

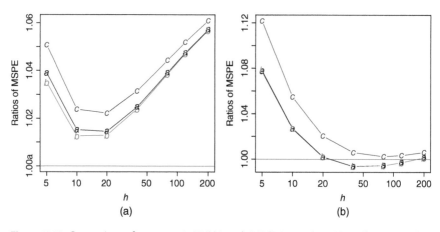

Figure 7.10 *Comparison of asymmetric EWMA and GARCH(1,1) using MSPE.* Shown are the ratios $\text{MSPE}^{ewma}(\eta, h, \lambda)/\text{MSPE}^{garch}(\eta)$. (a) $\eta = 1$; (b) $\eta = 10$. The skewness parameter λ takes values $\lambda = 0$ (black with "a"), $\lambda = 0.1$ (red with "b"), and $\lambda = 0.4$ (blue with "c").

where $\mathrm{MSPE}^{ewma}(\eta, h, \lambda)$ and $\mathrm{MSPE}^{garch}(\eta)$ are the means of squared predic-tion errors when prediction horizon is η (see (7.3) for the formula). Panel (a) has $\eta = 1$ and panel (b) has $\eta = 10$. The h-axis is logarithmic. The skewness parameter λ takes values $\lambda = 0$ (black with "a"), $\lambda = 0.1$ (red with "b"), and $\lambda = 0.4$ (blue with "c"). We can see that there is some improvement from using the asymmetric moving average when $\eta = 1$, but in the case $\eta = 10$ there is no improvement.

7.5 State Space Predictors

Let $R_1, \dots, R_t$ be the observed time series of financial returns. We consider the prediction of $R_{t+\eta}^2$, where $\eta \geq 1$ is the prediction horizon. The prediction of the squared return can be interpreted as the estimation of $E_t R_{t+\eta}^2$, where E_t is the conditional expectation, conditionally on the information available at time t. Thus, regression function estimators can be used to construct predictors.

Let us denote the response variable as

$$Y_i = R_{i+\eta}^2. \tag{7.13}$$

The predictive variables are the components of vector $X_i \in \mathbf{R}^d$, where $d \geq 1$. Vector X_i is computed from returns $R_i, \dots, R_{i-p+1}$, where $p \geq 1$ is the lag parameter. The initial observed time series $R_1, \dots, R_t$ leads to regression data

$$(X_i, Y_i), \quad i = p, \dots, t - \eta,$$

which is used to compute a regression function estimate. We transform the observations X_i by (4.36), which makes the marginals of X_i approximately stan-dard normal, but keeps the copula of X_i the same as the original copula.

The information available at time t is defined by the vector X_t of predictive variables. We apply two types of state variables. First, we choose the state vari-able as the vector of exponentially weighted moving average of past squared returns and past returns, as defined in (7.9) and (6.5):

$$X_t = \left(\sum_{i=1}^{t} p_i(t) R_i^2, \sum_{i=1}^{t} p_i(t) R_i \right). \tag{7.14}$$

Second, we choose the state variable as the vector of past squared returns:

$$X_t = \left(R_t^2, \dots, R_{t-p+1}^2 \right), \tag{7.15}$$

where $p \geq 1$.

We compute the means of prediction errors

$$\mathrm{MSPE}^{stsp}(\eta) = \frac{1}{T - t_0 - \eta + 1} \sum_{t=t_0}^{T-\eta} \left| \hat{f}(t) - R_{t+\eta}^2 \right|^2$$

for various predictors $\hat{f}(t)$, and compare this to the $\text{MSPE}^{garch}(\eta)$, which is the corresponding mean of the squared prediction errors of the GARCH(1, 1) predictor, as written in (7.3). We compute also the time series of cumulative sum of squared prediction errors

$$\text{CSPE}^{stsp}(\eta)_t = \sum_{i=t_0}^{t-\eta} \left| \hat{f}(i) - R^2_{i+\eta} \right|^2$$

for various predictors $\hat{f}(i)$, and compare this to the time series $\text{CSPE}^{garch}(\eta)_t$, which is the corresponding cumulative sum of the squared prediction errors of the GARCH(1, 1) predictor.

7.5.1 Linear Regression Predictor

The linear predictor is defined by

$$\hat{f}(t) = \hat{\alpha}_t + \hat{\beta}'_t X_t,$$

where $\hat{\alpha}_t \in \mathbf{R}$, $\hat{\beta}_t \in \mathbf{R}^d$, and $X_t \in \mathbf{R}^d$. In (7.14), $d = 2$ and in (7.15), $d = p$. The statistics $\hat{\alpha}_t$ and $\hat{\beta}_t$ minimize the least squares criterion

$$\sum_{i=t_0}^{t-\eta} (Y_i - \alpha - \beta' X_i)^2,$$

over $\alpha \in \mathbf{R}$ and $\beta \in \mathbf{R}^d$, where $t_0 \geq d$. The least squares solutions are given in (6.13)–(6.15).

7.5.1.1 Prediction with Volatility and Mean

We choose the state variable to be the vector with two elements. The elements are an exponentially weighted moving average of past squared returns and an exponentially weighted moving average of past returns, as defined in (7.14). The weights $p_i(t)$ involve the smoothing parameter h. We study the effect of $h > 0$ and the prediction horizon $\eta \geq 1$ to the sum of the squared prediction errors.

Figure 7.11 compares the means of squared prediction errors of the linear predictor to the GARCH(1, 1) predictor. Panel (a) shows functions

$$h \mapsto \frac{\text{MSPE}^{stsp}(\eta, h)}{\text{MSPE}^{garch}(\eta)}. \tag{7.16}$$

Panel (b) shows time series

$$\text{CSPE}^{stsp}(\eta, h)_t - \text{CSPE}^{garch}(\eta)_t \tag{7.17}$$

for smoothing parameter $h = 10$. The prediction horizon is $\eta = 1$ (black with "a") and $\eta = 10$ (red with "b"). We see that for prediction horizon $\eta = 1$ the GACRH(1, 1) predictor is better, but for prediction horizon $\eta = 10$ the linear

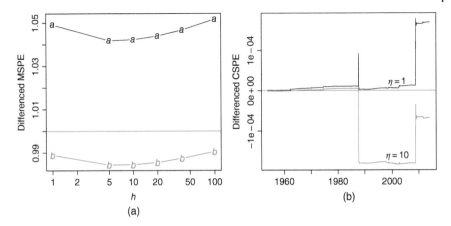

Figure 7.11 *Linear prediction with moving averages.* (a) Shown are functions $h \mapsto \text{MSPE}^{stsp}(\eta, h)/\text{MSPE}^{garch}(\eta)$. (b) Shown are time series $\text{CSPE}^{stsp}(\eta, h)_t - \text{CSPE}^{garch}(\eta)_t$ for smoothing parameter $h = 10$. The prediction horizon is $\eta = 1$ (black with "a") and $\eta = 10$ (red with "b").

predictor performs better. Panel (b) shows that autumns 1987 and 2008 give the most contribution to the means of squared prediction errors.

Figure 7.12 shows time series (7.17) for smoothing parameter $h = 10$. Panel (a) shows the beginning of the period and panel (b) shows the end of the period. The prediction horizon is $\eta = 1$ (black) and $\eta = 10$ (red). We see that the GARCH(1, 1) predictor performs better during other periods, but at autumn 1987 the linear predictor performs better.

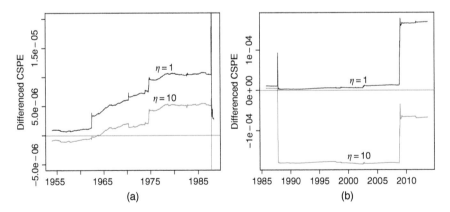

Figure 7.12 *Linear prediction with moving averages.* Shown are time series $\text{CSPE}^{stsp}(\eta, h)_t - \text{CSPE}^{garch}(\eta)_t$. Panel (a) shows the beginning of the period and panel (b) shows the end of the period. The prediction horizon is $\eta = 1$ (black) and $\eta = 10$ (red).

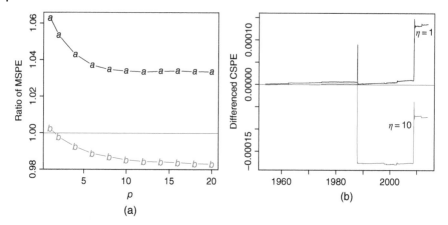

Figure 7.13 *Linear prediction with past squared returns.* (a) Shown are functions $p \mapsto$ $\mathrm{MSPE}^{stsp}(\eta,p)/\mathrm{MSPE}^{garch}(\eta)$. (b) Shown are time series $\mathrm{CSPE}^{stsp}(\eta,p)_t - \mathrm{CSPE}^{garch}(\eta)_t$ for smoothing parameter $h = 18$. The prediction horizon is $\eta = 1$ (black with "a") and $\eta = 10$ (red with "b").

We conclude that the linear predictor does not improve significantly GARCH(1, 1) predictor. We show in Section 7.5.2 that a corresponding kernel predictor improves GARCH(1, 1) predictor.

7.5.1.2 Prediction with Past Squared Returns

We choose the state variable to be the vector of the past squared returns, as defined in (7.15):

$$X_t = \left(R_t^2, \dots, R_{t-p+1}^2 \right).$$

There are two differences to the ARCH(p) predictor: (1) The linear function is fitted using least squares and not maximum likelihood. (2) Predictions with horizon $\eta > 1$ are done fitting the response variable $Y_i = R_{i+\eta}^2$, as explained in (7.13).

Figure 7.13 compares the performance of the linear predictor to the GARCH(1, 1) predictor. Panel (a) shows ratios $\mathrm{MSPE}^{stsp}(\eta,p)/\mathrm{MSPE}^{garch}(\eta)$. Panel (b) shows time series $\mathrm{CSPE}^{stsp}(\eta,p)_t - \mathrm{CSPE}^{garch}(\eta)_t$ for the lag parameter $p = 18$. The prediction horizon is $\eta = 1$ (black with "a") and $\eta = 10$ (red with "b"). We see that for a longer prediction horizon the linear predictor outperforms GARCH(1, 1) predictor, but the outperformance comes from a single time point: Autumn 1987.

7.5.2 Kernel Regression Predictor

We want to predict the squared return $Y_i = R_{i+\eta}^2$, where $\eta \geq 1$ is the prediction horizon. The prediction is based on the vector $X_i \in \mathbf{R}^d$, where $d \geq 1$. Vector

X_i is computed from returns $R_i, R_{i-1}, \ldots$. We apply the kernel regression estimator

$$\hat{f}(t) = \sum_{i=k}^{t-\eta} p_i(t) R_{i+\eta}^2,$$

where

$$p_i(t) = \frac{K_h(X_t - X_i)}{\sum_{j=k}^{t-\eta} K_h(X_t - X_j)}, \tag{7.18}$$

$K_h(x) = K(x/h)/h^d$ is the scaled kernel function, $K : \mathbf{R}^d \to \mathbf{R}$ is the kernel function, and $h > 0$ is the smoothing parameter. The kernel predictor was discussed in Section 6.1.2.

We choose the state variable as the vector of exponentially weighted moving average of past squared returns and past returns:

$$X_t = \left(\sum_{i=1}^{t} q_i(t) R_i^2, \sum_{i=1}^{t} q_i(t) R_i \right). \tag{7.19}$$

The exponentially weighted moving averages are defined in (7.9) and (6.5) for the squared returns and returns. Now we use the notation

$$q_i(t) = \frac{K((t-i)/g)}{\sum_{j=1}^{t} K((t-j)/g)},$$

where $K(x) = \exp(-x) I_{[0,\infty)}(x)$ is the kernel function and $g > 0$ is the smoothing parameter. The notation $q_i(t)$ and g is used to make a difference to the notation $p_i(t)$ and h in (7.18). Below we apply smoothing parameter $g = 40$.

Figure 7.14 compares the means of squared prediction errors of the kernel predictor to the GARCH(1, 1) predictor. Panel (a) shows functions

$$h \mapsto \frac{\mathrm{MSPE}^{stsp}(\eta, h)}{\mathrm{MSPE}^{garch}(\eta)}.$$

Panel (b) shows time series

$$\mathrm{CSPE}^{stsp}(\eta, h)_t - \mathrm{CSPE}^{garch}(\eta)_t \tag{7.20}$$

for smoothing parameter $h = 0.5$. The prediction horizon is $\eta = 1$ (black with "a") and $\eta = 10$ (red with "b"). We see that for prediction horizon $\eta = 1$, there is no big difference between the overall means of squared prediction errors, but for prediction horizon $\eta = 10$ the kernel predictor performs better, when smoothing parameter is suitably chosen. Panel (b) shows that autumns 1987 and 2008 give the most contribution to the means of squared prediction errors.

Figure 7.15 shows time series (7.20) for smoothing parameter $h = 0.5$. Panel (a) shows the beginning of the period and panel (b) shows the end of the period. The prediction horizon is $\eta = 1$ (black) and $\eta = 10$ (red).

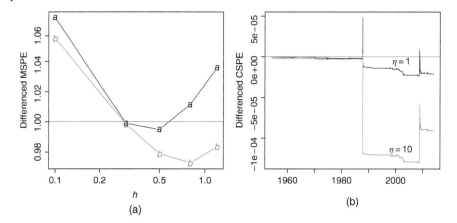

Figure 7.14 *Kernel prediction.* (a) Shown are functions $h \mapsto \text{MSPE}^{stsp}(\eta, h)/\text{MSPE}^{garch}(\eta)$. (b) Shown are time series $\text{CSPE}^{stsp}(\eta, h)_t - \text{CSPE}^{garch}(\eta)_t$ for smoothing parameter $h = 0.5$. The prediction horizon is $\eta = 1$ (black with "a") and $\eta = 10$ (red with "b").

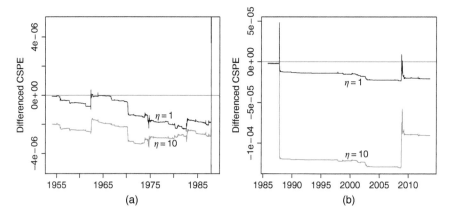

Figure 7.15 *Kernel prediction.* Shown are the time series $\text{CSPE}^{stsp}(\eta, h)_t - \text{CSPE}^{garch}(\eta)_t$ for smoothing parameter $h = 0.5$. Panel (a) shows the beginning of the period and panel (b) shows the end of the period. The prediction horizon is $\eta = 1$ (black) and $\eta = 10$ (red).

We see that the kernel predictor performs better during most periods until autumn 1987. At autumn 1987, the kernel predictor performs significantly better. After that the performance is about equal, until at autumn 2008 the GARCH(1, 1) predictor performs better.

Figure 7.16 shows the regression function, estimated using the complete data. Panel (a) shows a contour plot. The observations X_i are plotted as yellow points. Panel (b) shows a perspective plot. We see that the highest volatility is predicted when the moving average of squared returns is high and the moving average of

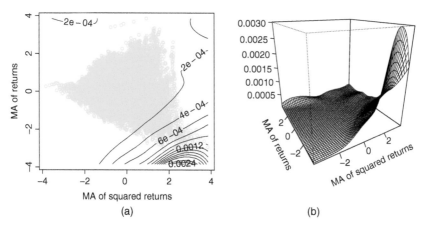

Figure 7.16 *Kernel prediction: Leverage effect.* The estimated regression function is visualized using (a) a contour plot and (b) a perspective plot.

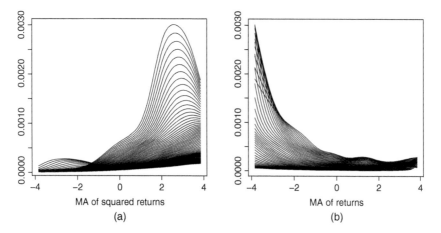

Figure 7.17 *Kernel prediction: Leverage effect.* (a) Slices $x_1 \mapsto \hat{f}(x_1, x_2)$ for several values of x_2. (b) Slices $x_2 \mapsto \hat{f}(x_1, x_2)$ for several values of x_1.

returns is low. This is called the leverage effect. Section 5.5 discusses the leverage effect as a stylized time series fact. The leverage effect is taken into account in the asymmetric GARCH model, defined in (5.46), and in the asymmetric moving average in (7.12).

Figure 7.17 shows slices of the estimated regression function $\hat{f}(x_1, x_2)$. Panel (a) shows slices $x_1 \mapsto \hat{f}(x_1, x_2)$ for several values of x_2, where x_1 is the moving average of squared returns and x_2 is the moving average of returns. Panel (b) shows slices $x_2 \mapsto \hat{f}(x_1, x_2)$ for several values of x_1.

8

Quantiles and Value-at-Risk

In quantile estimation, we study such questions as:

1) How much liquid capital a bank must possess in order that the probability of running out of cash during the next month is smaller than 1/10,000?
2) How much cash must be deposited in a margin account in order that the probability that cash does not cover the losses of a futures position during the next day is smaller than 1/10,000?

These questions can be formulated using the concepts of probability theory. Let Y be a real valued random variable. Let $0 < p < 1$ be a probability. The pth quantile is the smallest number $x \in \mathbf{R}$ so that $P(Y > x) \leq 1 - p$. In the first example random variable Y is the monthly loss of the investment portfolio of a bank. In the second example random variable Y is the daily loss of a futures position.

We want to estimate the pth quantile using previously observed data. In the first example, we do not have observations of the past losses (because the investment portfolio has changed), but we have observed the past returns of the components of the portfolio, and the probability distribution of the loss of the current portfolio can be deduced from an estimate of the joint distribution of its components. In the second example, we have observed the previous daily losses of the asset underlying the futures position.

The pth quantile is also called the quantile with confidence level p. The term value-at-risk (VaR) is used for the upper quantiles of the distribution of a loss of a portfolio.

What is special about quantile estimation? Quantiles can be defined as the values of the inverse of the distribution function, and thus quantile estimation is related to the estimation of the distribution function. For example, the population median is the pth quantile for $p = 0.5$: The population median is such $x \in \mathbf{R}$ that $P(Y \leq x) = 0.5 = P(Y \geq x)$. However, in quantile estimation we are interested in the cases where p is close to 0 or close to 1. For example, $p = 0.0001$ or $p = 0.9999$. By definition, there are few observations in the tail areas. This makes estimation of tail areas difficult. In order to estimate quantiles

Nonparametric Finance, First Edition. Jussi Klemelä.

when confidence level p is close to 0 or close to 1 we make special models for the tails of the distribution, and ignore the central area of the distribution.

One may question whether is reasonable even to try to estimate the extreme quantiles. Embrechts *et al.* (1997, p. VI) make the following comment: "Whatever theory can or cannot predict about extremal events, in practice the problems are there! As scientists, we cannot duck the question of the height of a sea dyke to be built in Holland, claiming that this is an inadmissible problem because, to solve it, we would have to extrapolate beyond the available data."

Section 8.1 discusses definitions of quantiles. Note that quantiles were defined already in Section 3.1.3. Section 8.2 discusses applications of quantile estimation. Section 8.3 discusses the measurement of the performance of quantile estimators.

Sections 8.4–8.7 study different types of quantile estimators. Section 8.4 studies nonparametric quantile estimators: empirical quantiles and kernel quantile estimators. Section 8.5 studies quantile estimators that are based on a volatility estimator. These estimators utilize a location–scale model, and can be considered as semiparametric quantile estimators. Section 8.6 studies quantile estimators which are based on fitting a parametric model to the excess distribution. We apply two techniques to make the fitting of the excess distribution useful in a time series setting. First, we apply a time localized parameter estimation. Second, we apply a quantile transformation to obtain residuals which are approximately uniformly distributed, and apply the empirical quantile estimator to the residuals. Section 8.7 uses some results of extreme value theory to derive quantile estimators.

Section 8.8 discusses the expected shortfall. Quantiles can be used as a risk measure, but the expected shortfall may be preferred as a risk measure. In fact, a quantile takes the number of exceedances of a threshold into account, but not the largeness of the exceedances, whereas an expected shortfall takes also the largeness of the exceedances into account.

There are many useful books related to quantile estimation. Embrechts *et al.* (1997) study quantile estimation with the emphasis on extreme value theory. The book by McNeil *et al.* (2005) is more practically oriented than Embrechts *et al.* (1997), but considers mostly purely parametric models. Malevergne and Sornette (2005) have analyzed financial data with nonparametric methods. Coles (2004) gives an introduction to statistical modeling of extreme values, describing classical asymptotic extreme value theory and models, threshold models, and a point process characterization of extremes.

8.1 Definitions of Quantiles

We have defined the pth quantile Q_p of the distribution of random variable Y as the smallest number $x \in \mathbf{R}$ such that $P(Y > x) \leq 1 - p$, where $0 < p < 1$.

The definition can be written as

$$Q_p = \inf\{x \in \mathbf{R} : P(Y > x) \le 1 - p\}. \tag{8.1}$$

We use both the capital letter and the small letter to denote a quantile, and sometimes we include the random variable in the notation:

$$Q_p = Q_p(Y) = q_p.$$

Since $P(Y > x) = 1 - F(x)$, where $F(x) = P(Y \le x)$ is the distribution function of Y, we can write

$$Q_p = \inf\{x \in \mathbf{R} : F(x) \ge p\}.$$

The generalized inverse of F is defined as $F^{\leftarrow}(p) = \inf\{x \in \mathbf{R} : F(x) \ge p\}$; see McNeil *et al.* (2005, p. 39). Thus, a quantile function is the generalized inverse of the distribution function. The distribution functions of discrete distributions are not monotonically increasing, and do not have a usual inverse. For example, the empirical distribution function does not have a usual inverse.

For practical purposes the returns and prices of stocks can be considered to have a continuous distribution: gross returns and prices can take almost any nonnegative value. Also, the loss of an investment portfolio can be considered for practical purposes to have a continuous distribution.[1] When Y has a continuous distribution, then there exists such $x \in \mathbf{R}$ that

$$P(Y > x) = 1 - p.$$

Thus,

$$P(Y \le x) = p.$$

The distribution function is $F(x) = P(Y \le x)$, so that the pth quantile x satisfies $F(x) = p$, and the pth quantile is

$$Q_p = F^{-1}(p),$$

where $F^{-1} : (0, 1) \rightarrow \mathbf{R}$ is the inverse of the distribution function.[2]

Figure 8.1 illustrates the definition of a quantile. Panel (a) shows a distribution function. The red vectors illustrate the inverting of the distribution function at level $p = 0.05$. The blue vectors illustrate the inverting at level $p = 0.99$. Panel (b) shows a density function corresponding to the distribution function. The red area has probability mass 5%. The right boundary of the red area indicates the location of the pth quantile for $p = 0.05$. The blue area has probability mass 1%. The left boundary of the blue area indicates the location of the pth quantile for $p = 0.99$.

---

1 Note, however, that the loss of a bond can be a discrete random variable. The value of a zero-coupon bond takes only two values at the maturity: the value is 0 in the case of the default and the value is 1 when the default does not happen.
2 We leave 0 and 1 out of the domain, because for the distributions with support $\mathbf{R}$ we have symbolically that $Q(0) = -\infty$ and $Q(1) = \infty$.

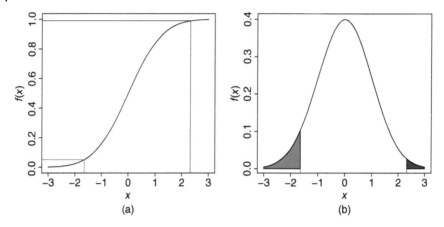

Figure 8.1 *Definition of quantiles.* (a) Definition by distribution function; (b) definition by density function.

In a time series setting, we can consider the estimation of conditional quantiles. Let $Y_t, t \in \{\ldots, -1, 0, 1, \ldots\}$, be a time series. Let

$$F_{Y_t|Y_{t-1}, Y_{t-2}, \ldots}(x) = P(Y_t \leq x | Y_{t-1}, Y_{t-2}, \ldots)$$

be the conditional distribution function of Y_t, given $Y_{t-1}, Y_{t-2}, \ldots$. Then the conditional pth quantile can be defined as

$$Q_p(Y_t | Y_{t-1}, Y_{t-2}, \ldots) = F^{-1}_{Y_t|Y_{t-1}, Y_{t-2}, \ldots}(p)$$

for continuous distributions.

Sometimes it is convenient to use the concepts of return level.

1) We call m-observation return level such level x_m which satisfies

$$P(Y > x_m) = 1/m.$$

Level x_m is exceeded on average once every m observations. It is convenient to express return levels on an annual scale. The N-year return level is the level expected to be exceeded once every N years. If there are n observations every year, then the N-year return level is the m-observation return level with $m = Nn$; see Coles (2004, p. 81).

2) We call return level x_p associated with return period $1/p$ the number satisfying

$$P(Y > x_p) = p.$$

Now x_p is expected to be exceeded on average once every $1/p$ years. Level x_p is exceeded by the annual maximum in any particular year with probability p.

8.2   Applications of Quantiles

Regulatory officials want to ensure that systemic financial institutions do not fall into a liquidity crisis. The companies want to control the probability of bankruptcy. Also, futures exchanges want to ensure that clients are able to meet the obligations of the derivatives they possess. For these purposes, estimation of quantiles is useful. Quantiles can also be used as a risk measure to characterize intuitively the riskiness of a portfolio.

8.2.1   Reserve Capital

Regulatory officials impose capital requirements on systemic financial institutions, such as large banks and large insurance companies. Regulatory officials want to ensure that systemic financial institutions do not fall into a liquidity crisis, because such crisis could have a negative impact on the whole economy. The regulators require that a bank has enough liquid reserves to cover the losses caused by adverse movements of the markets. See Rebonato (2007, Chapter 9) for more about a description of economic capital.

8.2.1.1   Value-at-Risk of a Portfolio

Value-at-risk can be used to determine the reserve capital of a bank or an enterprise. The term "value-at-risk" is used to denote upper quantiles of the loss distribution of a financial asset. Value-at-risk at level $0.5 < p < 1$ is such value that the probability of losing more during a given period has a smaller probability than $1 - p$. Thus, value-at-risk can be directly related to the amount of reserves. The loss of a portfolio at time $t + 1$, over one period, is defined as

$$L_{t+1} = -(V_{t+1} - V_t), \tag{8.2}$$

where V_t is the value of the portfolio at time t. The value-at-risk at the confidence level $p \in (0, 1)$ of the portfolio is defined as the smallest value x such that the probability that the loss L_{t+1} exceeds x is smaller or equal to $1 - p$: This is the p-quantile of the loss, when the p-quantile is defined in (8.1). We use both of the following notations:

$$\text{VaR}_p(L_{t+1}) = Q_p(L_{t+1}). \tag{8.3}$$

Typically p takes such values as 0.95 or 0.99. A larger value of value-at-risk indicates that the portfolio is more risky. The loss over $\eta \geq 1$ periods is

$$L_{t+\eta}^{(\eta)} = -(V_{t+\eta} - V_t),$$

and the value-at-risk over the horizon of $\eta \geq 1$ periods is

$$\text{VaR}_p\left(L_{t+\eta}^{(\eta)}\right).$$

Thus, the value-at-risk has two parameters: the risk horizon (daily, weekly, and 20-day horizon) and the confidence level p.

8.2.1.2 Decomposition of the Loss of a Portfolio

We can write the loss of a portfolio, as defined in (8.2), as

$$L_{t+1} = -V_t R_{t+1},$$

where R_{t+1} is the net return of the portfolio:

$$R_{t+1} = \frac{V_{t+1} - V_t}{V_t}.$$

Thus, the value-at-risk is obtained from the quantile $Q_p(R_{t+1})$ of the return distribution by the formula

$$\text{VaR}_p(L_{t+1}) = -V_t \, Q_p(R_{t+1}). \tag{8.4}$$

When the portfolio has initial value zero, then we cannot compute the return, and (8.4) cannot be used.

For example, the portfolio of stocks has the value

$$V_t = \sum_{i=1}^{d} \xi^i S_t^i,$$

where d is the number of stocks in the portfolio, S_t^i is the value of the ith stock at time t, and ξ^i is the quantity of the ith stock in the portfolio. Thus, for the portfolio of stocks the loss is

$$L_{t+1} = -(V_{t+1} - V_t) = -\sum_{i=1}^{d} \xi^i S_t^i R_{t+1}^i,$$

where

$$R_{t+1}^i = \frac{S_{t+1}^i - S_t^i}{S_t^i}$$

is the net return of the ith stock. Thus, the calculation of the value-at-risk requires knowledge of the distribution of a linear combination of the returns of the portfolio components.

8.2.1.3 Losses over Several Periods

We estimate the value-at-risk for the risk horizon, which is the same as the sampling frequency. However, often it is needed to estimate the value-at-risk for a longer risk horizon than the sampling period. For example, we might want to estimate the value-at-risk for the risk horizon of 10 days using daily observations. The loss over $\eta \geq 1$ periods is defined as

$$L_{t+\eta}^{(\eta)} = -(V_{t+\eta} - V_t).$$

We show in (8.30) that

$$Q_p\left(L_{t+\eta}^{(\eta)}\right) = \mu + \sigma \, Q_p(\epsilon),$$

where

$$\mu = E_t L_{t+\eta}^{(\eta)}, \quad \sigma^2 = \text{Var}_t\left(L_{t+\eta}^{(\eta)}\right), \quad \epsilon = \left(L_{t+\eta}^{(\eta)} - \mu\right)/\sigma.$$

We can write

$$L_{t+\eta}^{(\eta)} = -V_t \frac{V_{t+\eta} - V_t}{V_t} \approx -V_t \log \frac{V_{t+\eta}}{V_t} = -V_t \sum_{i=0}^{\eta-1} \log \frac{V_{t+i+1}}{V_{t+i}}. \tag{8.5}$$

The approximation

$$\sigma^2 \approx V_t^2 \sum_{i=0}^{\eta-1} E_t R_{t+i+1}^2, \quad R_{t+i+1} = \log \frac{V_{t+i+1}}{V_{t+i}}$$

could be used, and the techniques of volatility prediction of Chapter 7 lead to a quantile estimate.

Assuming that R_t are i.i.d., leads to

$$Q_p\left(L_{t+\eta}^{(\eta)}\right) = -V_t \eta m + V_t \sqrt{\eta} s \, Q_p(\epsilon),$$

where $m = ER_i$, $s^2 = \text{Var}(R_i)$. Assume that the logarithmic returns $R_{t+i+1} = \log(V_{t+i+1}/V_{t+i})$ are i.i.d. with the normal distribution $N(m, s^2)$. Then, $\epsilon \sim N(0, 1)$, and $Q_p(\epsilon) = \Phi^{-1}(p)$, where Φ is the distribution function of the $N(0, 1)$ distribution.

8.2.2 Margin Requirements

Value-at-risk can be applied to determine the safety deposit (margin) that a holder of a futures position or a writer of an option has to hold. The exchanges and brokers require that the investors save a deposit in the margin account. The amount of the deposit can be determined by the value-at-risk of the futures or option position.

Let us assume that the margin account earns the yearly rate $r > 0$. The gross return of a portfolio whose components are the risk-free rate and a risky asset is

$$R_{t+1} = (1 - b)(1 + r\Delta t) + b \frac{S_{t+1}}{S_t},$$

where $b \in \mathbf{R}$ is the weight of the risky asset, S_t is the price of the risky asset, and Δt is the length of the period in fractions of a year. The gross return $R_{t+1} \leq 0$ means that bankruptcy occurs. We want to choose b so that

$$P(R_{t+1} \leq 0) \approx p,$$

where $0 < p < 0.5$ is small. We have that

$$R_{t+1} \leq 0 \Leftrightarrow b\left(\frac{S_{t+1}}{1 + r\Delta t} - S_t\right) \leq -S_t.$$

Let us consider separately the case $b > 0$ and $b < 0$.

1) When $b > 0$, then we are buying the risky asset, and

$$R_{t+1} \leq 0 \Leftrightarrow \frac{S_{t+1}}{1 + r\Delta t} - S_t \leq -\frac{S_t}{b}.$$

Let $\hat{Q}_p$ be an estimate of the pth quantile of

$$\frac{S_{t+1}}{1 + r\Delta t} - S_t.$$

When $\hat{Q}_p < 0$, we can choose

$$b = -\frac{S_t}{\hat{Q}_p},$$

because $S_t > 0$.

2) When $b > 0$, then we are selling the risky asset, and

$$R_{t+1} \leq 0 \Leftrightarrow \frac{S_{t+1}}{1 + r\Delta t} - S_t \geq -\frac{S_t}{b}.$$

Let $\hat{Q}_{1-p}$ be an estimate of the $(1 - p)$th quantile of

$$\frac{S_{t+1}}{1 + r\Delta t} - S_t.$$

When $\hat{Q}_{1-p} > 0$, we can choose

$$b = -\frac{S_t}{\hat{Q}_{1-p}}.$$

8.2.3 Quantiles as a Risk Measure

Quantiles can be used as a risk measure. For example, if the 1% quantile of a monthly loss of S&P 500 returns is 20%, then an investor who owns S&P 500 index could expect to suffer 20% monthly loss once in every 8 years, roughly speaking, since $100 \times 1/12 \approx 8$. This kind of statement gives an investor some understanding of the riskiness of the position.

It can be argued that a reasonable risk measure satisfies the axioms of a coherent risk measure. Coherent risk measures were defined by Artzner *et al.* (1999). Coherent risk measures ρ satisfy the following properties, where X and Y are random variables interpreted as portfolio losses.

1) Monotonicity: if $X \leq Y$, then $\rho(X) \leq \rho(Y)$. If the outcome of an investment dominates the outcome of an other investment, then the risk must be greater.
2) Subadditivity: $\rho(X + Y) \leq \rho(X) + \rho(Y)$. Diversification reduces risk.
3) Positive homogeneity: $\rho(\lambda Y) = \lambda\rho(Y)$ for $\lambda \geq 0$. If the investor doubles his position for every asset, then he doubles also the risk.

4) Translation invariance: $\rho(Y + a) = \rho(Y) - a$ for $a \in \mathbf{R}$. If the amount a is added to the portfolio, then the capital requirement is reduced by the same amount.

The value-at-risk (an upper quantile of the loss) is not a coherent risk measure, but the expected shortfall is a coherent risk measure. Quantiles do not satisfy the subadditivity like the expected shortfall.[3]

Föllmer and Schied (2002, Section 4.1) define a monetary measure of risk as satisfying the condition of monotonicity and translation invariance. A monetary measure of risk is called a convex measure of risk if it satisfies convexity:

$$\rho(\lambda X + (1 - \lambda)Y) \leq \lambda\rho(x) + (1 - \lambda)\rho(Y),$$

for $0 \leq \lambda \leq 1$. A convex measure or risk is called a coherent risk measure if it satisfies the condition of positive homogeneity. Under the assumption of positive homogeneity, convexity is equivalent to subadditivity. However, Föllmer and Schied (2002, p. 155) note that risk may grow in a nonlinear ways as the size of the position increases, and thus it makes sense to study convex measures of risk, instead of coherent measures of risk.

To learn more about risk measures, see McNeil *et al.* (2005, Section 6.1) and Föllmer and Schied (2002, Section 4.1).

8.3 Performance Measures for Quantile Estimators

We can measure the performance of a quantile estimator by studying how well the quantile estimator captures the probability of an exceedance. Secondly, we can measure the performance by using a suitable loss function.

In this section, we only define the performance measures, and they are illustrated in Section 8.5.1.

---

3 Let X and Y be independent and identically distributed with the distribution $P(Y = 1) = 0.95$ and $P(Y = 0) = 0.05$. Random variables Y and X can be interpreted as pay-offs of zero-coupon bonds paying one at the maturity when there is no default, and the probability of default being 0.05. The quantiles of X and Y are

$$Q_p(Y) = Q_p(X) = \begin{cases} 0, & 0 < p \leq 0.95, \\ 1, & 0.95 < p \leq 1. \end{cases}$$

Let $p_1 = 0.95^2$ and $p_2 = 2 \cdot 0.95 \cdot 0.05$. The quantiles of $X + Y$ are

$$Q_p(X + Y) = \begin{cases} 0, & 0 < p \leq p_1, \\ 1, & p_1 < p \leq p_1 + p_2, \\ 2, & p_1 + p_2 < p \leq 1. \end{cases}$$

Thus, $Q_p(X + Y) > Q_p(X) + Q_p(Y)$ when $p_1 < p \leq 0.95$.

8.3.1 Measuring the Probability of Exceedances

To measure the performance of a quantile estimator for continuous distributions, we use the fact that the pth quantile q_p satisfies

$$P(Y \leq q_p) = p.$$

Let quantile estimator $\hat{q}_p$ be constructed using observations $Y_1, \ldots, Y_T$. If the estimator is good, then

$$\frac{\#\{Y_i : Y_i \leq \hat{q}_p\}}{T} \approx p.$$

However, we should not use the same data both for constructing the estimator and for evaluating the estimator, because this would give a too optimistic impression of the performance of the estimator.

8.3.1.1 Cross-Validation

We consider the case of time series observations and estimation of the conditional quantiles. There are many ways to choose the conditioning information. Let us observe time series $Y_1, \ldots, Y_T$. Let the conditional quantile estimator be

$$\hat{q}_t = \hat{Q}_p(Y_t | Y_{t-1}, Y_{t-2}, \ldots),$$

which is constructed using data $Y_1, \ldots, Y_{t-1}$. The cross-validation quantity is

$$\hat{p} = \frac{1}{T - t_0} \sum_{t=t_0+1}^{T} I_{(-\infty, \hat{q}_t]}(Y_t), \tag{8.6}$$

where $1 \leq t_0 \leq T - 1$. We start to evaluate the performance of the estimator after t_0 observations are available, because any estimator can behave erratically when it is constructed using only a couple of observations.

We might also observe vector time series $(X_1, Y_1), \ldots, (X_T, Y_T)$ and use the conditional quantile estimator

$$\hat{q}_t = \hat{Q}_p(Y_t | (X_{t-1}, Y_{t-1}), \ldots).$$

The same cross-validation estimate as in (8.6) can be used.[4]

---

4 Let us observe i.i.d. data $Y_1, \ldots, Y_T$. Let $\hat{q}_i$ be a quantile estimator, constructed using the other data but not the ith observation. The cross-validation quantity is

$$\hat{p} = \frac{1}{T - 1} \sum_{j=1, j \neq i}^{T} I_{(-\infty, \hat{q}_i]}(Y_j).$$

Note that the estimator could also be an estimator of a conditional quantile. In this case we observe $(Y_1, X_1), \ldots, (Y_T, X_T)$ and want to estimate $Q_p(Y | X = x)$. Then the estimator without the ith observation is $\hat{q}_i(x)$ and the cross-validation quantity is

$$\hat{p} = \frac{1}{T - 1} \sum_{j=1, j \neq i}^{T} I_{(-\infty, \hat{q}_i(X_j)]}(Y_j).$$

8.3.1.2 Probability Differences

Finally, the performance is measured using the difference

$$p - \hat{p}. \tag{8.7}$$

If $\hat{p} > p$, this means that the quantile estimates were in average larger than the true quantiles. When we are estimating the left tail, so that p is close to 0, then the relation $\hat{p} > p$ means that the true distribution has a heavier left tail than the quantile estimates would indicate. When we are estimating the right tail, so that p is close to 1, then this relation reverses, and the relation $\hat{p} > p$ means that the true distribution has a lighter right tail than the quantile estimates would indicate.

We will show the performance of quantile estimators by plotting the difference

$$R(p, \hat{p}) = \begin{cases} p - \hat{p}, & \text{when } p < 0.5, \\ \hat{p} - p, & \text{when } p > 0.5. \end{cases} \tag{8.8}$$

Thus, the difference $R(p, \hat{p})$ being negative means that the true distribution has a heavier tail than the quantile estimates would indicate. The difference $R(p, \hat{p})$ being positive means that the true distribution has a lighter tail than the quantile estimates would indicate.

An alternative definition replaces the differences with the ratios, so that

$$R_r(p, \hat{p}) = \begin{cases} p/\hat{p}, & \text{when } p < 0.5, \\ \hat{p}/p, & \text{when } p > 0.5. \end{cases} \tag{8.9}$$

The ratios are more informative for p very close to 0 or to 1.

Note that the absolute difference $|p - \hat{p}|$ is not as good performance measure as $p - \hat{p}$ or $\hat{p} - p$, because the absolute difference looses the information about the sign. Information about the sign of $p - \hat{p}$ is useful because it tells whether the quantile estimator tends to be too small or too large.

8.3.1.3 Confidence of the Performance Measure

Even when we would know the true quantiles, there is random fluctuation in the numbers $\hat{p}$. The random variables

$$Z_t = I_{(-\infty, q_p]}(Y_{t+1}), \quad t = t_0, \dots, T - 1,$$

are Bernoulli random variables with $P(Z_t = 1) = p$, where q_p is the true quantile. If random variables Y_t are independent, then random variables Z_t are independent, and

$$M = \sum_{t=t_0}^{T-1} Z_t$$

is a binomial random variable with the distribution $\text{Bin}(n, p)$, where $n = T - t_0$. The probability mass function of M is

$$P(M = i) = \binom{n}{i} p^i (1 - p)^{n-i}, \quad \text{for } i = 0, \dots, n.$$

Difference We want to bound the difference $p - \tilde{p}$, where $\tilde{p} = M/n$. We choose numbers c_0 and c_1 so that

$$P(c_0 \leq p - \tilde{p} \leq c_1) \geq 1 - \alpha, \tag{8.10}$$

where $0 < \alpha < 1$. We have

$$c_0 = p - n^{-1} z_{1-\alpha/2}, \quad c_1 = p - n^{-1} z_{\alpha/2}, \tag{8.11}$$

where $z_{\alpha/2}$ and $z_{1-\alpha/2}$ are such that $P(z_{\alpha/2} \leq M \leq z_{1-\alpha/2}) \geq 1 - \alpha$.

Ratio We can also use analogous bounds for the ratio $p/\tilde{p}$, where $\tilde{p} = M/n$. We choose numbers d_0 and d_1 so that

$$P(d_0 \leq p/\tilde{p} \leq d_1) \geq 1 - \alpha,$$

where $0 < \alpha < 1$. We have

$$d_0 = \frac{np}{z_{1-\alpha/2}}, \quad d_1 = \frac{np}{z_{\alpha/2}},$$

where $z_{\alpha/2}$ and $z_{1-\alpha/2}$ are such that $P(z_{\alpha/2} \leq M \leq z_{1-\alpha/2}) \geq 1 - \alpha$.

8.3.1.4 Probability Differences Over All Time Intervals

The estimate $\hat{p}$ in (8.6) is computed over time interval $[t_0, T]$. It is of interest to compute the estimate over a large collection of subintervals $[t_1, t_2]$, where $t_0 \leq t_1 < t_2 \leq T$. We denote

$$\hat{p}(t_1, t_2) = \frac{1}{t_2 - t_1} \sum_{t=t_1+1}^{t_2} I_{(-\infty, \hat{q}_t]}(Y_t).$$

When we study a large collection of time intervals, then we can find whether a quantile estimator is uniformly better than another quantile estimator, or whether there are time periods where the first estimator is better and time periods where the second estimator is better.

We can plot the two-variate function $(t_1, t_2) \mapsto R(p, \hat{p}(t_1, t_2))$. This two-variate function is rather unsmooth, so that perspective plots and contour plots are difficult to interpret. It can be easier to look at the slices

$$t_1 \mapsto R(p, \hat{p}(t_1, t_2)) \tag{8.12}$$

for a fixed t_2, and the slices

$$t_2 \mapsto R(p, \hat{p}(t_1, t_2)) \tag{8.13}$$

for a fixed t_1.

It is intuitive to plot the vectors joining the points $p_1, p_2 \in \mathbf{R}^2$, where

$$p_1 = (t_1, R(p, \hat{p}(t_1, t_2))), \quad p_2 = (t_2, R(p, \hat{p}(t_1, t_2))), \tag{8.14}$$

for a collection of pairs (t_1, t_2). These vectors are intuitive because they visualize the interval $[t_1, t_2]$ together with the value $R(p, \hat{p}(t_1, t_2))$. We divide time interval

$[t_0, T]$ into K subintervals of equal length, where $K = 1, 2, \ldots$. That is, we choose the collection of pairs (t_1, t_2), where

$$t_1 = t_0 + (k - 1)\delta, \quad t_2 = t_0 + k\delta, \quad \delta = [(T - t_0)/K], \tag{8.15}$$

for $k = 1, \ldots, K$.

8.3.2 A Loss Function for Quantile Estimation

A quantile can be characterized as a minimizer of

$$R(\theta) = E\rho_p(Y - \theta),$$

over $\theta \in \mathbf{R}$, where

$$\rho_p(x) = x\,[p - I_{(-\infty,0)}(x)] = \begin{cases} x(p - 1), & \text{if } x < 0, \\ xp, & \text{if } x \geq 0, \end{cases} \tag{8.16}$$

for $0 < p < 1$. That is,

$$Q_p(Y) = \mathrm{argmin}_{\theta \in \mathbf{R}} E\rho_p(Y - \theta). \tag{8.17}$$

The same holds for the conditional quantiles:

$$Q_p(Y|X = x) = \mathrm{argmin}_{g \in \mathcal{G}} E\rho_p(Y - g(X)),$$

where $\mathcal{G}$ is the class of measurable functions $\mathbf{R}^d \to \mathbf{R}.$[5] Figure 8.2 shows the loss function in (8.16) with $p = 0.5$ (black solid line) and with $p = 0.1$ (red dashed line).

8.3.2.1 Cross-Validation
We can measure the quality of a quantile estimator $\hat{q}_p$ by estimating

$$R(\hat{q}_p) = E\rho_p(Y - \hat{q}_p).$$

The estimation can be done using a sample mean, but we have to be careful not to use the same observations in the estimation of $R(\hat{q}_p)$ that were used in the construction of $\hat{q}_p$.

Let $Y_1, \ldots, Y_T$ be time series data. Let the conditional quantile estimator be

$$\hat{q}_t = \hat{Q}_p(Y_t|Y_{t-1}, \ldots),$$

---

5 We show that if the distribution function F_Y is strictly monotonic, then (8.17) holds. Note that

$$E\rho_p(Y - \theta) = (p - 1) \int_{-\infty}^{\theta} (y - \theta)\, dF_Y(y) + p \int_{\theta}^{\infty} (y - \theta)\, dF_Y(y)$$

and thus

$$\frac{\partial}{\partial \theta} E\rho_p(Y - \theta) = (1 - p) \int_{-\infty}^{\theta} dF_Y(y) - p \int_{\theta}^{\infty} dF_Y(y) = F_Y(\theta) - p.$$

Setting $\partial E\rho_p(Y - \theta)/\partial\theta = 0$, we get (8.17), when F_Y is strictly monotonic.

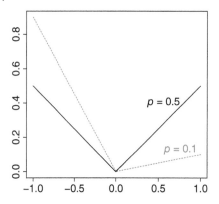

Figure 8.2 *Loss functions for quantile estima-tion.* Loss function in (8.16) with $p = 0.5$ (black solid line) and with $p = 0.1$ (red dashed line).

which is constructed using data $Y_1, \ldots, Y_{t-1}$, and calculate

$$\hat{R} = \frac{1}{T - t_0} \sum_{t=t_0+1}^{T} \rho_p(Y_t - \hat{q}_t), \tag{8.18}$$

where $1 \le t_0 \le T - 1$. We begin to evaluate the performance of the estimator after t_0 observations are available, because any estimator can behave erratically when only a couple of observations are used for its construction.[6]

8.3.2.2 Performance Over All Time Intervals

We can compare the performance of two conditional quantile estimators in a time series setting using the cumulative sums of the losses. This is similar to the use of cumulative sums of squared prediction errors to compare two predictors, in (6.26).

Let $Y_1, \ldots, Y_T$ be time series data. Let $\hat{q}_i^1$ and $\hat{q}_i^2$ be two conditional quantile estimators, which are constructed using data $Y_1, \ldots, Y_{i-1}$, and define

$$C_t = \sum_{i=t_0+1}^{t} \rho_p\left(Y_i - \hat{q}_i^1\right) - \sum_{i=t_0+1}^{t} \rho_p\left(Y_i - \hat{q}_i^2\right), \tag{8.19}$$

where $1 \le t_0 \le T - 1$. When $C_{t_1} < C_{t_2}$, then estimator $\hat{q}_i^1$ performs better on time period $[t_1, t_2]$ than estimator $\hat{q}_i^2$. Thus, a single time series plot of C_t reveals

---

6 Let $Y_1, \ldots, Y_T$ be i.i.d. data. Let $\hat{q}_i$ be a quantile estimate constructed using the other data but not the ith observation. The cross-validation quantity is

$$\hat{R} = \frac{1}{T - 1} \sum_{j=1, j \ne i}^{T} \rho_p(Y_j - \hat{q}_i).$$

Note that the quantile estimator could also be an estimator of the conditional quantile if we observe $(Y_1, X_1), \ldots, (Y_T, X_T)$, and want to estimate the conditional quantile $Q_p(Y|X = x)$.

all time periods where the first estimator is better than the second estimator, as explained in the connection of (6.26).

8.3.2.3   A Comparison to Probability Differences

Note that there is some resemblance to the performance measure $p - \hat{p}$ of (8.7). We can write

$$p - \hat{p} = \frac{1}{T - t_0} \sum_{t=t_0+1}^{T} [p - I_{(-\infty,0)}(Y_t - \hat{q}_t)].$$

On the other hand,

$$\hat{R} = \frac{1}{T - t_0} \sum_{t=t_0+1}^{T} (Y_t - \hat{q}_t)[p - I_{(-\infty,0)}(Y_t - \hat{q}_t)].$$

The probability difference $p - \hat{p}$ takes into account only whether there is an exceedance. The mean loss $\hat{R}$ takes also the largeness of the exceedance into account. Thus, $\hat{R}$ punishes more from the outliers in the values of $\hat{q}$.

8.3.2.4   Empirical Risk Minimization

Loss function ρ_p can be used in empirical risk minimization. Let us have regression data $(X_1, Y_1), \ldots, (X_T, Y_T)$. We can use function ρ_p to construct an estimator for the conditional quantile. The estimator is defined as the minimizer of the empirical risk. The estimator of the conditional quantile $f(x) = Q_p(Y|X = x)$ is

$$\hat{f} = \mathrm{argmin}_{g \in \mathcal{G}} \sum_{i=1}^{n} \rho_p(Y_i - g(X_i)),$$

where $\mathcal{G}$ is a suitable subset of the class of all measurable functions $g : \mathbf{R}^d \to \mathbf{R}$.

8.4   Nonparametric Estimators of Quantiles

For continuous distributions the quantile can be defined as the inverse of the distribution function:

$$Q_p(Y) = F^{-1}(p),$$

where Y is a random variable with a continuous distribution, $F(x) = P(Y \leq x)$ is the distribution function, and $p \in (0, 1)$. We obtain an estimate of the quantile by inverting the empirical distribution function.

Section 8.4.1 defines the empirical quantiles, which are obtained by inverting the empirical distribution function. Also, we define modifications of empirical quantiles that apply smoothing. Section 8.4.2 defines empirical conditional quantile estimators. The empirical quantile can be combined with time space or state space smoothing to obtain an estimator for a conditional quantile.

8.4.1 Empirical Quantiles

The basic empirical quantiles are obtained by inverting the empirical distribution function. The basic empirical quantiles can be modified by kernel smoothing methods.

8.4.1.1 Basic Empirical Quantile Estimator

We give three equivalent definitions for the empirical quantiles. The empirical distribution function, based on the observations $Y_1, \ldots, Y_T$, is defined in (3.30) as

$$\hat{F}(x) = \frac{1}{T} \sum_{i=1}^{T} I_{(-\infty, x]}(Y_i), \quad x \in \mathbf{R}. \tag{8.20}$$

The empirical distribution is the discrete distribution with the probability mass function $P(\{Y_i\}) = 1/T$ for $i = 1, \ldots, T$. Now we can define an estimate of the quantile by

$$\hat{Q}_p = \inf\{x : \hat{F}(x) \geq p\}, \tag{8.21}$$

where $0 < p < 1$. It holds that

$$\hat{Q}_p = \begin{cases} Y_{(1)}, & 0 < p \leq 1/T, \\ Y_{(2)}, & 1/T < p \leq 2/T, \\ \vdots & \\ Y_{(T-1)}, & 1 - 2/T < p \leq 1 - 1/T, \\ Y_{(T)}, & 1 - 1/T < p < 1, \end{cases} \tag{8.22}$$

where the ordered sample is denoted by $Y_{(1)} \leq Y_{(2)} \leq \cdots \leq Y_{(T)}$. A third way of writing the empirical quantile was given in (3.23):

$$\hat{Q}_p = Y_{(\lceil pT \rceil)}, \tag{8.23}$$

where we denote by $\lceil x \rceil$ the smallest integer $\geq x$. We can write equivalently $\hat{Q}_p = Y_{(\lfloor pT \rfloor + 1)}$, where $\lfloor x \rfloor$ is the largest integer $\leq x$.

Figure 8.3 illustrates the empirical quantiles. We use monthly S&P 500 data, described in Section 2.4.3. Panel (a) shows the empirical distribution function. Panel (b) shows the first half of the function, indicated by the black vectors in panel (a). The red vectors indicate the location of the pth empirical quantile for $p = 0.01$. The location is determined by the seventh largest observation.

When $p < 1/T$, then the quantile estimator based on the empirical distribution is equal to the smallest observation, no matter how small p is. When

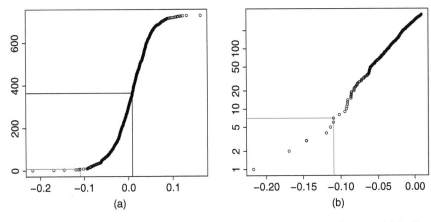

Figure 8.3 *Definition of empirical quantiles.* (a) The empirical distribution function; (b) the first half of the function, indicated by the black vectors in panel (a). The red vectors indicate the location of the pth empirical quantile for $p = 0.01$.

$p < 1/T$, we have to extrapolate outside the range of observations in order to estimate the quantile at level p. This can be done with parametric or semiparametric methods.

8.4.1.2   Smooth Empirical Quantiles

The smooth empirical quantiles apply kernel weights to modify the basic empirical quantiles. The distribution function can be estimated smoothly, or the quantiles can be directly estimated by a weighted sum of ordered observations.

Kernel Estimation of the Distribution Function   The kernel density estimator $\hat{f}$ is defined in Section 3.2.2. The corresponding estimator of the distribution function is

$$\hat{F}(x) = \int_{-\infty}^{x} \hat{f}(y)\, dy = \frac{1}{T} \sum_{i=1}^{T} \int_{-\infty}^{x} K_h(y - Y_i)\, dy, \tag{8.24}$$

where $x \in \mathbf{R}$, $K : \mathbf{R} \to \mathbf{R}$ is the kernel function, $K_h(x) = K(x/h)/h$, and $h > 0$ is the smoothing parameter. The corresponding quantile estimator is

$$\hat{Q}_p = \hat{F}^{-1}(p),$$

where $p \in (0, 1)$. Azzalini (1981) has studied this quantile estimator.

Figure 8.4 illustrates the extrapolation outside the range of data and interpolation between the data points. We use monthly S&P 500 data, described in Section 2.4.3.

In Figure 8.4(a) our purpose is to estimate the pth quantile for $p = 0.1\%$. We have less than 1000 observations, and the pth empirical quantile is the smallest

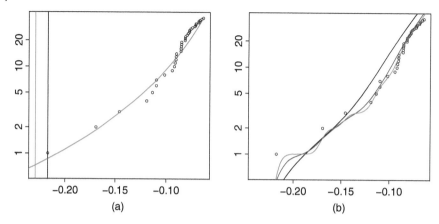

Figure 8.4 *Extrapolating and interpolating the empirical distribution.* (a) The black vertical line shows the empirical *p*th quantile for $p = 0.1\%$. The red vertical line shows the Pareto quantile, and the red curve shows the fitted Pareto distribution. (b) Kernel distribution function estimates. The blue curve has smoothing parameter *h* chosen by the normal reference rule. The black curve is oversmoothing and the green curve is undersmoothing.

observation, indicated by the black vertical line. The extrapolation can be done by fitting the Pareto model to the data, and taking the quantile estimate to be the *p*th quantile of the fitted Pareto distribution. The fitted Pareto distribution is shown by the red curve, and the *p*th quantile of the Pareto distribution is indicated by the red vertical line. Fitting of Pareto distributions is discussed in Section 3.4.2.

In Figure 8.4(b) we show distribution function estimates (8.24). The blue curve has smoothing parameter *h* chosen by the normal reference rule. The black curve is oversmoothing and the green curve is undersmoothing. We see that extrapolation outside the range of data is not possible with kernel distribution function estimator, but interpolation between the extreme observations is possible.

Kernel Quantile Estimators We have defined the empirical quantile in (8.21)–(8.23). Equation (8.23) shows that the empirical quantile depends on a single order statistics. *L*-estimators of quantiles are weighted averages of several order statistics. The use of several order statistics in the estimation of a quantile can improve the efficiency. A kernel quantile estimator is a special case of *L*-estimators of quantiles. A kernel quantile estimator can be defined for $0 < p < 1$ as

$$\hat{Q}_p^K = \frac{\sum_{i=1}^{T} K_h(i/T - p) \, Y_{(i)}}{\sum_{i=1}^{T} K_h(i/T - p)}, \tag{8.25}$$

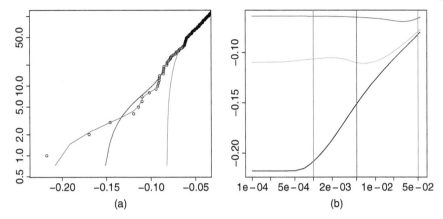

Figure 8.5 *Kernel quantile estimator.* (a) The implied distribution function estimates. The smoothing parameter is $h = 0.05$ (red), $h = 0.005$ (blue), and $h = 0.001$ (green). (b) The quantile estimates as a function of smoothing parameter. We estimate the pth quantile for $p = 0.05$ (purple), $p = 0.01$ (violet), and $p = 0.001$ (black).

where $K_h(x) = K(x/h)/h$ is the scaled kernel function, $K : \mathbf{R} \to \mathbf{R}$ is the kernel function, and $h > 0$ is the smoothing parameter. The estimator is a weighted average of order statistics $Y_{(i)}$ for which i/T is close to p.[7] Sheather and Marron (1990) studies the asymptotic mean squared error of (8.25) and show that the kernel quantile estimator has the same first-order asymptotics as the empirical quantile, but the higher order asymptotics reveal improvement.[8]

Figure 8.5 studies the kernel quantile estimator in (8.25). Panel (a) shows the implied distribution function estimates. The smoothing parameter is $h = 0.05$

---

7 Alternative definitions for the kernel quantile estimator are $T^{-1} \sum_{i=1}^{T} K_h(i/T - p) \, Y_{(i)}$, $T^{-1} \sum_{i=1}^{T} K_h((i - 0.5)/T - p) \, Y_{(i)}$, and $T^{-1} \sum_{i=1}^{T} K_h(i/(T + 1) - p) \, Y_{(i)}$. Unlike (8.25), these estimators are not normalized so that the weights sum up to 1. An additional form of a kernel quantile estimator can be defined as $\sum_{i=1}^{T} \left(\int_{i-1/T}^{i/T} K_h(t - p) \, dt \right) Y_{(i)}$, where the kernel function K is a density function. Sheather and Marron (1990) shows that these estimators are asymptotically equivalent with (8.25), under certain conditions.

8 The asymptotic variance of the empirical quantile is $T^{-1} p(1 - p)[Q'(p)]^2$, where $Q'(p)$ is the derivative of the quantile function. The asymptotic mean squared error of the kernel quantile estimator is under certain assumption

$$T^{-1} p(1 - p)[Q'(p)]^2 - 2T^{-1} h[Q'(p)]^2 \int_{-\infty}^{\infty} uK(u)K^{(-1)}(u) \, du$$

$$+ 4^{-1} h^4 [Q''(p)]^2 \left[\int_{-\infty}^{\infty} u^2 K(u) \, du \right]^2 + o(T^{-1}h) + o(h^4),$$

where $K^{(-1)}$ is the antiderivative of K. We have that $\int_{-\infty}^{\infty} uK(u)K^{(-1)}(u) \, du > 0$, which shows that the higher order asymptotics is better for the kernel quantile estimator than for the empirical quantile, because it reasonable to choose $h = o(T^{-1/4})$.

(red), $h = 0.005$ (blue), and $h = 0.001$ (green). Panel (b) shows the quantile estimates as a function of smoothing parameter. We estimate the pth quantile for $p = 0.05$ (purple), $p = 0.01$ (violet), and $p = 0.001$ (black). The vertical lines indicate the smoothing parameters that are used in panel (a). We see that when the smoothing parameter approaches zero, then the estimates approach the basic empirical quantiles.

8.4.2 Conditional Empirical Quantiles

The conditional empirical quantiles can be defined both for time space smoothing and state space smoothing.

8.4.2.1 Time Space Smoothing

Let $Y_1, \ldots, Y_T$ be stationary time series data. We can define one-sided moving average estimator of the conditional quantile by inverting the one-sided moving average estimator $\hat{F}_{Y_t}$ of the distribution function, defined by

$$\hat{F}_{Y_t}(y) = \sum_{i=1}^{t} p_i(t) \, I_{(-\infty, y]}(Y_i), \quad t = 1, \ldots, T,$$

where the weights $p_i(t)$ are defined in (6.4). This gives

$$\hat{Q}_p(Y_t | Y_{t-1}, \ldots) = \begin{cases} Y_{(1)}, & 0 < p \leq p_1(t), \\ Y_{(2)}, & p_1(t) < p \leq p_1(t) + p_2(t), \\ \vdots & \\ Y_{(t-1)}, & \sum_{i=1}^{t-2} p_i(t) < p \leq \sum_{i=1}^{t-1} p_i(t), \\ Y_{(t)}, & \sum_{i=1}^{t-1} p_i(t) < p < 1, \end{cases} \quad (8.26)$$

where the ordered sample is denoted by $Y_{(1)} \leq Y_{(2)} \leq \cdots \leq Y_{(t)}$.

8.4.2.2 State Space Smoothing

Let $(X_1, Y_1), \ldots, (X_T, Y_T)$ be identically distributed regression data, where $X_i \in \mathbf{R}^d$ is a predictive variable. An estimator of a conditional quantile of Y can be defined with the help of the estimator of the conditional distribution function $\hat{F}_{Y|X=x}(y)$, defined by

$$\hat{F}_{Y|X=x}(y) = \sum_{i=1}^{T} p_i(x) I_{(-\infty, y]}(Y_i), \quad (8.27)$$

where the kernel weights $p_i(x)$ are defined in (6.21). We get the conditional quantile estimator by taking the generalized inverse of the estimator of the

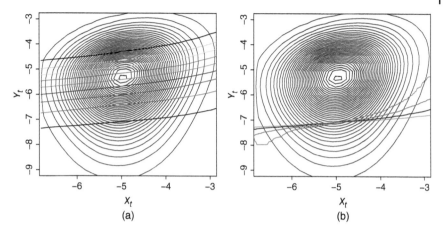

Figure 8.6 *Kernel estimates of conditional quantiles.* (a) Conditional quantile estimates for the levels $p = 0, 1, 0, 2, \ldots, 0.9$, when the smoothing parameter is $h = 0.7$; (b) estimates for the level $p = 0.1$ when the smoothing parameters are $h = 0.3, 0.5, 0.7, 0.9$. A contour plot of a kernel estimate of the density of (X_t, Y_t) is also shown.

conditional distribution function:

$$\hat{Q}_p(Y|X = x) = \inf\{y : \hat{F}_{Y|X=x}(y) \ge p\}. \tag{8.28}$$

The estimator can be written as

$$\hat{Q}_p(Y|X = x) = \begin{cases} Y_{(1)}, & 0 < p \le p_1(x), \\ Y_{(2)}, & p_1(x) < p \le p_1(x) + p_2(x), \\ \vdots & \\ Y_{(T-1)}, & \sum_{i=1}^{T-2} p_i(x) < p \le \sum_{i=1}^{T-1} p_i(x), \\ Y_{(T)}, & \sum_{i=1}^{T-1} p_i(x) < p < 1, \end{cases} \tag{8.29}$$

where the ordered sample is denoted by $Y_{(1)} \le Y_{(2)} \le \cdots \le Y_{(T)}$.

Figure 8.6 shows conditional quantile estimates when kernel weights are used. We apply the daily S&P 500 data of Section 2.4.1. Panel (a) shows estimates for the levels $p = 0, 1, 0, 2, \ldots, 0.9$, when the smoothing parameter is $h = 0.7$. Panel (b) shows estimates for the level $p = 0.1$ when the smoothing parameters are $h = 0.3, 0.5, 0.7, 0.9$. The standard normal kernel is used in both panels. The explanatory and the response variables as

$$X_t = \log_e \sqrt{\frac{1}{k} \sum_{i=1}^{k} R_{t-i}^2}, \quad Y_t = \log_e |R_t|,$$

where $k = 10$, and $R_t = (S_t - S_{t-1})/S_{t-1}$ are the net returns. We show also a contour plot of a kernel estimate of the density of (X_t, Y_t).

8.5 Volatility Based Quantile Estimation

Volatility based quantile estimators are build on a location–scale model. The location and the scale parameters are estimated, and the corresponding quantile estimate is derived. The scale parameter can be estimated using the sample standard deviation. Conditional quantile estimators are obtained from estimators of the conditional mean and the conditional standard deviation.

8.5.1 Location–Scale Model

Consider a location–scale model

$$Y = \mu + \sigma \, \epsilon,$$

where $\mu \in \mathbf{R}$, $\sigma > 0$, and ϵ is a real valued random variable. The quantile of Y is obtained from the quantile of ϵ by

$$Q_p(Y) = \mu + \sigma \, Q_p(\epsilon). \tag{8.30}$$

Indeed, for $x \in \mathbf{R}$,

$$P(Y \le x) = P\left(\epsilon \le \frac{x - \mu}{\sigma}\right) = F_\epsilon\left(\frac{x - \mu}{\sigma}\right),$$

where F_ϵ is the distribution function of ϵ. If ϵ has a continuous distribution, then F_ϵ is strictly increasing and the inverse function F_ϵ^{-1} exists. The pth quantile $Q_p(Y)$ of Y satisfies $P(Y \le Q_p(Y)) = p$, and we can solve this equation to get $Q_p(Y) = \mu + \sigma \, F_\epsilon^{-1}(p)$, which implies (8.30).

For a known F_ϵ, we get from estimators $\hat{\mu}$ and $\hat{\sigma}$ the quantile estimator

$$\hat{Q}_p(Y) = \hat{\mu} + \hat{\sigma} \, F_\epsilon^{-1}(p). \tag{8.31}$$

8.5.1.1 Examples of Location–Scale Quantile Estimators
For example, the Gaussian quantile estimator is

$$\hat{Q}_p(Y) = \hat{\mu} + \hat{\sigma}\Phi^{-1}(p), \tag{8.32}$$

where $\hat{\mu}$ is the sample mean, $\hat{\sigma}$ is the sample standard deviation, and Φ is the distribution function of the standard normal distribution. The Student quantile estimator is[9]

$$\hat{Q}_p(Y) = \hat{\mu} + \sqrt{\frac{v - 2}{v}} \, \hat{\sigma} t_v^{-1}(p), \tag{8.33}$$

---

9 If $X \sim t_v$, then $\mathrm{Var}(X) = v/(v - 2)$, so that $\sqrt{(v - 2)/v}\, t_v^{-1}(p)$ is the p-quantile of such t-distribution which is standardized to have unit variance.

where $v > 2$ is the degrees of freedom and t_v is the distribution function of the Student distribution with degrees of freedom v.

In practice the estimators are conditional in the sense that they are calculated sequentially: estimator computed at time t uses data $Y_1, \dots, Y_t$. That is, $\hat{\mu} = \hat{\mu}_t$ is the sample mean and $\hat{\sigma} = \hat{\sigma}_t$ is the sample standard deviation computed from $Y_1, \dots, Y_t$,

8.5.1.2 Estimation of S&P 500 Quantiles

We estimate quantiles of the S&P 500 returns Y_t using S&P 500 daily data, described in Section 2.4.1. We estimate the quantiles using three estimators:

1) Empirical quantile

$$\hat{Q}_p(Y) = Y_{(pt)}, \tag{8.34}$$

where $Y_{(1)} \le \cdots \le Y_{(t)}$ is the ordered sample. Note that the sequential empirical quantiles are a special case of (8.26), when we take $p_i(t) = 1/t$, $i = 1, \dots, t$.
2) The Gaussian quantile estimator, as defined in (8.32).
3) The Student quantile estimator, as defined in (8.33). We use $v = 6$ degrees of freedom.

We apply the performance measures that are defined in Section 8.3.

Probability Differences Figure 8.7 plots the probability differences $p \mapsto R(p, \hat{p})$, defined in (8.8). Panel (a) plots $p - \hat{p}$ for p close to 0, and panel (b) plots $\hat{p} - p$ for p close to 1. The performance of empirical quantiles is shown in red, that of the

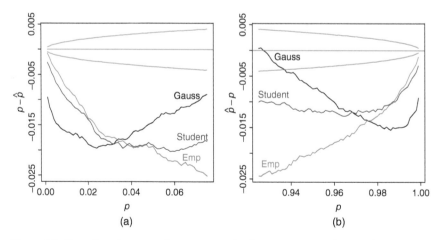

Figure 8.7 *Quantile estimator performance.* We plot function $p \mapsto R(p, \hat{p})$, as defined in (8.8). Empirical quantiles (red), Gaussian quantiles (black), and Student quantiles (blue). The green lines show level $\alpha = 0.05$ fluctuation bands. (a) Level p is close to zero; (b) p is close to one.

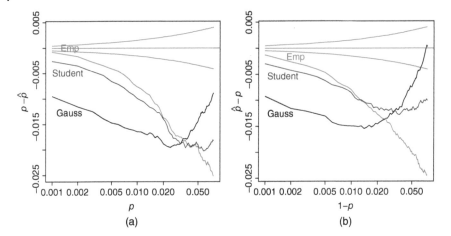

Figure 8.8 *Quantile estimator performance: Logarithmic x-axis.* We plot function $p \mapsto R(p, \hat{p})$, as defined in (8.8). Empirical quantiles (red), Gaussian quantiles (black), and Student quantiles (blue). The green lines show level $\alpha = 0.05$ fluctuation bands. (a) Level p is close to zero; (b) p is close to one.

Gaussian quantiles is shown in black, and that of the t-distribution quantiles is in blue. A green line is drawn at level 0, and it is accompanied by the level $\alpha = 0.05$ fluctuation bands, defined in (8.10) and (8.11).

Figure 8.8 plots the probability differences when the x-axes have a logarithmic scale, and in panel (b) x-axis takes values $1 - p$.

Figures 8.7 and 8.8 indicate that the true distribution has heavier tails than the quantile estimates would indicate. The empirical quantiles have the best performance when $p \leq 4\%$ or $p \geq 96\%$, but the Student quantiles are almost as good as empirical quantiles in this range. The Gaussian quantiles have the best performance when $4\% \leq p \leq 96\%$.

Figure 8.9 shows slices of two-variate probability differences when $p = 1\%$. Panel (a) shows slices in (8.12) for $t_2 = T$ and for $t_2 = 1979$. Panel (b) shows slices in (8.13) for $t_1 = t_0$ and for $t_1 = 1979$. The red curves show $p - \hat{p}$ for the empirical quantiles and the black curves show $p - \hat{p}$ for the Gaussian quantile estimates. We see that the empirical quantiles are better for all shown time intervals. The performance becomes worse for the time periods close to the end. The fluctuation is large when the time periods are short.

Figure 8.10 shows vectors joining points p_1 and p_2, as defined in (8.14), when the time intervals $[t_1, t_2]$ are defined in (8.15). We take the number $K = 4$ of intervals and $p = 1\%$. Panel (a) shows the case of empirical quantiles (red) and the case of Gaussian quantiles (black). Panel (b) shows the case of empirical quantiles (red) and the case of Student quantiles (blue). The green lines indicate the fluctuation bands for confidence 95%. The empirical quantiles seem to be uniformly better than Gaussian or Student quantiles.

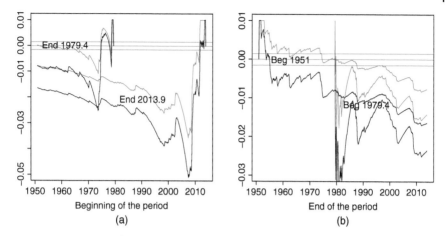

Figure 8.9 *Quantile estimator performance: Slices of probability differences.* We show $p - \hat{p}$ when $p = 1\%$ for the empirical quantiles (red) and for the Gaussian quantile estimates (black). Panel (a) shows slices in (8.12) for $t_2 = T$ and for $t_2 = 1979$. Panel (b) shows slices in (8.13) for $t_1 = t_0$ and for $t_1 = 1979$.

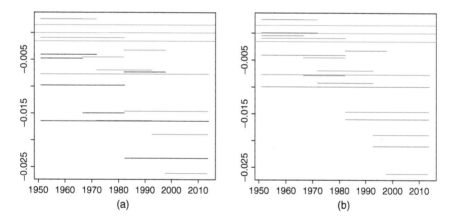

Figure 8.10 *Quantile estimator performance: Vectors over time intervals.* We show vectors joining points p_1 and p_2, as defined in (8.14), when the time intervals $([t_1, t_2]$ are defined in (8.15) for $K = 4$ and $p = 1\%$. (a) Empirical quantiles (red) and Gaussian quantiles (black); (b) empirical quantiles (red) and Student quantiles (blue).

The Loss Function We illustrate performance measurement using the three estimators defined in (8.32)–(8.34): the Gaussian quantile estimator, Student quantile estimator, and empirical quantile estimator. These estimators are conditional in the sense that they are computed sequentially: estimator $\hat{q}_t$ uses data $Y_1, \dots, Y_t$.

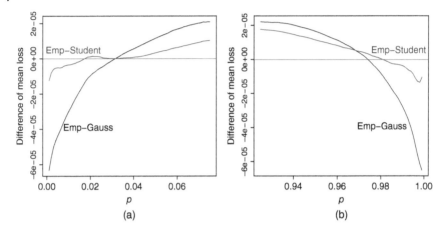

Figure 8.11 *Performance of quantile estimators: Expected loss.* We show functions $p \mapsto \hat{R}(\hat{q}^{emp}) - \hat{R}(\hat{q}^{gau})$ (black) and $p \mapsto \hat{R}(\hat{q}^{emp}) - \hat{R}(\hat{q}^{stu})$ (blue). (a) Range $p \in [0.001, 0.075]$ and (b) $p \in [0.925, 0.999]$.

Figure 8.11 compares the estimated expected losses of the three estimators. We show functions

$$p \mapsto \hat{R}(\hat{q}^{emp}) - \hat{R}(\hat{q}^{gau}) \text{ (black)}, \quad p \mapsto \hat{R}(\hat{q}^{emp}) - \hat{R}(\hat{q}^{stu}) \text{ (blue)},$$

where $\hat{R}(\hat{q})$ is defined in (8.18) when the estimator is $\hat{q}$. We denote by $\hat{q}^{emp}$ the empirical quantile estimator, by $\hat{q}^{gau}$ the Gaussian quantile estimator, and by $\hat{q}^{stu}$ the Student quantile estimator. Panel (a) shows range $p \in [0.001, 0.075]$. Panel (b) shows range $p \in [0.925, 0.999]$.

We obtain partially the same message as in Figures 8.7 and 8.8: The empirical quantiles are best in the far away tails but the Gaussian quantile estimator is best when we are closer to the center. Furthermore, the Student quantile estimator is better than the Gaussian quantile estimator in the far away tails. Note that we obtain less information than in Figures 8.7 and 8.8: We do not see whether the estimators overestimate or underestimate the true quantiles.

Figure 8.12 shows the time series of differences

$$C_t = \sum_{i=t_0+1}^{t} \rho_p \left(Y_i - \hat{q}_i^{emp}\right) - \sum_{i=t_0+1}^{t} \rho_p(Y_i - \hat{q}_i) \tag{8.35}$$

of cumulative sums, where $\hat{q}_i^{emp}$ is the empirical quantile estimator, and $\hat{q}_i$ is the Gaussian quantile estimator (black) or the Student quantile estimator (blue). Panel (a) considers pth quantiles for $p = 0.1\%$. Panel (b) considers pth quantiles for $p = 5\%$. We see that for $p = 0.1\%$ the empirical quantile estimator is better than the Gaussian quantile estimator, but the empirical quantile estimator is worse than the Student estimator at the beginning of the time period. For $p = 5\%$ the empirical quantile estimator is equally good as the Gaussian and

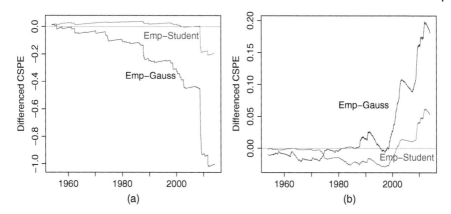

Figure 8.12 *Performance of quantile estimators: Cumulative losses.* We show time series C_t in (8.35). (a) $p = 0.1\%$ and (b) $p = 5\%$.

Student quantile estimator at the beginning of the period, but better at the end of the period.

8.5.2 Conditional Location–Scale Model

The conditional heteroskedasticity model

$$Y_t = \mu_t + \sigma_t \epsilon_t,$$

was defined in (5.29) for $\mu_t \equiv 0$. For the financial returns the signal (the expected return) is typically of a lower order than the noise, and thus in quantile estimation the location μ_t can usually be ignored. We do not ignore μ_t but use only the sample mean to estimate μ_t, instead of using any more sophisticated methods.

We use the conditional quantile estimator

$$\hat{Q}_p(Y_t | \mathcal{F}_{t-1}) = \hat{\mu}_t + \hat{\sigma}_t \, \hat{Q}_p(\epsilon_t), \tag{8.36}$$

where $\hat{\mu}_t$ is the prediction of Y_{t+1}, $\hat{\sigma}_t$ is the predicted volatility (an estimator of the conditional standard deviation of Y_{t+1}), and $\hat{Q}_p(\epsilon_t)$ is an estimator of the pth quantile of the distribution of $\epsilon_t = (Y_t - \mu_t)/\sigma_t$.

8.5.2.1 Examples of Conditional Location–Scale Quantile Estimators

We choose $\hat{\mu}_t$ to be the sequential sample mean. We choose the estimator $\hat{\sigma}_t$ of the conditional standard deviation to be either the GARCH(1, 1) predictor, the square root of an exponentially weighted moving average of squared returns, or a state space kernel predictor of volatility. These predictors are discussed in Chapter 7.

We choose the estimator $\hat{Q}_p(\epsilon_t)$ of the quantile of the residuals to be either a fixed quantile function, or the empirical quantile of the residuals. We can choose $\hat{Q}_p(\epsilon_t) = \Phi^{-1}(p)$, where Φ is the distribution function of the standard normal distribution. This leads to the estimator

$$\hat{Q}_p(Y_t|F_{t-1}) = \hat{\mu}_t + \hat{\sigma}_t \, \Phi^{-1}(p). \tag{8.37}$$

Second, we can choose $\hat{Q}_p(\epsilon_t) = \sqrt{(v-2)/v} \, t_v^{-1}(p)$, where t_v is the distribution function of the t-distribution with v degrees of freedom, $v > 2$. This leads to the quantile estimator

$$\hat{Q}_p(Y_t|F_{t-1}) = \hat{\mu}_t + \sqrt{\frac{v-2}{v}} \, \hat{\sigma}_t \, t_v^{-1}(p). \tag{8.38}$$

We can choose $\hat{Q}_p(\epsilon_t)$ to be the empirical quantile of the residuals $Y_t/\hat{\sigma}_t$. This leads to the quantile estimator

$$\hat{Q}_p(Y_t|F_{t-1}) = \hat{\mu} + \hat{\sigma}_t \, \hat{Q}^{res}(p), \tag{8.39}$$

where $\hat{Q}^{res}(p)$ is the pth empirical quantile, computed from

$$(Y_1 - \hat{\mu}_1)/\hat{\sigma}_1, \dots, (Y_t - \hat{\mu}_t)/\hat{\sigma}_t.$$

Empirical quantiles are defined in (8.21). Note that when $\sigma_1 = \cdots = \sigma_t$, then $\hat{\sigma}_t \, \hat{Q}^{res}(p)$ is the empirical quantile computed from $Y_1, \dots, Y_t$. Thus, the use of an empirical quantile of residuals makes sense only in the conditional quantile estimation. The method of using empirical quantiles of residuals was suggested in Fan and Gu (2003).

The distribution of the GARCH(1, 1) residuals $Y_t/\hat{\sigma}_t$ is studied in Figure 5.10, which shows the tail plots of the residuals. The maximum likelihood estimator of the GARCH(1, 1) model is defined with the assumption of standard normal innovations, but Figure 5.10 indicates that the tails of the residuals are better fitted with the t-distribution and thus it makes sense to try to use quantiles from the t-distribution.

8.5.2.2 Estimation of S&P 500 Quantiles

We study the performance of the GARCH(1, 1) volatility estimator in quantile estimation and compare its performance to the performance of moving average estimators. We use the daily S&P 500 data, described in Section 2.4.1.

The sequential GARCH(1, 1) volatility predictor $\hat{\sigma}_t^{garch}$ is defined in (7.7), where a different notation is used (here we use subindex t instead of $t + 1$). Exponentially weighted moving average $\hat{\sigma}_t^{ewma}$ for the estimation of conditional variance was defined in (7.9). We also use the name "EWMA(h) estimator" to refer to the exponentially weighted moving average estimator with smoothing parameter h.

The performance of the exponentially weighted moving average and GARCH(1, 1) volatility estimator is compared in Section 7.4.2.

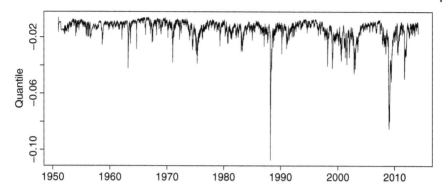

Figure 8.13 *GARCH(1, 1) quantiles.* The figure shows the time series of estimated pth quantiles with $p = 0.05$ for the S&P 500 returns data. The quantiles are estimated with the GARCH(1, 1) method with the residual distribution being the standard normal.

GARCH(1, 1)-Based Quantile Estimators GARCH(1, 1)-based quantile estimator is defined by (8.36), when $\hat{\sigma}_t$ is estimated with the GARCH(1, 1) method. The residual quantile is determined with one of the three methods in (8.37)–(8.39).

Figure 8.13 shows the time series of estimated conditional quantiles with the level $p = 0.05$. The distribution of the residuals is the standard normal distribution: we use the method (8.37).

Figure 8.14 compares the three methods in (8.37)–(8.39). Panel (a) plots the function $p \mapsto p - \hat{p}$ in the range $p \in [0.001, 0.075]$ and panel (b) plots the

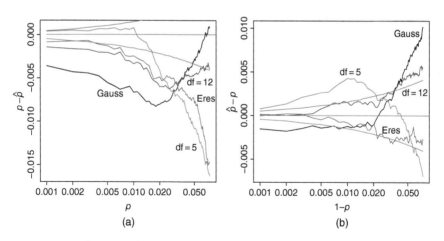

Figure 8.14 *Performance of GARCH(1, 1) quantile estimators: Probability differences.* (a) Functions $p \mapsto p - \hat{p}$ are shown for $p \in [0.001, 0.075]$. (b) Functions $p \mapsto \hat{p} - p$ are shown for $p \in [0.925, 0.999]$. The residuals are the standard normal (black) and the t-distribution with degrees of freedom 12 (blue), degrees of freedom 5 (red), and the empirical distribution (purple).

function $p \mapsto \hat{p} - p$ in the range $p \in [0.925, 0.999]$. Four cases are shown: the residual distribution is the standard normal distribution (black), the standardized t-distribution with degrees of freedom 5 (red) and 12 (blue), and the empirical distribution (purple). The green lines show the level $\alpha = 0.05$ fluctuation bands, defined in (8.10) and (8.11).

We see from Figure 8.14(a) that the Gaussian residuals perform well for level $p = 0.05$, but for level $p = 0.01$ using a t-distribution or the empirical distribution gives better estimates. The GARCH(1, 1)-based quantile estimates are estimating the left tail of the S&P 500 return distribution too light, except when the residuals are from the t-distribution with degrees of freedom 5, in which case the tail is estimated too heavy for levels $p < 0.01$. We see from Figure 8.14(b) that for the right tail the quantile estimates are more accurate than for the left tail. The standard t-distribution with degrees of freedom 12 gives a good overall performance.

Figure 8.15 shows slices of two-variate probability differences when $p = 0.1\%$. Panel (a) shows slices in (8.12) for $t_2 = T$ and for $t_2 = 1979$. Panel (b) shows slices in (8.13) for $t_1 = t_0$ and for $t_1 = 1979$. The purple curves show $p - \hat{p}$ when the quantiles of the residuals are the empirical quantiles and the black curves show $p - \hat{p}$ when the quantiles of the residuals are from the standard normal distribution. We see that the empirical quantiles are better for all shown time intervals. The performance becomes worse for the time periods close to the end. The fluctuation is large when the time periods are short.

Figure 8.16 shows functions

$$p \mapsto \hat{R}(\hat{q}^{eres}) - \hat{R}(\hat{q}),$$

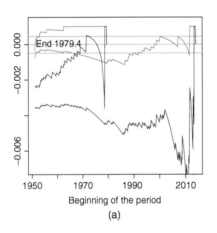
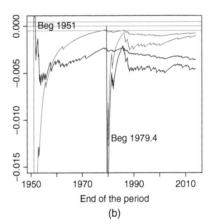

(a) (b)

Figure 8.15 *Performance of GARCH(1, 1) quantile estimators: Slices of probability differences.* We show $p - \hat{p}$ when $p = 0.1\%$ for the empirical residuals (purple) and for the Gaussian residuals (black). Panel (a) shows slices in (8.12) for $t_2 = T$ and for $t_2 = 1979$. Panel (b) shows slices in (8.13) for $t_1 = t_0$ and for $t_1 = 1979$.

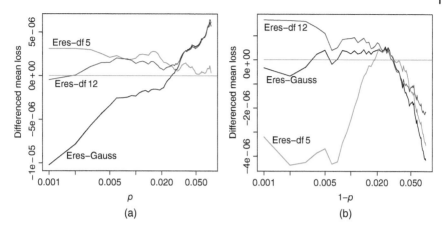

Figure 8.16 *Performance of GARCH(1, 1) quantile estimators: Loss function.* Functions $p \mapsto \hat{R}(\hat{q}^{eres}) - \hat{R}(\hat{q})$ (a) $p \in [0.001, 0.075]$; (b) $p \in [0.925, 0.999]$. The residuals in $\hat{q}$ are the standard normal (black) and the t-distribution with degrees of freedom 12 (blue) and degrees of freedom 5 (red).

where $\hat{R}(\hat{q})$ is the mean loss, defined in (8.18). Panel (a) shows range $p \in [0.001, 0.075]$. Panel (b) shows range $p \in [0.925, 0.999]$. We denote by $\hat{q}^{eres}$ the quantile estimator when the residual distribution is the empirical distribution. The estimator $\hat{q}$ is the quantile estimator with the standard normal distribution (black), the standardized t-distribution with degrees of freedom 5 (red) and 12 (blue). We see that for p close to 0, the t-distribution with degrees of freedom 5 gives the best result. When p is close to 1, the t-distribution with degrees of freedom 12 gives the best result.

Figure 8.17 shows time series of differences

$$C_t = \sum_{i=t_0+1}^{t} \rho_p \left(Y_i - \hat{q}_i^{eres} \right) - \sum_{i=t_0+1}^{t} \rho_p(Y_i - \hat{q}_i) \tag{8.40}$$

of cumulative sums, where $\hat{q}_i^{eres}$ is the quantile estimator with empirical residuals, and $\hat{q}_i$ has the Gaussian residuals (black), the Student residuals with degrees of freedom 5 (red), and degrees of freedom 12 (blue). Panel (a) considers pth conditional quantiles for $p = 0.1\%$. Panel (b) considers pth conditional quantiles for $p = 5\%$. We see that for $p = 0.1\%$ the Student distribution with five degrees of freedom leads to best results, for almost all time intervals. For $p = 1\%$ the Student distribution with 5 degrees of freedom and 12 degrees of freedom leads to almost equal performance, for almost all time intervals.

EWMA-Based Quantile Estimators Exponentially weighted moving average (EWMA) based quantile estimators are defined by (8.36), where $\hat{\sigma}_t$ is

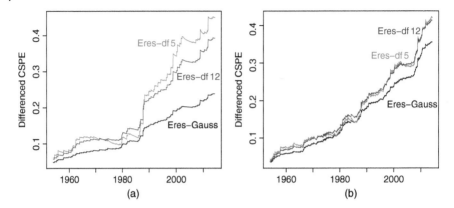

Figure 8.17 *Performance of GARCH(1, 1) quantile estimators: Cumulative losses.* We show time series C_t in (8.40). (a) $p = 0.1\%$; (b) $p = 1\%$. The residuals of $\hat{q}_i$ are the standard normal (black) and the t-distribution with degrees of freedom 12 (blue) and degrees of freedom 5 (red).

calculated with the EWMA method, and the residual quantile is determined with one of the three methods in (8.37)–(8.39).

Figure 8.18 shows functions

$$h \mapsto \hat{R}(\hat{q}_h),$$

where $\hat{R}(\hat{q})$ is the mean loss, defined in (8.18), and h is the smoothing parameter of the EWMA estimator of $\hat{\sigma}_t$. Panel (a) estimates pth conditional quantiles for $p = 0.1\%$. Panel (b) has $p = 1\%$. The quantile estimators $\hat{q}_h$ have the Gaussian residuals (black with "2"), the Student residuals with degrees of freedom 5 (red

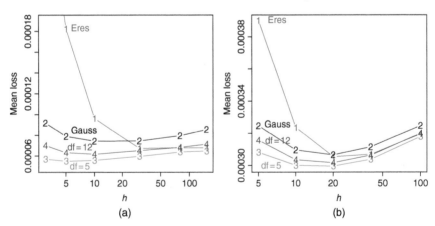

Figure 8.18 *EWMA(h) quantile estimator: Mean losses.* The mean loss is shown as a function of smoothing parameter h. (a) $p = 0.1\%$; (b) $p = 1\%$. The residual distributions are the standard Gaussian (black with "2"), the Student with degrees of freedom 5 (red with "3"), degrees of freedom 12 (blue with "4"), and the empirical (purple with "1").

with "3"), degrees of freedom 12 (blue with "4"), and the empirical residuals (purple with "1"). We see that the Student residual with degrees of freedom 5 gives the best results, for smoothing parameter $h = 5$ for $p = 0.1\%$ and $h = 10$ for 1%. The optimal smoothing parameter is larger in the case of empirical residuals than the optimal smoothing parameter in the case of Gaussian or Student residuals.

Figure 8.19 shows functions

$$h \mapsto \hat{R}(\hat{q}_h),$$

where $\hat{R}(\hat{q})$ is the mean loss, defined in (8.18), and h is the smoothing parameter of the EWMA estimator of $\hat{\sigma}_t$. Panel (a) estimates pth conditional quantiles for $p = 0.1\%$. Panel (b) has $p = 1\%$. The quantile estimators $\hat{q}_h$ have the Student residuals with degrees of freedom 4 (black with "1"), 5 (red with "2"), and 6 (blue with "3"). The horizontal lines show the mean losses when $\hat{\sigma}_t$ is the GARCH(1, 1) volatility. We see that for $p = 0.1\%$ the EWMA can perform better than GARCH(1, 1), but for $p = 1\%$ GARCH(1, 1) is better.

Figure 8.20 shows the performance of exponentially weighted moving average for four smoothing parameters: $h = 2$ (black), $h = 5$ (red), $h = 20$ (blue), and $h = 40$ (purple). Panel (a) plots the function $p \mapsto p - \hat{p}$ in the range $p \in [0.001, 0.075]$, and panel (b) plots the function $p \mapsto \hat{p} - p$ in the range $p \in [0.925, 0.999]$. The distribution of the residuals is Student with degrees of freedom equal to 4. The green horizontal line is drawn at level zero, and it is accompanied with the level $\alpha = 0.05$ fluctuation bands. The smoothing

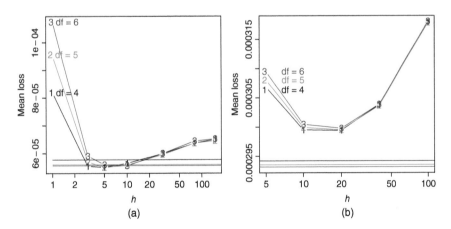

Figure 8.19 *EWMA(h) quantile estimator: Mean losses for Student residuals.* The mean loss is shown as a function of smoothing parameter h. (a) $p = 0.1\%$ and (b) $p = 1\%$. The residual distributions are the Student distributions with degrees of freedom 4 (black with "1"), 5 (red with "2"), and 6 (blue with "3"). The horizontal lines show the mean losses for GARCH(1, 1) volatility.

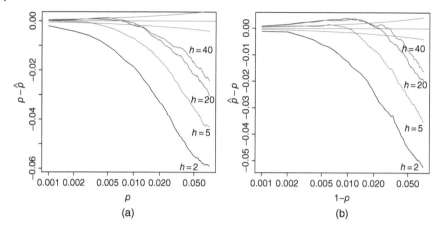

Figure 8.20 *EWMA(h) quantile estimator: Smoothing parameter selection for Student residuals.* Panel (a) shows the curves $p \mapsto p - \hat{p}$ for $p \in [0.001, 0.075]$ and panel (b) shows the curves $p \mapsto \hat{p} - p$ for the cases $p \in [0.925, 0.999]$. The distribution of the residuals is Student with degrees of freedom equal to 4. The smoothing parameters $h = 2, 5, 20, 40$ are shown with the colors black, red, blue, and purple.

parameters $h = 10$ and $h = 30$ give the best results for large p. However, for small p the smoothing parameter $h = 100$ gives the best results.

Figure 8.21 shows the performance of exponentially weighted moving average for four smoothing parameters: $h = 5$ (black), $h = 10$ (red), $h = 30$ (blue), and $h = 100$ (purple). Panel (a) plots the function $p \mapsto p - \hat{p}$ in the range $p \in [0.001, 0.075]$, and panel (b) plots the function $p \mapsto \hat{p} - p$ in the range $p \in [0.925, 0.999]$. The distribution of the residuals is the standard normal. The green horizontal line is drawn at level zero, and it is accompanied with the level $\alpha = 0.05$ fluctuation bands. The smoothing parameters $h = 10$ and $h = 30$ give the best results for large p. However, for small p the smoothing parameter $h = 100$ gives the best results.

Figure 8.22 shows the performance of the exponentially weighted moving average estimator with the smoothing parameter $h = 30$ for four residual distributions. Panel (a) shows the curves $p \mapsto p - \hat{p}$ for $p \in [0.001, 0.075]$ and panel (b) shows the curves $p \mapsto \hat{p} - p$ for the cases $p \in [0.925, 0.999]$. The blue curve shows the standard normal residual distribution, the black curve shows the standard t-distribution with degrees of freedom 12, the red curve shows degrees of freedom 5, and the purple curve shows the case of using empirical distribution. For the left tail the empirical residuals give the best result, except when $p \geq 0.05$, when the Gaussian residual give the best result. For the right tail the empirical residuals and the standard t-distribution with degrees of freedom 12 give the best results.

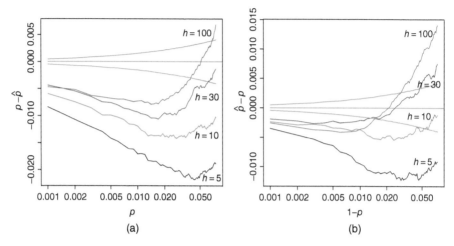

Figure 8.21 *EWMA(h) quantile estimator: Smoothing parameter selection for Gaussian residuals.* Panel (a) shows the curves $p \mapsto p - \hat{p}$ for $p \in [0.001, 0.075]$ and panel (b) shows the curves $p \mapsto \hat{p} - p$ for the cases $p \in [0.925, 0.999]$. The distribution of the residuals is the standard normal. The smoothing parameters $h = 5, 10, 30, 100$ are shown with the colors black, red, blue, and purple.

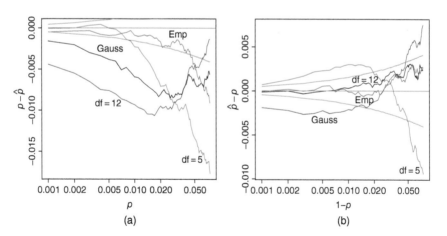

Figure 8.22 *EWMA(h) quantile estimator: Selection of residual distribution using probability differences.* (a) The curves $p \mapsto p - \hat{p}$ for $p \in [0.001, 0.075]$. (b) The curves $p \mapsto \hat{p} - p$ for $p \in [0.925, 0.999]$. The residual distributions standard normal, standard t-distribution with degrees of freedom 12, degrees of freedom 5, and the empirical distribution are shown with the colors blue, red, black, and green.

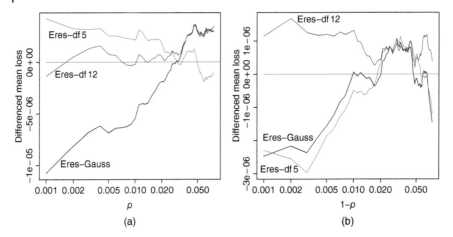

Figure 8.23 *Performance of EWMA(h) quantile estimators: Selection of residual distribution using loss function. Functions $p \mapsto \hat{R}(\hat{q}^{eres}) - \hat{R}(\hat{q})$ for (a) $p \in [0.001, 0.075]$ and (b) $p \in [0.925, 0.999]$. The residuals in $\hat{q}$ are the standard normal (black) and the t-distribution with degrees of freedom 12 (blue) and degrees of freedom 5 (red).*

Figure 8.23 shows functions

$$p \mapsto \hat{R}(\hat{q}^{eres}) - \hat{R}(\hat{q})$$

where $\hat{R}(\hat{q})$ is the mean loss, defined in (8.18). Panel (a) shows range $p \in [0.001, 0.075]$. Panel (b) shows range $p \in [0.925, 0.999]$. We denote by $\hat{q}^{eres}$ the quantile estimator when the residual distribution is the empirical distribution. The estimator $\hat{q}$ is the quantile estimator with the the standard normal distribution (black), the standardized t-distribution with degrees of freedom 5 (red) and 12 (blue). We see that for p close to 0, the Student distribution with degrees of freedom 5 gives the best result. When p is close to 1, the Student distribution with degrees of freedom 12 gives the best result.

Figure 8.24 shows time series of differences

$$C_t = \sum_{i=t_0+1}^{t} \rho_p \left(Y_i - \hat{q}_i^{emp} \right) - \sum_{i=t_0+1}^{t} \rho_p (Y_i - \hat{q}_i) \tag{8.41}$$

of cumulative sums, where $\hat{q}_i^{emp}$ is the empirical quantile estimator, and $\hat{q}_i$ is the Gaussian quantile estimator (black) or the Student quantile estimator (blue). Panel (a) considers pth conditional quantiles for $p = 0.1\%$. Panel (b) considers pth conditional quantiles for $p = 5\%$. We see that for $p = 0.1\%$ the empirical quantile estimator is better than the Gaussian quantile estimator, but the empirical quantile estimator is worse than the Student estimator at the beginning of the time period. For $p = 5\%$ the empirical quantile estimator is equally good as the Gaussian and Student quantile estimators at the beginning of the period, but better at the end of the period.

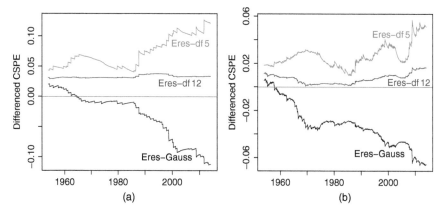

Figure 8.24 *Performance of EWMA quantile estimators: Cumulative losses. We show time series C_t in (8.41). (a) $p = 0.1\%$ and (b) $p = 5\%$. The residuals are the standard normal (black) and the t-distribution with degrees of freedom 12 (blue), degrees of freedom 5 (red), and the empirical distribution (purple).*

Kernel Smoothing-Based Quantile Estimators We define in Section 7.5.2 a kernel estimator $\hat{\sigma}_t^{ker}$ for the conditional standard deviation, which uses as predictors the exponentially weighted moving averages of past squared returns and past returns, as defined in (7.19). The smoothing parameter g involved in the definition of the predictors is taken to be $g = 40$. Estimator $\hat{\sigma}_t^{ker}$ is now applied to quantile estimation.

Section 7.5.2 shows that the kernel estimator is better than GARCH(1, 1) for volatility estimation, when the performance is measured in the terms of the sum of squared prediction errors. In this section we see that the kernel estimator leads to a better quantile estimator than GARCH(1, 1). Looking at the aggregated mean loss it seems that the GARCH(1, 1) would perform better for the estimation of the pth conditional quantiles when p is close to 0. However, looking at the cumulative losses in Figure 8.29 reveals that kernel estimator is better for all other time periods, but autumn 1987 makes the aggregate mean loss better for GARCH(1, 1), when p is close to 0.

Figure 8.25 shows functions

$$h \mapsto \hat{R}(\hat{q}_h),$$

where $\hat{R}(\hat{q}_h)$ is the mean loss, defined in (8.18), and h is the smoothing parameter of the kernel estimator $\hat{\sigma}_t^{ker}$. Panel (a) estimates the pth conditional quantiles for $p = 0.1\%$. Panel (b) has $p = 1\%$. The quantile estimators $\hat{q}_h$ have the Gaussian residuals (black with "2"), the Student residuals with degrees of freedom 5 (red with "3"), degrees of freedom 12 (blue with "4"), and the empirical residuals (purple with "1"). We see that the Student residuals with degrees of freedom 5 give the best results.

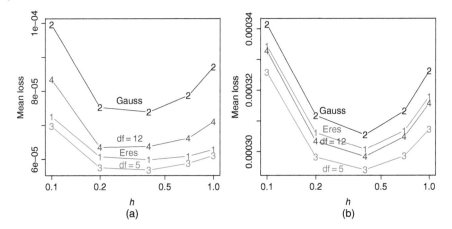

Figure 8.25 *Kernel quantile estimator: Mean losses.* The mean loss is shown as a function of smoothing parameter h. (a) $p = 0.1\%$ and (b) $p = 1\%$. The residual distributions are the standard Gaussian (black with "2"), the Student with degrees of freedom 5 (red with "3"), degrees of freedom 12 (blue with "4"), and the empirical (purple with "1").

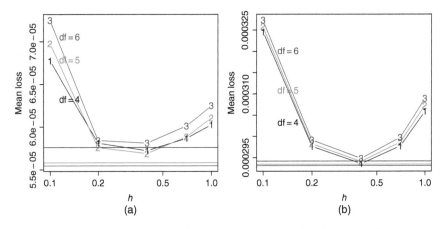

Figure 8.26 *Kernel quantile estimator: Mean losses for Student residuals when p is close to 0.* The mean loss is shown as a function of smoothing parameter h. (a) $p = 0.1\%$ and (b) $p = 1\%$. The residual distributions are the Student distributions with degrees of freedom 4 (black with "1"), 5 (red with "2"), and 6 (blue with "3"). The horizontal lines show the mean losses for GARCH(1, 1) volatility.

Figure 8.26 shows functions $h \mapsto \hat{R}(\hat{q}_h)$, where $\hat{R}(\hat{q}_h)$ is the mean loss, defined in (8.18), and h is the smoothing parameter of the kernel estimator $\hat{\sigma}_t^{ker}$. Panel (a) estimates pth conditional quantiles for $p = 0.1\%$. Panel (b) has $p = 1\%$. The quantile estimators $\hat{q}_h$ have the Student residuals with degrees of freedom 4 (black with "1"), 5 (red with "2"), and 6 (blue with "3"). The horizontal lines show

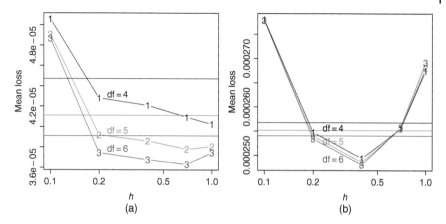

Figure 8.27 *Kernel quantile estimator: Mean losses for Student residuals when p is close to 1. The mean loss is shown as a function of smoothing parameter h. (a) p = 0.999 and (b) p = 0.99. The residual distributions are the Student distributions with degrees of freedom 4 (black with "1"), 5 (red with "2"), and 6 (blue with "3"). The horizontal lines show the mean losses for GARCH(1, 1) volatility.*

the mean losses when $\hat{\sigma}_t$ is the GARCH(1, 1) volatility. We see that GARCH(1, 1) gives better results than the kernel estimator, and the degrees of freedom 4 gives the best results.

Figure 8.27 shows functions $h \mapsto \hat{R}(\hat{q}_h)$, where $\hat{R}(\hat{q}_h)$ is the mean loss, defined in (8.18), and h is the smoothing parameter of the kernel estimator $\hat{\sigma}_t^{ker}$. Panel (a) estimates pth conditional quantiles for $p = 0.999$. Panel (b) has $p = 99$. The quantile estimators $\hat{q}_h$ have the Student residuals with degrees of freedom 4 (black with "1"), 5 (red with "2"), and 6 (blue with "3"). The horizontal lines show the mean losses when $\hat{\sigma}_t$ is the GARCH(1, 1) volatility. We see that kernel estimator gives better results than GARCH(1, 1), and the degrees of freedom 6 gives the best results.

Figure 8.28 shows the performance of the kernel estimator for the empirical residual distribution. Panel (a) shows the curves $p \mapsto p - \hat{p}$ for $p \in [0.001, 0.075]$, and panel (b) shows the curves $p \mapsto \hat{p} - p$ for the cases $p \in [0.925, 0.999]$. The smoothing parameter of the kernel estimator is $h = 0.2$ (black), $h = 0.5$ (red), and $h = 1$ (blue).

Figure 8.29 shows time series of differences

$$C_t = \sum_{i=t_0+1}^{t} \rho_p \left(Y_i - \hat{q}_i^{garch} \right) - \sum_{i=t_0+1}^{t} \rho_p \left(Y_i - \hat{q}_i^{ker} \right) \tag{8.42}$$

of cumulative sums, where $\hat{q}_i^{garch}$ uses the GARCH(1, 1) volatility and $\hat{q}_i^{ker}$ uses the kernel estimator of volatility. Panel (a) considers pth conditional quantiles for $p = 0.1\%$. Panel (b) considers pth conditional quantiles for $p = 5\%$. The distribution of the residuals is the Student distribution with degrees of freedom 5.

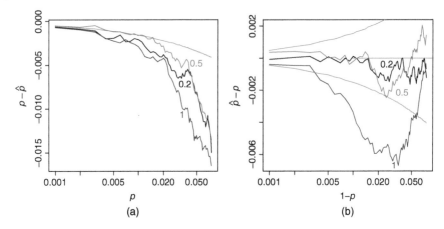

Figure 8.28 *Kernel quantile estimator: Empirical residual for many h.* (a) The curves $p \mapsto p - \hat{p}$ for $p \in [0.001, 0.075]$. (b) The curves $p \mapsto \hat{p} - p$ for $p \in [0.925, 0.999]$. The smoothing parameter of the kernel estimator is $h = 0.2$ (black), $h = 0.5$ (red), and $h = 1$ (blue).

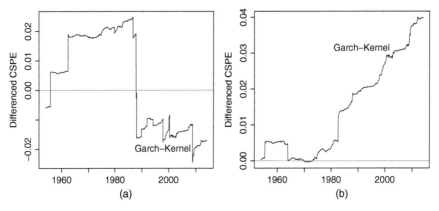

Figure 8.29 *Kernel versus GARCH quantile estimator: Cumulative losses.* We show time series C_t in (8.42). (a) $p = 0.1\%$ and (b) $p = 0.999$.

The smoothing parameter of the kernel estimator is $h = 0.4$. We see that for $p = 0.1\%$ the kernel estimator is better or equally good as GARCH estimator, expect at autumn 1987. For $p = 0.999$ the kernel estimator is better for almost all time periods.

8.6 Excess Distributions in Quantile Estimation

Let $Y \in \mathbf{R}$ be a random variable with a continuous distribution, let $F(x) = P(Y \leq x)$ be the distribution function of Y, and f the density function.

8.6.1 The Excess Distributions

We define the upper excess distribution and the lower excess distribution. We can use the term "excess distribution" without a further qualification when it is clear from the context whether we mean the upper or the lower excess distribution.

8.6.1.1 The Upper Excess Distribution

The distribution function of the excess distribution with threshold $u \in \mathbf{R}$ is

$$F_u(x) = P(Y - u \le x | Y > u) = \frac{F(x + u) - F(u)}{1 - F(u)}, \qquad (8.43)$$

where $x \ge 0$. The density function is

$$f_u(x) = \frac{f(x + u)}{1 - F(u)} \, I_{[0,\infty)}(x). \qquad (8.44)$$

Now it holds that

$$f(x)I_{[u,\infty)}(x) = (1 - F(u))f_u(x).$$

This implies that we can model the right tail of f by giving a parametric model for f_u. These definitions were given in (3.59) and (3.60) and the estimation of the excess distribution is studied in Section 3.4.

The quantiles can be written in terms of the excess distribution. If the quantile satisfies $Q_p(Y) > u$, then

$$Q_p(Y) = u + F_u^{-1}\left(1 - \frac{1 - p}{1 - F(u)}\right). \qquad (8.45)$$

Indeed, let $x > u$. Now

$$P(Y > x | Y > u) = \frac{P(Y > x)}{P(Y > u)}.$$

Thus,

$$\begin{aligned}
P(Y > x) &= P(Y > u)\, P(Y > x | Y > u) \\
&= P(Y > u)\, P(Y - u > x - u | Y > u) \\
&= P(Y > u)\, [1 - F_u(x - u)].
\end{aligned}$$

Thus,

$$F_u(x - u) = 1 - \frac{P(Y > x)}{P(Y > u)}$$

and

$$x = u + F_u^{-1}\left(1 - \frac{P(Y > x)}{P(Y > u)}\right).$$

Now we choose $x = Q_p(Y)$ and use the fact that if the distribution of Y is continuous, then $P(Y > Q_p(Y)) = 1 - p$ to get (8.45).

8.6.1.2 The Lower Excess Distribution

The distribution function of the lower excess distribution with threshold $u \in \mathbf{R}$ is

$$F_u(x) = P(u - Y \le x | Y < u) = \frac{F(u) - F(u - x)}{F(u)}, \tag{8.46}$$

where $x \ge 0$. The density function is

$$f_u(x) = \frac{f(u - x)}{F(u)} I_{[0,\infty)}(x).$$

Now it holds that

$$f(x) I_{(-\infty, u]}(x) = F(u) f_u(x).$$

This implies that we can model the left tail of f by giving a parametric model for f_u. The quantiles can be written in terms of the lower excess distribution. We have that

$$Q_p(Y) = u - F_u^{-1}\left(1 - \frac{p}{F(u)}\right), \tag{8.47}$$

when the quantile satisfies $Q_p(Y) < u$.[10]

8.6.1.3 Quantile Estimators Using Excess Distributions

Now we obtain semiparametric estimators of the quantiles. Let $0 < p < p_0 < 0.5$, or $0.5 < p_0 < p < 1$. First we estimate the p_0th quantile using empirical quantiles. Let the estimate be $u = \hat{Q}_{p_0}$. Then we model F_u parametrically,

---

10 Indeed, let $x < u$. Now

$$P(Y < x | Y < u) = \frac{P(Y < x)}{P(Y < u)}.$$

Thus,

$$
\begin{aligned}
P(Y < x) &= P(Y < u)\, P(Y < x | Y < u) \\
&= P(Y < u)\, P(Y - u < x - u | Y < u) \\
&= P(Y < u)\, P(u - Y > u - x | Y < u) \\
&= P(Y < u)\, [1 - F_u(u - x)].
\end{aligned}
$$

Thus,

$$F_u(u - x) = 1 - \frac{P(Y < x)}{P(Y < u)}$$

and

$$x = u - F_u^{-1}\left(1 - \frac{P(Y < x)}{P(Y < u)}\right).$$

Now we choose $x = Q_p(Y)$ and use the fact that if the distribution of Y is continuous, then $P(Y < Q_p(Y)) = p$ to get (8.47).

estimate the parameters, and use (8.45) and (8.47) to estimate the pth quantile. This is reasonable because estimation of the p_0th quantile is easier than the estimation of the pth quantile.

8.6.1.4 A Connection to Location–Scale Model

We can obtain the formula

$$Q_p(Y) = \mu + \sigma\, F_\epsilon^{-1}(p). \tag{8.48}$$

of location–scale modeling as a special case of the excess distribution modeling. Formula (8.48) was given in (8.30).

Let the distribution of ϵ be symmetric around zero. Then the density function of ϵ satisfies $f_\epsilon(x) = f_\epsilon(-x)$. Define

$$f_u(x) = \frac{2}{\sigma} f_\epsilon\left(\frac{x}{\sigma}\right) I_{[0,\infty)}(x), \qquad F_u(x) = 2\left(F_\epsilon\left(\frac{x}{\sigma}\right) - 1/2\right) I_{[0,\infty)}(x).$$

We have that

$$F_u^{-1}(p) = \sigma F_\epsilon^{-1}\left(\frac{p+1}{2}\right).$$

Choose $u = \mu$ and $F(u) = 1/2$. Then for $1/2 < p < 1$

$$Q_p(Y) = u + F_u^{-1}\left(1 - \frac{1-p}{1-F(u)}\right) = \mu + \sigma F_\epsilon^{-1}(p).$$

For $0 < p < 1/2$

$$Q_p(Y) = u - F_u^{-1}\left(1 - \frac{p}{F(u)}\right) = \mu - \sigma F_\epsilon^{-1}(-p) = \mu + \sigma F_\epsilon^{-1}(p).$$

8.6.2 Unconditional Quantile Estimation

We study the estimation of the lower quantiles using formula

$$Q_p(Y) = u - F_u^{-1}\left(1 - \frac{p}{F(u)}\right), \tag{8.49}$$

as derived in (8.47), where $0 < p < F(u) < 0.5$. We choose

$$u = \hat{F}^{-1}(p_0), \tag{8.50}$$

where $\hat{F}$ is the empirical distribution function, $\hat{F}^{-1}$ means the generalized inverse, and $0 < p < p_0 < 0.5$. The excess distribution F_u is modeled parametrically, and the parameters of F_u are estimated.

The study is made using the daily S&P 500 data, described in Section 2.4.1. The quantile estimator is unconditional, but the performance is measured sequentially: the estimator is updated at each time t using the previous data.

Section 3.4.2 describes the one-parameter exponential and Pareto distributions, and the two-parameter gamma, generalized Pareto, and Weibull distributions. We apply these distributions to model excess distributions.

We study the performance of quantile estimators for various values of p_0/p. Since $p_0 = p$ means that the estimator is the empirical quantile estimator, we are comparing the estimators to the empirical quantile estimator.

8.6.2.1 Exponential Excess Distribution

Let F_u be the excess distribution of Y. When F_u is an exponential distribution, then $F_u^{-1}(p) = -\beta \log(1 - p)$, where $\beta > 0$ is the parameter of the exponential distribution. When $0 < p < F(u) < 0.5$,

$$Q_p(Y) = u + \beta \log\left(\frac{p}{F(u)}\right),\tag{8.51}$$

and the maximum likelihood estimator of parameter β is

$$\hat{\beta}_{left} = \frac{1}{\#\mathcal{L}} \sum_{Y_i \in \mathcal{L}} (u - Y_i), \quad \mathcal{L} = \{Y_i : Y_i \le u\}.\tag{8.52}$$

When $0.5 < F(u) < p < 1$, then analogous formulas hold.[11]

Figures 8.30 and 8.31 show that the exponential excess distribution leads to a better quantile estimator than the empirical quantile estimator. The ratio $p_0/p \approx 3$ leads to best estimates, for small p.

Figure 8.30 studies the performance of the exponential quantile estimator as a function of p_0/p, for several values of p. We estimate quantiles with levels $p = 0.1\%$ (black), $p = 1\%$ (red), and $p = 5\%$ (blue). Panel (a) shows the ratios $p/\hat{p}$ as a function of the ratio p_0/p, where $\hat{p}$ is the implied estimate of the probability p, as defined in (8.6), and p_0 is the probability in (8.50). Panel (b) shows the estimates of the expected loss as a function of ratio p_0/p: we show functions

$$k \mapsto \frac{\hat{R}(k) - \min_k \hat{R}(k)}{\max_k \hat{R}(k) - \min_k \hat{R}(k)},\tag{8.53}$$

where $k = p_0/p \ge 1$ is the multiplier and the estimated expected loss $\hat{R}$ is defined in (8.18). The value of $\hat{R}$ depends on p and p_0 through the quantile estimates. We indicate by vertical lines the values p_0/p (a) minimizing $|p/\hat{p} - 1|$ (b) minimizing $\hat{R}$. We see that the ratio $p_0/p \approx 3$ gives the best results for all values of p. Note that the ratio $p_0/p = 1$ implies that the estimator is the empirical quantile.

---

11 The pth quantile of Y is

$$Q_p(Y) = u - \beta \log\left(\frac{1-p}{1-F(u)}\right),$$

and the parameter estimate is

$$\hat{\beta}_{right} = \frac{1}{\#\mathcal{R}} \sum_{Y_i \in \mathcal{R}} (Y_i - u), \quad \mathcal{R} = \{Y_i : Y_i \ge u\}.$$

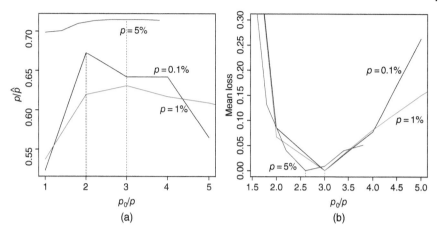

Figure 8.30 *Exponential model.* (a) Ratios $p/\hat{p}$ as a function of the multiplier $k = p_0/p$. (b) The expected loss $\hat{R}$ as a function of multiplier k. The quantile level p takes values $p = 0.1\%$ (black), $p = 1\%$ (red), and $p = 5\%$ (blue).

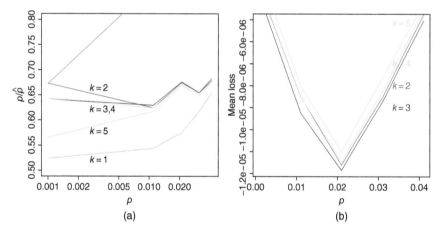

Figure 8.31 *Exponential model.* (a) Function $p \mapsto p/\hat{p}$, where $\hat{p}$ is the estimate of the exceedance probability. (b) Function $p \mapsto \hat{R}(k) - \hat{R}(1)$, where $\hat{R}$ is the estimated loss when the multiplier of the quantile estimator is $k = p_0/p$. The multiplier is $k = 1$ (violet), $k = 2$ (purple), $k = 3$ (dark green), $k = 4$ (pink), and $k = 5$ (yellow).

Figure 8.31 studies the performance of the exponential quantile estimator as a function of p, for several values of $k = p_0/p$. We consider values $k = 1$ (violet), $k = 2$ (purple), $k = 3$ (dark green), $k = 4$ (pink), and $k = 5$ (yellow). Panel (a) shows functions $p \mapsto p/\hat{p}$, where $\hat{p}$ is the implied estimate of the probability p, as defined in (8.6). The green lines show the fluctuation bands for $\alpha = 5\%$.

Panel (b) shows functions

$$p \mapsto \hat{R}(k) - \hat{R}(1),$$

where the estimated expected loss $\hat{R}$ is defined in (8.18), and $\hat{R}$ depends on $k = p_0/p$ through the quantile estimates. Note that the multiplier $k = 1$ implies that the quantile estimator is the empirical quantile, because then $Q_p(Y) = u$. Thus, negative values of $\hat{R}(k) - \hat{R}(1)$ imply that the quantile estimator with multiplier k is better than the empirical quantile. We see from panels (a) and (b) that the exponential quantiles are better than the empirical quantiles. We see from panel (a) that the true distribution seems to have a heavier tail than the exponential quantiles indicate, and that multiplier $k = 2$ leads to best results, at least for small p. We see from panel (b) that multiplier $k = 3$ leads to best results.

8.6.2.2 Pareto Excess Distribution

When the excess distribution F_u of Y is a Pareto distribution, then $F_u^{-1}(p) = u(1 - p)^{-1/\alpha} - u$ for $u > 0$ and $F_u^{-1}(p) = -u(1 - p)^{-1/\alpha} + u$ for $u < 0$, where $\alpha > 0$ is the parameter of the Pareto distribution. Here, we use the Pareto distribution with the support $[0, \infty)$, as defined in (3.74). When $0 < p < F(u) < 0.5$, then

$$Q_p(Y) = u \left(\frac{F(u)}{p} \right)^{1/\alpha}. \tag{8.54}$$

Note that when $u < 0$, then $Q_p(Y) < u$, because $\alpha > 0$ and $F(u)/p > 1$. The maximum likelihood estimator of parameter α is

$$\hat{\alpha}_{left} = \left(\frac{1}{\#\mathcal{L}} \sum_{Y_i \in \mathcal{L}} \log(Y_i/u) \right)^{-1}, \quad \mathcal{L} = \{Y_i : Y_i \le u\}.$$

Analogous formulas hold when $0.5 < F(u) < p < 1$.[12]

Figures 8.32 and 8.33 show that the Pareto excess distribution leads to better quantile estimates than the empirical quantiles, and ratio p_0/p can be even 6–10 for $p = 1$–5%. For $p = 0.1\%$ modeling with the Pareto excess distribution does not lead to a better performance than the performance of the empirical quantile estimator.

---

12 When p and u are such that $0.5 < F(u) < p < 1$, then the pth quantile of the distribution of Y is

$$Q_p(Y) = u \left(\frac{1 - F(u)}{1 - p} \right)^{1/\alpha}.$$

Note that when $u > 0$, then $Q_p(Y) > u$, because $\alpha > 0$ and $(1 - F(u))/(1 - p) > 1$. The maximum likelihood estimator of parameter α is

$$\hat{\alpha}_{right} = \left(\frac{1}{\#\mathcal{R}} \sum_{Y_i \in \mathcal{R}} \log(Y_i/u) \right)^{-1}, \quad \mathcal{R} = \{Y_i : Y_i \ge u\}.$$

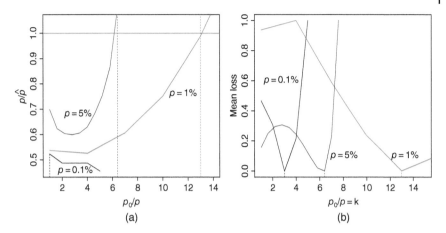

Figure 8.32 *Pareto model.* (a) Ratios $p/\hat{p}$ as a function of the multiplier $k = p_0/p$. (b) The expected loss $\hat{R}$ as a function of multiplier k. The quantile level p takes values $p = 0.1\%$ (black), $p = 1\%$ (red), and $p = 5\%$ (blue).

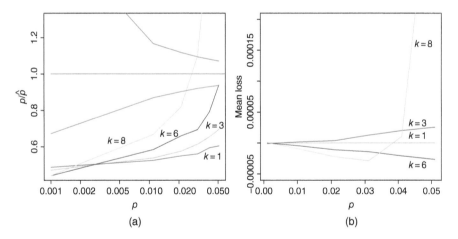

Figure 8.33 *Pareto model.* (a) Function $p \mapsto p/\hat{p}$, where $\hat{p}$ is the estimate of the exceedance probability. (b) Function $p \mapsto \hat{R}(k) - \hat{R}(1)$, where $\hat{R}$ is the estimated loss when the multiplier of the quantile estimator is $k = p_0/p$. The multiplier is $k = 1$ (violet), $k = 3$ (purple), $k = 6$ (dark green), and $k = 8$ (pink).

Figure 8.32 studies the performance of the Pareto quantile estimator as a function of p_0/p, for several values of p. We estimate quantiles with levels $p = 0.1\%$ (black), $p = 1\%$ (red), and $p = 5\%$ (blue). The setting is the same as in Figure 8.30. We see that the threshold $u = 2.5 \times p$ is close to optimum, and the expected loss does not change much when u is larger than that.

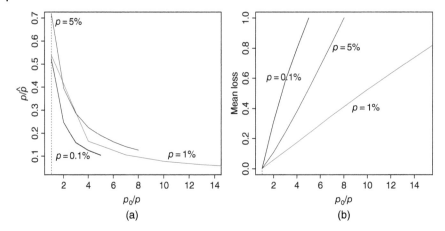

Figure 8.34 *Gamma model.* (a) Ratios $p/\hat{p}$ as a function of the multiplier $k = p_0/p$. (b) The expected loss $\hat{R}$ as a function of multiplier k. The quantile level p takes values $p = 0.1\%$ (black), $p = 1\%$ (red), and $p = 5\%$ (blue).

Figure 8.33 studies the performance of the Pareto quantile estimator as a function of p, for several values of $k = p_0/p$. We consider values $k = 1$ (violet), $k = 3$ (purple), $k = 6$ (dark green), and $k = 8$ (pink). The setting is the same as in Figure 8.31.

8.6.2.3 The Gamma Excess Distribution

Le the excess distribution F_u be a gamma distribution. The gamma distribution has parameters $\kappa > 0$ and $\beta > 0$. The quantile function $F_u^{-1}(p)$ cannot be given in a closed form, and the maximum likelihood estimators of the parameters do not have a closed-form expression.

Figures 8.34 and 8.35 show that the gamma excess distribution leads to worse quantile estimates than the empirical quantiles, for $p = 0.1$–5%.

Figure 8.34 studies the performance of the gamma quantile estimator as a function of p_0/p, for several values of p. We estimate quantiles with levels $p = 0.1\%$ (black), $p = 1\%$ (red), and $p = 5\%$ (blue). The setting is the same as in Figure 8.30.

Figure 8.35 studies the performance of the gamma quantile estimator as a function of p, for several values of $k = p_0/p$. We consider values $k = 1$ (violet), $k = 2$ (purple), and $k = 3$ (dark green). The setting is the same as in Figure 8.31.

8.6.2.4 The Generalized Pareto Excess Distribution

The parameters of the generalized Pareto distributions are $\xi \geq 0$ and $\beta > 0$. For $\xi = 0$ the distributions are exponential. When the excess distribution is a generalized Pareto distribution, then for $\xi > 0$, $F_u^{-1}(p) = (\beta/\xi)[(1-p)^{-\xi} - 1]$.

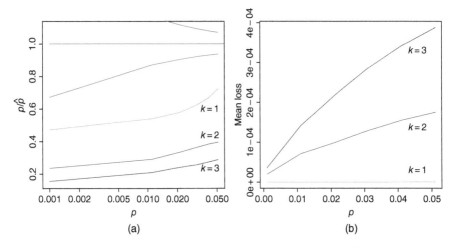

Figure 8.35 *Gamma model.* (a) Function $p \mapsto p/\hat{p}$, where $\hat{p}$ is the estimate of the exceedance probability. (b) Function $p \mapsto \hat{R}(k) - \hat{R}(1)$, where $\hat{R}$ is the estimated loss when the multiplier of the quantile estimator is $k = p_0/p$. The multiplier is $k = 1$ (violet), $k = 2$ (purple), and $k = 3$ (dark green).

For $\xi = 0$, $F_u^{-1}(p) = -\beta \log(1 - p)$. For $0 < p < F(u) < 0.5$ the pth quantile is

$$Q_p(Y) = u - (\beta/\xi)\left[\left(\frac{p}{F(u)}\right)^{-\xi} - 1\right].$$

For $0 < p < F(u) < 0.5$ we have analogous expressions.[13] The maximum likelihood estimators of parameters do not have a closed-form expression.

Figures 8.36 and 8.37 show that the generalized Pareto excess distribution leads to better quantile estimates than the empirical quantiles: for $p = 5\%$ we can have $p_0/p = 10$, and for $p = 1\%$ we can have $p_0/p = 20$. For $p = 0.1\%$ modeling with the generalized Pareto excess distribution does not lead to a better performance than the performance of the empirical quantile estimator.

Figure 8.36 studies the performance of the generalized Pareto quantile estimator as a function of p_0/p, for several values of p. We estimate quantiles with levels $p = 0.1\%$ (black), $p = 1\%$ (red), and $p = 5\%$ (blue). The setting is the same as in Figure 8.30.

Figure 8.37 studies the performance of the generalized Pareto quantile estimator as a function of p, for several values of $k = p_0/p$. We consider values $k = 1$ (violet), $k = 2$ (purple), and $k = 10$ (dark green). The setting is the same as in Figure 8.31.

---

13 The pth quantile with $0.5 < F(u) < p < 1$ is

$$Q_p(Y) = u + (\beta/\xi)\left[\left(\frac{1-p}{1-F(u)}\right)^{-\xi} - 1\right].$$

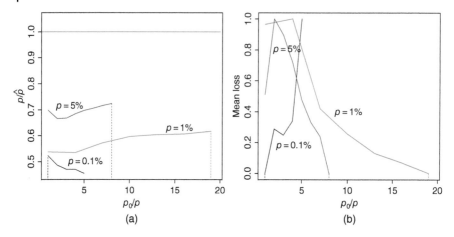

Figure 8.36 *Generalized Pareto model.* (a) Ratios $p/\hat{p}$ as a function of the multiplier $k = p_0/p$. (b) The expected loss $\hat{R}$ as a function of multiplier k. The quantile level p takes values $p = 0.1\%$ (black), $p = 1\%$ (red), and $p = 5\%$ (blue).

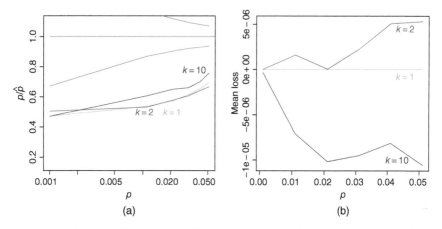

Figure 8.37 *Generalized Pareto model.* (a) Function $p \mapsto p/\hat{p}$, where $\hat{p}$ is the estimate of the exceedance probability. (b) Function $p \mapsto \hat{R}(k) - \hat{R}(1)$, where $\hat{R}$ is the estimated loss when the multiplier of the quantile estimator is $k = p_0/p$. The multiplier is $k = 1$ (violet), $k = 2$ (purple), and $k = 10$ (dark green).

8.6.2.5 The Weibull Excess Distribution

The parameters of the Weibull distributions are $\xi \geq 0$ and $\beta > 0$. For $\xi = 0$ the distributions are exponential. When the excess distribution is a generalized Pareto distribution, then for $\xi > 0$, $F_u^{-1}(p) = (\beta/\xi)[(1-p)^{-\xi} - 1]$. For $\xi = 0$, $F_u^{-1}(p) = -\beta \log(1-p)$. For $0 < p < F(u) < 0.5$ the pth quantile is

$$Q_p(Y) = u - (\beta/\xi) \left[\left(\frac{p}{F(u)} \right)^{-\xi} - 1 \right].$$

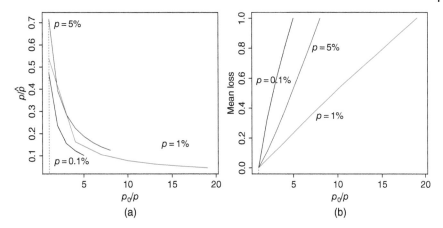

Figure 8.38 *Weibull model.* (a) Ratios $p/\hat{p}$ as a function of the multiplier $k = p_0/p$. (b) The expected loss $\hat{R}$ as a function of multiplier k. The quantile level p takes values $p = 0.1\%$ (black), $p = 1\%$ (red), and $p = 5\%$ (blue).

For $0 < p < F(u) < 0.5$ we have analogous expressions.[14] The maximum likelihood estimators of parameters do not have a closed-form expression.

Figures 8.38 and 8.39 show that the Weibull excess distribution leads to worse quantile estimates than the empirical quantiles, for $p = 0.1\text{–}5\%$.

Figure 8.38 studies the performance of the Weibull quantile estimator as a function of p_0/p, for several values of p. We estimate quantiles with levels $p = 0.1\%$ (black), $p = 1\%$ (red), and $p = 5\%$ (blue). The setting is the same as in Figure 8.30.

Figure 8.39 studies the performance of the Weibull quantile estimator as a function of p, for several values of $k = p_0/p$. We consider values $k = 1$ (violet), $k = 2$ (purple), and $k = 3$ (dark green). The setting is the same as in Figure 8.31.

8.6.3 Conditional Quantile Estimators

First we study time varying parameter of the excess distribution, and then we study the use of the empirical quantiles of the residual distribution.

We apply exponential, gamma, generalized Pareto, and Weibull excess distributions. The Pareto distribution is not studied because it is a special case of the generalized Pareto distribution.

First, we see that the generalized Pareto distribution leads to the best results when the time varying parameter is used. Second, we see that when the empirical quantiles of the residual distribution are used, then the performance is rather similar for all models.

---

14 The pth quantile with $0.5 < F(u) < p < 1$ is

$$Q_p(Y) = u + (\beta/\xi)\left[\left(\frac{1-p}{1-F(u)}\right)^{-\xi} - 1\right].$$

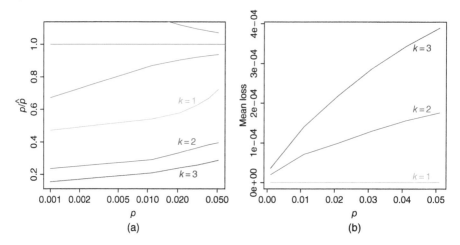

Figure 8.39 *Weibull model.* (a) Function $p \mapsto p/\hat{p}$, where $\hat{p}$ is the estimate of the exceedance probability. (b) Function $p \mapsto \hat{R}(k) - \hat{R}(1)$, where $\hat{R}$ is the estimated loss when the multiplier of the quantile estimator is $k = p_0/p$. The multiplier is $k = 1$ (violet), $k = 2$ (purple), and $k = 3$ (dark green).

8.6.3.1 Time Varying Parameter of the Excess Distribution

We have discussed time varying estimators for the excess distribution in Section 5.2.3. We apply the local likelihood estimator defined in (5.12) for the left tail as

$$\hat{\theta}_{left,t} = \text{argmax}_{\theta \in \Theta} \sum_{i:Y_i \in \mathcal{L}_t} p_i(t) \log f_u(u - Y_i, \theta),$$

where $u = \hat{q}_{p_0}$ is the empirical quantile computed from $Y_1, \dots, Y_t$, $0 < p < p_0 < 0.5$,

$$\mathcal{L}_t = \{Y_i : Y_i \leq u, \quad i = 1, \dots, t\}, \tag{8.55}$$

and the time space localized weights are defined in (5.13) as

$$p_i(t) = \frac{K((t-i)/h)}{\sum_{j:Y_j \in \mathcal{L}_t} K((t-j)/h)}, \tag{8.56}$$

where $h > 0$ is the smoothing parameter and $K : [0, \infty) \to \mathbf{R}$ is a kernel function.

Then we obtain the quantile estimator from (8.49) by inserting the parameter estimate:

$$\hat{q}_t = u - F_u^{-1}\left(1 - \frac{p}{p_0}, \hat{\theta}_{left,t}\right).$$

It turns out that the generalized Pareto distribution leads to the best results. The exponential model leads to almost as good results. The gamma model leads to worse results, and the Weibull model leads to the worst results.

Exponential Excess Distribution The time varying estimator for the left tail index is obtained from (8.52) as

$$\hat{\beta}_{left,t} = \sum_{i:Y_i \in \mathcal{L}_t} p_i(t)(u - Y_i),$$

where $\mathcal{L}_t$ is the set of observations in the left tail, as defined in (8.55), and $p_i(t)$ is the time space localized weight, as in (8.56). The conditional quantile estimator is obtained from (8.51) as

$$\hat{q}_t = u + \hat{\beta}_{left,t} \log\left(\frac{p}{p_0}\right),$$

where u is the p_0th empirical quantile and $0 < p < p_0 < 0.5$.

Figures 8.40–8.42 show that the exponential quantiles are better than the empirical quantiles. The smoothing parameter $h = 100$ leads to the best results.

Figure 8.40 studies the performance as a function of smoothing parameter h in estimating the pth quantile for $p = 1\%$. Panel (a) shows functions $h \mapsto p/\hat{p}$ and panel (b) shows functions $h \mapsto \hat{R}$, where $\hat{p}$ is the implied estimate of the probability p as defined in (8.6), and $\hat{R}$ is the estimated expected loss as defined in (8.18). The value of $\hat{R}$ depends on p and p_0 through the quantile estimates. The values of p_0 are $p_0 = 5\%$ (black with "1"), $p_0 = 10\%$ (blue with "2"), $p_0 = 15\%$ (violet with "3"), and $p_0 = 20\%$ (dark green with "4"). The red horizontal lines show the performance of the empirical quantile.

Figure 8.41 considers estimation of the pth quantile for $p = 0.1\%$, but otherwise the setting is similar to Figure 8.40.

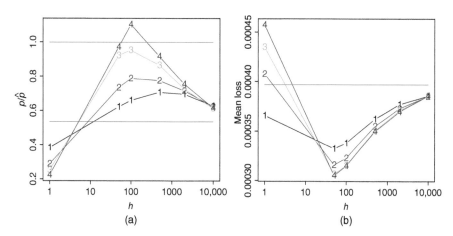

Figure 8.40 *Exponential model:* $p = 1\%$. (a) Function $h \mapsto p/\hat{p}$, where $\hat{p}$ is the estimate of the exceedance probability. (b) Function $h \mapsto \hat{R}$, where $\hat{R}$ is the estimated loss. The values of p_0 are $p_0 = 5\%$ (black with "1"), $p_0 = 10\%$ (blue with "2"), $p_0 = 15\%$ (violet with "3"), and $p_0 = 20\%$ (dark green with "4").

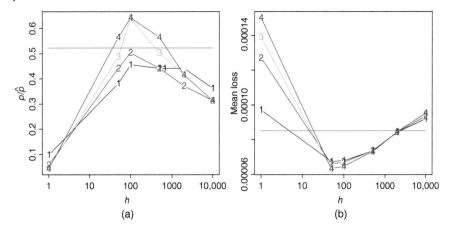

Figure 8.41 *Exponential model: $p = 0.1\%$.* (a) Function $h \mapsto p/\hat{p}$, where $\hat{p}$ is the estimate of the exceedance probability. (b) Function $h \mapsto \hat{R}$, where $\hat{R}$ is the estimated loss. The values of p_0 are $p_0 = 5\%$ (black with "1"), $p_0 = 10\%$ (blue with "2"), $p_0 = 15\%$ (violet with "3"), and $p_0 = 20\%$ (dark green with "4").

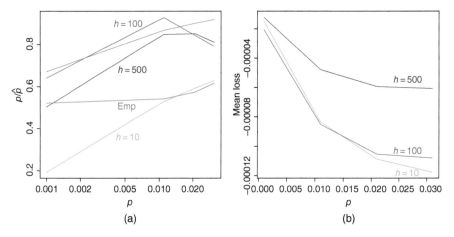

Figure 8.42 *Exponential model.* (a) Function $p \mapsto p/\hat{p}$, where $\hat{p}$ is the estimate of the exceedance probability. (b) Function $p \mapsto \hat{R}(k) - \hat{R}(1)$, where $\hat{R}(k)$ is the estimated loss when the multiplier of the quantile estimator is $k = p_0/p$, and $k = 1$ corresponds to the empirical quantile. The smoothing parameter is $h = 10$ (orange), $h = 100$ (purple), and $h = 500$ (dark green). In panel (a) the red curve corresponds to the empirical quantile.

Figure 8.42 studies the performance of the exponential quantile estimator as a function of p, for several values of smoothing parameter h. We take $p_0 = 15\%$. We consider values $h = 10$ (orange), $h = 100$ (purple), and $h = 500$ (dark green). Furthermore, we show the empirical quantiles (red). Panel (a) shows functions $p \mapsto p/\hat{p}$, where $\hat{p}$ is the implied estimate of the probability p, as defined in (8.6). The green lines show the fluctuation bands for $\alpha = 5\%$. Panel (b) shows functions

$$p \mapsto \hat{R}(k) - \hat{R}(1),$$

where the estimated expected loss $\hat{R}$ is defined in (8.18), and $\hat{R}$ depends on $k = p_0/p$ through the quantile estimates. Note that the multiplier $k = 1$ implies that the quantile estimator is the empirical quantile, because then $Q_p(Y) = u$. Thus, negative values of $\hat{R}(k) - \hat{R}(1)$ imply that the quantile estimator with multiplier k is better than the empirical quantile.

Gamma Excess Distribution The gamma densities are defined in (3.80). Parameter $\kappa > 0$ is the shape parameter, and $\beta > 0$ is the scale parameter. The logarithmic likelihood is written in (3.81). The time varying estimators for the parameters are maximizers of

$$l_{loc}(\kappa, \beta) = -\kappa \log \beta - \log \Gamma(\kappa) + (\kappa - 1) \sum_{\{Y_i \in \mathcal{L}_t\}} p_i(t) \log Y_i$$

$$- \frac{1}{\beta} \sum_{\{Y_i \in \mathcal{L}_t\}} p_i(t) Y_i,$$

where $\mathcal{L}_t$ is the set of observations in the left tail, as defined in (8.55), and $p_i(t)$ is the time space localized weight, as in (8.56). When $\kappa > 0$ is given, the time varying estimator of parameter β is

$$\beta_t(\kappa) = \frac{1}{\kappa} \sum_{\{Y_i \in \mathcal{L}_t\}} p_i(t) Y_i.$$

The localized maximum likelihood estimators are

$$\hat{\kappa}_t = \text{argmax}_{\kappa > 0} \, l_{loc}(\kappa, \beta_t(\kappa)), \quad \hat{\beta}_t = \beta_t(\hat{\kappa}_t).$$

We can write

$$\hat{\kappa}_t = \text{argmax}_{\kappa > 0} \left[-\kappa(1 + \log \beta) - \log \Gamma(\kappa) + (\kappa - 1) \sum_{\{Y_i \in \mathcal{L}_t\}} p_i(t) \log Y_i \right].$$

Figures 8.43–8.45 show that the gamma quantiles are better than the empirical quantiles for the estimation of the pth quantile with $p = 1\%$. When $p = 0.1\%$, then the gamma quantiles hardly beat the empirical quantiles. The smoothing parameter $h = 100$ leads to the best results.

Figure 8.43 studies the performance as a function of smoothing parameter h in estimating the pth quantile for $p = 1\%$. Panel (a) shows functions $h \mapsto p/\hat{p}$ and panel (b) shows functions $h \mapsto \hat{R}$, where $\hat{p}$ is the implied estimate of the probability p and $\hat{R}$ is the estimated expected loss. The setting is the same as in Figure 8.40.

Figure 8.44 studies the estimation of the pth quantile for $p = 0.1\%$. Otherwise the setting is the same as in Figure 8.43.

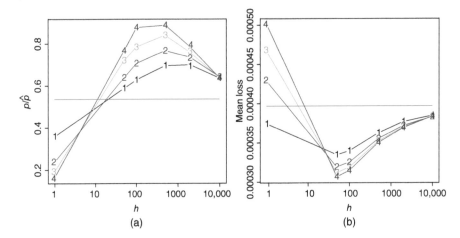

Figure 8.43 *Gamma model: p = 1%.* (a) Function $h \mapsto p/\hat{p}$, where $\hat{p}$ is the estimate of the exceedance probability; (b) function $h \mapsto \hat{R}$, where $\hat{R}$ is the estimated loss. The values of p_0 are $p_0 = 5\%$ (black with "1"), $p_0 = 10\%$ (blue with "2"), $p_0 = 15\%$ (violet with "3"), and 20% (dark green with "4"). The red horizontal lines show the performance of empirical quantiles.

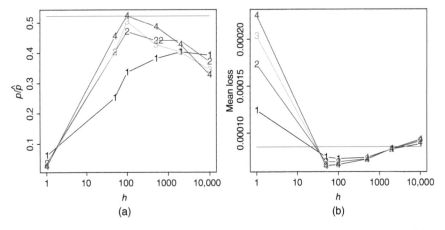

Figure 8.44 *Gamma model: p = 0.1%.* (a) Function $h \mapsto p/\hat{p}$, where $\hat{p}$ is the estimate of the exceedance probability. (b) Function $h \mapsto \hat{R}$, where $\hat{R}$ is the estimated loss. The values of p_0 are 5% (black with "1"), 10% (blue with "2"), 15% (violet with "3"), and 20% (dark green with "4"). The red horizontal lines show the performance of empirical quantiles.

Figure 8.45 studies the performance of the exponential quantile estimator as a function of p, for several values of smoothing parameter h. We take $p_0 = 15\%$. We consider values $h = 10$ (orange), $h = 100$ (purple), and $h = 500$ (dark green). Furthermore, we show the empirical quantiles (red). The setting is the same as in Figure 8.42.

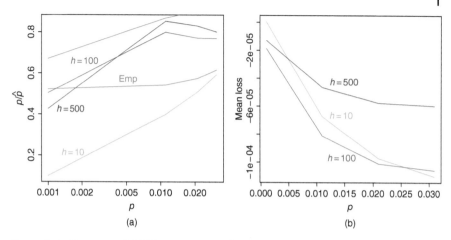

Figure 8.45 *Gamma model.* (a) Function $p \mapsto p/\hat{p}$, where $\hat{p}$ is the estimate of the exceedance probability. (b) Function $p \mapsto \hat{R}(k) - \hat{R}(1)$, where $\hat{R}(k)$ is the estimated loss when the multiplier of the quantile estimator is $k = p_0/p$, and $k = 1$ corresponds to the empirical quantile. The smoothing parameter is $h = 10$ (orange), $h = 100$ (purple), and $h = 500$ (dark green). In panel (a) the red curve corresponds to the empirical quantile.

Generalized Pareto Excess Distribution The generalized Pareto densities are defined in (3.82). Parameter $\xi > 0$ is the shape parameter, and $\beta > 0$ is the scale parameter. The generalized Pareto densities are written in (3.84) using the shape parameter $\alpha = 1/\xi$ and the scale parameter $\sigma = \beta/\xi$. The logarithmic likelihood is written in (3.85). The time varying estimators for the parameters are maximizers of

$$l_{loc}(\alpha, \sigma) = \log\left(\frac{\alpha}{\sigma}\right) - (1 + \alpha) \sum_{\partial Y_i \in \mathcal{L}_t} p_i(t) \log\left(1 + \frac{Y_i}{\sigma}\right),$$

where $\mathcal{L}_t$ is the set of observations in the left tail, as defined in (8.55), and $p_i(t)$ is the time space localized weight, as in (8.56). When $\sigma > 0$ is given, the time varying estimator of parameter α is

$$\alpha_t(\sigma) = \left[\sum_{\partial Y_i \in \mathcal{L}_t} p_i(t) \log\left(1 + \frac{Y_i}{\sigma}\right)\right]^{-1}.$$

The localized maximum likelihood estimators are

$$\hat{\sigma}_t = \operatorname{argmax}_{\sigma > 0} l_{loc}(\alpha_t(\sigma), \sigma), \quad \hat{\alpha}_t = \alpha_t(\hat{\sigma}_t).$$

We can write

$$\hat{\sigma}_t = \operatorname{argmax}_{\sigma > 0} \left[\log\left(\frac{\alpha_t(\sigma)}{\sigma}\right) - \left(1 + \frac{1}{\alpha_t(\sigma)}\right)\right].$$

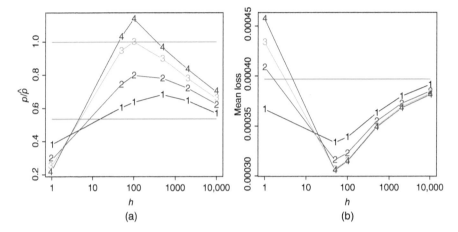

Figure 8.46 *Generalized Pareto model:* $p = 1\%$. (a) Function $h \mapsto p/\hat{p}$, where $\hat{p}$ is the estimate of the exceedance probability. (b) Function $h \mapsto \hat{R}$, where $\hat{R}$ is the estimated loss. The values of p_0 are $p_0 = 5\%$ (black with "1"), $p_0 = 10\%$ (blue with "2"), $p_0 = 15\%$ (violet with "3"), and 20% (dark green with "4"). The red horizontal lines show the performance of empirical quantiles.

The localized maximum likelihood estimators for ξ and β are

$$\hat{\xi}_t = 1/\hat{\alpha}_t, \quad \hat{\beta}_t = \hat{\xi}_t \hat{\sigma}_t.$$

Figures 8.46–8.48 show that the generalized Pareto quantiles are better than the empirical quantiles. The smoothing parameter $h = 100$ leads to the best results.

Figure 8.46 studies the performance as a function of smoothing parameter h in estimating the pth quantile for $p = 1\%$. Panel (a) shows functions $h \mapsto p/\hat{p}$ and panel (b) shows functions $h \mapsto \hat{R}$, where $\hat{p}$ is the implied estimate of the probability p and $\hat{R}$ is the estimated expected loss. The setting is the same as in Figure 8.40.

Figure 8.47 studies the estimation of the pth quantile for $p = 0.1\%$. Otherwise the setting is the same as in Figure 8.46.

Figure 8.48 studies the performance of the exponential quantile estimator as a function of p, for several values of smoothing parameter h. We take $p_0 = 15\%$. We consider values $h = 10$ (orange), $h = 100$ (purple), and $h = 500$ (dark green). Furthermore, we show the empirical quantiles (red). The setting is the same as in Figure 8.42.

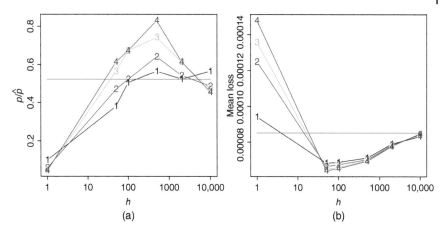

Figure 8.47 *Generalized Pareto model*: $p = 0.1\%$. (a) Function $h \mapsto p/\hat{p}$, where $\hat{p}$ is the estimate of the exceedance probability. (b) Function $h \mapsto \hat{R}$, where $\hat{R}$ is the estimated loss. The values of p_0 are 5% (black with "1"), 10% (blue with "2"), 15% (violet with "3"), and 20% (dark green with "4"). The red horizontal lines show the performance of empirical quantiles.

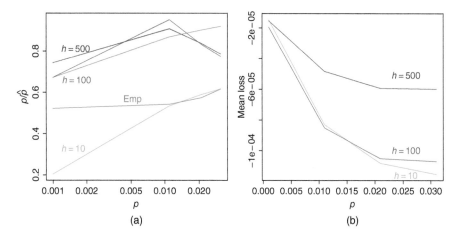

Figure 8.48 *Generalized Pareto model*. (a) Function $p \mapsto p/\hat{p}$, where $\hat{p}$ is the estimate of the exceedance probability. (b) Function $p \mapsto \hat{R}(k) - \hat{R}(1)$, where $\hat{R}(k)$ is the estimated loss when the multiplier of the quantile estimator is $k = p_0/p$, and $k = 1$ corresponds to the empirical quantile. The smoothing parameter is $h = 10$ (orange), $h = 100$ (purple), and $h = 500$ (dark green). In panel (a) the red curve corresponds to the empirical quantile.

Weibull Excess Distribution The Weibull densities are defined in (3.86). Parameter $\kappa > 0$ is the shape parameter and $\beta > 0$ is the scale parameter. The logarithmic likelihood is written in (3.87). The time varying estimators for the parameters are maximizers of

$$l_{loc}(\kappa, \beta) = \log(\kappa/\beta) + (\kappa - 1) \sum_{\{Y_i \in \mathcal{L}_t\}} p_i(t) \log(Y_i/\beta)$$
$$- \sum_{\{Y_i \in \mathcal{L}_t\}} p_i(t)(Y_i/\beta)^\kappa,$$

where $\mathcal{L}_t$ is the set of observations in the left tail, as defined in (8.55), and $p_i(t)$ is the time space localized weight, as in (8.56). When $\kappa > 0$ is given, the time varying estimator of parameter β is

$$\beta_t(\kappa) = \left(\sum_{\{Y_i \in \mathcal{L}_t\}} p_i(t) Y_i^\kappa \right)^{1/\kappa}.$$

The localized maximum likelihood estimators are

$$\hat{\kappa}_t = \mathrm{argmax}_{\kappa > 0}\, l_{loc}(\kappa, \beta_t(\kappa)), \quad \hat{\beta}_t = \beta_t(\hat{\kappa}_t).$$

Figures 8.49–8.51 show that the Weibull quantiles are better than the empirical quantiles for the estimation of the pth quantile with $p = 1\%$, but when $p = 0.1\%$, then the empirical quantile is better. The smoothing parameter $h = 100$ leads to the best results.

Figure 8.49 studies the performance as a function of smoothing parameter h in estimating the pth quantile for $p = 1\%$. Panel (a) shows functions $h \mapsto p/\hat{p}$ and panel (b) shows functions $h \mapsto \hat{R}$, where $\hat{p}$ is the implied estimate of the probability p and $\hat{R}$ is the estimated expected loss. The setting is the same as in Figure 8.40.

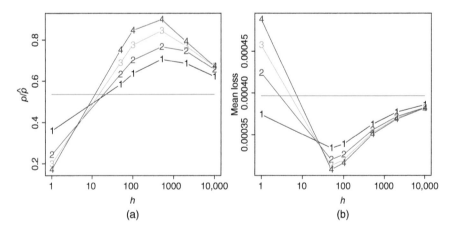

Figure 8.49 *Weibull model:* $p = 1\%$. (a) Function $h \mapsto p/\hat{p}$, where $\hat{p}$ is the estimate of the exceedance probability. (b) Function $h \mapsto \hat{R}$, where $\hat{R}$ is the estimated loss. The values of p_0 are $p_0 = 5\%$ (black with "1"), $p_0 = 10\%$ (blue with "2"), $p_0 = 15\%$ (violet with "3"), and 20% (dark green with "4"). The red horizontal lines show the performance of empirical quantiles.

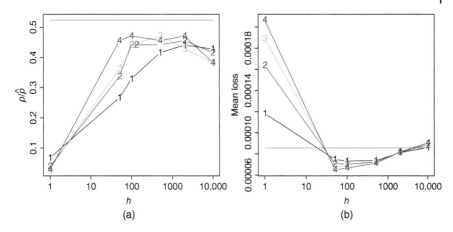

Figure 8.50 *Weibull model:* $p = 0.1\%$. (a) Function $h \mapsto p/\hat{p}$, where $\hat{p}$ is the estimate of the exceedance probability. (b) Function $h \mapsto \hat{R}$, where $\hat{R}$ is the estimated loss. The values of p_0 are 5% (black with "1"), 10% (blue with "2"), 15% (violet with "3"), and 20% (dark green with "4"). The red horizontal lines show the performance of empirical quantiles.

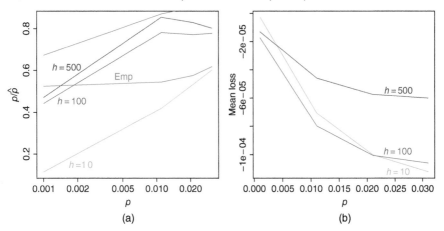

Figure 8.51 *Weibull model.* (a) Function $p \mapsto p/\hat{p}$, where $\hat{p}$ is the estimate of the exceedance probability. (b) Function $p \mapsto \hat{R}(k) - \hat{R}(1)$, where $\hat{R}(k)$ is the estimated loss when the multiplier of the quantile estimator is $k = p_0/p$, and $k = 1$ corresponds to the empirical quantile. The smoothing parameter is $h = 10$ (orange), $h = 100$ (purple), and $h = 500$ (dark green). In panel (a) the red curve corresponds to the empirical quantile.

Figure 8.50 studies the estimation of the pth quantile for $p = 0.1\%$. Otherwise the setting is the same as in Figure 8.49.

Figure 8.51 studies the performance of the Weibull quantile estimator as a function of p, for several values of smoothing parameter h. We take $p_0 = 15\%$. We consider values $h = 10$ (orange), $h = 100$ (purple), and $h = 500$ (dark green). Furthermore, we show the empirical quantiles (red). The setting is the same as in Figure 8.42.

8.6.3.2 The Empirical Residuals

We apply the idea of using the empirical quantiles of the residuals, as in (8.39). Let $F : \mathbf{R} \to \mathbf{R}$ be the continuous distribution function of the returns, and $F_u : [0, \infty) \to \mathbf{R}$ the lower excess distribution, as defined in (8.46) for $u \in \mathbf{R}$. At time t we observe returns $Y_1, \dots, Y_t$. The left tail is

$$\mathcal{L}_t = \{Y_i : Y_i \le u_t, \quad i = 1, \dots, t\},$$

where $u_t = \hat{q}_{p_0}$ is the empirical quantile from $Y_1, \dots, Y_t$, and $0 < p < p_0 < 0.5$. Let $n_t = \#\mathcal{L}_t$ be the number of observations in $\mathcal{L}_t$. Let us apply notation

$$\mathcal{L}_t = \left\{ Y^t_{t_1,1} < \cdots < Y^t_{t_{n_t}, n_t} \right\},$$

where $Y^t_{t_i, i}$ is observed at time $t_i \in \{1, \dots, t\}$. The residuals are defined as

$$U_i = F_{u_t}\left(u_t - Y^t_{t_i, i}, \hat{\theta}_{t_i}\right), \quad i = 1, \dots, n_t.$$

The residuals are approximately uniformly distributed, if the distribution $F_u(\cdot, \hat{\theta}_t)$ provides a good approximation of the true excess distribution $F_u(\cdot, \theta)$. Let us assume for notational convenience that the residuals are ordered:

$$U_1 < \cdots < U_{n_t}.$$

Let $\hat{e}_{1-p/p_0}$ be the empirical quantile with level $1 - p/p_0$. That is,

$$\hat{e}_{1-p/p_0} = U_{[n_t(1-p/p_0)]}.$$

The estimator of the conditional quantile is given by

$$\hat{q}_t = u_t - F_u^{-1}(\hat{e}_{1-p/p_0}, \hat{\theta}_t).$$

Note that when the estimator θ_t would not depend on t, then the quantile estimator would be the empirical quantile. In our definition the ith residual is defined using the ith estimate $\hat{\theta}_i$, but the quantile estimator $\hat{q}_t$ is defined using the current parameter estimate $\hat{\theta}_{tn_t}$.

It turns out that with the empirical residuals the results are robust with respect to the choice of the excess distribution. On the other hand, we are not able to improve the results which were obtained using the time varying generalized Pareto distribution.

The empirical quantiles of the residuals are better than the empirical quantiles. The smoothing parameter $h = 100$ leads to the best results.

Exponential Excess Function Figure 8.52 studies the performance as a function of smoothing parameter h in estimating the pth quantile for $p = 1\%$. Panel (a) shows functions $h \mapsto p/\hat{p}$ and panel (b) shows functions $h \mapsto \hat{R}$, where $\hat{p}$ is the implied estimate of the probability p as defined in (8.6), and $\hat{R}$ is the estimated expected loss as defined in (8.18). The value of $\hat{R}$ depends on p and p_0 through

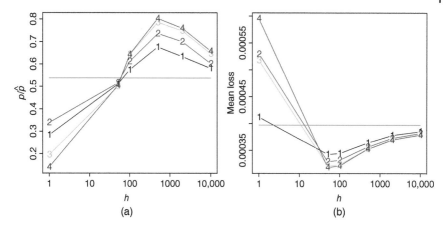

Figure 8.52 *Exponential model:* $p = 1\%$. (a) Function $h \mapsto p/\hat{p}$, where $\hat{p}$ is the estimate of the exceedance probability. (b) Function $h \mapsto \hat{R}$, where $\hat{R}$ is the estimated loss. The values of p_0 are $p_0 = 5\%$ (black with "1"), $p_0 = 10\%$ (blue with "2"), $p_0 = 15\%$ (violet with "3"), and 20% (dark green with "4"). The red horizontal lines show the performance of empirical quantiles.

the quantile estimates. The values of p_0 are $p_0 = 5\%$ (black with "1"), $p_0 = 10\%$ (blue with "2"), $p_0 = 15\%$ (violet with "3"), and $p_0 = 20\%$ (dark green with "4"). The red horizontal lines show the performance of the empirical quantile.

Figure 8.53 considers estimation of the pth quantile for $p = 0.1\%$, but otherwise the setting is similar to Figure 8.52.

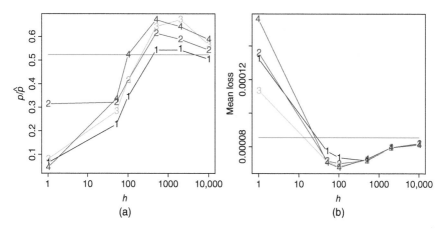

Figure 8.53 *Exponential model:* $p = 0.1\%$. (a) Function $h \mapsto p/\hat{p}$, where $\hat{p}$ is the estimate of the exceedance probability. (b) Function $h \mapsto \hat{R}$, where $\hat{R}$ is the estimated loss. The values of p_0 are $p_0 = 5\%$ (black with "1"), $p_0 = 10\%$ (blue with "2"), $p_0 = 15\%$ (violet with "3"), and 20% (dark green with "4"). The red horizontal lines show the performance of empirical quantiles.

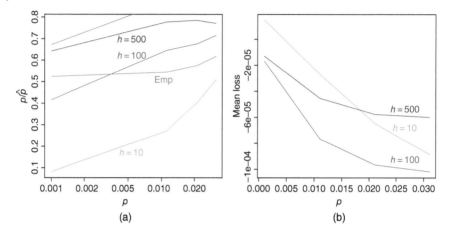

Figure 8.54 *Exponential model.* (a) Function $p \mapsto p/\hat{p}$, where $\hat{p}$ is the estimate of the exceedance probability. (b) Function $p \mapsto \hat{R}(k) - \hat{R}(1)$, where $\hat{R}(k)$ is the estimated loss when the multiplier of the quantile estimator is $k = p_0/p$, and $k = 1$ corresponds to the empirical quantile. The smoothing parameter is $h = 10$ (orange), $h = 100$ (purple), and $h = 500$ (dark green). In panel (a) the red curve corresponds to the empirical quantile.

Figure 8.54 studies the performance of the exponential quantile estimator as a function of p, for several values of smoothing parameter h. We take $p_0 = 15\%$. We consider values $h = 10$ (orange), $h = 100$ (purple), and $h = 500$ (dark green). Panel (a) shows functions $p \mapsto p/\hat{p}$, where $\hat{p}$ is the implied estimate of the probability p, as defined in (8.6). We also show the performance of empirical quantiles (red). The green lines show the fluctuation bands for $\alpha = 5\%$. Panel (b) shows functions

$$p \mapsto \hat{R}(k) - \hat{R}(1),$$

where the estimated expected loss $\hat{R}$ is defined in (8.18), and $\hat{R}$ depends on $k = p_0/p$ through the quantile estimates. Note that the multiplier $k = 1$ implies that the quantile estimator is the empirical quantile. Thus, negative values of $\hat{R}(k) - \hat{R}(1)$ imply that the quantile estimator with multiplier k is better than the empirical quantile.

Gamma Excess Function Figure 8.55 studies the performance as a function of smoothing parameter h in estimating the pth quantile for $p = 1\%$. Panel (a) shows functions $h \mapsto p/\hat{p}$ and panel (b) shows functions $h \mapsto \hat{R}$, where $\hat{p}$ is the implied estimate of the probability p and $\hat{R}$ is the estimated expected loss. The setting is the same as in Figure 8.52.

Figure 8.56 considers estimation of the pth quantile for $p = 0.1\%$, but otherwise the setting is similar to Figure 8.55.

Figure 8.57 studies the performance of the Weibull quantile estimator as a function of p, for several values of smoothing parameter h. We take $p_0 = 15\%$.

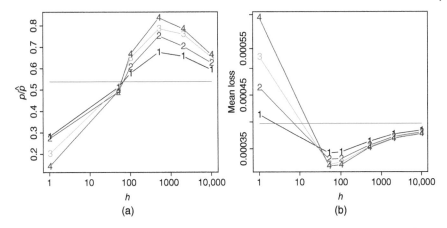

Figure 8.55 *Gamma model*: $p = 1\%$. (a) Function $h \mapsto p/\hat{p}$, where $\hat{p}$ is the estimate of the exceedance probability. (b) Function $h \mapsto \hat{R}$, where $\hat{R}$ is the estimated loss. The values of p_0 are $p_0 = 5\%$ (black with "1"), $p_0 = 10\%$ (blue with "2"), $p_0 = 15\%$ (violet with "3"), and 20% (dark green with "4"). The red horizontal lines show the performance of empirical quantiles.

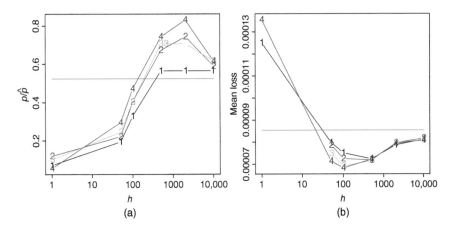

Figure 8.56 *Gamma model*: $p = 0.1\%$. (a) Function $h \mapsto p/\hat{p}$, where $\hat{p}$ is the estimate of the exceedance probability. (b) Function $h \mapsto \hat{R}$, where $\hat{R}$ is the estimated loss. The values of p_0 are $p_0 = 5\%$ (black with "1"), $p_0 = 10\%$ (blue with "2"), $p_0 = 15\%$ (violet with "3"), and 20% (dark green with "4"). The red horizontal lines show the performance of empirical quantiles.

We consider values $h = 10$ (orange), $h = 100$ (purple), and $h = 500$ (dark green). Furthermore, we show the empirical quantiles (red). The setting is the same as in Figure 8.54

Generalized Pareto Excess Function Figure 8.58 studies the performance as a function of smoothing parameter h in estimating the pth quantile for $p = 1\%$.

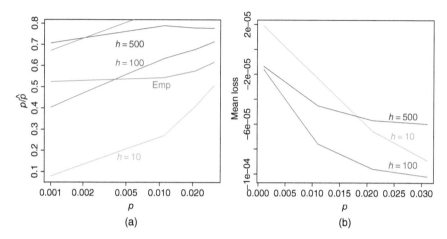

Figure 8.57 *Gamma model.* (a) Function $p \mapsto p/\hat{p}$, where $\hat{p}$ is the estimate of the exceedance probability. (b) Function $p \mapsto \hat{R}(k) - \hat{R}(1)$, where $\hat{R}(k)$ is the estimated loss when the multiplier of the quantile estimator is $k = p_0/p$, and $k = 1$ corresponds to the empirical quantile. The smoothing parameter is $h = 10$ (orange), $h = 100$ (purple), and $h = 500$ (dark green). In panel (a) the red curve corresponds to the empirical quantile.

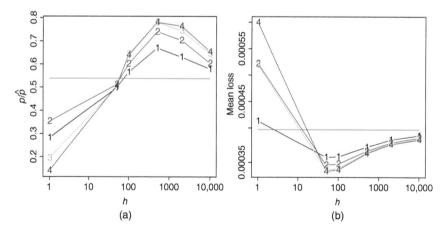

Figure 8.58 *Generalized Pareto model:* $p = 1\%$. (a) Function $h \mapsto p/\hat{p}$, where $\hat{p}$ is the estimate of the exceedance probability. (b) Function $h \mapsto \hat{R}$, where $\hat{R}$ is the estimated loss. The values of p_0 are $p_0 = 5\%$ (black with "1"), $p_0 = 10\%$ (blue with "2"), $p_0 = 15\%$ (violet with "3"), and 20% (dark green with "4"). The red horizontal lines show the performance of empirical quantiles.

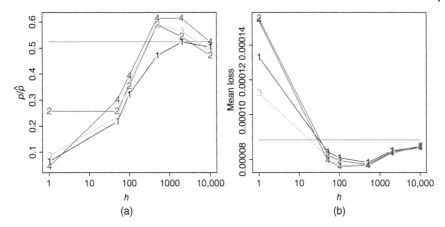

Figure 8.59 *Generalized Pareto model: $p = 0.1\%$. Gamma model: $p = 0.1\%$. (a)* Function $h \mapsto p/\hat{p}$, where $\hat{p}$ is the estimate of the exceedance probability. (b) Function $h \mapsto \hat{R}$, where $\hat{R}$ is the estimated loss. The values of p_0 are $p_0 = 5\%$ (black with "1"), $p_0 = 10\%$ (blue with "2"), $p_0 = 15\%$ (violet with "3"), and 20% (dark green with "4"). The red horizontal lines show the performance of empirical quantiles.

Panel (a) shows functions $h \mapsto p/\hat{p}$ and panel (b) shows functions $h \mapsto \hat{R}$, where $\hat{p}$ is the implied estimate of the probability p and $\hat{R}$ is the estimated expected loss. The setting is the same as in Figure 8.52.

Figure 8.59 considers estimation of the pth quantile for $p = 0.1\%$, but otherwise the setting is similar to Figure 8.58.

Figure 8.60 studies the performance of the generalized Pareto quantile estimator as a function of p, for several values of smoothing parameter h. We take $p_0 = 15\%$. We consider values $h = 10$ (orange), $h = 100$ (purple), and $h = 500$ (dark green). Furthermore, we show the empirical quantiles (red). The setting is the same as in Figure 8.54.

Weibull Excess Function Figure 8.61 studies the performance as a function of smoothing parameter h in estimating the pth quantile for $p = 1\%$. Panel (a) shows functions $h \mapsto p/\hat{p}$ and panel (b) shows functions $h \mapsto \hat{R}$, where $\hat{p}$ is the implied estimate of the probability p and $\hat{R}$ is the estimated expected loss. The setting is the same as in Figure 8.52.

Figure 8.62 considers estimation of the pth quantile for $p = 0.1\%$, but otherwise the setting is similar to Figure 8.61.

Figure 8.63 studies the performance of the Weibull quantile estimator as a function of p, for several values of smoothing parameter h. We take $p_0 = 15\%$. We consider values $h = 10$ (violet), $h = 100$ (purple), and $h = 500$ (dark green). Furthermore, we show the empirical quantiles (red). The setting is the same as in Figure 8.54.

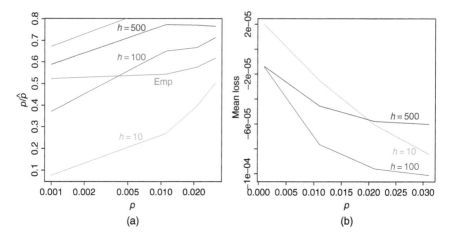

Figure 8.60 *Generalized Pareto model.* (a) Function $p \mapsto p/\hat{p}$, where $\hat{p}$ is the estimate of the exceedance probability. (b) Function $p \mapsto \hat{R}(k) - \hat{R}(1)$, where $\hat{R}(k)$ is the estimated loss when the multiplier of the quantile estimator is $k = p_0/p$, and $k = 1$ corresponds to the empirical quantile. The smoothing parameter is $h = 10$ (orange), $h = 100$ (purple), and $h = 500$ (dark green). In panel (a) the red curve corresponds to the empirical quantile.

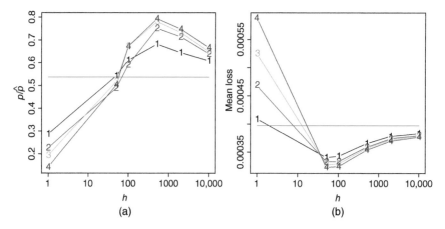

Figure 8.61 *Weibull model:* $p = 1\%$. (a) Function $h \mapsto p/\hat{p}$, where $\hat{p}$ is the estimate of the exceedance probability. (b) Function $h \mapsto \hat{R}$, where $\hat{R}$ is the estimated loss. The values of p_0 are $p_0 = 5\%$ (black with "1"), $p_0 = 10\%$ (blue with "2"), $p_0 = 15\%$ (violet with "3"), and 20% (dark green with "4"). The red horizontal lines show the performance of empirical quantiles.

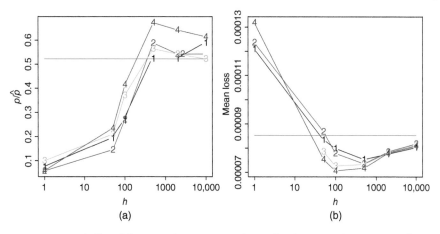

Figure 8.62 *Weibull model*: $p = 0.1\%$. (a) Function $h \mapsto p/\hat{p}$, where $\hat{p}$ is the estimate of the exceedance probability. (b) Function $h \mapsto \hat{R}$, where $\hat{R}$ is the estimated loss. The values of p_0 are $p_0 = 5\%$ (black with "1"), $p_0 = 10\%$ (blue with "2"), $p_0 = 15\%$ (violet with "3"), and 20% (dark green with "4"). The red horizontal lines show the performance of empirical quantiles.

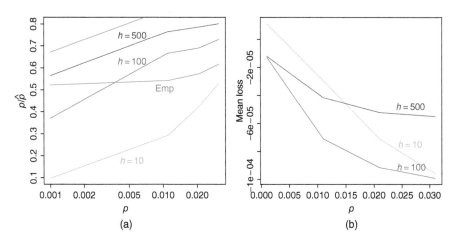

Figure 8.63 *Weibull model*. (a) Function $p \mapsto p/\hat{p}$, where $\hat{p}$ is the estimate of the exceedance probability. (b) Function $p \mapsto \hat{R}(k) - \hat{R}(1)$, where $\hat{R}(k)$ is the estimated loss when the multiplier of the quantile estimator is $k = p_0/p$, and $k = 1$ corresponds to the empirical quantile. The smoothing parameter is $h = 10$ (violet), $h = 100$ (purple), and $h = 500$ (dark green). In panel (a) the red curve corresponds to the empirical quantile.

8.7 Extreme Value Theory in Quantile Estimation

First, we describe the block maxima method and then the method of threshold exceedances. We consider now the estimation of the quantiles in the right tail, so that the level of the estimated quantile is $0.5 < p < 1$.

8.7.1 The Block Maxima Method

Let the real valued random variables $X_1, \ldots, X_n \in \mathbf{R}$ be independent and identically distributed. We use now the notation $X_1, \ldots, X_n$ instead of $Y_1, \ldots, Y_T$, because later the observations $Y_1, \ldots, Y_T$ will be divided into blocks of size n. Denote the maximum

$$M_n = \max\{X_1, \ldots, X_n\}.$$

We make the assumption that weak convergence holds for the maximum of the observations. This assumption was discussed in Section 3.5.2; see (3.97). We assume that

$$P\left(\frac{M_n - d_n}{c_n} \leq x\right) \xrightarrow{d} H_\xi(x), \tag{8.57}$$

as $n \to \infty$, for all x, where $d_n \in \mathbf{R}$, $c_n > 0$, and H_ξ is the distribution function of a generalized extreme value distribution with parameter $\xi \in \mathbf{R}$, defined in (3.100).

8.7.1.1 An Expression for the Quantiles

The convergence in (8.57) suggests that we have for large n the approximation

$$P\left(\frac{M_n - d_n}{c_n} \leq x\right) = H_\xi(x),$$

for all x. Then,

$$P(M_n \leq x) = H_\xi\left(\frac{x - d_n}{c_n}\right). \tag{8.58}$$

Since $X_1, \ldots, X_n$ is an i.i.d. sample from the distribution of X,

$$\begin{aligned} P(M_n \leq x) &= P(X_1 \leq x, \ldots, X_n \leq x) \\ &= P(X_1 \leq x) \cdots P(X_n \leq x) \\ &= [P(X \leq x)]^n. \end{aligned} \tag{8.59}$$

Let $q_p = Q_p(X)$ be the pth quantile. Then,

$$P(X \leq q_p) = p, \tag{8.60}$$

when X has a continuous distribution. Thus, combining (8.59) and (8.60),

$$P(M_n \leq q_p) = [P(X \leq q_p)]^n = p^n. \tag{8.61}$$

Thus, combining (8.58) and (8.61),

$$H_\xi\left(\frac{q_p - d_n}{c_n}\right) = p^n$$

and we get

$$Q_p(X) = d_n + c_n H_\xi^{-1}(p^n). \tag{8.62}$$

8.7.1.2 Estimation of the Parameters

The expression (8.62) for a quantile contains unknown parameters ξ, d_n, and c_n, which we have to estimate. Let us denote

$$H_{\xi,\mu,\sigma}(x) = H_\xi\left(\frac{x - \mu}{\sigma}\right).$$

We consider the family of distributions $H_{\xi,\mu,\sigma}$, where ξ is the shape parameter, μ is the location parameter, and σ is the scale parameter. The parameters can be estimated using the block maxima method. Let $Y_1, \dots, Y_T$ be i.i.d. observations. Since (8.58) holds, we could estimate the parameters if we would have several observations of the maxima. This can be achieved when we divide the observations into m blocks of size n, assuming for simplicity that $T = nm$. Denote by M_{ni}, $i = 1, \dots, m$, the maximum of the ith block:

$$M_{ni} = \max\{Y_{(i-1)n+1}, \dots, Y_{in}\}, \quad i = 1, \dots, m.$$

The maxima $M_{n1}, \dots, M_{nm}$ are independent[15] and we define the likelihood function

$$L(\xi, \mu, \sigma; M_{n1}, \dots, M_{nm}) = \prod_{i=1}^{m} h_{\xi,\mu,\sigma}(M_{ni}),$$

where $h_{\xi,\mu,\sigma}$ is the density function corresponding to the distribution function $H_{\xi,\mu,\sigma}$. We define the estimators $\hat{\xi}$, $\hat{\mu}$, and $\hat{\sigma}$ to be maximizers of the likelihood function. From (8.62) we get the estimator for a quantile

$$\hat{Q}_p(Y) = \hat{\mu} + \hat{\sigma} H_{\hat{\xi}}^{-1}(p^n). \tag{8.63}$$

Note that in (8.63) the sample size is T and n is the block size.

8.7.2 Threshold Exceedances

Section 8.6 is devoted to the quantile estimation on the basis of excess distributions. We summarize this approach and point out the connection to the asymptotics of threshold exceedances.

The excess distribution F_u with threshold $u > 0$ of random variable Y is defined in (8.43). It is stated in (3.102) that the limit distribution of the excess

---

15 Even when the original observations are not independent, the block maxima are approximately independent, for large block sizes.

distribution is a generalized Pareto distribution:

$$\lim_{u \to x_F} \sup_{0 \le x < x_F - u} |F_u(x) - G_{\xi, \beta(u)}(x)| = 0 \tag{8.64}$$

for some positive function $\beta(u)$ if and only if F belongs to the maximum domain of attraction of H_ξ, where $\xi \in \mathbb{R}$. We denote $X_F = \sup\{x : F(x) < 1\}$ and $G_{\xi, \beta}$ is the distribution function of the generalized Pareto distribution. The density functions are

$$g_{\xi, \beta}(x) = \begin{cases} \frac{1}{\beta}(1 + \xi x/\beta)^{-1/\xi - 1}, & \xi \neq 0, \\ \frac{1}{\beta} \exp\{-x/\beta\}, & \xi = 0, \end{cases} \tag{8.65}$$

where $\beta > 0$, $x \geq 0$, when $\xi \geq 0$, and $0 \leq x \leq -\beta/\xi$, when $\xi < 0$.

8.7.2.1 An Expression for the Quantiles

The convergence in (8.64) suggests the approximation

$$F_u(x) = G_{\xi, \beta}(x). \tag{8.66}$$

The pth quantile of Y was expressed in (8.45) as

$$Q_p(Y) = u + F_u^{-1}\left(1 - \frac{1 - p}{1 - F(u)}\right). \tag{8.67}$$

The inverse of the generalized Pareto distribution function is

$$G_{\xi, \beta}^{-1}(p) = \begin{cases} \frac{\beta}{\xi}[(1 - p)^{-\xi} - 1], & \xi \neq 0, \\ -\beta \log(1 - p), & \xi = 0. \end{cases}$$

8.7.2.2 Choosing and Estimating the Parameters

The expression (8.67) for the pth quantile contains the unknown probability $P(Y > u) = 1 - F(u)$, and the unknown parameters of the excess distribution $F_u = G_{\xi, \beta}$. Let $Y_1, \ldots, Y_T$ be an i.i.d. sample from the distribution of Y.

Choosing Threshold Threshold u is the parameter which is chosen by the user. Threshold u has to be sufficiently large so that approximation in (8.66) holds. On the other hand, when u is large then estimators of the parameters have a large variance. We can look at choosing of threshold u in two ways.

1) Choose first u and then estimate $P(Y > u)$.
   For example, we can use the estimate.

$$P(Y > u) = \frac{N_u}{T}, \tag{8.68}$$

   where $N_u = \#\{Y_i > u\}$ is the number of observations larger than u.
2) Choose first such p_0 that $0.5 < p_0 < p < 1$, and then choose u as an estimate of the p_0th quantile. Estimation of the p_0th quantile is easier than estimating

the pth quantile. For example, let

$$u = \hat{F}^{-1}(p_0),$$

where $\hat{F}$ is the empirical distribution function, and we take the generalized inverse. Now u is such that $N_u/T = 1 - p_0$, and we end up to the same estimate as in (8.68).

There are several ways to choose threshold u (or the level p_0 of the preliminary quantile estimator).

1) We can study the performance measures of quantile estimation, and choose u so that the performance measures are optimized. This was done in Section 8.6, where the probability of exceedances and the loss function for quantile estimation were used to measure performance.

2) We can choose the threshold u by studying the stability of parameter estimates. The shape parameter ξ is the same for all thresholds $u > u_0$ when the excesses over threshold u_0 follow a generalized Pareto distribution. Let $\sigma = \beta/\xi$. The scale parameter σ depends on $u > u_0$. Furthermore, if the excesses over threshold u_0 follow a generalized Pareto distribution and $u > u_0$, then

$$\sigma_u = \sigma_{u_0} - \xi u; \tag{8.69}$$

see Coles (2004, p. 79). We can define

$$\sigma^* = \sigma_u - \xi u.$$

When $u_1, u_2 > u_0$, then $\sigma_{u_1} - \xi u_1 = \sigma_{u_0} = \sigma_{u_2} - \xi u_2$. Thus, σ^* does not depend on u. Then, estimates of ξ and σ^* should be constant above u_0, if u_0 is a sufficiently large threshold so that the excesses follow the generalized Pareto distribution.

3) We can choose the threshold by studying the linearity of the mean residual plot. For a generalized Pareto distribution $EY = \sigma/(1 - \xi)$, when $\xi < 1$. When $\xi \geq 1$, then the expectation does not exist. Thus, if the excesses over threshold u follow a generalized Pareto distribution, then

$$E(Y - u|Y > u) = \frac{\sigma_u}{1 - \xi},$$

when $\xi < 1$; see Coles (2004, p. 79). Furthermore, if the excesses over threshold u_0 follow a generalized Pareto distribution and $u > u_0$, then (8.69) holds and $u \mapsto E(Y - u|Y > u)$ is a linear function. The mean residual life plot is the plot of points

$$\left(u, \frac{1}{N_u} \sum_{i=1}^{N_u} (X_i - u) \right)$$

for $u < \max\{Y_1, \ldots, Y_T\}$, where $\{X_1, \ldots, X_{N_u}\} = \{Y_i > u\}$ are the observations that exceed u. The level should be such that the points in the plot over level u can be approximately fitted by a linear function.

Estimating the Parameters We estimate the parameters ξ and β of the generalized Pareto distribution with the maximum likelihood method. Let us denote

$$\{X_1, \dots, X_{N_u}\} = \{Y_i : Y_i > u\}.$$

Now $X_1, \dots, X_{N_u}$ is a sample from the distribution F_u. Since $F_u \approx G_{\xi,\beta}$, we define the likelihood function

$$L(\xi, \beta; X_1, \dots, X_{N_u}) = \prod_{i=1}^{N_u} g_{\beta,\xi}(X_i),$$

where $g_{\beta,\xi} = G'_{\beta,\xi}$ is the density function of the generalized Pareto distribution. Define $\hat{\xi}$ and $\hat{\beta}$ as the maximizers of the likelihood function.

Finally, the quantile estimator is

$$\hat{Q}_p(Y) = u + G^{-1}_{\hat{\xi},\hat{\beta}}\left(1 - \frac{1-p}{N_u/n}\right).$$

8.8 Expected Shortfall

The expected shortfall for the right tail is defined as

$$ES_p(Y) = E(Y|Y \geq Q_p(Y)),$$

where Y is a random variable with a continuous distribution, $Q_p(Y)$ is the pth quantile, and $p \in (0, 1)$. The expected shortfall for the left tail and the expected shortfall for noncontinous distributions are defined in Section 3.1.3. Section 8.2.3 discusses the use of expected shortfall as a risk measure.

8.8.1 Performance of Estimators of the Expected Shortfall

In measuring the performance of quantile estimators in Section 8.3.1, we used the fact the for continuous distributions

$$P(Y > Q_p(Y)) = 1 - p,$$

which implies that

$$\frac{1}{T}\sum_{i=1}^{T} I_{(Q_p(Y),\infty)}(Y_i) \approx 1 - p.$$

Similarly, in the case of measuring the performance of estimators of the expected shortfall we use the fact that for continuous distributions

$$ES_p(Y) = \frac{1}{1-p} E[Y\, I_{(Q_p(Y),\infty)}(Y)],$$

which implies that

$$\frac{1}{T}\sum_{i=1}^{T} Y_i I_{(Q_p(Y),\infty)}(Y_i) \approx (1-p)ES_p(Y).$$

In measuring the performance, we have to be careful not to use the same data for estimation and for measuring the performance. Let us observe time series $Y_1, \ldots, Y_T$. Let $\hat{q}_t$ the estimator of the quantile $Q_p(Y)$ (or the estimator of the conditional quantile) at time t, and let $\hat{e}_t$ be the estimator of the expected shortfall $ES_p(Y)$ (or the estimator of the conditional expected shortfall) at time t. The estimators at time t use data $Y_1, \ldots, Y_t$. To measure the performance, we look closeness to zero of the difference

$$\frac{1}{T - t_0} \sum_{t=t_0}^{T-1} [Y_{t+1} I_{(\hat{q}_t, \infty)}(Y_{t+1}) - (1 - p)\hat{e}_t],$$

where $1 < t_0 < T$ is the time point which starts the measuring period.

8.8.2 Estimation of the Expected Shortfall

Estimation is done using identically distributed random variables $Y_1, \ldots, Y_T$. The different types of quantile estimators of Sections 8.4–8.6 lead to the corresponding estimators of the expected shortfall.

8.8.2.1 Empirical Expected Shortfall
Empirical quantiles are discussed in Section 8.4. The empirical expected shortfall can be derived from formula

$$ES_p(Y) = \frac{1}{1 - p} E[Y \, I_{(Q_p(Y),\infty)}(Y)], \tag{8.70}$$

where it is assumed that the distribution of Y is continuous. In this formula the expectation can be estimated by the sample mean:

$$E[Y \, I_{(Q_p(Y),\infty)}(Y)] \approx \frac{1}{T} \sum_{i=1}^{T} [Y_i \, I_{(Q_p(Y),\infty)}(Y_i)],$$

and $(1 - p)T \approx T - m$, where $m = pT$.

Thus, the empirical expected shortfall for the right tail is

$$\widehat{ES}_p(Y) = \frac{1}{T - m + 1} \sum_{i=m}^{T} Y_{(i)},$$

where $Y_{(1)} \leq \cdots \leq Y_{(T)}$ and $m = \lceil pT \rceil$, with $0.5 < p < 1$. Note that $Y_{(m)}$ is the empirical quantile, as defined in (8.21)–(8.23).

8.8.2.2 Expected Shortfall in a Location–Scale Model
The volatility based quantile estimators are discussed in Section 8.5. These estimators are based on a location–scale model. Let us consider the location–scale model

$$Y = \mu + \sigma \epsilon,$$

where $\mu \in \mathbf{R}$, $\sigma > 0$, and ϵ is a random variable with a continuous distribution. Now,

$$ES_p(Y) = \mu + \sigma ES_p(\epsilon).$$

In fact,

$$ES_p(Y) = E(Y|Y \geq Q_p(Y))$$

$$= \mu + \sigma E\left(\left.\frac{Y-\mu}{\sigma}\right| \frac{Y-\mu}{\sigma} \geq Q_p(\epsilon)\right)$$

$$= \mu + \sigma E(\epsilon|\epsilon \geq Q_p(\epsilon)),$$

where we used the fact $Q_p(Y) = \mu + \sigma Q_p(\epsilon)$, noted in (8.30), so that

$$Y \geq Q_p(Y) \Leftrightarrow \frac{Y-\mu}{\sigma} \geq Q_p(\epsilon).$$

Thus, the estimate for the expected shortfall can be obtained as

$$\widehat{ES}_p(Y) = \hat{\mu} + \hat{\sigma} ES_p(\epsilon),$$

where $\hat{\mu}$ is an estimate of μ and $\hat{\sigma}$ is an estimate of σ.

For example, if $\epsilon \sim N(0, 1)$ then the expected shortfall for the right tail is

$$ES_p(\epsilon) = \frac{\phi(\Phi^{-1}(p))}{1-p},$$

where ϕ is the density function of the standard normal distribution and Φ is the distribution function of the standard normal distribution.[16] If $\epsilon \sim t_\nu$, where t_ν is the t-distribution with ν degrees of freedom, then

$$ES_p(\epsilon) = \frac{g_\nu\left(t_\nu^{-1}(p)\right)}{1-p} \frac{\nu + \left(t_\nu^{-1}(p)\right)^2}{\nu-1},$$

where g_ν is the density function of the t-distribution with ν degrees of freedom and t_ν is the distribution function of the t-distribution with ν degrees of freedom; see McNeil *et al.* (2005, p. 46).

8.8.2.3 Excess Distributions in Expected Shortfall Estimation

Section 8.6 discusses estimation of quantiles when the excess distribution is modeled parametrically. The distribution function of the (upper) excess distribution is given in (8.43). The density function of the excess distribution is given in (8.44) as

$$f_u(x) = \frac{f(x+u)}{1-F(u)} I_{[0,\infty)}(x),$$

---

16 We have that $ES_p(\epsilon) = (1-p)^{-1} \int_{q_p}^{\infty} x\phi(x)\,dx = (1-p)^{-1}\phi(q_p)$, where $q_p = Q_p(\epsilon)$ and we use the fact that $\phi'(x) = -x\phi(x)$.

where f and F are the density and distribution function of the original distribution.

Let Y be the random variable whose density and distribution functions are $f : \mathbf{R} \to \mathbf{R}$ and $F : \mathbf{R} \to \mathbf{R}$. Let X be the random variable distributed as the excess distribution, whose density and distribution functions are $f_u : [0, \infty) \to \mathbf{R}$ and $F_u : [0, \infty) \to \mathbf{R}$. Then,

$$\mathrm{ES}_p(Y) = u + \frac{1 - F(u)}{1 - p} \, \mathrm{ES}_{q-u}(X), \tag{8.71}$$

where $q = Q_p(Y) \geq u$. Note that when $u = Q_p(u)$, then

$$\mathrm{ES}_p(Y) = Q_p(Y) + EX, \tag{8.72}$$

because $\mathrm{ES}_0(X) = EX$.

Let us prove (8.71). We have for $r \geq 0$ that

$$E[X \, I_{(r,\infty)}(X)] = \int_r^\infty x f_u(x) \, dx$$

$$= \frac{1}{1 - F(u)} \int_r^\infty x f(x + u) \, dx$$

$$= \frac{1}{1 - F(u)} \left(\int_{r+u}^\infty y f(y) \, dy - u(1 - F(r + u)) \right).$$

Thus,

$$\int_{r+u}^\infty y f(y) \, dy = (1 - F(u)) \int_r^\infty x f_u(x) \, dx + u(1 - F(r + u)).$$

Choose $r = q - u$, multiply both sides of the equation by $1/(1 - p)$, apply $F(q) = p$, and apply (8.70) to obtain (8.71).

Let us explain the difference between (8.71) and (8.72) in the case when $Q_p(Y)$ is estimated using the estimation of the excess distribution F_u, where $u < Q_p(Y)$. In the case of (8.71), we can use the same fitted F_u to estimate the expected shortfall $\mathrm{ES}_{q-u}(X)$, where $X \sim F_u$. In the case of (8.72) we have to choose a second threshold $u' > u$, estimate $F_{u'}$, and estimate EX for $X \sim F_{u'}$. When u is large, it can happen that F_u and $F_{u'}$ are close. In fact, the limit theorem in (3.102) says that the excess distribution is a generalized Pareto distribution for large u.

There exist closed-form expressions for $\mathrm{ES}_{q-u}(X)$ in some cases. These expressions are convenient to give in terms of the mean excess function

$$e(v) = E(X - v | X > v).$$

We can write

$$\mathrm{ES}_{q-u}(X) = e(v) + v, \quad v = Q_{q-u}(X).$$

1) The exponential distribution is defined in (3.65). Let $X \sim \exp(\beta)$, where $\beta > 0$ is the scale parameter. Then

$$e(v) = \beta.$$

The quantile is $Q_p(X) = -\beta \log(1 - p)$.

2) The gamma distribution is defined in (3.80). Let X follow the gamma distribution with parameters κ and β, where $\kappa > 1$ is the shape parameter and $\beta > 0$ is the scale parameter. Then

$$e(v) = \beta \left(1 + \frac{\beta(\kappa - 1)}{v} + o\left(\frac{1}{v}\right) \right),$$

as $v \to \infty$. The quantile does not have a closed-form expression.

3) The generalized Pareto distribution is defined in (3.82). Let X follow the generalized Pareto distribution with parameters ξ and β, where $0 < \xi < 1$ is the shape parameter, and $\beta > 0$ is the scale parameter. Then

$$e(v) = \frac{\beta + \xi v}{1 - \xi}.$$

The quantile is $Q_p(X) = \frac{\beta}{\xi}[(1 - p)^{-\xi} - 1]$.

4) The Weibull distribution is defined in (3.86). Let $X \sim$ Weibull(κ, β), where where $\kappa > 0$ is the shape parameter and $\beta > 0$ is the scale parameter. Then

$$e(v) = \frac{v^{1-\kappa} \beta^\kappa}{\kappa}(1 + o(1)),$$

as $v \to \infty$. The quantile is $Q_p(X) = \beta(-\log(1 - p))^{1/\kappa}$.

The formulas for the mean excess function can be found in Embrechts *et al.* (1997, Table 3.4.7, p. 161).

Part III

Portfolio Management

9

Some Basic Concepts of Portfolio Theory

Portfolio theory studies two related problems: (1) how to construct a portfolio with desirable properties and (2) how to evaluate the performance of a portfolio. In this chapter, we concentrate on the concepts related to the construction of portfolios. A portfolio is constructed by allocating the available wealth among some basic assets. The return of a portfolio is a weighted average of the returns of the basic assets, the weights expressing the proportion of wealth allocated to each basic assets. There exist also portfolios that require zero initial wealth. Such portfolios are constructed using borrowing or option writing.

A main topic of the chapter is to introduce concepts related to the comparison of return and wealth distributions, and this topic is addressed in Section 9.2. In order to study portfolio construction we need to define what it means that a wealth distribution or a return distribution is better than another such distribution. (Here wealth distribution means the probability distribution of wealth, when wealth is considered as a random variable, and we do not mean the distribution of wealth in the sense of allocation of wealth among different people.) In portfolio selection we try to select the weights of basic assets so that the distribution of the return of the portfolio is in some sense optimal.

The optimal distribution of the return is such that the expected return is high but the risk of negative returns is small. The expected return of a portfolio is determined by the expected returns of the basic assets, but the risk of the return distribution depends on the joint distribution of the returns of the basic assets. The two main ways to compare returns is the use of the mean–variance criterion and the use of the expected utility.

The issue of multiperiod portfolio selection is an important and interesting research topic. However, we do not address this topic in any depth, but only in Section 9.3. The bypassing of multiperiod portfolio selection can be justified by the fact that for the logarithmic utility function there is no difference between the one period and multiperiod portfolio selection. Thus, when we ignore the effect of varying risk aversion and restrict ourselves to the logarithmic utility, then we can ignore the issues related to multiperiod portfolio selection. Note

Nonparametric Finance, First Edition. Jussi Klemelä.
© 2018 John Wiley & Sons, Inc. Published 2018 by John Wiley & Sons, Inc.

that we discuss certain aspects of multiperiod portfolio in the connection of pricing of options, because prices of options are related to the initial wealth of a trading strategy, which approximately replicates the payoff of the option.

Section 9.1 discusses some basic concepts related to portfolios and their returns. These concepts include the concept of a trading strategy, wealth process, self-financing, portfolio weight, shorting, and leveraging. Section 9.2 discusses the comparison of return and wealth distributions. Section 9.3 discusses issues related to multiperiod portfolio selection.

9.1 Portfolios and Their Returns

The components of a portfolio can be stocks, bonds, commodities, currencies, or other financial assets. The risk-free bond (bank account) can also be included in the portfolio. The price of the risk-free bond is denoted by B_t. Let us have d risky portfolio components and let

$$S_t = \left(S_t^1, \dots, S_t^d\right)$$

be the vector of the prices of the risky portfolio components at time t. Prices satisfy $0 < B_t < \infty$ and $0 \le S_t^i < \infty$. The price vector which includes the risk-free bond is denoted by

$$\bar{S}_t = (B_t, S_t) = \left(B_t, S_t^1, \dots, S_t^d\right).$$

Sometimes it is convenient to denote

$$B_t = S_t^0.$$

The time series of the prices of the riskless bond, the vector time series of the prices of the risky assets, and the combined time series are denoted by

$$B = (B_t)_{t=0,\dots,T}, \qquad S = (S_t)_{t=0,\dots,T}, \qquad \bar{S} = (\bar{S}_t)_{t=0,\dots,T}.$$

As an example, the bond price could be defined as $B_t = (1+r)^t$, where $r > -1$ is the risk-free rate. To take changing rates into account we could define $B_0 = 1$ and $B_t = \prod_{k=1}^{t}(1+r_k)$ for $t \ge 1$, where $r_k > -1$ are the risk-free rates for one period. The risk-free rate r_k is different depending on the length of the period. For the 1-day period the risk-free rate could be the Eonia rate. For the 1-month period the risk-free rate could be the rate of a 1-month government bond.

9.1.1 Trading Strategies

A trading strategy is vector time series $\bar{\xi} = (\bar{\xi})_{t=1,\dots,T}$, where

$$\bar{\xi}_t = (\beta_t, \xi_t), \qquad \xi_t = \left(\xi_t^1, \dots, \xi_t^d\right), \qquad t = 1, \dots, T.$$

The value β_t expresses the number of bonds held between $t-1$ and t. The value ξ_t^i expresses the number of shares of the ith risky asset held between $t-1$ and

t. Vector $\bar{\xi}_t$ is chosen at time $t - 1$, using information which is available at time $t - 1$. Since the values $\bar{\xi}_t$ are known (chosen) at time $t - 1$, it is said that $\bar{\xi}_t$ is a predictable random vector. In our setting, components of $\bar{\xi}_t$ can be any real numbers and not just integers.

A portfolio is typically chosen using available relevant information. We assume that the relevant information is expressed with the state vector $Z_t \in \mathbf{R}^m$, where $m \geq 1$ is the length of Z_t. The vector $\bar{\xi}_t \in \mathbf{R}^{d+1}$ is obtained with a function

$$w : \mathbf{R}^m \to \mathbf{R}^{d+1}$$

and we have

$$\bar{\xi}_t = w(Z_{t-1}).$$

More generally, the function w may be time dependent, and the definition of the relevant information Z_t may be time dependent. In the time dependent case, we define $Z_t \in \mathbf{R}^{m(t)}$ and

$$w_t : \mathbf{R}^{m(t)} \to \mathbf{R}^{d+1}, \qquad t = 0, \ldots, T - 1,$$

which maps at each time $t - 1$ the relevant information to a portfolio vector. Now

$$\bar{\xi}_t = w_{t-1}(Z_{t-1}).$$

The relevant information for portfolio selection may include the following constituents:

1) The relevant information used in choosing the portfolio vector $\bar{\xi}_t$ can include the vector time series of the previous gross returns: $Z_t = (\bar{R}_1, \ldots, \bar{R}_t)$, where $\bar{R}_t = (B_t/B_{t-1}, S_t^1/S_{t-1}^1, \ldots, S_t^d/S_{t-1}^d)$. Since $\bar{R}_t \in \mathbf{R}^{d+1}$, we have that $Z_t \in \mathbf{R}^{t(d+1)}$.

   According to a version of efficient market hypothesis, the historical stock prices contain all relevant information. In this case, we use only the information in the past asset prices to choose the portfolio.

2) The relevant information can include information about the state of the economy, or about the state of individual companies. For example, Z_t can contain macroeconomic information like default spreads and term spreads. Also, Z_t can contain information about the individual companies like dividend yields and earnings.

9.1.2 The Wealth and Return in the One-Period Model

The one-period model has a special interest for portfolio selection, whereas for option pricing the multiperiod model is more interesting. In particular, for the logarithmic utility function the multiperiod portfolio selection reduces to the one-period portfolio selection (see Section 9.3).

We use the following notation for the inner product:

$$\bar{\xi}_t \cdot \bar{S}_t = \beta_t B_t + \xi_t \cdot S_t = \beta_t B_t + \sum_{i=1}^{d} \xi_t^i S_t^i.$$

Sometimes it is convenient to use the notation

$$\bar{\xi}_t' \bar{S}_t$$

for the inner product, where A' denotes the transpose of matrix A, and the vectors are taken as column vectors.

9.1.2.1 The Wealth and Self-financing

The wealth at time t is

$$W_t = \bar{\xi}_{t+1} \cdot \bar{S}_t. \tag{9.1}$$

At time $t + 1$ the wealth is equal to

$$W_{t+1} = \bar{\xi}_{t+1} \cdot \bar{S}_{t+1}.$$

We interpret (9.1) in the following way. We take $W_t > 0$ to be the total wealth available for investment at time t. The total wealth is allocated among the port-folio components. This self-financing condition states that no wealth is reserved for consumption and no wealth is inserted from outside into the portfolio. We could also interpret (9.1) to be the definition of the initial wealth, but in the multiperiod model the self-financing condition is applied at the beginning of each period.

9.1.2.2 Portfolio Weights

Let us assume $W_t > 0$. The portfolio weights are defined as

$$b_t^0 = \frac{\beta_{t+1} B_t}{W_t}, \qquad b_t^i = \frac{\xi_{t+1}^i S_t^i}{W_t}, \qquad i = 1, \ldots, d.$$

Note that we use time index t for the portfolio weights b_t^i but time index $t + 1$ for the portfolio quantities ξ_{t+1}^i, to follow the typical practice in the literature. We define the weight vector by

$$\bar{b}_t = \left(b_t^0, b_t\right), \qquad b_t = \left(b_t^1, \ldots, b_t^d\right).$$

The weight vector satisfies

$$\sum_{i=0}^{d} b_t^i = 1. \tag{9.2}$$

The number b_t^i determines the proportion of the total wealth invested in asset i at time t. The self-financing condition (9.1) leads to (9.2), when $W_t > 0$.

9.1.2.3 Portfolio Returns

The gross return of the portfolio is obtained as a weighted average of the gross returns of the portfolio components. Indeed, the gross return of the portfolio is equal to

$$R^p_{t+1} = \frac{W_{t+1}}{W_t} = \frac{\beta_{t+1}B_t}{W_t}\frac{B_{t+1}}{B_t} + \sum_{i=0}^{d} \frac{\xi^i_{t+1}S^i_t}{W_t}\frac{S^i_{t+1}}{S^i_t}$$

$$= \sum_{i=0}^{d} b^i_t R^i_{t+1} = \bar{b}_t \cdot \bar{R}_{t+1}, \tag{9.3}$$

where

$$\bar{R}_{t+1} = \left(R^0_{t+1}, R_{t+1}\right), \qquad R_{t+1} = \left(R^1_{t+1}, \ldots, R^d_{t+1}\right)$$

is the vector of the gross returns of the portfolio components. The gross returns of the portfolio components are defined by

$$R^0_{t+1} = \frac{B_{t+1}}{B_t}, \qquad R^i_{t+1} = \frac{S^i_{t+1}}{S^i_t}, \qquad i = 1, \ldots, d.$$

9.1.2.4 The Product and Additive Forms of Wealth

The wealth can be written either in the product form or in the additive form. These two ways of writing the wealth will be applied in Section 9.1.3 to write the wealth process.

Wealth in the Product Form   We can write the wealth at time $t + 1$ as

$$W_{t+1} = W_t \bar{b}_t \cdot \bar{R}_{t+1}, \tag{9.4}$$

where $\bar{b}_t$ satisfies restriction (9.2), which can be written as

$$\bar{b}_t \cdot 1_{d+1} = 1, \tag{9.5}$$

where 1_{d+1} is the vector of length $d + 1$ whose components are ones. Second, the wealth can be written using only the unrestricted weight vector b_t. Indeed, the restriction can be written as

$$b^0_t = 1 - 1_d \cdot b_t.$$

Thus,

$$\bar{b}_t \cdot \bar{R}_{t+1} = (1 - 1_d \cdot b_t)R^0_{t+1} + b_t \cdot R_{t+1}$$

$$= R^0_{t+1} + b_t \cdot \left(R_{t+1} - R^0_{t+1}\right), \tag{9.6}$$

where

$$R_{t+1} - R^0_{t+1}$$

is called the excess return. We arrive at

$$W_{t+1} = W_t \left[R_{t+1}^0 + b_t \cdot \left(R_{t+1} - R_{t+1}^0 \right) \right], \tag{9.7}$$

which expresses the wealth at time $t + 1$ in terms of the unrestricted weight vector b_t.

Wealth in the Additive Form   We can write the wealth at time $t + 1$ as

$$W_{t+1} = W_t + \bar{\xi}_{t+1} \cdot (\bar{S}_{t+1} - \bar{S}_t), \tag{9.8}$$

where $\bar{\xi}_{t+1}$ satisfies restriction (9.1) :

$$\bar{\xi}_{t+1} \cdot \bar{S}_t = W_t.$$

Second, the wealth can be written using only the unrestricted vector ξ_{t+1}. Indeed, the restriction can be written as

$$\beta_{t+1} = \frac{W_t}{B_t} - \xi_{t+1} \cdot \frac{S_t}{B_t}.$$

Thus,

$$W_{t+1} = W_t + \frac{W_t}{B_t}(B_{t+1} - B_t) + \xi_{t+1} \cdot \left(S_{t+1} - S_t - \frac{S_t}{B_t}(B_{t+1} - B_t) \right)$$

$$= B_{t+1} \frac{W_t}{B_t} + \xi_{t+1} \cdot B_{t+1} \left(\frac{S_{t+1}}{B_{t+1}} - \frac{S_t}{B_t} \right).$$

We arrive at

$$W_{t+1} = B_{t+1} V_{t+1}, \tag{9.9}$$

where

$$V_{t+1} = V_t + \xi_{t+1} \cdot (X_{t+1} - X_t),$$

and

$$V_t = \frac{W_t}{B_t}, \qquad X_t = \frac{S_t}{B_t}, \qquad X_{t+1} = \frac{S_{t+1}}{B_{t+1}}.$$

We have expressed the wealth at time $t + 1$ in terms of the unrestricted vector ξ_{t+1}.

9.1.3   The Wealth Process in the Multiperiod Model

The wealth process $W = (W_t)_{t=0,\ldots,T}$ can be written either multiplicatively or additively. Furthermore, we can write the wealth either so that the self-financing restrictions are implicitly assumed, or so that the self-financing conditions are eliminated by moving from the gross returns to the excess returns (product form) or by moving from the prices to the discounted prices (additive form). In the case of the product form the elimination of the self-financing restrictions does not bring essential simplifications but in the case of the additive form the elimination of the self-financing conditions simplifies the dynamic optimization algorithm for the maximization of the expected wealth.

9.1.3.1 The Wealth in the Product Form

We assume that $W_0 > 0$ and self-financing holds at each of the T periods (wealth W_{t+1} is obtained from wealth W_t only through the changes in asset prices and through the changes in wealth allocation). We can write

$$W_T = W_0 \prod_{t=0}^{T-1} \frac{W_{t+1}}{W_t}.$$

We get from (9.4) that

$$W_T = W_0 \prod_{t=0}^{T-1} \bar{b}_t \cdot \bar{R}_{t+1}, \tag{9.10}$$

where $\bar{b}_t \in \mathbf{R}^{d+1}$ satisfies restriction

$$\bar{b}_t \cdot 1_{d+1} = 1.$$

The wealth process can be written in terms of only the weights b_t of the risky assets. We obtain from (9.7) that

$$W_T = W_0 \prod_{t=0}^{T-1} \left[R_{t+1}^0 + b_t \cdot \left(R_{t+1} - R_{t+1}^0 \right) \right], \tag{9.11}$$

where $b_t \in \mathbf{R}^d$ is unrestricted.

When the sequence $b = (b_t)_{t=0,\dots,T-1}$ of portfolio vectors is constant, not changing with t, then we call the portfolios "constant weight portfolios." Note that when using a constant weight portfolio strategy there is a need to make a rebalancing at each period because the prices of the portfolio components are changing, and to keep the weights constant we have to decrease the weight of those assets whose price has increased and to increase the weights of those assets whose price has declined. In this sense a constant weight portfolio strategy is a counter trend strategy.

9.1.3.2 The Wealth in the Additive Form

The additive wealth process is applied more in option pricing than in portfolio management, but it is useful also in the portfolio selection, especially when the exponential utility is used. We summarize the definitions related to the additive wealth process, but the detailed explanations are given in Section 13.2.2, where option pricing is studied.

We can write

$$W_T = W_0 + \sum_{t=0}^{T-1}(W_{t+1} - W_t).$$

We get from (9.8) that

$$W_T = W_0 + \sum_{t=0}^{T-1}\bar{\xi}_{t+1} \cdot (\bar{S}_{t+1} - \bar{S}_t),$$

where $\bar{\xi}_{t+1}$ satisfy restrictions

$$\bar{\xi}_t \cdot \bar{S}_t = \bar{\xi}_{t+1} \cdot \bar{S}_t, \qquad t = 1, \ldots, T-1. \tag{9.12}$$

We say that a trading strategy $\bar{\xi}$ is self-financing if (9.12) holds.

We define the value process, which is useful because it involves only the numbers ξ_{t+1} of risky assets. The discounted price process is defined by

$$X_t^i = \frac{S_t^i}{B_t}, \qquad t = 0, \ldots, T, \qquad i = 1, \ldots, d.$$

We denote

$$X_t = \left(X_t^1, \ldots, X_t^d\right), \qquad \bar{X}_t = (1, X_t).$$

The value process is defined as

$$V_t = \frac{W_t}{B_t}, \qquad t = 0, \ldots, T.$$

We obtain from (9.9) that

$$W_T = B_T V_T, \tag{9.13}$$

where

$$V_T = V_0 + \sum_{t=0}^{T-1}\xi_{t+1} \cdot (X_{t+1} - X_t).$$

9.1.4 Examples of Portfolios

The collection of possible portfolios is determined by the collection of possible portfolio weights. The most general collection of portfolio weights consists of all weights satisfying the constraint (9.2):

$$B = \left\{ \left(b_t^0, \ldots, b_t^d\right) : \sum_{i=0}^{d} b_t^i = 1 \right\}.$$

We can impose various restrictions on portfolio weights and obtain smaller collections of weights. For example, we can allow leveraging but forbid shorting of stocks, or we can restrict ourselves to long only portfolios.

9.1.4.1 Shorting

A portfolio is described by giving weights for the portfolio components. The weights are such that they sum to one, as stated in (9.2). Without any further constraints, borrowing and short selling are allowed. When shorting is allowed, then the elements of portfolio vectors can take negative values. Borrowing is interpreted as selling short the risk-free rate. Thus, when borrowing is allowed, the weight of the risk-free rate can take negative values. When short selling or borrowing occurs, then some weights are larger than one.

Selling a stock short means that we sell a stock that we do not own. Typically the stock that is sold short is first borrowed from somebody who owns the stock. If the stock is sold without first borrowing it, the short selling is called naked short selling. Short selling a stock changes the character of the portfolio: a short position on a stock has an unlimited downside risk, but only a limited upside potential. In contrast, a long position on a stock can lose only the invested capital but has an unlimited upside potential.

A return that is obtained when being short a stock is

$$(1 - b)r_{t+1} + bR_{t+1}, \tag{9.14}$$

where $b < 0$, $R_{t+1} = S_{t+1}/S_t$ is the gross return of the stock to be shorted, and r_{t+1} is the gross return of another asset. For example, r_{t+1} can be the return of the risk-free investment. The return $2r_{t+1} - R_{t+1}$ arises when the available wealth is invested in the risk-free rate, the stock is shorted with the amount of the total wealth, and the proceedings obtained from shorting the stock are invested in the risk-free rate.

It can happen that $(1 - b)r_{t+1} + bR_{t+1} < 0$, because R_{t+1} is not bounded from above. Gross returns less or equal to zero can be interpreted as leading to bankruptcy, but they can also be interpreted as leading to debt.

Figure 9.1 shows functions $S_{t+1} \mapsto R_{t+1} = (1 - b) + bS_{t+1}/s$, where $s = S_t$ is the previous value of the stock. The case $b = 1$ (black) means that we are long the stock (we have bought the stock). The case $b = 2$ (blue) means that we

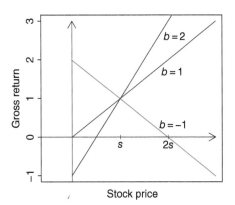

Figure 9.1 *Being long and short a stock.* The blue lines show the gross return of being long a stock for $b = 1$ and $b = 2$ as a function of the stock price. The red line shows the gross return of being short a stock. Shown are the functions $S_{t+1} \mapsto R_{t+1} = (1 - b) + bS_{t+1}/s$, where $s = S_t$ is the previous value of the stock.

are leveraged. The case $b = -1$ (red) means that we are short the stock. We have taken the gross return of the risk-free investment as $r_{t+1} = 1$.

9.1.4.2 Long Only Portfolios

In a long only portfolio borrowing and short selling are excluded. In the case of long only portfolios the portfolio weights are nonnegative. Thus, the weights satisfy

$$b_t^j \geq 0$$

for $j = 0, \ldots, d$.

The nonnegativity constraint together with the condition $\sum_{j=0}^{d} b_t^j = 1$ imply that

$$0 \leq b_t^j \leq 1$$

for $j = 0, \ldots, d$.

9.1.4.3 Leveraged Portfolios

A portfolio allowing leveraging but forbidding short selling is such that the weight of the risk-free rate can be negative but the weights of the other assets are nonnegative. In a leveraged portfolio it is allowed to borrow money and invest the borrowed money to stocks or other assets. Borrowing money is interpreted as shorting the risk-free rate. Let $B_t = S_t^0$ be the bank account. The portfolio vectors of a leveraged portfolio satisfy, in addition to the constraint $\sum_{j=0}^{d} b_t^j = 1$, the additional constraint

$$b_t^j \geq 0$$

for $j = 1, \ldots, d$.

We allow negative values for the portfolio weight b_t^0 of the bank account, but the other portfolio weights $b_t^j, j = 1, \ldots, d$, are nonnegative.

9.1.4.4 Restrictions on Short Selling

In practice investors have a constraint on the amount of short selling. It is natural to make a constraint on the amount of short selling by requiring that the portfolio weights satisfy

$$\sum_{j=0}^{d} |b_t^j| \leq L, \tag{9.15}$$

where $L \geq 1$. Under the constraint $\sum_{j=0}^{d} b_t^j = 1$, the constraint (9.15) is equivalent to any of the following two constraints:

$$\sum_{j=0}^{d} \left(b_t^j \right)_- \leq \frac{L-1}{2}, \qquad \sum_{j=0}^{d} \left(b_t^j \right)_+ \leq \frac{L+1}{2},$$

where we denote by $(b)_+ = \max\{0, b\}$ the positive part of $b \in \mathbf{R}$ and by $(b)_- = -\min\{0, b\}$ the negative part of b.[1] Thus, $C = (L-1)/2$ is such factor that we are allowed to short sell C times the current wealth.

9.1.4.5 Portfolios Used in Trading

There are several reasons to define very restricted finite collections of the allowed portfolio weights. The use of restricted collections of weights brings computational advantages, and restricted collections are often used in such trading strategies as market timing and stock selection.

1) *Computational advantages.* For computational reasons, we might prefer to search the portfolio vector from a rather small collection of the allowed portfolio weights. When the collection of the allowed portfolio weights is small, we do not have to use involved optimization techniques to find the portfolio weights.

2) *Market timing.* Some market timing strategies require only the choice between two different assets. These market timing strategies might be such that we have two available assets, and at the beginning of every month we choose to invest everything into the one asset and nothing into the other asset, or we might go long the one asset and go short the other asset. The two assets might both be risky assets, or the one asset might be the risk-free rate and only the other asset would be risky. Market timing strategies are often trend following strategies, which are discussed in Section 12.1.1.

3) *Stock selection.* Sometimes a mutual fund uses a strategy where a search is made for an optimal subset of the stocks in the index that is the benchmark for the performance. For example, a mutual fund whose aim is to beat the performance of S&P 500 index might try to select a subset of the stocks in S&P 500 index, invest equal weights to this subset, and allocate zero weights to the remaining stocks of S&P 500 index. For instance, the mutual fund might look for a subset of 20 companies whose price to earnings ratio is the smallest, and to invest 5% to each of the companies with the smallest P/E ratios. More involved stock selection methods might use regression on economic indicators to estimate the expected returns or the expected utility, as discussed in Sections 12.1.1 and 12.1.3.

---

1 We have $b = (b)_+ - (b)_-$ and $|b| = (b)_+ + (b)_-$. Thus,

$$\sum_{j=0}^{d} b_t^j = 1 \Leftrightarrow \sum_{j=0}^{d} \left(b_t^j\right)_+ = 1 + \sum_{j=0}^{d} \left(b_t^j\right)_-.$$

Then,

$$\sum_{j=0}^{d} |b_t^j| = \sum_{j=0}^{d} \left(b_t^j\right)_+ + \sum_{j=0}^{d} \left(b_t^j\right)_- \le L \Leftrightarrow \sum_{j=0}^{d} \left(b_t^j\right)_- \le \frac{L-1}{2}.$$

Let us have basis assets $S^0, \ldots, S^d$ and predictions $m(S^0), \ldots, m(S^d)$ for the performance of the basis assets. The performance predictions might be estimates for the expected return, estimates for the expected utility, estimates for the Markowitz criterion, or the price to earnings ratio (which could be considered as an estimate for the expected return) . These performance predictions are discussed in Section 12.1.

1) Let us consider the case $d = 1$, so that we have two basis assets S^0 and S^1. A possible strategy is to put weight one to the first asset and to put weight zero to the second asset, when $m(S^0) > m(S^1)$. Otherwise, when $m(S^0) \leq m(S^1)$, then we put weight zero to the first asset and weight one to the second asset. Now the set of the allowed portfolio vectors is

$$B = \{(1,0), (0,1)\}. \tag{9.16}$$

2) Let us secondly consider the case $d = 2$, so that we have three basis assets. As examples, we consider two strategies.
   a) We put weight one to the asset with the highest value for the performance measure $m(S^i)$, $i = 0, 1, 2$, and zero weight to the two other assets. Now the set of the allowed portfolio vectors is

   $$B = \{(1,0,0), (0,1,0), (0,0,1)\}. \tag{9.17}$$

   b) We put the equal weight $1/2$ to the two assets with the highest value for the performance measure $m(S^i)$, $i = 0, 1, 2$, and the zero weight to the remaining asset. Now the set of the allowed portfolio vectors is

   $$B = \{(1/2, 1/2, 0), (1/2, 0, 1/2), (0, 1/2, 1/2)\}. \tag{9.18}$$

3) Let us thirdly consider the general case of $d \geq 1$. We consider the strategy where we choose from $d + 1 = N$ basis assets a subset of $M < N$ assets with the highest values for the performance measure $m(S^i)$, $i = 0, \ldots, d$, and put equal weights to the M selected assets. Now the set of the allowed portfolio vectors is

$$B = \left\{ \left(\frac{1}{M} I_J(j) \right)_{j=0,\ldots,d} : J \subset \{0, \ldots, d\}, \#J = M \right\}, \tag{9.19}$$

where $I_J(j) = 1$ if $j \in J$ and otherwise $I_J(j) = 0$, we use the notation $(b_0, \ldots, b_d) = (b_j)_{j=0,\ldots,d}$, and $\#J$ means the number of elements in set J. We get (9.17) as a special case by choosing $d + 1 = N = 3$ and $M = 1$. We get (9.18) as a special case by choosing $d + 1 = N = 3$ and $M = 2$.

The previous collections of portfolio weights defined long only portfolios. We can define in an analogous way collections of portfolio weights that allow shorting.

1) Let us consider the case $d = 1$, so that we have two basis assets, and we assume that the first basis asset is the risk-free rate and the second asset is

a risky asset. Let the risky asset have return R_{t+1} and let the risk-free rate be r_{t+1}. Taking

$$B = \{(0,1),(2,-1)\} \tag{9.20}$$

means that we are either long of the stock, which gives return R_{t+1}, or we are short of the stock, which gives return $2r_{t+1} - R_{t+1}$. Taking

$$B = \{(1,0),(-1,2)\} \tag{9.21}$$

means that we are either not invested, which gives return r_{t+1}, or we are leveraged, which gives return $2R_{t+1} - r_{t+1}$.

2) When the number $d + 1 = N$ of the basis assets increases, the cardinality of the set of possible and reasonable portfolio vectors increases rapidly. As an example, let us consider the case $N = 3$, where the first asset is the risk-free rate and two basis assets are risky. Now,

$$B = \{(0,1,0),(2,-1,0),(0,0,1),(2,0,-1),(1,0,0)\} \tag{9.22}$$

describes the choices of being long of one of the stocks, being short of one of the stocks, and staying out of the market.

9.1.4.6 Pairs Trading

In pairs trading we have two risky assets and typically two alternatives are considered: (1) go long of the first asset and short of the second asset or (2) go short of the first asset and long of the second asset. Then the return of the portfolio is

$$R_{t+1}^p = (1-b)\frac{S_{t+1}^1}{S_t^1} + b\frac{S_{t+1}^2}{S_t^2}, \tag{9.23}$$

where (1) $b = -1$, or (2) $b = 2$. More generally, we can consider pairs trading with other values for b. Choosing the weights from set

$$B = \{(1+a,-a),(-a,1+a)\}, \tag{9.24}$$

where $a > 0$, means that we are leveraged of the first asset and short of the second asset. We can include the risk-free rate and consider returns

$$(1 - b_1 - b_2)(1+r) + b_1\frac{S_{t+1}^1}{S_t^1} + b_2\frac{S_{t+1}^2}{S_t^2}. \tag{9.25}$$

Sometimes a strategy for pairs trading is defined in terms of asset prices. The strategy could be such that coefficients $c_1, c_2 \in \mathbf{R}$ are determined so that the linear combination

$$c_1 S_t^1 + c_2 S_t^2$$

of prices satisfies certain conditions. For example the aim could be to choose c_1 and c_2 so that the linear combination is stationary. This is possible when the prices S_t^1 and S_t^2 are colinear. When $c_1 S_t^1 + c_2 S_t^2 > 0$, the return of the portfolio is

$$R_{t+1}^p = \frac{c_1 S_{t+1}^1 + c_2 S_{t+1}^2}{c_1 S_t^1 + c_2 S_t^2},$$

and the weight in (9.23) is

$$b = \frac{c_2 S_t^2}{c_1 S_t^1 + c_2 S_t^2}.$$

9.2 Comparison of Return and Wealth Distributions

In order to study portfolio selection and performance measurement we need to define what it means that a wealth distribution or a return distribution is better than another such distribution. Let the initial wealth be W_0 and the wealth at time T be W_T. Terminal wealth W_T is a random variable. When $W_0 > 0$ then we can define the gross return $R_T = W_T / W_0$. The gross return R_T is a random variable. We can use either the distribution of W_T or the distribution of R_T to study portfolio selection and performance measurement.

In portfolio selection, we need to choose the portfolio weights so that the return R_T or the terminal wealth W_T of the portfolio is optimized. To measure the performance of asset managers we need to define what it means that a return distribution (or the distribution of the terminal wealth) generated by an asset manager is better than the distribution generated by another asset manager.[2]

To compare return and wealth distributions, we make a mapping from a class of distributions to the set of real numbers. This mapping assigns to each distribution a number that can be used to rank the distributions.

It might seem reasonable to compare return and wealth distributions using only the expected returns and expected wealths: we would prefer always the distribution with the highest (estimated) expectation. However, this would lead to the preference of investment strategies with extremely high risk. Thus, the comparison of distributions has to take into account not only the expectation but also the risk associated with the distribution.

A classical idea to rank the return distributions is to use the variance penalized expected return. This idea is discussed in Section 9.2.1, and it is related to the Markowitz portfolio selection.

---

2 Note that in portfolio selection there is the additional problem that the distributions of the returns of the individual assets, their cross-sectional dependencies, and their time series properties are unknown and have to be estimated using historical data. Similarly, to measure the performance of fund managers we have to collect information of the past returns obtained by the fund managers. Then we estimate the return distributions and compare the estimated return distributions.

Figure 9.2 *Comparison of distributions.* Shown are two return densities, where the red distribution has a higher risk and a higher return than the black distribution. It is not obvious which return distribution should be preferred.

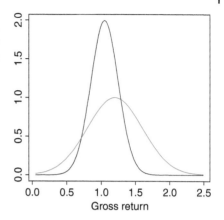

Gross return

The expected utility is discussed in Section 9.2.2. The Markowitz criterion uses only the first two moments of the distribution; it uses only the mean and the variance. However, the expected utility takes into account the higher order moments of the distribution. A Taylor expansion of the expected utility shows that all the moments make a contribution to the expected utility. Conversely, a Taylor expansion of the expected utility can be used to justify the mean–variance criterion, and various other criteria that involve a collection of moments of various degrees, such as the third and the fourth-order moments.

Figure 9.2 shows densities of two gross return distributions whose comparison is not obvious. The distributions are Gaussian, and the expected return of the red distribution is higher, but also the variance of the red distribution is higher.[3] Thus, the red return distribution has a higher risk and a higher expected return. There exists no universal or objective way to compare these two distributions. Instead, the comparison depends on the risk aversion of the investor. An investor with a high-risk aversion would prefer the black distribution, but an investor with a low-risk aversion would prefer the red distribution.

9.2.1 Mean–Variance Preferences

Portfolio choice with mean–variance preferences was proposed by Markowitz (1952, 1959). This method ranks the distributions of the portfolio return R^p_{t+1}

---

3 Strictly speaking, Gaussian distributions cannot be return distributions, because a Gaussian random variable can take negative values, whereas a gross return is larger or equal to zero, and a net return is larger or equal to -1. However, since Gaussian distributions have light tails, the probability of a negative value can be negligible, and we can use them to model return distributions well, although a log-normal distribution would be more appropriate, for example. In fact, to model a return distribution, distributions with unbounded support are typically used, like t-distributions (see Section 3.3).

according to

$$E_t R^p_{t+1} - \frac{\gamma}{2} \operatorname{Var}_t \left(R^p_{t+1} \right), \tag{9.26}$$

where $\gamma \geq 0$ is the risk aversion parameter, and E_t and Var_t mean the conditional expectation and conditional variance, respectively. The expected return is penalized by subtracting the variance of the return. Parameter γ measures the investor's risk aversion, or more precisely, absolute risk aversion, as defined in (9.30).

We consider now basically one-period model, with time points t and $t + 1$. We could apply the notations used in Section 9.1, and denote $T = 1$, and replace (9.26) by $E_0 R^p_1 - \frac{\gamma}{2} \operatorname{Var}_0(R^p_1)$. However, it is convenient to denote the time points by t and $t + 1$, because in practice we will use the sequence of one-period models with $t = 0, \dots, T - 1$.

Remember that the gross return of a portfolio was written in (9.3) as

$$R^p_{t+1} = \sum_{i=0}^{d} b^i_t R^i_{t+1} = \bar{b}'_t \bar{R}_{t+1},$$

where $\bar{R}_{t+1} = (R^0_{t+1}, \dots, R^d_{t+1})'$ is the column vector of the gross returns of the portfolio components, the gross return of a single portfolio component is $R^i_{t+1} = S^i_{t+1}/S^i_t$, and $\bar{b}_t = (b^0_t, \dots, b^d_t)'$ is the vector of the portfolio weights. Here $S^0_t = B_t$ is the risk-free bond and R^0_{t+1} is the risk-free gross return.

In order to calculate the conditional variance of R^p_{t+1} it is convenient to separate the risk-free rate. This was done in (9.6), where we wrote

$$\bar{b}'_t \bar{R}_{t+1} = \left(1 - 1'_d b_t\right) R^0_{t+1} + b'_t R_{t+1} = R^0_{t+1} + b'_t \left(R_{t+1} - R^0_{t+1} \right),$$

where $b_t = (b^1_t, \dots, b^d_t)'$ and $R_{t+1} = (R^1_{t+1}, \dots, R^d_{t+1})'$ are the weights and the returns of the risky assets.

We can write

$$E_t R^p_{t+1} = R^0_{t+1} + b'_t \left(E_t R_{t+1} - R^0_{t+1} \right),$$

and

$$\operatorname{Var}_t \left(R^p_{t+1} \right) = b'_t \operatorname{Var}_t(R_{t+1}) b_t,$$

where $E_t R_{t+1}$ is the d-vector of the expected returns of the risky assets and $\operatorname{Var}_t(R_{t+1})$ is the $d \times d$ covariance matrix of R_{t+1}. Note that the risk-free rate R^0_{t+1} is known at time t, and therefore it does not affect the conditional variance.[4]

Section 9.2.2 discusses the use of the expected utility to rank the distributions. The Markowitz ranking is related to the use of the quadratic

---

4 The unconditional variance can be written as $\operatorname{Var}(R^p_{t+1}) = \bar{b}'_t \operatorname{Var}(\bar{R}_{t+1}) \bar{b}_t$, because the risk-free rate R^0_{t+1} is a random variable.

utility function

$$U(x) = x - 1 - \frac{1}{2}(x-1)^2,$$

because the Markowitz criterion (9.26) with $\gamma = 1$ is approximately equal to $EU(R_{t+1}^p)$, the difference being due to the the fact that the expected quadratic utility involves the squared return but the Markowitz criterion in (9.26) involves variance.

Chapter 11 discusses portfolio selection when the Markowitz criterion is used. Next, we give two examples that illustrate how the variance of the portfolio can be decreased by a skillful choice of the portfolio weights. The first example considers uncorrelated basis assets and the second example considers correlated assets. In practice, it is difficult to find uncorrelated basis assets and it is even more difficult to find anticorrelated basis assets. However, even when the basis assets are correlated it is possible to decrease the risk of the portfolio by allocating the portfolio weights skillfully among the basis assets.

9.2.1.1 A Large Number of Uncorrelated Assets

The variance of the portfolio return can be close to zero, when we have a large number of uncorrelated basis assets. Consider d risky assets $S^1, \dots, S^d$, whose gross returns are $R_{t+1}^i = S_{t+1}^i / S_t^i$, $i = 1, \dots, d$. We denote $E_t R_{t+1}^i = \mu$, $\mathrm{Var}_t(R_{t+1}^i) = \sigma^2$, and we assume that the returns are uncorrelated. Let the portfolio vector be $b = (1/d, \dots, 1/d) \in \mathbf{R}^d$. Then,

$$E_t(b'R_{t+1}) = \mu, \qquad \mathrm{Var}_t(b'R_{t+1}) = \frac{\sigma^2}{d}.$$

Thus, when the number d of assets in the portfolio is large, the variance of the portfolio return is close to zero.

9.2.1.2 Two Correlated Assets

In the case of two risky basis assets, the variance of the portfolio return can be close to zero when the two assets are anticorrelated. Let R_{t+1}^1 and R_{t+1}^2 be the gross returns of two basis assets. Let us assume that the $\mathrm{Var}_t(R_{t+1}^1) = \mathrm{Var}_t(R_{t+1}^2) = \sigma^2$ and $\mathrm{Cor}_t(R_{t+1}^1, R_{t+1}^2) = \rho$. Then the variance of the portfolio return is

$$\mathrm{Var}_t\left(bR_{t+1}^1 + (1-b)R_{t+1}^2\right) = b^2\sigma^2 + (1-b)^2\sigma^2 + 2b(1-b)\sigma^2\rho,$$

where $b \in \mathbf{R}$ is the weight of the first asset. Figure 9.3 shows the function $(\rho, b) \mapsto b^2 + (1-b)^2 + 2b(1-b)\rho$, where we have chosen the variance of the portfolio components to be $\sigma^2 = 1$. The variance of the portfolio becomes smaller when $\rho \to -1$. When $0 \le b \le 1$, then variance of the portfolio is smaller than one, otherwise it is larger than one. Thus, the variance of the portfolio is smaller than the variance of the components when $0 \le b \le 1$, and the reduction in the variance is greatest when portfolio components are anticorrelated.

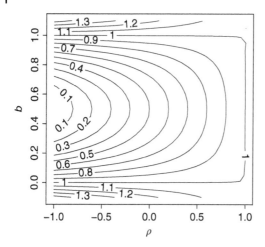

Figure 9.3 *Two correlated assets. A contour plot of function $(\rho, b) \mapsto b^2 + (1 - b)^2 + 2b(1 - b)\rho$ is shown. The function is equal to the variance of the portfolio when the portfolio components have variance one, correlation ρ, and the weight of the portfolio components are b and $1 - b$.*

9.2.2 Expected Utility

We can order distributions according to the value of the expected utility. Introducing the utility function U and ranking the distributions according to the expected utility $E_0 U(R_T)$ brings in the element of risk aversion, whereas ranking the return distributions solely according to the expected returns $E_0 R_T$ does not take risk into account.

The expected utility can be calculated either from the wealth or from the return. The expected utility calculated from the wealth is

$$E_0 U(W_T),$$

where W_T is the wealth (in Euros, Dollars, etc.), and $U : \mathbf{R} \rightarrow \mathbf{R}$ is a utility function. The negative wealth means that more is borrowed than owned. The expected utility calculated from the gross returns is

$$E_0 U(R_T),$$

where $U : (0, \infty) \rightarrow \mathbf{R}$ is a utility function and $R_T = W_T / W_0$ is the gross return. The gross return R_T is always nonnegative. It is natural to define $U(0) = -\infty$, because the gross return of zero means bankruptcy.

Sometimes it is equivalent to calculate the expected utility from the wealth and calculate it from the return. Consider the logarithmic utility $U(x) = \log x$. Now $E_0 \log W_T = E \log W_0 + E \log R_T$. This issue is discussed in Section 9.3.

Figure 9.4 illustrates the ranking of distributions according to the expected utility, when the densities have the same shape but different locations. Panel (a) shows four densities of gross return distributions. The distribution with the black density is the best because its expectation is the largest, and the distribution with the red density is the worst, because its expectation is the smallest.

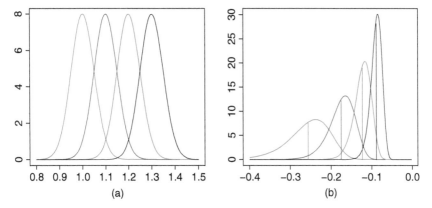

Figure 9.4 *Ranking distributions with the expected utility: Different means.* (a) Four density functions of gross return R_T; (b) the four density functions of $U(R_T)$. The expectations $E_0U(R_T)$ are marked with vertical vectors.

Panel (b) shows the densities of $U(R_T)$, where U is the power utility function with risk aversion $\gamma = 5$ and R_T is the return.[5] The power utility functions are defined in (9.28). The expectations $E_0U(R_T)$ are marked with vertical lines. We can see that although the densities of returns R_T are symmetrical, the densities of $U(R_T)$ are skewed to the left, so that the expectations $EU(R_T)$ are smaller than the modes of the distributions.

Figure 9.5 illustrates the ranking of distributions according to the expected utility, when the densities have the same location but different variances. The utility function is the power utility function with risk aversion $\gamma = 5$. The power utility functions are defined in (9.28). Panel (a) shows four densities of return distributions. The distribution with the black density is the best because its spread is the smallest, and the distribution with the red density is the worst, because its spread is the largest. Panel (b) shows the densities of $U(R_T)$, where U is the utility function and R_T is the return. The expectations $E_0U(R_T)$ are marked with vertical lines. We can see that although the mode of the red density is located furthest to the right, its expected value is furthest to the left.

---

5 The density function of $U(R)$ is $f_{U(R)}(x) = f_R(U^{-1}(x))/|U'(U^{-1}(x))|$, where f_R is the density function of return R. In fact, let $X \in \mathbf{R}$ be a random variable and $A : \mathbf{R} \to \mathbf{R}$ be a monotonic function. Denote with f_X the density of X and with $f_{A(X)}$ the density of $A(X)$. We have

$$f_{A(X)}(x) = \frac{\partial}{\partial x} P(A(X) \le x) = \frac{\partial}{\partial x} P(X \le A^{-1}(x))$$

$$= f_X(A^{-1}(x)) \cdot \left|\frac{\partial}{\partial x} A^{-1}(x)\right| = f_X(A^{-1}(x)) \cdot \frac{1}{|A'(A^{-1}(x))|},$$

where $A'(x) = \frac{\partial}{\partial x} A(x)$.

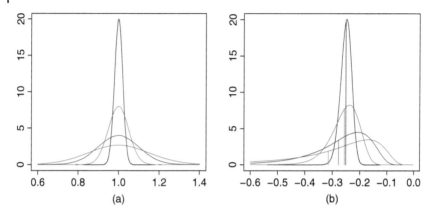

Figure 9.5 *Ranking of distributions with expected utility: Different variances.* (a) Four density functions of return R_T; (b) the density functions of $U(R_T)$. The expectations $E_0 U(R_T)$ are marked with vertical vectors.

9.2.2.1 Basic Properties of Utility Functions

In our examples a utility function can have as its domain either the positive real axis or the real line. When the argument is a gross return, then utility function $U : (0, \infty) \to \mathbf{R}$ is defined on the positive real axis.[6] When the argument is the wealth which can take negative values, then utility function $U : \mathbf{R} \to \mathbf{R}$ is defined on the real line.

It is natural to require that a utility function is strictly increasing and strictly concave.

1) A strictly increasing function $U : \mathbf{R} \to \mathbf{R}$ satisfies $U'(x) > 0$ for all $x \in \mathbf{R}$, when the function is differentiable. A function is strictly increasing if $U(x_2) > U(x_1)$ for all $x_2 > x_1$.
   A utility function should be increasing because investors prefer a larger wealth to a lesser wealth.

2) A strictly concave function $U : \mathbf{R} \to \mathbf{R}$ satisfies $U''(x) < 0$ for all $x \in \mathbf{R}$, when the function is two times differentiable. A concave function is such that the rate of increase decreases.
   A utility function should be concave since increasing the wealth makes the value of additional wealth decline: The marginal value of additional consumption is declining. The concavity of a utility function is a consequence of risk aversion: The curvature of the utility function captures the subjective aversion to risk.

Concavity can also be defined in the case where the function is not two times differentiable. A function $U : \mathbf{R} \to \mathbf{R}$ is strictly concave, when

$$pU(x_1) + (1 - p)U(x_2) < U(px_1 + (1 - p)x_2) \tag{9.27}$$

for all $0 \leq p \leq 1$ and for all $x_1, x_2 \in \mathbf{R}$.

---

6 When we use a net return, then the utility function should have domain $(-1, \infty)$.

In addition, sometimes it is assumed that utility function $U : (0, \infty) \rightarrow \mathbf{R}$ is continuously differentiability with $\lim_{x \to \infty} U'(x) = 0$, $\lim_{x \to 0} U'(x) = \infty$.

9.2.2.2 Power and Exponential Utility Functions

The power utility functions are defined as

$$U(x) = \begin{cases} \dfrac{x^{1-\gamma}}{1-\gamma}, & \text{if } \gamma > 0, \quad \gamma \neq 1, \\ \log x, & \text{if } \gamma = 1, \end{cases} \qquad x > 0, \tag{9.28}$$

where $\gamma > 0$ is the risk aversion parameter. Note that $U'(x) = x^{-\gamma}$ for $\gamma > 0$, $\gamma \neq 1$ and $\partial/\partial x \log x = x^{-1}$, which can be used to explain why the logarithmic function is obtained as the limit when $\gamma \rightarrow 1$. The power utility functions are constant relative risk aversion (CRRA) utility functions, as defined in (9.31).

The exponential utility functions are defined as

$$U(x) = 1 - e^{-\alpha x}, \qquad x \in \mathbf{R}, \tag{9.29}$$

where $\alpha > 0$ is the risk aversion parameter. The exponential utility functions are constant absolute risk aversion (CARA) utility functions, as defined in (9.30).

The power utility functions are defined on $(0, \infty)$, but the exponential utility functions are defined on the whole real line. Thus, the exponential utility functions can be applied in the case of negative wealth. The exponential utility functions are useful when we consider portfolios of derivatives (selling of options), because in these cases the wealth can become negative. There exists also other than power and exponential utility functions.[7]

Figure 9.6 plots normalized utility functions with different risk aversion parameters. Panel (a) shows power utility functions (9.28) and panel (b) shows exponential utility functions (9.29). The normalized utility functions are defined by

$$u(x) = \frac{U(x) - U(1)}{U(2) - U(1)}.$$

The normalization is such that $u(1) = 0$ and $u(2) = 1$. Note that the ordering of the distributions according to the expected utility is not affected by linear transformations $aU(x) + c$, $a > 0$, $c \in \mathbf{R}$, because

$$E[aU(R_{t+1}) + c] = aEU(R_{t+1}) + c.$$

Figure 9.6 shows that larger values of γ or α are used when one is more risk averse, because the curvature of the utility functions increases when γ or α are increased.

---

7 The utility functions $U(x) = I_{[a,\infty)}(x)$, $x \in \mathbf{R}$, where $a > 0$, are used when one wants to choose a portfolio maximizing the probability of reaching the given amount of capital. Note that these utility functions are not concave.

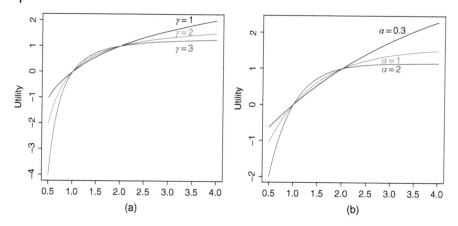

Figure 9.6 *Utility functions.* (a) Power utility functions (9.28) for risk aversion values $\gamma = 1$, $\gamma = 2$, and $\gamma = 3$; (b) exponential utility functions (9.29) for risk aversion values $\alpha = 0.3$, $\alpha = 1$, and $\alpha = 2$.

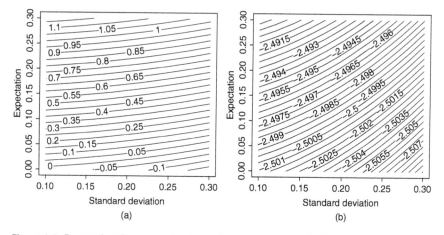

Figure 9.7 *Expected utility as a function of mean and standard deviation.* We show contour plots of functions $(\sigma, \mu) \mapsto EU(R)$, where R follows a normal distribution $N(\mu, \sigma^2)$. (a) $U(x) = \log x$; (b) $U(x) = x^{1-\gamma}/(1 - \gamma)$ with $\gamma = 5$.

Figure 9.7 shows contour plots of functions $(\sigma, \mu) \mapsto EU(R)$, where R follows distribution $R \sim 1 + r$, where $r \sim N(\mu_0, \sigma_0^2)$, where $\mu_0 = \mu/250$, $\sigma_0 = \sigma/\sqrt{250}$. In panel (a) the utility function is logarithmic $U(x) = \log x$ and in panel (b) $U(x) = x^{1-\gamma}/(1 - \gamma)$ with $\gamma = 5$.[8] The expected utility is maximized when the mean is high and the standard deviation is low, which happens in the upper left corner. We see that for the logarithmic utility the expected utility is determined

---

8 In panel (a) we have multiplied the values of $E(U(R))$ with 1000 and in panel (b) with 10.

by the expectation, but increasing the risk aversion to $\gamma = 5$ makes the expected utility sensitive both to mean and to standard deviation. When risk aversion is increased more, then the expected utility becomes sensitive only to standard deviation.

9.2.2.3 Taylor Expansion of the Utility

A Taylor expansion of a utility function can be used to gain insight into the differences between the use of the mean–variance criterion and the use of the expected utility, because the use of the mean–variance criterion is approximately equal to the use of the second-order Taylor expansion to approximate the utility function. Also, we can use a Taylor expansion to replace the expected utility with a series containing higher than the second-order moment, which leads to a useful tool in portfolio selection.

We restrict ourselves to the fourth-order Taylor expansion, because the extension to higher order expansions is obvious. For a utility function $U : \mathbf{R} \to \mathbf{R}$ that has fourth-order continuous derivatives we have the approximation

$$U(x + h) \approx U(x) + hU'(x) + \frac{1}{2}h^2 U''(x) + \frac{1}{6}h^3 U^{(3)}(x) + \frac{1}{24}h^4 U^{(4)}(x),$$

where $x, h \in \mathbf{R}$. This approximation holds for a power utility function $U : (0, \infty) \to \mathbf{R}$, when $x > 0$ and $x + h > 0$. We can write

$$b'_t R_{t+1} = b'_t R^e_{t+1} + R^0_{t+1},$$

where R^0_{t+1} is the risk-free rate and

$$R^e_{t+1} = R_{t+1} - R^0_{t+1}$$

is the vector of the excess gross returns. We can choose also $R^0_{t+1} = 1$, so that R^e_{t+1} is the net return instead of the excess return. When we take $x = R^0_{t+1}$ and $h = b'_t R^e_{t+1}$, then we obtain the approximation

$$U\left(b'_t R_{t+1}\right) \approx a_0 + a_1 b'_t R^e_{t+1} + \frac{a_2}{2}\left(b'_t R^e_{t+1}\right)^2 + \frac{a_3}{6}\left(b'_t R^e_{t+1}\right)^3 + \frac{a_4}{24}\left(b'_t R^e_{t+1}\right)^4,$$

where $a_0 = U(R^0_{t+1})$, $a_1 = U'(R^0_{t+1})$, $a_2 = U''(R^0_{t+1})$, $a_3 = U^{(3)}(R^0_{t+1})$, and $a_4 = U^{(4)}(R^0_{t+1})$.[9] As a special case, when $U(x) = \log(x)$, then the fourth-order Taylor expansion leads to the approximation

$$\log\left(b'_t R_{t+1}\right) \approx b'_t R^e_{t+1} - \frac{1}{2}\left(b'_t R^e_{t+1}\right)^2 + \frac{1}{3}\left(b'_t R^e_{t+1}\right)^3 - \frac{1}{4}\left(b'_t R^e_{t+1}\right)^4.$$

Figure 9.8 shows the first four Taylor approximations to the logarithmic utility. The black curve shows log-utility $x \mapsto \log x$, the blue curve shows

---

9 For example, when U is a power utility function with $\gamma \neq 1$ and $R^0_{t+1} = 1$, then $a_0 = U(1) = 1/(1 - \gamma)$, $a_1 = U'(1) = 1$, $a_2 = U''(1) = -\gamma$, $a_3 = U^{(3)}(1) = \gamma(\gamma + 1)$, and $a_4 = U^{(4)}(1) = -\gamma(\gamma + 1)(\gamma + 2)$.

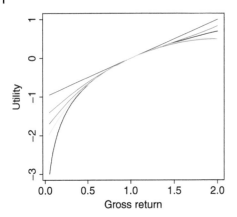

Figure 9.8 *Approximation of logarithmic utility.* The black curve shows log-utility $x \mapsto \log x$, the blue curve shows the linear approximation, the red curve shows the quadratic approximation, the green curve shows the third-order approximation, and the yellow curve shows the fourth-order approximation.

the linear function $x \mapsto x - 1$, the red curve shows the quadratic function $x \mapsto x - 1 - (x-1)^2/2$, the green curve shows the third-order polynomial $x \mapsto x - 1 - (x-1)^2/2 + (x-1)^3/3$, and the yellow curve shows the fourth-order polynomial $x \mapsto x - 1 - (x-1)^2/2 + (x-1)^3/3 - (x-1)^4/4$. The approximations are accurate when the gross return is close to one. A gross return close to one means that the asset price has not changed much. However, when the gross return is close to zero or much larger than one, then even the fourth-order approximation is not accurate. Thus, using the logarithmic utility in portfolio selection leads to taking large fluctuations into account, and in particular, the logarithmic utility is better than any finite approximation when we consider portfolios with extreme tail risk: the logarithmic utility approaches $-\infty$ when the gross return approaches zero.

9.2.2.4 Risk Aversion
We can classify utility functions using measures of risk aversion.

CARA Utility Functions The coefficient of absolute risk aversion of utility function U at point x is defined as

$$-\frac{U''(x)}{U'(x)}. \tag{9.30}$$

Utility functions with constant absolute risk aversion are called CARA utility functions. For example, the exponential utility functions, defined in (9.29), are CARA utility functions and have the coefficient of absolute risk aversion α, whereas the power utility functions, defined in (9.28), are not CARA utility functions because they have the coefficient of absolute risk aversion γx^{-1}. When an investor whose wealth is 100 is willing to risk 50, and after reaching wealth 1000, is still willing to risk 50, then the investor has constant absolute risk aversion. Most investors have decreasing absolute risk aversion (so that after reaching wealth 1000, the investor is willing to risk more than 50).

CRRA Utility Functions The coefficient of relative risk aversion of utility function U at point x is defined as

$$-x \frac{U''(x)}{U'(x)}. \tag{9.31}$$

Utility functions with constant relative risk aversion are called CRRA utility functions. For example, the power utility functions are CRRA utility functions and have the coefficient of relative risk aversion γ, whereas the exponential utility functions are not CRRA utility functions because they have the coefficient of relative risk aversion αx. When an investor whose wealth is 100 is willing to risk 50, and after reaching wealth 1000, is willing to risk 500, then the investor has constant relative risk aversion.

Expected Utility and Portfolio Weights It is helpful to plot a curve that shows estimates of the expected utility for a scale of risk aversion parameters. The expected utility curve is the function

$$\gamma \mapsto Eu_\gamma(R_{t+1}), \qquad \gamma > 0, \tag{9.32}$$

where

$$u_\gamma(x) = \frac{U_\gamma(x) - U_\gamma(1)}{U_\gamma(e) - U_\gamma(1)},$$

and U_γ is the power utility function, defined in (9.28). Since the expected value is unknown, we have to estimate it using a sample average of historical values.

In the case of two basis assets, it is helpful to look at functions

$$b \mapsto Eu_\gamma\left((1-b)R_{t+1}^1 + bR_{t+1}^2\right) \tag{9.33}$$

for various values of b, where R_{t+1}^1 and R_{t+1}^2 are the gross returns of the two basis assets.

Figure 9.9 considers daily S&P 500 and Nasdaq-100 data, described in Section 2.4.2. Panel (a) shows functions (9.32) for S&P 500 (black) and Nasdaq-100 (red). We see that Nasdaq-100 is better for a risky investor but S&P 500 is better for a risk averse investor. Panel (b) shows functions (9.33) for $\gamma = 1$ (blue) and $\gamma = 2$ (green). Here R_{t+1}^1 is the return of S&P 500 and R_{t+1}^2 is the return of Nasdaq-100. The optimal value of weight b is indicated by vertical lines. We see that when risk aversion γ increases, then the weight b of Nasdaq-100 decreases.

Figure 9.10 considers monthly S&P 500 data, described in Section 2.4.3. Panel (a) shows functions

$$\gamma \mapsto Eu_\gamma((1-b) + bR_{t+1}), \qquad \gamma > 0, \tag{9.34}$$

where R_{t+1} is the gross return of S&P 500. Thus, $(1-b) + bR_{t+1}$ is the gross return of a portfolio whose components are the risk-free rate with gross return

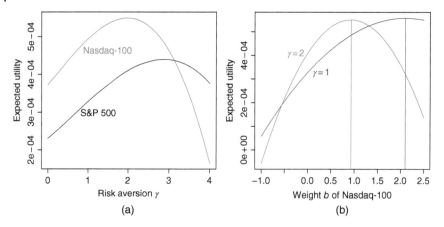

Figure 9.9 *Portfolio selection: S&P 500 and Nasdaq-100.* (a) Functions (9.32) for S&P 500 (black) and Nasdaq-100 (red); (b) functions (9.33) for $\gamma = 1$ (blue) and $\gamma = 2$ (green).

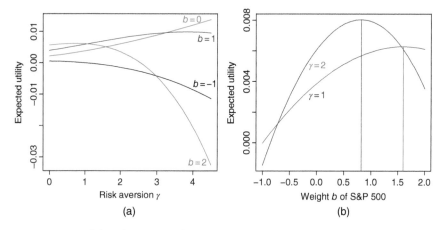

Figure 9.10 *Portfolio selection: Risk-free rate and S&P 500.* (a) Functions (9.34) for $b = -1$ (black), $b = 0$ (red), $b = 1$ (blue), and $b = 2$ (green); (b) functions (9.35) for $\gamma = 1$ (purple) and $\gamma = 2$ (dark green).

one and the S&P 500. We show cases $b = -1$ (black), $b = 0$ (red), $b = 1$ (blue), and $b = 2$ (green). Panel (b) shows functions

$$b \mapsto Eu_\gamma((1-b) + bR_{t+1}), \qquad b \in \mathbf{R} \tag{9.35}$$

for $\gamma = 1$ (purple) and $\gamma = 2$ (dark green). The optimal value of weight b is indicated by vertical lines.

9.2.3 Stochastic Dominance

Sometimes a return distribution stochastically dominates another return distribution, so that it is preferred regardless of the chosen utility function. However, stochastic dominance occurs rarely in practice.

The distribution of $X \in \mathbf{R}$ stochastically dominates the distribution of $Y \in \mathbf{R}$, if

$$F_X(t) \leq F_Y(t) \tag{9.36}$$

for all $t \in \mathbf{R}$, where $F_X(t) = P(X \leq t)$ and $F_Y(t) \leq P(Y \leq t)$ are the distribution functions. Inequality (9.36) is equivalent to

$$P(X \geq t) \geq P(Y \geq t) \tag{9.37}$$

for all $t \in \mathbf{R}$.

Stochastic dominance is also called first-order stochastic dominance to distinguish it from second-order stochastic dominance. The distribution of $X \in \mathbf{R}$ second-order dominates stochastically the distribution of $Y \in \mathbf{R}$, if

$$\int_{-\infty}^{x} F_X(t) \, dt \leq \int_{-\infty}^{x} F_Y(t) \, dt$$

for all $x \in \mathbf{R}$.

Figure 9.11 shows an example of first-order stochastic dominance. Panel (a) shows the densities of two distributions, and the distribution of the black density stochastically dominates the distribution of the red density. The densities have the same shape but the black density is located to the right of the red density. Panel (b) shows the distribution functions.

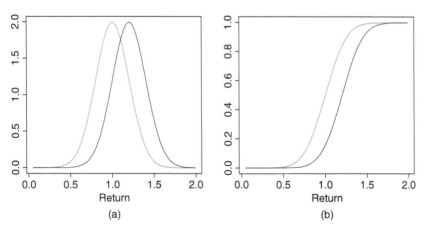

Figure 9.11 *First-order stochastic dominance.* The black distribution dominates the red distribution. (a) Density functions; (b) distribution functions.

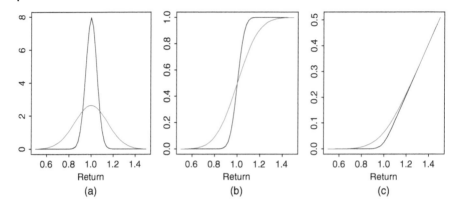

Figure 9.12 *Second-order stochastic dominance.* The black distribution dominates the red distribution. (a) Density functions; (b) distribution functions; (c) functions $G_X(x) = \int_{-\infty}^{x} F_X(t)\, dt$, $G_Y(x) = \int_{-\infty}^{x} F_Y(t)\, dt$, where F_X and F_Y are the distribution functions.

Figure 3.8(b) shows two empirical distribution functions, which are such that neither of the distribution functions dominates the other.

Figure 9.12 shows an example of second-order stochastic dominance. The distribution of the black density dominates the distribution of the red density. Panel (a) shows the densities of the two distributions, panel (b) shows the distribution functions, and panel (c) shows the functions $G_X(x) = \int_{-\infty}^{x} F_X(t)\, dt$, $G_Y(x) = \int_{-\infty}^{x} F_Y(t)\, dt$, where F_X and F_Y are the distribution functions. The black and the red densities have the same location, but the red distribution has a larger variance than the black distribution.

When a return distribution second-order dominates stochastically another return distribution, then it is preferred, regardless of risk aversion. In fact, it holds that the distribution of X second-order dominates the distribution of Y if and only if

$$EU(X) \geq EU(Y)$$

for every increasing and concave utility function $U : \mathbf{R} \to \mathbf{R}$, which is two times continuously differentiable. First-order stochastic dominance occurs if and only if the dominant distribution has a higher expected utility for all increasing and continuously differentiable utility functions.

9.3 Multiperiod Portfolio Selection

In the multiperiod model the wealth of the portfolio is obtained from (9.10) as

$$W_T = W_0 \prod_{t=0}^{T-1} \bar{b}_t \cdot \bar{R}_{t+1}, \tag{9.38}$$

where $W_0 > 0$ is the initial wealth at time 0, $\bar{b}_t \in \mathbf{R}^{d+1}$ is the vector of the portfolio weights, and $\bar{R}_{t+1} \in \mathbf{R}^{d+1}$ is the vector of gross returns of the $d+1$ portfolio components. We write

$$\bar{b}_t = (b_t^0, \dots, b_t^d), \qquad \bar{R}_{t+1} = (R_{t+1}^0, \dots, R_{t+1}^d),$$

where $R_{t+1}^i = S_{t+1}^i / S_t^i$, $i = 0, \dots, d$. The portfolio weights satisfy

$$\sum_{i=0}^d b_t^i = 1 \qquad \Leftrightarrow \qquad \bar{b}_t \cdot 1_{d+1} = 1. \tag{9.39}$$

Now $\bar{b}_t \cdot \bar{R}_{t+1} = W_{t+1}/W_t$ is the one period gross return. We can write

$$\bar{b}_t \cdot \bar{R}_{t+1} = R_{t+1}^0 + b_t \cdot (R_{t+1} - R_{t+1}^0), \tag{9.40}$$

$b_t \in \mathbf{R}^d$. In this way we do not have to worry about the restriction (9.39).

The wealth of the portfolio is obtained in additive form from (9.13) as

$$W_T = S_T^0 V_T, \tag{9.41}$$

where

$$V_T = V_0 + \sum_{t=0}^{T-1} \xi_{t+1} \cdot (X_{t+1} - X_t).$$

Here

$$V_t = \frac{W_t}{S_t^0}, \qquad X_t = \frac{S_t}{S_t^0}, \qquad S_t = (S_t^1, \dots, S_t^d).$$

The vector $\xi_{t+1} = (\xi_{t+1}^1, \dots, \xi_{t+1}^d)$ gives the numbers of risky assets in the portfolio. Vector $\xi_{t+1} \in \mathbf{R}^d$ is unrestricted. Note that the time indexing is such that ξ_{t+1} and b_t both describe the portfolio for the period $[t, t+1]$. Note that we have assumed in (9.41) that $S_t^0 > 0$ almost surely. This holds when S^0 is a risk-free investment.

The multiplicative way of writing the wealth presupposes a positive wealth, whereas the additive way of writing the wealth allows for a nonpositive wealth. The multiplicative way of writing the wealth is convenient for the power utility functions, whereas the additive way of writing the wealth is convenient for the exponential utility functions, because factoring the wealth as a product makes the writing of the backward induction convenient.

At time $t = 0$ we want to find the portfolio vector b_0 or ξ_1 so that

$$E_0 U(W_T) \tag{9.42}$$

is maximized, where the rebalancing of the portfolio will be made at times $1, \dots, T-1$. The maximization of (9.42) is over the sequence of weights $b_0, \dots, b_{T-1}$ or over the sequence of numbers $\xi_1, \dots, \xi_T$, although at time $t = 0$ we need to choose only b_0 or ξ_1.

We can summarize the results in the following way:

1) For the logarithmic utility function the multiperiod portfolio selection reduces to the one-period portfolio selection.
2) For the power utility functions (which include the logarithmic utility) and for the exponential utility functions the optimal portfolio vector does not depend on the initial wealth.

The power utility functions need a positive wealth as the argument, and they can be applied when the wealth process is written in the product form. The exponential utility functions can take a negative wealth as the argument, and they lead to tractable dynamic programming when the wealth process is written additively.

We describe first the one-period optimization in Section 9.3.1, and then the multiperiod optimization in Section 9.3.2. The understanding of the multiperiod optimization is easier when it is contrasted with the single period optimization. We describe first the case of the logarithmic utility function. After that we describe the solution for the power utility functions. Third, we describe the case of the exponential utility functions. In the multiperiod model we give also the formulas for arbitrary utility functions.

9.3.1 One-Period Optimization

We want to maximize at time 0 the expected utility of the wealth at time 1:

$$E_0 U(W_1).$$

We discuss the cases where U is the logarithmic utility function, a power utility function, and an exponential utility function.

9.3.1.1 The Logarithmic Utility Function

The logarithmic utility function is $U(x) = \log x$. We have

$$U(W_0 \bar{b}_0 \cdot \bar{R}_1) = \log W_0 + \log(\bar{b}_0 \cdot \bar{R}_1).$$

Thus, we need to maximize

$$E_0 \log(\bar{b}_0 \cdot \bar{R}_1)$$

over $\bar{b}_0$, under restriction $\bar{b}_0 \cdot 1_{d+1} = 1$. Thus, the optimal $\bar{b}_0$ does not depend on the initial wealth W_0.

The maximization can be done unrestricted when we apply (9.40), so that we need to maximize

$$E_0 \log \left[R_1^0 + b_0 \cdot (R_1 - R_1^0) \right]$$

over $b_0 \in \mathbf{R}^d$.

9.3.1.2   The Power Utility Functions

The power utility functions are $U(x) = x^{1-\gamma}/(1-\gamma)$ for $x > 0$, where $\gamma > 0$, $\gamma \neq 1$. We have

$$U(W_0 \bar{b}_0 \cdot \bar{R}_1) = \frac{W_0^{1-\gamma}}{1-\gamma} (\bar{b}_0 \cdot \bar{R}_1)^{1-\gamma}.$$

Thus, we need to maximize

$$E_0 \left[(\bar{b}_0 \cdot \bar{R}_1)^{1-\gamma} \right] \tag{9.43}$$

over $\bar{b}_0$, under restriction $\bar{b}_0 \cdot 1_{d+1} = 1$. Thus, the optimal $\bar{b}_0$ does not depend on the initial wealth W_0.

The maximization can be done unrestricted when we apply (9.40), so that we need to maximize

$$E_0 \left[\left(R_1^0 + b_0 \cdot (R_1 - R_1^0) \right)^{1-\gamma} \right]$$

over $b_0 \in \mathbf{R}^d$.

9.3.1.3   The Exponential Utility Functions

The exponential utility functions are $U(x) = 1 - e^{-\alpha x}$ for $x \in \mathbf{R}$, where $\alpha > 0$. The maximization of $E_0 U(W_1)$ is equivalent to the minimization of

$$E_0 \exp\{-\alpha W_1\}.$$

We apply the additive form in (9.41) to obtain

$$\exp\{-\alpha W_1\} = \exp\{-\alpha S_1^0 (V_0 + \xi_1 \cdot (X_1 - X_0))\}$$
$$= \exp\{-\alpha S_1^0 V_0\} \exp\{-\alpha S_1^0 \xi_1 \cdot (X_1 - X_0)\}.$$

Thus, we need to minimize

$$E_0 \exp\{-\alpha S_1^0 \xi_1 \cdot (X_1 - X_0)\}$$

over $\xi_1 \in \mathbf{R}^d$. This is unrestricted minimization. The optimal ξ_1 does not depend on the initial wealth W_0.

9.3.2   The Multiperiod Optimization

We want to maximize at time 0 the expected utility of the wealth at time T:

$$E_0 U(W_T).$$

We discuss the cases where U is the logarithmic utility function, a power utility function, an exponential utility function, and a general utility function.

9.3.2.1 The Logarithmic Utility Function

For the logarithmic utility $U(x) = \log x$ we have from (9.38) that

$$\log(W_T) = \log(W_0) + \sum_{t=0}^{T-1} \log(\bar{b}_t \cdot \bar{R}_{t+1}),$$

where

$$\bar{b}_t \cdot 1_{d+1} = 1.$$

We want to find portfolio vector $\bar{b}_0$ maximizing

$$E_0 \log(W_T),$$

under restriction $\bar{b}_0 \cdot 1_{d+1} = 1$. We see that for the logarithmic utility the optimal portfolio vector at time $t = 0$ is the maximizer over $\bar{b}_0$ of the single period expected logarithmic return

$$E_0 \log(\bar{b}_0 \cdot \bar{R}_1),$$

when the maximization is done under restriction $\bar{b}_0 \cdot 1_{d+1} = 1$. In particular, the initial wealth W_0 does not affect the solution.

We can use (9.40) to note that the maximization can be done without the restriction: we need to maximize

$$E_0 \log \left[R_1^0 + b_0 \cdot \left(R_1 - R_1^0 \right) \right]$$

over $b_0 \in \mathbf{R}^d$.

We have shown that in the case of the logarithmic utility the multiperiod optimization reduces to the single period optimization, since b_0 can be found by ignoring the time points $t = 1, \ldots, T - 1$.

9.3.2.2 The Power Utility Functions

Let $U : (0, \infty) \to \mathbf{R}$ be the power utility function

$$U(x) = \frac{x^{1-\gamma}}{1 - \gamma},$$

where $\gamma > 0$ is the risk aversion parameter. For $\gamma = 1$ we define $U(x) = \log x$. For $\gamma \neq 1$ we get from (9.38) that

$$U(W_T) = \frac{W_0^{1-\gamma}}{1 - \gamma} \prod_{t=0}^{T-1} (\bar{b}_t \cdot \bar{R}_{t+1})^{1-\gamma}.$$

Thus, for $\gamma \neq 1$ we need to maximize

$$E_0 \prod_{t=0}^{T-1} u(\bar{b}_t \cdot \bar{R}_{t+1}) \tag{9.44}$$

under restrictions

$$\bar{b}_t \cdot 1_{d+1} = 1, \qquad t = 0, \dots, T-1,$$

where

$$u(x) = x^{1-\gamma}.$$

Thus, the optimal portfolio vector $\bar{b}_t$ does not depend on the initial wealth W_0.
Using (9.40) we see that we can maximize

$$E_0 \prod_{t=0}^{T-1} u\left(R_{t+1}^0 + b_t \cdot \left(R_{t+1} - R_t^0\right)\right)$$

over $b_t \in \mathbf{R}^d$, $t = 0, \dots, T-1$.

The Case T = 2 Let us consider the maximization of (9.44) for the case $T = 2$.
We present first the case $T = 2$ because the structure of the argument is visible
already in the two period case but this case is notationally more transparent
than the general case $T \geq 2$. Define function

$$F(\bar{b}_0) = \max_{\bar{b}_1 \in B} E_0 \left[u\left(\bar{b}_0 \cdot \bar{R}_1\right) u\left(\bar{b}_1 \cdot \bar{R}_2\right)\right],$$

where

$$B = \left\{\bar{b} \in \mathbf{R}^{d+1} : \bar{b} \cdot 1_{d+1} = 1\right\}.$$

The optimal portfolio vector at time $t = 0$ is the maximizer of function F: Our
purpose is to find

$$\operatorname*{argmax}_{\bar{b}_0 \in B} F(\bar{b}_0).$$

We can write, using the law of the iterated expectations,

$$E_0 \left[u\left(\bar{b}_0 \cdot \bar{R}_1\right) u\left(\bar{b}_1 \cdot \bar{R}_2\right)\right] = E_0 E_1 \left[u\left(\bar{b}_0 \cdot \bar{R}_1\right) u\left(\bar{b}_1 \cdot \bar{R}_2\right)\right]$$
$$= E_0 \left[u\left(\bar{b}_0 \cdot \bar{R}_1\right) E_1 u\left(\bar{b}_1 \cdot \bar{R}_2\right)\right].$$

Thus,

$$F(\bar{b}_0) = E_0 \left[u(\bar{b}_0 \cdot \bar{R}_1) \max_{\bar{b}_1 \in B} E_1 u(\bar{b}_1 \cdot \bar{R}_2)\right].$$

Comparing to (9.43) we see the difference between the one- and two-period
portfolio selections: In the two-period case, we have the additional multiplier
$\max_{\bar{b}_1 \in B} E_1 u(\bar{b}_1 \cdot \bar{R}_2)$.
We can use (9.40) to note that the maximization can be done without the
restrictions. Write

$$\bar{b}_0 \cdot \bar{R}_1 = R_1^0 + b_0 \cdot \left(R_1 - R_1^0\right), \qquad \bar{b}_1 \cdot \bar{R}_2 = R_2^0 + b_1 \cdot \left(R_2 - R_2^0\right),$$

where $b_0 \in \mathbf{R}^d$ and $b_1 \in \mathbf{R}^d$ are unrestricted.

The Case $T \geq 2$ Let us consider the maximization of (9.44). Define

$$F(\bar{b}_0) = \max_{(\bar{b}_1,\ldots,\bar{b}_{T-1}) \in C} E_0 \prod_{t=0}^{T-1} u(\bar{b}_t \cdot \bar{R}_{t+1}),$$

where , $u(x) = x^{1-\gamma}$,

$$C = B^{T-1}, \qquad B = \left\{ \bar{b} \in \mathbf{R}^{d+1} : \bar{b} \cdot 1_{d+1} = 1 \right\}.$$

Here B^{T-1} means the $T - 1$ fold product $B \times \cdots \times B$. The optimal portfolio vector a time $t = 0$ is the maximizer of function F: Our purpose is to find

$$\hat{b}_0 = \underset{\bar{b}_0 \in B}{\operatorname{argmax}} F(\bar{b}_0).$$

We give a recursive formula for $\hat{b}_0$.

1) Denote

$$F_{T-1}(\bar{b}_{T-1}) = E_{T-1} u(\bar{b}_{T-1} \cdot \bar{R}_T)$$

and let $\hat{b}_{T-1}$ be the maximizer of $F_{T-1}(\bar{b}_{T-1})$ over $\bar{b}_{T-1} \in B$.
2) For $t = T - 2, \ldots, 0$, define

$$F_t(\bar{b}_t) = E_t \left[u(\bar{b}_t \cdot \bar{R}_{t+1}) F_{t+1}(\hat{b}_{t+1}) \right]$$

and let $\hat{b}_t$ be the maximizer of $F_t(\bar{b}_t)$ over $\bar{b}_t \in B$.

We can use (9.40) to note that the maximization can be done without the restrictions: we can define the functions to be maximized as

$$F_{T-1}(b_{T-1}) = E_{T-1} u \left(R_T^0 + b_T \cdot \left(R_T - R_T^0 \right) \right)$$

and

$$F_t(b_t) = E_t \left[U \left(R_{t+1}^0 + b_t \cdot \left(R_{t+1} - R_{t+1}^0 \right) \right) F_{t+1}(\hat{b}_{t+1}) \right],$$

$t = T - 2, \ldots, 0$. The maximization is done over $b_{T-1} \in \mathbf{R}^d$ in the one before the previous function, and over $b_t \in \mathbf{R}^d$ in the previous function.

9.3.2.3   The Exponential Utility Functions

Let $U : (0, \infty) \to \mathbf{R}$ be the exponential utility function

$$U(x) = 1 - \exp\{-\alpha x\},$$

where $\alpha > 0$ is the risk aversion parameter. Maximizing $EU(W_T)$ is equivalent to minimizing $E \exp\{-\alpha W_T\}$. We get from (9.41) that

$$\exp\{-\alpha W_T\} = \exp\left\{-\alpha S_T^0 V_0\right\} \prod_{t=0}^{T-1} \exp\left\{-\alpha S_T^0 \xi_{t+1} \cdot (X_{t+1} - X_t)\right\}.$$

Thus, we need to minimize

$$E_0 \prod_{t=0}^{T-1} u(\xi_{t+1} \cdot (X_{t+1} - X_t))$$

over $\xi_1, \dots, \xi_T \in \mathbf{R}^d$, where

$$u(x) = \exp\left\{-\alpha S_T^0\right\}.$$

Thus, the optimal portfolio vector ξ_1 does not depend on the initial wealth W_0.
Define

$$F(\xi_1) = \min_{\xi_2, \dots, \xi_T \in \mathbf{R}^d} E_0 \prod_{t=0}^{T-1} u(\xi_{t+1} \cdot (X_{t+1} - X_t)).$$

The optimal portfolio vector a time $t = 0$ is the minimizer of function F: Our purpose is to find

$$\hat{\xi}_1 = \operatorname*{argmin}_{\xi_1 \in \mathbf{R}^d} F(\xi_1).$$

We give a recursive formula for $\hat{\xi}_1$.

1) Denote

$$F_{T-1}(\xi_T) = E_{T-1} u(\xi_T \cdot (X_T - X_{T-1}))$$

and let $\hat{\xi}_T$ be the minimizer of $F_{T-1}(\xi_T)$ over $\xi_T \in \mathbf{R}^d$.
2) For $t = T - 2, \dots, 0$, define

$$F_t(\xi_{t+1}) = E_t \left[u(\xi_{t+1} \cdot (X_{t+1} - X_t)) F_{t+1}(\hat{\xi}_{t+2}) \right]$$

and let $\hat{\xi}_{t+1}$ be the minimizer of $F_t(\xi_{t+1})$ over $\xi_{t+1} \in \mathbf{R}^d$.

9.3.2.4 General Utility Functions

When the utility function is not the logarithmic function, of a power form, or of an exponential form, then the maximizer of the expected wealth can depend on the initial wealth. Also, for the general utility functions we do not obtain such factorization as for the logarithmic and power utility functions (when the product form is used) or such factorization as for the exponential utility functions (when the additive form is used). However, we can obtain recursive formulas for the maximization of the expected wealth.

The Product Form The optimal portfolio vector at time $t = 0$ is defined as the maximizer of the expected utility $EU(W_T)$. Thus, the optimal portfolio vector maximizes function F, defined as

$$F(\bar{b}_0) = \max_{(\bar{b}_1, \dots, \bar{b}_{T-1}) \in C} E_0 U \left(W_0 \prod_{t=0}^{T-1} \bar{b}_t \cdot \bar{R}_{t+1} \right),$$

where

$$C = B^{T-1}, \qquad B = \{\bar{b} \in \mathbf{R}^{d+1} : \bar{b} \cdot 1_{d+1}\}.$$

We can define the optimal portfolio vector recursively as follows:

1) Define

$$F_{T-1}(\bar{b}_0, \dots, \bar{b}_{T-1}) = E_{T-1}U\left(W_0 \prod_{t=0}^{T-1} \bar{b}_t \cdot \bar{R}_{t+1} \right)$$

and let $\hat{b}_{T-1} = \hat{b}_{T-1}(\bar{b}_0, \dots, \bar{b}_{T-2})$ be the maximizer of $F_{T-1}(\bar{b}_0, \dots, \bar{b}_{T-1})$ over $\bar{b}_{T-1} \in B$.

2) For $t = T - 2, \dots, 0$, define

$$F_t(\bar{b}_0, \dots, \bar{b}_t) = E_t U\left(W_0 \prod_{i=0}^{t} \bar{b}_i \cdot \bar{R}_{i+1} \prod_{i=t+1}^{T-1} \hat{b}_i \cdot \bar{R}_{i+1} \right),$$

where

$$\hat{b}_i = \hat{b}_i(\bar{b}_0, \dots, \bar{b}_t, \hat{b}_{t+1}, \dots, \hat{b}_{i-1}).$$

Let $\hat{b}_t = \hat{b}_t(\bar{b}_0, \dots, \bar{b}_{t-1})$ be the maximizer of $F_t(\bar{b}_0, \dots, \bar{b}_t)$ over $\bar{b}_t \in B$.

This gives a recursive definition of $\hat{b}_0$.[10]

The Additive Form The optimal portfolio vector at time $t = 0$ is defined as the maximizer of the expected utility $EU(W_T)$. Thus, the optimal portfolio vector maximizes function F, defined as

$$F(\xi_1) = \max_{\xi_2, \dots, \xi_{T-1} \in \mathbf{R}^d} E_0 U\left(W_0 + B_T \sum_{t=0}^{T-1} \xi_{t+1} \cdot (X_{t+1} - X_t) \right).$$

---

10 For example, in the two-period case

$$F(b_t) = \max_{b_{t+1}} E_t U\left(W_t \cdot b_t' R_{t+1} \cdot b_{t+1}' R_{t+2} \right).$$

We can write, using the law of the iterated expectations,

$$E_t U\left(W_t \cdot b_t' R_{t+1} \cdot b_{t+1}' R_{t+2} \right) = E_t E_{t+1} U\left(W_t \cdot b_t' R_{t+1} \cdot b_{t+1}' R_{t+2} \right).$$

Thus, in the two-period case,

$$F(b_t) = E_t \left[\max_{b_{t+1}} E_{t+1} U\left(W_t \cdot b_t' R_{t+1} \cdot b_{t+1}' R_{t+2} \right) \right].$$

We can define the optimal portfolio vector recursively as follows:

1) Define

$$F_{T-1}(\xi_1, \dots, \xi_T) = E_{T-1} U \left(W_0 + B_T \sum_{t=0}^{T-1} \xi_{t+1} \cdot (X_{t+1} - X_t) \right)$$

and let $\hat{\xi}_T = \hat{\xi}_T(\xi_1, \dots, \xi_{T-1})$ be the maximizer of $F_{T-1}(\xi_1, \dots, \xi_T)$ over $\xi_T \in \mathbf{R}^d$.

2) For $t = T - 2, \dots, 0$, define

$$F_t(\xi_1, \dots, \xi_t) = E_t U \left(W_0 + B_T \sum_{i=0}^{t} \xi_{i+1} \cdot (X_{i+1} - X_i) \right.$$

$$\left. + \sum_{i=t+1}^{T-1} \hat{\xi}_{i+1} \cdot (X_{i+1} - X_i) \right),$$

where

$$\hat{\xi}_{i+1} = \hat{\xi}_{i+1}(\xi_1, \dots, \xi_{t+1}, \hat{\xi}_{t+2}, \dots, \hat{\xi}_i).$$

Let $\hat{\xi}_{t+1} = \hat{\xi}_{t+1}(\xi_1, \dots, \xi_t)$ be the maximizer of $F_t(\xi_1, \dots, \xi_{t+1})$ over $\xi_{t+1} \in \mathbf{R}^d$.

This gives a recursive definition of $\hat{\xi}_1$.

10

Performance Measurement

When the performance of a fund is measured, there is a temptation to look only at the past return on the investment. However, it is important to measure the performance by taking the risk into account. An investor can increase both the expected return and the risk by leveraging, so that it is of interest to find the inherent quality of the fund, and leave the choice of the leveraging factor to the investor. The Sharpe ratio is defined as the ratio of the expected excess return to the standard deviation of the excess return. This is an example of a performance measure that penalizes the expected return with the risk.

The measures of performance are usually single numbers, but we cannot hope to completely reduce the characteristics of a fund into a single number. For example, the Sharpe ratio is a single number, but we obtain more information by giving separately the expected excess return and the standard deviation of the excess return, instead of giving only their ratio.

Section 9.2 discussed the ranking of return distributions from the point of view of portfolio selection. This discussion is relevant for the performance measurement. For example, in portfolio selection we could be interested in the conditional expected utility

$$E[U(R_{t+1}) \mid X_t = x], \tag{10.1}$$

where U is a utility function, R_{t+1} is a return, and the expectation is taken conditionally on the state variable X_t having value x. If we would have information about the conditional expected utility of all portfolios, then we could choose an optimal portfolio. When R_{t+1} is the return of a fund, then the conditional expected utility characterizes the properties of the fund. However, since the conditional expected utility is a function of x, this is a very complicated way to describe the properties of a fund. We can summarize the performance using a single number, and the unconditional expected utility

$$EU(R_{t+1})$$

Nonparametric Finance, First Edition. Jussi Klemelä.
© 2018 John Wiley & Sons, Inc. Published 2018 by John Wiley & Sons, Inc.

is a single number summarizing the performance. The unconditional expected utility averages over all states x.

Has the past performance of a fund been due to a luck or to a skill? This question could be answered by testing the null hypothesis, which states that the long-time performance of a fund is equal to the performance of a market index. Testing is related to the construction of confidence bands to a performance measure. Hypothesis testing and confidence bands can be used to try to answer the question whether the better performance of a fund, as compared to the performance of another fund, is due to a luck or skill. We address the issue of hypothesis testing and confidence bands only when the Sharpe ratio is the performance measure. We note that the confidence bands tend to be quite wide.

Performance measures are computed using a time period $[t_0, t_1]$ of historical returns. One has to address the issue whether the choice of the time period of historical returns affects the performance measures. This question is considered in Section 10.5, where we provide tools to simultaneously look at the all possible time intervals contained in $[t_0, t_1]$. These tools are an alternative to looking at the conditional expectation in (10.1). A problem with the conditional expectations in (10.1) is that we have to choose the conditioning state variables X_t, which is a difficult task, as there is a huge number of potentially useful conditioning variables. When we look at the performance of a fund over all subintervals of $[t_0, t_1]$, then we get clues about which conditioning variables are relevant for the performance of the fund. For example, we could find answers to the questions: Does this fund perform well only in the bull markets? Does this fund perform well only when the inflation is high?

Section 10.1 considers Sharpe ratio. Section 10.2 considers certainty equivalent. Section 10.3 discusses drawdown. Section 10.4 discusses alpha. Section 10.5 presents graphical tools to help the performance measurement.

10.1 The Sharpe Ratio

First, we give the definition of the Sharpe ratio. Then, we derive confidence intervals for the Sharpe ratio and test the equality of two Sharpe ratios. Finally, we give examples of some other measures of risk-adjusted return.

10.1.1 Definition of the Sharpe Ratio

The Sharpe ratio of a financial asset is defined as the expected excess return divided by the standard deviation of the excess return:[1]

$$ \text{Sh} = \frac{E\left(R_{t+1} - R_{t+1}^0\right)}{\text{sd}\left(R_{t+1} - R_{t+1}^0\right)}, \tag{10.2} $$

---

1 The returns of the risky asset and the risk-free returns can be either gross returns or net returns. The gross return is $R_{t+1} = S_{t+1}/S_t$ and the net return is $R_{t+1} - 1$, but the subtractions in the definition of the Sharpe ratio cancel the term -1. Sometimes logarithmic returns are used.

where R_{t+1} is the return of the asset, and R_{t+1}^0 is the return of a risk-free investment.

Note that we have not defined the Sharpe ratio as $E_t(R_{t+1} - R_{t+1}^0)/\mathrm{sd}_t(R_{t+1} - R_{t+1}^0)$, where the conditional expectation and the conditional standard deviation are used. Under this definition, we could write the Sharpe ratio as $(E_t R_{t+1} - R_{t+1}^0)/\mathrm{sd}_t(R_{t+1})$, since the risk-free rate for the period $t \mapsto t+1$ is not a random variable at time t, and thus it can be dropped from the conditional standard deviation. Instead, we have defined the Sharpe ratio in (10.2) using the unconditional expectation and the unconditional standard deviation, so that the risk-free rate is a random variable.

The return period can be, for example, 1 day, 1 month, or 1 year. The annualized Sharpe ratio is defined as

$$(\Delta t)^{-1/2} \frac{E\left(R_{t+1} - R_{t+1}^0\right)}{\mathrm{sd}\left(R_{t+1} - R_{t+1}^0\right)},$$

where Δt is the return horizon: $\Delta t = 1/250$ for daily returns, $\Delta t = 1/12$ for monthly returns, and so on.[2] The Sharpe ratio was defined by Sharpe (1966).

An estimator of the Sharpe ratio is obtained from historical returns $R_1, \dots, R_T$ and from historical risk-free rates $R_1^0, \dots, R_T^0$ by replacing the population mean and the population standard deviation by the sample mean and the sample standard deviation:

$$\widehat{\mathrm{Sh}} = \frac{\bar{X}}{s}, \tag{10.3}$$

where

$$\bar{X} = \frac{1}{T}\sum_{t=1}^{T} X_t, \qquad s = \left(\frac{1}{T}\sum_{t=1}^{T} X_t^2 - \bar{X}^2\right)^{1/2},$$

where $X_t = R_t - R_t^0$ is the excess return.

We can increase as much as we like the expected return of a given asset by leveraging. However, leveraging increases also the risk. The Sharpe ratio is invariant with respect to leveraging. Consider an asset with the expected excess return μ and the variance of the excess return $\sigma^2 > 0$:

$$E\left(R_{t+1} - R_{t+1}^0\right) = \mu, \qquad \mathrm{Var}\left(R_{t+1} - R_{t+1}^0\right) = \sigma^2,$$

where R_{t+1} is the time $t+1$ return of the risky asset and R_{t+1}^0 is the risk-free rate. Consider a portfolio of the risky asset and the risk-free rate, where the weight of the original asset is b and the weight of the risk-free asset is $1 - b$, where $b \in \mathbf{R}$. The return of the portfolio is

$$R_{t+1}^p = bR_{t+1} + (1-b)R_{t+1}^0.$$

---

2 The annualized mean is $(\Delta t)^{-1} \times E(R_{t+1} - R_{t+1}^0)$ and the annualized standard deviation is $(\Delta t)^{-1/2}\,\mathrm{sd}(R_{t+1} - R_{t+1}^0)$. The ratio of the annualized mean and the annualized standard deviation gives the annualized Sharpe ratio.

The excess return of the portfolio is

$$R^p_{t+1} - R^0_{t+1} = bR_{t+1} + (1-b)R^0_{t+1} - R^0_{t+1} = b\left(R_{t+1} - R^0_{t+1}\right).$$

Thus, $E(R^p_{t+1} - R^0_{t+1}) = b\mu$ and $\mathrm{Var}(R^p_{t+1} - R^0_{t+1}) = b^2\sigma^2$. Thus,

$$\frac{E\left(R^p_{t+1} - R^0_{t+1}\right)}{\mathrm{sd}\left(R^p_{t+1} - R^0_{t+1}\right)} = \begin{cases} \mathrm{sign}(b)\mu/\sigma, & b \neq 0, \\ 0, & b = 0. \end{cases}$$

10.1.2 Confidence Intervals for the Sharpe Ratio

Assume we observe historical excess returns $X_t = R_t - R^0_t$ for $t = 1, \ldots, T$. The estimator (10.3) of the Sharpe ratio, when multiplied by the annualizing factor, can be written as

$$\widehat{\mathrm{Sh}} = h(S_T),$$

where

$$h(x_1, x_2) = (\Delta t)^{-1/2} \frac{x_1}{\left(x_2 - x_1^2\right)^{1/2}},$$

$$S_T = \left(\bar{X}, T^{-1} \sum_{t=1}^{T} X_t^2\right)', \tag{10.4}$$

and $\bar{X} = T^{-1}\sum_{t=1}^{T} X_t$. Let us assume that

$$T^{1/2}(S_T - \theta) \xrightarrow{d} N(0, \Psi), \tag{10.5}$$

as $T \to \infty$, where $\theta = (EX, EX^2)'$ and Ψ is a 2×2 covariance matrix. The central limit theorem (10.5) holds at least when the observed excess returns are independent and identically distributed, and the fourth moment of the excess returns is finite. Independence does not hold for financial returns, but the central limit theorem holds when the returns are weakly dependent, as discussed in Section 3.5.1. An application of the delta-method gives[3]

$$T^{1/2}(h(S_T) - h(\theta)) \xrightarrow{d} N(0, \nabla h(\theta)' \Psi \nabla h(\theta)),$$

as $T \to \infty$, where $\nabla h(x) = (\partial h(x)/\partial x_1, \partial h(x)/\partial x_2)'$ is the gradient, and $h(\theta)$ is the Sharpe ratio. We have that

$$\nabla h(x) = (\Delta t)^{-1/2} \left(\frac{x_2}{\left(x_2 - x_1^2\right)^{3/2}}, -\frac{1}{2}\frac{x_1}{\left(x_2 - x_1^2\right)^{3/2}} \right).$$

---

3 The multivariate delta-method for statistics $T_n \in \mathbf{R}^d$ assumes that (1) function $h : \mathbf{R}^d \to \mathbf{R}$ has a gradient ∇h that is continuous at $\mu \in \mathbf{R}^d$, (2) statistics T_n satisfies $n^{1/2}(T_n - \mu) \xrightarrow{d} N(0, \Sigma)$, as $n \to \infty$. Then, $n^{1/2}(h(T_n) - h(\mu)) \xrightarrow{d} N(0, \nabla h(\mu)'\Sigma\nabla h(\mu))$ when $n \to \infty$.

The boundaries of the confidence interval for the Sharpe ratio $\text{Sh} = h(\theta)$ with the confidence level $0 < \alpha < 1$ are

$$\widehat{\text{Sh}} \pm z_{1-\alpha/2} T^{-1/2}\hat{\sigma},$$

where $z_{1-\alpha/2}$ is the $1 - \alpha/2$-quantile of the standard normal distribution,

$$\hat{\sigma} = \sqrt{\nabla h(S_T)'\hat{\Psi}\nabla h(S_T)}, \tag{10.6}$$

and $\hat{\Psi}$ is an estimator of Ψ. Indeed, $P(|\text{Sh} - \widehat{\text{Sh}}| \leq z_{1-\alpha/2} T^{-1/2}\hat{\sigma}) \approx P(|Z| \leq z_{1-\alpha/2}) = 1 - \alpha$, where $Z \sim N(0, 1)$.

10.1.2.1 Independent Returns

The central limit theorem (10.5) holds when the observed excess returns are independent and identically distributed, and the fourth moment of the excess returns is finite. In this case the asymptotic covariance matrix is

$$\Psi = \begin{bmatrix} \text{Var}(X) & \text{Cov}(X, X^2) \\ \text{Cov}(X, X^2) & \text{Var}(X^2) \end{bmatrix},$$

and estimator $\hat{\Psi}$ is obtained by using the sample variances and the sample covariance. We can write

$$\hat{\Psi} = T^{-1} \sum_{t=1}^{T} (Z_t - \bar{Z})(Z_t - \bar{Z})', \tag{10.7}$$

where $Z_t = (X_t, X_t^2)'$ and $\bar{Z} = S_T$ is defined in (10.4).

10.1.2.2 Dependent Returns

A central limit theorem holds when the dependence is weak. Let $(Z_t)_{t\in\mathbf{Z}}$ be a vector time series, where $Z_t \in \mathbf{R}^d$. A central limit theorem states that

$$T^{-1/2} \sum_{t=1}^{T} (Z_t - EZ_t) \xrightarrow{d} N(0, \Sigma), \tag{10.8}$$

where

$$\Sigma = \sum_{j=-\infty}^{\infty} \Gamma(j) = \Gamma(0) + \sum_{j=1}^{\infty} (\Gamma(j) + \Gamma(j)'),$$

and the autocovariance matrix $\Gamma(j)$ is defined as

$$\Gamma(j) = \text{Cov}(Z_t, Z_{t+j}).$$

Note that we used the property $\Gamma(j) = \Gamma(-j)'$. Weak dependence can be defined in terms of mixing coefficients.[4]

---

4 If the time series $(a'Z_t)_{t\in\mathbf{Z}}$ satisfies the conditions for the univariate central limit theorem for all $a \in \mathbf{R}^d$, then the multivariate central limit theorem holds.

To estimate Σ in (10.8) we use

$$\hat{\Sigma} = \hat{\Gamma}(0) + \sum_{j=1}^{T-1} w(j)(\hat{\Gamma}(j) + \hat{\Gamma}(j)'), \tag{10.9}$$

where

$$\hat{\Gamma}(j) = \frac{1}{T} \sum_{t=1}^{T-j} (Z_t - \bar{Z})(Z_{t+j} - \bar{Z})',$$

for $j = 0, \ldots, T-1$, and the weights are defined as

$$w(j) = K(j/h), \tag{10.10}$$

where $K : [0, \infty) \to [0, 1]$ is a kernel function satisfying $K(0) = 1$ and $|K(x)| \le 1$ for all $x > 0$. We get the estimator (10.7) by choosing $K(x) = I_{[0,1)}(x)$ and $h = 1$. We get the estimator

$$\hat{\Psi} = \frac{1}{T} \sum_{t=1}^{T} \sum_{s=1}^{T} (Z_t - \bar{Z})(Z_s - \bar{Z})',$$

by choosing $K(x) = I_{[0,1)}(x)$ and $h = T$. A further example is $K(x) = \max\{1 - x, 0\}$ and $1 \le h \le T$. The idea of using weights in asymptotic covariance estimation can be found in Newey and West (1987).

10.1.2.3 Confidence Intervals for the S&P 500 Sharpe Ratio

Figure 10.1(a) shows the confidence intervals for the Sharpe ratio of S&P 500 index, when the coverage probability is in the range $[0.01, 0.99]$. The S&P 500 monthly data is described in Section 2.4.3. We have used the estimator (10.7) of the asymptotic covariance matrix. The x-axis shows the range of possible values of the Sharpe ratio and the y-axis shows the coverage probabilities of the confidence intervals. The yellow vector shows the point estimate of the Sharpe ratio and the red vectors show the confidence interval with 0.95 coverage.

Figure 10.1(b) studies how the confidence intervals change when we use the autocorrelation robust estimator (10.9) of the asymptotic covariance matrix. We show the ratios $\hat{\sigma}_1/\hat{\sigma}_0$ as a function of smoothing parameter, where $\hat{\sigma}_0$ is defined by (10.6) and by the estimator (10.7) for the covariance matrix, whereas $\hat{\sigma}_1$ is defined by (10.6) and by the estimator (10.9) for the covariance matrix. These ratios are equal to the ratios of the lengths of the corresponding confidence intervals. We use the kernel function $K = I_{[0,1)}$ and try values $h = 1, 6, 11, \ldots, 216$. For $h = 1$ the ratio is equal to one, because the estimators for the covariance matrix are equal. We see that taking the autocorrelation into account makes the confidence bands eventually shorter, but for moderate h the confidence intervals can be longer, too.

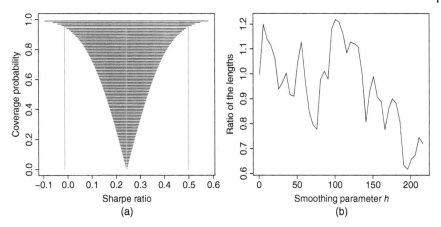

Figure 10.1 *Confidence intervals for the S&P 500 Sharpe ratio.* (a) Confidence intervals corresponding to a coverage probability in $[0.01, 0.99]$. The x-axis shows the range of possible values of the Sharpe ratio and the y-axis shows the coverage probabilities of the confidence intervals. The yellow vertical vector indicates the point estimate of the Sharpe ratio and the red vertical vectors show the confidence interval with 0.95 coverage. (b) The ratios $\hat{\sigma}_1/\hat{\sigma}_0$ as a function of smoothing parameter, where $\hat{\sigma}_0$ is the estimator assuming zero autocorrelation, whereas $\hat{\sigma}_1$ assumes autocorrelation.

10.1.3 Testing the Sharpe Ratio

Let us have two portfolios a and b with excess returns X and Y. We want to test the equality of the Sharpe ratios, so that the null hypothesis is

$$H_0 : \text{Sh}_a = \text{Sh}_b,$$

where $\text{Sh}_a = (\Delta t)^{-1/2} EX/\text{sd}(X)$ and $\text{Sh}_b = (\Delta t)^{-1/2} EY/\text{sd}(Y)$. Portfolio a could typically be an actively managed portfolio and portfolio b could be the benchmark index.

Let us have historical returns $X_1, \dots, X_T$ of portfolio a and historical returns $Y_1, \dots, Y_T$ of portfolio b. We use the test statistics

$$U_T = \widehat{\text{Sh}}_a - \widehat{\text{Sh}}_b,$$

where $\widehat{\text{Sh}}_a = (\Delta t)^{-1/2} \bar{X}/(T^{-1} \sum_{t=1}^{T} X_t^2 - \bar{X}^2)^{1/2}$ is the estimate of the Sharpe ratio of portfolio a, and $\widehat{\text{Sh}}_b$ is the estimate of the Sharpe ratio of portfolio b. The test statistics can be written as

$$U_T = h(S_T),$$

where

$$h(x_1, x_2, x_3, x_4) = (\Delta t)^{-1/2} \left(\frac{x_1}{\left(x_2 - x_1^2\right)^{1/2}} - \frac{x_3}{\left(x_4 - x_3^2\right)^{1/2}} \right),$$

and

$$S_T = \left(\bar{X}, T^{-1} \sum_{t=1}^{T} X_t^2, \bar{Y}, T^{-1} \sum_{t=1}^{T} Y_t^2 \right)'. \tag{10.11}$$

Let us assume that

$$T^{1/2}(S_T - \theta) \xrightarrow{d} N(0, \Psi), \tag{10.12}$$

as $T \to \infty$, where $\theta = (EX, EX^2, EY, EY^2)'$ and Ψ is a 4×4 covariance matrix. An application of the delta-method gives

$$T^{1/2}(h(S_T) - h(\theta)) \xrightarrow{d} N(0, \nabla h(\theta)' \Psi \nabla h(\theta)),$$

as $T \to \infty$, where $\nabla h(x)$ is the gradient, and $h(\theta)$ is the difference of the Sharpe ratios. We have that

$$\nabla h(x) = (\Delta t)^{-1/2}$$

$$\times \left(\frac{x_2}{(x_2 - x_1^2)^{3/2}}, -\frac{1}{2} \frac{x_1}{(x_2 - x_1^2)^{3/2}}, -\frac{x_4}{(x_4 - x_3^2)^{3/2}}, \frac{1}{2} \frac{x_3}{(x_4 - x_3^2)^{3/2}} \right).$$

When the alternative hypothesis is

$$H_1 : \mathrm{Sh}_a > \mathrm{Sh}_b,$$

then the null hypothesis is rejected for large values of the test statistics U_T, and thus the p-value for the one-sided test is

$$1 - \Phi(T^{1/2}\hat{\sigma}^{-1}U_T),$$

where Φ is the distribution function of the standard normal distribution and

$$\hat{\sigma} = \sqrt{\nabla h(S_T)' \hat{\Psi} \nabla h(S_T)}.$$

The estimator $\hat{\Psi}$ of Ψ can be defined similarly as in (10.7) or (10.9). Indeed, under the null hypothesis $P(T^{1/2}\hat{\sigma}^{-1}U > u_0) \approx P(Z > u_0) = 1 - \Phi(u_0)$, where $Z \sim N(0, 1)$. When the alternative hypothesis is

$$H_1 : \mathrm{Sh}_a \neq \mathrm{Sh}_b,$$

then the null hypothesis is rejected for large values of the absolute values of test statistics U_T, and thus the p-value for the two-sided test is $2(1 - \Phi(T^{1/2}\hat{\sigma}^{-1}|U_T|))$. Indeed, under the null hypothesis $P(T^{1/2}\hat{\sigma}^{-1}|U| > u_0) \approx P(|Z| > u_0) = 2(1 - \Phi(u_0))$. These tests were defined in Ledoit and Wolf (2008).

10.1.3.1 A Test Under Normality
A test for the equality of Sharpe ratios is presented in Jobson and Korkie (1981), with the corrected formula in Memmel (2003). They use the test statistics

$$U_T = \theta^{-1/2}T^{1/2}(\hat{\mu}_a\hat{\sigma}_b - \hat{\mu}_b\hat{\sigma}_a),$$

where $\hat{\mu}_a$ is the sample mean calculated from $X_1, \ldots, X_T$, $\hat{\sigma}_a$ is the corresponding sample standard deviation, $\hat{\mu}_b$ and $\hat{\sigma}_b$ are the sample mean and standard deviation calculated from $Y_1, \ldots, Y_T$, and

$$\theta = 2\hat{\sigma}_a^2\hat{\sigma}_b^2 - 2\hat{\sigma}_a\hat{\sigma}_b\hat{\sigma}_{ab} + \frac{1}{2}\hat{\mu}_a^2\hat{\sigma}_b^2 + \frac{1}{2}\hat{\mu}_b^2\hat{\sigma}_a^2 - \frac{\hat{\mu}_a\hat{\mu}_b}{\hat{\sigma}_a\hat{\sigma}_b}\hat{\sigma}_{ab}^2,$$

where $\hat{\sigma}_{ab}$ is the sample covariance. Under the assumption that the returns are normally distributed, the distribution of the test statistics under the null hypothesis can be approximated by the standard normal distribution: $U_T \sim N(0, 1)$. For the one sided alternative, the null hypothesis is rejected for the large values of the test statistics U_T and thus the p-value for the one-sided test is $1 - \Phi(U_T)$, where Φ is the distribution function of the standard normal distribution.

10.1.4 Other Measures of Risk-Adjusted Return

There exist several performance measures that resemble the Sharpe ratio. These performance measures are defined by dividing a measure for the expected return by a measure for the risk.

10.1.4.1 Information Ratio

The information ratio is defined as

$$\frac{E\left(R_{t+1} - R_{t+1}^b\right)}{\text{sd}\left(R_{t+1} - R_{t+1}^b\right)},$$

where R_{t+1} is the return of the portfolio and R_{t+1}^b is the return of a benchmark portfolio. Thus, the information ratio is like the Sharpe ratio, but the risk-free rate in the Sharpe ratio is replaced by the return of a benchmark.

The benchmark return is the return of an asset that is chosen as the benchmark for the asset manager. S&P 500 could be chosen as a benchmark for a US equity fund and MSCI World could be chosen as a benchmark for a global equity fund investing in developed markets. (MSCI is an acronym for Morgan Stanley Capital International.)

10.1.4.2 Sortino ratio

The Sortino ratio is otherwise similar to Sharpe ratio but the standard deviation is replaced by a lower partial moment, and the risk-free rate is replaced by a constant $r > 0$. The Sortino ratio is defined as

$$\frac{ER_{t+1} - r}{\text{LPM}_{r,2}(R_{t+1})},$$

where the lower partial moment of order 2 is defined in (3.15) as

$$\text{LPM}_{r,2} = E(r - R_{t+1})_+^2.$$

Another version of the Sortino ratio is defined as

$$\frac{E\left(R_{t+1} - R^0_{t+1}\right)}{E\left(R^0_{t+1} - R_{t+1}\right)^2_+},$$

where R^0_{t+1} is a risk-free rate.

10.1.4.3 Omega Ratio

The Omega ratio is the ratio of the upper partial moment of order one to the lower partial moment of order one:

$$\Omega(r) = \frac{E(R_{t+1} - r)_+}{E(r - R_{t+1})_+},$$

where r is the chosen threshold (the target rate). We have defined the upper and lower partial moments in (3.14) and (3.15). The definition was made in Shadwick and Keating (2002). Note that the Omega ratio is written often using the expressions

$$E(R_{t+1} - r)_+ = \int_r^\infty (1 - F(y)) \, dy$$

and

$$E(r - R_{t+1})_+ = \int_{-\infty}^r F(y) \, dy,$$

where $F : \mathbf{R} \to \mathbf{R}$ is the distribution function of R_{t+1}. The sample Omega ratio is

$$\frac{\sum_{i=1}^T (R_i - r)_+}{\sum_{i=1}^T (r - R_i)_+}.$$

Another version of the Omega ratio is defined as

$$\frac{E\left(R_{t+1} - R^0_{t+1}\right)}{E\left(R^0_{t+1} - R_{t+1}\right)_+},$$

where R^0_{t+1} is a risk-free rate. Note that $E(R_{t+1} - R^0_{t+1})$ and $E(R_{t+1} - R^0_{t+1})_+$ are close to each other, because probability $P(R_{t+1} - R^0_{t+1} < 0)$ is small.

10.2   Certainty Equivalent

The certainty equivalent of a return distribution is defined as

$$U^{-1}(EU(R_t)),$$

where R_t is a gross return and $U : (0, \infty) \to \mathbf{R}$ is a utility function. The certainty equivalent is the minimal risk-free rate that is preferred to the rate R_t.

As an example, let us consider return R_t that takes only two values. The distribution of the return is defined by

$$P(R_t = 1 + \alpha) = p, \qquad P(R_t = 1 - \alpha) = 1 - p$$

for some $0 < \alpha < 1$ and for some probability $0 \le p \le 1$. Then, for a concave utility function $U : (0, \infty) \to \mathbf{R}$, using (9.27),

$$
\begin{aligned}
EU(R_t) &= pU(1 + \alpha) + (1 - p)U(1 - \alpha) \\
&< U[p(1 + \alpha) + (1 - p)(1 - \alpha)] \\
&= U(1 + \alpha(2p - 1)).
\end{aligned}
$$

Thus, one would prefer always the certainly received amount $1 + \alpha(2p - 1)$ to the lottery. In particular, in the case $p = 1/2$, one would prefer to preserve the current wealth to the lottery with equal probabilities $1/2$ of winning and losing the amount α. Thus, the number $U^{-1}(EU(R_t))$ is called the certainty equivalent, since this is the minimal risk-free rate which is preferred to the rate R_t. Similarly, $U^{-1}(EU(W_t))$ is the minimum amount of wealth, guaranteed preservation of which allows the investor to decline the proposed game.

The certainty equivalent can be estimated using a time series of historical returns $R_1, \ldots, R_T$. The sample certainty equivalent is

$$
U^{-1}\left(\frac{1}{T} \sum_{t=1}^{T} U(R_t) \right). \tag{10.13}
$$

For example, when $U(x) = x^{1-\gamma}/(1 - \gamma)$ is the power utility, then $U^{-1}(x) = [(1 - \gamma)x]^{1/(1-\gamma)}$, where $\gamma \ge 0$, $\gamma \ne 1$. For $\gamma = 1$, $U(x) = \log x$ and $U^{-1}(x) = \exp(x)$.

10.3 Drawdown

Drawdown is a new time series constructed from the time series $S_0, \ldots, S_T$ of asset prices. Define the return for the period $u \mapsto t$ as

$$
R_{u,t} = \frac{S_t}{S_u},
$$

where $0 \le u \le t \le T$. The drawdown at time t is

$$
D_t = 1 - \min_{u=0,\ldots,t} R_{u,t},
$$

where $t = 0, \ldots, T$. Thus, drawdown at time t is one minus the minimum gross return. We can write

$$
D_t = 1 - \frac{S_t}{\max_{u=0,\ldots,t} S_u},
$$

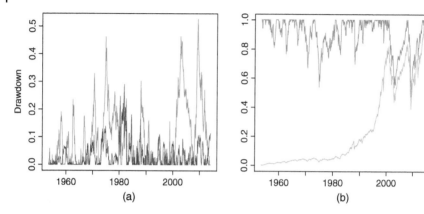

Figure 10.2 *Drawdown*. (a) Drawdown time series D_t for S&P 500 (red) and 10-year bond (blue); (b) time series $1 - D_t$ (red) and the cumulative wealth (orange) for S&P 500.

because

$$\frac{S_t}{S_{u_0}} = R_{u_0,t},$$

where

$$u_0 = \operatorname*{argmin}_{u=0,\dots,t} R_{u,t} = \operatorname*{argmax}_{u=0,\dots,t} S_u.$$

Large values of drawdown indicate that the asset has a high level of riskiness, just like a high value of variance indicates a high level of riskiness. Also, when $r_{u,t} = S_t/S_u - 1$ is the net return, then

$$D_t = - \min_{u=0,\dots,t} r_{u,t}.$$

Sometimes drawdown is defined as $\max_{u=0,\dots,t} S_u - S_t$, but this definition is not in terms of returns.

Interesting statistics are the maximum drawdown, the mean drawdown, and the variance of drawdowns.

Figure 10.2 shows drawdown time series for the monthly S&P 500 and 10-year bond data, described in Section 2.4.3. Panel (a) shows drawdown time series for S&P 500 (red) and 10-year bond (blue). Panel (b) shows time series $1 - D_t$ (red) and the cumulative wealth (orange) for S&P 500. The original time series of cumulative wealth starts with value one, but we have normalized the time series to take values on $[0, 1]$.

10.4 Alpha and Conditional Alpha

Linear regression can be used to describe assets and portfolios. A beta of an asset describes the exposure of a portfolio to a risk factor and the alpha of a portfolio can be used to measure the performance of the portfolio. The beta is the coefficient of the linear regression and the alpha is the intercept of the linear regression.

The alpha as a performance measure was proposed in Jensen (1968), and therefore the term Jensen's alpha is sometimes used. The alpha has been used to evaluate portfolio performance, for example, in Carhart (1997), Kosowski *et al.* (2006), and Fama and French (2010).

10.4.1 Alpha

First, we consider the case of a single risk factor. The single risk factor is usually the return of a market index. Second, we consider the case of several risk factors. The arbitrage pricing model is an example of using several risk factors.

10.4.1.1 A Single Risk Factor

Efficient Markets In the framework of Markowitz theory of portfolio selection, it can be shown that the optimal portfolios in the Markowitz sense are a combination of the market portfolio and the risk-free investment; see Section 11.3, where the concepts of the efficient frontier and the tangency portfolio are explained.[5] Thus, the returns of the optimal portfolios for the period $t \mapsto t+1$ are

$$R_{t+1} = (1 - \beta)R_{t+1}^0 + \beta R_{t+1}^M, \tag{10.14}$$

where R_{t+1}^0 is the return of the risk-free investment and R_{t+1}^M is the return of the market portfolio, both returns being for the investment period ending at time $t + 1$. The coefficient $\beta \geq 0$ is the proportion invested in the market portfolio. When $0 \leq \beta \leq 1$, then the portfolio is investing available wealth; but if $\beta > 1$, then amount $(\beta - 1)W_t$ is borrowed and amount $(\beta + 1)W_t$ is invested in the market portfolio, where W_t is the investment wealth at the beginning of the period.

The coefficient β is determined by the risk aversion of the investor. For an investor whose portfolio returns are R_{t+1} we do not know the coefficient β, but we obtain from (10.14) that

$$R_{t+1} - R_{t+1}^0 = \beta \left(R_{t+1}^M - R_{t+1}^0 \right).$$

We can collect past returns $R_{t+1}, t = 0, \ldots, T-1$, and use these, together with the past returns R_{t+1}^0 of the risk-free return and the past returns R_{t+1}^M of the

---

5 The Markowitz theory of portfolio selection defines the optimal stock portfolios as portfolios maximizing expected return for a given upper bound on the standard deviation of the portfolio return or, equivalently, as portfolios minimizing the standard deviation of the portfolio return for a given lower bound on the expected return of the portfolio. This defines single period portfolio choice, for a given investment period.

market portfolio, to estimate the coefficient β in the linear model

$$R_{t+1} - R_{t+1}^0 = \beta \left(R_{t+1}^M - R_{t+1}^0 \right) + \epsilon_{t+1}, \tag{10.15}$$

where ϵ_{t+1} is an error term. Now $R_{t+1} - R_{t+1}^0$ is the response variable and $R_{t+1}^M - R_{t+1}^0$ is the explanatory variable. The returns R_{t+1}^M of the market portfolio are approximated with the returns of a wide market index, like S&P 500 index, Wilshire 5000 index, or DAX 30 index. The risk-free rate R_{t+1}^0 can be taken to be the rate of return of a government bond. This model is called the capital asset pricing model, or CAP model.

Alpha of a Portfolio In (10.15), we have a regression model without a constant term. The exclusion of the intercept can be justified by arguments based on efficient markets. However, we can include the intercept, in order to study whether it is positive in some cases.

We extend model (10.15) to the model

$$R_{t+1} - R_{t+1}^0 = \alpha + \beta \left(R_{t+1}^M - R_{t+1}^0 \right) + \epsilon_{t+1}, \tag{10.16}$$

where R_{t+1} is the return of the actively managed portfolio, R_{t+1}^M is the return of the market portfolio, R_{t+1}^0 is the risk-free rate, and ϵ_{t+1} is an error term. The excess return of a market index is chosen as the explanatory variable, and the excess return of the actively managed portfolio is chosen as the response variable. The estimated constant $\hat{\alpha}$ is taken as the measure of the performance, so that larger values of $\hat{\alpha}$ indicate a better performance of the portfolio.

Denote the response variable $Y_t = R_t - R_t^0$ and the explanatory variable $X_t = R_t^M - R_t^0$. We have that

$$\alpha = EY - \beta EX, \qquad \beta = \frac{\text{Cov}(X, Y)}{\text{Var}(X)}, \tag{10.17}$$

when $E\epsilon_t = 0$ and $E(X_t \epsilon_t) = 0$. This follows from (10.22) and (10.23), by specializing to the one-dimensional case $d = 1$. Note that

$$\beta = \frac{\text{sd}(Y)}{\text{sd}(X)} \, \text{Cor}(X, Y),$$

where $\text{sd}(Y)$ and $\text{sd}(X)$ are the standard deviations, and $\text{Cor}(X, Y)$ is the correlation.

Given a sample (X_t, Y_t), $t = 1, \dots, T$, the estimators are

$$\hat{\alpha} = \bar{Y} - \hat{\beta}\bar{X}, \qquad \hat{\beta} = \frac{\sum_{i=1}^T (X_i - \bar{X})(Y_i - \bar{Y})}{\sum_{i=1}^T (X_i - \bar{X})^2}, \tag{10.18}$$

where $\bar{X}$ and $\bar{Y}$ are the sample means. The formulas are special cases of (10.25) and (10.26), for the case $d = 1$.

The beta of an asset gives information about the volatility of the stock in relation to the volatility of the benchmark. If $\beta < 0$, the asset tends to move in the opposite direction as the benchmark; if $\beta = 0$, the asset is uncorrelated with the benchmark; if $0 < \beta < 1$, the asset tends to move in the same direction as the benchmark but it tends to move less; and if $\beta > 1$, the asset tends to move in the same direction as the benchmark but it tends to move more.

We see from (10.18) that the alpha of an asset is not equal to the sample mean of the excess returns $Y_t = R_t - R_t^0$, but we have subtracted term $\hat{\beta}\bar{X}$. Thus, the assets that are negatively correlated with the market index have alpha larger than the sample mean of the excess returns, whereas the assets that are positively correlated with the market index have alpha smaller than the sample mean of the excess returns, when we assume that the sample mean $\bar{X}$ is positive.

Figure 10.3 shows alphas and betas for the S&P 500 components. S&P 500 components daily data is defined in Section 2.4.5. Panel (a) shows a scatter plot of $(\hat{\alpha}_i, \hat{\beta}_i)$, when i runs over the S&P 500 components which are included in the data. Panel (b) shows the linear functions $x \mapsto \hat{\alpha}_i + \hat{\beta}_i x$, when x-axis is the S&P 500 excess return, and the y-axis shows the excess returns of S&P 500 components. We see that almost all alphas are positive, and betas range between 0.2 and 0.8.

10.4.1.2 Several Risk Factors

Instead of one risk factor, we can consider several risk factors whose returns are $R_{t+1}^1, \ldots, R_{t+1}^d$. These risk factors should ideally be such that the returns R_{t+1} of all reasonable portfolios can be represented as

$$R_{t+1} = \left(1 - \sum_{i=1}^{d} \beta_i\right) R_{t+1}^0 + \beta_1 R_{t+1}^1 + \cdots + \beta_d R_{t+1}^d. \tag{10.19}$$

Since this relation can hold only approximately we need an error term ϵ_{t+1}. Since we want to allow for the possibility of abnormal returns we need the

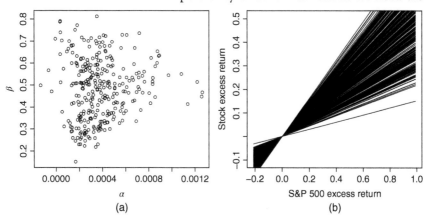

Figure 10.3 *Alphas and betas of S&P 500 components.* (a) A scatter plot of $(\hat{\alpha}_i, \hat{\beta}_i)$; (b) linear functions $x \mapsto \hat{\alpha}_i + \hat{\beta}_i x$.

intercept α. This leads to the extension of the one-dimensional model (10.16) into the model

$$R_{t+1} - R^0_{t+1} = \alpha + \sum_{i=1}^{d} \beta_i \left(R^i_{t+1} - R^0_{t+1} \right) + \epsilon_{t+1}, \tag{10.20}$$

where R_{t+1} is the return of the actively managed portfolio, R^i_{t+1}, $i = 1, \ldots, d$, are the returns of the risk factors, R^0_{t+1} is the risk-free rate, and ϵ_{t+1} is an error term.

Note that in (10.19) we have ensured that the weights of the assets sum to one with the help of a risk-free rate. There are other ways to make the portfolio weights sum to one. For example, we could have

$$R_{t+1} = (1 - \beta_1)R^0_{t+1} + \beta_1 R^1_{t+1} + \beta_2 \left(R^2_{t+1} - R^3_{t+1} \right). \tag{10.21}$$

This kind of construction is used in the Fama–French model (see (10.34)).

Least Squares Formulas Denote $Y = R_{t+1} - R^0_{t+1}$ and $X^i = R^i_{t+1} - R^0_{t+1}$. Now we can write the model (10.20) as

$$Y = \alpha + \beta'X + \epsilon,$$

where $\alpha \in \mathbf{R}$, $\beta \in \mathbf{R}^d$, and $X = (X^1, \ldots, X^d)'$. Note that in the case of construction (10.21) we would choose $X^1 = R^1_{t+1} - R^0_{t+1}$ and $X^2 = R^2_{t+1} - R^3_{t+1}$.
If $E(X\epsilon) = 0$, then

$$\alpha = EY - \beta'EX \tag{10.22}$$

and

$$\beta = \mathrm{Cov}(X)^{-1} E[(X - EX)(Y - EY)], \tag{10.23}$$

where

$$\mathrm{Cov}(X) = E[(X - EX)(X - EX)'],$$

and we assume additionally that $\mathrm{Cov}(X)$ is invertible.[6]
In the two-dimensional case $d = 2$ we have $\alpha = EY - \beta'_1 EX_1 - \beta_2 EX_2$,

$$\beta_1 = \frac{1}{\sigma_1^2\sigma_2^2 - \sigma_{12}^2} \left(\sigma_2^2 \, \mathrm{Cov}(X_1, Y) - \sigma_{12} \, \mathrm{Cov}(X_2, Y) \right),$$

and

$$\beta_2 = \frac{1}{\sigma_1^2\sigma_2^2 - \sigma_{12}^2} \left(\sigma_1^2 \, \mathrm{Cov}(X_2, Y) - \sigma_{12} \, \mathrm{Cov}(X_1, Y) \right),$$

where $\sigma_1^2 = \mathrm{Var}(X_1)$, $\sigma_2^2 = \mathrm{Var}(X_2)$, and $\sigma_{12} = \mathrm{Cov}(X_1, X_2)$.

---

6 We get the same solution by minimizing $E(\alpha + \beta'X - Y)^2$. Derivation with respect to α and setting the derivative equal to zero gives $\alpha = EY - \beta'EX$. Then we find $\hat{\beta}$ by minimizing $E(\beta'(X - EX) - (Y - EY))^2$. Derivation with respect to elements of β and setting these derivatives to zero gives $E(X - EX)(X - EX)'\beta = E(X - EX)(Y - EY)$, which leads to (10.23).

The least squares estimates are $\hat{\alpha} \in \mathbf{R}$ and $\hat{\beta} \in \mathbf{R}^d$ are defined as the minimizers of the least squares criterion

$$\sum_{i=1}^{n} (Y_i - \alpha - \beta' X_i)^2. \tag{10.24}$$

The solution can be written as

$$\hat{\alpha} = \bar{Y} - \hat{\beta}' \bar{X}, \tag{10.25}$$

$$\hat{\beta} = \left[\sum_{i=1}^{n} (X_i - \bar{X})(X_i - \bar{X})' \right]^{-1} \sum_{i=1}^{n} (X_i - \bar{X})(Y_i - \bar{Y})', \tag{10.26}$$

where $\bar{X} = n^{-1} \sum_{i=1}^{n} X_i$ and $\bar{Y} = n^{-1} \sum_{i=1}^{n} Y_i$.

Further Least Squares Formulas   It is often convenient to use notation where the intercept is included in the vector β. This can be done by choosing the first component of the vector of explanatory variables as the constant one. Denote

$$X = (1, X_2, \ldots, X_{d+1}) \in \mathbf{R}^{d+1}.$$

We use below the notation

$$K = d + 1. \tag{10.27}$$

Write the regression model as

$$Y = \beta' X + \epsilon, \tag{10.28}$$

where $\beta \in \mathbf{R}^K$, $X = (X_1, \ldots, X_K)'$, $Y \in \mathbf{R}$, and $\epsilon \in \mathbf{R}$ is the scalar error term. Multiplying (10.28) with vector X, we get

$$XY = XX'\beta + X\epsilon.$$

If $E(X\epsilon) = 0$, then

$$E(XY) = E(XX')\beta. \tag{10.29}$$

If $E(XX')$ is invertible, then

$$\beta = [E(XX')]^{-1} E(XY). \tag{10.30}$$

Let us observe

$$(X_i, Y_i), \qquad X_i = (1, X_{i,2}, \ldots, X_{i,d+1}) \in \mathbf{R}^{d+1}, \quad Y_i \in \mathbf{R}, \tag{10.31}$$

where $i = 1, \ldots, T$. We assume that $(X_1, Y_1), \ldots, (X_T, Y_T)$ are identically distributed and have the same distribution as (X, Y). The least squares estimator of parameter β can be written as

$$\hat{\beta} = (\mathbf{X}'\mathbf{X})^{-1} \mathbf{X}'\mathbf{y}, \tag{10.32}$$

where $\mathbf{X} = (X_1, \ldots, X_T)'$ is the $T \times K$ matrix whose rows are X_i', and $\mathbf{y} = (Y_1, \ldots, Y_T)'$ is the $T \times 1$ vector. The estimator can be written as

$$\hat{\beta} = \left(\frac{1}{n} \sum_{i=1}^{n} X_i X_i' \right)^{-1} \frac{1}{n} \sum_{i=1}^{n} X_i Y_i. \tag{10.33}$$

This estimator is the same as the least squares estimator in (10.32), as can be seen by noting that

$$\mathbf{X}'\mathbf{X} = \sum_{i=1}^{n} X_i X_i', \qquad \mathbf{X}'\mathbf{y} = \sum_{i=1}^{n} X_i Y_i.$$

Note that (10.33) is obtained from (10.30) by replacing the expectations with the sample means.[7]

A Three-Factor Model Fama and French (1993) proposes a three-factor model, where the factors are the market return, size, and value versus growth. The model is related to the arbitrage pricing theory. Let R_{t+1}^S be the return of a diversified portfolio of small stocks, and let R_{t+1}^L be the return of a diversified portfolio of large stocks, where largeness and smallness is measured by the market capitalization. Let R_{t+1}^V be the return of a diversified portfolio of value stocks, and let R_{t+1}^G be the return of a diversified portfolio of growth stocks, where a value stock has a high book-to-market ratio, and a growth stock has a low book-to-market ratio.

Fama and French (1993, 2012) formulate the model as[8]

$$R_{t+1} - R_{t+1}^0 = \alpha + \beta_1 \left(R_{t+1}^1 - R_{t+1}^0 \right)$$
$$+ \beta_2 \left(R_{t+1}^S - R_{t+1}^B \right) + \beta_3 \left(R_{t+1}^V - R_{t+1}^G \right) + \epsilon_{t+1}. \tag{10.34}$$

Factors of Smart Alpha It can happen that a hedge fund achieves a large positive alpha, when the alpha is measured in the capital asset pricing model (10.20) or in the arbitrage pricing model (10.34). However, we can introduce models with

---

7 The solution (10.32) can be found by writing the least squares criterion (10.24) with the matrix notation as

$$(\mathbf{y} - \mathbf{X}\beta)'(\mathbf{y} - \mathbf{X}\beta) = \mathbf{y}'\mathbf{y} - 2\beta'\mathbf{X}'\mathbf{y} + \beta'\mathbf{X}'\mathbf{X}\beta.$$

Derivation of this with respect to β, and setting the gradient to zero, gives the equations

$$\mathbf{X}'\mathbf{X}\beta = \mathbf{X}'\mathbf{y},$$

which leads to the solution (10.32).

8 Note that an alternative model would be

$$R_{t+1} - R_{t+1}^0 = \alpha + \sum_{i=1}^{3} \beta_i \left(R_{t+1}^i - R_{t+1}^0 \right) + \epsilon_{t+1},$$

where R_{t+1}^1 is the return of a market index, R_{t+1}^i is the return of the risk-free rate, $R_{t+1}^2 = 2R_{t+1}^S - R_{t+1}^L$, and $R_{t+1}^3 = 2R_{t+1}^V - R_{t+1}^G$.

additional factors. The alpha defined by a model with some additional factors can be called smart alpha.

The momentum factor has been proposed to be an additional factor, which generates positive returns. Carhart (1997) defines the momentum factor for monthly returns as the difference

$$R_{t+1}^W - R_{t+1}^L,$$

where R_{t+1}^W is the return of a diversified portfolio of the winners of the past year, and R_{t+1}^L is the return of a diversified portfolio of losers of the past year.

Fung and Hsieh (2004) define seven risk factors: three trend-following risk factors, two equity-oriented risk factors, and two bond-oriented risk factors. The trend-following risk factors are a bond trend-following factor, a currency trend-following factor, and a commodity trend-following factor. The equity-oriented risk factors are the equity market factor, which is the S&P 500 index monthly total return, and the size spread factor, which can be defined as the Wilshire Small Cap 1750 minus the Wilshire Large Cap 750 monthly return or Russell 2000 index monthly total return minus the S&P 500 monthly total return. The bond-oriented risk factors are the bond market factor, which is the monthly change in the 10-year treasury constant maturity yield (month end-to-month end), and the credit spread factor, which is the monthly change in the Moody's Baa yield minus the 10-year treasury constant maturity yield (month end-to-month end).

Eurex provides futures on six factor indexes. The six factors include the size and value factors from the three-factor model, and the momentum factor. Additional factors are the low-risk factor (stocks with volatility below average), quality factor (stocks with solid financial background based on debt coverage, earnings and other metrics), and carry factor (stocks with high-growth potential based on earnings and dividends).[9]

10.4.2 Conditional Alpha

We have applied a linear model to the evaluation of portfolio performance. The performance was measured by the estimate $\hat{\alpha}$ of the constant term α of linear regression. We can use varying coefficient regression to estimate conditional alpha. It has been argued that the conditional alpha measures better hedge fund performance, since hedge funds do not use long only strategies but apply short selling, buying of options, and writing of options.

We choose a collection of risk factors $X_t^1, \dots, X_t^d$ and make a linear regression of hedge fund return Y_t on these risk factors, where $t = 1, \dots, T$. The unconditional alpha is defined as

$$\hat{\alpha} = \text{argmin}_\alpha \min_{\beta_1, \dots, \beta_d} \sum_{t=1}^T \left(Y_t - \alpha - \beta_1 X_t^1 - \cdots - \beta_d X_t^d \right)^2.$$

---

9 http://www.eurexchange.com/exchange-en/products/idx/istoxx/.

The conditional alpha, conditionally on the information $Z_t \in \mathbf{R}^p$ at time t, is defined as

$$\hat{\alpha}(Z_t) = \text{argmin}_\alpha \min_{\beta_1,\dots,\beta_d} \sum_{i=1}^{t} \left(Y_i - \alpha - \beta_1 X_i^1 - \dots - \beta_d X_i^d\right)^2 p_i(Z_t),$$

where

$$p_i(Z_t) = K_h(Z_i - Z_t),$$

where $K_h(x) = K(x/h)/h^p$ is the scaled-kernel function, $K : \mathbf{R}^p \to \mathbf{R}$ is the kernel function, and $h > 0$ is the smoothing parameter.

10.5 Graphical Tools of Performance Measurement

We describe how cumulative wealth, Sharpe ratios, and certainty equivalents can be used to evaluate a given return time series using graphical tools.

A central idea is to find the periods of good performance and the periods of bad performance. It occurs seldom that a return series would indicate good performance for every time period. Instead, a typical series of returns of a financial asset has some periods of good performance and some periods of bad performance. It is useful to to find during which periods the performance is good and during which it is bad, instead of looking only at the aggregate performance. It is also of interest to find characteristics of the type: "the return series is good in recession," or "the return series is good when the commodity prices are rising."

We describe methods to evaluate a return time series. The methods can be used to study the properties of any return time series, but it is of particular interest to study a time series created by historical simulation, as described in Section 12.2. We can study also a time series of historical returns of an asset manager. In this section, we use the monthly data of S&P 500, US Treasury 10-year bond, And US Treasury 1-month bill, described in Section 2.4.3.

Section 10.5.1 describes the use of wealth in evaluation, Section 10.5.2 describes the use of Sharpe ratio in evaluation, and Section 10.5.3 describes the use of certainty equivalent in evaluation.

10.5.1 Using Wealth in Evaluation

Given a time series of gross returns $R_1, \dots, R_T \in (0, \infty)$, we can construct the time series of cumulative wealth by

$$W_0 = 1, \qquad W_{t+1} = W_t \times R_t, \qquad t = 0, \dots, T-1.$$

Now,

$$W_t = W_0 \prod_{i=1}^{t} R_t.$$

Time series $\{W_t\}$ of wealth can be more instructive to find periods of good returns than looking at the original return time series. Plotting the logarithmic wealth $\log W_t$ can be helpful in cases where W_t increases exponentially.

Figure 10.4 shows cumulative wealths of monthly time series of S&P 500 (red), 10-year US Treasury bond (blue), and 1-month US Treasury bill (black). Panel (a) has wealth at the y-axis, and panel (b) has a logarithmic scale at the y-axis. Time series in Figure 10.4 have a concrete interpretation as the cumulative wealth, but they do not reveal the periods of relative outperformance and underperformance in such a detail than we are able to see in Figures 10.5 and 10.6.

To compare two return time series, we can use the relative cumulative wealth. Let us consider two return time series $R_1^1, \dots, R_T^1$ and $R_1^2, \dots, R_T^2$. The corresponding time series of cumulative wealths are $W_0^1, \dots, W_T^1$ and $W_0^2, \dots, W_T^2$. The time series

$$Z_t = W_t^2 / W_t^1, \qquad t = 0, \dots, T, \qquad (10.35)$$

can be used to compare the two return series. Indeed, for $u < t$,

$$\frac{Z_t}{Z_u} = \frac{W_t^2/W_t^1}{W_u^2/W_u^1} = \frac{W_t^2/W_u^2}{W_t^1/W_u^1} = \frac{\prod_{i=u+1}^{t} R_i^2}{\prod_{i=u+1}^{t} R_i^1}.$$

Thus, when $Z_t > Z_u$, then asset 2 is performing better than asset 1 over time period $[u, t]$. Conversely, when $Z_t < Z_u$, then asset 1 is performing better than asset 2 over time period $[u, t]$.

Time series

$$\log Z_t = \log\left(W_t^2 / W_t^1\right), \qquad t = 0, \dots, T,$$

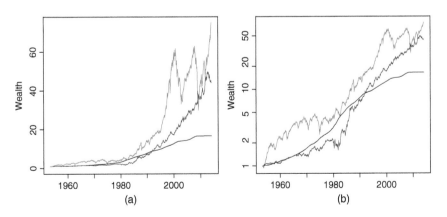

Figure 10.4 *Time series of cumulative wealths.* (a) The y-axis shows the cumulative wealth; (b) the y-axis has a logarithmic scale. We show the cumulative wealth of S&P 500 (red), 10-year bond (blue), and 1-month bill (black).

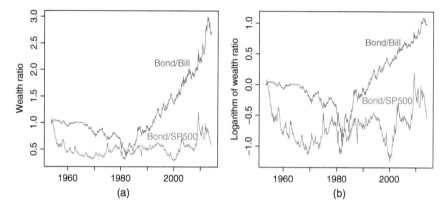

Figure 10.5 *Time series of relative cumulative wealth of 10-year bond.* We compare 10-year bond to S&P 500 and to 1-month bill. (a) The wealth ratio $Z_t = W_t^2/W_t^1$, where W_t^2 is the wealth of the 10-year bond, W_t^1 is the wealth of S&P 500 (green), and W_t^1 is the wealth of 1-month bill (purple). Panel (b) shows time series $\log Z_t$.

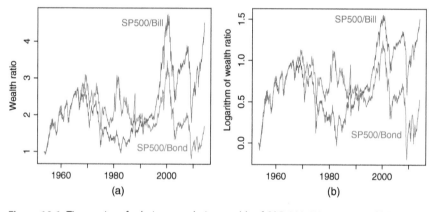

Figure 10.6 *Time series of relative cumulative wealth of S&P 500.* We compare S&P 500 to 10-year bond and to 1-month bill. (a) The wealth ratio $Z_t = W_t^2/W_t^1$, where W_t^2 is the wealth of S&P 500, W_t^1 is the wealth of 10-year bond (green), and W_t^1 is the wealth of 1-month bill (purple). Panel (b) shows time series $\log Z_t$.

can sometimes be more illustrative in comparing the two return series. Again, when $\log Z_t > \log Z_u$, then asset 2 is performing better than asset 1 over period $[u, t]$, where $u < t$. Conversely, when $\log Z_t < \log Z_u$, then asset 1 is performing better than asset 2 over period $[u, t]$.

Note that this graphical method is analogous to the looking at the time series (6.26), which shows the periods of good prediction performance, in terms of the sum of squared prediction errors.

Figure 10.5 compares monthly time series of US Treasury 10-year bond returns to the S&P 500 returns, and to 1-month US Treasury bill rates. Panel (a) shows the wealth ratio $Z_t = W_t^2/W_t^1$, when asset 2 is 10-year bond and asset 1 is S&P 500 (green), or asset 1 is 1-month bill (purple). Panel (b) shows time series $\log Z_t$. We can see a clear pattern in the purple curves (ratio of 10-year bond to 1-month bill): it is near to monotonically decreasing until about 1985, after that it is near to monotonically increasing. This means that 10-year bond performs worse than 1-month bill in practically all time periods before 1985, and better in practically all time periods after 1985. Such a clear pattern cannot be seen in the green curves (ratio of 10-year bond to S&P 500). However, look-ing at the details, we can detect the time periods where 10-year bond has better returns than S&P 500, unlike in Figure 10.4, where such details cannot be seen.

Figure 10.6 compares monthly time series of S&P 500 returns to US Treasury 10-year bond returns, and to US Treasury 1-month bill rates. Panel (a) shows the wealth ratio $Z_t = W_t^2/W_t^1$, where asset 2 is S&P 500. Asset 1 is 10-year bond (green), or asset 1 is 1-month bill (purple). Panel (b) shows time series $\log Z_t$. The green curves are mirror images of the green curves in Figure 10.5. The purple curve (ratio of S&P 500 to 1-month bill) does not express such a clear pattern as the purple curve in Figure 10.5 (ratio of 10-year bond to 1-month bill). However, we can see that the purple curves increase almost monotonically from 1953 until about 1970, and from about 1985 until about 2000. The pur-ple curves decrease almost monotonically from about 1970 until about 1985. After 2000 there are several periods of increase and decrease. Purple and green curves have somewhat similar periods of increase and decrease, but the moves in the purple curves are more profound.

10.5.2 Using the Sharpe Ratio in Evaluation

It is not enough to compute the Sharpe ratio of a return time series $R_1, \ldots, R_T$, but it is important to study the Sharpe ratios for any time periods $[u, t]$, where $1 \leq u < t \leq T$, instead of just computing the Sharpe ratio for the complete time period $[1, T]$.

10.5.2.1 Sharpe Ratios over All Possible Time Periods

We were able to compare graphically two return time series over all possible time periods by looking at the single time series of wealth ratios, defined in (10.35). However, when we want to compare the Sharpe ratios of two return time series over all time periods, such a simple tool does not seem to be avail-able. Instead, we define function $S(u, t)$ of two variables whose value is the annualized Sharpe ratio of the return series $R_u, R_{u+1}, \ldots, R_t$, where $u < t$. Given a time series $R_1, \ldots, R_T$, we define

$$S(u, t) = \frac{1}{\sqrt{\Delta t}} \frac{\hat{\mu}_u^t}{\hat{\sigma}_u^t}, \qquad 1 \leq u + \delta < t \leq T, \tag{10.36}$$

where Δt is equal to the time step between two observations of the time series (for monthly data $\Delta t = 1/12$), $\hat{\mu}_u^t$ is the sample mean over time period $[u, t]$ of the excess return, and $\hat{\sigma}_u^t$ is the sample standard deviation over time period $[u, t]$ of the excess return.[10] In addition, we have introduced parameter $0 < \delta < T$ to guarantee that there are at least two observations to calculate the Sharpe ratio. In fact, we need several observations to guarantee that the estimate of the Sharpe ratio has some accuracy.

To compare two return time series, we calculate function S for both of these time series: call these functions S_1 and S_2. Then we can study difference

$$Z(u, t) = S_2(u, t) - S_1(u, t).$$

Note that ratio $S_2(u, t)/S_1(u, t)$ is useful only when $S_2(u, t) > 0$ and $S_1(u, t) > 0$, but this is not always the case, because a return time series of a risky asset can have a smaller mean than the mean of the returns of the risk-free rate.

Figure 10.7 shows a contour plot of function S. In panel (a) function S is calculated from the monthly returns of of S&P 500. In panel (b) the returns are of US Treasury 10-year bond. Parameter δ in the definition of the domain of

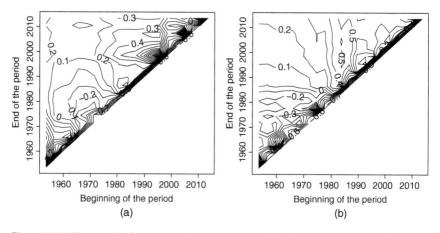

Figure 10.7 *Sharpe ratios for every period: Contour plots.* We show contour plots of function $S(u, t)$, defined in (10.36). (a) Sharpe ratios of S&P 500; (b) Sharpe ratios of US Treasury 10-year bond.

---

10 Let $X_i = R_i - r_i$ be the excess return, where r_i is the risk-free rate. Then,

$$\hat{\mu}_u^t = (t - u + 1)^{-1} \sum_{i=u}^{t} X_i, \tag{10.37}$$

and

$$\hat{\sigma}_u^t = \left((t - u + 1)^{-1} \sum_{i=u}^{t} X_i^2 - \left(\hat{\mu}_u^t \right)^2 \right)^{1/2}. \tag{10.38}$$

Note that the sample Sharpe ratio is defined in (10.3).

S is equal to 36 months, and furthermore, function S is evaluated only at the points $\{(u, t) : u, t \in \{1, \delta, 2\delta, \dots, T\}, 1 \le u + \delta < t \le T\}$. Function has lots of fluctuation near the diagonal, because near the diagonal u and t are close to each other, and thus the Sharpe ratio is computed over a short time period.

Figure 10.8 shows an image plot corresponding to the contour plot in Figure 10.7. The bright yellow shows the time periods where the Sharpe ratio is high and the red color shows the time periods where the Sharpe ratio is low. The image plot can be useful in showing more details than the contour plot. Parameter δ in the definition of the domain of S is equal to 12 months.

Functions $S(u, t)$ and $Z(u, t)$ are often quite unsmooth, which makes contour plots, perspective plots, or image plots inconvenient to interpret. However, we can plot few individual level sets of these functions. This shows for which time periods the performance, or the relative performance, is good. A level set of $S(u, t)$, for level $\lambda \in \mathbf{R}$, is defined by

$$\Lambda(S, \lambda) = \{(u, t) \in \mathcal{D} : S(u, t) \ge \lambda\}, \tag{10.39}$$

where the domain is

$$\mathcal{D} = \{(u, t) : u, t \in \{1, \dots, T\}, 1 \le u + \delta < t \le T\},$$

where $\delta > 0$.

Figure 10.9 shows a level set $\Lambda(S, \lambda)$ with blue color. The blue and red regions together show the domain of the function S. In panel (a) function S is calculated from the monthly returns of of S&P 500. In panel (b) the returns are of US Treasury 10-year bond. The level λ is the Sharpe ratio over the complete period. Parameter δ in the definition of the domain of S is equal to 36 months. Thus,

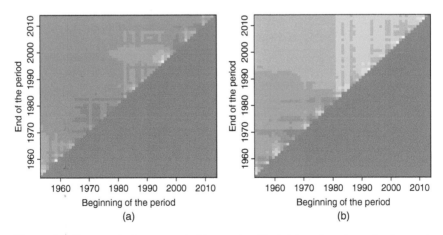

Figure 10.8 *Sharpe ratios for every period: Image plots.* The bright yellow shows the time periods where the Sharpe ratio is high and the red color shows the time periods where the Sharpe ratio is low. (a) S&P 500; (b) 10-year bond.

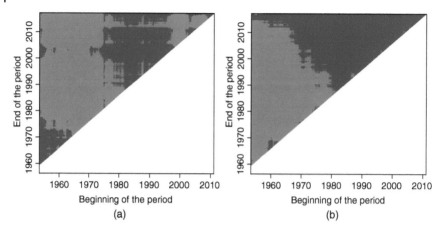

Figure 10.9 *Sharpe ratios for every period: Level sets.* We show a level set $\Lambda(S, \lambda)$ in (10.39) with blue color. (a) Function S is the Sharpe ratio of S&P 500; (b) function S is the Sharpe ratio of US Treasury 10-year bond. The blue regions show the time periods $[u, t]$ for which the Sharpe ratio is above the usual value, and the red color shows when it is below the usual value.

the blue region shows the time periods $[u, t]$ for which the Sharpe ratio is above the usual value, and the red region shows the time periods $[u, t]$ for which the Sharpe ratio is below the usual value.

10.5.2.2 Sharpe Ratios Over a Sequence of Intervals
A useful way to visualize function S, defined in (10.36), is to draw slices of this function. Slices are univariate functions

$$t \mapsto S(u_0, t), \qquad u \mapsto S(u, t_0),$$

where u_0 and t_0 are fixed. When u_0 is fixed, then we are looking at Sharpe ratios over periods with a fixed starting point u_0. When t_0 is fixed, then we are looking at Sharpe ratios over periods with a fixed end point t_0. For function $t \mapsto S(u_0, t)$ we choose u_0 so that $1 \leq u_0 \leq T - \delta$, and then t satisfies $u_0 + \delta \leq t \leq T$. For function $u \mapsto S(u, t_0)$ we choose t_0 so that $\delta \leq t_0 \leq T$, and then u satisfies $1 \leq u \leq t - \delta$.

Figure 10.10 shows slices of function S. Panel (a) shows slices $u \mapsto S(u, t_0)$, where $t_0 = 1953$ (red), $t_0 = 1986$ (blue), $t_0 = 1995$ (green), and $t_0 = 2014$ (black). Panel (b) shows slices $t \mapsto S(u_0, t)$, where $u_0 = 1954$ (black), $u_0 = 1986$ (red), $u_0 = 1995$ (blue), and $u_0 = 2014$ (green).

Figure 10.11 shows function $u \mapsto S(u, T)$ as black curves. In panel (a) we use S&P 500 monthly returns and in panel (b) we use monthly returns of US Treasury 10-year bond. Parameter δ is equal to 120 months. The green curves show time series of means: $u \mapsto 10 \times 12 \times \hat{\mu}_u^T$, and the blue curves show time series of standard deviations: $u \mapsto 4 \times \sqrt{12} \times \hat{\sigma}_u^T$, where $\hat{\mu}_u^t$ is defined in (10.37) and $\hat{\sigma}_u^t$ is defined in (10.38). Note that the upper borders of the level sets in Figure 10.9

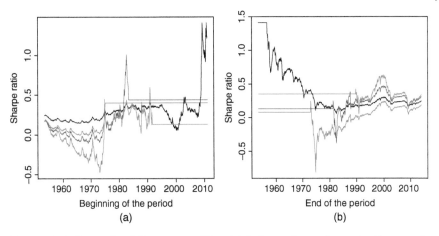

Figure 10.10 *Time series of Sharpe ratios: Slices.* (a) A slice at time u shows the Sharpe ratio computed with the data starting at u and ending t_0, where $t_0 = 1953$ (red), $t_0 = 1986$ (blue), $t_0 = 1995$ (green), and $t_0 = 2014$ (black). (b) A slice at time t shows the Sharpe ratio computed with the data starting at u_0 and ending t, where $u_0 = 1954$ (black), $u_0 = 1986$ (red), $u_0 = 1995$ (blue), and $u_0 = 2014$ (green).

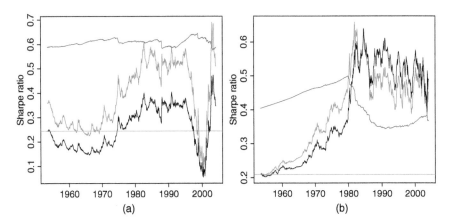

Figure 10.11 *Time series of Sharpe ratios.* (a) Sharpe ratios of S&P 500; (b) Sharpe ratios of US 10-year bond. The black curves show the Sharpe ratios, the green curves show the means of the excess returns, and the blue curves show the standard deviations of the excess returns. The time series at time t show Sharpe ratios computed with the data starting at t and ending T. The violet horizontal lines show the Sharpe ratios over the complete time period.

show the level sets of black functions in Figure 10.11. The violet horizontal lines show the Sharpe ratios over the complete time period.

Both time series of Sharpe ratios of S&P 500 and 10-year bond show a similar pattern: The Sharpe ratios make a jump at the end of 1970s. This pattern is more profound for 10-year bond than for S&P 500. We can see that the changes in the time series of Sharpe ratios are caused mainly by the changes in the time series of the arithmetic means of the returns.

Note that the Sharpe ratios in Figure 10.11 are relevant in the case when we choose time point t_0 to divide the historical data to the estimation part and to the testing part. After that we calculate the Sharpe ratio from the historically simulated returns $R_{t_0+1}, \ldots, R_T$, and compare the Sharpe ratio to the Sharpe ratio of a benchmark (S&P 500 or 10-year bond). The choice of time point t_0 affects the Sharpe ratio of the benchmark, as shown by the black curve in Figure 10.11, and in a sense, we study the robustness of the performance measures of the benchmarks to the choice of the time point t_0, which sets the beginning of the testing period.

10.5.3 Using the Certainty Equivalent in Evaluation

The certainty equivalent can be used in much the same way as the Sharpe ratio. Sample certainty equivalent is defined in (10.13). Given a time series $R_1, \ldots, R_T$ of gross returns, we define the sample certainty equivalent over interval $[u, t]$ as

$$ C(u, t) = U^{-1} \left(\frac{1}{t - u + 1} \sum_{i=u}^{t} U(R_i) \right), \qquad 1 \le u + \delta < t \le T, $$

where $U : (0, \infty) \to \mathbf{R}$ is a utility function. Parameter $0 < \delta < T$ guarantees that there are at least two observations to calculate the certainty equivalent. The slices of function C are often more informative than contour plots or perspective plots. Slices are univariate functions

$$ t \mapsto C(u_0, t), \qquad u \mapsto C(u, t_0), $$

where u_0 and t_0 are fixed. For function $t \mapsto S(u_0, t)$ we choose u_0 so that $1 \le u_0 \le T - \delta$, and then t satisfies $u_0 + \delta \le t \le T$. For function $u \mapsto S(u, t_0)$ we choose t_0 so that $\delta \le t_0 \le T$, and then u satisfies $1 \le u \le t - \delta$.

Figure 10.12 shows function $u \mapsto C(u, T)$: the sample mean is taken only over the last part of the original return time series. We have chosen $\delta = 120$. In panel (a) the returns R_i are monthly gross returns of S&P 500 index, and in panel (b) the returns R_i are monthly gross returns of US Treasury 10-year bond. The utility function $U : (0, \infty) \to \mathbf{R}$ is the power utility function, defined in (9.28). The green curve shows certainty equivalents when the risk aversion parameter of the utility function is $\gamma = 0$ (plain gross returns), the red curve shows the case $\gamma = 1$ (logarithmic utility), and the purple curve shows the case $\gamma = 5$. Note that the green curves in Figure 10.11 show the means of the excess returns, whereas the green curves in Figure 10.12 show the means of the gross returns.

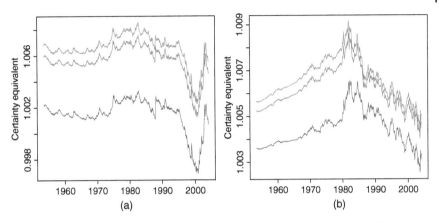

Figure 10.12 *Time series of certainty equivalents.* (a) Certainty equivalents of S&P 500; (b) certainty equivalents of US Treasury 10-year bond. The green curves show the case of risk aversion $\gamma = 0$, the red curves have $\gamma = 1$, and the purple curves have $\gamma = 5$. Time series at time t show certainty equivalents computed with the data starting at t and ending T.

We can see that the certainty equivalents of S&P 500 are rather stable when the testing period starts before the mid-1990s, whereas the certainty equivalents of 10-year bond are rather unstable even when the testing period starts early. The risk aversion parameter γ does not change qualitatively time series but affects only the level: a lower risk aversion leads to a larger certainty equivalent.

11

Markowitz Portfolios

Portfolio choice with mean–variance preferences was proposed by Markowitz (1952, 1959). The approach introduced the idea of balancing risk and return to find an optimal portfolio.

Markowitz approach can be used in the single period portfolio selection. Let R^p_{t+1} be the portfolio return. In the Markowitz approach there exists three ways to choose the portfolio:

1) Maximize the variance penalized expected return

$$E_t R^p_{t+1} - \frac{\gamma}{2} \operatorname{Var}_t \left(R^p_{t+1} \right),$$

   where $\gamma \geq 0$ is the risk aversion coefficient. Parameter γ measures the investor's relative risk aversion, as defined in (9.31).
2) Minimize the variance $\operatorname{Var}_t(R^p_{t+1})$ under a minimal requirement for the expected return: $E_t R^p_{t+1} \geq \mu_0$, where $\mu_0 \in \mathbf{R}$ is the minimal requirement for the expected return.
3) Maximize the expected return $E_t R^p_{t+1}$ under a condition that the variance is not too large: $\operatorname{Var}_t(R^p_{t+1}) \leq \sigma_0^2$, where $\sigma_0 > 0$ is the largest allowed standard deviation for the return.

Here E_t and Var_t mean the conditional expectation and the conditional variance, conditional on the information available at time t.

The variance penalized expected return was already discussed in Section 9.2.1. The variance penalized expected return is convenient because it involves explicitly the risk aversion parameter γ, which makes it possible to find a connection to the maximization of an expected utility. The other two approaches involve risk aversion more implicitly. When variance is minimized under a minimal requirement μ_0 for the expected return, then the minimal requirement μ_0 is a risk aversion parameter, because smaller values of μ_0 are associated with more risk aversion. When the expected return is maximized under a condition that the variance is less than or equal to σ_0^2, then σ_0 is a risk aversion parameter, because smaller values of σ_0 indicate more risk aversion.

Nonparametric Finance, First Edition. Jussi Klemelä.
© 2018 John Wiley & Sons, Inc. Published 2018 by John Wiley & Sons, Inc.

We explain with the help of Markowitz bullets the concepts of the minimum variance portfolio, the tangency portfolio, and the efficient frontier. This is done in Section 11.3.

The method of Lagrange multipliers appears in Section 11.1.2 and in Section 11.2. The method of Lagrange multipliers is a useful general method of optimization. The method of Lagrange multipliers helps to cope with the restriction that the sum of portfolio weights have to be equal to one. Further complications appear when we want to restrict ourselves to long-only portfolios, or to make some other additional restrictions on the portfolio weights. We do not consider these additional complications.

We use the notations of Section 9.1. The portfolio return was defined in (9.3) as

$$R^p_{t+1} = \bar{b}'_t \bar{R}_{t+1} = \sum_{i=0}^{d} b^i_t R^i_{t+1},$$

where

$$\bar{b}_t = \left(b^0_t, b_t \right), \quad b_t = \left(b^1_t, \dots, b^d_t \right)$$

is the vector of the portfolio weights, $\bar{b}'_t$ is the transpose of the column vector $\bar{b}_t$, and

$$\bar{R}_{t+1} = \left(R^0_{t+1}, R_{t+1} \right), \quad R_{t+1} = \left(R^1_{t+1}, \dots, R^d_{t+1} \right)$$

is the vector of the gross returns of the portfolio components. The portfolio weights satisfy the constraint

$$\bar{b}'_t 1_{d+1} = \sum_{i=0}^{d} b^i_t = 1.$$

Thus, $b^0_t = 1 - 1'_d b_t$, and

$$R^p_{t+1} = \left(1 - 1'_d b_t \right) R^0_{t+1} + b'_t R_{t+1} = R^0_{t+1} + b'_t \left(R_{t+1} - R^0_{t+1} \right),$$

where $R_{t+1} - R^0_{t+1}$ is called the excess return. Since we consider only single period portfolio selection, we do not need the time subscript in the notation. Thus, we denote the portfolio vector of risky assets by

$$b = b_t.$$

Also, since the expectations and variances are conditional on t, the risk-free rate is a constant (known at time t). We will denote the risk-free rate by

$$r = R^0_{t+1}.$$

The vector of means and the covariance matrix of the risky assets is denoted by

$$\mu = E\, R_{t+1}, \quad \mathrm{Var}(R_{t+1}) = \Sigma,$$

where $\mu = (\mu_1, \dots, \mu_d)'$ and Σ is the $d \times d$ matrix with elements σ_{ij}. Now,

$$E(b'R_{t+1}) = b'\mu, \quad \mathrm{Var}(b'R_{t+1}) = b'\Sigma b.$$

Section 11.1 considers the maximization of the variance penalized expected return. Section 11.2 considers minimization of the variance under a minimal requirement for the expected return. Section 11.3 considers concepts related to the Markowitz portfolio theory, such as the minimum variance portfolio and the tangency portfolio. Section 11.4 considers further topics related to Markowitz portfolio theory. Section 11.5 applies Markowitz formulas to portfolio selection.

11.1 Variance Penalized Expected Return

We consider the maximization of the variance penalized expected return. Section 11.1.1 considers portfolios where the risk-free rate is included. Section 11.1.2 considers portfolios without the risk-free rate.

11.1.1 Variance Penalization with the Risk-Free Rate

Let us consider the maximization of the variance penalized expected return when the risk-free rate is included. We consider first the general case of d risky asset and then the special cases of one risky asset and two risky assets.

11.1.1.1 Several Risky Assets and the Risk-Free Rate

The portfolio components are d risky assets and the risk-free rate. Let the return of the risk-free investment be r. We allocate the proportion $1 - b'1_d$ into the risk-free investment. Then the portfolio return is

$$R^p_{t+1} = b'R_{t+1} + (1 - b'1_d)r.$$

We choose the weight vector b as maximizing

$$b'\mu + (1 - b'1_d)r - \frac{\gamma}{2}\, b'\Sigma b. \tag{11.1}$$

Derivating with respect to b and setting the partial derivatives to zero gives

$$\mu - 1_d r - \gamma\Sigma b = 0.$$

Thus,

$$b = \gamma^{-1}\Sigma^{-1}(\mu - r1_d).$$

11.1.1.2 One Risky Asset and the Risk-Free Rate

Let us invest the proportion b to a stock and $1 - b$ to the risk-free rate whose gross return is $r > 0$. Now the gross return of the portfolio is

$$R^p_{t+1} = bR_{t+1} + (1 - b)r,$$

where R_{t+1} is the return of the stock. Let the expected return of the stock be $E_t R_{t+1} = \mu$ and the variance $\text{Var}_t(R_{t+1}) = \sigma^2$. Then,

$$E_t R_{t+1}^p - \frac{\gamma}{2} \text{Var}_t\left(R_{t+1}^p\right) = r + b(\mu - r) - \frac{\gamma}{2} b^2 \sigma^2. \tag{11.2}$$

Setting the derivative with respect to b to zero and solving for b gives the maximizer of (11.2) as

$$b = \frac{1}{\gamma} \frac{\mu - r}{\sigma^2}. \tag{11.3}$$

Let b_{rest} be the optimal weight of the long-only portfolio. The maximizer b_{rest} of (11.2) under the restriction that $b \in [0, 1]$ is obtained by projecting the unrestricted solution on $[0, 1]$. Thus,

$$b_{rest} = \begin{cases} 0, & \text{when } b \le 0, \\ b, & \text{when } 0 \le b \le 1, \\ 1, & \text{when } b \ge 1, \end{cases} \tag{11.4}$$

where b is given in (11.3).

11.1.1.3 Two Risky Assets and the Risk-Free Rate

Let us have two stocks and the risk-free rate and let us invest the proportion b_1 in the first stock, proportion b_2 in the second stock, and proportion $1 - b_1 - b_2$ in the risk-free rate. Now the portfolio return is

$$R_{t+1}^p = b_1 R_{t+1}^1 + b_2 R_{t+1}^2 + (1 - b_1 - b_2)r,$$

where R_{t+1}^1 is the return of the first stock and R_{t+1}^2 is the return of the second stock. Let the expected returns of the stocks be $E_t R_{t+1}^1 = \mu_1$, $E_t R_{t+1}^2 = \mu_2$ and let the variances of the returns be $\text{Var}_t(R_{t+1}^1) = \sigma_1^2$, $\text{Var}_t(R_{t+1}^2) = \sigma_2^2$. Denote the covariance of the returns by $\text{Cov}_t(R_{t+1}^1, R_{t+1}^2) = \sigma_{12}$. We have

$$E_t R_{t+1}^p - \frac{\gamma}{2} \text{Var}_t\left(R_{t+1}^p\right)$$
$$= b_1 \mu_1 + b_2 \mu_2 + (1 - b_1 - b_2)r - \frac{\gamma}{2}\left(b_1^2 \sigma_1^2 + b_2^2 \sigma_2^2 + 2b_1 b_2 \sigma_{12}\right).$$

Setting derivatives with respect to b_1 and b_2 to zero gives

$$\begin{cases} \mu_1 - r - \gamma \sigma_1^2 b_1 - \gamma \sigma_{12} b_2 = 0, \\ \mu_2 - r - \gamma \sigma_2^2 b_2 - \gamma \sigma_{12} b_1 = 0. \end{cases}$$

Thus,[1]

$$b_1 = \frac{1}{\gamma} \frac{(\mu_1 - r)\sigma_2^2 - \sigma_{12}(\mu_2 - r)}{\sigma_1^2 \sigma_2^2 - \sigma_{12}^2}$$

---

1 Solving b_2 gives

$$b_2 = \frac{\mu_2 - r - \gamma \sigma_{12} b_1}{\gamma \sigma_2^2}.$$

and

$$b_2 = \frac{1}{\gamma} \frac{(\mu_2 - r)\sigma_1^2 - \sigma_{12}(\mu_1 - r)}{\sigma_1^2 \sigma_2^2 - \sigma_{12}^2}.$$

11.1.2 Variance Penalization without the Risk-Free Rate

Let us consider the maximization of the variance penalized expected return when the risk-free rate is excluded. We solve first the case of d risky assets and then the case of two risky assets.

11.1.2.1 Several Risky Assets

The maximization of the variance penalized expected return chooses the weight vector b as maximizing

$$b'\mu - \frac{\gamma}{2} b'\Sigma b \quad \text{under} \quad b'1_d = 1, \tag{11.5}$$

where 1_d is the vector of length d whose all elements are equal to one, so that the constraint is

$$b'1_d = \sum_{i=1}^{d} b^i = 1.$$

Let us maximize

$$b'\mu - \frac{\gamma}{2} b'\Sigma b$$

under the constraint $b'1_d = 1$. We use the method of Lagrange multipliers and maximize the Lagrange function

$$b'\mu - \frac{\gamma}{2} b'\Sigma b + \lambda(b'1_d - 1),$$

where $\lambda \in \mathbf{R}$ is the Lagrange multiplier. Derivating with respect to b and λ and setting the partial derivatives to zero we get

$$\begin{cases} \mu - \gamma\Sigma b + \lambda 1_d = 0, \\ b'1_d - 1 = 0. \end{cases}$$

Thus,

$$b = \gamma^{-1}\Sigma^{-1}(\mu + \lambda 1_d).$$

Let us solve λ from $b'1_d = 1$, which leads to

$$\gamma^{-1}1'_d\Sigma^{-1}\mu + \lambda\gamma^{-1}1'_d\Sigma^{-1}1_d = 1,$$

This leads to

$$\mu_1 - r - \frac{\sigma_{12}}{\sigma_2^2}(\mu_2 - r) + b_1\gamma\left(\frac{\sigma_{12}^2}{\sigma_2^2} - \sigma_1^2\right) = 0.$$

and finally

$$\lambda = \frac{1 - \gamma^{-1} 1_d' \Sigma^{-1} \mu}{\gamma^{-1} 1_d' \Sigma^{-1} 1_d}.$$

11.1.2.2 Two Risky Assets

Let us have two stocks and put the proportion $1 - b$ to the first stock and proportion b to the second stock. Now,

$$R_{t+1}^p = (1 - b)R_{t+1}^1 + bR_{t+1}^2.$$

Let the expected returns of the stocks be $E_t R_{t+1}^1 = \mu_1$, $E_t R_{t+1}^2 = \mu_2$ and let the variances of the returns be $\mathrm{Var}_t(R_{t+1}^1) = \sigma_1^2$, $\mathrm{Var}_t(R_{t+1}^2) = \sigma_2^2$. Denote the covariance of the returns by $\mathrm{Cov}_t(R_{t+1}^1, R_{t+1}^2) = \sigma_{12}$. We have

$$E_t R_{t+1}^p - \frac{\gamma}{2} \, \mathrm{Var}_t \left(R_{t+1}^p \right)$$

$$= \mu_1 + b(\mu_2 - \mu_1) - \frac{\gamma}{2} \left[(1 - b)^2 \sigma_1^2 + b^2 \sigma_2^2 + 2(1 - b)b\sigma_{12} \right]$$

$$= \mu_1 - \frac{\gamma}{2} \, \sigma_1^2 + b \left[\mu_2 - \mu_1 - \gamma \left(\sigma_{12} - \sigma_1^2 \right) \right] - b^2 \, \frac{\gamma}{2} \left(\sigma_1^2 + \sigma_2^2 - 2\sigma_{12} \right).$$

Setting the derivative with respect to b to zero and solving for b gives

$$b = \frac{1}{\gamma} \frac{\mu_2 - \mu_1 - \gamma \left(\sigma_{12} - \sigma_1^2 \right)}{\sigma_1^2 + \sigma_2^2 - 2\sigma_{12}}. \tag{11.6}$$

Note that the maximizer b_{rest} under the restriction that $b \in [0, 1]$ is obtained by projecting the unrestricted solution:

$$b_{rest} = \min\{\max\{0, b\}, 1\}, \tag{11.7}$$

where b is given in (11.6).

11.2 Minimizing Variance under a Sufficient Expected Return

We consider the minimization of the variance under a condition that the expected return should be sufficiently large. Section 11.2.1 considers portfolios where the risk-free rate is included. Section 11.2.2 considers portfolios without the risk-free rate.

11.2.1 Minimizing Variance with the Risk-Free Rate

We consider first the case of d risky assets and the risk-free investment, and then the case of one risky assets and the risk-free investment.

11.2.1.1 Several Risky Assets and the Risk-Free Rate

Let us consider the case of d risky assets and a risk-free investment. We want to find the weight vector minimizing

$$b'\Sigma b \quad \text{under} \quad b'\mu + (1 - b'1_d)r = \mu_0, \tag{11.8}$$

where $\mu_0 \in \mathbf{R}$. We should choose $\mu_0 \geq r$, so that the required expected return is not smaller than the risk-free return.

The return vector of the risky investments is denoted by R_{t+1}, the expectation vector is μ, the covariance matrix is Σ, and the risk-free return is r. The proportion $1 - b'1_d$ is invested in the risk-free asset. The return of the portfolio is

$$R^p_{t+1} = b'R_{t+1} + (1 - b'1_d)r.$$

The expected return of the portfolio is

$$b'\mu + (1 - b'1_d)r.$$

Let us find b minimizing

$$\frac{1}{2} b'\Sigma b$$

under the constraint

$$b'\mu + (1 - b'1_d)r = \mu_0,$$

where $\mu_0 \in \mathbf{R}$ is a constant. Define the Lagrange function

$$L(b, \lambda) = \frac{1}{2} b'\Sigma b + \lambda[\mu_0 - b'\mu - (1 - b'1_d)r],$$

where $\lambda \in \mathbf{R}$ is the Lagrange multiplier. We solve the equation

$$0 = \frac{\partial}{\partial b} L(b, \lambda) = \Sigma b + \lambda(1_d r - \mu)$$

to get

$$b = \lambda \Sigma^{-1}(\mu - 1_d r).$$

The constraint can be written as

$$b'(\mu - r1_d) = \mu_0 - r,$$

which implies

$$\lambda(\mu - r1_d)'\Sigma^{-1}(\mu - r1_d) = \mu_0 - r$$

and

$$\lambda = \frac{\mu_0 - r}{(\mu - r1_d)'\Sigma^{-1}(\mu - r1_d)}.$$

Thus, the vector of the weights of the risky investments is

$$b = \frac{\mu_0 - r}{(\mu - r1_d)'\Sigma^{-1}(\mu - r1_d)} \Sigma^{-1}(\mu - r1_d).$$

11.2.1.2 One Risky Asset and the Risk-Free Rate

Let us consider the case where we have one risky asset with return R_{t+1} and a risk-free investment with return r. Let the expected return of the risky asset be $E_t R_{t+1} = \mu$ and the variance $\mathrm{Var}_t(R_{t+1}) = \sigma^2$. Let us invest the proportion b to the risky asset and the proportion $1 - b$ to the risk-free asset. The return of the portfolio is

$$R^p_{t+1} = bR_{t+1} + (1 - b)r.$$

The expected return of the portfolio is

$$E_t R^p_{t+1} = b\mu + (1 - b)r = r + b(\mu - r)$$

and the variance of the portfolio is

$$\mathrm{Var}_t\left(R^p_{t+1}\right) = b^2\sigma^2.$$

We want that the expected return should be at least $\mu_0 \in \mathbf{R}$ and we minimize the variance under this condition. Thus, we want to find b minimizing

$$\frac{1}{2} b^2\sigma^2$$

under the constraint

$$r + b(\mu - r) = \mu_0.$$

Define the Lagrange function

$$L(b, \lambda) = \frac{1}{2} b^2\sigma^2 + \lambda[\mu_0 - r - b(\mu - r)],$$

where $\lambda \in \mathbf{R}$ is the Lagrange multiplier. The solution of the equation

$$0 = \frac{\partial}{\partial b} L(b, \lambda) = \sigma^2 b - \lambda(\mu - r)$$

is

$$b = \lambda \frac{\mu - r}{\sigma^2}.$$

The constraint $b(\mu - r) = \mu_0 - r$ implies $\lambda = \sigma^2(\mu_0 - r)/(\mu - r)^2$. Thus, the weight of the risky investment is

$$b = \frac{\mu_0 - r}{\mu - r}.$$

When $r \le \mu_0 \le \mu$, then $0 \le b \le 1$.

11.2.2 Minimizing Variance without the Risk-Free Rate

We want to choose the weight vector b minimizing

$$b'\Sigma b \quad \text{under} \quad b'\mu = \mu_0, \quad b'1_d = 1, \tag{11.9}$$

where $\mu_0 \in \mathbf{R}$, and we should choose $\mu_0 \ge r$, so that the required expected return is not smaller than the risk-free return.

Let us consider portfolios of d risky assets and exclude the risk-free investment. The return vector of the risky investments is denoted by R_{t+1}. Let us denote $\mu = E_t R_{t+1}$ and $\Sigma = \text{Cov}_t(R_{t+1})$. Then,

$$E_t(b'R_{t+1}) = b'\mu, \quad \text{Var}_t(b'R_{t+1}) = b'\Sigma b.$$

We minimize

$$\frac{1}{2} b'\Sigma b$$

under the constraints

$$b'\mu = \mu_0, \quad b'1_d = 1,$$

where 1_d is the column vector of length d whose elements are equal to 1, and $\mu_0 \in \mathbf{R}$ is a constant. The Lagrange function is

$$L(b, \lambda_1, \lambda_2) = \frac{1}{2} b'\Sigma b + \lambda_1(\mu_0 - b'\mu) + \lambda_2(1 - b'1_d),$$

where $\lambda_1, \lambda_2 \in \mathbf{R}$ are the Lagrange multipliers. The solution of the equation[2]

$$0 = \frac{\partial}{\partial b} L(b, \lambda_1, \lambda_2) = \Sigma b - \lambda_1 \mu - \lambda_2 1_d$$

is

$$b = \Sigma^{-1}(\lambda_1 \mu + \lambda_2 1_d).$$

To get λ_1 and λ_2 we need to solve the equations

$$\begin{cases} \mu_0 = b'\mu = \lambda_1 \mu'\Sigma^{-1}\mu + \lambda_2 \mu'\Sigma^{-1}1_d, \\ 1 = b'1_d = \lambda_1 1'_d\Sigma^{-1}\mu + \lambda_2 1'_d\Sigma^{-1}1_d. \end{cases}$$

Denoting $\alpha = \mu'\Sigma^{-1}\mu$, $\beta = 1'_d\Sigma^{-1}1_d$, and $\delta = 1'_d\Sigma^{-1}\mu$, we get

$$\lambda_1 = \frac{\beta\mu_0 - \delta}{\alpha\beta - \delta^2}, \quad \lambda_2 = \frac{\alpha - \delta\mu_0}{\alpha\beta - \delta^2}.$$

Then, the vector of the portfolio weights is

$$b = \frac{1}{e} \Sigma^{-1}[(\alpha 1_d - \gamma\mu) + \mu_0 (\beta\mu - \gamma 1_d)],$$

where $e = \alpha\beta - \delta^2$.

11.3 Markowitz Bullets

A Markowitz bullet is a scatter plot of points, where each point corresponds to a portfolio, the x-coordinate of a point is the standard deviation of the return of the portfolio, and the y-coordinate of a point is the expected return of the

---

2 We have that $\frac{\partial}{\partial b} b'\Sigma b = (\Sigma + \Sigma')b$, and for symmetric matrices $(\Sigma + \Sigma')b = 2\Sigma b$.

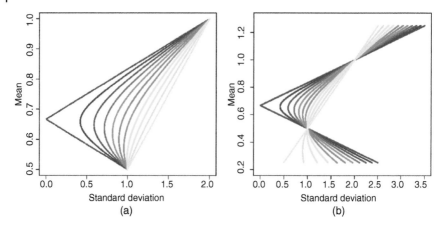

Figure 11.1 *Markowitz bullets: Portfolios of two risky assets when correlation varies.* (a) Shown are long-only portfolios that can be obtained from two risky assets when correlation between the risky assets varies between −1 and 1. (b) Shorting is allowed.

portfolio. The scatter plot is called a bullet because the boundary of the scatter plot is a part of a hyperbola, and thus its shape resembles the shape of a bullet.[3]

Figure 11.1 plots a collection of portfolios which are obtained from two risky assets. The expected net returns of the assets are 1 and 0.5. The standard deviations are 2 and 1. The correlation between the returns of the risky assets varies from −1 to 1. Panel (a) shows long-only portfolios. The blue wedge on the left shows all portfolios that can be obtained when correlation is −1. The orange vector on the right shows all portfolios that can be obtained when correlation is 1. When correlation is −1, then there exists a portfolio with zero variance. The portfolio with zero variance should have the same return as the risk-free rate, to exclude arbitrage. Panel (b) shows portfolios that can be obtained from the two risky assets when shorting is allowed. The weight of an asset varies between −0.5 and 1.5.

Figure 11.2 shows portfolios obtained from three risky assets as a blue area. The three risky assets are shown as orange points. The correlations between the risky assets are 0.2, 0.5, and 0.6. Panel (a) shows all long-only portfolios and panel (b) shows portfolios when shorting is allowed with restrictions. The shapes of the blue areas are irregular but the left boundaries are parts of hyperbolas.

Figure 11.3 shows a Markowitz bullet of long-only portfolios, when the risk-free rate is included and the borrowing is allowed. Panel (a) shows as a blue curve long-only portfolios whose components are two risky assets with

---

3 A hyperbola should not be confused with a parabola, which can be obtained as a graph of a polynomial of degree two. For example, the graph of $x \mapsto 1/x$ is a hyperbola, and the points (x, y) which satisfy $x^2/a^2 - y^2/b^2 = 1$ constitute a hyperbola, when $a, b > 0$ are constants.

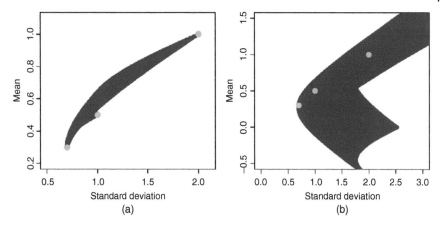

Figure 11.2 *Markowitz bullets: Portfolios of three risky assets.* (a) Long-only portfolios that can be obtained from three risky assets. (b) Portfolios when shorting is allowed.

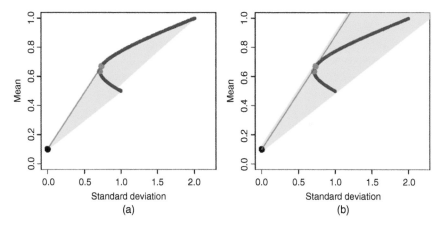

Figure 11.3 *Markowitz bullet: Long-only portfolios and leveraging.* Panel (a) shows long-only portfolios for two risky assets and the risk-free investment. Panel (b) shows portfolios for two risky assets and the risk-free investment when the weight of the risk-free investment is allowed to be negative, which means the borrowing is allowed.

correlation −0.4. The green point shows the minimum variance portfolio. The black point shows the risk-free investment whose net return is 0.1 and the variance is zero. The red point shows the tangency portfolio. The red line joining the risk-free investment and the tangency portfolio corresponds to the long-only portfolios whose components are the risk-free investment and the tangency portfolio. The yellow area corresponds to the long-only portfolios whose components are the risk-free investment and one of the portfolios on the blue curve; these are all possible long-only portfolios. Panel (b) shows portfolios from two risky assets and the risk-free investment when the weight of the risk-free investment is allowed to be negative, which amounts to allowing leveraging by borrowing.

We can use Figure 11.3 to define the concepts of the minimum variance portfolio, the tangency portfolio, and the efficient frontier.

1) The minimum variance portfolio is the portfolio of risky assets whose variance is the smallest among all portfolios of risky assets. When the risk-free rate is included, then the risk-free investment has the minimum variance zero.

2) Efficient frontier is the collection of those portfolio vectors that have the expected return greater than or equal to the expected return of the minimum variance portfolio:

   In Figure 11.3(a) the efficient frontier without the risk-free rate is the part of the blue curve going upward from the red point, but when the risk-free rate is included, then the red vector from the risk-free asset to the tangency portfolio, followed by the blue curve shows the efficient frontier. The efficient frontier consists of possible portfolios a rational investor should consider, because these portfolios have a higher expected return with the same variance than other portfolios. Adding the risk-free rate gives the possibility to get portfolios with a smaller standard deviation than any of the pure stock portfolios: some of the portfolios on the red vector are such that the standard error is smaller than the standard deviation of any of the pure stock portfolios.

   In Figure 11.3(b) borrowing is allowed. The borrowed money is invested in the stocks. Now the efficient frontier is the red half line starting from the risk-free investment and passing the tangency portfolio. We see that a rational investor chooses only portfolios that are a combination of the risk-free investment and the tangency portfolio. The other portfolios have a smaller expected return for the same variance.

3) The tangency portfolio is a portfolio which has the largest Sharpe ratio. Indeed, the tangency portfolio maximizes the slope of the vector drawn from the risk-free asset to a pure stock portfolio. The slope of the vector from the point $(0, R_f)$ to the point $(\mathrm{sd}(R_t^p), ER_t^p)$ is equal to the Sharpe ratio $(ER_t^p - R_f)/\mathrm{sd}(R_t^p)$, where R_f is the return of the risk-free asset and R_t^p is the return of a portfolio.

4) It can be argued that the tangency portfolio, shown as the red point in the blue curve, is in fact the market portfolio, because the rational investor buys only a combination of the tangency portfolio and the risk-free asset, and thus the price of the tangency portfolio is in the equilibrium equal to the price of the market portfolio.

Figure 11.4 plots standard deviations and means for a collection of portfolios when shorting of a stock is allowed. Panel (a) shows portfolios from two risky

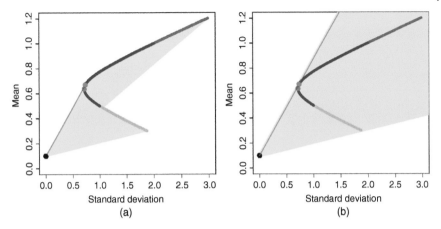

Figure 11.4 *Markowitz bullet: Shorting and leveraging.* Panel (a) shows portfolios from two risky assets, and from two risky assets and the risk-free investment. Panel (b) shows portfolios from two risky assets and a risk-free investment when the weight of the risk-free investment is allowed to be negative, so that borrowing is possible.

assets. The blue part shows the long-only portfolios, the orange part shows the portfolios where the less risky stock is shorted, and the purple part shows the portfolios where the more risky stock is shorted. The green bullet shows the minimum variance portfolio, the black bullet shows the risk-free investment, and the red bullet shows the tangency portfolio. Panel (b) shows portfolios of two risky assets and a risk-free investment when the weight of the risk-free investment is allowed to be negative, which amounts to allowing leveraging by borrowing.

Figure 11.5 shows how increasing the number of basis assets makes the Markowitz bullet larger. The blue hyperbola shows portfolios that can be obtained from two risky assets, the green area shows portfolios that can be

Figure 11.5 *Markowitz bullet: Uncorrelated assets.* Markowitz bullets are shown for an increasing number of assets: blue curve shows portfolios from two risky assets, the green area portfolios from three risky assets, and the yellow area portfolios from four risky assets, when the risky assets are uncorrelated.

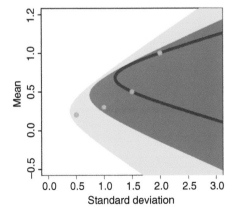

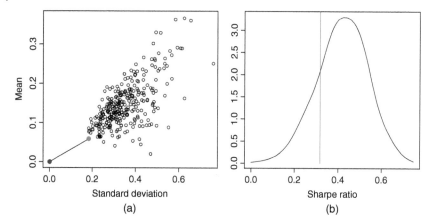

Figure 11.6 *Markowitz bullet: S&P 500 components.* (a) A scatter plot of annualized sample standard deviations and means of excess returns of a collection of stocks in the S&P 500 index. (b) A kernel density estimate of the distribution of the Sharpe ratios.

obtained from three risky assets, and the yellow area shows portfolios that can be obtained from four risky assets. The orange points show the risky assets. The covariances between the returns of the risky assets are zero.

Figure 11.6 studies a Markowitz bullet of the daily returns of S&P 500 components. The data is described in Section 2.4.5. Panel (a) shows a scatter plot of the annualized sample standard deviations and annualized sample means of the excess returns of the stocks included in the S&P 500 components data. The red bullet shows the location of the S&P 500 index. The blue bullet is at the origin: we take the risk-free rate equal to zero because the Markowitz bullet is computed from the excess returns.[4] Panel (b) shows a kernel density estimate of the distribution of the Sharpe ratios of the stocks included in S&P 500 components data. The red vertical line indicates the Sharpe ratio of the S&P 500 index. We see that the S&P 500 is not a tangent portfolio, since its Sharpe ratio is smaller than the most Sharpe ratios of the individual stocks.

11.4 Further Topics in Markowitz Portfolio Selection

11.4.1 Estimation

In order to apply Markowitz formulas, we have to estimate the vector μ of expected returns and the covariance matrix Σ of the returns of the risky assets.

---

4 The risk-free rate is computed form the yields of the US 1-month bill: the monthly yield is divided by 22 to obtain a risk-free rate for the one day period. The 1-month bill data is described in Section 2.4.3.

The sample means, sample variances, and sample covariances could be applied. However, we have discussed many other methods. Chapter 6 discusses various prediction methods that could be applied to estimate (predict) $\mu = E_t R_{t+1}$. Chapter 7 discusses various methods for volatility prediction that could be used to estimate $\sigma_i^2 = \text{Var}_t(R_{t+1}^i)$, $i = 1, \ldots, d$. Analogous methods can be used to estimate the covariances $\sigma_{ij} = \text{Cov}_t(R_{t+1}^i, R_{t+1}^j)$, $i, j = 1, \ldots, d$, $i \neq j$. For example, Section 5.4 considers multivariate time series models which are relevant for covariance prediction.

In the estimation of the covariance matrix Σ we have to take the curse of dimensionality into account, since the number d of risky assets can be high relative to the sample size. Note that the covariance matrix involves only the pairwise covariances, so that high dimensionality does not make it difficult to estimate any single component of the matrix Σ. However, there are $d(d-1)/2$ covariances, and a simultaneous estimation of such a large number of parameters is difficult.

11.4.2 Penalizing Techniques

Let us consider minimization of the variance of the portfolio return under a minimal requirement for the expected return. Let R_{t+1} be the return vector of risky assets with $E_t R_{t+1} = \mu$ and $\text{Var}_t(R_{t+1}) = \Sigma$. Then the expected return of the portfolio is $b'\mu$ and the variance is $b'\Sigma b$, where b is the vector of portfolio weights. We want to find weights b such that

$$b'\Sigma b$$

is minimized under the constraints

$$b'\mu = \mu_0, \quad b'1_d = 1,$$

where $\mu_0 \in \mathbf{R}$ is the requirement for the expected return of the portfolio. The minimization problem is equivalent to finding b such that

$$E_t(b'R_{t+1} - \mu_0)^2$$

is minimized under the same constraints. Let us assume to have observed historical returns $R_1, \ldots, R_T$ of the basis assets. The empirical version of the minimization problem is to find b such that

$$\sum_{t=0}^{T-1} (b'R_{t+1} - \mu_0)^2$$

is minimized under the constraints

$$b'\hat{\mu} = \mu_0, \quad b'1_d = 1,$$

where $\hat{\mu} = T^{-1} \sum_{t=0}^{T-1} R_{t+1}$. Brodie *et al.* (2009) proposed to add a penalization term and find b minimizing

$$\sum_{t=0}^{T-1} (b'R_{t+1} - \mu_0)^2 + \tau \sum_{i=1}^{d} |b_i|,$$

under the same constraints, where $\tau > 0$ is the regularizing parameter. The approach is similar to the approach in Lasso regression of Tibshirani (1996).

DeMiguel *et al.* (2009) showed that it is difficult to significantly or consistently outperform the naive strategy in which each available asset is given an equal weight in the portfolio.

11.4.3 Principal Components Analysis

Let μ be the d vector of the expected returns of the d risky assets. Given the d vector of portfolio weights b, the return of the portfolio is $b'R_{t+1}$. Let Σ be the $d \times d$ covariance matrix of the returns of the risky assets. We can make the principal component analysis of the covariance matrix and write

$$\Sigma = U\Lambda U',$$

where U is the $d \times d$ matrix whose columns are the eigenvectors of Σ and Λ is the $d \times d$ diagonal matrix, whose diagonal elements are the eigenvalues of Σ. We get d uncorrelated principal portfolios whose return vector is $U'R_{t+1}$. We can think of these principal portfolios as new basic assets and write any portfolio in terms of the principal components. If the original weights are b, then the new weights are $\tilde{b} = U'b$. Now we can calculate the variance of the portfolio as

$$\text{Var}(b'R_{t+1}) = b'\Sigma b = \sum_{i=1}^{d} \tilde{b}_i^2 \lambda_i,$$

where $\lambda_1, \ldots, \lambda_d$ are the eigenvalues of Σ. We can define the diversification distribution

$$p_i = \frac{\tilde{b}_i^2 \lambda_i}{\text{Var}_t(b'R_{t+1})},$$

where $i = 1, \ldots, d$. We can say that a portfolio is better diversified, if the diversification distribution is closer to the uniform distribution. This can be measured by

$$\exp\left\{ -\sum_{i=1}^{d} p_i \log_e p_i \right\}.$$

Partovi and Caputo (2004) used principal portfolios in their discussion of efficient frontier, and Meucci (2009) presented the idea of the diversification distribution.

11.5 Examples of Markowitz Portfolio Selection

We illustrate Markowitz portfolio selection using as the basic assets the S&P 500 and Nasdaq-100 indexes. The daily data set of S&P 500 and Nasdaq-100 is described in Section 2.4.2.

We consider portfolio selection without the risk-free rate. We maximize the variance penalized expected return (11.5) both without restrictions and with the restriction to the long-only weights.

Figure 11.7 shows the time series of the Markowitz weights of S&P 500. Panel (a) shows the unrestricted Markowitz weights and panel (b) shows the long-only Markowitz weights. The risk aversion parameter takes values $\gamma = 2, 5, 10, 50, 100$ (black, red, blue, green, and orange). When the weight of S&P 500 is denoted by b, then the weight of Nasdaq-100 is $1 - b$. We have estimated the mean vector and the covariance matrix sequentially, using the sample means and the sample covariance matrices. We start when there are 1000 observations (4 years of data). The weight of S&P 500 increases when the risk aversion parameter γ increases. After year 2000, the weight of S&P 500 jumps higher.

Figure 11.8 shows the Sharpe ratios, annualized means and annualized standard deviations, as a function of risk aversion parameter $\gamma = 2, 5, 10, 50, 100$. Panel (a) shows the Sharpe ratios. The black line with labels "1" is obtained when the unrestricted weights are used and the green line with labels "2" is obtained when the long-only weights are used. The horizontal lines show the Sharpe ratios for S&P 500 (blue) and Nasdaq-100 (red). The risk-free rate is deduced from the 1-month US bill rates, described in Section 2.4.3. The highest value of the Sharpe ratio is obtained for small risk aversion. When the risk

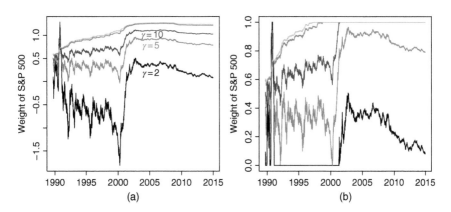

Figure 11.7 *S&P 500 and Nasdaq-100: Markowitz weights.* The time series of the weights for the S&P 500. (a) The unrestricted weights and (b) the long-only weights. The risk aversion parameter takes values $\gamma = 2, 5, 10, 50, 100$ (black, red, blue, green, and orange).

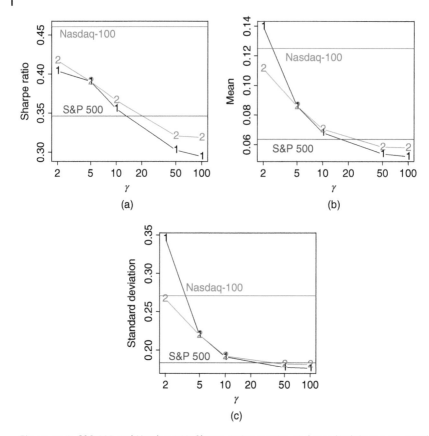

Figure 11.8 *S&P 500 and Nasdaq-100: Sharpe ratios, means, and standard deviations.* (a) The Sharpe ratios as a function of γ; (b) the annualized means; (c) the annualized standard deviations. The black line with labels "1" is obtained when the unrestricted weights are used and the green line with labels "2" is obtained when the long-only weights are used.

aversion increases, the Sharpe ratios of the Markowitz portfolio approach the Sharpe ratio of S&P 500. Panel (b) shows the annualized means as a function of the risk aversion and panel (c) shows the annualized standard deviations. Both means and standard deviations increase sharply when the risk aversion parameter decreases.

12

Dynamic Portfolio Selection

Dynamic portfolio selection means here such portfolio selection in a multi-period model where the portfolio weights are repeatedly rebalanced at the beginning of every period, using information available at the beginning of each period. We apply the methods of Chapter 6, where prediction of asset returns is studied, and the methods of Chapter 7, where prediction of volatility is studied.

The return of a portfolio is a linear combination of the returns of the portfolio components. Let us exclude the risk-free rate for a moment, so that the return of the portfolio is given by

$$R^p_{t+1} = b'_t R_{t+1},$$

where R^p_{t+1} is the gross return the portfolio, $b_t = (b^1_t, \ldots, b^d_t)$ is the vector of portfolio weights, and $R_{t+1} = (R^1_{t+1}, \ldots, R^d_{t+1})$ is the vector of the gross returns of the risky assets. We can consider the following approaches to choose the portfolio weights b_t.

1) Maximize the expected return

$$E_t R^p_{t+1} = b'_t \mu_t, \tag{12.1}$$

where $\mu_t = E_t R_{t+1}$.

2) Maximize the variance penalized expected returns

$$E_t R^p_{t+1} - \frac{\gamma}{2} \mathrm{Var}_t \left(R^p_{t+1} \right) = b'_t \mu_t - \frac{\gamma}{2} b'_t \Sigma_t b_t, \tag{12.2}$$

where $\gamma \geq 0$ is the risk-aversion parameter, $\mu_t = E_t R_{t+1}$, and $\Sigma_t = \mathrm{Cov}_t(R_{t+1})$ is the covariance matrix.

3) Maximize the expected utility

$$E_t U \left(R^p_{t+1} \right) = E_t U \left(b'_t R_{t+1} \right), \tag{12.3}$$

where U is a utility function.

Nonparametric Finance, First Edition. Jussi Klemelä.
© 2018 John Wiley & Sons, Inc. Published 2018 by John Wiley & Sons, Inc.

The use of solely the expected returns in portfolio selection is contrary to the intuition that both risk and the expected return should play a role in portfolio selection. However, it turns out that we obtain useful benchmarks using strategies that are based solely on the expected returns. The strategies that are based solely on the expected returns are often "trend following" or "momentum" strategies. A partial explanation for the success of the trend following and related strategies comes from the fact that even when the risk is not explicitly involved, these strategies make severe restrictions on the allowed portfolio weights. On the other hand, trend following strategies often provide "derived assets" or "factors," which are included as an asset into an otherwise diversified portfolio.

An important message of the chapter is that a two asset trend following brings better results than a one asset trend following. In a two asset trend following we follow the trend of the better asset, whereas in a one asset trend following we move to the risk-free rate, or short the asset, in those times where the trend is downwards. Thus, the better success of the two asset trend following can be explained by the fact that in the two asset trend following we do not have to exit the market, or short the market, but we can always stay invested.

A second important message of the chapter is that we can obtain almost as good results with return prediction as using trend following. Our return prediction model is not optimized, since we use only the dividend yield and term spread as the predicting variables of the linear regression. It is known that the dividend yield has lost some of its predicting power, since companies have substantially increased stock buy backs, instead of paying dividends. Thus, the approach of return prediction can be considered as as a fundamentally validated approach, although it is nontrivial to beat trend following with a return prediction model.

In (12.2) and (12.3) the risk aversion is incorporated in the maximization problem. In (12.1) we have a linear objective function, whereas in (12.2) we have a quadratic objective function. Section 11.1 describes how the maximization of (12.2) can be done. The maximization of (12.3) seems to be very difficult, since the portfolio vector b_t is inside the utility function, and we have to predict $U(b_t' R_{t+1})$ separately for each portfolio vector. However, the set of the allowed portfolio vectors can be chosen to be quite small, so that the maximization problem is tractable. Also, the numerical methods of convex optimization can be used, as is done in Györfi *et al.* (2006).

In our examples, we concentrate on the cases with one risky asset and two risky assets. Studying only these simplest cases gives insight into the structure of the problem.

Section 12.1 reviews prediction methods related to the utility maximization and Markowitz criterion. Section 12.2 explains how backtesting is used to evaluate trading strategies. Section 12.3 reviews trading strategies using one risky asset. Section 12.4 reviews trading strategies using two risky assets.

12.1   Prediction in Dynamic Portfolio Selection

We consider single period portfolio selection with d risky assets. We describe how the portfolio weights can be chosen by maximizing the expected returns, by maximizing the expected utility, or by maximizing the Markowitz criterion. The examples are studied in Sections 12.3 and 12.4, but in these examples we study only the maximization of the expected returns and the maximization of the Markowitz criterion. Examples of the maximization of the expected utility are studied in Klemelä (2014, Section 3.12.1).

12.1.1   Expected Returns in Dynamic Portfolio Selection

A portfolio can be selected using return forecasting. Tactical return forecasting approaches are timing approaches that can be divided into two categories: (1) single asset approaches that choose an exposure to a single asset at each time point and (2) cross-asset selection approaches that select portfolio weights for a collection of assets at each time point. Note that single asset approaches include trading with a single asset pair, like the yield spread between corporate credits and government bonds.

In timing approaches the decision is made based on indicators. For example, we can consider a single asset approach that goes long one unit of Treasuries, if equity markets have performed poorly, otherwise goes short one unit of Treasuries. Multiple indicators can be aggregated in several ways. For example, one can take an average of ± 1 trading signals. Another way to aggregate multiple indicators is to use regression analysis, for example, linear regression. The return of the asset, or the excess return of the asset, is predicted with a regression method and the size of the position is decided by the predicted value.

In cross-asset selection approaches, we can rank each asset based on a criterion or an aggregation of criterions. Then we buy the top-ranked assets and sell the bottom-ranked assets.

See Ilmanen (2011, Chapter 24) about tactical return forecasting approaches.

We describe strategies where the portfolio weights are chosen solely on the basis of the estimated expected returns. The expected returns can be estimated using moving averages and using regression on economic indicators.

12.1.1.1   Trend Following

We call trend following such strategies where moving averages are used to estimate the expected returns. Trend following strategies are also called momentum strategies. Momentum strategies are mentioned in Section 10.4, where they appear as factors to define the alpha of a portfolio. There are many variants of trend following and momentum strategies; see Ilmanen (2011, Chapter 14), where the focus is on the commodity trend following strategies.

In the simplest trend following strategy the expected return is estimated solely by the previous period return. The simplest trend following can be generalized to the use of moving averages to estimate the expected returns.

One Step Trend Following   We define one step trend following so that if the previous period return of the risky asset is larger than the risk-free rate, then the weight of the risky asset is one and the weight of the risk-free rate is zero. Otherwise, the weight of the risky asset is zero and the weight of the risk-free rate is one.

More formally, the set of the allowed portfolio vectors is

$$\{(1,0),(0,1)\}, \tag{12.4}$$

the first asset is the risk-free rate, and the second asset is the risky asset. The prediction of the next period return of the risky asset is $\hat{f}(t) = R_t$, where R_t is the current return of the risky asset. Let r_{t+1} be the risk-free rate for the period $t \mapsto t + 1$. The weight of the risky asset is

$$\hat{b}_t = \begin{cases} 1, & \text{if } \hat{f}(t) > r_{t+1}, \\ 0, & \text{if } \hat{f}(t) \le r_{t+1}. \end{cases} \tag{12.5}$$

Trend Following with Moving Averages   We choose the allowed portfolio vectors as in (12.4) and the portfolio selection rule as in (12.5), but the prediction of $\hat{f}(t)$ is now a moving average estimator.

Moving averages were discussed in Section 6.1.1. A moving average estimator of the expected return of a risky asset is

$$\hat{f}(t) = \sum_{i=1}^{t} p_i R_i, \tag{12.6}$$

where $R_1, \ldots, R_t$ are the historical returns and $p_1, \ldots, p_t$ are weights satisfying $p_1 \le \cdots \le p_t$ and $\sum_{i=1}^{t} p_i = 1$, as in (6.2).

One step trend following can be obtained as a special case of using weighted averages. When the weights are such that $p_t = 1$ and $p_1, \ldots, p_{t-1} = 0$, then $\hat{f}(t) = R_t$.

12.1.1.2   Regression to Estimate the Expected Returns

The first generalization of the trend following consists of replacing a moving average with a regression function estimate. We can choose the allowed portfolio vectors as in (12.4), and the portfolio selection rule as in (12.5), but the prediction $\hat{f}(t)$ is now given by a regression estimator.

We use regression data (X_i, Y_i), where $Y_i = R_{i+1}$, and X_i is a vector of economic or technical indicators, for $i = 1, \ldots, t - 1$. Linear and kernel estimators are defined in Section 6.1.2, but we give in the following section a summary of the definitions, specialized for the current setting.

Linear Regression A linear regression function estimator is

$$\hat{g}_t(x) = \hat{\alpha} + \hat{\beta}'x, \tag{12.7}$$

where $\hat{\alpha}$ and $\hat{\beta}$ are the least squares estimates, calculated from the regression data (X_i, Y_i), where $Y_i = R_{i+1}$, $i = 1, \ldots, t - 1$. Now $\hat{g}_t(x)$ is the prediction for the time $t + 1$ return, when $X_t = x$ is observed. We choose the prediction

$$\hat{f}(t) = \hat{g}_t(X_t). \tag{12.8}$$

Kernel Regression In kernel regression the regression function estimate is

$$\hat{g}_t(x) = \sum_{i=1}^{t-1} p_t(x)\, Y_i, \tag{12.9}$$

where

$$p_t(x) = \frac{K_h(X_t - x)}{\sum_{u=1}^{t-1} K_h(X_u - x)},$$

$K_h(x) = K(x/h)$ is the scaled kernel function, $K : \mathbf{R}^d \to \mathbf{R}$ is the kernel function, and $h > 0$ is the smoothing parameter. We choose the prediction

$$\hat{f}(t) = \hat{g}_t(X_t). \tag{12.10}$$

12.1.1.3 A General Setting of Trend Following and Related Strategies

The second generalization of the trend following consists of replacing the collection of the allowed portfolio vectors (12.4) with a more general collection. We allow now more than two basic assets. However, the portfolio weights are chosen solely on the basis of the estimated expected returns,

Let us have $d + 1$ basic assets, with expected one period returns

$$\mu_0 = E_t R_{t+1}^0, \ldots, \mu_d = E_t R_{t+1}^d,$$

and their estimates

$$\hat{\mu}_0, \ldots, \hat{\mu}_d.$$

Note that if R_{t+1}^0 is the risk-free rate, then the return of this asset is already known at time t. We choose the portfolio weight

$$\hat{b}_t = \underset{\bar{b} \in B}{\mathrm{argmax}}\ \bar{b}'\hat{\mu}, \tag{12.11}$$

where $\hat{\mu} = (\hat{\mu}_0, \ldots, \hat{\mu}_d)'$ and B is the set of the allowed portfolio vectors.

Various sets of the allowed portfolio vectors are described in (9.16)–(9.22). For example, in the case of three basic assets the set of all long-only weights is given by

$$B = \{(1 - b^1 - b^2, b^1, b^2) : (b^1, b^2) \in [0, 1]^2\}.$$

Note that since we are maximizing a linear functional in (12.11), the optimal solution lies in the corners of the simplex, and the same solution is obtained when we choose

$$B = \{(1,0,0),(0,1,0),(0,0,1)\}.$$

In both cases everything is invested into the asset with the best predicted return. To get different solutions, we can choose, for example,

$$B = \{(0.5,0.5,0),(0.5,0,0.5),(0,0.5,0.5)\}.$$

Now equal weights 0.5 are put into the two assets with the highest predicted returns.

12.1.2 Markowitz Criterion in Dynamic Portfolio Selection

The Markowitz criterion is defined as the variance penalized expected return. We want to maximize

$$E_t R^p_{t+1} - \frac{\gamma}{2} \mathrm{Var}_t \left(R^p_{t+1} \right)$$

over the portfolio vectors, where $\gamma \geq 0$ is the risk-aversion parameter. The notations E_t and Var_t mean that we take the expectation and the variance conditionally on the information available at time t. Section 11.1 considered this problem.

The return of a portfolio is a linear combination of the returns of the portfolio components:

$$R^p_{t+1} = \left(1 - 1'_d b_t \right) R^0_{t+1} + b'_t R_{t+1},$$

where R^p_{t+1} is the gross return of the portfolio, $b_t = (b^1_t, \ldots, b^d_t)$ is the vector of portfolio weights, R^0_{t+1} is the risk-free rate, and $R_{t+1} = (R^1_{t+1}, \ldots, R^d_{t+1})$ is the vector of the gross returns of the risky assets. We can write

$$E_t R^p_{t+1} - \frac{\gamma}{2} \mathrm{Var}_t \left(R^p_{t+1} \right) = \left(1 - 1'_d b_t \right) R^0_{t+1} + b'_t \mu_t - \frac{\gamma}{2} b'_t \Sigma_t b_t,$$

where $\mu_t = E_t R_{t+1}$ and $\Sigma_t = \mathrm{Cov}_t(R_{t+1})$ is the covariance matrix.

The expected returns μ_t can be estimated using moving averages as in (12.6), linear regression as in (12.8), or kernel regression as in (12.10).

The diagonal of Σ_t contains the conditional variances. Estimation of the conditional variances can be made using the methods of Chapter 7.

The off-diagonal elements of Σ_t are the conditional covariances. The conditional covariances can be estimated analogously as conditional variances. The moving average estimate of the covariance matrix Σ_t is defined by

$$\hat{\Sigma}_t = \hat{g}(t) - \hat{f}(t)\hat{f}(t)',$$

where

$$\hat{g}(t) = \sum_{i=1}^t p_i R_i R'_i, \qquad \hat{f}(t) = \sum_{i=1}^t p_i R_i,$$

$R_1, \ldots, R_t$ are the historical returns and $p_1, \ldots, p_t$ are weights satisfying $p_1 \leq \cdots \leq p_t$ and $\sum_{i=1}^{t} p_i = 1$, as in (6.2).

12.1.3 Expected Utility in Dynamic Portfolio Selection

Our purpose is to find a good approximation to the portfolio vector b_t^o that maximizes the expected utility:

$$b_t^o = \underset{b_t \in B}{\operatorname{argmax}} \, E_t U\left(b_t' R_{t+1}\right), \qquad (12.12)$$

where $U : \mathbf{R} \to \mathbf{R}$ is a utility function and $R_{t+1} = (R_{t+1}^0, \ldots, R_{t+1}^d)'$ is the vector of the single period returns of the basis assets. Collection B is the set of the allowed portfolio vectors. Set B is a subset of the collection of all vectors whose components sum to one:

$$B \subset \left\{ (b^0, \ldots, b^d) : \sum_{i=0}^{d} b^i = 1 \right\}. \qquad (12.13)$$

Utility functions are discussed in Section 9.2.2. The notation E_t means that we take the expectation conditionally on the information available at time t.

In order to approximate b_t^o in (12.12), we have to estimate the expected utility transformed return $E_t U(b_t' R_{t+1})$ for each portfolio vector $b_t \in B$. Estimation of the conditional expectation $E_t U(b_t' R_{t+1})$ is equivalent to the prediction of $U(b_t' R_{t+1})$, when the best prediction is defined as the minimizer of the mean squared prediction error (see Section 5.3.1).

12.1.3.1 Time Space Prediction
The formula for the weighted average in (6.1) gives an estimator for $E_t U(b' R_{t+1})$. The estimator is

$$\hat{f}_b(t) = \sum_{i=1}^{t} p_i U(b' R_i), \qquad (12.14)$$

where $p_1, \ldots, p_t$ are the weights summing to one and we have historical data $R_1, \ldots, R_t$ of the returns of the portfolio components. The portfolio vector at time t is

$$\hat{b}_t = \underset{b \in B}{\operatorname{argmax}} \, \hat{f}_b(t).$$

12.1.3.2 State Space Prediction
Let the available information be described by the vector X_t at time t. Possible choices for X_t are discussed in Section 6.3. The expectation E_t can be taken as the conditional expectation

$$E_t U(b_t' R_{t+1}) = E\left[U(b_t' R_{t+1}) \mid X_t\right].$$

Define, for a fixed portfolio vector $b \in \mathbf{R}^N$ with $\sum_{i=1}^{N} b^i = 1$, the response variable

$$Y_{b,t} = U(b'R_{t+1}).$$

We assume that $(Y_{b,i}, X_i), i = 1, \dots, t-1$, are identically distributed, and denote by (Y_b, X) a random vector that has the same distribution as $(Y_{b,i}, X_i)$. Define the regression function

$$f_b(x) = E(Y_b | X = x), \qquad x \in \mathbf{R}^p.$$

The regression function is estimated by

$$\hat{f}_{b,t} : \mathbf{R}^p \to \mathbf{R}, \tag{12.15}$$

using data $(Y_{b,i}, X_i), i = 1, \dots, t-1$. The function

$$\hat{b}_t(x) = \underset{b \in B}{\operatorname{argmax}} \hat{f}_{b,t}(x)$$

can be considered as an estimate of the theoretical weight function $b : \mathbf{R}^p \to B$, defined by

$$b(x) = \underset{b \in B}{\operatorname{argmax}} f_b(x).$$

At time t we choose the portfolio vector

$$\hat{b}_t(X_t).$$

The regression function estimator in (12.15) can be a linear estimator or a kernel estimator, for example. Linear and kernel estimators are defined in Section 6.1.2, but a summary of the definitions, specialized for the current setting, is given in the following sections.

Linear Regression   A linear regression function estimator is

$$\hat{f}_{b,t}(x) = \hat{\alpha} + \hat{\beta}'x, \tag{12.16}$$

where $\hat{\alpha}$ and $\hat{\beta}$ are the least squares estimates, calculated with the regression data $(X_i, Y_{b,i})$, where $Y_{b,i} = U(b'R_{i+1})$, $i = 1, \dots, t-1$. Now $\hat{f}_{b,t}(x)$ is the prediction for time $t + 1$ utility transformed return, when $X_t = x$ is observed. We choose the portfolio weights $b_t = \operatorname{argmax}_b \hat{f}_{b,t}(X_t)$.

Kernel Regression   In kernel regression the regression function estimate is

$$\hat{f}_{b,t}(x) = \sum_{i=1}^{t-1} p_t(x) \, Y_{b,i}, \tag{12.17}$$

where

$$p_t(x) = \frac{K_h(X_t - x)}{\sum_{u=1}^{t-1} K_h(X_u - x)},$$

$K_h(x) = K(x/h)$ is the scaled kernel function, $K : \mathbf{R}^d \to \mathbf{R}$ is the kernel function, and $h > 0$ is the smoothing parameter.

12.2 Backtesting Trading Strategies

Backtesting uses historical data to answer the question: what would have the performance of a trading strategy been, if it would have been applied in the past. Backtesting is called sometimes historical simulation, to highlight the fact that we could measure the performance of a trading strategy by a Monte-Carlo experiment, where the data is simulated from a model of the asset price dynamics, instead of using historical data. We describe how historical data can be used to create a return time series of a trading strategy.

Typically historical data is divided into two periods. The first period is the estimation period, and data covering the estimation period is used solely for choosing the parameters of the trading strategy. The second period is the testing period, where the performance of the trading strategy is measured. Note, however, that the parameters of the trading strategy are typically updated during the testing period, analogously as in the definition of the sequential out-of-sample sum of squares of prediction errors in (6.22).

Let us have historical data over time points $1, \ldots, T$, and let us consider a trading strategy where rebalancing is done at the beginning of each period (e.g., daily or monthly). Time point $1 < t_0 < T$ divides the data to the estimation part and to the testing part. At time points $t = t_0, \ldots, T-1$ we make the trading decisions, using data over time points $1, \ldots, t$ to determine the parameters of the trading strategy. Note that the estimation is sequential in the sense that the parameters of the trading strategy are constantly updated at time points $t = t_0, \ldots, T-1$. The decision at time point t leads to the return $R_{t+1} = W_{t+1}/W_t$, where W_t is the wealth generated by the trading strategy. Thus, we get time series $R_{t_0+1}, \ldots, R_T$ of historically simulated returns. This time series is used to evaluate the trading strategy. For example, we calculate a performance measure (Sharpe ratio and certainty equivalent), and compare the performance measure of the trading strategy to the performance measure of a benchmark.

For example, consider the following strategy, introduced in Section 12.1.1. We have historical data (X_t, R_t^1, R_t^2), $t = 1, \ldots, T$, where X_t is a vector of economic indicators (dividend yield, term spread), R_t^1 is the return of S&P 500 index, and R_t^2 is the return of US Treasury 10-year bond. We use economic indicators X_t as predicting variables for the S&P 500 return and for the 10-year bond return. We use the data $(X_1, R_2^1), \ldots, (X_{t-1}, R_t^1)$ to fit a linear regression, obtaining the coefficients $(\hat{\alpha}_t^1, \hat{\beta}_t^1)$. Return R_{t+1}^1 is predicted by $\hat{R}_{t+1}^1 = \hat{\alpha}_t^1 + X_t'\hat{\beta}_t^1$. Return R_{t+1}^2 is predicted similarly. We invest everything in S&P 500, when $\hat{R}_{t+1}^1 > \hat{R}_{t+1}^2$. Otherwise, everything is invested in the bond. Thus, the return of the trading strategy is

$$
R_{t+1} = \begin{cases} R_{t+1}^1, & \text{when } \hat{R}_{t+1}^1 > \hat{R}_{t+1}^2, \\ R_{t+1}^2, & \text{otherwise.} \end{cases}
$$

This is done at times $t = t_0, \ldots, T-1$, to obtain historically simulated returns $R_{t_0+1}, \ldots, R_T$. We compute the Sharpe ratio and certainty equivalents using historically simulated returns, and compare these to the Sharpe ratios of S&P 500 and 10-year bond.

12.3 One Risky Asset

We study portfolio selection when the portfolio components are one risky asset and the risk-free rate. We consider both the case when the risky asset is S&P 500 index and the case when the risky asset is US Treasury 10-year bond. The rebalancing will be done monthly, and thus the risk-free rate is 1-month Treasury bill. The data is described in Section 2.4.3.

12.3.1 Using Expected Returns with One Risky Asset

We consider portfolio strategies where the portfolio weight is chosen solely on the basis of expected returns. We call trend following strategies such strategies, where the expected return is estimated by guessing that the near future is similar to the near past. The guessing is done by using some type of a moving average. However, expected returns can also be estimated by using regression on economic indicators, and by using regression on previous returns. Regression based approaches search for more complicated patterns that relate the past observations to the expected returns, instead of just guessing that the near future is similar to the near past.

12.3.1.1 One Step Trend Following
One step trend following is described in (12.4)–(12.5), where one step trend following is defined so that if the previous period return of the risky asset is larger than the risk-free rate, then the weight of the risky asset is one and the weight of the risk-free rate is zero. Otherwise, the weight of the risky asset is zero and the weight of the risk-free rate is one.

Figure 12.1 shows the cumulative wealth of the trend following portfolios. Panel (a) shows the wealth on the y-axis and panel (b) uses logarithmic scale on the y-axis. The purple curves show the portfolio using 10-year bond as the risky asset, and the black curves show the portfolio using S&P 500 as the risky asset. The red curves show the cumulative wealth of S&P 500 and the blue curves show the cumulative wealth of 10-year bond, when the initial wealth is equal to one. We can note that the trend following brings some increase into the cumulative wealth, when compared to S&P 500 index or 10-year bond.

Figure 12.2 shows time series of wealth ratios $Z_t = W_t^2 / W_t^1$, as defined in (10.35). Panel (a) compares the wealth of a trend following to the wealth of its benchmark. Panel (b) compares 10-year bond trend following to the S&P 500 trend following. In panel (a) the black curve shows the ratio, where the numerator W_t^2 is the wealth of trend following with S&P 500 as the risky asset,

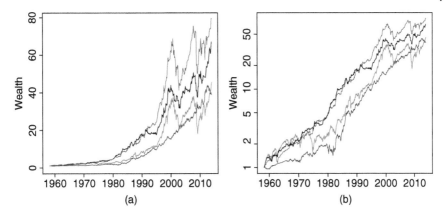

Figure 12.1 *Trend following with the previous 1-month return: Cumulative wealth.* (a) The cumulative wealth of the trend following portfolio with 10-year bond (purple) and S&P 500 (black); (b) logarithmic scale. The red curves show the cumulative wealth of S&P 500 and the blue curves show the cumulative wealth of 10-year bond.

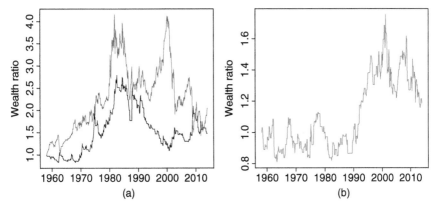

Figure 12.2 *Trend following with the previous 1-month return: Wealth ratios.* We show wealth ratios $Z_t = W_t^2/W_t^1$. (a) The ratio Z_t compares S&P 500 trend following to the benchmark (black) and 10-year bond trend following to the benchmark (purple); (b) the ratio Z_t compares 10-year bond trend following to the S&P 500 trend following.

and the denominator W_t^1 is the wealth of S&P 500. The purple curve shows the ratio, where the numerator W_t^2 is the wealth of trend following with US Treasury 10-year bond as the risky asset, and the denominator W_t^1 is the wealth of US Treasury 10-year bond. In panel (b) the numerator W_t^2 is the wealth of trend following with 10-year bond as the risky asset, and the denominator W_t^1 is the wealth of trend following with S&P 500 as the risky asset.

Figure 12.2(a) shows that trend following with 10-year bond performs better than trend following with S&P 500, relative to their benchmarks. The time series behave qualitatively similarly until the beginning of 1990s, when trend following with 10-year bond starts outperforming its benchmark, but trend following with S&P 500 is underperforming its benchmark. Figure 12.2(b) shows that trend following with 10-year bond outperforms trend following with S&P 500 significantly at the beginning of 1990s, but during other time periods the outperformance and underperformance are alternating.

Figure 12.3 studies Sharpe ratios of trend following with the previous 1-month return. In panel (a) we use S&P 500 monthly returns and in panel (b) we use monthly returns of US Treasury 10-year bond. Function $u \mapsto S(u, T)$ are shown as a black curve (S&P 500) and as a purple curve (10-year bond), where $S(u, T)$ is the Sharpe ratio computed from the returns on period $[u, T]$, and T is the time of the last observation. Note that Figure 10.11 shows the corresponding time series for the benchmarks. Parameter δ is equal to 120 months. The yellow curves show time series of means: $u \mapsto 10 \times 12 \times \hat{\mu}_u^T$, and the pink curves show time series of standard deviations: $u \mapsto 4 \times \sqrt{12} \times \hat{\sigma}_u^T$, where $\hat{\mu}_u^t$ is defined in (10.37) and $\hat{\sigma}_u^t$ is defined in (10.38). Note that the upper borders of the level sets in Figure 12.4 correspond to the level sets of black and purple functions in Figure 12.3. The black (S&P 500 trend) and the purple (10-year bond trend) horizontal lines show the Sharpe ratios over the complete time period. The Sharpe ratios of the indexes are shown as a red line (S&P 500) and as a blue line (10-year bond).

Figure 12.4 shows with the blue region the time periods $[u, t]$ for which the Sharpe ratio is above the benchmark, and with the red region the time periods $[u, t]$ for which the Sharpe ratio is below the benchmark. Panel (a) considers the monthly returns of S&P 500 and panel (b) considers the returns of US Treasury 10-year bond. The blue regions show the sets $\{(u, t) : S(u, t) \geq \lambda\}$, where λ is

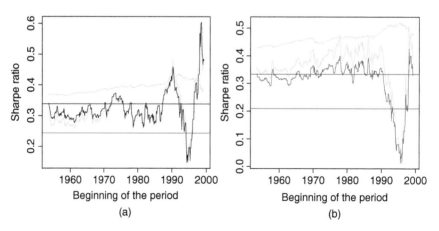

Figure 12.3 *Trend following with the previous 1-month return: Times series of Sharpe ratios.* (a) S&P 500 and (b) 10-year bond.

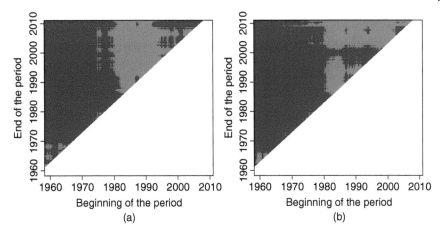

Figure 12.4 *Trend following with the previous 1-month return: Level sets of Sharpe ratios.* We show with blue the time periods where the Sharpe ratio of trend following was better than the Sharpe ratio of the benchmark. (a) S&P 500 and (b) 10-year bond.

equal to the Sharpe ratio of S&P 500 (panel (a)) and the Sharpe ratio of 10-year bond (panel (b)). Parameter δ in the definition of the domain of S is equal to 36 months (see (10.36)).

12.3.1.2 Trend Following with Moving Averages
Trend following with a moving average means that if a moving average of the returns of the risky asset is larger than the moving average of the risk-free rate, then the weight of the risky asset is one and the weight of the risk-free rate is zero. Otherwise, the weight of the risky asset is zero and the weight of the risk-free rate is one. We use the exponential moving average, defined in (6.7) (see also (12.6)). The 1-month trend following strategy of Figures 12.1–12.4 is obtained as a special case by choosing the window of the moving average so small that the moving average is equal to the previous 1-month return.

Figure 12.5 shows (a) the Sharpe ratios and (b) the certainty equivalents[1] of the moving average strategy, as a function of the smoothing parameter h. The black curve shows the performance when the risky asset is S&P 500 and the purple curve has 10-year bond as the risky asset. We can see that the performance measures of trend following with S&P 500 are highest for smaller smoothing parameters, whereas the performance measures of trend following with 10-year bond have a high level for all smoothing parameters. When the smoothing parameter becomes larger, then the Sharpe ratio of trend following with S&P 500 converges to the Sharpe ratio of S&P 500 index. This is due to

---

1 The sample certainty equivalent is defined in (10.13). Here, we use the logarithmic utility $U(x) = \log x$ and multiply by 100, so that the sample certainty equivalent is defined by $100 \times (\exp[T^{-1} \sum_{t=1}^{T} \log(R_t)] - 1)$.

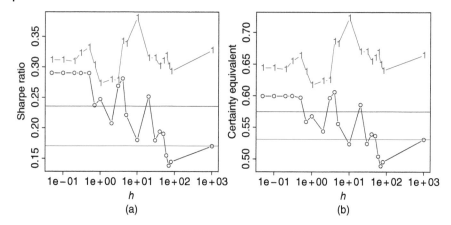

Figure 12.5 *Trend following with moving averages.* (a) Sharpe ratios of trend following strategies with S&P 500 (black curve) and 10-year bond (purple curve); (b) certainty equivalent is
shown for the same portfolios as in panel (a). The *x*-axis shows the smoothing parameter of
the moving average. The red horizontal line shows the Sharpe ratio of S&P 500 and the blue
horizontal line shows the Sharpe ratio of 10-year bond.

the fact that when the smoothing parameter is large, then trend following with
S&P 500 chooses often to invest everything to S&P 500, and seldom to the
risk-free rate (see Figure 6.1(a)). For trend following with 10-year bond the convergence toward the performance measure of the underlying 10-year bond does
not happen, because before 1980s a moving average with even a large smoothing parameter is smaller than the risk-free rate (see Figure 6.1(b)).

Figure 12.6 shows the time series of the ratios of the cumulative wealth of a
trend following strategy to the cumulative wealth of the benchmark. In panel (a)
we follow the trend of S&P 500 and the benchmark is S&P 500. In panel (b) we
follow the trend of US Treasury 10-year bond and the benchmark is US Treasury 10-year bond. The smoothing parameters of moving averages are $h = 0.1$
(black), $h = 1$ (red), $h = 10$ (blue), and $h = 100$ (green). For S&P 500 the smallest smoothing parameter gives the best result, but for 10-year bond a larger
smoothing parameter gives the best result.

12.3.1.3 Regression on Economic Indicators

We consider a portfolio strategy where the expected return of the risky asset
is estimated,[2] and if the estimated expected return is larger than the risk-free
rate, then the weight of the risky asset is one, and the weight of the risk-free rate
is zero. Otherwise, the weight of the risky asset is zero and the weight of the
risk-free rate is one. The expected returns are estimated using linear regression

---

2 The expression "estimating the expected return" means the same as "predicting the next period
return," because the theoretically optimal prediction (in the mean squared error sense) is equal to
the expected return (see Section 5.3.1).

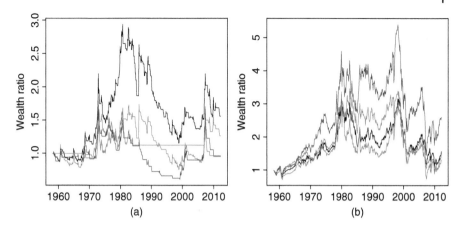

Figure 12.6 *Trend following with moving averages: Wealth ratios.* (a) Following the trend of S&P 500 and (b) following the trend of 10-year bond. We show the ratios of the cumulative wealth to cumulative wealth of the benchmark. The smoothing parameters of moving averages are $h = 0.1$ (black), $h = 1$ (red), $h = 10$ (blue), and $h = 100$ (green).

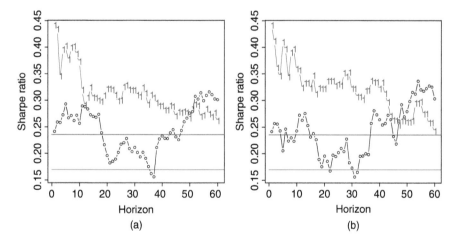

Figure 12.7 *Expected returns determined by economic indicators: Sharpe ratios.* Sharpe ratios of portfolios with S&P 500 (black curve) and 10-year bond (purple curve). (a) The prediction method of (6.35) and (b) the prediction method of (6.36). The x-axis shows the prediction horizon. The red horizontal line shows the Sharpe ratio of S&P 500 and the blue horizontal line shows the Sharpe ratio of 10 year bond.

on economic indicators, as explained in Section 6.4. Both dividend yield and term spread are used as predicting variables.

Figure 12.7 shows the Sharpe ratios as a function of the prediction horizon s. Note that we predict 1-month returns, but the 1-month return predictions are deduced from a prediction with horizon s. Panel (a) uses the prediction method

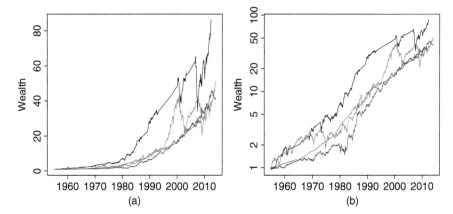

Figure 12.8 *Expected returns determined by economic indicators: Cumulative wealth.* (a) The cumulative wealth of the portfolio whose weights are chosen according to the expected returns, when the risky asset is 10-year bond (purple) and S&P 500 (black); (b) logarithmic scale. The red curves show the cumulative wealth of S&P 500 and the blue curves show the wealth of 10-year bond.

of (6.35), and panel (b) uses the prediction method of (6.36). We can see that the Sharpe ratios of portfolios with 10-year bond are highest for short prediction horizons, whereas the Sharpe ratios of portfolios with S&P 500 are highest for the long prediction horizons.

Figure 12.8 shows the cumulative wealth of the portfolios whose weights are chosen according to the expected returns. We use 12-month prediction horizon, so that $s = 12$ in (6.34). Panel (a) shows the wealth on the y-axis and panel (b) uses logarithmic scale on the y-axis. The purple curves show the portfolio using 10-year bond as the risky asset, and the black curves show the portfolio using S&P 500 as the risky asset. The red curves show the cumulative wealth of S&P 500 and the blue curves show the cumulative wealth of 10-year bond. The wealth is normalized to have value one at the beginning. We can note that trend following increases the cumulative wealth, when compared to S&P 500 index and to 10-year bond. The linear segments in the black curves indicate that during these periods the risk-free investment is chosen.

Figure 12.9 considers the same strategies as in Figure 12.8 but now we show wealth ratios. Panel (a) shows the ratio of the wealth of trend following with S&P 500 to the wealth of S&P 500 (black) and the ratio of the wealth of trend following with 10-year bond to the wealth of 10-year bond (purple). Panel (b) shows the ratio of the wealth of trend following with S&P 500 to the wealth of trend following with 10-year bond. Figure 12.2 shows the corresponding time series for 1-month trend following. We see that using economic indicators improves the trading with S&P 500, whereas 1-month trend following works better when trading with 10-year bond.

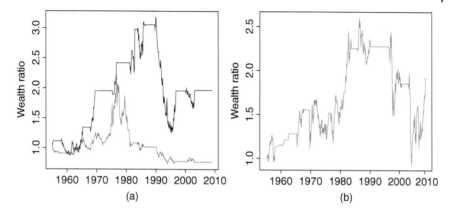

Figure 12.9 *Expected returns determined by economic indicators: Wealth ratios.* (a) The ratio of the wealth of trend following with S&P 500 to the wealth of S&P 500 (black), and the ratio of the wealth of trend following with 10-year bond to the wealth of 10-year bond (purple); (b) The ratio of the wealth of trend following with S&P 500 to the wealth of trend following with 10-year bond.

12.3.2 Markowitz Portfolios with One Risky Asset

Variance penalization with the risk-free rate was discussed in Section 11.1.1. Using this approach, when there is one risky asset and the risk aversion is $\gamma = 1$, we maximize

$$b(\mu - r) - \frac{1}{2} b^2 \sigma^2 \tag{12.18}$$

over $b \in \mathbf{R}$, where μ is the expected gross return of the risky asset, r is the risk-free rate, and σ^2 is the variance of the return of the risky asset. The maximizer was given in (11.3) as

$$b_0 = \frac{\mu - r}{\sigma^2}. \tag{12.19}$$

When the maximization is restricted to $b \in [0, 1]$, then the solution was given in (11.4) as

$$b = \begin{cases} 0, & \text{when } b_0 \le 0, \\ b_0, & \text{when } 0 \le b_0 \le 1, \\ 1, & \text{when } b_0 \ge 1. \end{cases} \tag{12.20}$$

We estimate the expected return using either moving averages or regression on economic indicators. The variance is estimated always using moving averages, which contains the sequentially computed sample variance as a special case. Variance estimation with moving averages is studied in Section 7.4.

12.3.2.1 Using Moving Averages

Figure 12.10 extends the approach of Figure 12.5. The expected return is again estimated using an exponentially weighted moving average, but now the weight of the risky asset is chosen as in (12.20). Variance σ^2 is the additional parameter to be estimated. Panel (a) shows the Sharpe ratios as a function of the smoothing parameter h of the moving average, which estimates the expected return. The variance is estimated by the sequentially calculated sample variance. We can see that the Sharpe ratio of the portfolio with S&P 500 as the risky asset is highest for smaller smoothing parameters, whereas the Sharpe ratio of the strategy with 10-year bond as the risky asset has a high level for all smoothing parameters. Panel (b) shows a contour plot: we show Sharpe ratio both as a function of the smoothing parameter of the moving average for the estimation of the expected return μ (x-axis), and as a function of the moving average for the estimation of volatility σ^2 (y-axis), when the portfolio has S&P 500 index as the risky asset.[3] We can see that the smoothing parameter of the volatility estimate has hardly any influence on the Sharpe ratio, and only the smoothing parameter of the estimate of the expected return has an influence.

Figure 12.11 shows wealth ratios of Markowitz strategies when the expected returns are estimated using moving averages and the variances are estimated using sequential sample variances. Panel (a) shows trading with S&P 500 and panel (b) shows trading with 10-year bond. The wealth of Markowitz strategies

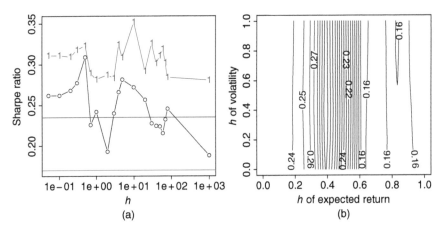

Figure 12.10 *Markowitz portfolios with moving averages: Sharpe ratios.* (a) Sharpe ratios of trend following strategies with S&P 500 (black curve) and 10-year bond (purple curve). The x-axis shows the smoothing parameter of the moving average. The red horizontal line shows the Sharpe ratio of S&P 500 and the blue horizontal line shows the Sharpe ratio of 10 year bond. (b) Sharpe ratios of trend following strategies with S&P 500 when both expected returns and volatility are estimated using moving averages.

---

3 The x-axis has values $0.05, 0.1, 0.2, 0.3, 0.5, 0.7, 1, 2, 3, 4, 5, 10, 20, 30, 40, 50, 60, 70, 80$ and the y-axis has values $1, 5, 10, 20, 30, 40$.

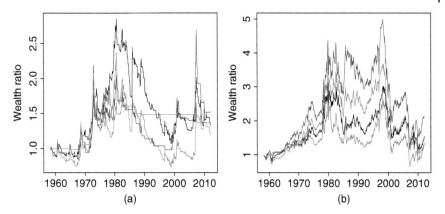

Figure 12.11 *Markowitz portfolios with moving averages: Wealth ratios.* (a) Trading with S&P 500 and (b) trading with 10-year bond. We show the ratios of the cumulative wealth to cumulative wealth of the benchmark. The smoothing parameters of moving averages are $h = 0.1$ (black), $h = 1$ (red), $h = 10$ (blue), and $h = 100$ (green).

is divided by the benchmark (S&P 500 index in (a) and 10-year bond in (b)). The smoothing parameters of moving averages are $h = 0.1$ (black), $h = 1$ (red), $h = 10$ (blue), and $h = 100$ (green). Note that Figure 12.6 shows the corresponding time series for the case where the variance is ignored. The Markowitz criterion does not seem to improve the wealth ratios, when compared to the simple trend following.

Figure 12.12 studies the effect of relaxing the restrictions on short selling and leveraging. Figure 12.10 studies the case where short selling and leveraging

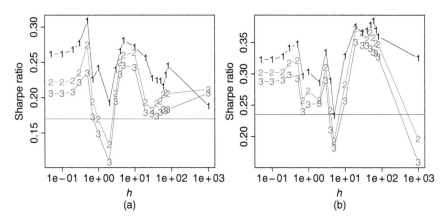

Figure 12.12 *Markowitz portfolios with moving averages: Sharpe ratios when short selling and leveraging are allowed.* Sharpe ratios of trend following strategies as a function of smoothing parameter h when the risky asset is (a) S&P 500; (b) 10-year bond. The red horizontal line show the Sharpe ratio of S&P 500 and the blue horizontal line show the Sharpe ratio of 10 year bond.

are not allowed. More generally, we can restrict the maximization in (12.18) to $b \in [-c, 1 + c]$, where $c > 0$. The weight that maximizes the Markowitz criterion is given by

$$b = \begin{cases} -c, & \text{when } b_0 \leq -c, \\ b, & \text{when } -c \leq b_0 \leq 1 + c, \\ 1 + c, & \text{when } b_0 \geq 1 + c, \end{cases}$$

where b_0 is defined in (12.19). Figure 12.12 shows Sharpe ratios of the portfolio as a function of smoothing parameter h when (a) the risky asset is S&P 500 index and (b) the risky asset is 10-year bond. The lines with label "1" show the case when $c = 0$ (these are the same lines as in Figure 12.10(a)). The lines with label "2" show the case when $c = 1$, and the lines with label "3" show the case when $c = 2$. We can see that the Sharpe ratios are decreasing when more short selling and leveraging are allowed.

12.3.2.2 Using Economic Indicators

Figure 12.13 shows the Sharpe ratios as a function of the prediction horizon s (in months). Note that we predict 1-month returns, and use the prediction horizon s in the construction of 1-month predictions. We use the Markowitz weight (12.20), estimate the expected return using linear regression on dividend yield and term spread, and estimate the variance with the sequential standard deviation. Panel (a) uses the prediction method of (6.35) and and panel (b) uses the

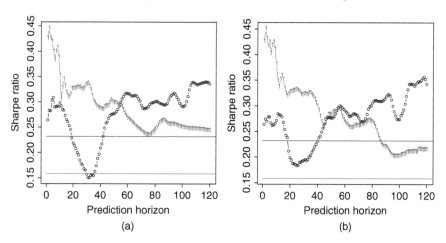

Figure 12.13 *Sharpe ratios as a function of prediction horizon when the expected returns are determined by economic indicators: Markowitz criterion.* Sharpe ratios of portfolios with S&P 500 (black curve) and 10-year bond (purple curve). (a) The prediction method of (6.35) and (b) the prediction method of (6.36). The x-axis shows the prediction horizon. The red horizontal line shows the Sharpe ratio of S&P 500 and the blue horizontal line shows the Sharpe ratio of 10 year bond.

prediction method of (6.36). The black curves show the cases where the risky asset is S&P 500 and the purple curves show the cases where the risky asset is the US 10-year bond. The blue horizontal lines show the Sharpe ratio of S&P 500 and the red horizontal lines show the Sharpe ratio of 10-year bond. Similar to Figure 12.7, we can see that the Sharpe ratios of portfolios with 10-year bond are highest for short prediction horizons, whereas the Sharpe ratios of portfolios with S&P 500 are highest for the long prediction horizons.

12.4   Two Risky Assets

Section 12.4.1 studies portfolio selection using only prediction of the returns and Section 12.4.2 studies the use of the Markowitz mean–variance criterion.

12.4.1   Using Expected Returns with Two Risky Assets

We study portfolio selection that uses only the expected returns to choose the weights (risk aversion is zero). This means that when the expected return of the bond is larger than the expected return of the stock, then everything is invested in the bond, and conversely.

We consider portfolio strategies where the portfolio weight is chosen solely on the basis of expected returns. We call trend following strategies such strategies, where the expected return is estimated solely on the basis of previous returns, by guessing that the near future is similar to the near past. However, expected returns can also be estimated by using regression on economic indicators and by using regression on previous returns.

12.4.1.1   One Step Trend Following

We have two risky assets and consider the decision rule that invests everything in the first asset if the previous period return of the first asset was bigger than the previous period return of the second asset. Otherwise, the second asset is chosen. Let us denote

$$\hat{f}_1(t) = R_t^1, \qquad \hat{f}_2(t) = R_t^2,$$

where R_t^1 and R_t^2 are the gross returns of assets 1 and 2. Let the weight of the first asset be $\hat{b}_t = 1$ if $\hat{f}_1(t) > \hat{f}_2(t)$ and otherwise $\hat{b}_t = 0$.

Figure 12.14 shows the cumulative wealth of the trend following portfolios. Panel (a) shows the wealth on the y-axis and panel (b) uses logarithmic scale on the y-axis. The black curves show the actively managed portfolio. The red curves show the cumulative wealth of S&P 500 and the blue curves show the wealth of 10-year bond. The wealth is normalized to be one at the beginning. We can note that the trend following increases the cumulative wealth, when compared to S&P 500 index or 10-year bond.

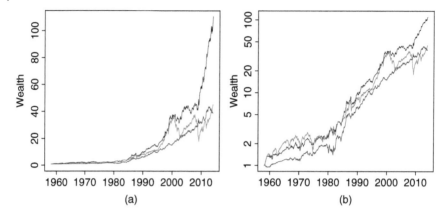

Figure 12.14 *Trend following with the previous 1-month return and two risky assets: Cumulative wealth.* (a) The cumulative wealth of the trend following portfolio (black); (b) logarithmic scale. The red curves show the cumulative wealth of S&P 500 and the blue curves show the wealth of 10-year bond.

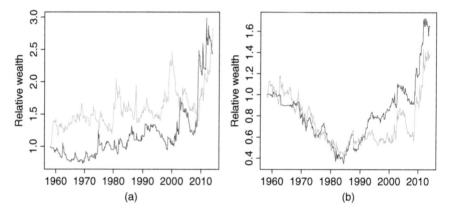

Figure 12.15 *Trend following with the previous 1-month return and two risky assets: Wealth ratios.* (a) The wealth of the two asset trend following is divided by the S&P 500 wealth (dark green) and 10-year bond wealth (orange); (b) the wealth of the two asset trend following is divided by the wealth of S&P 500 trend following (dark green) and the wealth of 10-year bond trend following (orange).

Figure 12.15 shows the ratios of cumulative wealth of the trend following portfolio. Panel (a) compares the trend following to the two benchmarks and panel (b) compares the two asset trend following to the two possible one asset trend following strategies. The one asset trend following is studied in Figures 12.1–12.4. In panel (a) the wealth of the two asset trend following is divided by the S&P 500 wealth (dark green) and 10-year bond wealth (orange). In panel (b) the wealth of the two asset trend following is divided by the wealth of S&P 500 trend following (dark green) and the wealth of 10-year bond trend following (orange). Panel (a) shows that the two asset trend following performs better than the two benchmarks in the most time periods. Panel (b) shows that the one asset trend following is better until 1980, but after that the two asset trend following is better.

12.4.1.2 Trend Following with Moving Averages

Figure 12.16 considers generalization of the simple 1-month trend following shown in Figure 12.1. We define trend following with a moving average so that if a moving average of the returns of the first risky asset is larger than the moving average of the second risky asset, then the weight of the first risky asset is one and the weight of the second risky asset is zero. Otherwise, the weight of the first risky asset is zero and the weight of the second risky asset is one. We use the exponential moving average, defined in (6.7). The 1-month trend following strategy of Figure 12.14 is obtained as a special case by choosing the window of the moving average so small that the moving average is equal to the previous 1-month return.

Figure 12.16 shows (a) the Sharpe ratios and (b) the certainty equivalents[4] of the moving average strategy as a function of the smoothing parameter h of the moving average. We can see that the performance measures of trend following with moving averages are highest for smaller smoothing parameters. When the

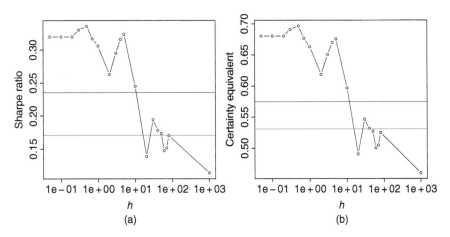

Figure 12.16 *Trend following with moving averages.* (a) Sharpe ratios of trend following strategies (black curve); (b) certainty equivalent is shown for the same portfolios as in panel (a). The *x*-axis shows the smoothing parameter of the moving average. The red horizontal line shows the Sharpe ratio of S&P 500 and the blue horizontal line shows the Sharpe ratio of 10 year bond.

---

4 The sample certainty equivalent is defined in (10.13). Here we use the logarithmic utility $U(x) = \log x$ and the normalization $100 \times (U^{-1}[T^{-1} \sum_{t=1}^{T} U(R_t)] - 1)$.

smoothing parameter becomes larger, then the Sharpe ratio of trend following with S&P 500 converges to the Sharpe ratio of S&P 500 index. This is because trend following with S&P 500 and with a large smoothing parameter chooses always to invest to S&P 500 and never to the risk-free rate. For the trend following with 10-year bond the convergence toward the performance measure of the underlying 10-year bond does not happen, because a moving average even with a large smoothing parameter is smaller than the risk-free rate during time periods before 1980s.

12.4.1.3 Regression on Economic Indicators

We consider a portfolio strategy where the expected returns of the risky assets are estimated, and if the estimated expected return of the first asset is larger than the estimated expected return of the second asset, then the weight of the first asset is one, and the weight of the second asset is zero. Otherwise, the weight of the first asset is zero, and the weight of the second asset is one. The expected returns are estimated using linear regression on economic indicators, as explained in Section 6.4. Both dividend yield and term spread are used as the predicting variables.

Figure 12.17 shows the Sharpe ratios as a function of the prediction horizon. Panel (a) uses the prediction method of (6.35) and and panel (b) uses the prediction method of (6.36). The two prediction methods seem to lead similar performances. The prediction horizon about $s = 12$ gives the best results.

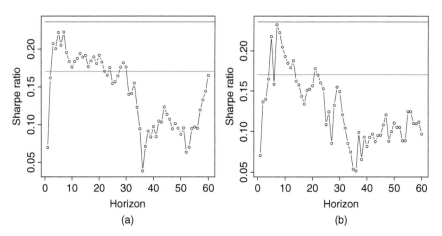

Figure 12.17 *Expected returns determined by economic indicators: Sharpe ratios.* Sharpe ratios of managed portfolios. (a) The prediction method of (6.35) and (b) the prediction method of (6.36). The x-axis shows the prediction horizon. The red horizontal line shows the Sharpe ratio of S&P 500 and the blue horizontal line shows the Sharpe ratio of 10 year bond.

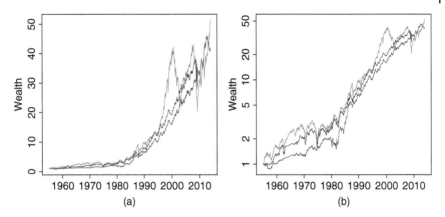

Figure 12.18 *Expected returns determined by economic indicators: Cumulative wealth.* (a) The cumulative wealth of the portfolio whose weights are chosen according to the expected return (black); (b) logarithmic scale. The red curves show the cumulative wealth of S&P 500 and the blue curves show the wealth of 10-year bond. The wealth is normalized to have value one at the beginning.

Note that we are making predictions of 1-month returns, but the predictions are constructed with the help of predictions with horizon s.

Figure 12.18 shows the cumulative wealth of the portfolio whose weights are chosen according to the expected returns. Panel (a) shows the wealth on the y-axis and panel (b) uses logarithmic scale on the y-axis. The prediction horizon is $s = 12$ months. The prediction method of (6.35) is used. The black curves show the actively managed portfolio. The red curves show the cumulative wealth of S&P 500 and the blue curves show the cumulative wealth of 10-year bond.

12.4.2 Markowitz Portfolios with Two Risky Assets

Variance penalization without a risk-free rate is discussed in Section 11.1.2. In this approach, when there are two risky assets and the risk aversion is $\gamma = 1$, we maximize

$$(1 - b)\mu_1 + b\mu_2 - \frac{1}{2}\left[(1 - b)^2\sigma_1^2 + b^2\sigma_2^2 + 2(1 - b)b\sigma_{12}\right] \qquad (12.21)$$

over $b \in \mathbf{R}$, where μ_1 and μ_2 are the expected gross returns of the risky assets, σ_1^2 and σ_2^2 are the variances of the returns of the risky assets, and σ_{12} is the covariance between the returns of the risky assets. The maximizer was given in (11.6) as

$$b_0 = \frac{\mu_2 - \mu_1 + \sigma_1^2 - \sigma_{12}}{\sigma_1^2 + \sigma_2^2 - 2\sigma_{12}}. \qquad (12.22)$$

When the maximization is restricted to $b \in [0,1]$, then the solution was given in (11.7) as

$$b = \begin{cases} 0, & \text{when } b_0 \leq 0, \\ b_0, & \text{when } 0 \leq b_0 \leq 1, \\ 1, & \text{when } b_0 \geq 1. \end{cases} \tag{12.23}$$

12.4.2.1 Moving Averages

The means, variances, and the covariance in (12.22) are estimated using the exponentially weighted moving averages. The smoothing parameter $h = \infty$ is interpreted as the sequential sample mean, variance, or covariance. Thus, we have up to five smoothing parameters to choose. However, we use always the same smoothing parameter for the two means, and the same smoothing parameter for the two variances. So, we have three smoothing parameters to choose.

Figure 12.19 shows Sharpe ratios as a function of smoothing parameters. The weight of the second risky asset is chosen as in (12.23). The means and variances are estimated with the exponentially weighted moving averages, and the covariance is the sequential sample covariance. Panel (a) shows the Sharpe ratios as a function of the smoothing parameter of the means μ_1 and μ_2, for the smoothing parameters of the variance in the range 1–1000. Panel (b) shows the Sharpe ratios as a function of the smoothing parameter of the variances σ_1^2 and σ_2^2, for the smoothing parameters of the means in the range 0.1–1000. The labels "1"–"6" and "1"–"5" correspond to the smoothing parameters in the increasing order. We can see that the Sharpe ratio of the portfolio is highest for

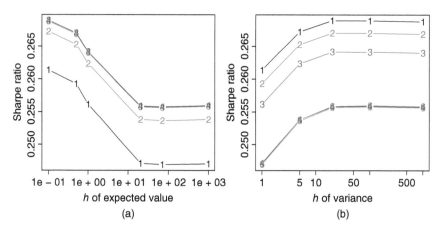

Figure 12.19 *Markowitz portfolios with two risky assets and moving averages: Sharpe ratios. (a) Sharpe ratios as a function of the smoothing parameter of the means μ_1 and μ_2, for the smoothing parameter of the variances in the range 1–1000; (b) Sharpe ratios as a function of the smoothing parameter of the variances σ_1^2 and σ_2^2, for the smoothing parameter of the means in the range 0.1–1000.*

the smallest smoothing parameters of the estimator of the expected return, but the variances require a large smoothing parameter.

Figure 12.20 shows Sharpe ratios as a function of smoothing parameters. The weight of the second risky asset is chosen as in (12.23). The means, variances, and the covariances are estimated with the exponentially weighted moving averages. The smoothing parameter of the estimator of the means is $h = 0.1$. Panel (a) shows the Sharpe ratios as a function of the smoothing parameter of the variances σ_1^2 and σ_2^2, for the smoothing parameters of the covariance in the range 1–1000. Panel (b) shows the Sharpe ratios as a function of the smoothing parameter of the covariance σ_{12}, for the smoothing parameters of the variances in the range 1–1000. The labels "1"–"5" correspond to the smoothing parameters in the increasing order. The Sharpe ratio of S&P 500 is shown with a red horizontal line, and the Sharpe ratio of 10 year bond is shown with a blue horizontal line. We can see that a large Sharpe ratio is obtained when the both smoothing parameters are about $h = 500$.

Figure 12.21 studies the effect of relaxing the restrictions on short selling and leveraging, whereas in Figure 12.19 short selling and leveraging are not allowed. Now, we restrict the maximization in (12.21) to $b \in [-c, 1 + c]$, where $c > 0$. The maximizing weight is

$$
b = \begin{cases} -c, & \text{when } b_0 \leq -c, \\ b_0, & \text{when } -c \leq b_0 \leq 1 + c, \\ 1 + c, & \text{when } b_0 \geq 1 + c, \end{cases}
$$

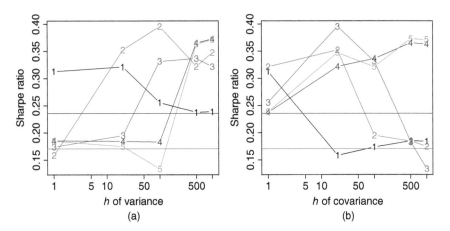

Figure 12.20 *Markowitz portfolios with two risky assets and moving averages: Sharpe ratios.* (a) Sharpe ratios as a function of the smoothing parameter of the variances σ_1^2 and σ_2^2, for the smoothing parameter of the covariance in the range 1–1000; (b) Sharpe ratios as a function of the smoothing parameter of the covariance σ_{12}, for the smoothing parameter of the variances in the range 1–1000.

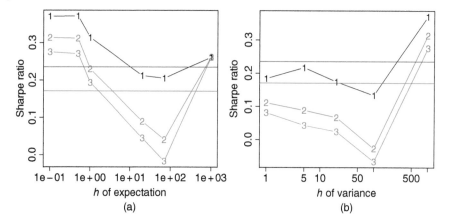

Figure 12.21 *Markowitz portfolios with two risky assets: Moving averages when short selling and leveraging are allowed.* (a) Sharpe ratios as a function of the smoothing parameter of the means μ_1 and μ_2; (b) Sharpe ratios as a function of the smoothing parameter of the variances σ_1^2 and σ_2^2. The curve with labels "1" $c = 0$ leverage, labels "2" indicate leverage $c = 1$, and labels "3" indicate leverage $c = 2$. The red horizontal line shows the Sharpe ratio of S&P 500 and the blue horizontal line shows the Sharpe ratio of 10-year bond.

where b_0 is defined in (12.22). Figure 12.21 shows Sharpe ratios as a function of smoothing parameters. The means and variances are estimated with the exponentially weighted moving averages, and the covariance is the sequential sample covariance. Panel (a) shows the Sharpe ratios as a function of the smoothing parameter of the means μ_1 and μ_2, for the smoothing parameter of the variances equal to 1000. Panel (b) shows the Sharpe ratios as a function of the smoothing parameter of the variances σ_1^2 and σ_2^2, for the smoothing parameter of the means equal to 0.1. The lines with label "1" show the case when $c = 0$ (these are the same lines as in Figure 12.19). The lines with label "2" show the case when $c = 1$ and the lines with label "3" show the case when $c = 2$. The red horizontal line shows the Sharpe ratio of S&P 500 and the blue horizontal line shows the Sharpe ratio of 10-year bond. We can see that the Sharpe ratios are smaller when more short selling and leveraging are allowed.

Figure 12.22 shows wealth and relative wealth. The smoothing parameter of the estimators of the expected returns is $h = 0.1$, for the variances and the covariance $h = 500$. Panel (a) shows the cumulative wealth of the Markowitz portfolio (black), S&P 500 (red), and 10-year bond (blue). Panel (b) shows the ratios of the wealth of the Markowitz portfolio to the wealth of S&P 500 (purple) and to the wealth of 10-year bond (orange). The wealth is normalized to have value one at the beginning. Panel (a) shows that the Markowitz portfolio beats the benchmarks. Panel (b) shows that the Markowitz portfolio beats the benchmarks during the most time periods.

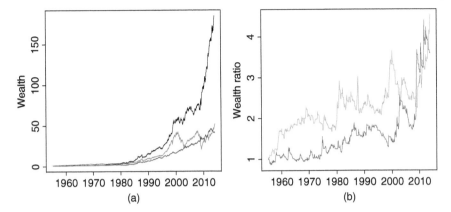

Figure 12.22 *Markowitz portfolios with two risky assets and moving averages: Wealth.* (a) The cumulative wealth of the Markowitz portfolio (black), S&P 500 (red), and 10-year bond (blue); (b) the ratios of the wealth of the Markowitz portfolio to the wealth of S&P 500 (purple) and to the wealth of 10-year bond (orange).

12.4.2.2 Using Economic Indicators

We use the Markowitz weight (12.23). The means are estimated using linear regression on dividend yield and term spread. The two methods of linear regression are defined in (6.35) and (6.36). The variances and the covariance are estimated with exponentially weighted moving averages. The smoothing parameter $h = \infty$ is interpreted as the sequential sample variance or covariance. Thus, we have up to five parameters to choose: two horizon parameters s for the estimation of the two means, two smoothing parameters h for the two variances, and one smoothing parameter h to estimate the covariance. However, we use always the same horizon parameter for the two means, and the same smoothing parameter for the two variances. So, we have three parameters to choose.

Figure 12.23 shows the Sharpe ratios of the Markowitz portfolios. The prediction method of (6.35) is used. The covariance is estimated by the sequential sample covariance. Panel (a) shows the Sharpe ratios as a function of the parameter s of the prediction horizon, for values $h = 5$–1000 of the smoothing parameter of the estimator of the variances. The symbols "1"–"4" correspond to the smoothing parameters of the variances in the increasing order. Panel (b) shows the Sharpe ratios as a function of the smoothing parameter of the estimator of the variances, for prediction horizons $s = 1$–120. Panel (a) shows that the prediction horizon around $s = 12$ gives the largest Sharpe ratio. Panel (b) shows that the smoothing parameter of the estimator of the variances should not be too small.

Figure 12.24 compares the two methods (6.35) and (6.36) of linear regression. The Sharpe ratios of (6.35) are divided by the Sharpe ratios of (6.36). Panel (a) shows the ratios of the Sharpe ratios as a function of the parameter s of the

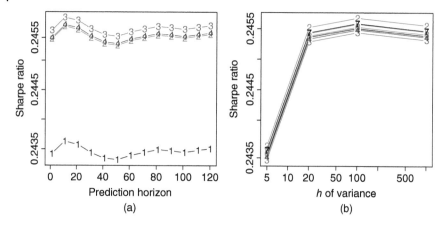

Figure 12.23 *Regression on economic indicators: Sharpe ratios.* (a) Sharpe ratios as a function of the prediction horizon, for values $h = 5$–1000 of the smoothing parameter of the variances; (b) Sharpe ratios as a function of the smoothing parameter of the variances, for prediction horizons $s = 1$–120.

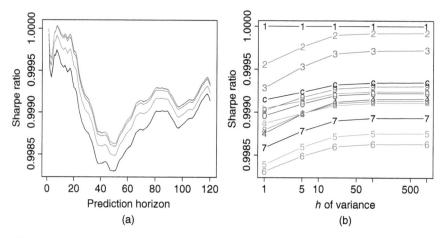

Figure 12.24 *Regression on economic indicators: Ratios of the Sharpe ratios.* The Sharpe ratios of (6.35) are divided by the Sharpe ratios of (6.36). (a) The ratios as a function of the prediction horizon, for values $h = 1$–1000 of the smoothing parameter of the variances; (b) the ratios as a function of the smoothing parameter of the variances, for prediction horizons $s = 1$–120.

prediction horizon, for values $h = 1$–1000 of the smoothing parameter of the estimator of the variances. Panel (b) shows the ratios of the Sharpe ratios as a function of the smoothing parameter of the estimator of the variances, for prediction horizons $s = 1$–120. We see that the ratios of the Sharpe ratios are smaller than one: the Sharpe ratios of (6.36) are larger than the Sharpe ratios of (6.35).

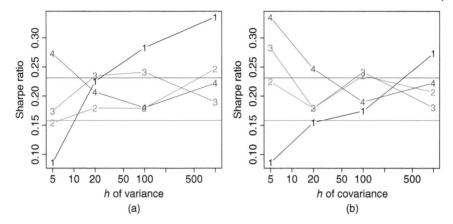

Figure 12.25 *Regression on economic indicators: Sharpe ratios.* (a) The Sharpe ratios as a function of the smoothing parameter of the estimator of the variances, for values $h = 5$–1000 of the smoothing parameter of the covariance; (b) the Sharpe ratios as a function of the smoothing parameter of the estimator of the covariance, for values $h = 5$–1000 of the smoothing parameter of the variances.

Figure 12.25 shows the Sharpe ratios of the Markowitz portfolios. The prediction method of (6.35) is used. The prediction horizon of the expected returns is $s = 12$. Panel (a) shows the Sharpe ratios as a function of the smoothing parameter of the estimator of the variances, for values $h = 5$–1000 of the smoothing parameter of the covariance. Panel (b) shows the Sharpe ratios as a function of the smoothing parameter of the estimator of the covariance, for values $h = 5$–1000 of the smoothing parameter of the variances. The symbols "1"–"4" correspond the smoothing parameters in the increasing order. The red horizontal lines show the Sharpe ratio of S&P 500 and the blue horizontal lines show the Sharpe ratio of 10-year bond. Panel (a) shows that the smoothing parameter of the variances should be large, especially when the smoothing parameter of the covariance is small. Panel (b) shows that when the smoothing parameter of the variances is large, then the smoothing parameter of the covariance should be small.

Figure 12.26 is otherwise similar to Figure 12.25, but now the the prediction method of (6.36) is used, instead of the prediction method of (6.35).

Figure 12.27 shows wealth and relative wealth. The prediction horizon of the estimator of the expected return is $s = 12$. The prediction method of (6.35) is used. The smoothing parameter of the variances is $h = 20$, and the smoothing parameter of the covariance is $h = 1$. Panel (a) shows the cumulative wealth of the Markowitz portfolio (black), S&P 500 (red), and 10-year bond (blue). Panel (b) shows the ratios of the wealth of the Markowitz portfolio to the wealth of S&P 500 (purple) and to the wealth of 10-year bond (orange). The wealth is normalized to have value one at the beginning. Panel (a) shows

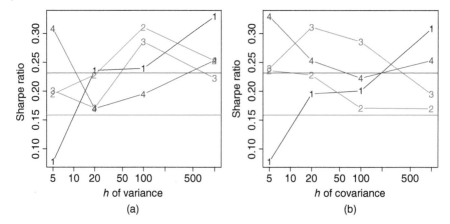

Figure 12.26 *Regression on economic indicators: Sharpe ratios of method (6.36).* (a) The Sharpe ratios as a function of the smoothing parameter of the estimator of the variances, for values $h = 5$–1000 of the smoothing parameter of the covariance; (b) the Sharpe ratios as a function of the smoothing parameter of the estimator of the covariance, for values $h = 5$–1000 of the smoothing parameter of the variances.

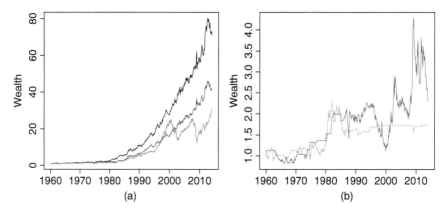

Figure 12.27 *Regression on economic indicators: Wealth.* (a) The cumulative wealth of the Markowitz portfolio (black), S&P 500 (red), and 10-year bond (blue); (b) the ratios of the wealth of the Markowitz portfolio to the wealth of S&P 500 (purple) and to the wealth of 10-year bond (orange).

that the Markowitz portfolio beats the benchmarks. Panel (b) shows that the Markowitz portfolio beats the benchmarks during the most time periods, and the performance of the Markowitz portfolio is close to the performance of the 10-year bond after about 1985.

Figure 12.28 shows wealth and relative wealth. The prediction horizon of the estimator of the expected return is $s = 12$. The prediction method of (6.36) is used. The smoothing parameter of the variances is $h = 40$, and the smoothing

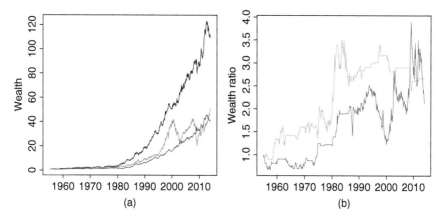

Figure 12.28 *Regression on economic indicators with the method (6.36): Wealth.* (a) The cumulative wealth of the Markowitz portfolio (black), S&P 500 (red), and 10-year bond (blue); (b) the ratios of the wealth of the Markowitz portfolio to the wealth of S&P 500 (purple) and to the wealth of 10-year bond (orange).

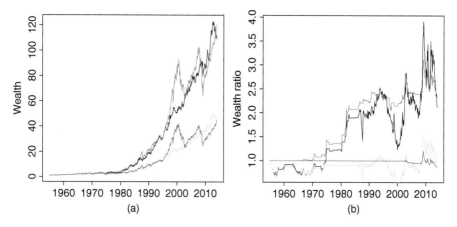

Figure 12.29 *Conditional versus unconditional estimation.* (a) Time series of the portfolio wealth when the first and second moments are estimated conditionally (black), first moments sequentially and second moments conditionally (green), first moments conditionally and second moments sequentially (yellow), and first and second moments are estimated sequentially (purple); (b) relative wealths with respect to the wealth of S&P 500.

parameter of the covariance is $h = 1$. Panel (a) shows the cumulative wealth of the Markowitz portfolio (black), S&P 500 (red), and 10-year bond (blue). Panel (b) shows the ratios of the wealth of the Markowitz portfolio to the wealth of S&P 500 (purple) and to the wealth of 10-year bond (orange). The wealth is normalized to have value one at the beginning. Panel (a) shows that the Markowitz portfolio beats the benchmarks, and the Markowitz portfolio of Figure 12.27. Panel (b) shows that the Markowitz portfolio beats the benchmarks during the most time periods.

Figure 12.29 studies the importance of conditional prediction of second moments. Panel (a) shows the wealths and panel (b) shows the relative wealths with respect to the wealth of S&P 500. The black curve is the same as in Figure 12.28, so that both the expectations and the second moments are estimated conditionally. The green curve corresponds to the case where only the second moments are estimated conditionally but the returns are predicted sequentially. The yellow curve corresponds to the case where only the returns are predicted conditionally but the second moments are predicted sequentially. The purple curve corresponds to the case where both the expectations and the second moments are estimated sequentially. The wealth is normalized to have value one at the beginning. We see that the black and green curves are much better, so that the conditional expectation of the second moments is more important than the conditional estimation of the expectations.

Part IV

Pricing of Securities

13

Principles of Asset Pricing

Asset pricing can be studied in two different settings: absolute pricing and relative pricing. Absolute pricing tries to explain the prices in terms of fundamental macroeconomic variables, applying utility functions and preferences. Relative pricing tries to explain the prices of a group of assets given the prices of a more fundamental group of assets.

We concentrate on relative pricing. Derivatives are assets whose payoffs are defined in terms of the payoffs of some basis assets. For example, an European call option gives the right to buy the underlying asset at the given expiration time T at the given strike price K. Thus, the payoff of the call option at time T is equal to

$$C_T = \max\{S_T - K, 0\},$$

where S_T is the value of the underlying asset. We want to find a "fair price" C_t for the call option, when $t < T$ is a previous time.

Derivatives are traded in exchanges just like stocks, and the price of a derivative is determined in an exchange by supply and demand. It can be argued that the pricing of the market is typically efficient. However, it is of interest to try to find fair prices by statistical and probabilistic methods at least for the following two reasons. (1) Sometimes options are bought and sold over the counter and not in exchanges. In this case, there is no information provided by the markets. (2) It is possible that the market prices are irrational. This can certainly happen in illiquid markets. In this case, a market participant can profit from the knowledge of scientific methods of pricing.

Besides pricing of options, it is of equal importance to hedge options. In fact, our main emphasis will be on the pricing by quadratic hedging. In this approach, the price of an option will be the initial investment of a trading strategy, which minimizes the quadratic error

$$E(W_T - C_T)^2,$$

Nonparametric Finance, First Edition. Jussi Klemelä.

where W_T is the wealth obtained by the hedging strategy, and C_T is the value of the derivative at the expiration. This approach will be developed more in detail in Chapter 16.

Section 13.1 studies general principles of pricing heuristically, discussing such concepts as absolute and relative pricing, arbitrage, the law of one price, and completeness of models. In addition, we introduce the idea of quadratic hedging.

Section 13.2 presents basics of mathematical asset pricing in discrete time. We describe the first and the second fundamental theorems of asset pricing (Theorems 13.1 and 13.3). The first fundamental theorem says that a market is arbitrage-free if and only if there exists an equivalent martingale measure. The second fundamental theorem says that every derivative can be replicated if and only if the martingale measure is unique. If every derivative can be replicated, then it is said that the market is complete. Theorem 13.2 states that the arbitrage-free prices of European options are expectations with respect to an equivalent martingale measure.

We give a proof of the first fundamental theorem of asset pricing. The proof is constructive: we construct an equivalent martingale measure for an arbitrage-free market. The most proofs of the first fundamental theorem of asset pricing found in the literature are not constructive, but apply abstract functional analysis. However, the construction of suitable equivalent martingale measures is useful for practical applications, because these measures lead to the collections of arbitrage-free prices. We do not give a proof of the second fundamental theorem of asset pricing. This is due to the fact that the general theory of complete markets seems to be less relevant from the practical point of view than the theory of incomplete markets, although the Black–Scholes model is useful in applications. Chapter 14 describes the theory of Black–Scholes pricing and Chapter 15 is devoted to incomplete models.

Section 13.3 discusses methods for the comparison of different pricing and hedging methods. The main method for the comparison is to use historical simulation to generate trajectories of prices, hedge the derivative through the trajectories, and then compute the sample mean of the squared hedging errors.

13.1 Introduction to Asset Pricing

Section 13.1.1 discusses absolute pricing with the help of coin tossing games. These examples show that utility functions can be useful in determining reasonable prices. Section 13.1.2 discusses how the principle of excluding arbitrage and the law of one price can be applied in relative pricing. The one period binary model is introduced. This model will be used to derive the Black–Scholes prices in Chapter 14. Section 13.1.3 discusses relative pricing in cases where arbitrage cannot be applied. The one period ternary model is an example of such case. In these cases, a fair price can be defined by minimizing the mean squared hedging error, for example.

13.1.1 Absolute Pricing

Let us consider a coin tossing game where a participant receives 1 € when heads occur and 0 € when tails occurs. The probability of getting heads is 1/2 and the probability of obtaining tails is 1/2. What is the fair price to participate in this game? It can be argued that the fair price is the expected gain:

$$0.5 \cdot 1 \,€ + 0.5 \cdot 0 \,€ = 0.5 \,€.$$

The fairness of the price can be justified by the law of large numbers. The law of large numbers implies that the gain from repeated independent repetitions of the game with price 0.5 € converges to zero with probability 1. A larger price than 0.5 € would give an almost sure profit to the organizer of the game in the long run and a smaller price than 0.5 € would give an almost sure profit to the player of the game in the long run.

It does not seem as clear what the price should be if we change the game so that a participant receives 1 million € when heads occur and 0 € when tails occur. Only few people would be willing to invest half a million € in order to participate in this game. The law of large numbers cannot be applied to justify a price because the probability of a bankruptcy is quite large when a player repeats the game.[1]

It can be argued that the price of the game should be equal to the expected utility: Let S be the random variable with $P(S = s) = 0.5$ and $P(S = 0) = 0.5$, where $s = 1$ million €. Then the expected utility is $EU(S)$, where U is a utility function.

The St. Petersburg paradox can be used to suggest that a utility function should be used. In the St. Petersburg paradox, the banker flips the coin until the heads come out the first time. The player receives 2^{k-1} coins when there are k tosses of the coin (1 coin if the heads come out in the first toss, 2 coins if the heads come out in the second toss, 4 coins if the heads come out in the third toss, and so on). What is the fair entrance fee to the game? We can calculate the expected gain. The probability that there are k tosses is $p_k = 2^{-k}$. Thus the expected payoff is

$$\sum_{k=1}^{\infty} p_k 2^{k-1} = \sum_{k=1}^{\infty} \frac{1}{2} = \infty.$$

Thus, it would seem that the entrance fee could be arbitrarily high. However, applying common sense, it does not seem reasonable to pay a high entrance fee.

---

1 Note also that a doubling strategy gives an almost sure win. A player who follows the classical doubling strategy doubles his bet until the first time he wins. If he starts with 1 €, his final gain is 1 € almost surely.

The paradox can be solved by using a utility function to measure the utility of the wealth. For example, the logarithmic utility function $x \mapsto \log_e(x)$ gives the expected utility of the game

$$\sum_{k=1}^{\infty} p_k \log_e(2^{k-1}) = \log_e 2,$$

which would give the price of two coins for the game.[2]

The St Petersburg paradox suggests that we could use the expected utility instead of the expected monetary payoff to determine fair prices. Utility maximization will be discussed in Section 15.2.5, as a method for derivative pricing, but otherwise we do not study further this approach.

13.1.2 Relative Pricing Using Arbitrage

Sometimes relative pricing can be done solely by applying the principle of excluding arbitrage. We illustrate this type of relative pricing using a coin tossing example.[3] After that, arbitrage and the law of one price are discussed more generally.

---

2 The usefulness of utility functions can also argued by considering a lottery. Consider a lottery where the player chooses seven numbers from the numbers $1, \ldots, 39$. Then seven numbers are drawn randomly from the numbers $1, \ldots, 39$ without replacement. If the player has chosen exactly the same numbers that were drawn, the player wins 2 million €. What is the fair price entrance price for the lottery? The probability of winning is

$$\binom{39}{7}^{-1} = \frac{1}{15,380,937},$$

and thus the expected gain is

$$\frac{1}{15,380,937} \cdot 2,000,000 \text{€} \approx 0.13 \text{€}.$$

However, a typical price for a Finnish lottery of this type is 0.8 € and playing the game with this entrance price results in the expected loss of 0.67 €. Using the logarithmic utility gives a positive utility

$$\frac{1}{15,380,937} \cdot \log_e(2,000,000) - \left(1 - \frac{1}{15,380,937}\right) \cdot \log_e(0.8) \approx 0.22,$$

which would give the price of $e^{0.22} \approx 1.25$ €. Thus, the market price of a lottery can be justified by pointing out that the market price is equal to the expected utility, for some utility function. Intuitively, a lottery is attractive because it provides an opportunity to a dramatic improvement of wealth, with a negligible price.

3 There are at least three differences in the setting of option pricing, as compared to the pricing of coin flipping games. (1) In the coin flipping games the payment is made and the payoff is received almost simultaneously, whereas the expiration time of the option can be several months or even years ahead. Thus, one has to take into account the cost of money in option pricing. (2) The probabilities of the outcomes are known in the coin flipping, whereas the distribution of stock prices is unknown and has to be estimated with statistical techniques. Arbitrage is based on "known knowns," statistical arbitrage in the coin flipping is based on "known unknowns," whereas statistical arbitrage in option pricing has to deal with "unknown unknowns." (3) In the coin flipping, we

13.1.2.1 Pricing in a One Period Binary Model

Let us consider two games related to the same tossing of a coin. The first game is such that the player receives u € when heads occur and d € when tails occurs, where $u > d \geq 0$. The participation to this game can be compared to buying a stock. We denote with S the random variable with $P(S = u) = 0.5$ and $P(S = d) = 0.5$.

The second game is such that the player receives 1 € when heads occur and 0 € when tails occurs. The participation to this game can be compared to buying a derivative. Indeed, the second game can be considered as a derivative because the payoff in the second game is random variable $C = f(S)$ for $f : \{u, d\} \to \mathbf{R}$, where $f(u) = 1$ and $f(d) = 0$. Random variable C has the distribution $P(C = 1) = 0.5$ and $P(C = 0) = 0.5$. The third asset is a bond with value $B = 1$. The price of bond is 1 and the price of stock is denoted with $\pi(S)$. We want to find the price of the derivative.

The derivative can be replicated with the bond and the stock: Consider the portfolio with β bonds and ξ stocks. We choose

$$\beta = \frac{-d}{u - d}, \quad \xi = \frac{1}{u - d}.$$

The replicating portfolio is $W = \beta B + \xi S$. Indeed, we have that $P(W = C) = 1$, because

$$\beta + \xi d = 0,$$
$$\beta + \xi u = 1.$$

By the law of one price, to exclude the possibility of arbitrage, the price of the derivative has to be equal to the price of the portfolio:[4]

$$\pi(C) = \pi(W).$$

The price of the portfolio is

$$\pi(W) = \beta + \xi \pi(S).$$

Thus, the price of the derivative is

$$\pi(C) = \frac{\pi(S) - d}{u - d}. \tag{13.1}$$

The price of the derivative is in general not equal to 0.5. If $\pi(S) = (u + d)/2$, then the price of the derivative is $\pi(C) = 0.5$. If $\pi(S) < (u + d)/2$, then the price of the derivative satisfies $\pi(C) < 0.5$.[5]

---

cannot usually invest in any games whose outcome is related to the game we are pricing, whereas in the case of options we can make transactions with the underlying stock.

4 If $\pi(C) < \pi(V)$, then buying C and selling V would give an almost sure profit. If $\pi(C) > \pi(V)$, then selling C and buying V would give an almost sure profit.

5 Exercise: Does the possibility for hedging always decrease the price of the derivative to a lower value than the expected value?

We have given the price of the derivative in (13.1) in terms of the price of the stock. This is an example of relative pricing: a price of an asset is expressed in terms of the prices of another asset.

13.1.2.2 Arbitrage

Arbitrage is both a term of everyday language and a technical term used in mathematical finance.

Arbitrage is used in everyday language to denote a financial operation where one obtains a profit with probability one by a simultaneous selling and buying of assets. We give two examples of this type of arbitrage.

1) The stock of Daimler is listed both in Frankfurt and Stuttgart stock exchanges. If the stock can be bought in Frankfurt with the price of 10 € and sold in Stuttgart with the price of 11 €, we obtain a risk free profit of 1 € (minus the transaction costs).

2) Suppose the price of a stock is 10 € and a call option with strike price $K = 8$ € with the expiration time in 1 week can be bought with the price of 1 €. Then, we can sell the stock short and buy the call option. The profit of the operation will be $-1 + 10 - 8 = 1$ € (buying the call costs 1 €, selling the stock short gives 10 €, and exercising the option costs 8 €).
   In general, we have a lower bound $S_t - K$ for the price of a call option, where S_t is the price of the stock at the time of buying the option, and K is the strike price. See (14.9) for a more precise lower bound.

In mathematical finance, an arbitrage is a financial operation whose payoff is always nonnegative and sometimes positive, that is, the probability of a non-negative payoff is one and the probability of a positive payoff is greater than zero. More formally, arbitrage portfolio W_t is such that its value at time t satisfies $W_t \le 0$ but its value W_T at a later time T satisfies $P_t(W_T \ge 0) = 1$ and $P_t(W_T > 0) > 0$. A reasonable system of prices should be such that arbitrage is excluded, so that there does not exist an arbitrage portfolio.

The absence of arbitrage implies the law of one price.

13.1.2.3 The Law of One Price and the Monotonicity Theorem

The law of one price states that if two financial assets have the same payoffs then they have the same price: If two portfolios satisfy

$$P_t \left(W_T^{(a)} = W_T^{(b)} \right) = 1,$$

then their prices are equal at a previous time t:

$$W_t^{(a)} = W_t^{(b)}.$$

The absence of arbitrage implies that the law of one price holds. Indeed, consider the case where the law of one price does not hold. Then we have two assets with different prices at time t, say $W_t^{(a)} < W_t^{(b)}$, and the prices of the assets are the same with probability 1 at a later time T: $P_t(W_T^{(a)} = W_T^{(b)}) = 1$. Then we can by the cheaper asset at time t and sell the more expensive asset at time t to obtain the amount $W_t^{(b)} - W_t^{(a)} > 0$. This amount can be put into a bank account. At time T the two assets have the same price, and thus we have locked the profit of time t. We have shown that there exists an arbitrage opportunity. Thus we have shown that the absence of arbitrage implies that the law of one price holds.

The monotonicity theorem states that if two financial assets satisfy

$$P_t\left(W_T^{(a)} \le W_T^{(b)}\right) = 1,$$

then their prices satisfy

$$W_t^{(a)} \le W_t^{(b)}$$

at time t. Furthermore, if $P_t(W_T^{(a)} < W_T^{(b)}) = 1$, then their prices satisfy $W_t^{(a)} < W_t^{(b)}$ at time t. This formulation of the monotonicity theorem is similar to the formulation of Blyth (2014, p. 48).

The law of one price implies the linearity of the pricing function. Let

$$W = \xi^1 S^1 + \cdots + \xi^d S^d$$

be a portfolio, and let $S_t^1, \ldots, S_t^d$ be the prices of the basis assets at time t. Then the price of the portfolio at time t is

$$W_t = \xi^1 S_t^1 + \cdots + \xi^d S_t^d. \tag{13.2}$$

13.1.2.4 Pricing using The Law of One Price
The law of one price can be used to price linear assets by replication.[6] Furthermore, the law of one price can be used to price all assets in complete markets. By a market we mean a collections of tradable assets together with assumptions about the probability distributions of the asset values. A complete market is such that any possible payoff can be obtained by a portfolio of assets. That is, assume that the market has tradable assets $S^1, \ldots, S^d$. Assume that an arbitrary payoff C_T can be obtained, so that $P_t(\xi^1 S_T^1 + \cdots + \xi^d S_T^d = C_T) = 1$. The law of one price implies that price of this payoff is

$$C_t = \xi^1 S_t^1 + \cdots + \xi^d S_t^d,$$

where we applied the linearity in (13.2).

Futures are linear derivatives, and thus the law of one price can be used to price futures; see Section 14.1. Futures can be priced by the law of one price because futures can be defined as a portfolio of the underlying asset and a bond:

---

6 We can use the arbitrage argument directly, but we have noted that the absence of arbitrage implies the law of one price and thus we use below the pricing with the replication.

the payoff of a futures contract is a linear combination of the payoffs of the underlying asset and a bond.

The payoff of an option is not a linear function of the payoff of the underlying. Thus options cannot be priced as easily as futures. The law of one price can be used to price options in the Black–Scholes model, because the Black–Scholes model is a complete model for the markets, so that all derivatives can be replicated (linearly).

The law of one price can be used to derive the put-call parity, which says that the prices of two options satisfy an equation. The law of one price can also be used to give bounds to option prices without assuming the Black–Scholes model, or any other market model. See Section 14.1.2 for the derivation of the put-call parity.

13.1.3 Relative Pricing Using Statistical Arbitrage

We have derived the price of the derivative in (13.1) using the replication of the derivative with a stock and a bond. The exact replication is possible only under special circumstances. It suffices to move from the binary model to a ternary model to make exact replication impossible, so that only approximate replication is possible. We use the term "statistical arbitrage" to mean quadratic hedging (variance optimal hedging), quantile hedging, and other similar methods for approximate replication.

13.1.3.1 Pricing in a One Period Ternary Model

Let us have two games related to the same tossing of a dice. The first game is such that the player receives d € when the dice shows 1 or 2, c € when the dice shows 3 or 4, and u € when the dice shows 5 or 6, where $0 \le d < c < u$. The participation to this game is an analogy to buying a stock and we denote with S the random variable with $P(S = d) = P(S = c) = P(S = u) = 1/3$.

The second game is such that the player receives 0 € when the dice shows 1, 2, 3, or 4 and 1 € when the dice shows 5 or 6. The participation to this game is an analogy to buying a derivative and we denote with C the random variable $C = f(S)$, where $f : \{1, \dots, 6\} \to \mathbf{R}$ is defined by $f(x) = 0$ when $x \in \{1, \dots, 4\}$ and $f(x) = 1$ when $x \in \{5, 6\}$. Now $P(C = 0) = 2/3$ and $P(C = 1) = 1/3$. The third asset is a bond with value $B = 1$. The price of the bond is 1 and the price of the stock is denoted with $\pi(S)$. We want to find the price $\pi(C)$ of the derivative.

The derivative cannot be replicated with the bond and the stock: Consider the portfolio with β bonds and ξ stocks. The portfolio is $W = \beta B + \xi S$. We have $P(W = C) = 1$ when β and ξ satisfy

$$\beta + \xi d = 0,$$
$$\beta + \xi c = 0,$$
$$\beta + \xi u = 1.$$

We can typically not find such β and ξ because, in general, two parameters cannot satisfy three equations simultaneously. To obtain an approximate replication we could choose β and ξ so that $E(W - C)^2$ is minimized. We have that

$$E(W - C)^2 = P(S = d)(\beta + \xi d)^2 + P(S = c)(\beta + \xi c)^2$$
$$+ P(S = u)(\beta + \xi u - 1)^2.$$

Since the probabilities are all equal to $1/3$, we get the least squares solution for $\bar{\xi} = (\beta, \xi)'$:

$$\bar{\xi} = (\mathcal{X}'\mathcal{X})^{-1}\mathcal{X}'\mathcal{Y},$$

where

$$\mathcal{X} = \begin{bmatrix} 1 & d \\ 1 & c \\ 1 & u \end{bmatrix}, \quad \bar{\xi} = \begin{bmatrix} \beta \\ \xi \end{bmatrix}, \quad \mathcal{Y} = \begin{bmatrix} 0 \\ 0 \\ 1 \end{bmatrix}.$$

The solution is

$$\beta = \frac{1}{3} - \frac{1}{3}(d + c + u)\xi, \quad \xi = \frac{u - (d + c + u)/3}{d^2 + c^2 + u^2 - (d + c + u)^2/3}.$$

We set the price of the derivative to be equal to the price of the approximately replicating portfolio:

$$\pi(C) = \pi(W) = \beta + \xi\pi(S).$$

If $\pi(S) = ES = (d + c + u)/3$, then $\pi(C) = EC = 1/3$.

13.1.3.2 Statistical Arbitrage and the Law of Approximate Price

Statistical arbitrage is a financial operation where a profit is obtained with a high probability. The principle of excluding statistical arbitrage is a pricing principle, which can be used when the principle of excluding arbitrage does not apply. However, the concept of statistical arbitrage can be defined in many ways. Let us compare the principle of excluding arbitrage to the idea of excluding statistical arbitrage.

1) *Excluding arbitrage.* The value of a derivative is C_T at time T. Let us have another asset whose value is W_T at time T. Assume that the values are equal with probability 1: $P_t(C_T = W_T) = 1$. Then it should hold that the value of the derivative and the other asset are equal at all previous times: $C_t = W_t$ for all previous times t. Otherwise, there would be an arbitrage opportunity: sell the more expensive instrument and buy the cheaper instrument to obtain a risk free profit at time T.

2) *Excluding statistical arbitrage.* The value of a derivative is C_T at time T. Let us have an other asset whose value is W_T at time T. If the random variables C_T and W_T are "close," then the prices C_t and W_t should be close at all previous times t. The closeness of random variables can be defined in many

ways. For example, we can say that two random variables C_T and W_T are close when $E_t(C_T - W_T)^2$ is small. A derivative can be priced with statistical arbitrage if we can construct an asset, which replicates the payoff of the derivative with high probability.

Pricing with statistical arbitrage requires that we define the best approximation W_T to a random payoff C_T. As an example, we can consider a call option written at time t, with the strike price K and with the expiration time T. The payout of the option at the expiration time is $C_T = \max\{S_T - K, 0\}$, where S_T is the price of the underlying instrument at time T. The best constant approximation of random variable C_T in the sense of mean squared error is its expectation:[7]

$$\operatorname*{argmin}_{P \in \mathbf{R}} E_t(C_T - P)^2 = E_t C_T,$$

where the minimization is taken with respect to all real numbers. Thus, expectation $E_t C_T$ can give a first approximation to the price of C_T. We can use the underlying asset to provide a better approximation. The best approximation of C_T with a function $f(S)$ of S is the conditional expectation:

$$\operatorname*{argmin}_{f} E_t(C_T - f(S))^2 = E_t(C_T|S),$$

where the minimization is taken with respect to functions $f : \mathbf{R} \to \mathbf{R}$, and function $f = E_t(C_T|S)$ takes values $f(s) = E_t(C_T|S = s)$. Thus, the conditional expectation $E_t(C_T|S = s)$ could be a candidate for the fair price. However, the conditional expectation is typically not a tradable asset, and we will make a further restriction to find such function $f(S)$, which is tradable, which leads to linear approximations.

13.2 Fundamental Theorems of Asset Pricing

Our intention is to describe the basic mathematical terminology and fundamental theorems of asset pricing in discrete time models. Our presentation follows Shiryaev (1999) and Föllmer and Schied (2002). The mathematics of asset pricing is a fascinating topic with elegant results and we hope that the presentation will inspire readers to study the subject in a greater detail.

The first fundamental theorem of asset pricing says that a market is arbitrage-free if and only if there exists an equivalent martingale measure. Furthermore, these martingale measures define the collection of arbitrage-free prices for a derivative. In a complete model, there is exactly one equivalent martingale measure, but in an incomplete model there are many equivalent martingale measures. Thus, the main problem will be to choose the martingale measure for pricing from a collection of available martingale measures. Our emphasis will be in incomplete models.

---

7 This follows from $E_t(C_T - P)^2 = E_t(C_T - E_t C_T)^2 + (E_t C_T - P)^2$.

13.2.1 Discrete Time Markets

Let $B = (B_t)_{t=0,\ldots,T}$ be the time series of prices of a riskless bond (bank account). Let $S = (S_t)_{t=0,\ldots,T}$ be the vector time series of prices of risky assets, where $S_t = (S_t^1, \ldots, S_t^d)$. The price vector that contains both the bond and the risky assets is denoted by

$$\bar{S}_t = (B_t, S_t) = \left(B_t, S_t^1, \ldots, S_t^d\right), \quad t = 0, 1, \ldots, T.$$

13.2.1.1 Filtered Probability Spaces

The underlying probability space $(\Omega, \mathcal{F}, P)$ is accompanied with a filtration of sigma-algebras $\mathcal{F}_0 \subset \mathcal{F}_1 \subset \cdots \subset \mathcal{F}_T$.[8] The price process of stocks is adapted with respect to the filtration: S_t is measurable with respect to $\mathcal{F}_t$.[9] The price process of the bond is predictable with respect to the filtration: B_t is measurable with respect to $\mathcal{F}_{t-1}$, $t = 1, \ldots, T$, and B_0 is measurable with respect to $\mathcal{F}_0$. Thus, the value of B_t is known at time $t - 1$, which makes B locally riskless. The prices are assumed to be nonnegative. We assume that[10]

$$\mathcal{F}_0 = \{\emptyset, \Omega\}, \quad \mathcal{F}_T = \mathcal{F}.$$

Thus, B_0 and elements of S_0 are constants (with probability 1).

13.2.1.2 Trading Strategies

A trading strategy is

$$\bar{\xi}_t = (\beta_t, \xi_t) = \left(\beta_t, \xi_t^1, \ldots, \xi_t^d\right), \quad t = 1, \ldots, T.$$

The values β_t and ξ_t^i express the quantity of the bond and the ith asset held between $t - 1$ and t. The trading strategy is predictable: β_t and ξ_t^i are measurable with respect to $\mathcal{F}_{t-1}$. This means that β_t and ξ_t^i are determined at time $t - 1$, using the information available at time $t - 1$.

13.2.1.3 Examples

Let us give examples of the locally riskless bond B. We can take $B_t = (1 + r)^t$, where $r > -1$ is a constant, or $B_t = \exp\{rt\}$, where $r \in \mathbf{R}$. In addition, we can take $B_0 = 1$, and

$$B_t = \prod_{k=1}^{t}(1 + r_k)$$

---

8 A sigma-algebra $\mathcal{F}$ is a set of subsets of Ω, which satisfies axioms (1) $\Omega \in \mathcal{F}$, (2) if $A \in \mathcal{F}$, then $A^c \in \mathcal{F}$, and (3) if $A_1, A_2, \ldots \in \mathcal{F}$, then $\cup_{i=1}^{\infty} A_i \in \mathcal{F}$.
9 Measurability of S_t with respect to $\mathcal{F}_t$ means that $\{\omega \in \Omega : S_t(\omega) \in A\} \in \mathcal{F}_t$ for each Borel set $A \subset \mathbf{R}$.
10 The case where $\mathcal{F}_T \subset \mathcal{F}$ and $\mathcal{F}_T \neq \mathcal{F}$ might arise in the following way. It might be known that in the near future an earnings report will be given. This knowledge increases implied volatility, and it increases the prices of the options, but the knowledge might not affect the price or the volatility of the underlying.

for $t \geq 1$, where $r_t > -1$ is predictable. We can also take $B_0 = 1$ and $B_t = \prod_{k=1}^{t} \exp\{r_k\}$, where $r_t \in \mathbf{R}$ is predictable.

Consider the two period binary model as an example of adaptability and predictability. Now $T = 2$ and $d = 1$. The initial stock price is $S_0 = s_0 = 1$. The next price is $S_1 = s_0(1 + w_1)$ and the final price is $S_2 = S_1(1 + w_2)$, where w_1 and w_2 are random variables. Random variable w_1 satisfies $P(w_1 = \epsilon) = p$ and $P(w_1 = -\epsilon) = 1 - p$ for $0 < p < 1$, where $\epsilon > 0$ is a fixed constant. Random variable w_2 has the same distribution as w_1, and is independent of w_1. Choose

$$\Omega = \{(d, d), (d, u), (u, d), (u, u)\},$$

where u and d refer to the upwards and downwards movements of the stock. Set Ω describes all possible trajectories of the process. Now,

$$S_1 = \begin{cases} 1 - \epsilon, & \text{when } \omega \in \{(d, d), (d, u)\}, \\ 1 + \epsilon, & \text{when } \omega \in \{(u, d), (u, u)\}. \end{cases}$$

Let $A \subset \mathbf{R}$. It follows that

$$S_1^{-1}(A) = \begin{cases} \varnothing, & \text{when } 1 - \epsilon \notin A \text{ and } 1 + \epsilon \notin A, \\ \{(d, d), (d, u)\}, & \text{when } 1 - \epsilon \in A \text{ and } 1 + \epsilon \notin A, \\ \{(u, d), (u, u)\}, & \text{when } 1 - \epsilon \notin A \text{ and } 1 + \epsilon \in A, \\ \Omega, & \text{when } 1 - \epsilon \in A \text{ and } 1 + \epsilon \in A. \end{cases}$$

In order for the stock prices to be adapted to the filtration we need

$$\{\{(d, d), (d, u)\}, \{(u, d), (u, u)\}\} \subset \mathcal{F}_1.$$

We have that

$$S_2 = \begin{cases} (1 - \epsilon)^2, & \text{when } \omega \in \{(d, d)\}, \\ (1 - \epsilon)(1 + \epsilon), & \text{when } \omega \in \{(d, u), (u, d)\}, \\ (1 + \epsilon)^2, & \text{when } \omega \in \{(u, u)\}. \end{cases}$$

It follows that in order for the stock prices to be adapted to the filtration we need

$$\{\{(d, d)\}, \{(d, u), (u, d)\}, \{(u, u)\}\} \subset \mathcal{F}_2.$$

Let the initial bond price be $B_0 = b_0$. Let the next bond price be $B_1 = b_0(1 + r_1)$, where $r_1 > -1$ is a constant. Let the final bond price be $B_2 = B_1(1 + r_2)$, where $r_2 = w_1/2$. Now bond prices are predictable with respect to the filtration. Bond price at time $t = 2$ depends only on the stock price at time $t = 1$. Thus, the bond price at time $t = 2$ is a random variable which is known at time $t = 1$.

13.2.2 Wealth and Value Processes

We define the wealth and value processes. The value process is obtained from the wealth process by dividing with the bond price. After that we derive an expression for the wealth and value processes under the assumption of self-financing.

13.2.2.1 The Wealth and Value Processes

We use the following notation for the inner product:

$$\bar{\xi}_t \cdot \bar{S}_t = \beta_t B_t + \xi_t \cdot S_t = \beta_t B_t + \sum_{i=1}^{d} \xi_t^i S_t^i.$$

At time 0 the initial wealth is $W_0 \in \mathbf{R}$ and after that,

$$W_t = \bar{\xi}_t \cdot \bar{S}_t, \quad t = 1, \dots, T.$$

Indeed, the portfolio vector $\bar{\xi}_t$ is chosen at time $t - 1$ and hold during the period $t - 1 \mapsto t$.

We assume that $P(B_t > 0) = 1$ for all t and choose the bond as a numéraire. The discounted price process is defined by

$$X_t^i = \frac{S_t^i}{B_t}, \quad t = 0, \dots, T, \quad i = 1, \dots, d.$$

We denote

$$X_t = \left(X_t^1, \dots, X_t^d \right), \quad \bar{X}_t = (1, X_t). \tag{13.3}$$

The value process is defined as $V_0 = W_0/B_0$ and

$$V_t = \frac{W_t}{B_t} = \bar{\xi}_t \cdot \bar{X}_t, \quad t = 1, \dots, T.$$

13.2.2.2 The Wealth Process under Self-financing

We assume in most cases that the trading strategy is self-financing. The local quadratic hedging without self-financing in Section 16.2.3 is a case where self-financing is not assumed.

Let us describe trading under the condition of self-financing. At time 0 the initial wealth is $W_0 \in \mathbf{R}$. The wealth is allocated among the available assets: the quantities $\bar{\xi}_1$ are chosen so that

$$W_0 = \bar{\xi}_1 \cdot \bar{S}_0.$$

The prices change from $\bar{S}_0$ to $\bar{S}_1$, and the wealth changes accordingly from W_0 to W_1. After that, wealth W_1 is allocated among the available assets. We obtain

$$W_1 = \bar{\xi}_1 \cdot \bar{S}_1 = \bar{\xi}_2 \cdot \bar{S}_1.$$

We continue in this way to obtain

$$W_t = \bar{\xi}_t \cdot \bar{S}_t = \bar{\xi}_{t+1} \cdot \bar{S}_t, \quad t = 1, \dots, T - 1.$$

The final wealth is

$$W_T = \bar{\xi}_T \cdot \bar{S}_T.$$

At time T we need not do the reallocation, because it is the last time instance.

We have described a process of trading, which is self-financing. We say that a trading strategy $\bar{\xi}$ is self-financing if

$$\bar{\xi}_t \cdot \bar{S}_t = \bar{\xi}_{t+1} \cdot \bar{S}_t, \quad t = 1, \ldots, T-1. \tag{13.4}$$

When the trading strategy is self-financing, then no external funds are received, and no funds are reserved for consumption.

Under the assumption (13.4) of self-financing, the change of wealth can be written as

$$W_t - W_{t-1} = \bar{\xi}_t \cdot \bar{S}_t - \bar{\xi}_{t-1} \cdot \bar{S}_{t-1} = \bar{\xi}_t \cdot (\bar{S}_t - \bar{S}_{t-1}),$$

for $t = 2, \ldots, T$. Thus, the wealth at time t can be written as[11]

$$W_t = W_0 + \sum_{k=1}^{t} (W_k - W_{k-1})$$

$$= \bar{\xi}_1 \cdot \bar{S}_0 + \sum_{k=1}^{t} \bar{\xi}_k \cdot (\bar{S}_k - \bar{S}_{k-1}), \tag{13.5}$$

where $0 \le t \le T$.

13.2.2.3   The Value Process Under Self-Financing

Let us assume that the rebalancing is made respecting the condition of self-financing: $\bar{\xi}_{t+1}$ is chosen so that the wealth W_t is allocated among the assets. Equation $W_t = \bar{\xi}_{t+1} \cdot \bar{S}_t$ sets a linear constraint on the vector $\bar{\xi}_{t+1}$. It is convenient to write the wealth process so that the quantity β_t of bonds is eliminated, and this can be done using the value process.

The self-financing condition in (13.4) implies that the discounted price process X_t in (13.3) satisfies

$$\bar{\xi}_t \cdot \bar{X}_t = \bar{\xi}_{t+1} \cdot \bar{X}_t, \quad t = 1, \ldots, T-1. \tag{13.6}$$

Similarly as for the wealth process, it holds that

$$V_0 = \bar{\xi}_1 \cdot \bar{X}_0,$$
$$V_t = \bar{\xi}_t \cdot \bar{X}_t = \bar{\xi}_{t+1} \cdot \bar{X}_t, \quad t = 1, \ldots, T-1,$$
$$V_T = \bar{\xi}_T \cdot \bar{X}_T. \tag{13.7}$$

Under the condition (13.6) of self-financing, an increment of the value process can be written as

$$V_t - V_{t-1} = \bar{\xi}_t \cdot \bar{X}_t - \bar{\xi}_{t-1} \cdot \bar{X}_{t-1} = \bar{\xi}_t \cdot (\bar{X}_t - \bar{X}_{t-1}) = \xi_t \cdot (X_t - X_{t-1}),$$

---

11 We can write the wealth process both in a multiplicative way and in an additive way. We have written the multiplicative wealth process in (9.10). This expression is valid for the cases where we have a positive initial wealth $W_0 > 0$. Now we write the wealth process W_t in an additive way. This way of writing the wealth process does not presuppose a positive initial wealth. For example, the writer of an option can start with zero initial wealth, when the hedging is done by borrowing the funds needed in hedging.

where the last equality follows because the first element of $\bar{X}_t$ is 1 for all t. Thus, the value at time t can be written as

$$V_t = V_0 + \sum_{k=1}^{t}(V_k - V_{k-1})$$

$$= V_0 + \sum_{k=1}^{t} \xi_k \cdot (X_k - X_{k-1}). \tag{13.8}$$

Note that the value process is written in terms of the quantity ξ_t of stocks. The quantity β_t of bonds is obtained from the equations

$$\beta_1 = V_0 - \xi_1 \cdot X_0,$$
$$\beta_{t+1} = \beta_t + (\xi_t - \xi_{t+1}) \cdot X_t, \quad t = 1, \dots, T - 1, \tag{13.9}$$

which follow from self-financing equations (13.7).

The gains process is defined as

$$G_0 = 0, \quad G_t = \sum_{k=1}^{t} \xi_k \cdot (X_k - X_{k-1}), \quad t = 1, \dots, T.$$

For a self-financing strategy

$$V_t = V_0 + G_t. \tag{13.10}$$

The gains process is a discrete stochastic integral. The gains process is a transformation of X_t by means of ξ_t.[12]

The value process can be used to derive some expressions for the wealth. For example, when $B_t = (1 + r)^t$, then[13]

$$W_t = (1+r)^t W_0 + \sum_{k=1}^{t} (1+r)^{t-k} \xi_k \cdot [S_k - (1+r)S_{k-1}].$$

---

12 When $X = (X_t, \mathcal{F}_t)$ is a stochastic sequence and $\xi = (\xi_t, \mathcal{F}_{t-1})$ is a predictable sequence, then the stochastic sequence $(\xi \cdot X) = ((\xi \cdot X)_t, \mathcal{F}_t)$ is called the transformation of X by means of ξ, where we denote

$$(\xi \cdot X)_t = \xi_0 \cdot X_0 + \sum_{k=1}^{t} \xi_k \cdot (X_k - X_{k-1}). \tag{13.11}$$

If X is a martingale, then $\xi \cdot X$ is called a martingale transformation. In our case $\xi_0 = 0$.

13 We have that $B_t X_k = B_{t-k} S_k$ and $B_t X_{k-1} = B_{t-k+1}(1+r)S_{k-1}$. Thus,

$$W_t = B_t V_0 + \sum_{k=1}^{t} B_t \xi_k \cdot (X_k - X_{k-1})$$

$$= B_t W_0 + \sum_{k=1}^{t} B_{t-k} \xi_k \cdot [S_k - (1+r)S_{k-1}].$$

13.2.3 Arbitrage and Martingale Measures

An arbitrage opportunity is a self-financing trading strategy $\bar{\xi}$ so that its value process satisfies

$$V_0 = 0, \quad P(V_T \geq 0) = 1, \quad P(V_T > 0) > 0.$$

This means that with an initial investment of zero it is possible to get a final wealth, which is always nonnegative and sometimes positive.

A martingale is a stochastic process $M = (M_t)_{t=0,\ldots,T}$ on a filtered probability space $(\Omega, F, (F_t), Q)$ if[14]

1) M_t it is adapted (M_t is F_t measurable),
2) $E_Q|M_t| < \infty$ for $0 \leq t \leq T$,
3) $E_Q(M_t|F_{t-1}) = M_{t-1}$ for $1 \leq t \leq T$.

A martingale difference satisfies conditions 1 and 2, but condition 3 takes the form $E_Q(M_t|F_{t-1}) = 0$ for $1 \leq t \leq T$. Thus, a martingale difference is a martingale if $P(M_0 = 0) = 1$.

A probability measure Q on $(\Omega, F, (F_t))$ is called a martingale measure, or a risk neutral measure, if the discounted price process X_t is a d-dimensional martingale. Then,

$$E_Q \left| \frac{S_t^i}{B_t} \right| < \infty$$

for $t = 0, \ldots, T$, and

$$E_Q \left(\frac{S_t^i}{B_t} \middle| F_{t-1} \right) = \frac{S_{t-1}^i}{B_{t-1}}$$

Q-almost surely for $t = 1, \ldots, T$, where $i = 1, \ldots, d$.

An equivalent martingale measure is a martingale measure, which is equivalent to the original measure P. Measures P and Q are equivalent, if $P(A) = 0$ if and only if $Q(A) = 0$. The equivalence of measures is denoted by $P \approx Q$. Let $\mathcal{P}$ be the set of equivalent martingale measures:

$$\mathcal{P} = \{Q : Q \text{ is a martingale measure with } Q \approx P\},$$

where P is the underlying probability measure of the market model.

The first fundamental theorem of asset pricing states that a market model is arbitrage-free if and only if there exists an equivalent martingale measure.

---

14 If $E|M_t| < \infty$, then the conditional expectation $E(M_t|F_{t-1})$ is a F_{t-1}-measurable random variable such that

$$\int_A E(M_t|F_{t-1})dP = \int_A M_t dP$$

for each $A \in F_{t-1}$.

Theorem 13.1   *The market model is arbitrage-free if and only if $P \neq \emptyset$.*

Theorem 13.1 was proved in Harrison and Kreps (1979) and Harrison and Kreps (1981) in the case of finite Ω. Dalang *et al.* (1990) proved it for arbitrary Ω. A proof of Theorem 13.1 can be found in Föllmer and Schied (2002, Theorem 5.17) and in Shiryaev (1999, p. 413). We proof Theorem 13.1 by first showing that the existence of an equivalent martingale measure implies no-arbitrage. After that, an equivalent martingale measure is constructed for an arbitrage-free model.

13.2.3.1 The Existence of a Risk Neutral Measure Implies No-Arbitrage

We think that it is instructive to prove the result first for the case $T = 1$, and after that for the general case $T \geq 1$.

A Proof in the One Period Model   Assume that there exists an equivalent martingale measure $Q \in P$. The martingale measure satisfies

$$S_0^i = E_Q \left(\frac{S_1^i}{B_1} \right)$$

for $i = 1, \ldots, d$. The value process is

$$V_0 = \frac{\bar{\xi}_1 \cdot \bar{S}_0}{B_0}, \quad V_1 = \frac{\bar{\xi}_1 \cdot \bar{S}_1}{B_1}.$$

Take a portfolio such that $\bar{\xi}_1 \cdot \bar{S}_0 = 0$. Then,

$$0 = \bar{\xi}_1 \cdot \bar{S}_0 = \beta_1 B_0 + \sum_{i=1}^{d} \xi_1^i S_0^i = \beta_1 B_0 + \sum_{i=1}^{d} E_Q \left(\frac{\xi_1^i S_1^i}{B_1} \right) = E_Q \left(\frac{\bar{\xi}_1 \cdot \bar{S}_1}{B_1} \right).$$

Thus, we cannot have $E_Q(\bar{\xi}_1 \cdot \bar{S}_1) > 0$. Thus, we cannot have $E_P(\bar{\xi}_1 \cdot \bar{S}_1) > 0$, and we cannot have $P(\bar{\xi}_1 \cdot \bar{S}_1 > 0) > 0$, and $\bar{\xi}_1$ cannot be an arbitrage opportunity.

A Proof in the Multiperiod Model   A proof can be found in Shiryaev (1999, p. 417). We assume that there exists a martingale measure Q, which is equivalent to P and such that $(X_t)_{t=0,\ldots,T}$ is a d-dimensional martingale with respect to Q, where $X_t = S_t/B_t$. We noted in (13.10) that the value process satisfies

$$V_t = V_0 + G_t,$$

where

$$G_0 = 0, \quad G_t = \sum_{k=1}^{t} \xi_k \cdot (X_k - X_{k-1}), \quad t = 1, \ldots, T.$$

Note that since X is a martingale with respect to Q, then sequence $G = (G_t)_{t=0,...,T}$ is a martingale transformation with respect to Q, when a martingale transformation is defined by (13.11).

Let ξ_t be a strategy with $V_0 = 0$, and $P(V_T \geq 0) = 1$, so that $P(G_T \geq 0) = 1$, and $Q(G_T \geq 0) = 1$. Let us assume that $|\xi_t| \leq C < \infty$ for $t = 1, \ldots, T$, where $C > 0$ is a constant. Then G is a martingale, and $E_Q G_T = G_0 = 0$. Since $Q(G_T \geq 0) = 1$, then $Q(G_T = 0) = 1$, which implies $P(G_T = 0) = 1$, and $P(V_T = 0) = 1$.

The case of unbounded ξ_t is handled in Shiryaev (1999, p. 98, Chapter II §1c)[15] and in Föllmer and Schied (2002, Theorem 5.15, p. 229).

13.2.3.2   A Construction of an Equivalent Martingale Measure

We have taken the proof from Shiryaev (1999, p. 413), which follows the ideas of Rogers (1994). The construction of equivalent martingale measures is based on the Esscher conditional transformations. The Esscher transforms were used also in Gerber and Shiu (1994) to construct an equivalent martingale measure. Note that the most proofs found in literature are not constructive, but apply a separation theorem in finite-dimensional Euclidean spaces, for example.[16] It is instructive to first consider the case of the one period model with one risky asset, second consider the case of the multiperiod model with one risky asset, and third consider the general case.

A Martingale Measure in the One Period Model with One Risky Asset   Let us consider the one period model ($T = 1$) with one risky asset ($d = 1$). We assume for simplicity that $B_0 = B_1 = 1$. Let

$$\Delta X = S_1 - S_0.$$

The absence of arbitrage implies that[17]

$$P(\Delta X > 0) > 0, \quad P(\Delta X < 0) > 0.$$

We need to construct measure Q so that

$$Q \approx P, \quad E_Q|\Delta X| < \infty, \quad E_Q(\Delta X) = 0.$$

Let

$$Z_a(x) = \frac{e^{ax}}{\phi(a)},$$

---

15 A theorem is proved in Shiryaev (1999, p. 98, Chapter II §1c) which states that the properties of being (1) a local martingale, (2) a generalized martingale, and (3) a martingale transformation are equivalent.

16 A separating hyperplane theorem can be stated as follows. Suppose that $C \subset \mathbf{R}^d$ is a convex set with $0 \notin C$. Then there exists $\eta \in \mathbf{R}^d$ with $\eta \cdot x \geq 0$ for all $x \in C$, and with $\eta \cdot x_0 > 0$ for at least one $x_0 \in C$; see Föllmer and Schied (2002, Proposition A.1).

17 If $P(\Delta X \geq 0) = 1$, then we know for sure that the stock price does not fall, and we should buy the stock as much as possible to induce arbitrage. If $P(\Delta X \leq 0) = 1$, then we know for sure that the stock price does not rise, and we should sell short the stock as much as possible to induce arbitrage.

for $a \in \mathbf{R}$, where

$$\phi(a) = E_P e^{a\Delta X}.$$

We can assume that $\phi(a) < \infty$ for each a such that $\phi(a) > 0$.[18] In addition, $Z_a(x) > 0$ and $E_P Z_a(\Delta X) = 1$. We define the probability measure

$$Q_a(dx) = Z_a(x)P(dx).$$

Now $Q_a \approx P$. We have that $\phi''(x) > 0$, and thus ϕ is strictly convex on $\mathbf{R}$. Let

$$\phi^* = \inf\{\phi(a) : a \in \mathbf{R}\}.$$

We have to prove that

$$\text{there exists such } a^* \text{ that } \phi(a^*) = \phi^*. \qquad (13.12)$$

If (13.12) holds, then we define

$$Q = Q_{a^*}.$$

Now Q is the required measure because $\phi'(a^*) = 0$ and

$$E_{Q_{a^*}}(\Delta X) = E_P\left(\frac{\Delta X e^{a^*\Delta X}}{\phi(a^*)}\right) = \frac{\phi'(a^*)}{\phi(a^*)} = 0.$$

Let us prove (13.12). Let us assume that (13.12) does not hold and derive a contradiction. Let $\{a_n\}$ be a sequence such that

$$\phi(a_n) > \phi^*, \quad \phi(a_n) \downarrow \phi^*. \qquad (13.13)$$

Then $a_n \to \infty$ or $a_n \to -\infty$. Otherwise, we can choose a convergent subsequence, the minimum is attained at a finite point, and (13.12) holds. Let

$$u_n = \frac{a_n}{|a_n|}, \quad u = \lim_{n\to\infty} u_n (= \pm 1).$$

We have that

$$Q(u\Delta X > 0) > 0.$$

Thus there exists $\delta > 0$ such that

$$Q(u\Delta X > \delta) = \epsilon > 0, \quad Q(u\Delta X = \delta) = 0.$$

Thus,

$$Q(a_n \Delta X > \delta |a_n|) = Q(u_n \Delta X > \delta) \to \epsilon,$$

---

18 Otherwise, we can move from P to measure

$$\tilde{P}(dx) = \frac{e^{-x^2} P(dx)}{E_P e^{-\Delta X^2}}, \quad x \in \mathbf{R}.$$

as $n \to \infty$. For sufficiently large n we have

$$\phi(a_n) = E_Q e^{a_n \Delta X} \geq E_Q \left(e^{a_n \Delta X} I_{(a_n \Delta X > \delta |a_n|)} \right) \geq \frac{1}{2} \epsilon e^{\delta |a_n|} \to \infty,$$

which contradicts (13.13).

A Martingale Measure in the Multiperiod Model with One Risky Asset Let us consider the multiperiod model ($T \geq 1$) with one risky asset ($d = 1$). We assume for simplicity that $B_0 = \cdots = B_T = 1$. Let

$$\Delta X_t = S_t - S_{t-1},$$

where $t = 1, \dots, T$. The absence of arbitrage implies that[19]

$$P(\Delta X_t > 0 | \mathcal{F}_{t-1}) > 0, \quad P(\Delta X_t < 0 | \mathcal{F}_{t-1}) > 0 \tag{13.14}$$

P-almost surely, for $t = 1, \dots, T$. We need to construct measure Q so that

$$Q \approx P,$$
$$E_Q |\Delta X_t| < \infty \quad \text{for } t = 1, \dots, T,$$
$$E_Q(\Delta X_t | \mathcal{F}_{t-1}) = 0 \quad \text{for } t = 1, \dots, T.$$

Then $\Delta X_1, \dots, \Delta X_T$ is a martingale difference with respect to Q, and $S_0, \dots, S_T$ is a martingale with respect to Q. Let

$$\phi_t(a, \omega) = E_P(e^{a \Delta X_t} | \mathcal{F}_{t-1})(\omega)$$

for $a \in \mathbf{R}$, We can assume that $\phi(a)$ is finite.[20] For a fixed ω function $\phi_t(a, \omega)$ is strictly convex, as follows from (13.14). There exists a unique finite

$$a_t^* = a_t^*(\omega)$$

such that $\inf_a \phi_t(a, \omega)$ is attained at a_t^*, which can be shown similarly as (13.12). We can show that $a^*(\omega)$ is $\mathcal{F}_{t-1}$-measurable.[21] Let $Z_0 = 1$ and

$$Z_t(\omega) = Z_{t-1}(\omega) \frac{\exp\left\{ a_t^*(\omega) \Delta X_t(\omega) \right\}}{E_P \left(\exp\left\{ a_t^* \Delta X_t \right\} | \mathcal{F}_{t-1} \right)(\omega)},$$

---

19 If $P(\Delta X_t \geq 0) = 1$, then we know for sure that the stock price does not fall for the period $t - 1 \mapsto t$, and we should buy the stock as much as possible at time $t - 1$ to obtain arbitrage. If $P(\Delta X_t \leq 0) = 1$, then we know for sure that the stock price does not rise during the period $t - 1 \mapsto t$, and we should sell short the stock as much as possible at time $t - 1$ to obtain arbitrage.
20 Otherwise, we can move from P to measure $\tilde{P}(d\omega) = c \exp\{ -\sum_{i=0}^{T} \Delta X_i^2(\omega) \} P(d\omega)$, where $c^{-1} = E_P \exp\{ -\sum_{t=0}^{T} \Delta X_i^2 \}$ is the normalizing constant.
21 In fact, for a closed interval $[A, B]$,

$$\{ \omega : a_t^*(\omega) \in [A, B] \} = \bigcap_{m=1}^{\infty} \bigcup_{a \in \mathbf{Q} \cap [A,B]} \left\{ \omega : \phi_t(a, \omega) < \phi_t^*(\omega) + \frac{1}{m} \right\} \in \mathcal{F}_{t-1},$$

where $\mathbf{Q}$ is the set of rational numbers and $\phi_t^*(\omega) = \phi_t(a^*, \omega)$.

for $t = 1, \ldots, T$. Now $Z_t(\omega) > 0$, $Z_t(\omega)$ are F_t-measurable, and they form a martingale:

$$E_P(Z_t|F_{t-1}) = Z_{t-1}$$

P-almost surely. We define the probability measure

$$Q(d\omega) = Z_T(\omega)P(d\omega).$$

Now $Q \approx P$, $E_Q|\Delta X_t| < \infty$, and $E_Q(\Delta X_t|F_{t-1}) = 0$, for $t = 1, \ldots, T$.

A Martingale Measure in the Multiperiod Model with Several Risky Assets Let $T \geq 1$ and $d \geq 1$. We assume for simplicity that $B_0 = \cdots = B_T = 1$. Let

$$\Delta X_t = S_t - S_{t-1},$$

where $t = 1, \ldots, T$. Now ΔX_t are vectors of length d. The portfolio vector ξ_t is a d-dimensional F_{t-1}-measurable vector. The components are bounded, so that $|\xi_t^i(\omega)| \leq C < \infty$ for $\omega \in \Omega$ and $i = 1, \ldots, d$. The absence of arbitrage implies that

$$P(\xi_t \cdot \Delta X_t > 0|F_{t-1}) > 0, \quad P(\xi_t \cdot \Delta X_t < 0|F_{t-1}) > 0$$

P-almost surely, for $t = 1, \ldots, T$. We need to construct measure Q so that

$$Q \approx P,$$
$$E_Q|\Delta X_t^i| < \infty \quad \text{for } t = 1, \ldots, T, \quad i = 1, \ldots, d,$$
$$E_Q\left(\Delta X_t^i|F_{t-1}\right) = 0 \quad \text{for } t = 1, \ldots, T, \quad i = 1, \ldots, d.$$

Then $\Delta X_1, \ldots, \Delta X_T$ is a martingale difference with respect to Q, and $S_0, \ldots, S_T$ is a martingale with respect to Q. Let

$$\phi_t(a, \omega) = E_P(e^{a \cdot \Delta X_t}|F_{t-1})(\omega)$$

for $a \in \mathbf{R}^d$. There exists a unique finite $a_t^* = a_t^*(\omega)$ such that $\inf_a \phi_t(a, \omega)$ is attained at a_t^*, and a_t^* is F_{t-1}-measurable.[22] Let $Z_0 = 1$ and

$$Z_t(\omega) = Z_{t-1}(\omega)\frac{\exp\left\{a_t^*(\omega) \cdot \Delta X_t(\omega)\right\}}{E_P\left(\exp\left\{a_t^* \cdot \Delta X_t\right\}|F_{t-1}\right)(\omega)},$$

for $t = 1, \ldots, T$. We define the probability measure

$$Q(d\omega) = Z_T(\omega)P(d\omega),$$

and Q is the required equivalent martingale measure.

---

22 If the support of $P(\Delta X_t|F_{t-1})(\omega)$ does not lie in a proper subspace of $\mathbf{R}^d$, then $\phi_n(a)$ are strictly convex, and we can find the unique minimizer. It the support is concentrated on a proper subspace of $\mathbf{R}^d$, then the proof is more complicated, but can be found in Rogers (1994).

13.2.3.3   Estimation of an Equivalent Martingale Measure

We estimate the Esscher martingale measures using S&P 500 daily data, described in Section 2.4.1. We consider both a one period model and a two period model.

A Martingale Measure for S&P 500: One Period   We consider the one period model where the period consists of 20 days. Let

$$\Delta X = S_t - S_{t-20}$$

be the price increment. Our S&P 500 data provides a sample of identically distributed observations of ΔX: we use data $\Delta X_i = S_0 R_i - S_0$, where $S_0 = 100$ and $R_i = S_i/S_{i-20}$ is the gross return over the period of 20 days. We use nonoverlapping increments. The risk free rate is $r = 0$.

The density dQ/dP of the Esscher martingale measure with respect to underlying physical measure of ΔX can be estimated as

$$Z_{a^*}(x), \quad x \in \mathbf{R},$$

where $Z_a(x) = e^{ax}/\hat{\phi}(a)$, $\hat{\phi}(a)$ is the sample average of $e^{a\Delta X}$, and a^* is the minimizer of $\hat{\phi}(a)$ over $a \in \mathbf{R}$. The underlying physical density f of ΔX with respect to the Lebesgue measure can be estimated using the kernel estimator $\hat{f}$. The kernel density estimator is defined in (3.43). The density q of the martingale measure with respect to the Lebesgue measure can be estimated as

$$\hat{q}(x) = Z_{a^*}(x)\hat{f}(x), \quad x \in \mathbf{R}.$$

Figure 13.1(a) shows the estimate Z_{a^*} of the density of the martingale measure with respect to the physical measure (red). The blue curve shows the density of the risk neutral log-normal density with respect to the estimated physical measure. We see that the measures put more probability mass on the negative increments than the physical measure. Fitting of a log-normal distribution is discussed in the connection of Figure 3.11.[23] Panel (b) shows the kernel estimate $\hat{f}$ of the density of the physical measure as a red curve, and the estimate $\hat{q}$ of the density of the Esscher martingale measure with respect to the Lebesgue measure as a red dashed curve. We apply the standard normal kernel and the normal reference rule to choose the smoothing parameter. The blue curves show the corresponding densities in the log-normal model.

Figure 13.2 shows density ratios. Panel (a) shows the ratio

$$\frac{Z_{a^*}(y)}{Z^{\log n}(y)},$$

---

[23] The physical measure is estimated as $S_T \sim \text{lognorm}(\log S_0 + \hat{m}, \hat{s}^2)$, where $\hat{m}$ is the sample mean and $\hat{s}^2$ is the sample variance of $\log(S_t/S_{t-20})$. The risk neutral measure is estimated as $S_T \sim \text{lognorm}(\log S_0, \hat{s}^2)$.

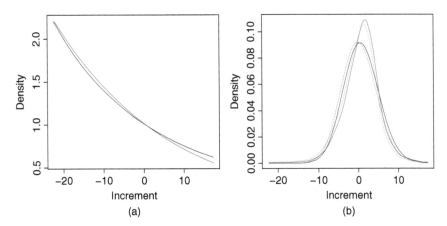

Figure 13.1 *Esscher martingale measure: One Period.* (a) The density of the Esscher martingale measure (red) and the density of the risk neutral log-normal measure (blue). The densities are with respect to the physical measure. (b) The kernel density estimate of the physical measure (red), and the corresponding Esscher martingale measure (red dashed). The log-normal physical measure (blue), and the corresponding risk neutral log-normal density (dashed blue).

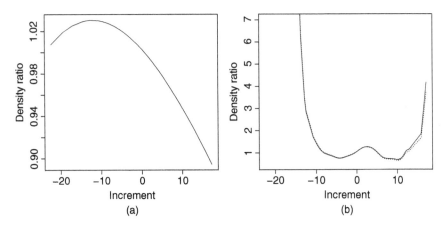

Figure 13.2 *Esscher martingale measure: Density ratios in one period.* (a) The density estimate of the Esscher martingale measure divided by the density estimate of the log-normal martingale measure. (b) The density estimate of the physical measure divided by the estimate log-normal physical measure (solid). The dashed line shows the ratio of the corresponding risk neutral densities.

and panel (b) shows the ratios

$$\frac{\hat{f}(y)}{\hat{f}^{\log n}(y)} \text{ (solid),} \qquad \frac{\hat{q}(y)}{\hat{q}^{\log n}(y)} \text{ (dashed),}$$

where $\hat{f}^{\log n}(y)$ is an estimate of the log-normal physical measure, $\hat{q}^{\log n}(y)$ is an estimate of the log-normal risk neutral measure, and $Z^{\log n}(y) = \hat{q}^{\log n}(y)/\hat{f}^{\log n}(y)$.

A Martingale Measure for S&P 500: Two Periods   Let us estimate the Esscher martingale measure for the two period model with two periods of 10 days. Let

$$\Delta X_1 = S_{t-10} - S_{t-20}, \quad \Delta X_2 = S_t - S_{t-10}$$

be the price increments. Our S&P 500 data provides a sample of identically distributed observations of $\Delta X = (\Delta X_1, \Delta X_2)$. The observations are

$$(\Delta X_{1i}, \Delta X_{2i}),$$

where

$$\Delta X_{1i} = S_0(S_{i-10}/S_{i-20} - 1), \quad \Delta X_{2i} = S_0(S_i - S_{i-10})/S_{i-20},$$

where $S_0 = 100$. We use nonoverlapping increments. Let

$$Z_1(x_1) = \frac{e^{a_1^* x_1}}{\hat{\phi}(a_1^*)},$$

where $\hat{\phi}(a)$ is the sample average of $e^{a \Delta X_1}$, and a_1^* is the minimizer of $\hat{\phi}(a)$ over $a \in \mathbf{R}$. Let

$$Z_2(x_1, x_2) = Z_1(x_1) \frac{\exp\{a_2^*(x_1)x_2\}}{\hat{\phi}_2(a_2^*(x_1), x_1)},$$

where $a_2^*(x_1)$ is the minimizer of $\hat{\phi}_2(a, x_1)$ over $a \in \mathbf{R}$, and $\hat{\phi}_2(a, x_1)$ is a regression estimate evaluated at x_1, when the response variable is $e^{a \Delta X_2}$ and the explanatory variable is ΔX_1. We apply a kernel regression estimate of (6.20) and (6.21) to define

$$\hat{\phi}_2(a, x_1) = \sum_{i=1}^{n} p_i(x_1) \, e^{a \Delta X_{2i}},$$

where $\Delta X_{1i}, i = 1, \ldots, n$, are the observation of ΔX_1,

$$p_i(x_1) = \frac{K((x_1 - \Delta X_{1i})/h)}{\sum_{j=1}^{n} K((x_1 - \Delta X_{1j})/h)}$$

are the kernel weights, $K : \mathbf{R} \to \mathbf{R}$ is the Gaussian kernel function and $h > 0$ is the smoothing parameter, chosen by the normal reference rule.

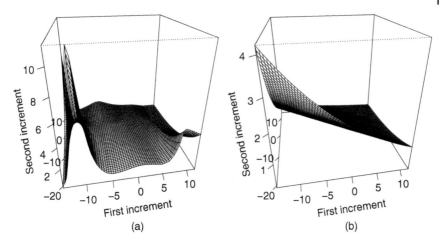

Figure 13.3 *Esscher martingale measure: Two Periods.* Estimates of the density of the Esscher measure with respect to the physical measure. (a) Increments are assumed to be dependent. (b) Increments are assumed to be independent.

The density dQ/dP of the martingale measure with respect to the underlying physical measure of $(\Delta X_1, \Delta X_2)$ can be estimated as

$$Z_2(x_1, x_2), \quad (x_1, x_2) \in \mathbf{R}^2. \tag{13.15}$$

We can also assume the independence of the increments and estimate dQ/dP by

$$Z_1(x_1)Z_1(x_2), \quad (x_1, x_2) \in \mathbf{R}^2. \tag{13.16}$$

Figure 13.3 shows estimates of the density of the Esscher measure with respect to the physical measure. In panel (a) we show estimate (13.15), which does not assume independence, and in panel (b) we show estimate (13.16), which assumes independence.

13.2.3.4 Examples of Equivalent Martingale Measures
We calculate the class of equivalent martingale measures in the one period binary model, in the one period ternary model, and in the one period model with a finite amount of states, which generalizes the two previous models.

The One Period Binary Model Let us have two assets: bond B and stock S. The value of the bond at time 1 is $1 + r$, where $r > -1$. The value of the stock at time 1 is u with probability p and d with probability $1 - p$, where $u > d$ and $0 < p < 1$. That is,

$$P(S = u) = p, \quad P(S = d) = 1 - p.$$

Let the price of the bond be $\pi^0 = 1$ and the price of the stock be π^1. Let us consider probability measure Q which is defined by

$$Q(S = u) = q, \quad Q(S = d) = 1 - q,$$

where $0 \leq q \leq 1$. If Q is a martingale measure, then it satisfies

$$E_Q S = qu + (1 - q)d = \pi^1(1 + r).$$

This holds if

$$q = \frac{\pi^1(1 + r) - d}{u - d}. \tag{13.17}$$

Thus, there exists an equivalent martingale measure if and only if

$$d < \pi^1(1 + r) < u. \tag{13.18}$$

Thus, the market is arbitrage-free if and only if (13.18) holds. The martingale measure is unique. The calculation will be repeated in (14.18), where derivative pricing is discussed.[24] Note that the pricing in the binary model was already studied in (13.1).

The One Period Ternary Model   Let us have two assets: bond B and stock S. The value of the bond at time 1 is $1 + r$, where $r > -1$. The value of the stock at time 1 is u with probability p_1, c with probability p_2, and d with probability $p_3 = 1 - p_1 - p_2$, where $d < c < u$, and $0 < p_1, p_2, p_3 < 1$. That is,

$$P(S = u) = p_1, \quad P(S = c) = p_2, \quad P(S = d) = 1 - p_1 - p_2.$$

Let the price of the bond be $\pi^0 = 1$ and the price of the stock be π^1. Let us consider probability measure Q, which is defined by

$$Q(S = u) = q_1, \quad Q(S = c) = q_2, \quad Q(S = d) = q_3,$$

where $0 \leq q_1, q_2, q_3 \leq 1$ and $q_3 = 1 - q_1 - q_2$. If Q is a martingale measure, then it satisfies

$$E_Q S = q_1 u + q_2 c + (1 - q_1 - q_2)d = \pi^1(1 + r).$$

This holds if $(q_1, q_2) \in A$, where

$$A = \{(q_1, q_2) \in [0, 1]^2 : q_1(u - d) + q_2(c - d) = \pi^1(1 + r) - d\}.$$

------
24 Let

$$C = \{E_Q Y : Q \in \mathcal{Q}\}, \tag{13.19}$$

where $\mathcal{Q}$ is the convex set of probability measures equivalent to P and $Y = S/(1 + r) - \pi$. Now $C = (d/(1 + r) - \pi^1, u/(1 + r) - \pi^1)$. If $0 \notin C$, then there is arbitrage. If $d \geq \pi^1(1 + r)$, then arbitrage is obtained by borrowing with the risk-free rate and buying the stock. If $u \leq (1 + r)\pi^1$, then the arbitrage is obtained by selling the stock short and investing in the risk-free rate.

We can write

$$A = \left\{ \left(q_1, -\frac{u-d}{c-d}q_1 + \frac{\pi^1(1+r)-d}{c-d} \right) \right.$$
$$\left. : \frac{\pi^1(1+r)-c}{u-d} \le q_1 \le \frac{\pi^1(1+r)-d}{u-d} \right\}.$$

Thus, there exists an equivalent martingale measure if and only if

$$d < \pi^1(1+r) < u+c-d. \tag{13.20}$$

Thus, the market is arbitrage-free if and only if (13.20) holds. There are several martingale measures.[25]

A Finite Amount of States Let us consider the one period model with a finite amount of states. Now the probability space Ω has a finite number of elements. We have $n = d+1$ basic securities and m possible states. In the binary model $n = m = 2$. In the ternary model $n = 2$ and $m = 3$.

The jth risky asset S^j takes m values, corresponding to the m different states. Let A be the $m \times d$ matrix whose elements are $a_{ij} = S^j(\omega_i)$, where ω_i is the ith state and S^j is the jth risky asset.

Let $p = (p_1, \dots, p_m)'$ be the $m \times 1$ vector of the probabilities of the m states. Let $\pi = (\pi^1, \dots, \pi^d)'$ be the $d \times 1$ vector of the prices of the risky assets at time 0. Let $S = (S^1, \dots, S^d)'$ be the vector of the risky assets.

Let $q = (q_1, \dots, q_m)'$ be a $m \times 1$ vector of probabilities of the m states. Vector q is a martingale measure if

$$\pi = \frac{1}{1+r}E_Q S = \frac{1}{1+r}A'q. \tag{13.21}$$

We can assume that $\text{rank}(A) = d$ and $m \ge d$, because the redundant basic assets can be removed. A redundant asset would correspond to a column of A which could be expressed as a linear combination of the other columns.

When $m = d$, then there exists a unique equivalent martingale measure, and this is the solution to (13.21):

$$q = (1+r)(A')^{-1}\pi. \tag{13.22}$$

When $m > d$, then the system (13.21) of d linear equations with m variables has many solutions.

---

25 Let C be defined in (13.19). Now $C = (d/(1+r) - \pi^1, (u+c-d)/(1+r) - \pi^1)$. If $0 \notin C$, then there is arbitrage. If $d \ge \pi^1(1+r)$, then arbitrage is obtained by borrowing with the risk-free rate and buying the stock. If $u+c-d \le (1+r)\pi^1$, then the arbitrage is obtained by selling the stock short and investing in the risk-free rate.

13.2.4 European Contingent Claims

13.2.4.1 The Definition of an European Continent Claim

We use the terms "contingent claim" and "derivative" interchangeably. However, these terms can have a different meaning.

1) A European contingent claim is a nonnegative random variable C defined on the probability space $(\Omega, \mathcal{F}, P)$.
2) A derivative of the underlying assets $\bar{S} = (S^0, \ldots, S^d)$ is a European contingent claim, which is measurable with respect to the σ-algebra $\mathcal{F}_T$ generated by the price process $\bar{S}_t$, $t = 0, \ldots, T$.

We assume that $\mathcal{F}_T = \mathcal{F}$, and thus in our case the two definitions lead to the same concept. Time T is called the maturity, or the expiration date.[26] The examples of European contingent claims include the following.

1) A call and a put on the ith asset are defined by

$$C^{call} = \left(S_T^i - K\right)_+, \quad C^{put} = \left(K - S_T^i\right)_+,$$

where $K > 0$ is the strike price and $(x)_+ = \max\{0, x\}$.
2) An Asian call and put option are defined by

$$C^{a-call} = \left(S_{av}^i - K\right)_+, \quad C^{a-put} = \left(K - S_{av}^i\right)_+,$$

where

$$S_{av}^i = \frac{1}{|\mathcal{T}|} \sum_{t \in \mathcal{T}} S_t^i,$$

$\mathcal{T} \subset \{0, \ldots, T\}$, and $|\mathcal{T}|$ is the cardinality of $\mathcal{T}$.
3) A knock-out option on the ith asset is defined by

$$C^{barrier} = \begin{cases} 0, & \text{when } \max_{0 \le t \le T} S_t^i \ge B, \\ \left(S_T^i - K\right)_+, & \text{otherwise}, \end{cases}$$

where $K > 0$ is the strike price, and $B > K$ is the barrier.

13.2.4.2 Arbitrage-Free Prices of European Continent Claims

A European contingent claim C is attainable (replicable, redundant), if there exists a self-financing trading strategy $\bar{\xi}$ whose terminal portfolio value is equal to C:

$$P(C = \bar{\xi}_T \cdot \bar{S}_T) = 1.$$

---

[26] The value of an European option is determined at the time T of the expiration, and in this respect an European option resembles bonds, because the price of a bond is fixed at the expiration, although the price of a bond at the expiration is not a random variable.

The trading strategy $\bar{\xi}$ is called a replicating strategy for C. A contingent claim is attainable if and only if the discounted claim

$$H = \frac{C}{B_T}$$

has the form

$$H = \bar{\xi}_T \cdot \bar{X}_T = V_T = V_0 + \sum_{t=1}^{T} \xi_t \cdot (X_t - X_{t-1}),$$

for a self-financing trading strategy $\bar{\xi} = (\beta, \xi)$ with value process V. Now it is natural to take the initial value

$$V_0 = \bar{\xi}_1 \cdot \bar{X}_0$$

to be the fair price of H, because a different price would lead to an arbitrage opportunity. The corresponding arbitrage-free price of the contingent claim C is

$$B_0 V_0 = \bar{\xi}_1 \cdot \bar{S}_0.$$

We need to define an arbitrage-free price also for those contingent claims which are not attainable. In fact, in typical market models a contingent claim cannot be replicated. Föllmer and Schied (2002, p. 238) formulate the following definition. An arbitrage-free price of a discounted claim H is a real number π^H, if there exists an adapted stochastic process X^{d+1} such that

1) $X_0^{d+1} = \pi^H$,
2) $X_t^{d+1} \geq 0$ for $t = 1, \ldots, T-1$,
3) $X_T^{d+1} = H$, and
4) the enlarged market model with price process $(1, X^1, \ldots, X^{d+1})$ is arbitrage-free.

According to this definition, an arbitrage-free price of a contingent claim is such that trading with this price at time 0 does not allow an arbitrage opportunity. A corresponding arbitrage-free price of the continent claim C is then

$$\pi^C = B_0 \pi^H.$$

We can express the class of arbitrage-free prices with the help of equivalent martingale measures. Föllmer and Schied (2002, Theorem 5.30, p. 239) formulate the following theorem.

Theorem 13.2 *Set $\Pi(H)$ of arbitrage-free prices is nonempty and given by*

$$\Pi(H) = \{E_Q H : Q \in P \text{ and } E_Q H < \infty\}.$$

Proof. Let us prove that $\Pi(H)$ is nonempty. We can find $\tilde{Q} \approx P$ such that $E_{\tilde{Q}}H < \infty$. For example, we can take $d\tilde{Q} = c(1+H)^{-1}dP$, where c is the normalizing constant. Under $\tilde{Q}$ the market model is arbitrage-free. Thus, Theorem 13.1 implies the existence of $Q \in \mathcal{P}$. In addition, this Q can be chosen such that $E_Q H < \infty$. Thus $E_Q H \in \Pi(H)$.

Let us first prove that inclusion $\subset$ holds. Let X^{d+1} satisfy the requirements. Theorem 13.1 implies that there exists an equivalent martingale measure Q for the extended market model. Measure Q satisfies

$$X_t^i = E_Q\left(X_T^i \mid \mathcal{F}_t\right),$$

for $t = 0, \dots, T$ and $i = 1, \dots, d+1$. In particular, $Q \in \mathcal{P}$ and $\pi^H = E_Q H$.

To prove inclusion $\supset$, let $\pi^H = E_Q H$ for some $Q \in \mathcal{P}$. Let

$$X_t^{d+1} = E_Q(H \mid \mathcal{F}_t)$$

for $t = 0, \dots, T$. Process X^{d+1} satisfies the requirements and Q is an equivalent martingale measure for the extended market model, which is thus arbitrage-free, by Theorem 13.1. Thus $\pi^H \in \Pi(H)$. □

The price of contingent claim C can now be written as

$$\pi^C = B_0 E_Q\left(\frac{C}{B_T}\right). \tag{13.23}$$

In the Black–Scholes model we use continuous compounding, where $B_0 = 1$, and $B_T = e^{rT}$, so that for a call option $\pi^C = e^{-rT} E_Q(S_T - K)_+$; see (14.47). In many cases we denote by t the time of writing the option and then $B_0 = e^{rt}$, so that $\pi^C = e^{-r(T-t)} E_Q(S_T - K)_+$.

13.2.4.3 Pricing Kernel

The pricing kernel (discount factor) z, related to the martingale measure Q, is defined as the discounted density of Q with respect to the physical measure P:

$$z = \frac{B_0}{B_T} \frac{dQ}{dP}.$$

The price of C is given in (13.23) as $\pi^C = B_0 E_Q(C/B_T)$. Now the price of derivative C can be written as

$$\pi^C = E_Q\left(\frac{B_0}{B_T} C\right) = E_P(zC).$$

The One Period Binary Model In the one period binary model, the martingale measure was defined as the measure

$$Q(S = u) = q, \quad Q(S = d) = 1 - q,$$

where the probability q is defined in (13.17). The pricing kernel is function $z :$ $\{d, u\} \to \mathbf{R}$, defined by

$$z(d) = \frac{1}{1+r}\frac{1-q}{1-p}, \qquad z(u) = \frac{1}{1+r}\frac{q}{p}.$$

Let $C = f(S)$ be a derivative, where $f : \mathbf{R} \to [0, \infty)$. The price of C is

$$E_Q f(S) = E_P[z(S)f(S)].$$

A Finite Amount of States   Let us continue to study the one period model with a finite amount of states. Let $p = (p_1, \dots, p_m)'$ be the $m \times 1$ vector of the probabilities of the m states. Let $\pi = (\pi^1, \dots, \pi^d)'$ be the $d \times 1$ vector of the prices of the risky assets at time 0. Let $S = (S^1, \dots, S^d)'$ be the vector of the risky assets. Let A be the $m \times d$ matrix whose elements are the values $a_{ij} = S^j(\omega_i)$ of the jth asset at the ith state.

Let $q = (q_1, \dots, q_m)'$ be an equivalent martingale measure, which is a solution to the equation in (13.21). In the case $m = d$ we may obtain q from (13.22) as $q = (1 + r)(A')^{-1}\pi$. Let

$$z(\omega_i) = \frac{1}{1+r}\frac{q_i}{p_i}$$

for $i = 1, \dots, m$. Let $C = f(S) = f(S^1, \dots, S^d)$ be a derivative. The price of C is

$$E_Q f(S) = \sum_{i=1}^{m} q_i f(a_{i1}, \dots, a_{id}) = \sum_{i=1}^{m} p_i z(\omega_i) f(a_{i1}, \dots, a_{id}) = E_P(zf(S)).$$

The Arrow–Debreu securities take value 1 in one state and value 0 in other states: for the jth Arrow–Debreu security S_j it holds that $S^j(\omega_j) = 1$, and $S^j(\omega_i) = 0$ for $i \neq j$. Then $A = I_d$ is the $d \times d$ identity matrix. Then $q = (1 + r)\pi$ and $z(\omega_i) = \pi_i/p_i$.[27]

13.2.5   Completeness

The second fundamental theorem of asset pricing says that every European contingent claim can be attained (replicated) if and only if there exists a unique equivalent martingale measure. If every contingent claim can be attained, then every contingent claim has a unique arbitrage-free price and every contingent claim can be hedged perfectly. The case that there is only one equivalent martingale measure occurs never in practice, but it is possible to be close to this situation.

---

27 Let $z = \pi'[E_P SS']^{-1}S$. We can see that z works as a pricing kernel for linear assets. Let $C = \xi'S$ be a linear asset. Now C is a derivative with price $\xi'\pi$. We have that

$$E_P(zC) = E_P(\pi'[E_P SS']^{-1}SS'\xi) = \pi'\xi.$$

Let $d = m$ and let S be the vector of d Arrow–Debreu securities Then, $E_P SS' = \text{diag}(p_1, \dots, p_d)$, and $z(\omega_i) = \pi_i/p_i$, where ω_i is the ith state, $i = 1, \dots, m$.

An arbitrage-free market model is called complete if every European contingent claim is attainable. Now we state the second fundamental theorem of asset pricing.

Theorem 13.3 *An arbitrage-free market model is complete if and only if there exists exactly one equivalent martingale measure:* $|\mathcal{P}| = 1$.

A proof can be found in Föllmer and Schied (2002, Theorem 5.38, p. 245), where an additional statement is proved: In a complete market, the number of atoms in $(\Omega, \mathcal{F}_T, P)$ is bounded above by $(d+1)^T$.[28]
Let us give examples related to completeness.

13.2.5.1 The One Period Binary Model
In the one period binary model, we have two assets: bond B and stock S. The bond satisfies $P(B = 1 + r) = 1$, where $r > -1$. The stock satisfies

$$P(S = u) = p, \quad P(S = d) = 1 - p,$$

where $u > d$ and $0 < p < 1$. Let the price of the bond be $\pi^0 = 1$ and the price of the stock be π^1. The space of attainable payoffs is

$$\mathcal{V} = \{\beta(1+r) + \xi S : \beta, \xi \in \mathbf{R}\}.$$

Let us consider contingent claim $C = f(S)$, where $f : \{u, d\} \to [0, \infty)$. To replicate the contingent claim, we need to choose β and ξ so that

$$P(f(S) = \beta B + \xi S) = 1.$$

This leads to equations

$$\beta(1+r) + \xi d = f(d),$$
$$\beta(1+r) + \xi u = f(u).$$

We have two equations and two free variables. The model is complete.

13.2.5.2 The One Period Ternary Model
Let us have two assets: bond B and stock S. The bond satisfies $P(B = 1 + r) = 1$, where $r > -1$. The stock satisfies

$$P(S = u) = p_1, \quad P(S = c) = p_2, \quad P(S = d) = 1 - p_1 - p_2,$$

where $d < c < u$, and $0 < p_1, p_2, p_3 < 1$. Let the price of the bond be $\pi^0 = 1$ and the price of the stock be π^1. The space of attainable payoffs is

$$\mathcal{V} = \{\beta(1+r) + \xi S : \beta, \xi \in \mathbf{R}\}.$$

---

28 Set $A \in \mathcal{F}$ is an atom of probability space $(\Omega, \mathcal{F}, P)$, if $P(A) > 0$ and if each $B \in \mathcal{F}$ with $B \subset A$ satisfies either $P(B) = 0$ or $P(B) = P(A)$.

Let us consider contingent claim $C = f(S)$, where $f : \{u, c, d\} \to [0, \infty)$. To replicate the contingent claim, we need to choose β and ξ so that

$$P(f(S) = \beta B + \xi S) = 1.$$

This leads to equations

$$\beta(1 + r) + \xi d = f(d),$$
$$\beta(1 + r) + \xi c = f(c),$$
$$\beta(1 + r) + \xi u = f(u).$$

We have three equations and two free variables. The model is not complete.

13.2.5.3 A Finite Amount of States

Let us continue to study the one period model with a finite amount of states. Let $p = (p_1, \ldots, p_m)'$ be the $m \times 1$ vector of the probabilities of the m states. Let $\pi = (\pi^1, \ldots, \pi^d)'$ be the $d \times 1$ vector of the prices of the risky assets at time 0. Let $S = (S^1, \ldots, S^d)'$ be the vector of the risky assets. Let A be the $m \times d$ matrix whose elements are the values $a_{ij} = S^j(\omega_i)$ of the jth asset at the ith state.

We can assume that rank$(A) = d$ and $m \geq d$, because the redundant basic assets can be removed. A redundant asset would correspond to a column of A which could be expressed as a linear combination of the other columns. A redundant basic asset could be considered as a derivative.

A derivative security is random variable C which takes m possible values. Let those values be in $m \times 1$ vector c. To replicate C, we need to find $d \times 1$ vector ξ so that $P(C = \xi \cdot S) = 1$. This leads to the matrix equation

$$c = A\xi.$$

When $m = d$, then

$$\xi = A^{-1}c.$$

When $m > d$, then we do not always have a solution, because there are d free variables and m equations.

We can choose an approximate replication by minimizing the sum of squared replication errors. Let the replication error be

$$\|A\xi - c\|,$$

where $\| \cdot \|$ is the Euclidean norm in $\mathbf{R}^m$. The minimizer is

$$x = (A'A)^{-1}A'c,$$

which is the same formula as the formula for the least squares coefficients in the linear regression $c = A\xi + \epsilon$. Note that $d \times d$ matrix $A'A$ has rank d, when A has rank d, and thus $A'A$ is invertible.

The Arrow–Debreu securities provide an example of derivatives. An Arrow–Debreu security has price 1 in one state and 0 in the other states. There are as many Arrow–Debreu securities as there are states. When $m = d$, the columns of A^{-1} give the portfolio weights for replicating the Arrow–Debreu securities.

13.2.6 American Contingent Claims

An American contingent claim is defined as a non-negative adapted process

$$C = (C_t)_{t=0,\dots,T},$$

on the filtered space $(\Omega, (F_t)_{t=0,\dots,T})$. The random variable C_t is the payoff if the American option is exercised at time t. For example, in the case of the American call option with strike price K, $C_t = \max\{0, S_t - K\}$, where S_t is the price of the underlying asset at time t.

The buyer of an American contingent claim has the right to choose the exercise time $\tau \in \{0, \dots, T\}$. The buyer receives the amount C_τ at time τ.

A stopping time is a random variable τ taking values in $\{0, \dots, T\} \cup \{+\infty\}$ such that $\{\tau = t\} \in F_t$ for $t = 0, \dots, T$. An exercise strategy is a stopping time taking values in $\{0, \dots, T\}$. The payoff obtained by using τ is equal to $C_{\tau(\omega)}(\omega)$. We denote with $\mathcal{T}$ the set of exercise strategies.

13.2.6.1 European and Bermudan Options

An European contingent claim is obtained as a special case of an American contingent claim, when we choose $C_t = 0$ for $t = 0, \dots, T - 1$. The value of the American option is larger or equal to the value of the corresponding European option.

A Bermudan option can be exercised at times $\mathbf{T} \subset \{0, \dots, T\}$. Formally we can define a Bermudan contingent claim as a non-negative adapted process $C = (C_t), t \in \mathbf{T} \subset \{0, \dots, T\}$, on the filtered space $(\Omega, (F_t)_{t=0,\dots,T})$. A Bermudan option can be obtained as an American option with $C_t = 0$ for $t \notin \mathbf{T}$.

On the other hand, an American option can be considered as a special case of a Bermudan option with $\mathbf{T} = \{0, \dots, T\}$. Also, from the point of view of the continuous time model with time space $[0, T]$, an American option in a discrete time model could be considered as a Bermudan option whose possible exercise times are $\{0, \dots, T\}$.

13.2.6.2 The Set of Arbitrage-Free Prices

Let H be a discounted American claim and let H_τ be the payoff which is obtained for a fixed exercise strategy $\tau \in \mathcal{T}$. Now H_τ can be considered as a discounted European contingent claim, whose set of arbitrage-free prices is given in Theorem 13.2 as

$$\Pi(H_\tau) = \{E_Q H_\tau : Q \in P, \ E_Q H_\tau < \infty\}.$$

Föllmer and Schied (2002, Definition 6.31) give the following definition for an arbitrage-free price. A number $\pi \in \mathbf{R}$ is called an arbitrage-free price of a discounted American claim H if

1) There exists some $\tau \in \mathcal{T}$ and $\pi' \in \Pi(H_\tau)$ such that $\pi \leq \pi'$.
   (The price π is not too high.)
2) There does not exist $\tau' \in \mathcal{T}$ such that $\pi < \pi'$ for all $\pi' \in \Pi(H_{\tau'})$.
   (The price π is not too low.)

The set of arbitrage-free prices is characterized in Föllmer and Schied (2002, Theorem 6.33). It is assumed that $H_t \in \mathcal{L}^1(Q)$ for all $Q \in \mathcal{P}$ and $t = 0, \dots, T$. The set $\Pi(H)$ of arbitrage-free prices is an interval with endpoints $\pi_{inf}(H)$ and $\pi_{sup}(H)$, where

$$\pi_{inf}(H) = \inf_{Q \in \mathcal{P}} \sup_{\tau \in \mathcal{T}} E_Q H_\tau = \sup_{\tau \in \mathcal{T}} \inf_{Q \in \mathcal{P}} E_Q H_\tau,$$

$$\pi_{sup}(H) = \sup_{Q \in \mathcal{P}} \sup_{\tau \in \mathcal{T}} E_Q H_\tau = \sup_{\tau \in \mathcal{T}} \sup_{Q \in \mathcal{P}} E_Q H_\tau.$$

The interval can be a single point, open, or half open.

13.2.6.3 Exercise Strategies for the Buyer

Let us call an exercise strategy τ^* optimal if

$$EH_{\tau^*} = \sup_{\tau \in \mathcal{T}} EH_\tau, \tag{13.24}$$

where

$$H_t = \frac{C_t}{B_t}, \quad t = 0, \dots, T.$$

Thus, an exercise strategy is defined to be optimal, if it maximizes the expectation among the class $\{H_\tau : \tau \in \mathcal{T}\}$ of the payoffs. Note that the definition can be generalized so that we maximize $E_Q U(H_\tau)$, where U is a utility function, and Q is a probability measure, possibly different from the physical measure P.

Föllmer and Schied (2002, Theorem 6.20) shows that the exercise strategy τ_{min} is optimal, when we define

$$\tau_{min} = \min\{t \geq 0 : U_t = H_t\},$$

where

$$U_T = H_T,$$
$$U_t = \max\{H_t, E[U_{t+1}|F_t]\}, \quad \text{for } t = T - 1, \dots, 0. \tag{13.25}$$

It is assumed that $H_t \in \mathcal{L}(\Omega, F_t, P)$ for $t = 0, \dots, T$. Process U_t is called a Snell envelope of H_t with respect to the measure P. Föllmer and Schied (2002, Proposition 6.22) notes that any optimal exercise strategy τ satisfies $\tau \geq \tau_{min}$, so that τ_{min} is the minimal optimal exercise strategy.

Föllmer and Schied (2002, Theorem 6.23) shows that τ_{max} is also an optimal exercise strategy, when we define

$$\tau_{max} = \inf\{t \geq 0 : E[U_{t+1} - U_t|F_t] \neq 0\} \wedge T,$$

where $a \wedge b$ means $\min\{a, b\}$. In addition, τ_{max} is the largest optimal exercise strategy in the sense that any optimal exercise strategy τ satisfies $\tau \leq \tau_{max}$.

13.2.6.4 American Options in Complete Models

Let us assume that the market model is complete, so that there exists the unique equivalent martingale measure Q. We defined the optimal exercise time in (13.24) using the physical measure P, and the construction of the optimal exercise time was made in (13.25) using the physical measure P. Let us define the Snell envelope $(U_t^Q)_{t=0,\dots,T}$ using the equivalent martingale measure Q. The value U_0^Q can be considered as the unique arbitrage-free price, because

$$H_\tau = U_0^Q + \sum_{k=1}^\tau \xi_k(X_k - X_{k-1})$$

holds Q-almost surely, when τ is optimal with respect to Q; see Föllmer and Schied (2002, Corollary 6.24).

13.3 Evaluation of Pricing and Hedging Methods

The evaluation of option pricing and hedging can be done either from the point of view of the writer or from the point of view of the buyer. The writer's point of view is to minimize the hedging error, or to optimize the return of the hedging portfolio. The buyer's point of view is to find fair prices for options. For example, the buyer could be interested whether the buying of the options leads to abnormal returns, as compared to the returns of the underlying.

13.3.1 The Wealth of the Seller

We assume that the seller (writer) of the option hedges the position, and thus the wealth of the seller of the option at the expiration is equal to the hedging error. The writer receives the option premium, makes self-financed trading to replicate the option, and pays the terminal value of the option. We consider the pricing to be fair and the hedging to be effective if the distribution of the hedging error (replication error) is as concentrated around zero as possible. However, we have to study separately the negative hedging errors (losses) and the positive hedging errors (gains).

13.3.1.1 Hedging Error
The hedging error e_T of the writer of the option is obtained from (13.10) as

$$e_T = C_0 + G_T(\xi) - C_T,$$

where

$$G_T(\xi) = \sum_{k=1}^T \xi_k(S_k - S_{k-1}).$$

Here the risk free rate is $r = 0$, C_0 is the price of the option, C_T is the terminal value of the option, $\xi = (\xi_k)_{k=1,\ldots,T}$ are the hedging coefficients, S_k are the stock prices, the current time is denoted by 0, the time to expiration is T days, and hedging is done daily.

13.3.1.2 Historical Simulation

We denote the time series of observed historical daily prices by $S_0, \ldots, S_N$. We construct $N - T$ sequences of prices:

$$S_i = (S_{i,i}, \ldots, S_{i,i+T}), \qquad i = 1, \ldots, N - T,$$

where

$$S_{i,i+j} = 100 \cdot S_{i+j}/S_i,$$

for $j = 0, \ldots, T$. Each sequence has length $T + 1$ and the initial price in each sequence is $S_{i,i} = 100$. We estimate the distribution of the hedging error e_T from the observations

$$e_T^1, \ldots, e_T^{N-T},$$

where e_T^i is computed from the prices S_i.

An example of a computation of hedging errors is given in (14.81), where Black-Scholes hedging is applied with sequential sample standard deviations as the volatility estimates. See also (14.80), where Black-Scholes hedging is applied with non-sequential sample standard deviations as the volatility estimates.

13.3.1.3 Comparison of the Error Distributions

We estimate the distribution of the hedging error e_T using data $e_T^1, \ldots, e_T^{N-T}$. A graphical summary of the error distribution is obtained by using tail plots and kernel density estimation, for example.

The Mean of Squared Hedging Errors In quadratic hedging the purpose is to minimize the mean squared hedging errors Ee_T^2; see Section 15.1 and Chapter 16. Thus it is natural to estimate the quality of a hedging strategy by the sample mean of squared hedging errors

$$\frac{1}{N-T} \sum_{i=1}^{N-T} \left(e_T^i\right)^2.$$

We can decompose the mean into the mean over negative hedging errors and over positive hedging errors:

$$\frac{1}{N-T} \left(\sum_{i \in I_-} \left(e_T^i\right)^2 + \sum_{i \in I_+} \left(e_T^i\right)^2 \right),$$

where

$$\mathcal{I}_- = \{i : e_T^i < 0\}, \qquad \mathcal{I}_+ = \{i : e_T^i \geq 0\}.$$

This composition is reasonable because the negative hedging errors are losses for the writer of the option and the positive hedging errors are gains for the writer of the option.

The Expected Utility of the Error Distribution   Even when the purpose of the hedging is typically to minimize the hedging error and not to maximize the wealth of the hedger, it is of interest to study the properties of the wealth distribution from the point of view of portfolio theory. This can be done by estimating the expected utility of the error distribution. We can use the exponential utility function $U(x) = 1 - e^{-\alpha x}$, where $\alpha > 0$ is the risk aversion parameter. Note that the hedging errors e_T can take any real value, and thus we cannot apply the power utility functions. The expected utility is estimated by

$$\frac{1}{N-T} \sum_{i=1}^{N-T} U\left(e_T^i\right).$$

See Section 9.2.2 for a discussion about expected utility.

13.3.2   The Wealth of the Buyer

We can ignore the hedging of the options and try to evaluate solely the fairness of the price. This approach can be considered to be the approach of the buyer of the option. We have at least the following possibilities.

1) *Comparison with the market prices.* The success of a pricing approach can be evaluated by testing whether the prices of the approach provide a good fit to the observed market prices of the derivatives.

    The comparison with the market prices is possible for liquid options. Note that a pricing method for illiquid options can be obtained by calibrating the parameters of the pricing method using liquid options. For example, the implied volatility of Black–Scholes pricing can be obtained from liquid options and then used as the volatility of the Black–Scholes formula to price illiquid options.

2) *Sharpe ratios.* We can estimate the Sharpe ratios of option strategies. The Sharpe ratios obtained by option buying should not be too far away from the Sharpe ratios of the underlying assets. This is illustrated in Section 17.2.3.

14

Pricing by Arbitrage

Pricing by arbitrage means that an asset is priced with the unique arbitrage-free price. Pricing by arbitrage can be applied in two different settings: (1) We can price by arbitrage linear securities, like forwards and futures, in any markets. (2) We can price by arbitrage nonlinear securities, like options, in complete markets. We discuss both of these cases in this chapter.

If two assets have the same terminal value with probability one, then the assets should have the same price. Otherwise, we could obtain a risk-free profit by selling the more expensive asset and by buying the cheaper asset. Linear assets, like futures, can be defined as a linear function of the underlying assets. Thus, they can be replicated, and their price is the initial value of the replicating portfolio. Nonlinear assets, like options, can be replicated only under restrictive assumptions on the markets (under the assumption of complete markets). However, the restrictive assumptions are often not too far away from the real properties of the markets.

The concepts of an arbitrage-free market and a complete market were studied in Chapter 13. The results of Chapter 13 imply that if the market is arbitrage-free and complete, then there is only one arbitrage-free price. In fact, Theorem 13.2 states that the arbitrage-free prices are obtained as expected values with respect to the equivalent martingale measures, and Theorem 13.3 (the second fundamental theorem of asset pricing) states that if the market is arbitrage-free and complete, then there is only one equivalent martingale measure.

Section 14.1 discusses pricing of futures, the put–call parity, and the American call options.

Section 14.2 studies binary models. In these models the price of a stock can at any time move only one step up, or one step down. The binary models are complete models, thus a derivative has a unique arbitrage-free price. These prices can be easily computed. The prices are called the Cox–Ross–Rubinstein prices, and they were introduced in Cox *et al.* (1979).

Section 14.2.3 studies asymptotics of multiperiod binary models, as the number of periods increases, and the length of the periods decreases. A multiperiod

Nonparametric Finance, First Edition. Jussi Klemelä.
© 2018 John Wiley & Sons, Inc. Published 2018 by John Wiley & Sons, Inc.

binary model converges to a log-normal model. In the log-normal model the stock price follows a geometric Brownian motion. The geometric Brownian motion is a market model that does not exactly describe the actual markets but it can be rather close to the actual markets. The Black–Scholes prices are obtained as the limits of the Cox–Ross–Rubinstein prices. The derivation of the Black–Scholes prices from the Cox–Ross–Rubinstein prices is both elegant and it leads to efficient numerical recipes, although there are many other derivations of the Black–Scholes price (see Section 14.3.3).

Section 14.3 studies the properties of the Black–Scholes prices. Section 14.4 studies Black–Scholes hedging. Section 14.5 studies combining the Black–Scholes hedging with time changing volatility estimates. This method of option pricing provides a benchmark against which we can evaluate competing pricing and hedging methods.

14.1 Futures and the Put–Call Parity

Futures can be priced by arbitrage regardless whether the market is complete or not. The payoff of a futures contract is a linear function of the underlying, and thus a futures contract can be replicated even in an incomplete market. The put–call parity provides a related example of pricing by arbitrage, which works whether the market is complete or not: the payoff of a put subtracted from the payoff of a call is a linear function of the underlying.

14.1.1 Futures

We consider pricing of stock futures, currency forwards, forward zero-coupon bonds, and forward rate agreements.

14.1.1.1 Stock Futures
A futures contract on a stock is made at time t_0. The contract specifies that the buyer of the contract has to buy the stock at a later time T with price K. We assume that the stock does not pay dividends during the time period from t_0 to T. Let us denote with S_t the price of the stock at time t. The value of the futures contract at the expiration time T is

$$F_T = S_T - K,$$

because the buyer of the futures contract gives away K and receives S_T.

The price of a zero-coupon bond is taken as $e^{-r(T-t)}$, where $t \in [t_0, T]$, r is the annualized risk-free rate, and $T - t$ is the time between t and T in fractions of a year. This is the method of continuous compounding.[1]

---

1 Assume that there are n payments of interest. Every payment increases the savings by the factor $1 + r(T - t)/n$. Then the total compounded savings at time T is $(1 + r(T - t)/n)^n P \to e^{r(T-t)} P$ as $n \to \infty$, where $P > 0$ is the initial savings.

What is the fair value F_t of the futures contract for $t_0 \le t < T$? We may replicate the futures contract buy buying the stock and borrowing the amount $e^{-r(T-t_0)}K$. At time $t \in [t_0, T]$ the value of this portfolio is

$$S_t - e^{-r(T-t)}K$$

with probability one. At time T the value of this portfolio is $F_T = S_T - K$ with probability one. Thus,

$$F_t = S_t - e^{-r(T-t)}K, \tag{14.1}$$

for $t \in [t_0, T]$, by the law of one price, to exclude arbitrage.

When an investor enters a futures contract in a futures exchange, this does not imply any cash flows, although the exchange requires from the investor a liquid collateral in order to secure a possible future payment. The fair forward price is called such value of K that makes the value F_t of the futures contract zero. To get $F_t = 0$ we need that

$$K = K_t = e^{r(T-t)}S_t. \tag{14.2}$$

The future prices that are quoted in a futures exchange are the values K_t (which are determined by the supply and demand). Numbers K_t are called futures prices or forward prices.

A buyer of a stock future or a stock index future saves the carrying costs but looses the possible stock dividends. When the annualized dividend rate is known to be d, then the fair future price is

$$K_t = e^{(r-d)(T-t)}S_t.$$

14.1.1.2 Currency Forwards

A currency forward is made at time t_0. We denote with e_t the exchange rate USD/euro at time t. Let $K > 0$ be an exchange rate. The contract stipulates that the buyer will buy V US dollars with euros at time T, using the USD/euro exchange rate K. The value of the contract for the buyer at time T is

$$F_T = V(e_T - K) \text{ euros.} \tag{14.3}$$

Indeed, the buyer uses KV euros to buy V dollars. Then the buyer exchanges the dollars to euros to obtain $e_T V$ euros. The profit of the buyer is $e_T V - KV$.

What is the price F_t of the currency forward for $t_0 \le t < T$? We can replicate the currency forward using zero-coupon bonds. Let $Z^f(t, T)$ be the USD price of the US zero-coupon bond and let $Z^d(t, T)$ be the euro price of an European zero-coupon bond. The zero-coupon bonds are such that $Z^f(T, T) = 1$ USD and $Z^d(T, T) = 1$ euro. Let us consider the portfolio with λ_1 units of US zero-coupon bonds and with λ_2 units of European zero-coupon bonds. The value of the portfolio at time t is

$$P_t = \lambda_1 Z^f(t, T) \text{ USD} + \lambda_2 Z^d(t, T) \text{ euro.}$$

We choose

$$\lambda_1 = V, \quad \lambda_2 = -KV.$$

The value of the portfolio a time T is

$$P_T = \lambda_1 \times 1 \text{ USD} + \lambda_2 \times 1 \text{ euro}$$
$$= (\lambda_1 e_T + \lambda_2) \text{ euro}$$
$$= V(e_T - K) \text{ euro.} \tag{14.4}$$

Since (14.3) and (14.4) are equal, the portfolio replicates the currency forward, and the price F_t of the currency forward is equal to P_t at time t.

The price is in euros

$$F_t = V\left(e_t Z^f(t, T) - K Z^d(t, T)\right) \text{ euro.}$$

To make $F_t = 0$ we need to choose the exchange rate

$$K = \frac{e_t Z^f(t, T)}{Z^d(t, T)}.$$

14.1.1.3 Forward Zero-Coupon Bonds

A time t_0 two parties make an agreement to exchange at a future time T_1 a zero-coupon bond whose maturity is at T_2, with a cash payment K. At time t_0 only the agreement is made, and at time T_1 the cash payment is made and the zero-coupon bond is received. This is called a forward zero-coupon bond. Forward zero-coupon bonds are considered in Section 18.2.1.

Let F_t be the price of the forward zero-coupon bond. Let $Z(t_0, T_1)$ and $Z(t_0, T_2)$ be the prices of zero-coupon bonds with maturities T_1 and T_2. Then,

$$F_{T_1} = Z(T_1, T_2) - K.$$

Consider a portfolio with one unit of $Z(t, T_2)$ and short of K units of $Z(t, T_1)$. The portfolio has value

$$V_t = Z(t, T_2) - K Z(t, T_1)$$

for $t \in [t_0, T_1]$. Thus, $V_{T_1} = F_{T_1}$ and we have replicated the forward zero-coupon bond. Thus, at time t_0 the prices are equal:

$$F_{t_0} = Z(t_0, T_2) - K Z(t_0, T_1).$$

To make $F_{t_0} = 0$, we need to choose the cash payment K as

$$K = P(t_0, T_1, T_2) = \frac{Z(t_0, T_2)}{Z(t_0, T_1)}. \tag{14.5}$$

14.1.1.4 Forward Rate Agreements

A forward rate agreement allows to change a floating Libor rate against a fixed rate. The contract is made at time t_0. The reset time of the Libor is $T_1 > t_0$.

At this time the Libor rate is settled to be $L(T_1, T_2)$, where $T_2 > T_1$ is the maturity of the Libor. The agreement stipulates that the buyer of the contract pays the payment with the fixed rate K, and receives the payment with the Libor rate $L(T_1, T_2)$. Thus, the payoff for the buyer at time T_2 is

$$P(L(T_1, T_2) - K)\tau, \tag{14.6}$$

where P is the notional, and τ is the time between T_1 and T_2. Forward rate agreements are considered in Section 18.2.2.

Let $Z(t, T_1)$ and $Z(t, T_2)$ be zero-coupon bonds. The forward rate agreement can be replicated by the trading strategy

$$P[V(t) - Z(t, T_2) - KZ(t, T_2)\tau],$$

where $V(t)$ is the strategy of buying $Z(t, T_1)$ and reinvesting the payoff at time T_1 with the prevailing τ-period Libor rate. Thus, the payoff of $V(t)$ at time T_2 is equal to $1 + L(T_1, T_2)\tau$. Thus, the trading strategy gives the payoff (14.6). The present value of the forward rate agreement is equal to

$$F_t = P[Z(t, T_1) - Z(t, T_2) - KZ(t, T_2)\tau]. \tag{14.7}$$

To make $F_t = 0$, we need to choose

$$K = \frac{1}{\tau}\left(1 - \frac{Z(t, T_1)}{Z(t, T_2)}\right).$$

14.1.1.5 Backwardation and Contango

Backwardation refers to the price relation where the spot price is higher than the forward price and contango refers to the case where the spot price is lower than the forward price:

$$\text{backwardation} \leftrightarrow K_t < S_t,$$
$$\text{contango} \leftrightarrow K_t > S_t.$$

At the expiration, the spot price and the futures price should be equal (to prevent arbitrage). Thus, the contango relationship means that distant delivery months trade at a greater price than near-term delivery months (the term structure is upward-sloping). Depending on the type of the underlying, either the contango or the backwardation is typical.

In the case of stock futures, or stock index futures, the theoretically fair price of a futures contract is greater than the spot price, according to (14.2). Thus, contango is typical for stock futures. However, futures prices K_t can differ from the theoretically fair price, and thus both contango and backwardation are possible. In the case of stock futures we can change the terminology in the following way:

$$\text{backwardation} \leftrightarrow K_t < e^{r(T-t)}S_t,$$
$$\text{contango} \leftrightarrow K_t > e^{r(T-t)}S_t.$$

According to this terminology we do not compare futures prices to the spot prices but to the theoretically fair prices.[2]

When the underlying is a commodity, then there are borrowing costs for the seller of the futures contract, similarly as in the case of stock futures. In addition, there are costs for storing the commodity, for the seller of the futures contract. Thus, contango is typical for futures on commodities, and contango is typically more profound than for stock futures.

The buyer of a treasury bond future saves carrying costs (determined by the short-term rate) but looses the interest received by the bond owners (determined by the long-term rate). If the short-term rate is significantly less than the long-term rate, then the distant delivery months may trade at a lesser price than the near-term delivery months: this is backwardation.

14.1.2 The Put–Call Parity

The price of a put can always be expressed in terms of the price of a call, and conversely. We have the put–call parity:

$$C_t - P_t = S_t - Ke^{-r(T-t)}, \tag{14.8}$$

where $K > 0$ is the common strike price of the call and put, r is the annualized interest rate, T is the expiration time, and $T - t$ is the time from t to T in fractions of a year. It is clear that at the expiration we have $C_T - P_T = S_T - K$. The put–call parity extends this relation for times t before the expiration time T. Thus, we do not need to know fair values for C_t and P_t in order to have an expression for their difference.[3]

14.1.2.1 A Derivation of the Put–Call Parity

Consider portfolio V_t obtained by buying the call and writing the put:

$$V_t = C_t - P_t.$$

At the expiration, we have $V_T = C_T - P_T = S_T - K$. Indeed,

$$C_T - P_T = \max\{0, S_T - K\} - \max\{0, K - S_T\}$$
$$= \begin{cases} S_T - K - 0, & \text{when } S_T \geq K, \\ 0 - (K - S_T), & \text{when } S_T < K. \end{cases}$$

Let F_t be the forward contract to buy stock at time T with price K. This forward contract was valued in (14.1), where we showed that

$$F_t = S_t - Ke^{-r(T-t)}.$$

---

2 Exercise: Discuss what happens to a rolling strategy when the prices are in contango. What problems are induced to the indexes whose underlying is a futures contract?
3 We can derive the Black–Scholes prices for the calls and puts using the put–call parity. This derivation is described in Section 14.3.3.

The replication was obtained by buying the stock and borrowing the amount $Ke^{-r(T-t)}$. At the expiration, we have $F_T = S_T - K$. Since $V_T = F_T$ with probability one, we have

$$V_t = F_t$$

for all times t before T, to exclude arbitrage. This is the statement (14.8) that we wanted to prove.

14.1.2.2 Bounds for the Call Price
We have bounds for the call price C_t:

$$\max\left\{S_t - e^{-r(T-t)}K, 0\right\} \le C_t \le S_t. \tag{14.9}$$

The upper bound $C_t \le S_t$ is obvious, since the right to buy a stock must be less valuable than the stock itself. To prove the lower bound, we can note that $C_t \ge 0$ is obvious, since the right to buy a stock involves no obligations.[4] The put–call parity and the fact that $P_t \ge 0$ gives

$$S_t - Ke^{-r(T-t)} = C_t - P_t \le C_t.$$

We have proved the lower bound.[5]

The lower bound in (14.9) implies that

$$\max\{0, S_t - K\} < C_t, \tag{14.10}$$

for $t < T$. This follows because $e^{-r(T-t)} < 1$ when $t < T$. The inequality (14.10) leads to the definition of the time value of the option; see (2.6). The lower bound says that the value of the option is larger than the intrinsic value $\max\{S_t - K, 0\}$. The difference of the price of the option and its intrinsic value is called the time value of the European option. For an European put option the lower bound fails unless $r = 0$. As a consequence of the put–call parity, the time-value of a put option whose intrinsic value is large (the option is in the money), is usually negative.

14.1.3 American Call Options

An American call option has the same price as the corresponding European call option, when the stock does not pay dividends. On the other hand, an

---

4 The claims follow from the monotonicity theorem stated in Section 13.1.2. Since $C_T = (S_T - K)_+ \le S_T$, we have $C_t \le S_t$. Since $C_T = (S_T - K)_+ \ge 0$, we have $C_t \ge 0$.
5 The lower bound can be deduced also by using Theorem 13.2, where the prices were expressed as expectations with respect to the risk-neutral measure. Let the numéraire be the riskless bond $B_u = e^{r(u-t)}$, where $t \le u \le T$. By the convexity of $x \mapsto (x)_+ = \max\{0, x\}$, the arbitrage-free price of an European call option is given for any risk-neutral measure $Q \in \mathcal{P}$ by

$$C_t = E_Q\left[\frac{(S_T - K)_+}{B_T}\right] \ge \left(E_Q\left[\frac{S_T - K}{B_T}\right]\right)_+ = \left(S_t - \frac{K}{B_T}\right)_+,$$

since $E_Q(S_T/B_T) = S_t$ for the risk-neutral measure Q.

American put option has a different price than the corresponding European put option.

The put–call parity was derived for European options. However, this parity can be used to show that

$$C_t^A = C_t^E, \tag{14.11}$$

when the stock does not pay dividends, where C_t^A is the price of an American call option and C_t^E is the price of the corresponding European call option. We know that $C_t^A \geq C_t^E$, because an American option has more rights than the corresponding European option. The lower bound in (14.9) implies

$$S_t - K \leq S_t - Ke^{-r(T-t)} \leq C_t^E \leq C_t^A,$$

where $S_t - K$ is the cash flow generated by the exercise of the American call option. Since $C_t^A \geq S_t - K$, an early exercise is always suboptimal and one should sell the American call option and not exercise it. Since it is not optimal to exercise the option, the possibility for the early exercise is not worth of anything, and the American call option has the same value as the European call option.[6]

However, an American put option is in general worth more than the corresponding European put option. For the American options we have

$$C_t^A - P_t^A < S_t - Ke^{-r(T-t)}.$$

The difference between the put and call options comes from the fact that the value of a put option increases as the price of the stock decreases. When the value of the stock decreases, then the absolute price changes become smaller. We can reach the point where the stock takes so small values that further decreases in the stock price would not give a better rate of return to the put option than the risk-free rate. At that point it is better to exercise the option and invest in the risk-free rate. With calls the situation reverses because as the stock price increases, absolute price changes increase. Pricing of American put options in the multiperiod binary model is discussed in Section 14.2.4.

14.2 Pricing in Binary Models

We consider pricing and hedging of derivatives in binary models. In a binary model the stock can at any given time go only one step higher or one step lower. Section 14.2.1 studies one-period binary models. Section 14.2.2 uses

---

6 We can use an arbitrage argument to show that $C_t^A \leq C_t^E$. Assume that $C_t^A > C_t^E$. Then buy the European option and sell the American option. We receive $C_t^A - C_t^E$, which can be put in the bank account. If the owner of the American option does not exercise the option, we receive $e^{r(T-t)}(C_t^A - C_t^E) > 0$ at the maturity, because $C_T^A = C_T^E$. If the owner of the American option exercises the option at time t_1, then we sell the European option and receive $C_{t_1}^E - (S_{t_1} - K) \geq 0$. In this case we receive at the maturity $e^{r(T-t)}(C_t^A - C_t^E) + e^{r(T-t)}(C_{t_1}^E - (S_{t_1} - K)) > 0$.

one-period binary models as building blocks for multiperiod binary models. A multiperiod binary model approximates the Black–Scholes model, as is shown in Section 14.2.3.

14.2.1 The One-Period Binary Model

First, we describe the one-period binary model. Second, we find the fair price and the optimal hedging coefficient. Third, we find the equivalent martingale measure. Fourth, we provide additional pricing and hedging formulas.

14.2.1.1 A Description of the One-Period Binary Model

In the single period model, there are time points $t = 0$ and $t = 1$. The market consist of stock S, bond B, and contingent claim C.

The bond takes value 1 at $t = 0$ and value $1 + r$ at $t = 1$:

$$B_0 = 1, \quad B_1 = 1 + r,$$

where $r > -1$ is the risk-free rate.

The initial value of the stock is s_0. The stock can take only two values at time 1. At $t = 1$ the value is $s_{1,0}$ with probability $1 - p$ and the value is $s_{1,1}$ with probability p:

$$P(S_0 = s_0) = 1, \quad P(S_1 = s_{1,0}) = 1 - p, \quad P(S_1 = s_{1,1}) = p,$$

where $0 < p < 1$. We assume that

$$s_{1,0} < s_0(1 + r) < s_{1,1}. \tag{14.12}$$

Note that if $s_{1,0} \geq s_0(1 + r)$, then it would not be rational to invest in the bond, and if $s_{1,1} \leq s_0(1 + r)$, then it would not be rational to invest in the stock. The "rationality" can be formalized as the absence of arbitrage.

The contingent claim C takes two possible values $c_{1,0}$ and $c_{1,1}$ at $t = 1$:

$$P(C_1 = c_{1,0}) = 1 - p, \quad P(C_1 = c_{1,1}) = p.$$

For example, in the case of a call option we have $C_1 = (S_1 - K)_+$ for some strike price $K > 0$. Then $c_{1,0} = (s_{1,0} - K)_+$ and $c_{1,1} = (s_{1,1} - K)_+$.

We want to find a fair value $C_0 \in \mathbf{R}$ for the derivative at $t = 0$. Also, we want to find the optimal hedging coefficient $\xi \in \mathbf{R}$, which is used to hedge the position of the writer of the option.

14.2.1.2 Pricing and Hedging in the One-Period Binary Model

We replicate the contingent claim C with a portfolio

$$W = \beta B + \xi S.$$

The portfolio consists of β units of the bond and ξ units of the stock. The portfolio takes values

$$W_0 = \beta + \xi s_0,$$
$$W_1 = \beta(1 + r) + \xi S_1.$$

To obtain $P(W_1 = C_1)$ we need to choose β and ξ so that

$$\beta(1 + r) + \xi s_{1,0} = c_{1,0},$$
$$\beta(1 + r) + \xi s_{1,1} = c_{1,1}.$$

The first equation leads to

$$\beta = (1 + r)^{-1}(c_{1,0} - \xi s_{1,0}). \tag{14.13}$$

Inserting this value of β to the second equation gives

$$\xi = \frac{c_{1,1} - c_{1,0}}{s_{1,1} - s_{1,0}}. \tag{14.14}$$

The law of one price implies that the arbitrage-free price C_0 of the contingent claim equals the value W_0 of the replicating portfolio at time $t = 0$: $C_0 = W_0$, and thus

$$C_0 = \beta + \xi s_0, \tag{14.15}$$

where β and ξ are defined in (14.13) and (14.14).

The number β of bonds can be written as $\beta = C_0 - \xi s_0$. Since $W = \beta B + \xi S$, we can write

$$P(C_0 + \xi(S_1 - S_0) = C_1) = 1. \tag{14.16}$$

We can interpret the replicating portfolio from the point of view of the writer of the option as the portfolio where the writer receives the option premium C_0, invests C_0 in the bank account, borrows the amount ξs_0 from the bank, and invests ξs_0 in the stock.

14.2.1.3 The Equivalent Martingale Measure

Price C_0 can be written as the expectation with respect to the equivalent martingale measure. We have

$$
\begin{aligned}
C_0 &= (1 + r)^{-1}\left[c_{1,0} - \xi s_{1,0} + \xi(1 + r)s_0\right] \\
&= (1 + r)^{-1}\left[\frac{s_{1,1} - s_0(1 + r)}{s_{1,1} - s_{1,0}}c_{1,0} + \frac{s_0(1 + r) - s_{1,0}}{s_{1,1} - s_{1,0}}c_{1,1}\right] \\
&= (1 + r)^{-1}\left[(1 - q)c_{1,0} + qc_{1,1}\right] \\
&= (1 + r)^{-1}E_Q C_1, \tag{14.17}
\end{aligned}
$$

where

$$q = \frac{s_0(1 + r) - s_{1,0}}{s_{1,1} - s_{1,0}}, \tag{14.18}$$

and E_Q means the expectation with respect to the probability measure Q with

$$Q(S_1 = s_{1,0}) = 1 - q, \quad Q(S_1 = s_{1,1}) = q.$$

Probability measure Q is called a risk-neutral measure because

$$E_Q S_1 = s_0(1 + r),$$

and it is called a martingale measure because

$$E_Q \left(\frac{S_1}{1+r} \right) = s_0.$$

Note that condition (14.12) guarantees that $0 < q < 1$, so that Q is equivalent to P.

14.2.1.4 Further Pricing and Hedging Formulas

It is of interest that the price can be written as

$$C_0 = (1+r)^{-1} \left[EC_1 - \xi E(S_1 - (1+r)s_0) \right]$$
$$= EH_1 - \xi E\Delta X_1, \tag{14.19}$$

where

$$H_1 = \frac{C_1}{1+r}, \quad \Delta X_1 = \frac{S_1}{1+r} - s_0.$$

Indeed, similarly to (14.13), we get

$$\beta = (1+r)^{-1}(c_{1,1} - \xi s_{1,1}),$$

which can be combined with (14.13) to get

$$\xi = p\beta + (1-p)\beta = (1+r)^{-1} \left[EC_1 - \xi ES_1 \right].$$

Combining this formula for β with the formula $C_0 = \beta + \xi s_0$ shows (14.19). In (14.19), we have written the derivative price as a discounted expectation of the derivative price, with an additional correction term.

The hedging coefficient can be written as

$$\xi = \frac{\text{Cov}(C_1, S_1)}{\text{Var}(S_1)}. \tag{14.20}$$

To derive (14.20), note that

$$\text{Cov}(C_1, S_1) = p(1-p)(s_{1,1} - s_{1,0})(c_{1,1} - c_{1,0}),$$
$$\text{Var}(S_1) = p(1-p)(s_{1,1} - s_{1,0})^2.$$

Thus, (14.20) is equal to (14.14). We can also write

$$\xi = \frac{\text{Cov}(H_1, \Delta X_1)}{\text{Var}(\Delta X_1)} = \frac{\text{Cov}(C_1, R_1)}{s_0 \text{Var}(R_1)}, \tag{14.21}$$

where $R_1 = S_1/s_0$ is the gross return. This way of writing the hedging coefficient appears in (16.10), where quadratic hedging is considered.

14.2.2 The Multiperiod Binary Model

We start with a description of the multiperiod binary model, and then proceed to the pricing formulas, derive the equivalent martingale measure, and give hedging formulas.

14.2.2.1 A Description of the Multiperiod Binary Model

The market consists of stock S, bond B, and contingent claim C. We define the discrete time processes

$$S = (S_k)_{k=0,\ldots,n}, \quad B = (B_k)_{k=0,\ldots,n}. \tag{14.22}$$

Note that in Section 13.2 we denoted the time steps of the discrete time markets by $t = 0, \ldots, T$. We have changed the notation, because we construct a time series that approximates the geometric Brownian motion on $[0, T]$. The approximation is done by dividing interval $[0, T]$ to n periods of equal lengths and letting $n \to \infty$.

The bond takes value $B_k = (1 + r\Delta t)^k$ at step k, where $r > -1$ is the annualized interest rate, and Δt is the time between two periods in fractions of a year.

At step $k = 0, \ldots, n$, the stock can take $k + 1$ values

$$s_{k,j} = u^j d^{k-j} s_0, \quad j = 0, \ldots, k, \tag{14.23}$$

where

$$0 < d < 1 + r\Delta t < u.$$

The stock price S_k is a random variable with

$$P(S_k = s_{k,j}) = \binom{k}{j} p^j (1 - p)^{k-j}, \quad j = 0, \ldots, k, \tag{14.24}$$

where $0 < p < 1$,

The stochastic process of stock prices can be described by a recombining binary tree, as in Figure 14.1, where $n = 3$. At step $k = 0$, the stock takes value

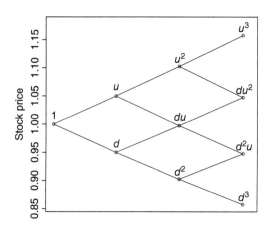

Figure 14.1 *A recombining binary tree. There are $n = 3$ periods and the initial value is $s_0 = 1$.*

s_0 (in the figure $s_0 = 1$). If the value of the stock at time $k - 1$ is s, then at step k the stock can take values ds and us, so that

$$s_{k,j} = ds_{k-1,j}, \quad s_{k,j+1} = us_{k-1,j},$$

The probabilities of the up and down movements are p and $1 - p$:

$$P(S_k = su \mid S_{k-1} = s) = p, \quad P(S_k = sd \mid S_{k-1} = s) = 1 - p.$$

The derivative can take at the expiration $n + 1$ values $c_{n,j}$, $j = 0, \ldots, n$. The random variable C_n takes the value $C_n = c_{n,j}$, when $S_n = s_{n,j}$. For example, when the contingent claim is the call option with $C_n = \max\{0, S_n - K\}$, where $K > 0$ is the strike price, then $c_{n,j} = \max\{0, s_{n,j} - K\}$. We have

$$P(C_n = c_{n,j}) = \binom{n}{j} p^j (1 - p)^{n-j}, \quad j = 0, \ldots, n. \tag{14.25}$$

We want to find the arbitrage-free price C_0 of the derivative at time $k = 0$.

14.2.2.2 Pricing in the Multiperiod Binary Model

The evolution of the stock price in a multiperiod binary model has been described using a recombining binary tree. The price in a multiperiod binary model can be found by backward induction. We know the price at the expiration, when $k = n$. We can use the single period model to calculate the price at step $k = n - 1$, and go backwards step by step to obtain the price at step $k = 0$. The price C_0 of the derivative is calculated using the backwards induction with the following steps.

1) At the expiration step $k = n$, the prices of the derivative are given by $c_{n,j}$, $j = 0, \ldots, n$.
2) Let the current step be $k - 1$ and the current node $s_{k-1,j}$. Then the two possible prices for the derivative are $c_{k,j}$ and $c_{k,j+1}$. We can use the single period model to calculate the price at step $k - 1$. We get the price from (14.17) as

$$c_{k-1,j} = (1 + r\Delta t)^{-1} \left[(1 - q)c_{k,j} + q c_{k,j+1} \right], \tag{14.26}$$

where

$$q = \frac{s_{k-1,j}(1 + r\Delta t) - s_{k,j}}{s_{k,j+1} - s_{k,j}} = \frac{(1 + r\Delta t) - d}{u - d},$$

and the second equality follows from $s_{k,j+1} = s_{k-1,j} u$ and $s_{k,j} = s_{k-1,j} d$.

We have described a recursive algorithm for the computation of the price, but we can also obtain the following explicit expression for the price.

Theorem 14.1 *The arbitrage-free price of the derivative at step $k = 0$ is*

$$C_0 = (1 + r\Delta t)^{-n} \sum_{j=0}^{n} \binom{n}{j} q^j (1 - q)^{n-j} c_{n,j}. \tag{14.27}$$

Proof. Let us make the induction hypothesis that the formula (14.27) holds for binary trees with $n-1$ levels. Then we can price the derivative at step $k = 1$. The price is different depending on whether we are at node $s_{1,0}$ or at node $s_{1,1}$:

$$c_{1,0} = (1 + r\Delta t)^{-n+1} \sum_{j=0}^{n-1} \binom{n-1}{j} q^j(1-q)^{n-j-1} c_{n,j},$$

$$c_{1,1} = (1 + r\Delta t)^{-n+1} \sum_{j=0}^{n-1} \binom{n-1}{j} q^j(1-q)^{n-j-1} c_{n,j+1}.$$

Applying the one-step binomial model we get the price at step $k = 0$:

$$C_0 = (1 + r\Delta t)^{-1} \left[(1-q)c_{1,0} + qc_{1,1} \right]$$

$$= (1 + r\Delta t)^{-n} \left[\binom{n-1}{0} q^0(1-q)^n c_{n,0} \right.$$

$$+ \sum_{j=1}^{n-1} \left[\binom{n-1}{j} + \binom{n-1}{j-1} \right] q^j(1-q)^{n-j} c_{n,j}$$

$$+ \left. \binom{n-1}{n-1} q^n(1-q)^0 c_{n,n} \right]$$

$$= (1 + r\Delta t)^{-n} \sum_{j=0}^{n} \binom{n}{j} q^j(1-q)^{n-j} c_{n,j},$$

where we used the fact

$$\binom{n-1}{j} + \binom{n-1}{j-1} = \binom{n}{j}.$$

We have proved Theorem 14.1. □

The arbitrage-free price of the derivative is obtained not only at step $k = 0$, but the price is obtained at all steps $k = 0, \dots, n-1$. In fact, the price of the derivative, under the condition that the stock price at step k is $S_k = s_{k,j}$, is given by the formula

$$c_{k,j} = (1 + r\Delta t)^{-(n-k)} \sum_{i=0}^{n-k} \binom{n-k}{i} q^i(1-q)^{n-k-i} c_{n,j+i}. \tag{14.28}$$

14.2.2.3 The Equivalent Martingale Measure
Let us define probability measure Q by

$$Q(S_k = s_{k,j}) = \binom{k}{j} q^j(1-q)^{k-j}, \quad j = 0, \dots, k,$$

where $k = 0, \dots, n$. Measure Q is obtained from the physical measure P, defined in (14.24), by replacing the probability p of an up-movement by probability q. The price in (14.27) can be written as the expectation

$$C_0 = (1 + r\Delta t)^{-n} E_Q C_n, \tag{14.29}$$

where C_n is the random variable taking value $c_{n,j}$, when S_n takes value $s_{n,j}$. The measure Q is called the equivalent martingale measure, or the risk-neutral measure.

The price $c_{k,j}$ of the derivative given in (14.28), under the condition that the stock price at step k is $S_k = s_{k,j}$, can be written as

$$c_{k,j} = (1 + r\Delta t)^{-(n-k)} E_Q(C_n \mid S_k = s_{k,j}) \tag{14.30}$$

Accordingly, we can define random variables

$$C_k = (1 + r\Delta t)^{-(n-k)} E_Q(C_n \mid S_k).$$

Defining

$$H_k = \frac{C_k}{(1 + r\Delta t)^k}.$$

we obtain a more elegant formula

$$H_k = E_Q(H_n \mid S_k). \tag{14.31}$$

14.2.2.4 Hedging in the Multiperiod Binary Model

The following theorem gives the hedging coefficients of the replicating portfolio. The replication means that the wealth process of the trading strategy obtains the value of the derivative with probability one, or equivalently, the value process of the trading strategy obtains the discounted value of the derivative with probability one. The value process of a self-financing trading strategy is given in (13.8).

Theorem 14.2 *Define the hedging coefficient for step $k = 0, \dots, n-1$ by*

$$\xi_{k+1} = \frac{c_{k+1,j+1} - c_{k+1,j}}{s_{k+1,j+1} - s_{k+1,j}}, \quad \text{when } S_k = s_{k,j}, \tag{14.32}$$

where $j = 0, \dots, k$, and $c_{k+1,j}$ and $c_{k+1,j+1}$ are the arbitrage-free prices of the derivative at step $k + 1$, given by (14.30). Then the derivative C_n is replicated: the hedging coefficients (14.32) satisfy

$$P\left(H_n = H_0 + \sum_{k=0}^{n-1} \xi_{k+1}(X_{k+1} - X_k) \right) = 1,$$

where $H_n = C_n/(1 + r\Delta t)^n$ and $X_k = S_k/(1 + r\Delta t)^k$.

Proof. Let us make the induction hypothesis that the claim holds for the multiperiod binary models with $n-1$ periods. According to the induction hypothesis

$$P\left(H_n = H_1 + \sum_{k=1}^{n-1} \xi_{k+1}(X_{k+1} - X_k) \,\bigg|\, S_1 = s_{1,0}\right) = 1,$$

$$P\left(H_n = H_1 + \sum_{k=1}^{n-1} \xi_{k+1}(X_{k+1} - X_k) \,\bigg|\, S_1 = s_{1,1}\right) = 1,$$

where H_k are defined by (14.31). We can apply the replication (14.16) in the one-period model to obtain

$$P(H_1 = H_0 + \xi_1(X_1 - X_0)) = 1,$$

where

$$\xi_1 = \frac{c_{1,1} - c_{1,0}}{s_{1,1} - s_{1,0}},$$

as given in (14.14). □

There are other ways to write the hedging coefficient. We have from the formula (14.21) of the one-period model that

$$\xi_{k+1} = \frac{\mathrm{Cov}_k(H_{k+1}, \Delta X_{k+1})}{\mathrm{Var}_k(\Delta X_{k+1})},$$

where $\Delta X_{k+1} = X_{k+1} - X_k$, $X_k = S_k/(1 + r\Delta t)^k$, and $H_k = C_k/(1 + r\Delta t)^k$. Here Var_k and Cov_k mean conditional variance and covariance, conditional on S_k, and under probability measure Q, which is the equivalent martingale measure.

We can also write the hedging coefficient as

$$\xi_{k+1} = \frac{\mathrm{Cov}_k(H_n, \Delta X_{k+1})}{\mathrm{Var}_k(\Delta X_{k+1})}. \tag{14.33}$$

Let us prove (14.33). We have from (14.31) that

$$H_{k+1} = E_{k+1}H_n.$$

Thus,

$$\mathrm{Cov}_k(H_{k+1}, \Delta X_{k+1}) = \mathrm{Cov}_k(E_{k+1}(H_n), \Delta X_{k+1})$$
$$= \mathrm{Cov}_k(H_n, \Delta X_{k+1}).$$

We have shown (14.33).

An additional formula for the hedging coefficient is

$$\xi_{k+1} = \frac{\mathrm{Cov}_k(H_n, X_n)}{(n-k)\mathrm{Var}_k(\Delta X_{k+1})}. \tag{14.34}$$

In fact, $X_n = X_k + \sum_{i=k+1}^{n}(X_i - X_{i-1})$. Thus,

$$\text{Cov}_k(H_n, X_n) = \sum_{i=k+1}^{n} \text{Cov}_k(H_n, \Delta X_i)$$

$$= \sum_{i=k+1}^{n} \text{Cov}_k(H_i, \Delta X_i)$$

$$= (n - k)\, \text{Cov}_k(H_{k+1}, \Delta X_{k+1}).$$

14.2.3 Asymptotics of the Multiperiod Binary Model

We start with the asymptotic normality of the logarithmic returns in the multiperiod binary model, then show the convergence of the arbitrage-free prices in the multiperiod binary model to the Black–Scholes prices, and finally show the convergence of the hedging coefficients in the multiperiod binary model to the Black–Scholes hedging coefficients.

14.2.3.1 Choice of the Parameters
Let

$$\Delta t = \frac{T}{n}.$$

We consider asymptotics when $n \to \infty$. We choose the up and down factors as

$$u = 1 + \sigma\sqrt{\Delta t}, \quad d = 2 - u = 1 - \sigma\sqrt{\Delta t}, \tag{14.35}$$

where $\sigma > 0$. We choose the probabilities of the up and down movements as

$$p = \frac{1}{2} + \frac{\mu_0\sqrt{\Delta t}}{2\sigma}, \quad 1 - p = \frac{1}{2} - \frac{\mu_0\sqrt{\Delta t}}{2\sigma},$$

where $\mu_0 \in \mathbf{R}$. With these choices the logarithmic return $\log(S_n/s_0)$ converges in distribution to the normal distribution $N(\mu T, \sigma^2 T)$, where $\mu = \mu_0 - \sigma^2/2$. Note that when we choose d and u as in (14.35), then the probability q of the up movement in the risk-neutral distribution is

$$q = \frac{1}{2} + \frac{r\sqrt{\Delta t}}{2\sigma}. \tag{14.36}$$

14.2.3.2 Asymptotic Normality in the Multiperiod Binary Model
We show that the distribution of the stock price in the multiperiod binary model converges in distribution to a log-normal distribution, as the number of steps increases.

In the n-period binary model the stock price S_k, $k = 0, \dots, n$, can be written as

$$S_k = s_0 \prod_{i=1}^{k}\left(1 + w_i\sigma\sqrt{\Delta t}\,\right), \tag{14.37}$$

where $w_1, w_2, \ldots$ are such i.i.d. random variables that $w_i = 1$ when S_i is a result of an up-movement and $w_i = -1$ when S_i is a result of a down-movement, so that

$$P(w_i = 1) = \frac{1}{2} + \frac{\mu_0 \sqrt{\Delta t}}{2\sigma}, \quad P(w_i = -1) = \frac{1}{2} - \frac{\mu_0 \sqrt{\Delta t}}{2\sigma}.$$

It holds that

$$\log_e \left(\frac{S_n}{S_0} \right) \xrightarrow{d} N(\mu T, \sigma^2 T), \tag{14.38}$$

as $n \to \infty$, where

$$\mu = \mu_0 - \frac{1}{2} \sigma^2.$$

We show a slightly more general result: For $0 < t \le T$

$$\log_e \left(\frac{S_{k_n}}{S_0} \right) \xrightarrow{d} N(\mu t, \sigma^2 t), \tag{14.39}$$

as $n \to \infty$, where k_n is such that $k_n \Delta t \to t$, as $n \to \infty$. In particular, we can choose $k_n = [t/\Delta t]$. Now (14.38) follows as a special case when we choose $k_n = n$ and $t = T$.

We denote below $k = k_n$. Let us denote $X_i = w_i \sigma \sqrt{\Delta t}$. We can write

$$\log_e \left(\frac{S_k}{S_0} \right) = \sum_{i=1}^{k} \log_e(1 + X_i) = Q_n - \frac{1}{2} R_n + S_n,$$

where

$$Q_n = \sum_{i=1}^{k} X_i, \quad R_n = \sum_{i=1}^{k} X_i^2, \quad S_n = \sum_{i=1}^{k} X_i^2 r(X_i),$$

with $r(x) = x^{-2}[\log(1 + x) - x - x^2/2]$. Now it holds that $\lim_{x \to 0} r(x) = 0$. We have that

1) $Q_n \xrightarrow{d} N(t\mu_0, t\sigma^2)$,
2) $R_n \xrightarrow{p} t\sigma^2$,
3) $S_n \xrightarrow{p} 0$,

as $n \to \infty$. Because $\mu_0 = \mu + \sigma^2/2$, the claim (14.39) follows from items 1–3.
To prove item 1, we note that

$$Ew_i = \frac{\mu_0 \sqrt{\Delta t}}{\sigma}, \quad Ew_i^2 = 1, \quad \text{Var}(w_i) = 1 - \left(\frac{\mu_0 \sqrt{\Delta t}}{\sigma} \right)^2.$$

Thus,

$$EX_i = \mu_0 \Delta t, \quad \text{Var}(X_i) = \sigma^2 \Delta t \left[1 - \left(\frac{\mu_0 \Delta t}{\sigma} \right)^2 \right].$$

By the choice of $k = k_n$, it holds $k_n \Delta t \to t$ as $n \to \infty$, and thus

$$kEX_i \to t\mu_0, \quad k\text{Var}(X_i) \to t\sigma^2,$$

as $n \to \infty$. Thus, item 1 follows by the central limit theorem.

To prove item 2, we use the fact that $EX_i^2 = \sigma^2 \Delta t$, which implies that $kEX_i^2 \to t\sigma^2$ as $n \to \infty$. Thus, the weak law of the large numbers implies item 2.

To prove item 3, we note that

$$|S_n| \le R_n \max_{i=1,\dots,k} |r(X_i)| \le R_n \max\{r(\sigma\Delta t), r(-\sigma\Delta t)\}.$$

Item 2 implies that $R_n = O_p(1)$ and since $\lim_{x\to 0} r(x) = 0$, we have that $\max\{r(\sigma\Delta t), r(-\sigma\Delta t)\} = o(1)$, so that $S_n = o_p(1)$.

14.2.3.3 Convergence of the Price

The arbitrage-free prices (14.27) in the multiperiod binomial model are called the Cox–Ross–Rubinstein prices. We show that the Cox–Ross–Rubinstein put and call prices converge to the Black–Scholes put and call prices, as $n \to \infty$. This is done in two steps. First, we show that the put and call prices in the multiperiod binary model converge to the expected values of the option pay-offs, when the expectation is taken with respect to a log-normal distribution. Second, we calculate closed-form expressions for the expected values.

Bounded Continuous Payoff Functions A fundamental theorem about weak convergence states that if

1) $X_n \xrightarrow{d} X$,
2) $f : \mathbf{R} \to \mathbf{R}$ is bounded and continuous,

then

$$Ef(X_n) \to Ef(X),$$

as $n \to \infty$; see Billingsley (2005, Theorem 25.8, p. 335).

1) First, we apply the weak convergence in (14.38) to obtain

$$S_n \xrightarrow{d} s_0 \exp\left\{(r - \sigma^2/2)T + \sigma\sqrt{T}\, Z\right\}, \tag{14.40}$$

where the stock price S_n is a random variable taking values $s_{n,j}$ with probabilities

$$P(S_n = s_{n,j}) = \binom{n}{j} q^j (1-q)^j, \quad j = 0, \dots, n,$$

and $Z \sim N(0, 1)$.

2) Second, we have noted in (14.29) that a Cox–Ross–Rubinstein price satisfies

$$C_0 = (1 + r\Delta t)^{-n} E_Q C_n, \tag{14.41}$$

where the expectation is with respect to the risk-neutral measure Q, and C_n is a random variable taking values $c_{k,j}$, defined by (14.25).

Let $f : \mathbf{R} \to \mathbf{R}$ be such function that $f(s_{k,j}) = c_{k,j}$. Then the option payoff can be written as

$$C_n = f(S_n).$$

If function f is continuous and bounded, then (14.40) and (14.41) imply that

$$\lim_{n \to \infty} C_0 = e^{-rT} E f(S_T), \tag{14.42}$$

where the distribution of S_T is defined by

$$\log_e \left(\frac{S_T}{S_0} \right) \sim N \left(T \left(r - \frac{\sigma^2}{2} \right), T\sigma^2 \right), \tag{14.43}$$

and we used the fact that $(1 + r\Delta t)^{-n} \to e^{-rT}$, because $\Delta t = T/n$.

Unbounded Continuous Payoff Functions The fundamental theorem about weak convergence applies for any sequence converging weakly. However, in our case we are interested in the special case of the convergence of a binomial distribution toward a log-normal distribution. In this special case Föllmer and Schied (2002, Proposition 5.39, p. 265) notes that the convergence of expectations can be proved also when we relax the condition of the boundedness. Let $f : (0, \infty) \to \mathbf{R}$ be measurable, almost everywhere continuous, and

$$|f(x)| \leq C(1 + x)^q \text{ for some } C \geq 0 \text{ and } 0 \leq q < 2. \tag{14.44}$$

Then,

$$E f(S_n) \to E f(S_T)$$

as $n \to \infty$, where S_n satisfies (14.40) and S_T is distributed as (14.43).

Put Prices The payoff function of a put option is

$$x \mapsto (K - x)_+,$$

where $K > 0$ is the strike price. Function $x \mapsto (K - x)_+$, $x \geq 0$, is bounded and continuous. Thus, the Cox–Ross–Rubinstein price P_0 of an European put option converges to the expectation:

$$\lim_{n \to \infty} P_0 = e^{-rT} E(K - S_T)_+. \tag{14.45}$$

The right hand side is equal to the Black–Scholes put price, as shown in (14.54).

Call Prices The payoff function of a call option is

$$x \mapsto (x - K)_+,$$

where $K > 0$ is the strike price. Function $x \mapsto (x - K)_+$, $x \geq 0$, is continuous but not bounded. Thus, the convergence of the Cox–Ross–Rubinstein price to the expected value cannot be inferred similarly as in the case of put options.

However, we can apply the put-call parity in (14.8) to conclude that the Cox–Ross–Rubinstein prices have to satisfy

$$C_0 - P_0 = s_0 - K(1 - r\Delta t)^{-n},$$

where C_0 is the Cox–Ross–Rubinstein call price, and P_0 the put price. Since we have shown (14.45), it holds that

$$\lim_{n \to \infty} C_0 = e^{-rT} E(K - S_T)_+ + s_0 - Ke^{-rT} = e^{-rT} E(S_T - K)_+. \tag{14.46}$$

The right hand side is equal to the Black–Scholes call price, as shown in (14.52).

Calculation of the Expectations We have proved in (14.45) and (14.46) that the arbitrage-free put and call prices in the multiperiod binary model approach the expected values of the option payoffs, when the expectations are with respect to the equivalent martingale measure (the risk neutral log-normal model). Thus, we want to calculate the expected values of the put and call payoffs. The expected values are the Black–Scholes prices, which we denote

$$C_0^{bs} = e^{-rT} E(S_T - K)_+ \tag{14.47}$$

and

$$P_0^{bs} = e^{-rT} E(K - S_T)_+, \tag{14.48}$$

where the expectations are taken with respect to the distribution of S_T defined by

$$S_T = s_0 \exp\left\{ \mu T + \sigma\sqrt{T}\, Z \right\}, \tag{14.49}$$

where $Z \sim N(0, 1)$ and

$$\mu = r - \frac{1}{2}\sigma^2.$$

Note that we have discussed the log-normal distribution in (3.50).

The density of the standard normal distribution is $\phi(z) = (2\pi)^{-1/2} e^{-z^2/2}$, where $z \in \mathbf{R}$. Then,

$$E(S_T - K)_+ = \int_w^\infty \left(s_0 \exp\{z\sigma\sqrt{T} + \mu T\} - K \right) \phi(z)\, dz,$$

where

$$w = \frac{\log_e(K/s_0) - \mu T}{\sigma\sqrt{T}}.$$

By writing $z\sigma\sqrt{T} - z^2/2 = -1/2(z - \sigma\sqrt{T})^2 + \sigma^2 T/2$ we have

$$\exp\left\{ z\sigma\sqrt{T} \right\} \phi(z) = \exp\left\{ \frac{1}{2}\sigma^2 T \right\} \phi\left(z - \sigma\sqrt{T} \right).$$

Thus,

$$E(S_T - K)_+ = s_0 e^{\mu T} \int_w^\infty e^{z\sigma\sqrt{T}} \, \phi(z) \, dz - K \int_w^\infty \phi(z) \, dz$$

$$= s_0 e^{\mu T + \sigma^2 T/2} \int_{w-\sigma\sqrt{T}}^\infty \phi(z) \, dz - K \int_w^\infty \phi(z) \, dz$$

$$= s_0 e^{\mu T + \sigma^2 T/2} \Phi(\sigma\sqrt{T} - w) - K\Phi(-w), \tag{14.50}$$

where $\Phi(x) = \int_{-\infty}^x \phi(t)dt$ is the distribution function of the standard normal distribution. Since $\mu = r - \sigma^2/2$,

$$E(S_T - K)_+ = s_0 e^{rT} \Phi(z_+) - K\Phi(z_-), \tag{14.51}$$

where

$$z_\pm = \frac{\log_e(s_0/K) + (r \pm \sigma^2/2)T}{\sigma\sqrt{T}},$$

because $\sigma\sqrt{T} - w = z_+$ and $-w = z_-$. This leads to the call price

$$C_0^{bs} = s_0 \Phi(z_+) - K e^{-rT} \Phi(z_-). \tag{14.52}$$

Similarly,

$$E(K - S_T)_+ = \int_{-\infty}^w \left(K - s_0 \exp\{z\sigma\sqrt{T} + \mu T\} \right) \phi(z) \, dz$$

$$= K\Phi(w) - s_0 e^{\mu T + \sigma^2 T/2} \Phi(w - \sigma\sqrt{T}), \tag{14.53}$$

which leads to the put price

$$P_0^{bs} = K e^{-rT} \Phi(-z_-) - s_0 \Phi(-z_+). \tag{14.54}$$

We have calculated explicit expressions for the call and put prices at time $t = 0$, when the stock price is $S_0 = s_0$. The Black–Scholes call and put prices at time $t \in [0, T)$, when the stock price is $S_t = s$, are given by

$$C_t(s) = e^{-r(T-t)} E(S_T - K)_+ \tag{14.55}$$

and

$$P_t(s) = e^{-r(T-t)} E(K - S_T)_+, \tag{14.56}$$

where the expectations are taken with respect to the distribution of S_T defined by

$$S_T = s \exp\left\{ \mu(T - t) + \sigma\sqrt{T - t} \, Z \right\},$$

where $Z \sim N(0, 1)$ and $\mu = r - \frac{1}{2}\sigma^2$. The corresponding explicit expressions are given in (14.58) and (14.59), and the Black–Scholes prices are discussed in Section 14.3.1.

14.2.3.4 Convergence of the Hedging Coefficient

We show that the hedging coefficients of puts and calls in the multiperiod binary model converge to the Black–Scholes hedging coefficients. The Black–Scholes hedging coefficients are called deltas, and they are obtained by differentiating the Black–Scholes prices with respect to the stock price.

Let the initial time and the initial stock price be

$$t \in [0, T), \quad s > 0.$$

Let $C_t(s)$ be either the Black–Scholes call price or the put price at time $t \in [0, T)$, when the stock price at time t is $S_t = s$, and the expiration is at time T. These prices are written as expectations in (14.55) and (14.56), and in a more explicit form in (14.58) and (14.59).

Let us consider step $k = 0, \ldots, n-1$ of the multiperiod binary model. Let $s_{k,j} = u^j d^{k-j} s_0$ be one of the possible stock prices at step k, where $j = 0, \ldots, k$. These prices were defined in (14.23). Let $k = k_n$ be such that

$$k_n \Delta t \to t,$$

as $n \to \infty$. In particular, we can choose $k_n = [t/\Delta t]$, where $[x]$ is the largest integer $\leq x$. Choose

$$j = j_n = \left\lceil \frac{\log(s/s_0) - k \log d}{\log(u/d)} \right\rceil.$$

Then

$$s_{k,j} \to s,$$

as $n \to \infty$.

The hedging coefficient of the multiperiod binary model was written in (14.32) as

$$\xi_{k+1}(s_{k,j}) = \frac{c_{k+1,j+1} - c_{k+1,j}}{s_{k+1,j+1} - s_{k+1,j}},$$

where $c_{k+1,j+1}$ and $c_{k+1,j}$ are the two possible prices of the derivative at step $k+1$, when the value of the stock at step k is $S_k = s_{k,j}$.[7] Now it holds that

$$\lim_{n \to \infty} \xi_{k+1}(s_{k,j}) = C'_t(s), \tag{14.57}$$

where $C'_t(s)$ is the derivative of the Black–Scholes price with respect to stock price.

---

7 These values are given in (14.28) and (14.30) as

$$c_{k+1,j+1} = (1 + r\Delta t)^{-(n-k-1)} E_Q(C_n \mid S_{k+1} = s_{k+1,j+1}),$$
$$c_{k+1,j} = (1 + r\Delta t)^{-(n-k-1)} E_Q(C_n \mid S_{k+1} = s_{k+1,j}).$$

To show (14.57), note that from (14.46) and (14.45), and the continuity of the functions $s \mapsto C_t(s)$ we have that

$$c_{k+1,j+1} \asymp C_t(s_{k+1,j+1}), \quad c_{k+1,j} \asymp C_t(s_{k+1,j}),$$

where $a_n \asymp b_n$ means that $\lim_{n \to \infty}(a_n/b_n) = 1$. Thus,

$$\xi_{k+1} \asymp \frac{C_t(s_{k+1,j+1}) - C_t(s_{k+1,j})}{s_{k+1,j+1} - s_{k+1,j}}.$$

Also, $s_{k+1,j+1} = s_{k,j}u = s_{k,j}(1 + \sigma\sqrt{\Delta t}) \to s$, $s_{k+1,j} = s_{k,j}d = s_{k,j}(1 - \sigma\sqrt{\Delta t}) \to s$, and $s_{k+1,j+1} - s_{k+1,j} = s_{k,j}(u - d) = 2s_{k,j}\sigma\sqrt{\Delta t} \to 0$, as $n \to \infty$.

14.2.3.5 Rate of Convergence

Figure 14.2 illustrates the convergence of (a) the ratio of the Cox–Ross–Rubinstein call price to the Black–Scholes call price, and (b) the ratio of the Cox–Ross–Rubinstein call hedging coefficient to the Black–Scholes call hedging coefficient. The ratios are plotted as a function of the number n of the steps in the multiperiod binary model, where $n = 100, \ldots, 1000$. The moneyness S/K is 0.9 (green), 0.95 (black), and 1 (red). The annualized volatility is $\sigma = 10\%$, the interest rate is $r = 0$, and the time to maturity is 1 month: $T - t = 1/12$. We see that the convergence of the at-the-money options is faster than the convergence of the out-of-the-money options.

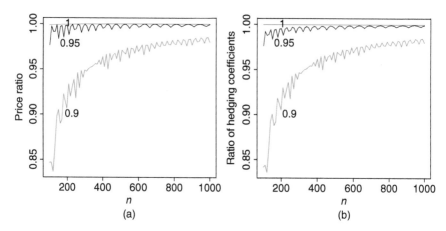

(a)   (b)

Figure 14.2 *Convergence of the Cox–Ross–Rubinstein price and hedging coefficient.* Plotted are (a) the ratios of the Cox–Ross–Rubinstein call price to the Black–Scholes price and (b) the ratios of the Cox–Ross–Rubinstein call hedging coefficient to the Black–Scholes hedging coefficient as a function of n. The three curves show the cases where the moneyness is 0.9 (green), 0.95 (black), and 1 (red).

14.2.3.6 Asian and Knock-Out-Options

We have proved the weak convergence at one point: Let $t \in (0, T]$ and $t_{k,n} = k_n \Delta t \to t$. Then

$$S_{k_n} \xrightarrow{d} S_t,$$

where S_{k_n} is the price of the stock at step k_n in the multiperiod binomial model, and S_t is the price of the stock at time t in the geometric Brownian motion model. If the option payoff is $f(S_{k_n})$, then the price in the multiperiod binomial model converges to the price of the option whose payoff is $f(S_t)$ in the geometric Brownian motion model. Asian options and knock-out options depend on values of stocks in more than one point.

Asian Options In the case of Asian options, the option payoffs can be written as

$$f(S_{k_1}, \dots, S_{k_m}), \quad f(S_{t_1}, \dots, S_{t_m}),$$

where $0 \leq k_1 < \cdots k_n \leq n$ are steps, and $0 \leq t_1 < \cdots < t_m \leq T$ are fixed time points. For example, in the case of an Asian call option

$$f(x_1, \dots, x_m) = \left(\frac{1}{m} \sum_{i=1}^{m} x_i - K \right)_+.$$

The corresponding prices converge for a suitable f, when

$$(S_{k_1}, \dots, S_{k_m}) \xrightarrow{d} (S_{t_1}, \dots, S_{t_m}).$$

Knock-Out Options The payoffs of knock-out options depend on the trajectory of the prices through

$$\max_{k=0,\dots,n} S_k, \quad \sup_{t \in [0,T]} S_t.$$

Let us divide the time interval $[0, T]$ into n subintervals. Denote the $n + 1$ boundaries of the intervals by

$$t_{k,n} = k\Delta t, \quad k = 0, \dots, n.$$

Points $t_{k,n}$ fill the interval $[0, T]$ asymptotically. We can define a continuous time process $W_{t,n}$, $t \in [0, T]$, by linearly interpolating S_k: define

$$W_{t,n} = S_{k-1,n} + \frac{S_{k,n} - S_{k-1,n}}{\Delta t} (t - t_{k-1,n}),$$

when $t \in [t_{k-1,n}, t_{k,n}]$. Geometric Brownian motion S_t is defined in (5.62). We can show that process $W_{t,n}$ converges weakly to the geometric Brownian motion

$$S_t = S_0 \exp \left\{ \left(\mu_0 - \frac{1}{2} \sigma^2 \right) t + \sigma B_t \right\}, \quad 0 \leq t \leq T,$$

where B_t is the standard Brownian motion. The weak convergence

$$(W_{t,n})_{t \in [0,T]} \xrightarrow{d} (S_t)_{t \in [0,T]},$$

as $n \to \infty$, happens in the metric space $C([0, T])$ of the continuous functions on $[0, T]$. The prices of the options whose payoffs are

$$f((W_{t,n})_{t \in [0,T]}), \qquad f((S_t)_{t \in [0,T]}),$$

converge for suitable $f : C[0, T] \to \mathbf{R}$.

14.2.4 American Put Options

An American put option has a different price than the corresponding European put option, and the price of an American put option does not have a closed-form expression in the multiperiod binary model. However, an American call option has the same price as the corresponding European call option, when the stock does not pay dividends (see Section 14.1.3). Thus, the American call options can be priced similarly as the European call options using the Black–Scholes prices or the recombining binary trees.

The American put options have to be priced by taking into account the possibility of an early exercise. We can use the recombining binary tree to price the American put options. At every node of the tree we consider whether it is better to exercise or to keep the option for a future exercise. We are not able to obtain a closed-form formula for the price of the American put options, but we obtain an algorithm for the computation of the price. First, the single period binary model is studied. Second, the multiperiod binary model leads to the final algorithm.

14.2.4.1 American Put Options in the One-Period Binary Model

In the one-period binary model, the American put option can be exercised at time $t = 0$ or at time $t = 1$. Let us denote with P_0^A the value of the American put option at time $t = 0$ and let us denote with P_0^E the value of the European put option at time $t = 0$. An arbitrage argument shows that

$$P_0^A = \max \left\{ K - s_0, P_0^E \right\}.$$

The value of an European put option can be obtained from (14.17) as

$$P_0^E = (1 + r\Delta t)^{-1} E_Q (K - S_1)_+,$$

where

$$E_Q(K - S_1)_+ = (1 - q)(K - s_{1,0})_+ + q(K - s_{1,1})_+$$

with

$$q = \frac{s_0(1 + r\Delta t) - s_{1,0}}{s_{1,1} - s_{1,0}}.$$

14.2.4.2 American Put Options in the Multiperiod Binary Model

The price of an American put option is determined in the n-step binomial model by recursion. Remember that in the n-step binomial model the possible prices at step k, $k = 0, \ldots, n$, are

$$s_{k,j} = u^j(2 - u)^{k-j}s_0, \quad j = 0, \ldots, k$$

with

$$u = 1 + \sigma\sqrt{\Delta t}.$$

The recursive steps are the following.

1) At time $T = t_{n,n}$ the prices of the American put option are given by

$$H_T(s_{n,j}) = \max\{K - s_{n,j}, 0\},$$

$j = 0, \ldots, n$.

2) At time $t_{k-1,n}$, $k = 1, \ldots, n$, when the stock has price $s_{k-1,j}$, we know from the previous steps of the algorithm that the two possible prices for the derivative at time $t_{k,n}$ are $H_{t_{k,n}}(s_{k,j})$ and $H_{t_{k,n}}(s_{k,j+1})$. We can use the single period model to calculate the price at time $t_{k-1,n}$. We get the price from the one-step binary model as

$$H_{t_{k-1,n}}(s_{k-1,j}) = \max\{K - s_{k-1,j}, E_q H_{t_{k,n}}(S_{t_{k,n}})\},$$

where

$$E_q H_{t_{k,n}}(S_{t_{k,n}}) = (1 + r\Delta t)^{-1}[(1 - q)H_{t_{k,n}}(s_{k,j}) + qH_{t_{k,n}}(s_{k,j+1})]$$

and

$$q = \frac{s_{k-1,j}(1 + r\Delta t) - s_{k,j}}{s_{k,j+1} - s_{k,j}} = \frac{1}{2} + \frac{r\sqrt{\Delta t}}{2\sigma},$$

with $\Delta t = (T - t)/n$.

14.3 Black–Scholes Pricing

First we describe the properties of Black–Scholes call and put prices, second we discuss implied volatility, third we describe various ways to derive the Black–Scholes prices, and fourth we give Black–Scholes formulas for options on forwards, for fixed income options, and for currency options.

14.3.1 Call and Put Prices

The Black–Scholes price of the call option at time t, with strike price K, and with the maturity date T, is equal to

$$C_t(S_t, K, T) = S_t\Phi(z_+) - Ke^{-r(T-t)}\Phi(z_-), \tag{14.58}$$

where S_t is the stock price at time t, $r > 0$ is the annualized risk-free rate,

$$z_\pm = \frac{\log_e(S_t/K) + (r \pm \sigma^2/2)(T - t)}{\sigma\sqrt{T - t}},$$

and Φ is the distribution function of the standard Gaussian distribution. The time $T - t$ to expiration is expressed in fractions of a year. The put price is equal to

$$P_t(S_t, K, T) = -S_t\Phi(-z_+) + Ke^{-r(T-t)}\Phi(-z_-). \tag{14.59}$$

Note that it can be convenient to write

$$z_\pm = \frac{\log_e\left(S_t e^{r(T-t)}/K\right) \pm (T - t)\sigma^2/2}{\sigma\sqrt{T - t}}.$$

The Black–Scholes price is derived under the assumption of a log-normal distribution of the stock price: It is assumed that at time $t < T$

$$S_T \sim S_t \exp\left\{\mu(T - t) + \sigma\sqrt{T - t}\, Z\right\},$$

where $Z \sim N(0, 1)$, $\mu \in \mathbf{R}$ is the drift, and $\sigma > 0$ is the volatility. Note that under the risk-neutral measure $\mu = r - \sigma^2/2$. The volatility σ is the only unknown parameter that need to be estimated, since μ does not appear in the price formula.

14.3.1.1 Computation of Black–Scholes Prices

For the application of the Black–Scholes formula the time $T - t$ is taken as the time in fractions of year. For example, when the time to expiration is 20 trading days, then $T - t = 20/252$. Alternatively, when time to expiration is 20 calendar days, then $T - t = 20/365$.

The risk-free rate r is expressed as the annualized rate.

The only unknown parameter σ has to be estimated. Let $S_{t_0}, \ldots, S_{t_n}$ be an equally spaced sample of stock prices and let us denote $\Delta t = t_i - t_{i-1}$, for $i = 1, \ldots, n$. We assume that $Y_i = \log(S_{t_i}/S_{t_{i-1}})$, $i = 1, \ldots, n$, are i.i.d. $N(m, s^2)$, so that the stock prices satisfy (3.50). We can estimate s^2 with the sample variance

$$\hat{s}^2 = \frac{1}{n}\sum_{i=1}^{n}(Y_i - \bar{Y})^2,$$

where $\bar{Y} = n^{-1}\sum_{i=1}^{n}Y_i$. Then an estimator of $\sigma = s(\Delta t)^{-1/2}$ is

$$\hat{\sigma} = \hat{s}\,(\Delta t)^{-1/2}. \tag{14.60}$$

For example, if we sample stock prices daily, then $\Delta t = 1/250$ and $\hat{\sigma} = \hat{s}\,\sqrt{250}$.[8] If we sample stock prices monthly, then $\Delta t = 1/12$ and $\hat{\sigma} = \hat{s}\,\sqrt{12}$.

---

8 The actual number of trading days in a year is between 250 and 252. There are 365 days in a year, but if we ignore the days, when there are no trading, then $\Delta t = 1/250$. Sampling of price data is discussed in Section 2.1.2.

The normalized sample standard deviation in (14.60) is called the annualized sample standard deviation.

14.3.1.2 Characteristics of Black–Scholes Prices

We study the qualitative behavior of the Black–Scholes prices as a function of five parameters σ, $T - t$, r, S_t, and K.

The prices of calls and puts increase as σ increases. We have

$$\lim_{\sigma \to \infty} C_t(S_t, K, T) = S_t$$

and

$$\lim_{\sigma \to 0} C_t(S_t, K, T) = \left(S_t - e^{-r(T-t)}K\right)_+,$$

which are the bounds derived from the put–call parity in (14.9). The prices of calls and puts increase as the time to maturity $T - t$ increases. The price of a call increases as S_t increases and the price of a call decreases as K increases, but for puts the relations reverse. The price of a call increases as the interest rate r increases but the price of a put decreases as the interest rate r increases.

Figure 14.3 shows Black–Scholes prices for calls and puts as a function of the call moneyness S/K. The call prices increase and the put prices decrease as a function of moneyness. Panel (a) shows the cases of annualized volatility 5% (black), 10% (red), and 20% (green). The time to maturity is 20 trading days and the interest rate is $r = 0$. Panel (b) shows the cases of interest rates $r = 0$ (black), $r = 10\%$ (red), and $r = 20\%$ (green). The time to maturity is 20 trading days and the annualized volatility is 10%. We see from panel (a) that increasing volatility

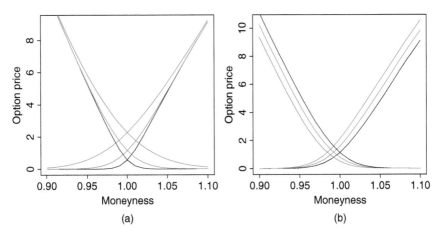

Figure 14.3 *Call and put prices as a function of call moneyness S/K. The call prices increase and the put prices decrease as a function of moneyness. (a) The annualized volatility is 5% (black), 10% (red), and 20% (green); (b) the interest rate is $r = 0$ (black), $r = 10\%$ (red), and $r = 20\%$ (green).*

increases the prices, both for call and puts. The effect of increasing time to maturity is similar. We see from panel (b) that for calls increasing interest rates increases prices, but for puts increasing interest rates decreases prices.

Figure 14.4 shows that the call and put prices are not symmetric. The ratios of call prices to put prices are shown as a function of the call moneyness S/K: We show the functions

$$S/K \mapsto \frac{C_t(S,K)}{P_t(S, 2S - K)},$$

where $S = 100$ is the current stock price. The strike prices for the calls take values $105, 104, \ldots, 95$, and the corresponding strike prices for the puts are $95, 96, \ldots, 105$. Panel (a) shows the cases of annualized volatility 5% (black), 10% (red), and 20% (green). The interest rate is $r = 0$. Panel (b) shows the cases of interest rates $r = 0$ (black), $r = 10\%$ (red), and $r = 20\%$ (green). The annualized volatility is 10%. The time to maturity is 20 trading days in both panels. We see that the call prices are higher than the put prices for out-of-the-money options. This is related to the asymmetry of the log-normal distribution (see Figure 3.11). For at-the-money options the difference between the call and the put prices is not large. Increasing the volatility makes the ratios of the call and put prices closer to one, and decreasing the interest rate makes the ratios of the call and put prices closer to one.

14.3.1.3 Black–Scholes Prices and Volatility

To apply Black–Scholes prices we need to estimate the volatility. We study how the Black–Scholes prices change when the volatility estimate changes. We apply the data of S&P 500 daily prices, described in Section 2.4.1.

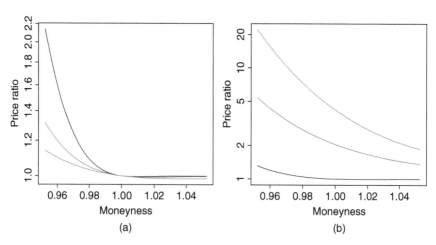

Figure 14.4 *Ratios of call prices to put prices as a function of moneyness S/K. (a) Annualized volatility is 5% (black), 10% (red), and 20% (green); (b) interest rate is $r = 0$ (black), $r = 10\%$ (red), and $r = 20\%$ (green).*

Figure 14.5 shows time series of Black–Scholes call prices. The volatility is equal to the sequential annualized sample standard deviation. Panel (a) shows call prices with moneyness $S/K = 1$ and panel (b) shows call prices with moneyness $S/K = 0.95$. The time to expiration is either 20 trading days (black curves) or 30 trading days (red curves). The risk-free rates are deduced from 1-month Treasury bill rates. We can see that the time series of Black–Scholes prices follow closely to the time series of sequential standard deviations in Figure 7.5.

Figure 14.6 studies Black–Scholes prices when the volatility is equal to the sequentially estimated GARCH(1, 1) volatility. Panel (a) shows time series of call prices with moneyness $S/K = 1$ (black) and $S/K = 0.95$ (green). The time to expiration is 20 trading days. Panel (b) shows kernel density estimates of the distributions of the prices. The horizontal and the vertical lines show the Black–Scholes prices when the volatility is the annualized sample standard deviation computed from the complete sample. The risk-free rate is zero.

14.3.1.4 The Greeks

The greeks are defined by differentiating the option price with respect to the price of the underlying, the time to the expiration, the interest rate, or the volatility. In this section, we denote the Black–Scholes call and put prices by

$$C_t(S, K, T, \sigma, r), \quad P_t(S, K, T, \sigma, r),$$

where t is the current time, S is the current price of the underlying, K is the strike price, T is the time of the expiration, σ is the volatility, and r is the interest rate. Sometimes we leave out some of the arguments and denote, for example, $C_t(S) = C_t(S, K, T, \sigma, r)$.

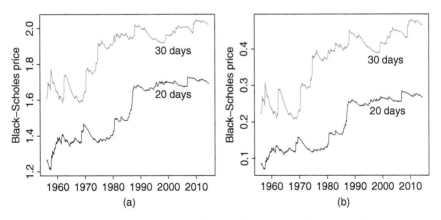

(a) (b)

Figure 14.5 *Time series of Black–Scholes call prices.* (a) Moneyness $S/K = 1$ and (b) moneyness $S/K = 0.95$. Call prices are computed using the sequential sample standard deviation. The time to expiration is 20 trading days (black) or 30 trading days (red).

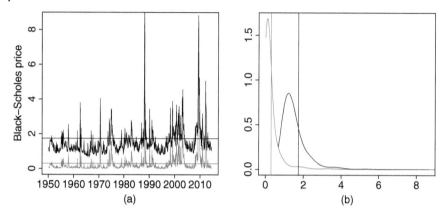

Figure 14.6 *Black–Scholes prices with GARCH-volatility.* (a) Time series of prices; (b) kernel estimates of the density of the collection of prices. The moneyness is $S/K = 1$ (black) and $S/K = 0.95$ (green).

The Delta The delta is the derivative of the price function with respect to the underlying. The delta is the hedging coefficient in the Black–Scholes hedging, as discussed in Section 14.4. The call and put delta are

$$C_s = \frac{\partial C_t(S)}{\partial S}, \quad P_s = \frac{\partial P_t(S)}{\partial S}. \tag{14.61}$$

We have that

$$0 < C_s < 1, \quad -1 < P_s < 0.$$

In fact, if S increases, then the price of the call increases and the price of the put decreases: $C_s > 0$ and $P_s < 0$. The absolute value of the change in the value of a call or a put cannot exceed the absolute value of the change in the underlying: $|C_s| < 1$ and $|P_s| < 1$.

The call delta is equal to

$$C_s = \Phi(z_+), \tag{14.62}$$

and the put delta equal to

$$P_s = -\Phi(-z_+). \tag{14.63}$$

Let us calculate the call delta. The delta of a call is given in (14.62) because

$$\frac{\partial C(S)}{\partial S} = \Phi(z_+) + e^{-r(T-t)} \left[Se^{r(T-t)} \frac{\partial \Phi(z_+)}{\partial S} - K \frac{\partial \Phi(z_-)}{\partial S} \right],$$

$$\frac{\partial \Phi(z_+)}{\partial S} = \phi(z_+) \cdot \frac{\partial z_+}{\partial S} = \phi(z_+) \cdot \frac{1}{\sigma\sqrt{T-t}} \cdot \frac{1}{S},$$

$$\frac{\partial \Phi(z_-)}{\partial S} = \phi(z_-) \cdot \frac{1}{\sigma\sqrt{T-t}} \cdot \frac{1}{S},$$

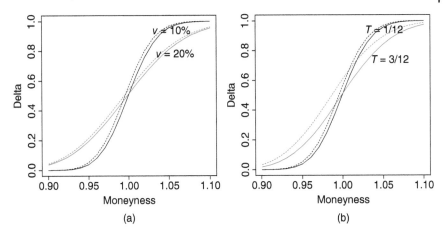

Figure 14.7 *Call deltas.* Call deltas are shown as a function of moneyness S/K. (a) The annualized volatility is 10% (black) and 20% (red) with time to expiration 1 month; (b) the time to expiration is 1 month (black) and 3 months (red) with volatility 10%. The interest rate is $r = 0$ (solid lines) and $r = 0.05$ (dashed lines).

and finally[9]

$$e^{r(T-t)}\phi(z_+) = \frac{K}{S}\,\phi(z_-).$$

Figure 14.7 shows the call delta as a function of moneyness S/K. In panel (a) the annualized volatility is 10% (black) and 20% (red) with time to expiration 1 month. In panel (b) the time to expiration is 1 month (black) and 3 months (red) with volatility 10%. The solid lines show the case of interest rate $r = 0$ and the dashed lines show the case of interest rates $r = 5\%$. The more interesting part is the moneyness ≤ 1. In this region the deltas are ≤ 0.5 in both panels. Panel (a) shows that when moneyness is ≤ 1, then a larger volatility leads to a larger delta. Panel (b) shows that increasing time to maturity has a similar qualitative effect as increasing volatility.

Figure 14.8 shows a time series of Black–Scholes deltas using the data of S&P 500 daily prices, described in Section 2.4.1. The daily prices are used to create a time series with the sampling frequency of 20 and 30 trading days. Panel (a) shows call prices with moneyness $S/K = 1$ and panel (b) shows call prices when moneyness $S/K = 0.95$. The time to expiration is either 20 trading days (black curves) or 30 trading days (red curves). The volatility is equal to the sequential sample standard deviation. The risk-free rates are deduced from 1-month

---

9 We have that $z_+\sigma\sqrt{T-t} - (T-t)\sigma^2/2 = \log(S/K) + r(T-t)$ and thus

$$\phi(z_-) = \phi(z_+ - \sigma\sqrt{T-t}) = \phi(z_+)\exp\left\{z_+\sigma\sqrt{T-t} - \frac{1}{2}\sigma^2(T-t)\right\}$$

$$= \phi(z_+)e^{r(T-t)}\frac{S}{K}.$$

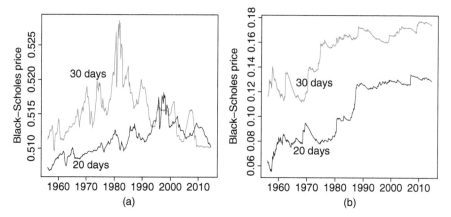

Figure 14.8 *Time series of Black–Scholes call deltas.* (a) Moneyness $S/K = 1$ and (b) moneyness $S/K = 0.95$. Call deltas are computed using the sequential sample standard deviation. The time to expiration is 20 trading days (black) or 30 trading days (red).

Treasury bill rates. The corresponding time series of Black–Scholes prices is given in Figure 14.5.

The Gamma The gamma is the second derivative of the price function with respect to the underlying:
$$C_{ss} = \frac{\partial^2 C_t(S)}{\partial S^2}, \quad P_{ss} = \frac{\partial^2 P_t(S)}{\partial S^2}.$$
The price functions are convex with respect to S and thus
$$C_{ss} > 0, \quad P_{ss} > 0.$$
The call gamma and the put gamma are given by
$$C_{ss} = P_{ss} = \frac{\Phi'(z_+)}{S\sigma\sqrt{T-t}}.$$
Figure 14.9 shows the call gamma as a function of moneyness S/K. In panel (a) the annualized volatility is 10% (black) and 20% (red) with time to expiration 1 month. In panel (b) the time to expiration is 1 month (black) and 3 months (red) with volatility 10%. The solid lines show the case of interest rate $r = 0$ and the dashed lines show the case of interest rates $r = 5\%$.

Theta The theta is the derivative of the price function with respect to time:
$$C_\theta = \frac{\partial C_t(S, K, T)}{\partial t}, \quad P_\theta = \frac{\partial P_t(S, K, T)}{\partial t}.$$
As t increases the value of the option decreases (everything else being equal) and thus
$$C_\theta < 0, \quad P_\theta < 0.$$

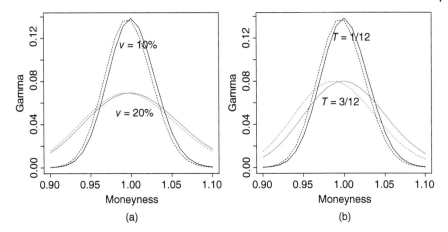

Figure 14.9 *Call gammas.* Call gammas are shown as a function of moneyness S/K. (a) The annualized volatility is 10% (black) and 20% (red) with the time to expiration 1 month; (b) the time to expiration is 1 month (black) and 3 months (red) with volatility 10%. The interest rate is $r = 0$ (solid lines) and $r = 0.05$ (dashed lines).

The call theta is equal to

$$C_\theta = -\frac{S\sigma\Phi'(z_+)}{2\sqrt{T-t}} - rKe^{-r(T-t)}\Phi(z_-),$$

and the put theta is equal to

$$P_\theta = -\frac{S\sigma\Phi'(z_+)}{2\sqrt{T-t}} + rKe^{-r(T-t)}\Phi(-z_-).$$

Figure 14.10 shows the call theta as a function of moneyness S/K. In panel (a) the annualized volatility is 10% (black) and 20% (red) with time to expiration 1 month. In panel (b) the time to expiration is 1 month (black) and 3 months (red) with volatility 10%. The solid lines show the case of interest rate $r = 0$ and the dashed lines show the case of interest rates $r = 1\%$.

Vega The vega is the derivative of the price function with respect to volatility:

$$C_\sigma = \frac{\partial C_t(S, K, T, \sigma)}{\partial \sigma}, \quad P_\sigma = \frac{\partial P_t(S, K, T, \sigma)}{\partial \sigma}.$$

The call vega and the put vega are equal to

$$C_\sigma = P_\sigma = S\sqrt{T-t}\,\Phi'(z_+).$$

Figure 14.11 shows the call vega as a function of moneyness S/K. In panel (a) the annualized volatility is 10% (black) and 20% (red) with time to expiration 1 month. In panel (b) the time to expiration is 1 month (black) and 3 months (red) with volatility 10%. The solid lines show the case of interest rate $r = 0$ and the dashed lines show the case of interest rates $r = 5\%$.

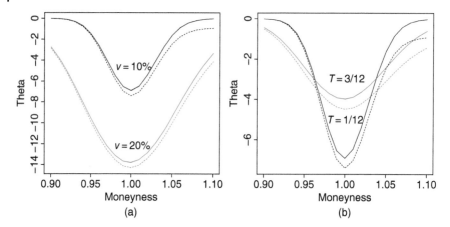

Figure 14.10 *Call thetas.* Call thetas are shown as a function of moneyness S/K. (a) The annualized volatility is 10% (black) and 20% (red) with the time to expiration 1 month; (b) the time to expiration is 1 month (black) and 3 months (red) with volatility 10%. The interest rate is $r = 0$ (solid lines) and $r = 0.01$ (dashed lines).

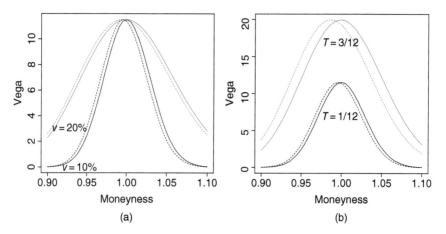

Figure 14.11 *Call vega.* Call vegas are shown as a function of moneyness S/K. (a) The annualized volatility is 10% (black) and 20% (red) with the time to expiration 1 month; (b) the time to expiration is 1 month (black) and 3 months (red) with volatility 10%. The interest rate is $r = 0$ (solid lines) and $r = 0.05$ (dashed lines).

Rho The rho is the derivative of the price function with respect to interest rate:

$$C_r = \frac{\partial C_t(S, K, T, \sigma, r)}{\partial r}, \quad P_r = \frac{\partial P_t(S, K, T, \sigma, r)}{\partial r}.$$

The call rho is equal to

$$C_r = (T - t)Ke^{-r(T-t)}\Phi(z_-),$$

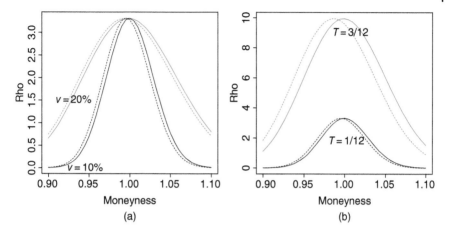

Figure 14.12 *Call rho.* Call rhos are shown as a function of moneyness S/K. (a) The annualized volatility is 10% (black) and 20% (red) with the time to expiration 1 month; (b) the time to expiration is 1 month (black) and 3 months (red) with volatility 10%. The interest rate is $r = 0$ (solid lines) and $r = 0.05$ (dashed lines).

and the put rho is equal to

$$P_r = -(T - t)Ke^{-r(T-t)}\Phi(-z_-).$$

Figure 14.12 shows the call rho as a function of moneyness S/K. In panel (a) the annualized volatility is 10% (black) and 20% (red) with time to expiration 1 month. In panel (b) the time to expiration is 1 month (black) and 3 months (red) with volatility 10%. The solid lines show the case of interest rate $r = 0$ and the dashed lines show the case of interest rates $r = 5\%$.

14.3.2 Implied Volatilities

Implied volatilities can be derived both from call prices or from put prices. Let $C_t(S, K, T, \sigma, r)$ be a Black–Scholes call price. The mapping

$$\sigma \mapsto C_t(S, K, T, \sigma, r). \tag{14.64}$$

is bijective and can be inverted. Let us denote by IV the inverse of mapping (14.64). When we observe a market price c for a call option, then

$$\sigma = \text{IV}(c)$$

is the implied volatility of the option. The implied volatility of a put option can be defined similarly.

14.3.2.1 Quoting Option Prices
It is helpful to quote option prices using implied volatilities. Option prices and implied volatilities are in a bijective correspondence, but it is easier to compare the prices of options with different maturities and strike prices using the

implied volatilities than using the market prices. This is analogous to expressing the bond prices with annualized rates instead of quoted prices.[10] The implied volatilities can be used to quote prices even when we do not think that the Black–Scholes prices are fair prices, similarly as the bond rates can be defined using various conventions.

14.3.2.2 The Volatility Surface

If the Black–Scholes model describes the true distribution of the asset prices, and if the market prices coincide with the Black–Scholes prices, then the implied volatilities of options with different strike prices and with different maturities are all equal. However, in practice the implied volatilities are different for the options with different strike prices and with different maturities.

The volatility surface gives for each strike price and for each maturity the corresponding implied volatility. Let c_{ij} be the market prices of call (put) options with strike prices K_i and expiration dates T_j, where $i = 1, \dots , m, j = 1, \dots , n$. The options are otherwise similar. The volatility surface is the function

$$\mathrm{VS}(K_i, T_i - t) = \mathrm{IV}(c_{ij}),$$

where $T_i - t$ is the time to the expiration.

The volatility surface is typically not a constant function. Instead, for a fixed maturity $T - t$, function

$$K \mapsto \mathrm{VS}(K, T - t)$$

is typically u-shaped (smile) or skew (one sided smile). Instead of the strike K one may take as the argument the moneyness S/K (or $S/Ke^{-r(T-t)}$), or the delta of the options.

Options written on equity indices yield often skews. This might be due to the fact that a crash in stock markets leads to increased volatility, whereas a rise in stock markets is not usually involved with an increased volatility. Options on various interest rates yield more monotonous one sided smiles than the equity indices.

Currency markets yield often symmetric smiles. Currency markets are more symmetric than stock markets because a big movement in either direction results in an increased volatility (it is always a crash for one of the currencies).

The smile tends to flatten out with maturity. This might be due to the better Gaussian approximation when the maturity is longer.

14.3.2.3 Pricing of Options Using Implied Volatilities

Out-of-the-money options can be priced in the following way. (1) Find the implied volatility of at-the-money options. (2) Adjust the implied volatility

---

10 For example, let P be the present value of 100 received in 3 months time. We define the Libor rate L as the solution of $P = 100/(1 + L/4)$, or as the solution L^* of $P = 100/(1 + L^*)^{1/4}$.

(using experience) to get a new volatility. (3) Calculate the price of the out-of-the-money option using the new volatility.

14.3.2.4 VIX Index

The VIX index of CBOE (Chicago Board Options Exchange) uses prices of options to derive the volatility that is expected by the markets. Section 6.3.1 contains a discussion of the VIX index and Figure 6.5 shows a time series of the VIX index. Let us derive the formula of the VIX index.

It is assumed that the stock price follows a geometric Brownian motion, as defined in (5.62). That is,

$$S_t = S_0 \exp\left\{\left(\mu - \frac{1}{2}\sigma^2\right)t + \sigma W_t\right\}, \quad 0 \le t \le T,$$

where W_t is the standard Brownian motion, $\mu \in \mathbf{R}$, and $\sigma > 0$. Under the equivalent martingale measure $\mu = r$, where r is the yearly risk-free interest rate. Thus, for the equivalent martingale measure we have

$$E \log \frac{S_T}{S_0} = T\left(r - \frac{1}{2}\sigma^2\right).$$

We can solve for σ^2 to get

$$\sigma^2 = \frac{2}{T}\left(rT - E \log \frac{S_T}{S_0}\right).$$

A Taylor expansion gives

$$\log \frac{S_T}{K_0} = \frac{S_T - K_0}{K_0} - \int_{K_0}^{\infty} \frac{1}{K^2}(S_T - K)_+ dK - \int_0^{K_0} \frac{1}{K^2}(K - S_T)_+ dK,$$

where $0 < K_0 < \infty$. Thus,

$$\log \frac{S_T}{S_0} = \log \frac{K_0}{S_0} + \frac{S_T - K_0}{K_0}$$
$$- \int_{K_0}^{\infty} \frac{1}{K^2}(S_T - K)_+ dK - \int_0^{K_0} \frac{1}{K^2}(K - S_T)_+ dK.$$

Theorem 13.2 implies that when the expectation is taken with respect to a risk-neutral measure, then

$$E(S_T - K)_+ = e^{rT} C_0(K), \quad E(K - S_T)_+ = e^{rT} P_0(K),$$

where $C_0(K)$ and $P_0(K)$ are the arbitrage-free prices of the call and the put. Under the risk-neutral measure

$$E \frac{S_T - K_0}{K_0} = \frac{e^{rT} S_0}{K_0} - 1 = \frac{F_0}{K_0} - 1,$$

where $F_0 = e^{rT} S_0$ is the futures price, as given in (14.2). Thus,

$$E \log \frac{S_T}{S_0} = \log \frac{K_0}{S_0} + \frac{F_0}{K_0} - 1$$

$$-e^{rT} \int_{K_0}^{\infty} \frac{1}{K^2} C_0(K) dK - e^{rT} \int_0^{K_0} \frac{1}{K^2} P_0(K) dK.$$

We arrive at the variance formula

$$\sigma^2 = \frac{2}{T} \left(rT - \left(\frac{F_0}{K_0} - 1 \right) - \log \frac{K_0}{S_0} \right.$$

$$\left. + e^{rT} \int_{K_0}^{\infty} \frac{1}{K^2} C_0(K) dK + e^{rT} \int_0^{K_0} \frac{1}{K^2} P_0(K) dK \right). \tag{14.65}$$

The variance formula (14.65) was derived in Demeterfi *et al.* (1999). CBOE uses the approximation

$$rT - \left(\frac{F_0}{K_0} - 1 \right) - \log \frac{K_0}{S_0} = \log \frac{F_0}{K_0} - \left(\frac{F_0}{K_0} - 1 \right) \approx -\frac{1}{2} \left(\frac{F_0}{K_0} - 1 \right)^2.$$

CBOE:s VIX index is defined as[11]

$$\text{VIX}^2 = \frac{2 e^{rT}}{T} \sum_i \frac{1}{K_i^2} Q(K_i) \Delta K_i - \frac{1}{T} \left(\frac{F_0}{K_0} - 1 \right)^2, \tag{14.66}$$

where $Q(K_i)$ is the midpoint of the bid-ask spread for the option, and $\Delta K_i = (K_{i+1} - K_{i-1})/2$. The calculation is done for two expiration dates, and the final index value is a weighted average of these. Note that the put–call parity (14.8) gives the equality $F_0 = K_A + e^{rT}(C_0(K_A) - P_0(K_A))$.

14.3.3 Derivations of the Black–Scholes Prices

We have derived the Black–Scholes prices as the limits of the prices in the multiperiod binary model. Now we describe shortly the martingale derivation of the Black–Scholes prices, derivation of the prices using the Black–Scholes differential equation, and the derivation of the prices using the put–call parity.

14.3.3.1 Martingale Derivation
The second fundamental theorem of asset pricing says that an arbitrage-free market model is complete if and only if there exists exactly one equivalent martingale measure. We have stated this theorem for the discrete time model in Theorem 13.3.

The Black–Scholes prices can be derived as a corollary of the second fundamental theorem of asset pricing: If the Black–Scholes market model is

---

11 See http://www.cboe.com/micro/volatility/introduction.aspx.

arbitrage-free and complete, then the Black–Scholes prices are the discounted expectations with respect to the equivalent martingale measure.

Shiryaev (1999, p. 710) states in a continuous time framework that if there exists a unique equivalent martingale measure, then the unique arbitrage-free price of the option is the discounted expected value of the payoff with respect to the unique equivalent martingale measure.

We give a sketch of some elements of the derivation of the Black–Scholes prices directly from the second fundamental theorem of asset pricing. Details can be found in Shiryaev (1999, p. 739).

The Black–Scholes model assumes that the stock price follows the geometric Brownian motion, as defined in (5.62). The stock price satisfies

$$dS_t = \mu S_t dt + \sigma S_t dW_t, \tag{14.67}$$

and the bank account satisfies

$$dB_t = rB_t dt,$$

where $0 \le t \le T$.[12] Thus,

$$S_t = S_0 \exp\left\{ \left(\mu - \frac{1}{2}\sigma^2\right) t + \sigma W_t \right\}, \quad B_t = B_0 e^{rt}.$$

The Girsanov's theorem was stated in (5.64). We apply Girsanov's theorem with the constant function $a_s(\omega, t) = (\mu - r)/\sigma$. Then,

$$\tilde{W}_t = W_t - \frac{\mu - r}{\sigma} t, \quad 0 \le t \le T, \tag{14.68}$$

is a Brownian motion with respect to measure Q_T, defined by $dQ_T = Z_T dP_T$, where

$$Z_T = \exp\left\{ -\frac{\mu - r}{\sigma} W_T - \frac{1}{2}\left(\frac{\mu - r}{\sigma}\right)^2 T \right\}.$$

Measure Q_T is the unique martingale measure that is equivalent to P_T; see Shiryaev (1999, p. 708). Thus, the price of the call option $C_T = (S_T - C)_+$ is

$$C_0 = e^{-rT} E_{Q_T} C_T.$$

We need to find the distribution of S_T under Q_T. From (14.68) we obtain that

$$\text{Law}(\mu T + \sigma W_T \mid Q_T) = \text{Law}(rT + \sigma \tilde{W}_T \mid Q_T)$$
$$= \text{Law}(rT + \sigma W_T \mid P_T).$$

---

12 The Black–Scholes model assumes that the market is idealized in the following sense. (1) There are no arbitrage opportunities, (2) selling of securities is possible at any time, (3) there are no transaction costs, (4) the market interest rate is constant, (5) there are no dividends during the life time of the option, (6) security trading is continuous, and (7) the stock price follows the log-normal Brownian motion.

Thus,

$$\mathrm{Law}(S_T \mid Q_T) = \mathrm{Law}(S_0 e^{(\mu - \sigma^2/2)T + \sigma W_T} \mid Q_T)$$
$$= \mathrm{Law}(S_0 e^{(r - \sigma^2/2)T + \sigma W_T} \mid P_T).$$

14.3.3.2 The Black–Scholes Differential Equation

Black and Scholes (1973) and Merton (1973) derived the Black–Scholes price by solving a differential equation. The Black–Scholes partial differential equation is

$$\frac{1}{2} C_{ss} \sigma^2 S_t^2 + r C_s S_t - r C_t + C_\theta = 0, \tag{14.69}$$

where $C_t = C(t, S_t)$ is the value of the option at time t, $C_\theta = \partial C(t, S_t)/\partial \partial t$ is the theta of the option, $C_s = \partial C(t, S_t)/\partial \partial S_t$ is the delta of the option, $C_{ss} = \partial^2 C(t, S_t)/\partial \partial S_t^2$ is the gamma of the option, and $0 \le t \le T$. When $C(t, S_t)$ is the value of a call option, then the solution is found under the boundary condition

$$C(T, S_T) = (S_T - K)_+.$$

The price of the option is $C_0 = C(0, S_0)$. The differential equation is solved, for example, in Shiryaev (1999, p. 746).

The Black–Scholes partial differential equation can be derived heuristically in the following way. Itô's lemma (5.61) applied to the function $C_t = C(t, S_t)$ gives

$$dC_t = \left(C_\theta + \mu S_t C_s + \frac{1}{2} \sigma^2 S_t^2 C_{ss} \right) dt + \sigma S_t C_s\, dW_t. \tag{14.70}$$

Assume that value C_t of the option is replicated by the portfolio

$$V_t = \xi_t S_t + \beta_t B_t,$$

that is,

$$dV_t = \xi_t\, dS_t + \beta_t\, dB_t.$$

The assumptions $dS_t = S_t \mu dt + S_t \sigma dW_t$ and $dB_t = r B_t dt$ imply

$$dV_t = (\beta_t r B_t + \xi_t \mu S_t)\, dt + \xi_t \sigma S_t\, dW_t$$
$$= \left[r(V_t - \xi_t S_t) + \xi_t \mu S_t \right] dt + \xi_t \sigma S_t\, dW_t. \tag{14.71}$$

Equating (14.70) and (14.71) gives $C_t - V_t = 0$, that is,

$$0 = \left[C_\theta - r(V_t - \xi_t S_t) + \mu S_t (C_s - \xi_t) + \frac{1}{2} \sigma^2 S_t^2 C_{ss} \right] dt$$
$$+ \sigma S_t (C_s - \xi_t) dW_t.$$

This means that we want a perfect replication of C_t. Choose $\xi_t = C_s$, which is called delta hedging. This makes dW_t to disappear and leads to the differential equation (14.69). Delta hedging has removed all uncertainty, and we have obtained a perfect hedge.

14.3.3.3 Derivation of the Prices Using the Put–Call parity

We can derive the Black–Scholes prices for the calls and puts using the put–call parity given in Section 14.1.2. The basic idea is that if we are willing to assume that the prices are expectations with respect to a log-normal distribution, then the put–call parity implies that the log-normal distribution should be the risk-neutral log-normal distribution.

We assume that the distribution of the stock price S_T is defined by (14.49). Let us denote by C_t the price of the call option at time t. We assume that the price of the call option is equal to

$$C_t = e^{-r(T-t)}E_t C_T = e^{-r(T-t)}E_t(S_T - K)_+$$

and the value of the put option is equal to

$$P_t = e^{-r(T-t)}E_t P_T = e^{-r(T-t)}E_t(K - S_T)_+.$$

Thus, using (14.50) and (14.53),

$$C_t - P_t = S_t e^{(\mu+\sigma^2/2-r)(T-t)} - e^{-r(T-t)}K,$$

because $\Phi(x) + \Phi(-x) = 1$ for all $x \in \mathbf{R}$. The put–call parity (14.8) implies that we have to take

$$\mu = r - \frac{1}{2}\sigma^2. \tag{14.72}$$

Inserting (14.72) to (14.50) and (14.53) leads to (14.58) and (14.59). This derivation was noted in Derman and Taleb (2005).

14.3.4 Examples of Pricing Using the Black–Scholes Model

We give some examples of Black–Scholes prices. The examples include pricing functions of options on a forward, caplets, swaptions, options on a foreign currency, and barrier options.

14.3.4.1 Options on a Forward

Let the underlying be a futures contract F_t with the maturity T'. Consider a call option with the expiration time $T \le T'$. The payoff is $(F_T - K)_+$, where $K > 0$ is the strike price. The price of the call option is obtained by replacing S_t in the Black–Scholes formula (14.58) by $e^{-r(T-t)}F_t$. This gives the price

$$C_t(F_t, K, T) = e^{-r(T-t)}[F_t\Phi(z_+) - K\Phi(z_-)], \tag{14.73}$$

where

$$z_\pm = \frac{\log_e(F_t/K) \pm (T-t)\sigma^2/2}{\sigma\sqrt{T-t}}.$$

The volatility σ is the volatility of the stock. This is called Black's formula, and it was introduced in Black (1976).

Why have we replaced S_t with $e^{-r(T-t)}F_t$? This is due to the fact that under the risk-neutral measure the stock price has the distribution

$$dS_t = rdt + \sigma S_t dW_t.$$

The pricing formula of the futures contract in (14.2) gives $F_t = e^{r(T-t)}S_t$. Thus, the distribution of F_t under the risk-neutral measure is

$$dF_t = \sigma S_t dW_t.$$

The distribution of F_T is

$$F_T \sim F_t \exp\left\{-\frac{1}{2}\sigma^2(T-t) + \sigma\sqrt{T-t}\,Z\right\},$$

where $Z \sim N(0, 1)$. The price of the call option is

$$C_t(F_t, K, T) = e^{-r(T-t)}E(F_T - K)_+,$$

and this expectation is calculated similarly as in (14.51).

The formula (14.73) holds when the option is subject to the stock type settlement. If the option is subject to the futures type settlement, then set $r = 0$ in (14.73). The futures type settlement means that the gains and losses are realized daily, whereas in the stock type settlement the gain or loss is realized at the time of liquidation.

14.3.4.2 Caplets

Caplets are discussed in Section 18.3.1. A caplet is a call option on the Libor rate $L(T_1, T_2)$. The payoff of a caplet at time T_1 is

$$\mathrm{CPL}_{T_1} = P \cdot (L(T_1, T_2) - K)_+(T_2 - T_1),$$

where P is the principal, and $K > 0$ is the strike. We assume that under the risk-neutral measure

$$L(T_1, T_2) = f(t, T_1, T_2)\exp\left\{-\frac{1}{2}\sigma^2(T_1 - 1) + \sigma\sqrt{T_1 - t}Z\right\}, \qquad (14.74)$$

where $Z \sim N(0, 1)$ and $f(t, T_1, T_2)$ is the forward rate, defined in (18.14). Note that at time T_1 the forward Libor rate is equal to the spot Libor rate: $f(T_1, T_1, T_2) = L(T_1, T_2)$. The Black's formula for the price of the caplet is

$$
\begin{aligned}
&C_t(K, T_1, T_2) \\
&\quad = P\left[f(t, T_1, T_2)\Phi(z_+) - K\Phi(z_-)\right] Z(t, T_2)(T_2 - T_1), \qquad (14.75)
\end{aligned}
$$

where

$$z_\pm = \frac{\log(f(t, T_1, T_2)/K) \pm \sigma^2(T_1 - t)/2}{\sigma\sqrt{T_1 - t}}.$$

The caplet price can be written as the expectation

$$C_t(K, T_1, T_2) = Z(t, T_2)E(\mathrm{CPL}_{T_1}),$$

where the expectation is with respect to the risk-neutral distribution in (14.74). The expectation is calculated similarly as in (14.51). The caplet price $C_t(K, T_1, T_2)$ is obtained from the Black–Scholes price of a call option on a stock when S_t is replaced by $f(t, T_1, T_2)Z(t, T_2)$ and $Ke^{-r(T-t)}$ is replaced by $KZ(t, T_2)$.

Caps are defined in Section 18.3.2. Let $T_0 < \cdots < T_m$ be the time points for the caplets on the Libor rates $L(T_{i-1}, T_i)$, where $i = 1, \dots, m$. A cap is priced by

$$P \sum_{i=1}^{m} C_t(K, T_{i-1}, T_i),$$

where $C_t(K, T_{i-1}, T_i)$ are the prices of the caplets.

14.3.4.3 Swaptions

Swaptions are discussed in Section 18.3.3. Let t be the current time, T_0 be the expiry time, and T_m be the maturity time of the swaption. The payoff of an European call option on a swap is given in (18.27) as

$$\mathrm{SWP}_{T_0} = (\mathrm{SR}(T_0, T_0, T_m) - K)_+ A(T_0, T_0, T_m),$$

where $\mathrm{SR}(t, T_0, T_m)$ is the equilibrium swap rate, K is the strike,

$$A(t, T_0, T_m) = P \sum_{i=1}^{m} (T_i - T_{i-1})Z(t, T_i),$$

P is the principal, and $Z(t, T)$ is a zero-coupon bond. We assume that under the risk-neutral measure

$$\mathrm{SR}(T_0, T_0, T_m) \sim \mathrm{SR}(t, T_0, T_m) \exp\left\{ -\frac{1}{2}\sigma^2(T_0 - t) + \sigma\sqrt{T_0 - t}\, Z \right\},$$

where $Z \sim N(0, 1)$. The Black's formula for the price of the swaption is

$$C_t(K, T_0, T_m) = [\mathrm{SR}(t, T_0, T_m)\Phi(z_+) - K\Phi(z_-)] \cdot A(t, T_0, T_m),$$

where

$$z_\pm = \frac{\log(\mathrm{SR}(t, T_0, T_m)/K) \pm \sigma^2(T_0 - t)/2}{\sigma\sqrt{T_0 - t}}.$$

The swaption price can be written as the expectation

$$C_t(K, T_0, T_m) = Z(t, T_0)E(\mathrm{SWP}_{T_0}),$$

where the expectation is with respect to the risk-neutral distribution in (14.74). The expectation is calculated similarly as in (14.51). The caplet price $C_t(K, T_0, T_m)$ is obtained from the Black–Scholes price of a call option on a stock when S_t is replaced by $\mathrm{SR}(t, T_0, T_m)A(t, T_0, T_m)$ and $Ke^{-r(T-t)}$ is replaced by $KA(t, T_0, T_m)$.

Note that the simultaneous log-normality for all caplets and all swaptions is not consistent because a swap rate is a linear combination of forward rates and cannot be log-normal if the underlying forward rates are.

14.3.4.4 Options on a Foreign Currency

Let S be the price of a foreign currency in the domestic currency units. The payoff of an European call option on a foreign currency is given by

$$C_T = (S_T - K)_+,$$

where K is the strike, and T is the expiration time. According the Garman–Kohlhagen model, under the risk-neutral measure,

$$S_T \sim S_t \exp\left\{ r_d - r_f - \frac{1}{2}\sigma^2(T-t) + \sigma\sqrt{T-t}\, Z \right\},$$

where $Z \sim N(0,1)$. Then the price of the call option is

$$C_t(S_t, K, T) = e^{-r_f(T-t)}S_t\Phi(z_+) - e^{-r_d(T-t)}K\Phi(z_-),$$

where

$$z_\pm = \frac{\log(S_t/K) + (r_d - r_f \pm \sigma^2/2)(T-t)}{\sigma\sqrt{T-t}},$$

where r_f is the risk-free rate in the foreign currency and r_d is the risk-free rate in the domestic currency. The put price is

$$P_t(S_t, K, T) = -e^{r_f(T-t)}S_t\Phi(-z_+) + e^{-r_d(T-t)}K\Phi(-z_-).$$

The price of the call option can be written as

$$C_t(S_t, K, T) = e^{-r_d(T-t)}E(S_T - K)_+,$$

where the expectation is with respect to the risk-neutral measure. The expectation is calculated similarly as in (14.51).

The call price $C_t(S_t, K, T)$ on the exchange rate is obtained from the Black–Scholes price of a call option on a stock when S_t is replaced by the exchange rate S_t and $Ke^{-r(T-t)}$ is replaced by $Ke^{-(r_d-r_f)(T-t)}$.

14.3.4.5 Down-and-Out Call

A down-and-out call on stock S has the payoff

$$C_T = \begin{cases} 0, & \text{when } \min_{0\le t\le T} S_t \le H, \\ \max\{0, S_T - K\}, & \text{otherwise,} \end{cases}$$

where $K > 0$ is the strike price and $H > 0$ is the barrier. We assume that the stock price has a log-normal distribution

$$S_T \sim S_t \exp\left\{ r - \frac{1}{2}\sigma^2(T-t) + \sigma\sqrt{T-t}\, Z \right\},$$

under the risk-neutral measure, where $Z \sim N(0,1)$. The price of the down-and-out call is

$$C_t(S_t, K, T, H) = C_t(S_t, K, T) - J_t,$$

where $C_t(S_t, K, T)$ is the Black–Scholes price of the vanilla call and

$$J_t = S_t \left(\frac{H}{S_t}\right)^{2(r-\sigma^2/2)/\sigma^2+2} \Phi\left(z_+^b\right)$$

$$-Ke^{-r(T-t)}\left(\frac{H}{S_t}\right)^{2(r-\sigma^2/2)/\sigma^2+2} \Phi\left(z_-^b\right),$$

where

$$z_\pm^b = \frac{\log(H^2/(S_t K)) + (r \pm \sigma^2/2)(T-t)}{\sigma\sqrt{T-t}}.$$

14.4 Black–Scholes Hedging

The hedging coefficients of calls and puts are given in (14.61). The Black–Scholes hedging coefficients are equal to the deltas of options. A delta is the derivative of the price of the option with respect to the price of the underlying: the call and put deltas are

$$C_s = \frac{\partial C_t(S)}{\partial S}, \quad P_s = \frac{\partial P_t(S)}{\partial S}.$$

This is shown in Section 14.2.3; see (14.57).[13] For the Black–Scholes pricing functions the call delta and the put delta are shown in (14.62) and (14.63) to be equal to

$$C_s = \Phi(z_+) \tag{14.76}$$

and

$$P_s = -\Phi(-z_+). \tag{14.77}$$

---

13 Let us explain informally the relation of an option delta to the hedging coefficient. We consider the wealth of the writer of the option at time $t + \Delta t$, when the option is written at time t. The wealth of the writer of the option is equal to the hedging error. We have

$$e_{t+\Delta t} = e^{r\Delta t}C_t(S_t) + \xi_{t+\Delta t}(S_{t+\Delta t} - e^{r\Delta t}S_t) - C_{t+\Delta t}(S_{t+\Delta t}),$$

where $\xi_{t+\Delta t}$ is the number of stocks that are bought at time t to hedge the position until $t + \Delta t$. The amount $\xi_{t+\Delta t}S_t$ is borrowed with the risk-free rate at time t. If the hedging gives a position without risk, we should have $e_{t+\Delta t} = 0$, which gives

$$\xi_{t+\Delta t} = \frac{C_{t+\Delta t}(S_{t+\Delta t}) - e^{r\Delta t}C_t(S_t)}{S_{t+\Delta t} - e^{r\Delta t}S_t} \approx \frac{C_{t+\Delta t}(S_{t+\Delta t}) - C_t(S_t)}{S_{t+\Delta t} - S_t} \approx C_t'(S_t),$$

when $\Delta t \to 0$. This gives the instantaneous optimal hedging coefficient as

$$\xi_t = \frac{\partial C_t(S)}{\partial S}\bigg|_{S=S_t}.$$

In this section, our purpose is to illustrate how hedging can be used to approximately replicate options, and to study how hedging frequency, expected stock returns, and the volatility of stock returns affect the replication. The study is made using Black–Scholes hedging, since Black–Scholes hedging provides a benchmark for comparing various hedging methods.

We take the purpose of hedging to be to make the probability distribution of the hedging error of the writer of the option as concentrated around zero as possible. We consider S&P 500 options and estimate the distribution of the hedging error of the writer using the S&P 500 daily data of Section 2.4.1.

Section 14.4.1 reviews historical simulation for the estimation of the distribution of the hedging error. Section 14.4.2 studies the effect of hedging frequency to the distribution of the terminal wealth. Section 14.4.3 studies the effect of the strike price, Section 14.4.4 studies the effect of the mean return, and Section 14.4.5 studies the effect of the return volatility.

14.4.1 Hedging Errors: Nonsequential Volatility Estimation

We have discussed the estimation of the hedging error using historical simulation in Section 13.3.1. Let us write the formulas for hedging error again, this time taking into account the varying hedging frequencies.

14.4.1.1 Hedging Errors
The hedging error e_T of the writer of the option is obtained from (13.10) by the formula

$$e_T = C_t + G_T(\xi) - C_T, \tag{14.78}$$

where

$$G_T(\xi) = \sum_{k=t+1}^{T} \xi_k (S_k - S_{k-1}),$$

when we take the risk-free rate $r = 0$. Here C_t is the price of the option, C_T is the terminal value of the option, ξ_k are the hedging coefficients, and S_k are the stock prices. In (14.78) the current time is denoted by t, the time to expiration is $T - t$ days, and hedging is done daily. The hedging can be done with a lesser frequency. When hedging is done n times during the period $[t, T]$, then

$$G_T(\xi) = \sum_{k=1}^{n} \xi_{t_{k,n}} (S_{t_{k,n}} - S_{t_{k-1,n}}), \tag{14.79}$$

where

$$t_{k,n} = t + k\Delta t, \quad \Delta t = \frac{T - t}{n}.$$

14.4.1.2 Historical Simulation

Let us denote the time series of observed historical daily prices by $S_0, \ldots, S_N$. Let us denote the time to the expiration by

$$U = T - t.$$

(The interpretation is that the last observation S_N is made at time t, and we are interested to write an option whose expiration time is $T > t$.) We construct $N - U$ sequences of prices:

$$S_i = (S_{i,i}, \ldots, S_{i,i+U}), \quad i = 1, \ldots, N - U,$$

where

$$S_{i,i+j} = 100 \cdot S_{i+j}/S_i,$$

for $j = 0, \ldots, U$. Each sequence has length $U + 1$ and the initial prices are always $S_{i,i} = 100$. We estimate the distribution of the hedging error e_T from the observations

$$e_T^1, \ldots, e_T^{N-U},$$

where e_T^i is computed from the prices S_i. For the estimation of the density we use both the histogram estimator and the kernel density estimator, defined in Section 3.2.2.

We consider hedging of call options. The Black–Scholes prices are given in (14.58) and the hedging coefficients are given in (14.62). The volatility σ is estimated using the annualized sample standard deviation, computed from the complete data. We study the effect of volatility estimation to hedging in Section 14.5, where sequential (out-of-sample) volatility estimation is studied. In this section, we use in-sample volatility estimation.

More precisely, the ith hedging error is

$$e_T^i = C_0 + \sum_{k=i+1}^{i+U} \xi_{i,k}(S_{i,k} - S_{i,k-1}) - C_{i,U}, \qquad (14.80)$$

where C_0 is the Black–Scholes price from (14.58), computed with the stock price $S = 100$, time to expiration U, and volatility $\sigma = \hat{\sigma}$. Here $\hat{\sigma}$ is the annualized sample standard deviation computed from return data

$$R_1, \ldots, R_N,$$

where

$$R_i = \frac{S_i}{S_{i-1}},$$

$i = 1, \ldots, N$. Thus, the volatility is estimated in-sample. The hedging coefficient $\xi_{i,k}$ is the Black–Scholes delta from (14.62), computed with the stock price

$S = S_{i,k-1}$, time to expiration $U + i - k + 1$, and volatility $\sigma = \hat{\sigma}$. The terminal value is $C_{i,U} = (S_{i,i+U} - K)_+$.

14.4.2 Hedging Frequency

In this section, we illustrate how a change in the hedging frequency changes the distribution of the hedging error. We study hedging of S&P 500 call options, using daily data of Section 2.4.1. The time to expiration is 20 days. The volatility in the Black–Scholes formula is the non-sequential annualized sample standard deviation.

14.4.2.1   Moneyness $S/K = 1$

Figure 14.13 shows time series of hedging errors. Panel (a) shows the case when there is no hedging, so that $e^i_T = C_0 - C_{i,U}$. Panel (b) shows the case when the hedging is done daily. Moneyness is $S/K = 1$.

Figure 14.14 shows tails plots of the hedging errors. Panel (a) shows the left tail plots and panel (b) shows the right tail plots. We show cases of no hedging (red), hedging once (black), hedging twice (blue), and hedging 20 times (green).

Figure 14.15 shows (a) histograms of hedging errors and (b) kernel density estimates of the distribution of hedging errors. Panel (a) shows cases of no hedging (red) and daily hedging (green). Panel (b) shows additionally the cases of hedging once (black) and hedging twice (blue). Note that the red kernel density estimate is very inaccurate, because the underlying distribution is such that a large part of the probability mass is concentrated at one point.

14.4.2.2   Moneyness $S/K = 0.95$

Figure 14.16 shows tails plots of the hedging errors. Panel (a) shows the left tail plots and panel (b) shows the right tail plots. We show cases of no hedging (red), hedging once (black), hedging twice (blue), and hedging 20 times (green).

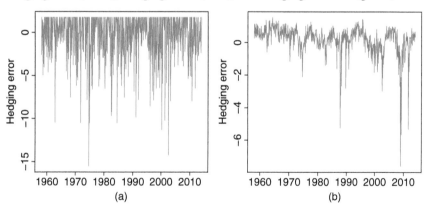

Figure 14.13 *Hedging frequency: Time series of hedging errors.* (a) There is no hedging; (b) hedging is done daily.

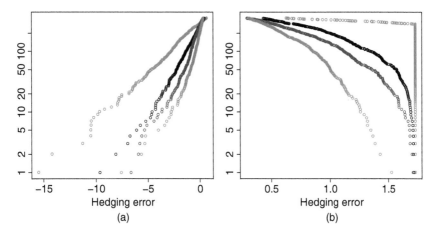

Figure 14.14 *Hedging frequency: Tail plots.* (a) Left tail plots and (b) right tail plots. We show cases of no hedging (red), hedging once (black), hedging twice (blue), and hedging 20 times (green).

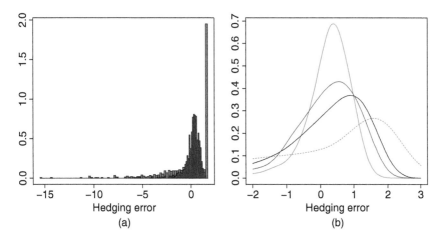

Figure 14.15 *Hedging frequency: Density estimates of hedging errors.* (a) Histograms; (b) kernel density estimates. Panel (a) shows cases of no hedging (red) and hedging 20 times (green). Panel (b) shows additionally the cases of hedging once (black) and hedging twice (blue).

Figure 14.17 shows (a) histograms of hedging errors and (b) kernel density estimates of the distribution of hedging errors when the moneyness is $S/K = 0.95$. Panel (a) shows cases of no hedging (red) and daily hedging (green). Panel (b) shows additionally the cases of hedging once (black) and hedging twice (blue). The red kernel density estimate is very inaccurate, because the underlying distribution is such that a large part of the probability mass is concentrated at one point.

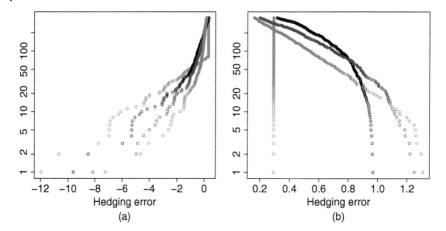

Figure 14.16 *Hedging frequency: Tail plots with moneyness* $S/K = 0.95$. (a) Left tail plots; (b) right tail plots. We show cases of no hedging (red), hedging once (black), hedging twice (blue), and hedging 20 times (green).

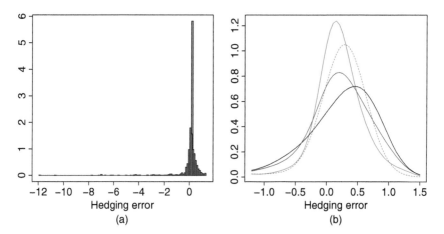

Figure 14.17 *Hedging frequency: Density estimates of hedging errors with moneyness* $S/K = 0.95$. (a) Histograms; (b) kernel density estimates. Panel (a) shows cases of no hedging (red) and hedging 20 times (green). Panel (b) shows additionally the cases of hedging once (black) and hedging twice (blue).

14.4.2.3 Expected Utility

Figure 14.18 shows the estimated expected utility as a function of the hedging frequency. In panel (a) the moneyness is $S/K = 1$. In panel (b) $S/K = 0.95$. We apply the exponential utility function $U(x) = 1 - e^{\alpha x}$, where the risk-aversion parameter takes values $\alpha = 0.001$ (black), $\alpha = 0.005$ (red), and $\alpha = 0.01$ (blue). The expected utilities are estimated using the sample averages. Increasing hedging frequency clearly increases the expected utility. We can see that when

Figure 14.18 *Expected utility as a function of hedging frequency.* (a) The moneyness is $S/K = 1$; (b) $S/K = 0.95$. The risk aversion parameter takes values $\alpha = 0.001$ (black), $\alpha = 0.005$ (red), and $\alpha = 0.01$ (blue).

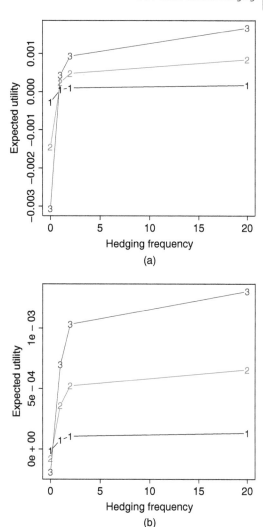

(a)

(b)

the risk aversion is small, then the increase in the expected utility as a function of hedging frequency is much smaller than the increase when the risk aversion is large.

14.4.3 Hedging and Strike Price

In this section, we illustrate how a change in the strike price changes the distribution of the hedging error. We study hedging of S&P 500 call options, using daily data of Section 2.4.1. The time to expiration is 20 days and hedging is done daily. The volatility in the Black–Scholes formula is the nonsequential annualized sample standard deviation.

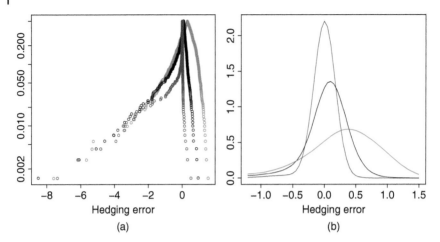

Figure 14.19 *Several strike prices.* (a) Tail plots; (b) kernel density estimates. The strike prices are $K = 100$ (red), $K = 105$ (black), and $K = 110$ (blue), when the stock price is $S = 100$.

Figure 14.19(a) shows tail plots of hedging errors and panel (b) shows kernel density estimates of the distribution of the hedging error. The strikes prices of calls are $K = 100$ (red), $K = 105$ (black), and $K = 110$ (blue), when the stock price is $S = 100$. For in-the-money options the distributions have a larger spread than for out-of-the-money options. Also, the distributions for at-the-money options have a center that is located to the right from the centers for out-of-the-money options.

14.4.4 Hedging and Expected Return

We study the effect of the mean return to the distribution of the hedging error. This is done by manipulating the S&P 500 data. We change observations so that the new net returns are

$$\tilde{r}_i = r_i - \bar{r} + \mu,$$

where $r_i = S_i/S_{i-1} - 1$ are the observed net returns, $\bar{r}$ is the sample mean of r_i, and μ is a value for the expected return that we choose. The new prices are obtained from the net returns by

$$\tilde{S}_i = \prod_{j=1}^{i}(\tilde{r}_j + 1).$$

For the S&P 500 returns the annualized mean return is about 8%. We try the annualized expected returns 50% and −8%. Thus, $\mu = 0.5/250$ and $\mu = -0.08/250$.

Figure 14.20 shows histogram estimates of the distribution of the hedging error when the expected annualized return is 50%. In panel (a) there is no

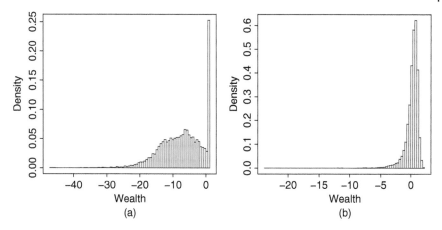

Figure 14.20 *No hedging versus daily hedging with a large positive drift.* (a) No hedging: A histogram from realizations of $C_0 - C_T$. (b) Daily hedging: A histogram from realizations of $C_0 + G_T(\xi) - C_T$.

hedging and in panel (b) there is daily hedging. That is, panel (a) shows a histogram made from realizations of the random variable $C_0 - C_T$, where $C_T = (S_T - K)_+$, and C_0 is the Black–Scholes price. Panel (b) shows a histogram made from realizations of the random variable $C_0 + G_T(\xi) - C_T$. The time to the expiration of the call option is $T = 61$ days, and the strike price is $K = 105$ with the initial stock price $S_t = 100$. Since μ is large the call option gives a profit to its owner with a large probability, as can be seen from panel (a). However, large μ does not change much the distribution of the hedging error when hedging is done daily, as can be seen from panel (b). In fact, the corresponding distribution of the hedging error when the expected return is moderate is shown in Figure 14.15.

Figure 14.21 shows the setting of Figure 14.20 when the annualized expected return is -8%, instead of 50%. Panel (a) shows the distribution of the hedging error of the writer when no hedging is done and panel (b) shows the hedging error when delta hedging is done daily. Since the expected return is negative, the writer of the call option gets a profit with a large probability, but this does not affect much the hedging error when the hedging is done daily, as can be seen from panel (b).

(1) When the drift is larger than the risk-free rate, then the Black–Scholes price C_0 is smaller than the expectation $e^{-r(T-t)}E(S_T - K)_+$. The possibility of hedging makes C_0 smaller than $e^{-r(T-t)}E(S_T - K)_+$. The expectation $E(S_T - K)_+$ increases when the drift increases but the possibility of hedging makes the price independent of the drift. (2) When the drift is equal to the risk-free rate, then $e^{-r(T-t)}E(S_T - K)_+$ is close to C_0, but hedging reduces the risk of the writer of the option, because it changes the wealth distribution of the writer of the option.

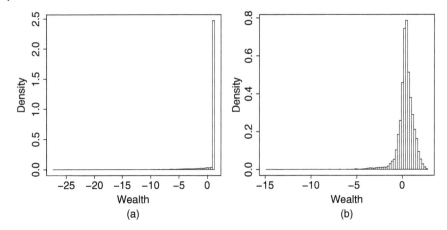

Figure 14.21 *No hedging versus daily hedging with a negative drift.* (a) No hedging: A histogram from realizations of $C_0 - C_T$. (b) Daily hedging: A histogram from realizations of $C_0 + G_T(\xi) - C_T$.

(3) When the drift is negative, then $e^{-r(T-t)}E(S_T - K)_+$ is smaller than C_0, and hedging reduces the expected profit of the writer of the option. However, the hedging reduces also the risk of the writer of the option, and thus hedging is reasonable even in the case of negative drift.

14.4.5 Hedging and Volatility

We study the effect of the return volatility to the distribution of the hedging error. We manipulate the S&P 500 data by changing observations so that the new net returns are

$$\tilde{r}_i = \frac{\sigma}{\hat{\sigma}}(r_i - \bar{r}) + \bar{r},$$

where $r_i = S_i/S_{i-1} - 1$ are the observed net returns, $\hat{\sigma}$ is the sample standard deviation of r_i, $\bar{r}$ is the sample mean of r_i, and σ is a value of our choice for the volatility. The new prices are obtained from the net returns by

$$\tilde{S}_i = \prod_{j=1}^{i}(\tilde{r}_j + 1).$$

For the S&P 500 returns the annualized sample standard deviation is about 15%. We try the annualized standard deviation 50%. Thus, $\sigma = 0.5/\sqrt{250}$.

Figure 14.22 shows histogram estimates of the distribution of the hedging error when the annualized volatility is 50%. In panel (a) there is no hedging, and in panel (b) there is daily hedging. That is, panel (a) shows a histogram made from realizations of the random variable $C_0 - C_T$, where $C_T = (S_T - K)_+$,

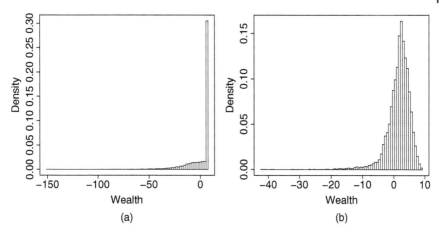

Figure 14.22 *Large volatility.* (a) No hedging: A histogram of realizations of $C_0 - C_T$. (b) Daily hedging: A histogram of realizations of $C_0 + G_T(\xi) - C_T$.

and C_0 is the Black–Scholes price. Panel (b) shows a histogram made from realizations of the random variable $C_0 + G_T(\xi) - C_T$. The time to the expiration is $T = 61$ days, and the strike price is $K = 105$ with the initial stock price $S_t = 100$. We see that the larger volatility makes the dispersion of the probability distribution of the hedging error larger.

14.5 Black–Scholes Hedging and Volatility Estimation

We continue to study the distribution of the hedging error e_T, as in Section 14.4. In this section, our aim is to study how the volatility estimation affects the distribution of the hedging error. The Black–Scholes prices and the hedging coefficients depend on the annualized volatility σ. We compare the performance of GARCH(1, 1) and exponentially weighted moving averages for the estimation of σ. The performance of Black–Scholes hedging will be used as a benchmark.

14.5.1 Hedging Errors: Sequential Volatility Estimation

We have discussed in Section 13.3.1 the estimation of the distribution of the hedging error using historical simulation. We write again the formulas for the hedging error and historical simulation, adapting to the current setting.

14.5.1.1 Hedging Errors
The hedging error e_T of the writer of the option is obtained from (13.10) as

$$e_T = C_0 + G_T(\xi) - C_T,$$

where

$$G_T(\xi) = \sum_{k=1}^{T} \xi_k(S_k - S_{k-1}).$$

Here the risk-free rate is $r = 0$, C_0 is the price of the option, C_T is the terminal value of the option, ξ_k are the hedging coefficients, S_k are the stock prices, the current time is denoted by 0, the time to expiration is T days, and hedging is done daily.

14.5.1.2 Historical Simulation

We denote the time series of observed historical daily prices by $S_0, \dots, S_N$. We construct $N - T$ sequences of prices:

$$S_i = (S_{i,i}, \dots, S_{i,i+T}), \quad i = 1, \dots, N - T,$$

where

$$S_{i,i+j} = 100 \cdot S_{i+j}/S_i,$$

for $j = 0, \dots, T$. Each sequence has length $T + 1$ and the initial price in each sequence is $S_{i,i} = 100$. We estimate the distribution of the hedging error e_T from the observations

$$e_T^1, \dots, e_T^{N-T},$$

where e_T^i is computed from the prices S_i.

More precisely, the ith hedging error is

$$e_T^i = C_{i,0} + \sum_{k=i+1}^{T} \xi_{i,k}(S_{i,k} - S_{i,k-1}) - C_{i,T}, \tag{14.81}$$

where $C_{i,0}$ is the Black–Scholes price from (14.58), computed with the stock price $S = 100$, time to expiration T, and volatility $\sigma = \hat{\sigma}_i$. Here $\hat{\sigma}_i$ is the annualized volatility estimate computed using return data

$$R_1, \dots, R_i,$$

where

$$R_i = \frac{S_i}{S_{i-1}}, \quad i = 1, \dots, N - T.$$

Thus, the volatility estimation is done sequentially (out-of-sample). The hedging coefficient $\xi_{i,k}$ is the Black–Scholes delta from (14.62), computed with the stock price $S = S_{i,k-1}$, time to expiration $T - k$, and volatility $\sigma = \hat{\sigma}_{i+k-1}$. The terminal value is $C_{i,T} = (S_{i,T} - K)_+$.

For the estimation of the density we use both the histogram estimator and the kernel density estimator, defined in Section 3.2.2.

14.5.2 Distribution of Hedging Errors

In the following examples the time to expiration is $T = 20$ trading days. We start pricing and hedging after 4 years of data has been collected. The risk-free rate is equal to zero. We consider S&P 500 call options and use the daily S&P 500 data, described in Section 2.4.1.

As a summary of the results, we can note that the GARCH(1, 1) and the exponential moving average (with a suitable smoothing parameter) improve the distribution of the hedging error from the point of view of the writer of the option, when compared to the sequential sample standard deviation. GARCH(1, 1) and the exponential moving average lead to a distribution whose left tail is lighter: with these volatility estimators the losses of the writer of the option are smaller. On the other hand, GARCH(1, 1) and the exponential moving average lead to larger positive hedging errors. The positive hedging errors are gains for the writer of the option.

Figure 14.23 shows (a) the means of negative hedging errors and (b) the means of positive hedging errors as a function of the moneyness S/K. The arithmetic means of positive and negative hedging errors are defined as

$$\bar{e}_+ = \text{mean}\left(\max\left\{e_T^i, 0\right\}\right), \quad \bar{e}_- = -\text{mean}\left(\min\left\{e_T^i, 0\right\}\right),$$

where the hedging errors e_T^i are defined in (14.81). The hedging is done with sequentially computed sample standard deviation (red with "s"), GARCH(1, 1) volatility (blue with "g"), the exponential moving average with $h = 1$ (yellow with "1"), and with smoothing parameter $h = 40$ (green with "2"). We see that

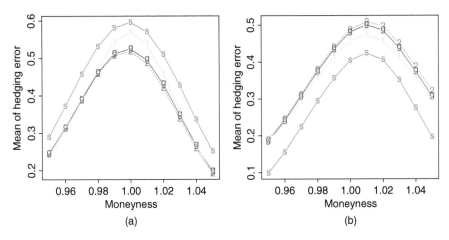

Figure 14.23 *Means of hedging errors.* The figure shows the means of (a) negative hedging errors as a function of moneyness and (b) positive hedging errors. The volatility is estimated by the sequentially computed sample standard deviation (red with "s"), GARCH(1,1) (blue with "g"), the exponential moving average with $h = 1$ (yellow with "1"), and with $h = 40$ (green with "2").

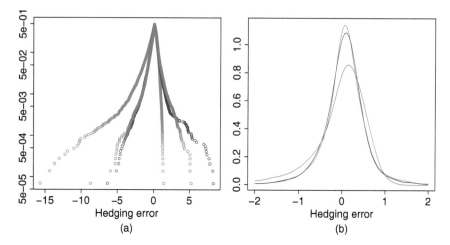

Figure 14.24 *Distribution of hedging errors.* The figure shows (a) tail plots of hedging errors and (b) kernel density estimates of the distribution of the hedging error. The volatility is estimated by the sequentially computed sample standard deviation (red), GARCH(1, 1) (blue), and the exponential moving average with $h = 40$ (green).

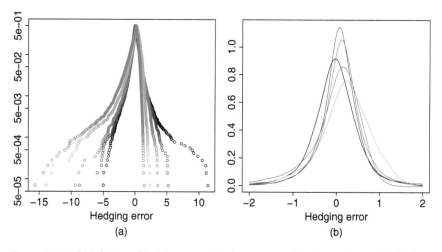

Figure 14.25 *Distribution of hedging errors: Moving averages.* The figure shows (a) tail plots of hedging errors and (b) kernel density estimates of the distribution of the hedging error. The volatility is estimated by the sequentially computed standard deviation (red), and by the exponentially weighted moving average with $h = 1000$ (purple), $h = 100$ (green), $h = 40$ (blue), and $h = 5$ (black).

the sample standard deviation is the best for the positive hedging errors and the worst for the negative hedging errors. The performance of GARCH(1, 1) and the exponentially weighted moving average with $h = 40$ are close to each other. The exponentially weighted moving average with $h = 1$ has the performance between the sample standard deviation and the GARCH(1, 1). We can see that hedging errors are larger for the at-the-money options than for the out-of-the-money options.

Figure 14.24 shows (a) tail plots of hedging errors and (b) kernel density estimates of the distribution of the hedging error. The moneyness of call options is $S/K = 1$. The volatility is estimated by the sequentially computed standard deviation (red), by GARCH(1, 1) (blue), and by the exponentially weighted moving average with $h = 40$ (green). Tail plots are defined in Section 3.2.1 and the kernel density estimator is defined in Section 3.2.2. We apply the standard normal kernel function and the smoothing parameter is chosen by the normal reference rule.

Figure 14.25 shows (a) tail plots of hedging errors and (b) kernel density estimates of the hedging error. The volatility is estimated by the sequentially computed standard deviation (red), and by the exponentially weighted moving average with $h = 1000$ (purple), $h = 100$ (green), $h = 40$ (blue), and $h = 5$ (black).

15

Pricing in Incomplete Models

We give an overview of various approaches to price derivatives in incomplete markets. In an incomplete market, there exists derivatives which cannot be exactly replicated. The second fundamental theorem of asset pricing (Theorem 13.3) states that if the market is arbitrage-free and complete, then there is only one equivalent martingale measure, and thus there is only one arbitrage-free price. When the arbitrage-free market is not complete, then there are many equivalent martingale measures, and thus there are many arbitrage-free prices.

In this chapter, we describe some approaches for choosing the equivalent martingale measure from a set of available equivalent martingale measures, in the case of an incomplete market. Chapter 16 is devoted to the study of quadratic hedging and pricing. In this chapter, we give only a short description of quadratic pricing, and concentrate to describe other methods.

Utility maximization provides a general method for the construction of an equivalent martingale measure. We show that the Esscher measure, which was used to prove the first fundamental theorem of asset pricing (Theorem 13.1), is related to the maximization of the expected utility, when the utility function is the exponential utility function. The concept of marginal rate of substitution provides a heuristic way to connect the utility maximization to the pricing of options. Minimizing the relative entropy between a martingale measure and the physical measure provides an equally natural way to construct an equivalent martingale measure. Again, we can show that minimizing the relative entropy is related to the maximizing the expected utility with the exponential utility function. It is of interest to compare the Esscher prices to the Black–Scholes prices: we see that the Esscher prices are close to the Black–Scholes prices for at-the-money calls, whereas for out-of-the-money calls the Esscher prices are lower.

We describe formulas for constructing an equivalent martingale measure by an absolutely continuous change of measure. These formulas are given for conditionally Gaussian returns, and for conditionally Gaussian logarithmic

Nonparametric Finance, First Edition. Jussi Klemelä.
© 2018 John Wiley & Sons, Inc. Published 2018 by John Wiley & Sons, Inc.

returns. An absolutely continuous change of measure shifts the market measure so that it becomes risk-neutral.

GARCH models provide good volatility predictions, and it is natural to ask whether GARCH models could be suitable for pricing options. We can apply the absolutely continuous change of measure to obtain an equivalent martingale measure in the GARCH market model. The standard GARCH(1, 1) model can be modified so that we obtain a model where the prices can be expressed almost in a closed form. We still need a numerical integration to compute the prices, but Monte Carlo simulation of price trajectories is not needed. The modified GARCH(1, 1) model was presented in Heston and Nandi (2000).

It is on interest to construct a nonparametric pricing method, and compare its properties to other methods. We construct a method which combines historical simulation, the Esscher measure, and conditioning on the current volatility.

An equivalent martingale measure can be deduced from the market prices of the options. This martingale measure could be called the implied martingale measure, because there is an analogy to the implied volatility. When the implied martingale measure is used, we have to assume that the market prices of the options are rational. This requires that we use option prices of liquid markets to estimate the implied martingale measure. The implied martingale measure which is deduced from the prices of liquid options can be used to price illiquid options.

Section 15.1 describes quadratic hedging and pricing. Section 15.2 describes pricing with the help of utility maximization. Section 15.3 considers pricing with the help of absolutely continuous changes of measures (Girsanov's theorem). Section 15.4 describes the use of a GARCH model in option pricing. Section 15.5 describes a method of nonparametric pricing which uses historical simulation. Section 15.6 discusses pricing with the help of estimating the risk-neutral density. Section 15.7 mentions quantile hedging.

Pricing in incomplete markets is studied in Duffie and Skiadas (1994), El Karoui and Quenez (1995), Karatzas (1996), and Gourioux *et al.* (1998). Bingham and Kiesel (2004, Chapter 7) discuss pricing in incomplete models, including mean–variance hedging and models driven by Lévy processes. Pricing of derivatives in the context of general econometric theory is presented in Magill and Quinzii (1996). Further references include Karatzas and Kou (1996).

15.1 Quadratic Hedging and Pricing

Quadratic hedging and pricing is discussed in detail in Chapter 16 (see also Föllmer and Schied, 2002, Definition 10.36, p. 393). At this point, we give a brief summary of the method.

Let us explain the idea of quadratic hedging using the case with one risky asset ($d = 1$) and two periods ($T = 2$). The initial wealth is W_0 and the wealth obtained by trading with a bond and a stock is

$$W_t = \beta_t B_t + \xi_t S_t, \quad t = 1, 2,$$

where B_t is the price of the bond, S_t is the price of the stock, β_t is the number of bonds in the portfolio, and ξ_t is the number of stocks in the portfolio. Our aim is to replicate the terminal value C_2 of the contingent claim. We measure the quality of the approximation by the quadratic hedging error

$$E(C_2 - W_2)^2.$$

The minimization is done over self-financing trading strategies and over the initial wealth W_0. The self-financing means that (β_1, ξ_1) and (β_2, ξ_2) satisfy

$$\beta_1 B_0 + \xi_1 S_0 = W_0, \quad \beta_2 B_1 + \xi_2 S_1 = W_1.$$

The self-financing restriction connects the initial wealth W_0 to the final wealth W_2. The quadratic price is the initial wealth W_0 that minimizes the quadratic hedging error.

The minimization is done easier when we use the value process, instead of the wealth process. Our final formulation for the general case $d \geq 1$ and $T \geq 1$ will be the following. In quadratic hedging, the quadratic hedging error

$$E(H_T - V_0 - G_T(\xi))^2$$

is minimized among strategies $\xi = (\xi_t)_{t=1,\dots,T}$[1] and among the initial investment $V_0 \in \mathbf{R}$, where $H_T = C_T/B_T$ is the discounted contingent claim. The terminal value of the gains process is defined by

$$G_T(\xi) = \sum_{t=1}^{T} \xi_t \cdot (X_t - X_{t-1}),$$

where $X_t = S_t/B_t$ is the discounted price vector.

15.2 Utility Maximization

It turns out that an equivalent martingale measure can be found by looking at the portfolios which maximize the expected utility. Section 15.2.1 shows that the Esscher martingale measure is related to the maximization of the expected utility with the exponential utility function. Section 15.2.2 considers other utility functions. Section 15.2.3 shows that the Esscher measure is the equivalent martingale measure which minimizes the relative entropy with respect

---

1 The minimizing is done over $\xi = (\xi_t)_{t=1,\dots,T} \in S$, where

$$S = \{\xi : \xi \text{ is predictable and } G_t(\xi) \in \mathcal{L}^2(P) \text{ for all } t\},$$

where $\mathcal{L}^2(P)$ is the set of square integrable random variables.

to the physical measure. Section 15.2.4 computes examples of Esscher prices. Section 15.2.5 discusses the heuristics of the marginal rate of substitution.

The use of Esscher transform and more general methods of utility maximization has to be combined with the estimation of the underlying distribution of the stock price process. This issue has been addressed in Bühlmann *et al.* (1996), Siu *et al.* (2004), Christoffersen *et al.* (2006), and Chorro *et al.* (2012), where parametric modeling was used. We address the issue in Section 15.5, where a nonparametric estimation is applied (see also Section 15.2.4).

15.2.1 The Exponential Utility

The Esscher transformation was applied in the proof of Theorem 13.1 (the first fundamental theorem of asset pricing) to construct an equivalent martingale measure in an arbitrage-free market. Let us recall the definition of the Esscher measure for the case of one risky asset. Let $X_t = S_t/B_t$ be the discounted stock price, where $t = 0, \dots, T$. Let

$$\Delta X_t = X_t - X_{t-1}, \quad t = 1, \dots, T.$$

Let

$$\phi_t(a, \omega) = E_P\left(e^{a\Delta X_t} \mid \mathcal{F}_{t-1}\right)(\omega),$$

where $a \in \mathbf{R}$. Let

$$a_t^* = a_t^*(\omega) \tag{15.1}$$

be the unique finite minimizer of $\phi_t(a, \omega)$ over $a \in \mathbf{R}$. Let $Z_0 = 1$ and

$$Z_t(\omega) = \prod_{i=1}^{t} z_i(\omega), \quad z_t(\omega) = \frac{\exp\left\{a_t^*(\omega)\Delta X_t(\omega)\right\}}{E_P\left(\exp\left\{a_t^*\Delta X_t\right\} \mid \mathcal{F}_{t-1}\right)(\omega)}$$

for $t = 1, \dots, T$. We define the probability measure

$$Q(d\omega) = Z_T(\omega)P(d\omega). \tag{15.2}$$

Now $Q \approx P$, $E_Q|\Delta X_t| < \infty$, and $E_Q(\Delta X_t \mid \mathcal{F}_{t-1}) = 0$ for $t = 1, \dots, T$. Hence $\Delta X_1, \dots, \Delta X_T$ is a martingale difference and $X = (X_t)_{t=0,\dots,T}$ is a martingale with respect to Q.

The Esscher martingale measure is related to the maximization of the expected utility with the exponential utility function. The exponential utility function is defined as

$$U(x) = 1 - e^{-ax}, \quad x \in \mathbf{R},$$

where $\alpha > 0$ is the parameter of risk aversion. We want to maximize

$$E_P U(V_T)$$

over self-financing trading strategies $\xi = (\xi_1, \ldots, \xi_T)$, where

$$V_T = V_0 + G_T, \quad G_T = \sum_{t=1}^{T} \xi_t \Delta X_t.$$

The maximization is equivalent to the minimization of

$$E \exp\{-\alpha G_T\}.$$

We can write

$$E \exp\{-\alpha G_T\}$$
$$= E \exp\{-\alpha \xi_1 \Delta X_1\} \times E_1 \exp\{-\alpha \xi_2 \Delta X_2\} \times \cdots \times E_{T-1} \exp\{-\alpha \xi_T \Delta X_T\},$$

where $E_t = E_P(\cdot \mid \mathcal{F}_t)$. Minimization of

$$E_{t-1} \exp\{-\alpha \xi_t \Delta X_t\}$$

over $\mathcal{F}_{t-1}$-measurable ξ_t is equivalent to the minimization of

$$\phi_t(a) = E_{t-1} \exp\{-a \Delta X_t\}$$

over $a \in \mathbf{R}$. Thus, we arrive at the minimizer a_t^* in (15.1), and we can define an equivalent martingale measure (15.2) with the help of these minimizers.

15.2.2 Other Utility Functions

The construction of the Esscher martingale measure can be generalized to cover other utility functions. For example, consider the one period model with d risky assets and let $\Delta X = (\Delta X^1, \ldots, \Delta X^d)$ be the vector of discounted net gains:

$$\Delta X^i = \frac{S_1^i}{1+r} - S_0^i, \quad i = 1, \ldots, d,$$

where $S_0^1, \ldots, S_0^d$ are the prices of the risky assets at time $t = 0$. Föllmer and Schied (2002, Corollary 3.10) states that if the market is arbitrage-free, and the utility function U and a maximizer ξ^* of $EU(\xi \cdot \Delta X)$ satisfy certain assumptions,[2] then

$$\frac{dQ}{dP} = \frac{U'(\xi^* \cdot \Delta X)}{EU'(\xi^* \cdot \Delta X)} \tag{15.3}$$

defines an equivalent martingale measure Q.

The proof is based on the fact that a martingale measure has to satisfy $E_Q \Delta X = 0$, and this follows for the measure defined in (15.3) because the maximizer satisfies the first-order condition $E[U'(\xi^* \cdot \Delta X)\Delta X] = 0$. Note

---

2 It is assumed that $U : D \to \mathbf{R}$ is a continuously differentiable utility function. Let $S(D) = \{\xi \in \mathbf{R}^d : P(\xi \cdot \Delta X \in D) = 1\}$. It is assumed that $EU(\xi \cdot \Delta X) < \infty$ for all $\xi \in S(D)$. Let ξ^* be a maximizer of $EU(\xi \cdot \Delta X)$. It is assumed that either (1) $D = \mathbf{R}$ and U is bounded from above, or (2) $D = [a, \infty)$ for some $a < 0$, and ξ^* is an interior point of $S(D)$.

that the existence of a maximizer is implied by Föllmer and Schied (2002, Theorem 3.3).

The Esscher density

$$\frac{dQ}{dP} = \frac{e^{a^* \cdot \Delta X}}{E e^{a^* \cdot \Delta X}} \tag{15.4}$$

is a special case of (15.3). Indeed, when the utility function is the exponential utility function $U(x) = 1 - e^{-\alpha x}$, where $x \in \mathbf{R}$ and $\alpha > 0$ is the risk aversion, then (15.3) gives the density

$$\frac{dQ}{dP} = \frac{e^{-\alpha \xi^* \cdot \Delta X}}{E e^{-\alpha \xi^* \cdot \Delta X}}.$$

Portfolio ξ^* maximizes the expected utility $1 - E e^{-\alpha \xi \cdot \Delta X}$ if and only if $a^* = -\alpha \xi^*$ minimizes the moment generating function $E e^{a \cdot \Delta X}$. Thus, we obtain the density in (15.4), and the martingale measure Q is independent of the risk aversion α.

Note that we have maximized $EU(\xi \cdot \Delta X)$ over ξ, which is not the same as maximizing the expected utility of the wealth. However, in the one-period case the wealth is written in (9.9) as

$$W_1 = (1 + r)(W_0 + \xi \cdot \Delta X).$$

Let u be a strictly increasing, strictly concave, and continuous utility function. We want to find ξ which maximizes $Eu(W_1)$ over all $\xi \in \mathbf{R}^d$ such that W_1 is P-almost surely in the domain of u. Define

$$U(x) = u((1 + r)(W_0 + x)).$$

The original utility maximization is equivalent to the maximization of

$$EU(\xi \cdot \Delta X),$$

among all $\xi \in \mathbf{R}^d$ such that $\xi \cdot \Delta X \in D$, where D is the domain of U.

15.2.3 Relative Entropy

The Esscher martingale measure was shown to be related to the maximization of the expected utility with the exponential utility function. We can also show that the Esscher measure can be obtained by minimizing the Kullback–Leibler distance to the physical market measure.

The closeness of probability distributions can be measured by the relative entropy (the Kullback–Leibler distance). The relative entropy of a probability measure Q with respect to a probability measure P is defined as

$$H(Q|P) = E\left(\frac{dQ}{dP} \log \frac{dQ}{dP}\right),$$

when P dominates Q. When P does not dominate Q, then we define $H(Q|P) = \infty$.

Föllmer and Schied (2002, Corollary 3.25) states the following result for the one-period model with d risky assets: When the market model is arbitrage-free, then there exists a unique equivalent martingale measure $Q \in \mathcal{P}$, which minimizes the relative entropy $H(P'\,|\,P)$ over all $P' \in \mathcal{P}$, where $\mathcal{P}$ is the set of equivalent martingale measures. Furthermore, the density of Q is the Esscher density

$$\frac{dQ}{dP} = \frac{e^{a^* \cdot \Delta X}}{E e^{a^* \cdot \Delta X}},$$

where a^* is the minimizer of the moment generating function $E e^{a \cdot \Delta X}$, and ΔX is the vector of discounted net gains: $\Delta X = S_1/(1+r) - S_0$, where $S_1 = (S_1^1, \ldots, S_1^d)$ is the vector of prices at time $t = 1$, and $S_0 = (S_0^1, \ldots, S_0^d)$ is the vector of prices at time $t = 0$.

15.2.4 Examples of Esscher Prices

We apply the data of S&P 500 daily prices, described in Section 2.4.1. We estimate the Esscher call prices when the time to expiration is $T = 20$ trading days. The estimation is done for the T-period model. We apply nonsequential estimation: the Esscher measure is estimated using the complete time series, and then the prices are estimated using the complete time series together with the estimated Esscher measure. We take the risk-free rate $r = 0$.

We denote the observed historical prices by $S_0, \ldots, S_N$. We construct $M = N - T$ sequences of prices:

$$S_i = (S_{i,0}, \ldots, S_{i,T}), \quad i = 1, \ldots, M,$$

where

$$S_{i,j} = 100 \frac{S_{i+j}}{S_i}, \quad j = 0, \ldots, T.$$

Each sequence has length $T + 1$, and the initial prices are $S_{i,0} = 100$. We apply nonoverlapping sequences, and restrict ourselves to the values $i = (l-1)T + 1$, $l = 1, \ldots, [N/T]$.

Let us compute differences

$$Y_{i,k} = S_{i,k} - S_{i,k-1} \tag{15.5}$$

for $i = 1, \ldots, M$ and $k = 1, \ldots, T$. These differences are a sample of identically distributed observations of price increments $S_t - S_{t-1}$. Note that price increments are not a stationary time series when the time period is long; see Figure 5.2(a). However, in our construction we have made M sequences of prices, each price sequence starts at 100, and thus the price differences make an approximately stationary sequence; see Figure 5.2(b) and (c).

Let us assume that the price increments are independent. Let $\hat{\phi}(a)$ be the sample average of $e^{aY_{i,k}}$. Let a^* be the minimizer of $\hat{\phi}(a)$ over $a \in \mathbf{R}$. Let

$$z(y) = \frac{e^{a^* y}}{\hat{\phi}(a^*)}$$

and

$$\hat{Z}_T(y_1, \ldots, y_T) = \prod_{i=1}^{T} z(y_i).$$

The density of the martingale measure Q with respect to underlying physical measure P of the price increments $(\Delta S_1, \ldots, \Delta S_T)$ is estimated by

$$\frac{Q(dy)}{P(dy)} \approx \hat{Z}_T(y_1, \ldots, y_T), \quad (y_1, \ldots, y_T) \in \mathbf{R}^T.$$

Let C_T be the payoff of a contingent claim. The price implied by the measure Q is

$$C_0 = E_Q C_T \approx E(C_T \hat{Z}_T). \tag{15.6}$$

We have constructed nonoverlapping price increments. The Esscher martingale measure was estimated using these price increments. Next we estimate the price (15.6) using a sample average. Let the contingent claim be $f(S_T)$. The estimate of price C_0 is

$$\hat{C}_0 = \frac{1}{M} \sum_{i=1}^{M} (f(S_{i,T}) \hat{Z}_T(Y_{i,1}, \ldots, Y_{i,T})),$$

where $Y_{i,k}$ are the price differences in (15.5).

Figure 15.1 compares the Esscher call prices to the Black–Scholes prices. The time to maturity is 20 trading days. Panel (a) shows the call prices as a function of moneyness S/K. The Esscher prices are shown with a red curve. The Black–Scholes prices are shown with a black curve. The volatility of the

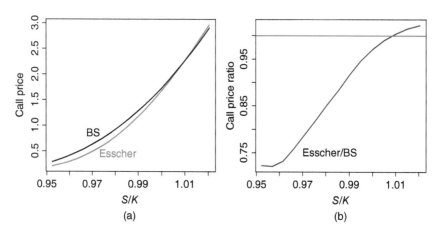

Figure 15.1 *Esscher call prices compared to Black–Scholes prices.* (a) Esscher prices (red) and Black–Scholes prices (black) as a function of S/K. (b) The ratio of Black–Scholes prices to Esscher prices.

Black–Scholes prices is taken as the annualized sample standard deviation over the complete sample. Panel (b) shows the ratio of Black–Scholes prices to Esscher prices as a function of S/K. We see that the Black–Scholes prices are less than the Esscher prices, except for the in-the-money calls. The result confirms with Figure 13.2(a), which shows that the Esscher density takes smaller values than the density of the Black–Scholes martingale measure for large increments.

15.2.5 Marginal Rate of Substitution

A martingale measure can be derived by using an argument based on marginal rate of substitution, as presented in Davis (1997) or in Cochrane (2001).

The value at time T, obtained by a self-financing trading, is written in (13.8) as

$$V_T = V_0 + \sum_{t=1}^{T} \xi_t \cdot (X_t - X_{t-1}).$$

Let us denote by

$$V_T(x, \xi)$$

the value which is obtained when the initial value is $V_0 = x$, and the self-financing trading strategy is $\xi = (\xi_t)_{t=1,...,T}$. The objective is to maximize

$$EU(V_T(x, \xi))$$

over self-financing trading strategies ξ, where $U : \mathbf{R} \to \mathbf{R}$ is a utility function. Utility functions are discussed in Section 9.2.2. The fair price of a derivative could be defined to be such that diverting a little of funds into the derivative has a neutral effect on the investor's achievable utility. Let H_T be a discounted contingent claim. Let us denote

$$\mathcal{W}(\delta, x, p) = \sup_{\xi} EU \left(V_T(x - \delta, \xi) + \frac{\delta}{p} H_T \right),$$

where p is the price of H_T. We define the fair price of H_T at time $t = 0$ to be the solution $p(x)$ of the equation

$$\frac{\partial}{\partial \delta} \mathcal{W}(\delta, x, p) \bigg|_{\delta=0} = 0.$$

It can be proved that

$$p(x) = \frac{E[U'(V_T(x, \xi^*))H_T]}{\tilde{U}'(x)},$$

where ξ^* is the maximizer of $E[U(V_T(x, \xi))]$, and

$$\tilde{U}(x) = E[U(V_T(x, \xi^*))] = \sup_{\xi} E[U(V_T(x, \xi))].$$

Here we assume that $\tilde{U}$ is differentiable at each $x \in (0, \infty)$ and that $\tilde{U}'(x) > 0$. Now

$$\frac{U'(V_T(x, \xi^*))}{\tilde{U}'(x)}$$

is called the stochastic discount factor (pricing kernel, change of measure, state-price density). We have a single discount factor which is pricing each different asset. The price reflects riskiness and the riskiness depends on the covariance of H_T with the pricing kernel, and not the variance. An asset that does badly in recession is less desirable than an asset that does badly in boom, when the assets are otherwise similar.

15.3 Absolutely Continuous Changes of Measures

Girsanov's theorem gives a formula for changing the physical market measure to an equivalent martingale measure. First, we describe formulas for changing the measure when the returns are conditionally Gaussian. Second, we consider conditionally Gaussian logarithmic returns. Note that a version of Girsanov's theorem in continuous time is given in (5.64).

15.3.1 Conditionally Gaussian Returns

Let us assume that the excess returns

$$R_t = \frac{S_t}{S_{t-1}} - \frac{B_t}{B_{t-1}}$$

are conditionally Gaussian. We assume that they satisfy

$$R_t = \mu_t + \sigma_t \epsilon_t$$

for $t = 1, \ldots, T$, where

$$\mathcal{L}(\epsilon_t \mid \mathcal{F}_{t-1}; P) = N(0, 1), \tag{15.7}$$

ϵ_t is $\mathcal{F}_t$-measurable, and μ_t and σ_t are predictable. The notation in (15.7) means that the conditional distribution of ϵ_t, conditional on $\mathcal{F}_{t-1}$, under probability measure P, is the standard normal distribution. Here we assume that $\mathcal{F}_0 = \{\emptyset, \Omega\}$, and $\mathcal{F}_T = \mathcal{F}$. We assume that $\sigma_t > 0$. The assumption on $\epsilon = (\epsilon_1, \ldots, \epsilon_T)$ implies that ϵ is a sequence of independent random variables with $\epsilon_t \sim N(0, 1)$.

15.3.1.1 The Martingale Measure

The equivalent martingale measure Q is such that under Q, the excess returns satisfy

$$R_t = \sigma_t \epsilon_t, \tag{15.8}$$

where $\epsilon = (\epsilon_t)_{t=1,\ldots,T}$ are i.i.d. with $N(0, 1)$.

15.3.1.2 Construction of the Martingale Measure
Let

$$Z_t = \exp\left\{ -\sum_{k=1}^{t} \frac{\mu_k}{\sigma_k}\epsilon_k - \frac{1}{2}\sum_{k=1}^{t}\left(\frac{\mu_k}{\sigma_k}\right)^2 \right\}$$

for $t = 1, \ldots, T$. Now $Z_t > 0$ and $EZ_T = 1$. We define the probability measure Q on $(\Omega, \mathcal{F})$ by

$$Q(d\omega) = Z_T(\omega)P(d\omega). \tag{15.9}$$

Measure Q is an equivalent martingale measure: the process X of discounted prices is a martingale.

The fact that Q is a martingale measure follows from

$$\mathcal{L}(R_t \mid \mathcal{F}_{t-1}; Q) = N(0, \sigma_t^2), \tag{15.10}$$

which is equivalent to

$$\mathcal{L}(\Delta X_t \mid \mathcal{F}_{t-1}; Q) = N\left(0, \left(\frac{S_{t-1}}{B_t}\right)^2 \sigma_t^2\right).$$

The proof can be found in Shiryaev (1999, p. 443). Let us give the main steps of the proof. Let $\lambda \in \mathbf{R}$ and i be the imaginary unit. Then,[3]

$$E_Q(e^{i\lambda R_t} \mid \mathcal{F}_{t-1}) = \frac{1}{Z_{t-1}}E_P(Z_t e^{i\lambda R_t} \mid \mathcal{F}_{t-1})$$

$$= E_P\left(\exp\left\{ i\lambda R_t - \frac{\mu_t}{\sigma_t}\epsilon_t - \frac{1}{2}\left(\frac{\mu_t}{\sigma_t}\right)^2 \right\} \Big| \mathcal{F}_{t-1}\right)$$

$$= \exp\left\{ -\frac{1}{2}\lambda^2\sigma_t^2 \right\}, \tag{15.11}$$

---

3 Let us justify the first equality. Denote $Y = e^{i\lambda R_t}$. For $A \in \mathcal{F}_{t-1}$,

$$E_Q(I_A Y) = E(I_A YZ_t) = E(I_A E(YZ_t \mid \mathcal{F}_{t-1})) = E_Q\left(I_A \frac{1}{Z_{t-1}}E_P(YZ_t \mid \mathcal{F}_{t-1})\right),$$

Q-almost surely. Thus, the first equality follows from the definition of the conditional expectation. For more details, see Föllmer and Schied (2002, Proposition A.12, p. 405) and Shiryaev (1999, p. 438), where terms "conversion lemma" and "generalized Bayes' formula" are used for this equality.

Q-almost surely, since

$$i\lambda R_t - \frac{\mu_t}{\sigma_t}\epsilon_t - \frac{1}{2}\left(\frac{\mu_t}{\sigma_t}\right)^2$$

$$= \left(i\lambda\sigma_t - \frac{\mu_t}{\sigma_t}\right)\epsilon_t + i\lambda\mu_t - \frac{1}{2}\left(\frac{\mu_t}{\sigma_t}\right)^2$$

$$= \left(i\lambda\sigma_t - \frac{\mu_t}{\sigma_t}\right)\epsilon_t - \frac{1}{2}\left(i\lambda\sigma_t - \frac{\mu_t}{\sigma_t}\right)^2$$

$$+ \frac{1}{2}\left(i\lambda\sigma_t - \frac{\mu_t}{\sigma_t}\right)^2 + i\lambda\mu_t - \frac{1}{2}\left(\frac{\mu_t}{\sigma_t}\right)^2$$

$$= \left(i\lambda\sigma_t - \frac{\mu_t}{\sigma_t}\right)\epsilon_t - \frac{1}{2}\left(i\lambda\sigma_t - \frac{\mu_t}{\sigma_t}\right)^2 - \frac{1}{2}\lambda^2\sigma_t^2,$$

and

$$E_P\left(\exp\left[\left(i\lambda\sigma_t - \frac{\mu_t}{\sigma_t}\right)\epsilon_t - \frac{1}{2}\left(i\lambda\sigma_t - \frac{\mu_t}{\sigma_t}\right)^2\right]\Bigg| \mathcal{F}_{t-1}\right) = 1.$$

Equation (15.11) shows that the characteristic function of $R_t \,|\, \mathcal{F}_{t-1}$ under Q is the characteristic function of $N(0, \sigma_t^2)$, which leads to (15.10).

15.3.1.3 The Relation to the Esscher Measure

The Esscher transformation leads to the same equivalent martingale measure as the absolutely continuous change of measure, in the special case of Gaussian returns with unit variance. We consider the case of one risky asset. Let

$$\Delta X_t = \mu_t + \epsilon_t, \quad t = 0, \dots, T,$$

where $X_t = S_t/B_t$ is the discounted stock price, $\Delta X_t = X_t - X_{t-1}$, and $\epsilon_t \sim N(0, 1)$. Now

$$\phi_t(a) = E(e^{a\Delta X_t} \,|\, \mathcal{F}_{t-1}) = \exp\left\{\frac{1}{2}a^2 + a\mu_t\right\},$$

where $a \in \mathbf{R}$. Then $a_t^* = -\mu_t$ minimizes $\phi_t(a)$ over $a \in \mathbf{R}$. We have

$$z_t = \frac{\exp\left\{a_t^*\Delta X_t\right\}}{\phi\left(a_t^*\right)} = \exp\left\{-\mu_t\Delta X_t + \frac{1}{2}\mu_t^2\right\} = \exp\left\{-\mu_t\epsilon_t - \frac{1}{2}\mu_t^2\right\}.$$

The Esscher martingale measure is $Q(d\omega) = Z_T(\omega)P(d\omega)$, where $Z_T = \prod_{i=1}^T z_i$, which is the same measure as (15.9) for $\sigma_t \equiv 1$.

15.3.2 Conditionally Gaussian Logarithmic Returns

Let us assume that the excess logarithmic returns are conditionally Gaussian:

$$h_t = \log\frac{X_t}{X_{t-1}} = \mu_t + \sigma_t\epsilon_t,$$

where μ_t, σ_t, and ϵ_t satisfy the same assumptions as in the case of conditionally Gaussian excess returns. This implies that ϵ is a sequence of independent random variables with $\epsilon_t \sim N(0,1)$. Here $X_t = S_t/B_t$ is the discounted stock price, so that h_t is the excess logarithmic return:

$$h_t = \log \frac{S_t}{S_{t-1}} - \log \frac{B_t}{B_{t-1}}.$$

15.3.2.1 The Martingale Measure

The equivalent martingale measure Q is such that under this measure the excess logarithmic returns satisfy

$$h_t = -\frac{1}{2}\sigma_t^2 + \sigma_t\epsilon_t, \tag{15.12}$$

where $\epsilon = (\epsilon_t)_{t=1,\dots,T}$ are i.i.d. with $N(0,1)$.

15.3.2.2 Construction of the Martingale Measure

Let

$$Z_t = \exp\left\{ -\sum_{k=1}^{t}\left(\frac{\mu_k}{\sigma_k} + \frac{\sigma_k}{2}\right)\epsilon_k - \frac{1}{2}\sum_{k=1}^{t}\left(\frac{\mu_k}{\sigma_k} + \frac{\sigma_k}{2}\right)^2 \right\}$$

for $t = 1, \dots, T$. Let us define measure Q by

$$Q(d\omega) = Z_T(\omega)P(d\omega).$$

It can be proved that Q is an equivalent martingale measure.

We derive measure Q using the approach of Shiryaev (1999, p. 449). Let us assume that

$$Z_T = \prod_{t=1}^{T} z_t,$$

where z_t has the form

$$z_t = \frac{e^{a_t h_t}}{E(e^{a_t h_t} \mid \mathcal{F}_{t-1})},$$

and a_t are $\mathcal{F}_{t-1}$-measurable. Let us choose a_t so that the discounted price $(X_t)_{t=1,\dots,T}$ is a martingale, where

$$X_t = X_0 \exp\{H_t\}, \quad H_t = \sum_{i=1}^{t} h_t.$$

Variables a_t must be such that

$$E\left(e^{(a_t+1)h_t} \mid \mathcal{F}_{t-1}\right) = E\left(e^{a_t h_t} \mid \mathcal{F}_{t-1}\right). \tag{15.13}$$

Indeed, we need to have

$$E_Q(X_t \mid \mathcal{F}_{t-1}) = X_{t-1}.$$

We have

$$
\begin{aligned}
E_Q(X_t \mid \mathcal{F}_{t-1}) &= X_0 e^{H_{t-1}} E_Q(e^{h_t} \mid \mathcal{F}_{t-1}) \\
&= X_0 e^{H_{t-1}} E(e^{h_t} z_t \mid \mathcal{F}_{t-1}) \\
&= \frac{X_0 e^{H_{t-1}}}{E(e^{a_t h_t} \mid \mathcal{F}_{t-1})} E(e^{h_t} e^{a_t h_t} \mid \mathcal{F}_{t-1}).
\end{aligned}
$$

Equality in (15.13) is equivalent to

$$
\exp\left\{ \frac{1}{2}(a_t + 1)^2 \sigma_t^2 + (a_t + 1)\mu_t \right\} = \exp\left\{ \frac{1}{2} a_t^2 \sigma_t^2 + a_t \mu_t \right\},
$$

which is equivalent to

$$
a_t^2 \sigma_t^2 + \frac{1}{2}\sigma_t^2 + \mu_t = 0 \quad \Leftrightarrow \quad a_t = -\frac{\mu_t}{\sigma_t^2} - \frac{1}{2}.
$$

Thus,

$$
E(e^{a_t h_t} \mid \mathcal{F}_{t-1}) = \exp\left\{ -\frac{\mu_t^2}{2\sigma_t^2} + \frac{\sigma_t^2}{8} \right\}
$$

and

$$
z_t = \exp\left\{ -\left(\frac{\mu_t}{\sigma_t} + \frac{\sigma_t}{2} \right) \epsilon_t - \frac{1}{2}\left(\frac{\mu_t}{\sigma_t} + \frac{\sigma_t}{2} \right)^2 \right\}.
$$

15.4 GARCH Market Models

When the market model is a GARCH model, then we can apply the absolutely continuous changes of measures of Section 15.3 to derive an equivalent martingale measure. We consider first the Heston–Nandi method, which applies numerical integration to compute the expectation with respect to the equivalent martingale measure. Second, we consider the method of Monte Carlo simulation for the computation of the expectation. Third, we compare the risk-neutral densities of the Heston–Nandi model and GARCH(1, 1) model.

The GARCH(1, 1) model is defined in (5.38). Pricing under GARCH models was considered in Duan (1995). The Heston–Nandi model was presented in Heston and Nandi (2000).

A related approach was followed in Aït-Sahalia *et al.* (2001), where a complete continuous time diffusion model was postulated for the stock price, the volatility was estimated, and the Girsanov's theorem was applied to obtain the risk-neutral measure, which can be used for pricing.

Chorro *et al.* (2012) consider GARCH models where the innovations follow a generalized hyperbolic distribution, instead of the standard normal distribution. They construct the equivalent martingale measure as in Gerber and Shiu

(1994), by choosing the density of the martingale measure (with respect to the physical measure) to have an exponential affine parametrization.

15.4.1 Heston–Nandi Method

We follow Heston and Nandi (2000) and consider model (5.50). We assume that the logarithmic returns satisfy

$$\log S_t - \log S_{t-1} = r + \mu_t + \sigma_t \epsilon_t,$$

where $r = \log(B_t/B_{t-1})$ is the constant risk-free rate, μ_t is predictable, ϵ_t are i.i.d. $N(0,1)$, and

$$\sigma_t^2 = \alpha_0 + \alpha_1(\epsilon_{t-1} - \gamma\sigma_{t-1})^2 + \beta\sigma_{t-1}^2,$$

where $\gamma \in \mathbf{R}$ is the skewness parameter. Because we assume conditionally Gaussian logarithmic returns, then (15.12) implies that there exists an equivalent martingale measure Q which is such that under Q

$$\log S_t - \log S_{t-1} = r - \frac{1}{2}\sigma_t^2 + \sigma_t \epsilon_t.$$

15.4.1.1 Prices of Call Options
The prices of call options have an expression which can be computed using numerical integration. Let

$$C_T = \max\{S_T - K, 0\}$$

be the payoff of a call option, and

$$H_T = e^{-rT}\max\{S_T - K, 0\}$$

be the discounted payoff of the call option. It follows from Theorem 13.2 that an arbitrage-free price of the call option is

$$E_Q H_T.$$

The expectation $E_Q H_T$ can be written in terms of the characteristic function. Let $f(i\phi)$ be the characteristic function of $\log S_T$, under the condition that the stock price at time 0 is S_0, where

$$f(\phi) = E_Q\left(S_T^\phi\right) = E_Q \exp\{\phi \log S_T\}$$

is the moment generating function, as defined in (5.53). The formula of f involves the stock price S_0, the interest rate r, and time T to expiration. Then

$$
\begin{aligned}
E_Q H_T = S_0 &\left(\frac{1}{2} + \frac{1}{\pi} \int_0^\infty \mathrm{Re}\left[\frac{K^{-i\phi} f(i\phi+1)}{i\phi f(1)} \right] d\phi \right) \\
&- e^{-rT} K \left(\frac{1}{2} + \frac{1}{\pi} \int_0^\infty \mathrm{Re}\left[\frac{K^{-i\phi} f(i\phi)}{i\phi} \right] d\phi \right),
\end{aligned}
\tag{15.14}
$$

where $\mathrm{Re}(z)$ denotes the real part of a complex number z. The expression in (15.14) is not a closed form expression, but we need numerical integration to compute the values. Also the values of the moment generating function are computed using a recursive formula. The price (15.14) is analogous to the Black–Scholes price in (14.58), where the cumulative distribution function Φ of the standard normal distribution does not have a closed form expression.

Let us prove (15.14). Let q be the density of Q with respect to Lebesgue measure. Let

$$q^*(x) = \frac{\exp(x)q(x)}{f(1)}.$$

Now q^* is a density function because $q^*(x) \geq 0$ and $\int_{-\infty}^{\infty} q^*(x)dx = 1$. Note that

$$f(1) = E_Q \exp\{\log S_T\} = e^{rT} S_0.$$

The moment generating function of q^* is

$$\int_{-\infty}^{\infty} \exp(\phi x)q^*(x)\,dx = \frac{1}{f(1)} \int_{-\infty}^{\infty} \exp((\phi+1)x)q(x)\,dx = \frac{f(\phi+1)}{f(1)}.$$

Now,

$$E_Q C_T = \int_{-\infty}^{\infty} \max\{\exp(x) - K, 0\}q(x)\,dx$$

$$= \int_{\log K}^{\infty} \exp(x)q(x)\,dx - K \int_{\log K}^{\infty} q(x)\,dx$$

$$= f(1) \int_{\log K}^{\infty} q^*(x)\,dx - K \int_{\log K}^{\infty} q(x)\,dx.$$

The characteristic function of q is $f(i\phi)$. Thus,

$$\int_{\log K}^{\infty} q(x) = \frac{1}{2} + \frac{1}{\pi} \int_0^{\infty} \mathrm{Re}\left[\frac{e^{-i\phi \log K} f(i\phi)}{i\phi}\right] d\phi;$$

see Billingsley (2005, Theorem 26.2, p. 346). The characteristic function of q^* is $f(i\phi + 1)/f(1)$, and a similar formula is obtained for $\int_{\log K}^{\infty} q^*(x)\,dx$. We have proved (15.14).

Figure 15.2 shows Heston–Nandi GARCH(1, 1) call prices divided by the Black–Scholes call prices as a function of the moneyness. Parameters α_0, α_1, β, and γ are estimated from S&P 500 daily data, described in Section 2.4.1. In panel (a), time to expiration is 20 trading days, and the annualized volatility takes values 10% (black), 15% (red), and 20% (blue). The solid lines have $\gamma = 7$, which is about equal to the value estimated from the S&P 500 data. The dashed lines have $\gamma = 0$. In panel (b), the annualized volatility is 15%, and the expiration time takes values 5 days (black), 20 days (red), and 40 days (blue). The risk-free rate is $r = 0$.

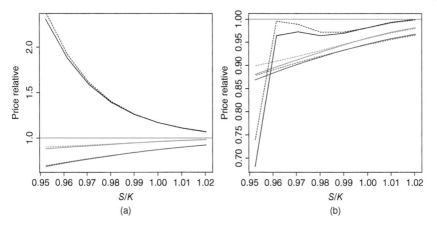

Figure 15.2 *The ratios of Heston–Nandi to Black–Scholes prices.* Shown are the Heston–Nandi call prices divided by the Black–Scholes prices as a function of moneyness S/K. (a) Time to expiration is 20 trading days. The annualized volatility takes values 10% (black), 15% (red), and 20% (blue). (b) The annualized volatility is 15%. The time to expiration takes values 5 days (black), 20 days (red), and 40 days (blue). The solid lines have $\gamma = 7$ and the dashed lines have $\gamma = 0$.

Panel (a) shows that the Heston–Nandi prices are lower than the Black–Scholes prices when the volatility is high, but when the volatility is low, then the Heston–Nandi prices are higher than the Black–Scholes prices. When the moneyness increases then the ratio of prices approaches one. Panel (b) shows that when the time to expiration becomes shorter, then the ratio of Heston–Nandi prices to the Black–Scholes prices increases. The skewness parameter γ has a smaller influence than the volatility and the time to expiration.

15.4.1.2 Hedging Coefficients of Call Options
The hedging coefficient of a call option is

$$\xi = \frac{1}{2} + \frac{1}{\pi} \int_0^\infty \mathrm{Re}\left[\frac{K^{-i\phi}f(i\phi + 1)}{i\phi f(1)}\right] d\phi,$$

where $K > 0$ is the strike price and $f(\phi) = E_Q(\exp\{\phi \log S_T\} \mid S_0)$ is the moment generating function of $\log S_T$, as defined in (5.53). The formula of f involves the stock price S_0, the interest rate r, and time T to expiration. Here T is the number of trading days to the expiration.

Figure 15.3 plots the ratios of Heston–Nandi GARCH(1, 1) call hedging coefficients to Black–Scholes call hedging coefficients as a function of the moneyness. We have the same setting as in Figure 15.2. We see from panel (a) that for a moderate and large volatility the Heston–Nandi deltas are smaller for out-of-the-money options, and larger for in-the-money options, than the

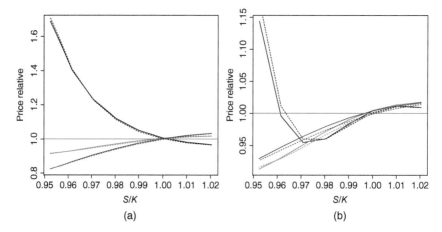

Figure 15.3 *Ratios of Heston–Nandi deltas to Black–Scholes deltas.* Shown are Heston–Nandi call hedging coefficients divided by the Black–Scholes hedging coefficients as a function of moneyness S/K. (a) Time to expiration is 20 trading days. The annualized volatility takes values 10% (black), 15% (red), and 20% (blue). (b) The annualized volatility is 15%. The time to expiration takes values 5 days (black), 20 days (red), and 40 days (blue). The solid lines have $\gamma = 7$ and the dashed lines have $\gamma = 0$.

Black–Scholes deltas. For small volatility, the behavior is opposite. We see from panel (b) that the time to expiration has a similar kind of effect as the volatility.

15.4.1.3 Hedging Errors of Call Options

Figure 15.4 shows hedging errors for hedging a call option with moneyness $S/K = 1$.[4] We use S&P 500 daily data, described in Section 2.4.1. Panel (a) shows tail plots and panel (b) shows kernel density estimates. We consider three cases: (1) The red plots show the case where the volatility in the Heston–Nandi formula is taken to be the current GARCH volatility in the Heston–Nandi model. (2) The green plots show the case where the volatility is the stationary volatility in (5.49). (3) The blue plots show the case of Black–Scholes hedging

---

4 The hedging error e_T of the writer of the option is obtained from (13.10) as

$$e_T = C_0 + G_T(\xi) - C_T,$$

where

$$G_T(\xi) = \sum_{k=1}^{T} \xi_k (S_k - S_{k-1}),$$

the risk-free rate is $r = 0$, C_0 is the price of the option, C_T is the terminal value of the option, ξ_k are the hedging coefficients, S_k are the stock prices, the current time is denoted by 0, the time to expiration is T days, and hedging is done daily. When hedging is done with a lesser frequency, then we use formula (14.79) for $G_T(\xi)$.

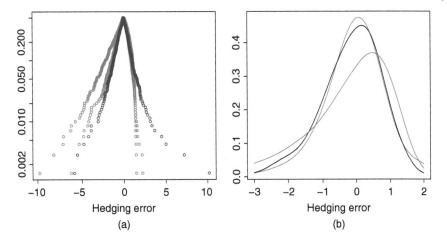

Figure 15.4 *Heston–Nandi hedging errors.* (a) Tail plots; (b) kernel density estimates of hedging errors. We show cases (1) the Heston–Nandi hedging with Heston–Nandi volatility (red), (2) the Heston–Nandi hedging with stationary volatility (green), and (3) the Black–Scholes hedging with GARCH(1, 1) volatility (blue).

with GARCH(1, 1) volatility. The parameters are estimated sequentially. Time to expiration is 20 days. The hedging is done twice: at the beginning and at the 10th day. The data is divided into 20 days periods using nonoverlapping sequences. The risk-free rate is $r = 0$. The hedging is started after obtaining 8 years (2000 days) of observations. We see that the Black–Scholes hedging leads to a better tail distribution of the hedging errors: the losses are smaller and the gains are larger. Note that the hedging errors of Black–Scholes hedging look different than in Figure 14.24, because in Figure 14.24 the hedging is done daily, and overlapping sequences are used.

15.4.2 The Monte Carlo Method

We assume that the excess returns follow a shifted GARCH(1, 1) model. The assumption that the excess logarithmic returns $h_t = \log(X_t/X_{t-1})$ follow a shifted GARCH(1, 1) model leads to similar prices, and we do not show results for this case.

It is assumed that

$$R_t = \frac{S_t}{S_{t-1}} - \frac{B_t}{B_{t-1}} = \mu_t + \sigma_t \epsilon_t,$$

where

$$\sigma_t^2 = \alpha_0 + \alpha_1 R_{t-1}^2 + \beta \sigma_{t-1}^2,$$

$\alpha_0 > 0$, $\alpha_0, \beta \geq 0$, and $\{\epsilon_t\}$ are i.i.d. with the standard normal distribution $N(0, 1)$. Under the martingale measure Q, obtained by an absolutely continuous change of measure, the excess returns can be written as

$$R_t = \sigma_t \epsilon_t. \tag{15.15}$$

Pricing under measure Q can be done by estimating

$$H_0 = E_Q H_T,$$

where H_T is the discounted contingent claim. We simulate n sequences

$$S_i = \left(S_0^{(i)}, \dots, S_T^{(i)} \right), \quad i = 1, \dots, n, \tag{15.16}$$

and use the estimate

$$\hat{H}_0 = \frac{1}{n} \sum_{i=1}^{n} H_T(S_i),$$

where $H_T(S_i)$ is the value of the discounted contingent claim for the trajectory S_i. We consider call options, so that $H_T(S_i) = \max\{0, S_T^{(i)} - K\}$.

We need to simulate trajectories $S_0, \dots, S_T$ under the dynamics in (15.15). This can be done by simulating sequences $R_1, \dots, R_T$ of excess returns and sequences $r_1, \dots, r_T$ of risk-free returns. The sequence of stock prices is obtained as

$$S_0, S_1 = S_0(R_1 + r_1), \dots, S_T = S_{T-1}(R_T + r_T).$$

In our simulation, we assume that the risk-free rates are zero, so that the risk-free gross returns are $r_i = 1$. To start the simulation we choose σ_1 to be the current GARCH(1, 1) volatility, and $R_1 = \sigma_1 \epsilon_1$.

Figure 15.5 shows Monte Carlo approximations of call prices divided by the Black–Scholes price as a function of the number n of Monte Carlo samples. In panel (a), the moneyness is $S/K = 1$, and in panel (b), the moneyness is $S/K = 0.95$. The black curves show the GARCH(1, 1) prices, and the red curves show the Heston–Nandi GARCH(1, 1) prices. The red horizontal line shows the Heston–Nandi prices computed using the closed form expression (15.14). Time to expiration is 20 trading days. We have applied data of S&P 500 daily prices, described in Section 2.4.1. The initial standard deviation is the sample standard deviation of S&P 500 returns, and the GARCH(1, 1) parameters are estimated using S&P 500 data. The Black–Scholes volatility is the annualized sample standard deviation. The risk-free rate is $r = 0$. We see that the GARCH(1, 1) price is lower than the Black–Scholes price when the moneyness is one, and the GARCH(1, 1) price is higher than the Black–Scholes price when the moneyness is 0.95. The Heston–Nandi prices are lower than the prices in the standard GARCH(1, 1) model.

Figure 15.6 shows standard GARCH(1, 1) call prices divided by the Black–Scholes call prices as a function of the moneyness. Parameters $\alpha_0, \alpha_1,$

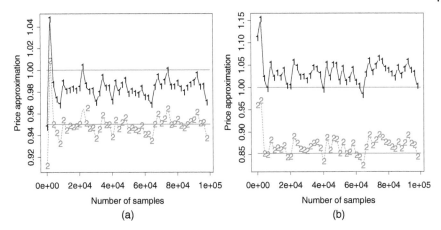

Figure 15.5 *Monte Carlo approximation of GARCH prices.* Approximations of call prices divided by the Black–Scholes price as a function of the number of Monte Carlo samples. (a) Moneyness is $S/K = 1$. (b) Moneyness is $S/K = 0.95$. We show the GARCH(1, 1) price ratios (black) and the Heston–Nandi GARCH(1, 1) price ratios (red), and the red horizontal line shows the Heston–Nandi price ratio computed using (15.14).

and β are estimated from S&P 500 daily data, described in Section 2.4.1. In panel (a), time to expiration is 20 trading days, and the annualized volatility takes values 10% (black), 15% (red), and 20% (blue). In panel (b), the annualized volatility is 15%, and the expiration time takes values 5 days (black), 20 days (red), and 40 days (blue). The risk-free rate is $r = 0$.

Panel (a) shows that the standard GARCH(1, 1) prices are lower than the Black–Scholes prices when the volatility is high, but when the volatility is low, then the standard GARCH(1, 1) prices are higher than the Black–Scholes prices. When the moneyness increases then the ratio of prices approaches one. Panel (b) shows that when the time to expiration becomes shorter, then the ratio of the standard GARCH(1, 1) prices to the Black–Scholes prices increases.

Figure 15.7 shows standard GARCH(1, 1) call prices divided by the Heston–Nandi call prices as a function of the moneyness. The setting is the same as in Figure 15.6.

15.4.3 Comparison of Risk-Neutral Densities

We study the risk-neutral distributions of the Heston–Nandi model when the parameters change. Also, we compare the risk-neutral distributions of the Heston–Nandi model to the risk-neutral distributions of the standard GARCH(1, 1) model.

In the GARCH models, the physical distribution of stock prices is given by

$$\log S_t = \log S_{t-1} + r + \mu_t + \sigma_t \epsilon_t,$$

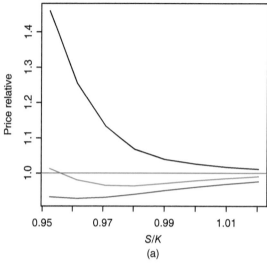

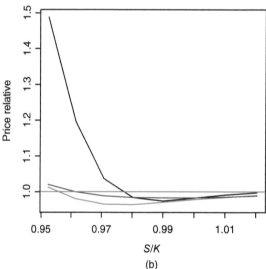

Figure 15.6 *The ratios of standard GARCH(1, 1) prices to Black–Scholes prices.* Shown are the standard GARCH(1, 1) call prices divided by the Black–Scholes price as a function of moneyness S/K. (a) Time to expiration is 20 trading days. The annualized volatility takes values 10% (black), 15% (red), and 20% (blue). (b) The annualized volatility is 15%. The time to expiration takes values 5 days (black), 20 days (red), and 40 days (blue).

where $t = 1, \ldots, T$. The volatility σ_t is defined differently in the standard GARCH(1, 1) model and in the Heston–Nandi model. A risk-neutral distribution of stock prices is given by

$$\log S_t = \log S_{t-1} + r - \frac{1}{2}\sigma_t^2 + \sigma_t \epsilon_t,$$

where $t = 1, \ldots, T$. We can estimate the distribution of S_T by first estimating the parameters of the model, second generating M Monte Carlo trajectories which

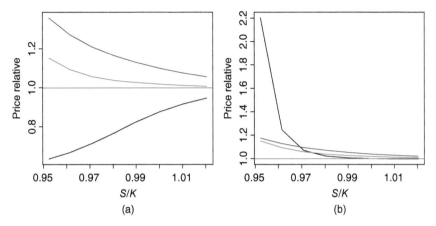

Figure 15.7 *The ratios of standard GARCH(1, 1) prices to Heston–Nandi prices.* Shown are the standard GARCH(1, 1) call prices divided by the Heston–Nandi prices as a function of money-ness S/K. (a) Time to expiration is 20 trading days. The annualized volatility takes values 10% (black), 15% (red), and 20% (blue). (b) The annualized volatility is 15%. The time to expiration takes values 5 days (black), 20 days (red), and 40 days (blue).

give M observations $S_T^1, \ldots, S_T^M$, and finally using a kernel density estimator. We use S&P 500 daily data, described in Section 2.4.1.[5]

Figure 15.8 shows the estimated risk-neutral densities. Panel (a) compares the Heston–Nandi model with the standard GARCH(1, 1) model. The red curve shows the risk-neutral distribution of the Heston–Nandi model, the black curve shows the risk-neutral distribution of the standard GARCH(1, 1) model, and the green curve shows the risk-neutral distribution of the Black–Scholes model.[6] Panel (b) studies the effect of the skewness parameter γ. The red curve shows the case γ about seven, which is the value estimated from data. The orange curve shows the case $\gamma = 0$, but it is very close to the case $\gamma = 7$. The blue curve shows the case $\gamma = 50$. We see that the green density

---

5 Note that in Section 13.2.3, we estimated the risk-neutral distribution of $S_T - S_0$ (one-period model) and the risk-neutral distribution of $(S_T - S_{T/2}, S_{T/2} - S_0)$ (two-period model). Estimation of the risk-neutral distribution of S_T is enough for the pricing of European call and put options.

6 In the Black–Scholes model the physical and the risk-neutral distributions of S_T are given by

$$\log S_t = \log S_{t-1} + r + \mu + \sigma \epsilon_t, \quad \log S_t = \log S_{t-1} + r - \frac{1}{2}\sigma^2 + \sigma \epsilon_t,$$

where $\epsilon_t \sim N(0, 1)$ are i.i.d. Thus, the risk-neutral distribution is the log-normal distribution

$$S_T \sim \text{lognorm}\left(\log S_0 + T\left(r - \frac{1}{2}\sigma^2\right), T\sigma^2\right),$$

where T is the number of trading days to the expiration. Parameter σ is estimated by the sample standard deviation from the daily logarithmic returns.

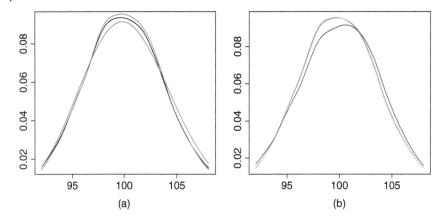

Figure 15.8 *Risk-neutral densities*. (a) Heston–Nandi model (red), standard GARCH(1, 1) model (black), and the Black–Scholes model (green). (b) Heston–Nandi risk-neutral densities for $\gamma = 7$ (red), $\gamma = 0$ (orange), and $\gamma = 50$ (blue).

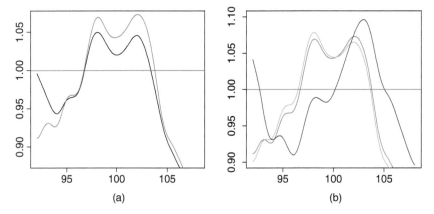

Figure 15.9 *Risk-neutral densities: Ratios*. (a) Heston–Nandi (red) and standard GARCH(1, 1) (black) risk-neutral density ratios. (b) Heston–Nandi risk-neutral density ratios for $\gamma = 7$ (red), $\gamma = 0$ (orange), and $\gamma = 50$ (blue).

is skewed so that large negative returns and moderate positive return are more probable than in the case of $\gamma = 0$.

Figure 15.9 shows the risk-neutral densities of Figure 15.8 divided by the Black–Scholes risk-neutral density, which is shown with green in Figure 15.8(a). Panel (a) shows the Heston–Nandi (red) and standard GARCH(1, 1) (black) risk-neutral densities divided by the Black–Scholes risk-neutral density. Panel (b) shows the Heston–Nandi risk-neutral densities for $\gamma = 7$ (red),

$\gamma = 0$ (orange), and $\gamma = 50$ (blue), divided by the Black–Scholes risk-neutral density. Panel (a) shows that the risk-neutral densities in the GARCH models take higher values at the center than the Black–Scholes risk-neutral density, and the ratio has a hat shape. Panel (b) shows that the densities are close to each other for $\gamma = 0$ and $\gamma = 7$, but when $\gamma = 50$, then the skewness is visible.

15.5 Nonparametric Pricing Using Historical Simulation

It is interesting to compare nonparametric pricing to the Black–Scholes pricing and to the GARCH pricing methods. We define nonparametric pricing combining three elements: (1) historical simulation, (2) Esscher transformation, (3) conditioning with the current volatility, which is taken to be the GARCH(1, 1) volatility.

We have used Monte Carlo simulation to create sequences of prices in (15.16). Analogously, historical prices can be used to create sequences of prices. The Esscher transformation was applied in the proof of Theorem 13.1 (the first fundamental theorem of asset pricing) to construct an equivalent martingale measure in an arbitrage-free market. In Section 15.2, the Esscher transformation was shown to be related to utility maximization.

We consider T-period model, with time to expiration being T trading days. Our price data is $S_0, \dots, S_N$. We construct $M = N - T$ sequences of prices:

$$S_i = (S_{i,0}, \dots, S_{i,T}), \quad i = 1, \dots, M,$$

where

$$S_{i,j} = 100 \, \frac{S_{i+j}}{S_i}.$$

Each sequence has length $T + 1$, and the initial prices are $S_{i,0} = 100$.

15.5.1 Prices

We define first the unconditional price and then the price which conditions on the current volatility.

The unconditional price is computed by

$$\hat{H}_0 = \frac{1}{M} \sum_{i=1}^{M} H_T(S_i) f(S_i),$$

where $H_T(S_i)$ is the value of the discounted contingent claim for the trajectory S_i, and f is the estimated Esscher density. For example, in the case of a call option, $H_T(S_i) = (S_{i,T} - K)_+$. The estimated Esscher density is $f : \mathbf{R}^{T+1} \to \mathbf{R}$ with

$$f(s_0, \ldots, s_T) = \prod_{i=1}^{T} g(\Delta s_i),$$

where $\Delta s_i = s_i - s_{i-1}$ and

$$g(\Delta s_i) = \frac{\exp\{a^* \Delta s_i\}}{\phi(a^*)}.$$

Value a^* is the minimizer of

$$\phi(a) = \frac{1}{MT} \sum_{i=1}^{M} \sum_{j=1}^{T} \exp\{a(S_{i,j} - S_{i,j-1})\}$$

over $a \in \mathbf{R}$.

Next we define the conditional price. Let σ be the current estimated GARCH(1, 1) volatility. The price is estimated as

$$\hat{H}_0 = \sum_{i=1}^{M} p_i(\sigma) H_T(S_i) f(S_i),$$

where $H_T(S_i)$ is the value of the discounted contingent claim for the trajectory S_i, and f is the Esscher density. The weight is defined as

$$p_i(\sigma) = \frac{K_h(\log(\sigma) - \log(\sigma_i))}{\sum_{j=1}^{M} K_h(\log(\sigma) - \log(\sigma_j))},$$

where σ_i is the estimated GARCH(1, 1) volatility at day i. Furthermore, K_h is the scaled kernel

$$K_h(x) = K\left(\frac{x}{h}\right),$$

where $K : \mathbf{R} \to \mathbf{R}$ is a kernel function. We choose

$$K(x) = I_{[-1,1]}(x).$$

We apply the S&P 500 daily data, described in Section 2.4.1.

Figure 15.10 shows nonparametric call prices divided by the Black–Scholes call price as a function of the smoothing parameter. In panel (a) moneyness $S/K = 1$, and in panel (b) $S/K = 0.95$. The annualized volatility takes values 10% (black), 15% (red), and 20% (blue). Time to expiration is $T = 20$ trading days, risk-free rate is $r = 0$. The volatility is the sample standard deviation. We see that the price ratios are higher for small volatility. The prices stabilize when $h \geq 1$.

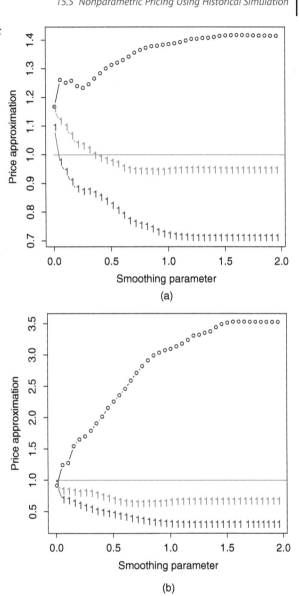

Figure 15.10 *Nonparametric prices.* Ratios of nonparametric call prices to the Black–Scholes price as a function of the smoothing parameter. (a) Moneyness is $S/K = 1$. (b) Moneyness is $S/K = 0.95$. The annualized volatility takes values 10% (black), 15% (red), and 20% (blue).

Figure 15.11 shows nonparametric call prices divided by the Black–Scholes call prices as a function of the moneyness S/K. In panel (a), time to expiration is 20 trading days, and the annualized volatility takes values 10% (black), 15% (red), and 20% (blue). In panel (b), the annualized volatility is 15%, and the expiration time takes values 5 days (black), 20 days (red), and 40 days (blue). The risk-free rate is $r = 0$.

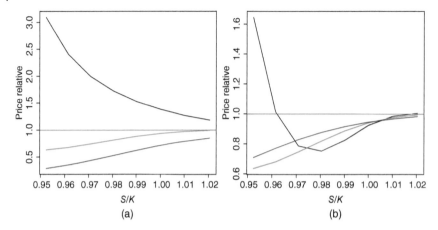

Figure 15.11 *The ratios of nonparametric prices to Black–Scholes prices.* Shown are the nonparametric call prices divided by the Black–Scholes prices as a function of moneyness S/K. (a) Time to expiration is 20 trading days. The annualized volatility takes values 10% (black), 15% (red), and 20% (blue). (b) The annualized volatility is 15%. The time to expiration takes values 5 days (black), 20 days (red), and 40 days (blue).

15.5.2 Hedging Coefficients

We compute the nonparametric hedging coefficients by approximating the derivative of the price numerically. The numerical approximation is done by the difference quotient. The price is computed at the stock price S and $S + \delta$, and the hedging coefficient is taken as

$$\frac{H_0(S + \delta) - H_0(S)}{\delta}, \tag{15.17}$$

where $\delta > 0$ is small.

Figure 15.12 shows nonparametric hedging coefficients divided by the Black–Scholes delta as a function of the smoothing parameter. In panel (a) moneyness $S/K = 1$, and in panel (b) $S/K = 0.95$. Time to expiration is $T = 20$ trading days, risk-free rate is $r = 0$. The volatility is the sample standard deviation. Parameter δ in (15.17) is taken as $\delta = 0.005, 0.01, 0.05$ (black, red, and blue). We see that when $S/K = 1$, then the nonparametric deltas are larger than the Black–Scholes deltas; when $S/K = 0.95$, then the nonparametric deltas are smaller than the Black–Scholes deltas.

Figure 15.13 shows nonparametric call deltas divided by the Black–Scholes call deltas as a function of the moneyness S/K. In panel (a), time to expiration is 20 trading days, and the annualized volatility takes values 10% (black), 15% (red), and 20% (blue). In panel (b), the annualized volatility is 15%, and the expiration time takes values 5 days (black), 20 days (red), and 40 days (blue). The risk-free rate is $r = 0$.

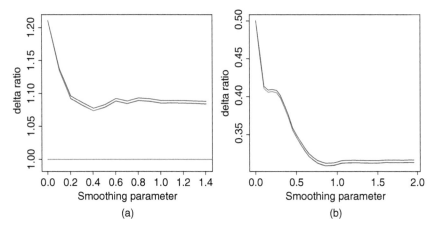

Figure 15.12 *Nonparametric deltas.* Ratios of nonparametric call deltas to the Black–Scholes delta as a function of the smoothing parameter. (a) Moneyness is $S/K = 1$. (b) Moneyness is $S/K = 0.95$. Parameter δ in (15.17) is taken as $\delta = 0.005, 0.01, 0.05$ (black, red, and blue).

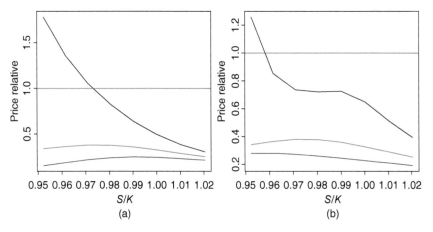

Figure 15.13 *The ratios of nonparametric deltas to Black–Scholes deltas.* Shown are the non-parametric call deltas divided by the Black–Scholes deltas as a function of moneyness S/K. (a) Time to expiration is 20 trading days. The annualized volatility takes values 10% (black), 15% (red), and 20% (blue). (b) The annualized volatility is 15%. The time to expiration takes values 5 days (black), 20 days (red), and 40 days (blue).

Figure 15.14 shows hedging errors of nonparametric hedging. Panel (a) shows tail plots of the empirical distribution function and panel (b) shows kernel density estimates. Time to expiration is $T = 20$ trading days. The blue curves show the case where hedging is done once, and the red curves show the case where hedging is done twice. We show also Black–Scholes hedging errors: in the green curves, hedging is done once and in the violet

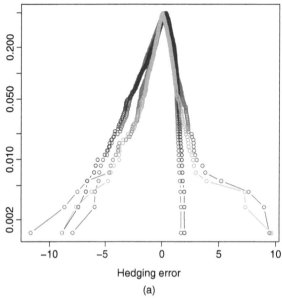

(a)

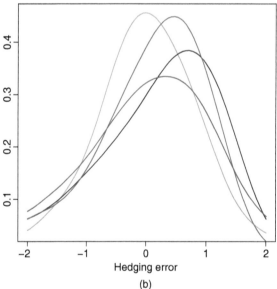

(b)

Figure 15.14 *Nonparametric hedging errors*. (a) Tail plots; (b) kernel density estimates. Nonparametric hedging is done once (blue) and twice (red). The Black–Scholes hedging is done once (green) and twice (violet).

curves, the hedging is done twice. The risk-free rate is $r = 0$. The volatility is the sample standard deviation. The smoothing parameter is $h = 0.5$. Panel (a) shows that the Black–Scholes hedging performs better in the tails, but panel (b) shows that the nonparametric hedging performs better in the central area.

15.6 Estimation of the Risk-Neutral Density

Let us consider an European call option

$$H = \max\{S_T - K, 0\},$$

where S_T is the price of the stock at the expiration, and $K > 0$ is the strike price. We assume that the risk-free rate is $r = 0$. Theorem 13.2 states that arbitrage-free prices of an European option H can be written as

$$E_Q H,$$

where the expectation is with respect to an equivalent martingale measure Q. We assume that the distribution of S_T under Q has density $f : [0, \infty) \to [0, \infty)$ with respect to the Lebesgue measure. The density depends on the initial stock price S_0. The price of the call option can be written as

$$E_Q H = \int_K^\infty f(s)(s - K)\, ds. \tag{15.18}$$

Differentiating with respect to K, we get

$$\frac{\partial}{\partial K} E_Q H = -\int_K^\infty f(s)\, ds,$$

where we used the fact $(\partial/\partial K) \int_K^\infty g(s)\, ds = -g(K)$. Differentiating second time, we get

$$\frac{\partial^2}{\partial K^2} E_Q H = f(K). \tag{15.19}$$

We can apply (15.19) to compute an approximation to the density f, when prices are observed for several strike prices K. Note that (15.18) shows that the problem can be considered as deconvolution problem, where the pricing function

$$K \mapsto E_Q H$$

should be inverted in order to obtain f.[7]

---

7 Hobson (1998, Lemma 2.3) proves that price functions and probability measures are in a one–one correspondence. It is assumed that there is a price function $K \mapsto h(K)$ for call options with strike price K, the interest rate is zero, h is decreasing and convex, $h(0) = S_0$, $\lim_{K \to \infty} h(K) = 0$, and at time $0 \le t < T$ the European call can be sold for at least $(S_t - K)_+$. Then there is a probability measure Q with $E_Q S_T = S_0$, and whose support is $(0, \infty)$. The probability measure is obtained by $Q((K, \infty)) = -h'(K)$, where h' is the right derivative. Also, $h(K) = \int_0^\infty (s - K)_+ Q(ds)$. Probability measure Q is a martingale distribution of S_T. The formula for Q is sometimes called Breeden–Lizenberger formula, since it can be found from Breeden and Lizenberger (1978).

The density f is a risk-neutral distribution for S_T. Similarly, we can consider estimating the risk-neutral distribution of $S_{T'}$ for $0 < T' < T$. Thus, we are able to estimate all marginal distributions of the prices process $S = (S_t)_{t=0,\ldots,T}$, but not the complete risk-neutral distribution Q.

15.6.1 Deducing the Risk-Neutral Density from Market Prices

Let us denote by $h(K) = E_Q H$ the option price when the strike price is K. The market prices provide values $h(K_1), \ldots, h(K_n)$ for strike prices $0 < K_1 < \cdots < K_n$. These observations can be used to deduce the risk-neutral density of S_T, implied by the market prices. The implied risk-neutral density can be estimated using liquid options, and the estimated density can be used to price illiquid options. It is possible that the market prices are not fair. On the other hand, market prices can incorporate information which is difficult to obtain by statistical procedures. For example, market prices can incorporate information about event risks, like information about the elections in the near future.

Aït-Sahalia and Lo (1998) considered the observations $h(K_1), \ldots, h(K_n)$ as coming from a regression model

$$h(K_i) = p(K_i) + \epsilon_i, \quad i = 1, \ldots, n,$$

where $p(K_i)$ is the true pricing function and ϵ_i is random noise. They used semiparametric regression to estimate the true pricing function, and then took the second derivative to obtain the risk-neutral density. They considered the pricing function to have five arguments:

$$p(K) = p(K, S, \tau, r, \delta),$$

where S is the stock price, τ is the time to expiration, r is the risk-free rate for that maturity, and δ is the dividend yield for the asset. They used two dimensional kernel regression to estimate the implied volatility, as a function of moneyness and time to maturity. Then they applied the Black–Scholes formula to obtain the pricing function.

Aït-Sahalia and Duarte (2003) estimated the pricing function using a combination of constrained univariate least squares regression and smoothing. The constrained regression is useful because the pricing function is increasing and convex as a function of the strike price. The convexity follows from (15.19), because $f(K) \geq 0$ implies that the second derivative of the pricing function is nonnegative, which implies convexity.

15.6.2 Examples of Estimation of the Risk-Neutral Density

We can use risk-neutral densities to give insight about a pricing method. Some pricing methods are such that the risk-neutral density can be expressed in a closed form (Black–Scholes model), or we can simulate observations from the risk-neutral density and estimate the risk-neutral density based on the simulated observations (Heston–Nandi and the standard GARCH(1, 1) model). See Figures 15.8 and 15.9 for risk-neutral densities in the Black–Scholes, Heston–Nandi and the standard GARCH(1, 1) model. The nonparametric pricing using historical simulation is described in Section 15.5. The nonparametric pricing is such that the risk-neutral density is not easy to find directly, but it can be deduced from the prices, by inverting the information in the prices.

We compute the risk-neutral densities for four methods: (1) the Heston–Nandi pricing described in Section 15.4.1, (2) the GARCH(1, 1) pricing described in Section 15.4.2, (3) the nonparametric pricing using historical simulation described in Section 15.5, and (4) the Black–Scholes pricing.

Let us denote by $h(K) = E_Q H$ the price when the strike price is K. Let us observe prices $h(K_1), \ldots, h(K_n)$ for strike prices $0 < K_1 < \cdots < K_n$. The first derivative is approximated by

$$\hat{h}'(K_i) = \frac{h(K_i) - h(K_{i-1})}{K_i - K_{i-1}}$$

for $i = 2, \ldots, n$. The approximations of the second derivative give estimates of the density:

$$\hat{f}(K_i) = \frac{\hat{h}'(K_i) - \hat{h}'(K_{i-1})}{K_i - K_{i-1}} \tag{15.20}$$

for $i = 3, \ldots, n$.

The numerical differentiation which leads to the density in (15.20) can lead to a unsmooth density. We can smooth the density by using two-sided moving averages. The smoothing is done below in the case of the nonparametric pricing. We apply below the S&P 500 daily data of Section 2.4.1.

Figure 15.15 shows the price functions. Panel (a) shows pricing functions as a function of strike price and panel (b) shows the ratios of the prices to the Black–Scholes prices. In panel (a), the prices are for the standard GARCH(1, 1) pricing (black), for nonparametric pricing (orange), for Heston–Nandi version of GARCH(1, 1) pricing (red), and for the Black–Scholes pricing (dark green). In panel (b), the prices are divided by the Black–Scholes prices. The number of Monte Carlo samples in GARCH pricing is $n = 10,000$. The smoothing parameter of nonparametric GARCH pricing is $h = 1$. The volatility is the sample standard deviation in all three cases.

Figure 15.16 shows densities of S_T under risk-neutral measures. Panel (a) shows densities with respect to the Lebesgue measure and panel (b) shows the ratios of the densities to the Black–Scholes density. In panel (a), the densities are for the standard GARCH(1, 1) pricing (black), for nonparametric pricing (orange), for Heston–Nandi version of GARCH(1, 1) pricing (red), and for the

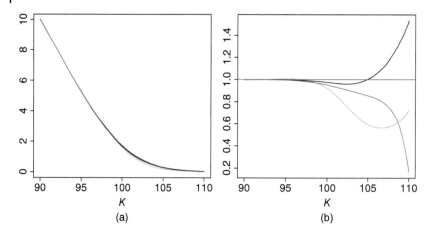

Figure 15.15 *Price functions.* (a) The standard GARCH(1, 1) pricing (black), nonparametric pricing (orange), Heston–Nandi pricing (red), and Black–Scholes pricing (dark green). (b) The prices divided by the Black–Scholes prices.

Black–Scholes pricing (dark green). In panel (b), the densities are divided by the Black–Scholes density.

15.7 Quantile Hedging

Let H_T be a discounted European contingent claim. A fair price of H_T could be considered as the initial investment V_0 such that there exists a trading strategy

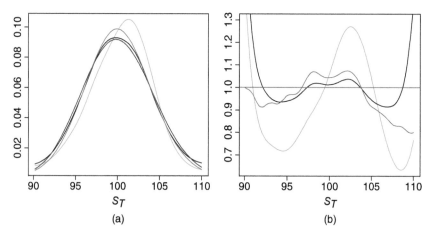

Figure 15.16 *Risk-neutral densities.* (a) The standard GARCH(1, 1) pricing (black), nonparametric pricing (orange), Heston–Nandi pricing (red), and the Black–Scholes pricing (dark green). (b) The densities are divided by the Black–Scholes density.

which leads to value V_T which is close to H_T. This is the basic idea of quadratic hedging, which minimizes $E(V_T - H_T)^2$ (see Section 15.1). However, from the point of view of the writer of the option it is desirable that $V_T \geq H_T$. This leads to the definition of quantile hedging, where the probability of $V_T \geq H_T$ is maximized.

Let v be a bound for the initial investment. We want to find a self-financing trading strategy whose value process maximizes the probability

$$P(V_T \geq H_T),$$

among all those strategies which satisfy

1) $V_0 \leq v$,
2) $V_t \geq 0$ for $t = 0, \ldots, T$.

We have to assume that v is not too large, since letting v to be large would allow very expensive strategies. We assume that $v \leq \sup \Pi(H)$, where $\Pi(H)$ is the set of arbitrage-free prices, as characterized in Theorem 13.2. Föllmer and Schied (2002, Section 8.1, p. 245) studies quantile hedging; see also Föllmer and Leukert (1999).

16

Quadratic and Local Quadratic Hedging

Quadratic hedging was introduced in Sections 13.1.3 and 15.1. In quadratic hedging we find the best approximation of the option in the sense of the mean-squared error. Quadratic hedging is related to the idea of statistical arbitrage: The fair price is defined as such price that makes the probability of gains and losses small for the writer of the option.

Quadratic hedging makes it possible to price and hedge options in a completely nonparametric way. In quadratic hedging we can derive prices and hedging coefficients without any modeling assumptions, making only some rather weak assumptions about square integrability and about a bounded mean–variance trade-off. There are many ways to implement quadratic hedging nonparametrically. We use kernel estimation in our implementation.

Let H_T be the discounted payoff of an European option. For example, for a call option $H_T = (1 + r)^{-T}(S_T - K)_+$. In quadratic hedging the mean-squared hedging error

$$E(H_T - V_0 - G_T(\xi))^2$$

is minimized among strategies $\xi = (\xi_t)_{t=1,\dots,T}$ and among the initial investment $V_0 \in \mathbf{R}$. The terminal value of the gains process is defined by

$$G_T(\xi) = \sum_{t=1}^{T} \xi_t \cdot (X_t - X_{t-1}),$$

where $X_t = S_t/(1 + r)^t$ is the discounted price vector. The problem resembles least-squares linear regression, where H_T is the response variable, $\Delta X_t = X_t - X_{t-1}$ are the explanatory variables, V_0 is the intercept, and ξ_t are the regression coefficients. However, now we have a time series setting, where the "explanatory" variable ΔX_t is observed at time t. We can use the knowledge of the observed values of $\Delta X_1, \dots, \Delta X_{t-1}$ in choosing ξ_t. In the usual linear regression all regression coefficients are chosen at the same time: $t = T$.

Quadratic hedging in discrete time is explained in monographs Föllmer and Schied (2002, p. 393), Bouchaud and Potters (2003), and Černý (2004b). Earlier

Nonparametric Finance, First Edition. Jussi Klemelä.

studies include Föllmer and Schweizer (1989), Bouchaud and Sornette (1994), Schweizer (1994), and Schäl (1994). Early continuous time studies include Föllmer and Sondermann (1986) and Duffie and Richardson (1991).

We study quadratic hedging and pricing in three steps: first for the one period model, then for the two period model, and finally the formulas are given for the general multiperiod model. The multiperiod model contains as special cases the one and two period models, but we think that it is helpful to study the one and two period models separately, because in these models the formulas are more transparent and notationally more convenient than in the multiperiod model. The generalization from the two period model to the multiperiod model is straightforward.

Local quadratic hedging simplifies the minimization problem of quadratic hedging, and it can achieve easier computations. In the one period model quadratic hedging and local quadratic hedging are equivalent, but in the multiperiod models they are different.

The price and the hedging coefficients of quadratic hedging do not have a closed-form expression, but only a recursive definition. This recursive definition can be used in computations, but the implementation is not trivial. We implement only the local quadratic hedging and pricing. We need to estimate various conditional expectations. We estimate the conditional expectations using historical simulation: The time series of previous returns is used to construct a large number of price sequences, and conditional expectations are estimated as sample means over price sequences. The observed volatility is used as the conditioning variable. Separate methods are used for the case of independent and dependent increments.

We evaluate quadratic hedging by studying the distribution of the hedging errors. The distribution should be concentrated around zero as well as possible. The main observation is that even the simplest setting of local quadratic hedging with independent increments leads to a distribution of the hedging errors that is better concentrated around zero than the distribution of the hedging errors when Black–Scholes hedging with GARCH$(1, 1)$ volatility is used.

Section 16.1 studies global quadratic hedging and pricing. Section 16.2 studies local quadratic hedging and pricing. Section 16.3 studies implementations of local quadratic hedging.

16.1 Quadratic Hedging

The exact solution for quadratic hedging can be given using backward induction. We present the solution in three steps: first for the one period model, second for the two period model, and third for the general model.

16.1.1 Definitions and Assumptions

We recall the notation from Section 13.2, and in particular from Section 13.2.2. We assume that there is only one risky asset: $d = 1$. The price process of the riskless bond is denoted by $B = (B_t)_{t=0,\ldots,T}$. We choose

$$B_t = (1+r)^t,$$

where $r > -1$. The notation is a short hand for $B_t = (1 + r\Delta t)^t$, where Δt is the time between two steps, expressed in fractions of a year, and r is the annual interest rate. The time series of prices of the risky asset is denoted by $S = (S_t)_{t=0,\ldots,T}$. The complete price vector is denoted by

$$\bar{S}_t = (B_t, S_t). \quad t = 0, \ldots, T.$$

A trading strategy is

$$\bar{\xi}_t = (\beta_t, \xi_t), \quad t = 1, \ldots, T,$$

where the values β_t and ξ_t express the quantity of the bond and the risky asset held between $t - 1$ and t.

16.1.1.1 Wealth and Value Processes
The wealth at time 0 is W_0, and after that

$$W_t = \beta_t B_t + \xi_t S_t = \bar{\xi}_t \cdot \bar{S}_t, \quad t = 1, \ldots, T.$$

Under the condition of self-financing the wealth at time t was written in (13.5) as

$$W_t = W_0 + \sum_{k=1}^{t} \bar{\xi}_k \cdot (\bar{S}_k - \bar{S}_{k-1}),$$

where $W_0 = \bar{\xi}_1 \cdot \bar{S}_0$.

The discounted price process is defined by

$$X_t = \frac{S_t}{B_t}, \quad t = 0, \ldots, T.$$

We denote $\bar{X}_t = (1, X_t)$. The value process was defined in (13.8) as

$$V_t = \frac{W_t}{B_t}, \quad t = 0, \ldots, T.$$

Under the condition of self-financing the value at time t can be written as

$$V_t = V_0 + \sum_{k=1}^{t} \xi_k (X_k - X_{k-1}), \tag{16.1}$$

where $V_0 = \bar{\xi}_1 \cdot \bar{X}_0$.

The gains process is defined as

$$G_0 = 0, \qquad G_t = \sum_{k=1}^{t} \xi_k (X_k - X_{k-1}), \qquad t = 1, \dots, T.$$

For a self-financing strategy

$$V_t = V_0 + G_t. \tag{16.2}$$

16.1.1.2 Quadratic Hedging

We use the terms "quadratic hedging" and "global quadratic hedging" to mean the same thing. The term "global quadratic hedging" is used when a distinction to "local quadratic hedging" is emphasized. We use the term "quadratic price" to mean the price that is implied by quadratic hedging.

Let C_T be the value of an European option. For example, $C_T = (S_T - K)_+$. Let $H_T = C_T / B_T$. A quadratic strategy (V_0, ξ) is a minimizer of

$$E(H_T - V_0 - G_T(\xi))^2$$

over $V_0 \in \mathbf{R}$ and over self-financing strategies $\xi = (\xi_t)_{t=1,\dots,T}$. We obtain the quadratic price

$$C_0 = V_0.$$

In the general case the quadratic price is $B_0 V_0$, but in our case $B_0 = 1$. The bond coefficients β_t are determined from the equations

$$\beta_1 = V_0 - \xi_1 X_0,$$
$$\beta_{t+1} = \beta_t + (\xi_t - \xi_{t+1}) X_t, \qquad t = 1, \dots, T - 1, \tag{16.3}$$

as noted in (13.9). The complete quadratic hedging strategy, which includes both the quantities of bond and stock, is given by

$$\bar{\xi}_t = (\beta_t, \xi_t), \qquad t = 1, \dots, T.$$

16.1.1.3 Mean Self-Financing

We have defined quadratic hedging as a hedging strategy that minimizes the mean-squared error among self-financing strategies. It is also possible to define a version of quadratic hedging where the mean-squared error is minimized among so-called mean self-financing strategies. We consider this approach only in Section 16.2.3, where local quadratic hedging without self-financing is discussed.

The self-financing condition in (13.4) states that

$$\xi_t S_t + \beta_t B_t = \xi_{t+1} S_t + \beta_{t+1} B_t, \qquad t = 1, \dots, T - 1. \tag{16.4}$$

We can dispose the restriction to the self-financing strategies, and assume only that the strategies are mean self-financing. Assumption

$$E_{t-1}(\xi_{t+1} S_t + \beta_{t+1} B_t - \xi_t S_t - \beta_t B_t) = 0$$

is equivalent with

$$E_{t-1}(\xi_{t+1}X_t + \beta_{t+1} - \xi_t X_t - \beta_t) = 0.$$

Let us make the mean self-financing assumption

$$E_{t-1}D_t = 0, \quad t = 1, \ldots, T, \tag{16.5}$$

where

$$D_t = \xi_{t+1}X_t + \beta_{t+1} - (\xi_t X_t + \beta_t), \quad t = 1, \ldots, T-1,$$
$$D_T = H_T - (\xi_T X_T + \beta_T).$$

Note that if we define the cumulative cost process by

$$C_0 = U_0,$$
$$C_t = U_t - G_t, \quad t = 1, \ldots, T-1,$$
$$C_T = H_T - G_T,$$

where

$$U_t = \xi_{t+1}X_t + \beta_{t+1}, \quad G_t = \sum_{i=1}^{t} \xi_t(X_t - X_{t-1}),$$

then

$$D_t = \Delta C_t = C_t - C_{t-1}.$$

Schäl (1994) considers quadratic hedging in discrete time with mean self-financing strategies. The definition of mean self-financing in continuous time was given in Föllmer and Sondermann (1986).[1]

We noted that the equality $V_T = V_0 + G_T$ in (16.2) holds only for self-financing strategies. Nevertheless, it is possible to minimize the mean-squared error

$$E_0(H_T - V_0 - G_T)^2,$$

over strategies ξ, which are not necessarily self-financing. Note that $V_0 = U_0$.
We have that

$$H_T - V_0 - G_T = C_T - C_0.$$

Under the condition (16.5) of mean self-financing we have that

$$E(C_T - C_0) = E\left(\sum_{t=1}^{T} \Delta C_t\right) = 0.$$

1 Note that Schäl (1994) defines self-financing so that not only conditions (16.4) are required, but also the condition that the final wealth equals the terminal value of the option: $C_T = \xi_T S_T + \beta_T B_T$, which is equivalent to $H_T = \xi_T X_T + \beta_T$. According to this terminology, self-financing requires replication.

Thus,

$$E(H_T - V_0 - G_T)^2 = \text{Var}(H_T - V_0 - G_T).$$

Thus, the term "variance optimal hedging" can be used in this case. Also,

$$V_0 = E(C_T) = E(H_T - G_T),$$

and it is natural to call V_0 the fair price.

16.1.1.4 Assumptions

We have to assume the square integrability of the relevant terms:

$$EH_T^2 < \infty, \qquad EX_t^2 < \infty, \quad t = 0, \dots, T.$$

The assumptions can be written as $H_T \in \mathcal{L}^2(\Omega, \mathcal{F}_T, P)$ and $X_t \in \mathcal{L}^2(\Omega, \mathcal{F}_t, P)$. In addition, we have to restrict ourselves to the square integrable trading strategies. That is, the value and the gains processes are assumed to satisfy

$$EV_t^2 < \infty, \qquad EG_t^2 < \infty, \quad t = 0, \dots, T. \tag{16.6}$$

We say that the bounded mean–variance trade-off holds if

$$[E(X_t - X_{t-1} \mid \mathcal{F}_{t-1})]^2 \le C \cdot \text{Var}(X_t - X_{t-1} \mid \mathcal{F}_{t-1}) \tag{16.7}$$

P-almost surely, for $t = 1, \dots, T$, where C is a constant. Föllmer and Schied (2002, Theorem 10.39, p. 395) consider the existence and uniqueness of a quadratic hedging strategy. They show that with $d = 1$ risky assets the bounded mean–variance trade-off guarantees the existence of a quadratic strategy (which they call a variance-optimal strategy). The strategy is unique up to modifications in the set $\{\text{Var}(X_t - X_{t-1} \mid \mathcal{F}_{t-1}) = 0\}$.

16.1.2 The One Period Model

We consider the pricing of an European option in the single period model. In the single period model the underlying security has value S_0 at the beginning of the period and value S_1 at the expiration of the option. The price S_0 is a fixed number and S_1 is a random variable. At time zero the option price is C_0. The value of the European option at the expiration is denoted by C_1. For example, in the case of a call option $C_1 = \max\{S_1 - K, 0\}$, where K is the strike price. In the single period model the option is hedged only once (at time 0).

16.1.2.1 Pricing in the One Period Model

In the one period model

$$\begin{aligned}
V_1 &= V_0 + \xi \Delta X_1, \\
\Delta X_1 &= X_1 - X_0 = (1+r)^{-1} S_1 - S_0, \\
H_1 &= (1+r)^{-1} C_1.
\end{aligned} \tag{16.8}$$

We want to find V_0 and ξ minimizing

$$E(V_0 + \xi \Delta X_1 - H_1)^2.$$

This is the population version of a linear least-squares regression with the explanatory variable ΔX_1 and the response variable H_1. We obtain the solutions[2]

$$\xi = \frac{\mathrm{Cov}(\Delta X_1, H_1)}{\mathrm{Var}(\Delta X_1)} \tag{16.10}$$

and

$$V_0 = EH_1 - \xi E \Delta X_1. \tag{16.11}$$

The hedging coefficient can be written as

$$\xi = \frac{\mathrm{Cov}(S_1, C_1)}{\mathrm{Var}(S_1)} \tag{16.12}$$

because constant S_0 can be removed from the covariance and the variance, and the discounting factors of the nominator and denominator cancel each other. Note that for a call option $\xi \in [0, 1]$ and for a put option $\xi \in [-1, 0]$.[3]

We see that the quadratic price

$$V_0 = EH_1 - \xi E \Delta X_1 = (1+r)^{-1}[EC_1 - \xi E(S_1 - (1+r)S_0)] \tag{16.13}$$

is obtained by subtracting a correction term from the expected value of the payoff of the option.

---

2 Derivating with respect to V_0 and ξ we get the equations

$$\begin{cases} E(V_0 + \xi \Delta X_1 - H_1) = 0, \\ E[(V_0 + \xi \Delta X_1 - H_1)\,\Delta X_1] = 0. \end{cases}$$

From the second equation we get

$$\xi = \frac{E[(H_1 - V_0)\Delta X_1]}{E\Delta X_1^2}. \tag{16.9}$$

From the first equation we get $V_0 = EH_1 - \xi E\Delta X_1$. Inserting this to (16.9) gives

$$\xi\left(1 - \frac{(E\Delta X_1)^2}{E\Delta X_1^2}\right) = \frac{E[(H_1 - EH_1)\Delta X_1]}{E\Delta X_1^2}.$$

It holds that

$$1 - \frac{(E\Delta X_1)^2}{E\Delta X_1^2} = \frac{\mathrm{Var}(\Delta X_1)}{E\Delta X_1^2}, \quad E[(H_1 - EH_1)\Delta X_1] = \mathrm{Cov}(\Delta X_1, H_1).$$

3 By the Cauchy–Schwarz inequality $|\mathrm{Cov}(S_1, C_1)| \le \sqrt{\mathrm{Var}(S_1)\mathrm{Var}(C_1)}$. For the call option $\mathrm{Var}(C_1) \le \mathrm{Var}(S_1 - K) = \mathrm{Var}(S_1)$ and for the put option $\mathrm{Var}(C_1) \le \mathrm{Var}(K - S_1) = \mathrm{Var}(S_1)$. For the call option $\mathrm{Cov}(S_1, C_1) \ge 0$ and for the put option $\mathrm{Cov}(S_1, C_1) \le 0$.

The value process is useful in the multiperiod model, but in the single period model we can use the wealth process as well. We can use the following equivalent formulation. The initial wealth is W_0. This amount is invested in the bank account. The amount ξS_0 is borrowed at the risk-free rate and this money is invested in stock, so that there are ξ stocks in the portfolio. The value of the portfolio at time 1 is

$$W_1 = (1 + r)W_0 + \xi(S_1 - (1 + r)S_0).$$

We want to find values of W_0 and ξ that minimize

$$E(W_1 - C_1)^2. \tag{16.14}$$

16.1.2.2 Scatter Plots of Stock Price Differences and Option Payoffs

Figure 16.1 shows a scatter plot of points $(S_1 - S_0, (S_1 - K)_+)$ together with linear fits. We use the daily data of S&P 500 prices, described in Section 2.4.1.

We take $S_0 = 100$ and $S_1 = 100R$, where R are the gross returns of S&P 500 over 10 trading days. Values $(S_1 - K)_+$ are the payoffs of call options with strike price K, and the expiration time 10 trading days. In panel (a) $K = 100$, and in panel (b) $K = 105$. The green lines show the least-squares linear fit $\Delta S_1 \mapsto V_0 + \xi \Delta S_1$, where $\Delta S_1 = S_1 - S_0$ and

$$\xi = \frac{\widehat{\mathrm{Cov}}(\Delta S_1, C_1)}{\widehat{\mathrm{Var}}(\Delta S_1)}, \quad V_0 = \widehat{EC_1} - \xi \cdot \widehat{E\Delta S_1},$$

where we use the sample means, variances, and covariances. The red lines show the linear fit $\Delta S_1 \mapsto a + b\Delta S_1$, where a is the Black–Scholes price, and b

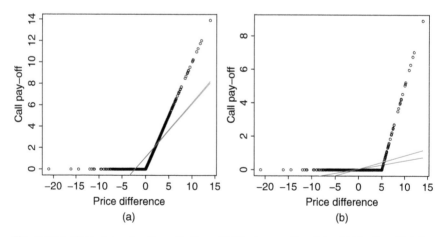

Figure 16.1 *Linear approximations of call payoffs.* We show scatter plots of $(\Delta S_1, (S_1 - K)_+)$ for (a) $K = 100$ and (b) $K = 105$. The green curves show the least-squares fit and the red curves show the Black–Scholes fit.

is the Black–Scholes delta. The volatility is estimated by the sample standard deviation. We see that when $K = 100$, there is hardly any difference between the least-squares fit and the Black–Scholes fit. When $K = 105$, then the least-squares hedging coefficient is higher than the Black–Scholes delta.

16.1.2.3 Pricing in the One Period Model Continued

We derive the solution (16.10)–(16.11) in a different way. The different way of deriving the solution is such that it can be generalized to multiperiod models. Furthermore, it helps us to find the equivalent martingale measure and the hedging error.

We need to find V_0 and ξ minimizing

$$E(V_1 - H_1)^2 = E(V_0 + \xi \Delta X_1 - H_1)^2,$$

where $\Delta X_1 = (1 + r)^{-1} S_1 - S_0$ and $H_1 = (1 + r)^{-1} C_1$. We can write

$$E(V_1 - H_1)^2 = E(V_0 - H_1)^2 + 2\xi E[(V_0 - H_1)\Delta X_1] + \xi^2 E \Delta X_1^2. \qquad (16.15)$$

For a fixed V_0 the minimizer over $\xi \in \mathbf{R}$ is

$$\xi = \frac{E[(H_1 - V_0)\Delta X_1]}{E \Delta X_1^2}. \qquad (16.16)$$

We have that

$$\min_{\xi} E(V_1 - H_1)^2$$

$$= E(V_0 - H_1)^2 - \frac{(E[(H_1 - V_0)\Delta X_1])^2}{E \Delta X_1^2}$$

$$= k_0 V_0^2 - 2V_0 k_0 H_0 + k_0 H_0^2 - k_0 H_0^2 + E H_1^2 - \frac{[E(H_1 \Delta X_1)]^2}{E \Delta X_1^2}$$

$$= k_0 (V_0 - H_0)^2 + \epsilon_0^2, \qquad (16.17)$$

where we denote

$$H_0 = \frac{1}{k_0} \left[E H_1 - \frac{E(\Delta X_1) E(\Delta X_1 H_1)}{E \Delta X_1^2} \right] \qquad (16.18)$$

and

$$k_0 = 1 - \frac{[E(\Delta X_1)]^2}{E \Delta X_1^2}, \qquad \epsilon_0^2 = E H_1^2 - k_0 H_0^2 - \frac{[E(H_1 \Delta X_1)]^2}{E \Delta X_1^2}.$$

We see from (16.17) that the mean-squared error is minimized by choosing $V_0 = H_0$. Equation (16.16) implies that the optimal hedging coefficient is

$$\xi = \frac{E[(H_1 - H_0)\Delta X_1]}{E \Delta X_1^2}, \qquad (16.19)$$

where H_0 is defined in (16.18). The formulas (16.18) and (16.19) are equivalent with the formulas (16.10) and (16.11).[4] Note that formulas (16.18) and (16.19) define ξ in terms of H_0, whereas (16.10) and (16.11) define H_0 in terms of ξ.

16.1.2.4 The Martingale Measure in the One Period Model

Let us assume that our one period model is arbitrage-free. Theorem 13.1 implies that there exists an equivalent martingale measure. Theorem 13.2 implies that any arbitrage-free price can be written as an expected value $E_Q H$ for some equivalent martingale measure Q. Let us find the martingale measure Q, which is implied by the quadratic hedging.

The density of Q with respect to the underlying physical measure is obtained from (16.18) as

$$\frac{dQ}{dP}(\omega) = \frac{1}{k_0} \left[1 - \frac{E(\Delta X_1)}{E\left(\Delta X_1^2\right)} \Delta X_1(\omega) \right] = a - b\Delta X_1(\omega), \tag{16.20}$$

where

$$k_0 = 1 - \frac{[E(\Delta X_1)]^2}{E\Delta X_1^2} = \frac{\mathrm{Var}(\Delta X_1)}{E\Delta X_1^2},$$

and

$$a = \frac{E\Delta X_1^2}{\mathrm{Var}(\Delta X_1)}, \quad b = \frac{E\Delta X_1}{\mathrm{Var}(\Delta X_1)}. \tag{16.21}$$

Now, we have found a measure Q such that

$$C_0 = H_0 = E_P \left(\frac{dQ}{dP} H_1 \right) = E_Q H_1 = \frac{1}{1+r} E_Q C_1.$$

Note that in our notation $E = E_P$.

The Martingale Measure for S&P 500 in the One Period Model Let us estimate the equivalent martingale measure associated with quadratic hedging using S&P 500 daily data of Section 2.4.1.

---

4 We can write

$$H_0 = \frac{1}{k_0} \left[EH_1 - \frac{E(\Delta X_1(H_1 - EH_1))}{E\Delta X_1^2} E\Delta X_1 - \frac{(E\Delta X_1)^2 EH_1}{E\Delta X_1^2} \right]$$

$$= EH_1 - \frac{\mathrm{Cov}(H_1, \Delta X_1)}{\mathrm{Var}(\Delta X_1)} E\Delta X_1.$$

Inserting this to (16.19) gives

$$\xi = \frac{\mathrm{Cov}(H_1, \Delta X_1)}{\mathrm{Var}(\Delta X_1)}.$$

A time series of increments $\Delta X_1 = S_1/(1+r) - S_0$ is not approximately stationary when the time series covers a long time period; see Figure 5.2. The time series of gross returns is nearly stationary, even when the time series extends over a long time period; see Figure 2.1(b). Thus, we can use historical simulation to create a time series of increments from the excess gross returns.

We take the interest rate $r = 0$. We consider a one-step model with the step of 20 days. The excess gross return is equal to the net return

$$R_t - 1 = \frac{S_t}{S_{t-20}} - 1.$$

Now,

$$\Delta X_t = S_t - S_{t-20} = S_{t-20}(R_t - 1).$$

Let

$$\Delta X_t = 100 \left(\frac{S_t}{S_{t-20}} - 1 \right)$$

be the increment. Our S&P 500 data provides a sample of identically distributed observations ΔX_t from the distribution of ΔX_1. We use non-overlapping increments.

Let us estimate the density

$$Z = \frac{dQ}{dP}$$

of the martingale measure with respect to underlying physical measure of ΔX_1. The estimate is

$$\hat{Z}(x) = \hat{a} - \hat{b}x, \quad x \in \mathbf{R},$$

where $\hat{a}$ and $\hat{b}$ are estimates of a and b in (16.21), obtained by replacing the expectations and variances by sample averages and sample variances.

The underlying physical density of ΔX_1 with respect to the Lebesgue measure can be estimated using the kernel estimate $\hat{f}(x)$ of (3.43). The density of the martingale measure with respect to the Lebesgue measure can be estimated as

$$\hat{q}(x) = \hat{Z}(x)\hat{f}(x), \quad x \in \mathbf{R}.$$

Figure 16.2(a) shows the estimate $\hat{Z}$ of the density of the martingale measure with respect to the physical measure (dark green). The red curve shows the case of the Esscher measure and the blue curve the case of the Black–Scholes measure. The blue curve shows the density of the risk-neutral log-normal density with respect to the estimated physical measure. These are taken from Figure 13.1. Panel (b) shows the density $\hat{q}$ (dashed dark green) and $\hat{f}$

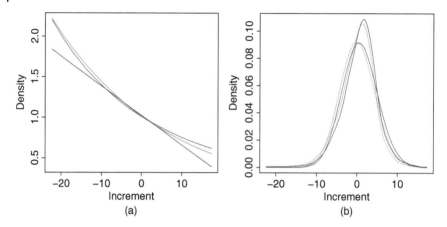

Figure 16.2 *Martingale measure in the one period model.* (a) The density of the quadratic martingale measure with respect to the physical measure (dark green), Esscher measure (red), and Black–Scholes measure (blue). (b) The kernel density estimate of the physical measure (solid dark green), and the corresponding quadratic martingale measure (dashed dark green). The log-normal physical measure and the corresponding risk-neutral log-normal density are depicted as solid blue and dashed blue curves, respectively.

(solid dark green). We show also the physical density (solid blue) and the risk-neutral density (dashed blue) in the Black–Scholes model.

The Martingale Measure in the One Period Binomial Model Let us study the one period binomial model, as defined in Section 14.2.1. In this model at time 0 the stock has value s_0, and at time 1 the stock can take values $s_{1,0}$ and $s_{1,1}$, where $s_{1,0} < (1 + r)s_0 < s_{1,1}$. The probability of the up movement is p and the probability of the down movement is $1 - p$ with $0 < p < 1$. Let us denote

$$D = s_{1,0}/(1 + r) - s_0, \quad U = s_{1,1}/(1 + r) - s_0.$$

We have that

$$P(\Delta X_1 = D) = 1 - p, \quad P(\Delta X_1 = U) = p.$$

From (16.20) we obtain that the martingale measure Q satisfies

$$Q(\Delta X_1 = x) = \left(\frac{\sigma^2 + \mu^2}{\sigma^2} - \frac{\mu}{\sigma^2} x \right) P(\Delta X_1 = x),$$

where $\mu = E\Delta X_1$, $\sigma^2 = \text{Var}(\Delta X_1)$, and $x = U$ or $x = D$. We have that

$$\mu = pU + (1 - p)D, \quad \sigma^2 = p(1 - p)(U - D)^2,$$

and

$$\sigma^2 + \mu^2 = pU^2 + (1 - p)D^2.$$

Thus,

$$Q(\Delta X_1 = U) = \frac{D}{D - U} = \frac{(1 + r)s_0 - s_{1,0}}{s_{1,1} - s_{1,0}},$$

$$Q(\Delta X_1 = D) = \frac{s_{1,1} - (1 + r)s_0}{s_{1,1} - s_{1,0}},$$

which is equal to the martingale measure already derived in (14.18). In fact, the binomial model is a complete model and there is only one equivalent martingale measure.

16.1.3 The Two Period Model

We consider pricing and hedging of an European option in the two period model. The general multiperiod model is considered in Section 16.1.4, and this presentation includes the two period model as a special case. However, we think that it is easier to read the presentation of the multiperiod model when the two period model is presented first.

In the two period model the underlying security takes values S_0, S_1, and S_2. The price S_0 is a fixed number and S_1 and S_2 are random variables. The option is written at time 0, and it expires at time 2. Hedging is done at times 0 and 1 by choosing the hedging coefficients ξ_1 and ξ_2. The value of the European option at the expiration is denoted by C_2. For example, in the case of a call option $C_2 = \max\{S_2 - K, 0\}$, where K is the strike price.

16.1.3.1 An Introduction to the Minimization Problem

The minimization problem can be solved either using the value process or by using the wealth process. The use of the value process is more convenient.

The Minimization Using the Value Process   In the two period model the value process and the discounted contingent claim are defined as

$$V_0 = W_0 = \text{initial wealth},$$
$$V_1 = V_0 + \xi_1 \Delta X_1,$$
$$V_2 = V_0 + \xi_1 \Delta X_1 + \xi_2 \Delta X_2,$$
$$H_2 = (1 + r)^{-2} C_2,$$

where $\Delta X_1 = X_1 - X_0$, $\Delta X_2 = X_2 - X_1$, $X_1 = (1 + r)^{-1} S_1$, and $X_2 = (1 + r)^{-2} S_2$. We want to find $V_0, \xi_1, \xi_2 \in \mathbf{R}$ minimizing

$$E_0(V_2 - H_2)^2. \tag{16.22}$$

Notation E_0 means the unconditional expectation with respect to the underlying measure P, and we denote by E_1, the conditional expectation, with respect to sigma-algebra $\mathcal{F}_1$:

$$E_1(\cdot) = E(\cdot \mid \mathcal{F}_1).$$

Unlike in the one period model this minimization problem cannot be considered as a usual population version of a linear least squares regression. We can consider ΔX_1 and ΔX_2 as explanatory variables and H_2 as the response variables, but now intercept V_0 and coefficient ξ_1 are chosen at time 0, and coefficient ξ_2 is chosen at time 1. In the usual regression problem all parameters are chosen at time 0.

The minimization problem can be solved in the following way. First, we find $\xi_2 \in \mathbf{R}$ minimizing

$$E_1(V_1 + \xi_2 \Delta X_2 - H_2)^2.$$

Let the minimizer be $\xi_2(V_0, \xi_1)$. The notation indicates that the minimizer depends on V_0 and ξ_1. Second, we find $V_0 \in \mathbf{R}$ and $\xi_1 \in \mathbf{R}$ minimizing

$$E(V_0 + \xi_1 \Delta X_1 + \xi_2(V_0, \xi_1)\Delta X_2 - H_2)^2.$$

The Minimization Using the Wealth Process The wealths at times 0, 1, and 2 are

$$W_0 = \bar{\xi}_1 \cdot \bar{S}_0,$$
$$W_1 = W_0 + \bar{\xi}_1 \cdot \Delta \bar{S}_1,$$
$$W_2 = W_0 + \bar{\xi}_1 \cdot \Delta \bar{S}_1 + \bar{\xi}_2 \cdot \Delta \bar{S}_2,$$

where $\Delta \bar{S}_1 = \bar{S}_1 - \bar{S}_0$, $\Delta \bar{S}_2 = \bar{S}_2 - \bar{S}_1$, $\bar{S}_1 = (B_1, S_1)$, $\bar{S}_2 = (B_2, S_2)$, $\bar{\xi}_1 = (\beta_1, \xi_1)$, and $\bar{\xi}_2 = (\beta_2, \xi_2)$. We want to find $\bar{\xi}_1, \bar{\xi}_2 \in \mathbf{R}^2$ so that

$$E_0(W_2 - C_2)^2$$

is minimized, under the self-financing constraints. The minimization problem can be solved in the following way. First, we find $\bar{\xi}_2$ minimizing

$$E_1(W_1 + \bar{\xi}_2 \cdot \Delta \bar{S}_2 - C_2)^2,$$

under the self-financing constraint

$$\bar{\xi}_2 \cdot \bar{S}_1 = \bar{\xi}_1 \cdot \bar{S}_1.$$

Let the minimizer be $\bar{\xi}_2(\bar{\xi}_1)$. The notation indicates that the minimizer depends on $\bar{\xi}_1$. Second, we find $\bar{\xi}_1$ minimizing

$$E(W_0 + \bar{\xi}_1 \cdot \Delta \bar{S}_1 + \bar{\xi}_2(\bar{\xi}_1) \cdot \Delta \bar{S}_2 - C_2)^2.$$

We can now see why it is easier to solve the problem using the value process: It is possible to apply unconstrained minimization when the value process is used.

16.1.3.2 Solving the Minimization Problem

Let us solve the problem of minimizing (16.22). We have that

$$\min_{V_0, \xi_1, \xi_2} E_0(V_2 - H_2)^2 = \min_{V_0, \xi_1} E_0 \min_{\xi_2} E_1(V_2 - H_2)^2.$$

Since

$$E_1(V_2 - H_2)^2 = E_1(V_1 - H_2)^2 + 2\xi_2 E_1[(V_1 - H_2)\Delta X_2] + \xi_2^2 E_1 \Delta X_2^2,$$

the minimizer over $\xi_2 \in \mathbf{R}$ is

$$\xi_2 = \frac{E_1[(H_2 - V_1)\Delta X_2]}{E_1 \Delta X_2^2}. \tag{16.23}$$

We have that

$$\min_{\xi_2} E_1(V_2 - H_2)^2$$

$$= E_1(V_1 - H_2)^2 - \frac{(E_1[(H_2 - V_1)\Delta X_2])^2}{E_1 \Delta X_2^2}$$

$$= k_1 V_1^2 - 2V_1 k_1 H_1 + k_1 H_1^2 - k_1 H_1^2 + E_1 H_2^2 - \frac{[E_1(H_2 \Delta X_2)]^2}{E_1 \Delta X_2^2}$$

$$= k_1(V_1 - H_1)^2 + \epsilon_1^2, \tag{16.24}$$

where we denote

$$k_1 = 1 - \frac{[E_1(\Delta X_2)]^2}{E_1\left(\Delta X_2^2\right)},$$

$$H_1 = \frac{1}{k_1}\left[E_1 H_2 - \frac{E_1(\Delta X_2)E_1(\Delta X_2 H_2)}{E_1\left(\Delta X_2^2\right)}\right], \tag{16.25}$$

and

$$\epsilon_1^2 = E_1 H_2^2 - k_1 H_1^2 - \frac{[E_1(H_2 \Delta X_2)]^2}{E_1 \Delta X_2^2}.$$

It holds that

$$\min_{\xi_1,\xi_2} E_0(V_2 - H_2)^2 = \min_{\xi_1} E_0[k_1(V_1 - H_1)^2] + \epsilon_1^2.$$

Similar calculations which lead to (16.23) show that the minimizer over $\xi_1 \in \mathbf{R}$ is

$$\xi_1 = \frac{E_0[(H_1 - V_0)k_1 \Delta X_1]}{E_0\left(k_1 \Delta X_1^2\right)}. \tag{16.26}$$

Finally, we have to find V_0 minimizing

$$\min_{\xi_1} E_0[k_1(V_1 - H_1)^2].$$

The minimizer over $V_0 \in \mathbf{R}$ is

$$H_0 = \frac{1}{k_0}\left[E_0(k_1 H_1) - \frac{E_0(k_1 \Delta X_1)E_0(k_1 \Delta X_1 H_1)}{E_0\left(k_1 \Delta X_1^2\right)}\right], \tag{16.27}$$

where

$$k_0 = E_0 k_1 - \frac{[E_0(k_1 \Delta X_1)]^2}{E_0\left(k_1 \Delta X_1^2\right)}.$$

Indeed, similar calculations which lead to (16.24) show that

$$\min_{\xi_1} E_0[k_1(V_1 - H_1)^2] = k_0(V_0 - H_0)^2 + \epsilon_0^2,$$

where

$$\epsilon_0^2 = E_0\left(k_1 H_1^2\right) - k_0 H_0^2 - \frac{[E_0(k_1 H_1 \Delta X_1)]^2}{E_0\left(k_1 \Delta X_1^2\right)}.$$

We can summarize the results in the following proposition.

Proposition 16.1 *In the two period model the fair price at time 0 is H_0 given in (16.27) and the optimal hedging coefficient is ξ_1 given in (16.26), when we define the fair price and the optimality of the hedging coefficient by the mean-squared error criterion.*

Note that hedging at time 1 is not done by coefficient ξ_2 in (16.23). Instead, at time 1 we consider the one period model between times 1 and 2, and choose the hedging coefficient of the one period model, as in (16.16).

16.1.3.3 The Martingale Measure in the Two Period Model

Let us find the martingale measure Q, which is implied by the quadratic hedging. We have to find such measure Q that the option price is the discounted expectation with respect to the measure Q:

$$C_0 = H_0 = E_Q H_2 = \frac{1}{(1+r)^2} E_Q C_2.$$

We obtain from (16.25) and (16.27) that the density of Q with respect to the underlying physical measure P is

$$\frac{dQ}{dP} = f_{1|0} \cdot f_{2|1},$$

where

$$f_{1|0}(\omega) = \frac{k_1(\omega)}{k_0}\left[1 - \frac{E_0(k_1 \Delta X_1)}{E_0\left(k_1 \Delta X_1^2\right)} \Delta X_1(\omega)\right],$$

$$f_{2|1}(\omega) = \frac{1}{k_1(\omega)}\left[1 - \frac{E_1(\Delta X_2)(\omega)}{E_1\left(\Delta X_2^2\right)(\omega)} \Delta X_2(\omega)\right],$$

$$k_0 = E_0 k_1 - \frac{[E_0(k_1 \Delta X_1)]^2}{E_0\left(k_1 \Delta X_1^2\right)}, \quad k_1(\omega) = 1 - \frac{[E_1(\Delta X_2)(\omega)]^2}{E_1\left(\Delta X_2^2\right)(\omega)}.$$

In fact,

$$H_0 = E_0(H_1 f_{1|0}), \quad H_1 = E_1(H_2 f_{2|1}),$$

which can be written as

$$H_0 = E_0 f_{1|0} E_1 f_{2|1} H_2 = E_P\left(\frac{dQ}{dP} H_2\right) = E_Q H_2,$$

where we use the notation $E_0 = E_P$.

When the increments are independent, then the martingale measure Q is defined by the density

$$\frac{dQ}{dP}(\omega) = f_1 \cdot f_2,$$

where

$$f_i = a - b\Delta X_i,$$

and

$$a = \frac{E(\Delta X_1)^2}{\text{Var}(\Delta X_1)}, \quad b = \frac{E(\Delta X_1)}{\text{Var}(\Delta X_1)};$$

compare this to a and b as defined in (16.21).

A Martingale Measure for S&P 500 in the Two Period Model Let us estimate the equivalent martingale measure associated with quadratic hedging using S&P 500 daily data of Section 2.4.1. Let us consider a two-step model with two steps of 10 days. We take interest rate $r = 0$. Let

$$\Delta X_1 = S_t - S_{t-10}, \quad \Delta X_2 = S_{t-10} - S_{t-20}$$

be the increments. Our S&P 500 data provides a sample of identically distributed observations from the distribution of $\Delta X = (\Delta X_1, \Delta X_2)$. We use non-overlapping increments.

Let us estimate the density

$$Z(x_1, x_2) = \frac{dQ}{dP}(x_1, x_2) = f_{1|0}(x_1) f_{2|1}(x_2, x_1)$$

of martingale measure Q with respect to the physical measure P.

First, we have to estimate $E_1 \Delta X_2$ and $E_1 \Delta X_2^2$ using nonparametric regression. Let us denote

$$g(x_1) = E(\Delta X_2 \mid \Delta X_1 = x_1), \quad h(x_1) = E((\Delta X_2)^2 \mid \Delta X_1 = x_1).$$

Let $\hat{g}$ and $\hat{h}$ be the kernel regression estimates.[5] Then, we obtain the estimate of k_1 as

$$\hat{k}_1(x_1) = 1 - \frac{\hat{g}^2(x_1)}{\hat{h}(x_1)}.$$

The estimate of $f_{2|1}$ is

$$\hat{f}_{2|1}(x_2, x_1) = \frac{1}{\hat{k}_1(x_1)} \left(1 - \frac{\hat{g}(x_1)}{\hat{h}(x_1)} x_2 \right).$$

Second, we have to estimate $\alpha = E_0(k_1 \Delta X_1)$, $\beta = E_0(k_1(\Delta X_1)^2)$, and $\gamma = E_0 k_1$. The estimates $\hat{\alpha}$, $\hat{\beta}$, and $\hat{\gamma}$ are the sample averages. Then, we obtain the estimate of k_0 as

$$\hat{k}_0 = \hat{\gamma} - \frac{\hat{\alpha}^2}{\hat{\beta}}.$$

The estimate of $f_{1|0}$ is

$$\hat{f}_{1|0}(x_1) = \frac{\hat{k}_1(x_1)}{\hat{k}_0} \left(\hat{\gamma} - \frac{\hat{\alpha}}{\hat{\beta}} x_1 \right).$$

Now, we have obtained the estimate

$$\hat{Z}(x_1, x_2) = \hat{f}_{1|0}(x_1)\hat{f}_{2|1}(x_2, x_1). \tag{16.28}$$

The density of the martingale measure with respect to the Lebesgue measure can be estimated as

$$\hat{q}(x_1, x_2) = \hat{Z}(x_1, x_2)\hat{f}(x_1, x_2), \quad (x_1, x_2) \in \mathbf{R}^2,$$

where $\hat{f}(y_1, y_2)$ is a two-dimensional kernel density estimate of the underlying physical measure of $(\Delta X_1, \Delta X_2)$. The kernel density estimator is defined in (3.43).

When the returns are assumed to be independent, then we use the estimate

$$\hat{Z}(x_1, x_2) = \hat{f}(x_1) \cdot \hat{f}(x_2), \tag{16.29}$$

---

5 We apply the kernel regression estimator of (6.20)–(6.21) to define

$$\hat{g}(x_1) = \sum_{i=1}^{n} p_i(x_1) \Delta X_{2i}, \quad \hat{h}(x_1) = \sum_{i=1}^{n} p_i(x_1)(\Delta X_{2i})^2,$$

where $(\Delta X_{1i}, \Delta X_{2i})$, $i = 1, \dots, n$, are the observations from the distribution of $(\Delta X_1, \Delta X_2)$,

$$p_i(x_1) = \frac{K((x_1 - \Delta X_{1i})/h)}{\sum_{j=1}^{n} K((x_1 - \Delta X_{1j})/h)},$$

$K : \mathbf{R} \to \mathbf{R}$ is a kernel function and $h > 0$ is a smoothing parameter.

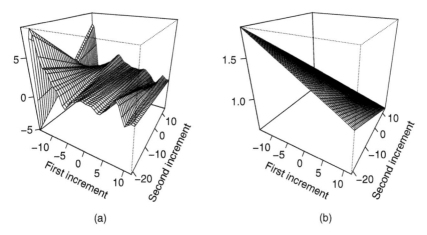

(a)   (b)

Figure 16.3 *The quadratic martingale measure: Two period model.* Estimates of the density of the quadratic martingale measure with respect to the physical measure. (a) Increments are not assumed independent. (b) Increments are assumed to be independent.

where

$$\hat{f}(x_i) = \hat{a} - \hat{b}x_i,$$

and $\hat{a}$ and $\hat{b}$ are the sample versions of

$$a = \frac{E(\Delta X_1)^2}{\text{Var}(\Delta X_1)}, \quad b = \frac{E(\Delta X_1)}{\text{Var}(\Delta X_1)}.$$

In the sample versions we replace the means and variances with the sample means and sample variances.

Figure 16.3 shows estimates of the density of the quadratic martingale measure with respect to the physical measure. In panel (a) we show estimate (16.28), which does not assume independence, and in panel (b) we show estimate (16.29), which assumes independence. In our setting regression estimation is difficult, and assuming independence leads to a more stable result. It is clear that the regression estimation could be improved by applying separate methods for the prediction of the first moment $g(x_1)$ and for the second moment $h(x_1)$.

16.1.4 The Multiperiod Model

We have derived the optimal hedging coefficient and the fair price in the mean-squared error sense for the two period model in Section 16.1.3. It is straightforward to generalize the results from the two period model to a general multiperiod model. The hedging coefficients and the fair price are derived using dynamic programming (backward induction).

16.1.4.1 Pricing in the Multiperiod Model

Let $V = (V_t)_{t=0,...,T}$ be the value process of a self-financing portfolio:

$$V_t = V_0 + \sum_{k=1}^{t} \xi_k \Delta X_k, \tag{16.30}$$

where

$$\Delta X_k = X_k - X_{k-1},$$

and $X_k = (1+r)^{-k} S_k$. We want to find $V_0, \xi_1, \ldots, \xi_T$ minimizing

$$E(V_T - H_T)^2, \tag{16.31}$$

where $H_T = (1+r)^{-T} C_T$ is the discounted value of the derivative at the expiration.[6]

When $V_0, \xi_1, \ldots, \xi_T$ minimize (16.31), then we say that V_0 is the fair price in the mean-squared error sense and ξ_1 is the optimal hedging coefficient in the mean squared error sense. The coefficients $\xi_2, \ldots, \xi_T$ are needed to derive ξ_1 in our backward induction, but they do not equal the optimal hedging coefficients at times $t = 2, \ldots, T$. Instead, at time $t = 2$ we need to make a new calculation of coefficients, say $\tilde{\xi}_1, \ldots, \tilde{\xi}_{T-1}$, where $\tilde{\xi}_1$ is the optimal hedging coefficient at time $t = 2$.

The following theorem is proved in Černý (2004b, Section 13.4), where $d \geq 1$, so that the number of risky assets is allowed to be larger than one. Černý (2004a) is an article with the same result, and Bertsimas *et al.* (2001) contains a similar kind of result. A similar kind of proof can be found in Schäl (1994), who considers the case of mean self-financing strategies. The case of independent increments was considered by Wolczyńska (1998) and Hammarlid (1998).

Theorem 16.2 *Let $H_T = (1+r)^{-T} C_T$ and let H_t be defined recursively for $t = 0, \ldots, T-1$ by*

$$H_t = \frac{1}{k_t} \left[E_t(k_{t+1} H_{t+1}) - \frac{E_t(k_{t+1} \Delta X_{t+1} H_{t+1})}{E_t \left[k_{t+1} \Delta X_{t+1}^2 \right]} E_t(k_{t+1} \Delta X_{t+1}) \right], \tag{16.32}$$

where $k_T = 1$ and

$$k_t = E_t k_{t+1} - \frac{(E_t[k_{t+1} \Delta X_{t+1}])^2}{E_t \left[k_{t+1} \Delta X_{t+1}^2 \right]}. \tag{16.33}$$

---

6 Equivalently, we want to find $W_0, \xi_1, \ldots, \xi_T$ minimizing

$$E(W_T - C_T)^2,$$

where $W_T = (1+r)^T V_T$ is the wealth, and $C_T = (1+r)^T H_T$ is the value of the option at the expiration.

Then, the fair price in the mean-squared error sense is H_0 and the optimal hedging coefficient is

$$\xi_1 = \frac{E_0[k_1(H_1 - H_0)\Delta X_1]}{E_0\left[k_1 \Delta X_1^2\right]}.$$ (16.34)

Proof. Define $J_T = (V_T - H_T)^2$ and

$$J_t = \min_{\xi_{t+1}} E_t J_{t+1}, \quad t = 0, \dots, T - 1.$$

Then,

$$\min_{V_0, \xi_1, \dots, \xi_T} E(V_T - H_T)^2 = \min_{V_0} J_0$$ (16.35)

and

$$\xi_{t+1} = \mathrm{argmin}_{\xi_{t+1}} E_t J_{t+1}.$$

Indeed,

$$\min_{V_0, \xi_1, \dots, \xi_T} E(V_T - H_T)^2 = \min_{V_0, \xi_1, \dots, \xi_T} E E_{T-1} J_T$$

$$= \min_{V_0, \xi_1, \dots, \xi_{T-1}} E \min_{\xi_T} E_{T-1} J_T$$

$$= \min_{V_0, \xi_1, \dots, \xi_{T-1}} E J_{T-1},$$

and we may continue recursively to obtain (16.35). Define $\epsilon_T = 0$ and

$$\epsilon_t^2 = E_t \left[\epsilon_{t+1}^2\right] + E_t \left[k_{t+1} H_{t+1}^2\right] - k_t H_t^2 - \frac{(E_t[k_{t+1} H_{t+1} \Delta X_{t+1}])^2}{E_t \left[k_{t+1} \Delta X_{t+1}^2\right]},$$ (16.36)

for $t = 0, \dots, T - 1$. We will prove that

$$J_t = k_t(V_t - H_t)^2 + \epsilon_t^2.$$ (16.37)

This holds for $t = T$ (with $k_T = 1$ and $\epsilon_T = 0$). Make the induction hypothesis (for backward induction) that

$$J_{t+1} = k_{t+1}(V_{t+1} - H_{t+1})^2 + \epsilon_{t+1}^2.$$

Write

$$V_{t+1} = V_t + \xi_{t+1} \Delta X_{t+1}.$$

Then,

$$J_{t+1} = k_{t+1} \left[(V_t - H_{t+1})^2 + 2\xi_{t+1}(V_t - H_{t+1})\Delta X_{t+1} + \xi_{t+1}^2 \Delta X_{t+1}^2\right] + \epsilon_{t+1}^2.$$

The solution ξ_{t+1} for $\min_{\xi_{t+1}} E_t J_{t+1}$ satisfies

$$E_t[k_{t+1}(V_t - H_{t+1})\Delta X_{t+1}] + \xi_{t+1} E_t \left[k_{t+1} \Delta X_{t+1}^2\right] = 0,$$

which gives

$$\xi_{t+1} = \frac{E_t[k_{t+1}(H_{t+1} - V_t)\Delta X_{t+1}]}{E_t[k_{t+1}\Delta X_{t+1}^2]}.$$

Then, we get

$$J_t = \min_{\xi_{t+1}} E_t J_{t+1}$$

$$= E_t[k_{t+1}(V_t - H_{t+1})^2] - \frac{(E_t[k_{t+1}(H_{t+1} - V_t)\Delta X_{t+1}])^2}{E_t[k_{t+1}\Delta X_{t+1}^2]} + E_t \epsilon_{t+1}^2,$$

which leads to (16.37). □

The Minimal Hedging Error The proof implies that the minimal hedging error is given by ϵ_0^2, defined recursively in (16.36). Indeed, from (16.35) and (16.37), we obtain that

$$\min_{W_0, \xi_1, \dots, \xi_T} E(V_T - H_T)^2 = \min_{W_0} J_0 = \min_{W_0} \left(k_0(W_0 - H_0)^2 + \epsilon_0^2\right) = \epsilon_0^2.$$

The Hedging Coefficients The proof implies that the sequence of quadratic hedging coefficients is given by

$$\xi_{t+1} = \frac{E_t[k_{t+1}(H_{t+1} - H_t)\Delta X_{t+1}]}{E_t[k_{t+1}\Delta X_{t+1}^2]},$$

where $t = 0, \dots, T - 1$. The coefficient ξ_1 is applied at time 0, and the coefficients $\xi_2, \dots, \xi_T$ will not be applied, because at time $t = 2$, we need to construct a model of $T - 1$ periods.

Excess Gross Returns and Quadratic Hedging The formulas for the price and the hedging coefficient are written using the increment

$$\Delta X_{t+1} = X_{t+1} - X_t = \frac{1}{(1 + r)^{t+1}}(S_{t+1} - (1 + r)S_t).$$

We can write the formulas as well using the excess gross return

$$R_{t+1} = \frac{S_{t+1}}{S_t} - (1 + r).$$

The formulas (16.32) and (16.34) can be written as

$$H_t = \frac{1}{k_t}\left[E_t(k_{t+1}H_{t+1}) - \frac{E_t(k_{t+1}R_{t+1})E_t(k_{t+1}R_{t+1}H_{t+1})}{E_t[k_{t+1}R_{t+1}^2]}\right], \qquad (16.38)$$

where $k_T = 1$ and

$$k_t = E_t k_{t+1} - \frac{(E_t[k_{t+1}R_{t+1}])^2}{E_t[k_{t+1}R_{t+1}^2]}.$$

The fair price in the mean-squared error sense is H_0 and the optimal hedging coefficient at time $t = 0$ is

$$\xi_1 = \frac{E_0[k_1(H_1 - H_0)R_1]}{S_0 E_0\left[k_1 R_1^2\right]}. \tag{16.39}$$

Indeed, we can multiply by $(1 + r)^{t+1}$ and divide by S_t both the nominators and the denominators, and these terms can be moved inside the conditional expectations $E_t(\,\cdot\,)$, because they are $\mathcal{F}_t$-measurable.

16.1.4.2 The Martingale Measure

Let us assume that the model is arbitrage-free. Theorem 13.1 says that there exists an equivalent martingale measure. Let us find the martingale measure Q associated with quadratic hedging. According to Theorem 13.2 the martingale measure is such that the option price is the discounted expectation with respect to the measure:

$$C_0 = H_0 = E_Q H_T = \frac{1}{(1+r)^T} E_Q C_T.$$

The density of Q with respect to the underlying physical measure is

$$\frac{dQ}{dP} = \prod_{t=0}^{T-1} f_{t+1|t},$$

where

$$f_{t+1|t} = \frac{k_{t+1}}{k_t}\left[1 - \frac{E_t(k_{t+1}\Delta X_{t+1})}{E_t\left[k_{t+1}\Delta X_{t+1}^2\right]}\Delta X_{t+1}\right].$$

In fact, (16.32) in Theorem 16.2 implies that

$$H_0 = E_0(H_1 f_{1|0}),$$
$$H_1 = E_1(H_2 f_{2|1}),$$
$$\vdots$$
$$H_{T-1} = E_{T-1}(H_T f_{T|T-1}),$$

which can be written as

$$H_0 = E_0 f_{1|0} E_1 f_{2|1} \cdots E_{T-1} f_{T|T-1} H_T = E_P\left(\frac{dQ}{dP}H_T\right) = E_Q H_T,$$

where we use the notation $E_0 = E_P$. We can derive a similar expression for the density using the excess gross returns R_{t+1} instead of increments ΔX_{t+1}: we can apply (16.38).

16.1.4.3 Simplifying Assumptions

We can simplify the price formula (16.32) and the hedging formula (16.34) making restrictive assumptions on the increments $\Delta X_1, \dots, \Delta X_T$. These assumptions are the martingale assumption, the assumption of a deterministic mean–variance ratio, the assumption of independence, and the assumption of independence and identical distribution.

Similar simplifications can be made to the formulas (16.38) and (16.39) when the assumptions are made on the process $R_1, \dots, R_T$ of the gross returns.[7]

Quadratic Hedging Under the Martingale Assumption Let us assume that $X = (X_t)_{t=0,\dots,T}$ is a martingale with respect to the underlying physical measure P. Then,

$$E_t X_{t+1} = X_t, \quad t = 0, \dots, T-1,$$

P-almost surely. Thus,

$$E_t \Delta X_{t+1} = 0, \quad t = 0, \dots, T-1.$$

Now, we have that $k_t = 1$, and

$$H_t = E_t H_{t+1}.$$

This implies that the option price is the expected value:

$$C_0 = H_0 = E H_T = \frac{1}{(1+r)^T} E C_T.$$

The first hedging coefficient is

$$\xi_1 = \frac{E[(H_1 - H_0)\Delta X_1]}{E\left(\Delta X_1^2\right)} = \frac{\mathrm{Cov}(H_1, \Delta X_1)}{\mathrm{Var}(\Delta X_1)} = \frac{\mathrm{Cov}(C_1, S_1)}{\mathrm{Var}(S_1)},$$

where $H_1 = E_1 H_T$, and $H_0 = E H_1$. The expression for ξ_1 is the same as in the one period model; see (16.19), (16.10), and (16.12). Using the rule of iterated expectations we can also write

$$\xi_1 = \frac{\mathrm{Cov}(H_T, \Delta X_1)}{\mathrm{Var}(\Delta X_1)}.$$

We can derive the result easily without using Theorem 16.2. Indeed,

$$E(V_0 + G_T(\xi) - H_T)^2 = E(V_0 - H_T)^2 + 2E[(V_0 - H_T)G_T(\xi)] + EG_T(\xi)^2,$$

---

7 For example, the assumption about independence of the increments and independence of gross returns are not equivalent. Consider the random walk model for the stock price: $S_t = S_0 + \sum_{i=1}^{t} \epsilon_i$, where $S_0 > 0$ is a constant, and ϵ_i are independent. Now, $S_{t+1} - S_t = \epsilon_{t+1}$, and thus the increments are independent. However, the gross returns S_{t+1}/S_t are not independent. Consider the geometric random walk model for the stock price: $S_t = S_0 \exp\left\{\sum_{i=1}^{t} \epsilon_i\right\}$. Now, $S_{t+1}/S_t = \exp\{\epsilon_{t+1}\}$, and thus the gross returns are independent. However, the increments $S_{t+1} - S_t$ are not independent.

where $G_T(\xi) = \sum_{t=1}^{T} \xi_t \Delta X_t$. Under the martingale assumption,

$$EG_T(\xi)^2 = E \sum_{t=1}^{T} \xi_t E_{t-1}(\Delta X_t)^2.$$

Also,

$$E[(V_0 - H_T)G_T(\xi)] = E \sum_{t=1}^{T} \xi_t E_{t-1}[(V_0 - H_t)\Delta X_t].$$

Thus, we obtain a sum of similar one period optimizations as in (16.15).

Deterministic Mean–Variance Ratio Let us denote

$$\mu_t = E_{t-1}\Delta X_t, \quad \sigma_t^2 = \text{Var}_{t-1}(\Delta X_t).$$

Let us assume that the ratio

$$\frac{\mu_t^2}{\sigma_t^2}$$

is deterministic for $t = 1, \ldots, T$. This assumption is made in Föllmer and Schied (2002, Proposition 10.40, p. 396) to derive an expression for the variance-optimal hedging strategy. Note that the mean–variance ratio was used in (16.7) to formulate a sufficient condition for the existence and uniqueness of the variance-optimal hedging strategy (the bounded mean–variance trade-off). Under the assumption of a deterministic mean–variance ratio it holds that k_t in (16.33) is deterministic. In fact, now $k_T = 1$, and

$$k_t = k_{t+1}\left(1 - \frac{(E_t[\Delta X_{t+1}])^2}{E_t\left[\Delta X_{t+1}^2\right]}\right) = k_{t+1}\frac{\sigma_{t+1}^2}{\mu_{t+1}^2 + \sigma_{t+1}^2}, \quad t = T - 1, \ldots, 0.$$

That is,

$$k_t = \prod_{i=t}^{T-1} \frac{\sigma_{i+1}^2}{\mu_{i+1}^2 + \sigma_{i+1}^2}, \quad t = 0, \ldots, T - 1.$$

Values H_t are defined recursively for $t = 0, \ldots, T - 1$ by

$$H_t = \frac{k_{t+1}}{k_t}\left[E_t(H_{t+1}) - \frac{E_t(\Delta X_{t+1}H_{t+1})}{E_t\left[\Delta X_{t+1}^2\right]}E_t(\Delta X_{t+1})\right], \tag{16.40}$$

where we start at $H_T = (1 + r)^{-T}C_T$. The fair price in the mean-squared error sense is H_0 and the optimal hedging coefficient is

$$\xi_1 = \frac{E[(H_1 - H_0)\Delta X_1]}{E\left[\Delta X_1^2\right]}. \tag{16.41}$$

Independent Increments Let us assume that the increments of discounted prices

$$\Delta X_t = X_t - X_{t-1} = \frac{1}{(1+r)^t}(S_t - (1+r)S_{t-1}), \quad t = 1, \dots, T,$$

are independent. Assume that the sigma-algebras are generated by the price process: $F_t = \sigma(S_1, \dots, S_t) = \sigma(X_1, \dots, X_t)$. Then, the independence of increments implies that the conditional expectations reduce to unconditional expectations, and

$$\mu_t = E_{t-1}\Delta X_t, \quad \sigma_t^2 = \text{Var}_{t-1}(\Delta X_t)$$

are deterministic. Thus, the ratio μ_t^2/σ_t^2 is deterministic, and we obtain the price and hedging formulas (16.40) and (16.41).

i.i.d. Increments Let us assume that the increments of discounted prices ΔX_t are independent and identically distributed. Let us denote

$$E_t \Delta X_{t+1} = \mu, \quad \text{Var}_t(\Delta X_{t+1}) = \sigma^2.$$

We have that

$$\frac{k_{t+1}}{k_t} = \frac{E_t\left[\Delta X_{t+1}^2\right]}{\text{Var}_t(\Delta X_{t+1})} = \frac{\sigma^2 + \mu^2}{\sigma^2}.$$

The price and hedging formulas are obtained from (16.40) and (16.41). Values H_t are defined recursively for $t = 0, \dots, T-1$ by

$$H_t = \frac{\sigma^2 + \mu^2}{\sigma^2} E\left[H_{t+1}\left(1 - \frac{\mu}{\sigma^2 + \mu^2}\Delta X_{t+1}\right)\right],$$

where we start at $H_T = (1+r)^{-T}C_T$. The fair price in the mean-squared error sense is H_0 and the optimal hedging coefficient is

$$\xi_1 = \frac{E[(H_1 - H_0)\Delta X_1]}{E\left[\Delta X_1^2\right]}.$$

The density of the martingale measure Q with respect to the underlying physical measure is

$$\frac{dQ}{dP} = \prod_{t=1}^{T} f_t,$$

where

$$f_t = \frac{\sigma^2 + \mu^2}{\sigma^2}\left[1 - \frac{\mu}{\sigma^2 + \mu^2}\Delta X_t\right].$$

16.2 Local Quadratic Hedging

Local quadratic hedging applies a much simpler recursive scheme for minimizing the quadratic hedging error than global quadratic hedging of Section 16.1. Local quadratic hedging solves the minimization only approximately. This numerical error could be compensated if a more accurate statistical estimation is possible.

Local quadratic hedging reduces the minimization of quadratic hedging error to a series of minimizations in one period models. Thus, in the one period model global and local quadratic hedging are identical. We introduce local quadratic hedging using the two period model, and after that cover the multiperiod model.

16.2.1 The Two Period Model

We introduce local quadratic hedging using the two period model. In the two period model the value process and the discounted contingent claim are defined as

$$V_0 = W_0 = \text{initial wealth,}$$
$$V_1 = V_0 + \xi_1 \Delta X_1,$$
$$V_2 = V_0 + \xi_1 \Delta X_1 + \xi_2 \Delta X_2,$$
$$H_2 = (1+r)^{-2} C_2,$$

where $\Delta X_1 = X_1 - X_0$, $\Delta X_2 = X_2 - X_1$, $X_1 = (1+r)^{-1} S_1$, and $X_2 = (1+r)^{-2} S_2$.

In local quadratic hedging the minimization is done in two steps.

1) First, we find $H_1 \in \mathbf{R}$ and $\xi_2 \in \mathbf{R}$ minimizing

$$E_1(H_1 + \xi_2 \Delta X_2 - H_2)^2.$$

This is the population version of a linear least-squares regression with the response variable H_2 and the explanatory variable ΔX_2. The minimizers are

$$\xi_2 = \frac{\text{Cov}_1(H_2, \Delta X_2)}{\text{Var}_1(\Delta X_2)}, \quad H_1 = E_1 H_2 - \xi_2 E_1 \Delta X_2.$$

2) Second, we find $H_0 \in \mathbf{R}$ and $\xi_1 \in \mathbf{R}$ minimizing

$$E(H_0 + \xi_1 \Delta X_1 - H_1)^2.$$

This is the population version of a linear least-squares regression with the response variable H_1 and the explanatory variable ΔX_1. The minimizers are

$$\xi_1 = \frac{\text{Cov}_0(H_1, \Delta X_1)}{\text{Var}_0(\Delta X_1)}, \quad H_0 = E_0 H_1 - \xi_1 E_0 \Delta X_1.$$

The minimization problems are easier to solve than in the case of global quadratic hedging. However, we are not able to minimize

$$E(V_2 - H_2)^2,$$

but only to minimize it approximately.

We can write the price of the discounted contingent claim obtained by local quadratic hedging as

$$H_0 = E_0 H_2 - E_0(\xi_2 E_1 \Delta X_2) - \xi_1 E_0 \Delta X_1. \tag{16.42}$$

The first hedging coefficient ξ_1 can be written as

$$\xi_1 = \frac{\text{Cov}_0(H_2, \Delta X_1)}{\text{Var}_0(\Delta X_1)} - \frac{\text{Cov}_0(\xi_2 \Delta X_2, \Delta X_1)}{\text{Var}_0(\Delta X_1)}. \tag{16.43}$$

The hedging coefficients ξ_1 and ξ_2 give the number of stocks in the hedging portfolio. The number of bonds β_1 and β_2 are obtained from the self-financing restrictions as in (16.3):

$$\begin{aligned}
\beta_1 &= H_0 - \xi_1 X_0 \\
&= E_0 H_2 - E_0(\xi_2 E_1 \Delta X_2) - \xi_1 E_0 X_1, \\
\beta_2 &= \beta_1 + (\xi_1 - \xi_2) X_1 \\
&= E_0 H_2 - E_0(\xi_2 E_1 X_2) + E_0(\xi_2 X_1) - \xi_2 X_1 + \xi_1(X_1 - E_0 X_1).
\end{aligned} \tag{16.44}$$

16.2.1.1 A Comparison to the Global Quadratic Hedging

To highlight the difference between the local and the global quadratic hedging, let us recall the global quadratic hedging of Section 16.1. In the global quadratic hedging we want to find $V_0, \xi_1, \xi_2 \in \mathbf{R}$ minimizing

$$E_0(V_2 - H_2)^2. \tag{16.45}$$

The minimization problem can be solved in two steps. First, we find $\xi_2 \in \mathbf{R}$ minimizing

$$E_1(V_1 + \xi_2 \Delta X_2 - H_2)^2.$$

Let the minimizer be $\xi_2(V_0, \xi_1)$. The minimizer depends on V_0 and ξ_1. Second, we find $V_0 \in \mathbf{R}$ and $\xi_1 \in \mathbf{R}$ minimizing

$$E_0(V_0 + \xi_1 \Delta X_1 + \xi_2(V_0, \xi_1) \Delta X_2 - H_2)^2.$$

The quadratic price is V_0.

16.2.1.2 The Martingale Measure

Let us study the martingale measure Q implied by the local hedging. The density of the martingale measure with respect to the physical measure P is

$$\frac{dQ}{dP} = f_{1|0} f_{2|1},$$

where

$$f_{t|t-1} = a_{t-1} - b_{t-1}\Delta X_t,$$

with

$$a_{t-1} = \frac{E_{t-1}\Delta X_t^2}{\mathrm{Var}_{t-1}(\Delta X_t)}, \quad b_{t-1} = \frac{E_{t-1}\Delta X_t}{\mathrm{Var}_{t-1}(\Delta X_t)},$$

for $t = 1, 2$. The derivation of the martingale measure is given in (16.55) for the multiperiod model.

We can write the density in terms of the excess gross return. Namely,

$$f_{t|t-1} = a_{t-1} - b_{t-1}R_t,$$

where

$$R_t = \frac{S_t}{S_{t-1}} - (1 + r),$$

and

$$a_{t-1} = \frac{E_{t-1}R_t^2}{\mathrm{Var}_{t-1}(R_t)}, \quad b_{t-1} = \frac{E_{t-1}R_t}{\mathrm{Var}_{t-1}(R_t)},$$

for $t = 1, 2$. This is possible, because the denominator and nominator can be multiplied by the square of $(1 + r)/S_{t-1}$, which is $\mathcal{F}_{t-1}$-measurable, and can be placed inside E_{t-1}.

A Martingale Measure for S&P 500: Two Steps Let us estimate the equivalent martingale measure associated with quadratic hedging using S&P 500 daily data of Section 2.4.1. Let us consider a two-step model with two steps of 10 days. We choose interest rate $r = 0$. Let

$$\Delta X_1 = S_t - S_{t-10}, \quad \Delta X_2 = S_{t-10} - S_{t-20}$$

be the price increments, where $t - 20$ is the current time. When t runs through a long time period the observations are not stationary, but we can use our S&P 500 data to provide a sample of identically distributed observations of $\Delta X = (\Delta X_1, \Delta X_2)$. We use non-overlapping increments. The observations are

$$\Delta X_{1t} = 100\left(\frac{S_t}{S_{t-10}} - 1\right), \quad \Delta X_{2t} = 100\left(\frac{S_{t-10}}{S_{t-20}} - 1\right).$$

Let us estimate the density

$$Z(y_1, y_2) = \frac{dQ}{dP}(y_1, y_2) = f_{1|0}(y_1)f_{2|1}(y_2, y_1)$$

of martingale measure Q with respect to the physical measure P.

First, we have to estimate $E_1 \Delta X_2$ and $E_1 (\Delta X_2)^2$ using nonparametric regression. Let us denote

$$g(y_1) = E(\Delta X_2 \,|\, \Delta X_1 = y_1), \quad h(y_1) = E((\Delta X_2)^2 \,|\, \Delta X_1 = y_1).$$

Let $\hat{g}$ and $\hat{h}$ be the kernel regression estimates.[8] Then, we obtain the estimates of a_1 and b_1 as

$$\hat{a}_1(y_1) = \frac{\hat{h}(y_1)}{\hat{h}(y_1) + \hat{g}^2(y_1)}, \quad \hat{b}_1(y_1) = \frac{\hat{g}(y_1)}{\hat{h}(y_1) + \hat{g}^2(y_1)},$$

The estimate of $f_{2|1}$ is

$$\hat{f}_{2|1}(y_2, y_1) = \hat{a}_1(y_1) + \hat{b}_1(y_1) y_2.$$

Second, we have to estimate $\alpha = E_0(\Delta X_1)$ and $\beta = E_0(\Delta X_1^2)$. The estimates $\hat{\alpha}$ and $\hat{\beta}$ are the sample averages. Then, we obtain the estimates of a_0 and b_0 as

$$\hat{a}_0 = \frac{\hat{\beta}}{\hat{\beta} + \hat{\alpha}^2}, \quad \hat{b}_0 = \frac{\hat{\alpha}}{\hat{\beta} + \hat{\alpha}^2}.$$

The estimate of $f_{1|0}$ is

$$\hat{f}_{1|0}(y_1) = \hat{a}_0 + \hat{b}_0 y_1.$$

Now, we have obtained the estimate

$$\hat{Z}(y_1, y_2) = \hat{f}_{1|0}(y_1)\hat{f}_{2|1}(y_2, y_1).$$

The density of the martingale measure with respect to the Lebesgue measure can be estimated as

$$\hat{q}(y_1, y_1) = \hat{Z}(y_1, y_2)\hat{f}(y_1, y_2), \quad (y_1, y_2) \in \mathbf{R}^2,$$

where $\hat{f}(y_1, y_2)$ is a two-dimensional kernel density estimate of the underlying physical measure of $(\Delta X_1, \Delta X_2)$. The kernel density estimator is defined in (3.43).

Figure 16.4 shows estimates of the density of the local quadratic martingale measure with respect to the physical measure. Panel (a) shows a contour plot and panel (b) shows a perspective plot.

---

8 We apply the kernel regression estimator of (6.20) and (6.21) to define

$$\hat{g}(y_1) = \sum_{i=1}^n p_i(y_1) Y_{2i}, \quad \hat{h}(y_1) = \sum_{i=1}^n p_i(y_1) Y_{2i}^2,$$

$$p_i(y_1) = \frac{K((y_1 - Y_{1i})/h)}{\sum_{j=1}^n K((y_1 - Y_{1j})/h)},$$

where $(Y_{1i}, Y_{2i}), i = 1, \dots, n$, are the observations of $(\Delta X_1, \Delta X_2), K : \mathbf{R} \to \mathbf{R}$ is a kernel function and $h > 0$ is a smoothing parameter.

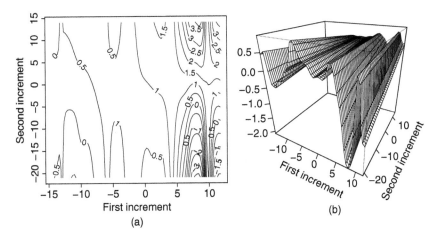

Figure 16.4 *A local quadratic martingale measure: Two period model.* (a) A contour plot; (b) a perspective plot. Estimates of the density of the local quadratic martingale measure with respect to the physical measure.

16.2.2 The Multiperiod Model

Let

$$H_T = \frac{1}{(1+r)^T} C_T$$

be the discounted value of the derivative at the expiration. We define recursively values H_{t-1} and ξ_t, $t = T, \dots, 1$, starting with the value H_T. Let H_{t-1} and ξ_t be the minimizers of

$$E_{t-1}(H_{t-1} + \xi_t \Delta X_t - H_t)^2,$$

for $t = T, \dots, 1$, over $H_{t-1}, \xi_t \in \mathbf{R}$, where

$$\Delta X_t = X_t - X_{t-1}, \quad X_t = \frac{S_t}{(1+r)^t}.$$

This is a conditional population least-squares linear regression problem with the response variable H_t and the explanatory variable ΔX_t. The problem is similar to the minimization problem in the one period model of Section 16.1.2, but now we are conditioning on $\mathcal{F}_{t-1}$. The solutions are

$$\xi_t = \frac{\text{Cov}_{t-1}(\Delta X_t, H_t)}{\text{Var}_{t-1}(\Delta X_t)}$$

and

$$H_{t-1} = E_{t-1} H_t - \xi_t E_{t-1} \Delta X_t.$$

Value H_0 is the price suggested by local quadratic hedging and ξ_1 is the hedging coefficient at time 0, which is suggested by local quadratic hedging.

We can write ξ_t using the undiscounted prices S_t and C_t as

$$\xi_t = \frac{\text{Cov}_{t-1}(S_t, C_t)}{\text{Var}_{t-1}(S_t)}, \tag{16.46}$$

where

$$C_{t-1} = (1+r)^{-1}E_{t-1}C_t + \xi_t\left(S_{t-1} - (1+r)^{-1}E_{t-1}S_t\right). \tag{16.47}$$

The price can be written as:[9]

$$H_0 = E_0 H_T - \sum_{t=0}^{T-1} E_0(\xi_{t+1}\Delta X_{t+1}). \tag{16.48}$$

The hedging coefficients can be written as[10]

$$\xi_t = \frac{1}{\text{Var}_{t-1}(\Delta X_t)}\left(\text{Cov}_{t-1}(H_T, \Delta X_t) - \sum_{u=t}^{T-1}\text{Cov}_{t-1}(\xi_{u+1}\Delta X_{u+1}, \Delta X_t)\right).$$

When F_t are independent, then

$$\xi_t = \frac{\text{Cov}_{t-1}(H_T, \Delta X_t)}{\text{Var}_{t-1}(\Delta X_t)}, \tag{16.49}$$

where $t = 1, \ldots, T$, and the price is

$$H_0 = E_0 H_T - \sum_{t=0}^{T-1} \xi_{t+1} E_0(\Delta X_{t+1}). \tag{16.50}$$

Also, similarly as in (14.34), when F_t are independent,

$$\xi_t = \frac{\text{Cov}_{t-1}(H_T, X_T)}{(T-t+1)\text{Var}_{t-1}(\Delta X_t)},$$

where $t = 1, \ldots, T$.

---

9 In fact, let us make the induction assumption that the formula holds for the $T-1$ period models. Then,

$$H_1 = E_1 H_T - \sum_{t=1}^{T-1} E_1(\xi_{t+1}\Delta X_{t+1}).$$

On the other hand, $H_0 = E_0 H_1 - \xi_1 E_0 \Delta X_1$.

10 We combine

$$\xi_t = \frac{\text{Cov}_{t-1}(\Delta X_t, H_t)}{\text{Var}_{t-1}(\Delta X_t)}$$

and the generalization of (16.48) to $0 \le t \le T-1$:

$$H_t = E_t H_T - \sum_{u=t}^{T-1} E_t(\xi_{u+1}\Delta X_{u+1}).$$

16.2.2.1 A Comparison to Black–Scholes

We compare the quadratic prices and hedging coefficients to the Black–Scholes prices and hedging coefficients. We assume the independence of increments and use formulas (16.49) and (16.50). We apply S&P 500 daily data of Section 2.4.1. The Black–Scholes prices and deltas are computed using the annualized standard deviation as the volatility.

Figure 16.5 compares quadratically optimal prices to Black–Scholes prices. Panel (a) shows the quadratically optimal prices (black) and the Black–Scholes prices (red) as a function of moneyness S/K. Time to expiration is 20 trading days. Panel (b) shows the ratios of the quadratically optimal prices to the Black–Scholes prices as a function of moneyness. Time to expiration is 20 days (black), 40 days (red), 60 days (blue), and 80 days (green). We see from panel (b) that when the moneyness is less than one, then the quadratic prices are less than the Black–Scholes prices. When the moneyness is about 0.95, then increasing the time to expiration makes the ratio of the quadratic prices to the Black–Scholes prices increase.

Figure 16.6 compares quadratic hedging coefficients to Black–Scholes hedging coefficients. Panel (a) shows the quadratic hedging coefficients (black) and the Black–Scholes hedging coefficients (red) as a function of moneyness S/K. Time to expiration is 20 trading days. Panel (b) shows the ratios of the quadratic hedging coefficients to the Black–Scholes hedging coefficients as a function of moneyness. Time to expiration is 20 days (black), 40 days (red), 60 days (blue), and 80 days (green). We see from panel (b) that when the moneyness is about one, then increasing the time to expiration makes the ratio of the quadratic hedging coefficient to the Black–Scholes hedging coefficient increase. When

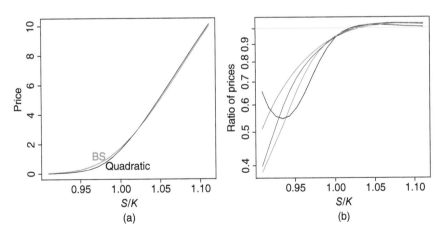

Figure 16.5 *Call prices.* (a) The quadratic prices (black) and the Black–Scholes prices (red) as a function of moneyness S/K. (b) The ratios of the quadratic prices to the Black–Scholes prices. Time to expiration is 20 days (black), 40 days (red), 60 days (blue), and 80 days (green).

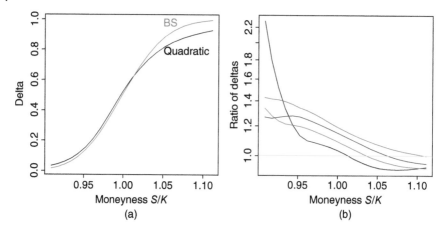

Figure 16.6 *Hedging coefficients.* (a) The quadratic hedging coefficients (black) and the Black–Scholes hedging coefficients (red) as a function of moneyness S/K. (b) The ratios of the quadratic hedging coefficients to the Black–Scholes hedging coefficients. Time to expiration is 20 days (black), 40 days (red), 60 days (blue), and 80 days (green).

the moneyness is less than 0.95 and the time to expiration is 20 days, then the quadratic hedging coefficient is much larger than the Black–Scholes hedging coefficient.

16.2.2.2 Square Integrability

The local quadratic trading strategy needs to be square integrable, in the sense of assumption (16.6). The square integrability is studied in Föllmer and Schied (2002, Proposition 10.10, p. 377). In fact, in order to guarantee the satisfaction of (16.6), it is enough to assume

$$\left(E_{t-1}\Delta X_t\right)^2 \leq C\, \mathrm{Var}_{t-1}(\Delta X_t) \qquad (16.51)$$

P-almost surely, for $t = 1, \dots, T$, for a constant C. Condition (16.51) of the bounded mean–variance trade-off appeared already in (16.7), where it was stated to guarantee the existence of a global quadratic trading strategy. Denote $\sigma_t^2 = \mathrm{Var}_{t-1}(\Delta X_t)$. Assumption (16.51) implies that $E_{t-1}(\Delta X_t)^2 \leq (1 + C)\sigma_t^2$. Thus we have for the local quadratic hedging coefficients ξ_t that

$$E(\xi_t \Delta X_t)^2 = E\left[I_{\{\sigma_t^2 \neq 0\}}\frac{\mathrm{Cov}_{t-1}(H_t, \Delta X_t)^2}{\sigma_t^4}E_{t-1}(\Delta X_t)^2\right]$$

$$\leq (1 + C)E\left[\frac{\mathrm{Cov}_{t-1}(H_t, \Delta X_t)^2}{\sigma_t^2}\right]$$

$$\leq (1 + C)E\,\mathrm{Var}_{t-1}(H_t),$$

where we used for the first equality the law of the iterated expectations, and for the third inequality the Cauchy–Schwarz inequality. Thus, the square integrability of H_T implies the square integrability of $\xi_T \Delta X_T$, which implies the square integrability of H_{T-1}. The backward induction shows that the square integrability of H_T implies the square integrability of $\xi_t \Delta X_t$ for $t = T, \ldots, 1$, under assumption (16.51).

16.2.2.3 The Equivalent Martingale Measure

Let us find the equivalent martingale measure Q implied by the local quadratic hedging. The density of the martingale measure with respect to the physical measure P is

$$\frac{dQ}{dP} = \prod_{t=1}^{T} f_{t|t-1}, \tag{16.52}$$

where

$$f_{t|t-1} = a_t - b_t \Delta X_t,$$

with

$$a_t = \frac{E_{t-1}\Delta X_t^2}{\mathrm{Var}_{t-1}(\Delta X_t)}, \quad b_t = \frac{E_{t-1}\Delta X_t}{\mathrm{Var}_{t-1}(\Delta X_t)}. \tag{16.53}$$

In order that density dQ/dP is positive we have to assume that

$$\Delta X_t E_{t-1}(\Delta X_t) < E_{t-1}\Delta X_t^2 \tag{16.54}$$

P-almost surely on $\{\mathrm{Var}_{t-1}(\Delta X_t) \neq 0\}$. Otherwise, Q would be a signed measure and not a probability measure.

A Derivation of the Equivalent Martingale Measure   Let us show that

$$H_0 = E_Q H_T.$$

This follows because we can write

$$H_{t-1} = E_{t-1}H_t - \frac{\mathrm{Cov}_{t-1}(H_t, \Delta X_t)}{\mathrm{Var}_{t-1}(\Delta X_t)} E_{t-1}\Delta X_t$$

$$= \left(\frac{E_{t-1}\Delta X_1^2}{\mathrm{Var}_{t-1}(\Delta X_1)} + \frac{(E_{t-1}\Delta X_t)^2}{\mathrm{Var}_{t-1}(\Delta X_t)} \right) E_{t-1}H_t$$

$$\quad - \frac{E_{t-1}(\Delta X_t(H_t - E_{t-1}H_t))}{\mathrm{Var}_{t-1}(\Delta X_t)} E_{t-1}\Delta X_t$$

$$= \frac{E_{t-1}\Delta X_1^2}{\mathrm{Var}_{t-1}(\Delta X_1)} E_{t-1}H_t - \frac{E_{t-1}(\Delta X_t H_t)}{\mathrm{Var}_{t-1}(\Delta X_t)} E_{t-1}\Delta X_t$$

$$= a_t E_{t-1}H_t - b_t E_{t-1}(\Delta X_t H_t)$$

$$= E_{t-1}[H_t(a_t - b_t \Delta X_t)]. \tag{16.55}$$

Now, we have

$$
\begin{aligned}
H_0 &= E_0(H_1 f_{1|0}) \\
H_1 &= E_1(H_2 f_{2|1}) \\
&\vdots \\
H_{T-1} &= E_{T-1}(H_T f_{T|T-1}),
\end{aligned}
\tag{16.56}
$$

which can be written as

$$
H_0 = E_0 f_{1|0} E_1 f_{2|1} \cdots E_{T-1} f_{T|T-1} H_T = E_P\left(\frac{dQ}{dP} H_T\right) = E_Q H_T,
$$

where we use the notation $E_0 = E = E_P$. Note that (16.56) implies that

$$
H_t = E_t f_{t+1|t} \cdots E_{T-1} f_{T|T-1} H_T = E_Q(H_T \mid \mathcal{F}_t).
\tag{16.57}
$$

Characterizations of the Equivalent Martingale Measure   Measure Q in (16.52) can be characterized as a minimal martingale measure. Föllmer and Schied (2002, Definition 10.21, p. 382) define a minimal martingale measure to be such measure Q which is equivalent to P, $E(dQ/dP)^2 < \infty$, and such that every square integrable P-martingale M which is strongly orthogonal to X is also a Q-martingale. The strong orthogonality of M and X means that

$$
\mathrm{Cov}_t(M_{t+1} - M_t, X_{t+1} - X_t) = 0
$$

P-almost surely, for $t = 0, \dots, T-1$.

Föllmer and Schied (2002, Theorem 10.22, p. 383) states that if Q is a minimal martingale measure, then (16.57) holds. Föllmer and Schied (2002, Corollary 10.28, p. 388) states that there exists at most one minimal martingale measure.

Föllmer and Schied (2002, Theorem 10.30, p. 390) proves the existence and the uniqueness of a minimal martingale measure, and gives formula (16.58) for the density of the minimal martingale measure. Let us assume the condition (16.54) of positivity and the condition (16.51) of the bounded mean–variance trade-off. Then there exists a unique minimal martingale measure Q with density

$$
\frac{dQ}{dP} = Z_T,
\tag{16.58}
$$

where $Z_0 = 1$ and

$$
Z_t = \prod_{i=1}^{t} (1 + \Lambda_i - \Lambda_{i-1}), \quad t = 1, \dots, T,
$$

with $\Lambda_0 = 1$ and

$$
\Lambda_t = 1 - \sum_{i=1}^{t} b_i (Y_i - Y_{i-1}), \quad t = 1, \dots, T.
$$

where b_i is defined in (16.53), and Y is the martingale part of the Doob decomposition of X. The Doob decomposition of X is

$$X = Y + B,$$

where Y is a martingale and B is predictable. The Doob decomposition is defined as

$$Y_t = X_0 + \sum_{i=1}^{t}(X_i - E_{i-1}X_i), \quad B_t = \sum_{i=1}^{t}(E_{i-1}X_i - X_{i-1}),$$

where $t = 1, \ldots, T$, $Y_0 = X_0$, and $B_0 = 0$; see Föllmer and Schied (2002, Proposition 6.1, p. 277).

Now we can show that the measure in (16.52) is the same as the minimal martingale measure in (16.58). Indeed,

$$1 + \Lambda_t - \Lambda_{t-1} = 1 - b_t(Y_t - Y_{t-1}) = 1 - b_t(X_t - E_{t-1}X_t)$$

and

$$\begin{aligned}
f_{t|t-1} &= a_t - b_t\Delta X_t \\
&= 1 - b_t\left(\frac{1 - a_t}{b_t} + \Delta X_t\right) \\
&= 1 - b_t(\Delta X_t - E_{t-1}\Delta X_t) \\
&= 1 - b_t(X_t - E_{t-1}X_t).
\end{aligned}$$

16.2.3 Local Quadratic Hedging without Self-Financing

It is of interest to note that when we define a local quadratic hedging without self-financing, then the price will be the same, the hedging coefficients of the stocks will be the same, and only the hedging coefficients of the bonds will be different. A local quadratic hedging without self-financing can be defined in a similar way as the local quadratic hedging with self-financing, but we replace the value process with the wealth process.

16.2.3.1 Backward Induction
Let us consider the two period model with $T = 2$. Let us describe local quadratic hedging when the wealth process is used. The wealth at times 0, 1, and 2 is equal to

$$\begin{aligned}
W_0 &= \bar{\xi}_1 \cdot \bar{S}_0 = \beta_1 + \xi_1 S_0, \\
W_1 &= \bar{\xi}_1 \cdot \bar{S}_1 = \beta_1(1 + r) + \xi_1 S_1, \\
W_2 &= \bar{\xi}_2 \cdot \bar{S}_2 = \beta_2(1 + r)^2 + \xi_2 S_2.
\end{aligned}$$

The self-financing condition would state that β_2 and ξ_2 should be chosen so that

$$\beta_2(1 + r) + \xi_2 S_1 = \beta_1(1 + r) + \xi_1 S_1.$$

In local quadratic hedging without self-financing we first find $\beta_2 \in \mathbf{R}$ and $\xi_2 \in \mathbf{R}$ minimizing

$$E_1(W_2 - C_2)^2 = E_1\big(\beta_2(1+r)^2 + \xi_2 S_2 - C_2\big)^2.$$

This is the population version of a linear least-squares regression with the response variable C_2 and the explanatory variable S_2. The minimizers are

$$\xi_2 = \frac{\mathrm{Cov}_1(C_2, S_2)}{\mathrm{Var}_1(S_2)}, \quad \beta_2 = (1+r)^{-2}(E_1 C_2 - \xi_2 E_1 S_2).$$

Let us denote

$$C_1 = \beta_2(1+r) + \xi_2 S_1.$$

Term C_1 is obtained by "discounting" term $\beta_2(1+r)^2 + \xi_2 S_2$. Second, we find $\beta_1 \in \mathbf{R}$ and $\xi_1 \in \mathbf{R}$ minimizing

$$E_0(W_1 - C_1)^2 = E_0\big(\beta_1(1+r) + \xi_1 S_1 - C_1\big)^2.$$

This is the population version of a linear least-squares regression with the response variable C_1 and the explanatory variable S_1. The minimizers are

$$\xi_1 = \frac{\mathrm{Cov}_0(C_1, S_1)}{\mathrm{Var}_0(S_1)}, \quad \beta_1 = (1+r)^{-1}(E_0 C_1 - \xi_1 E_0 S_1).$$

The optimal price in the local quadratic sense is

$$C_0 = \beta_1 + \xi_1 S_0.$$

Term C_0 is obtained by "discounting" term $\beta_1(1+r) + \xi_1 S_1$. The price can be written as

$$C_0 = E_0 H_2 - E_0(\xi_2 E_1 \Delta X_2) - \xi_1 E_0 \Delta X_1.$$

The price is equal to the price which is obtained with the self-financing condition, as can be seen from (16.42). The first hedging coefficient can be written as

$$\xi_1 = \frac{\mathrm{Cov}_0(C_2, S_1)}{\mathrm{Var}_0(S_1)} - \frac{\mathrm{Cov}_0(\xi_2((1+r)^{-1}S_2 - S_1), S_1)}{\mathrm{Var}_0(S_1)}.$$

The hedging coefficient is equal to the hedging coefficient in (16.43), which is obtained with the self-financing restriction.

16.2.3.2 A Comparison with the Case of Self-Financing

We have seen that the price C_0 and the hedging coefficients ξ_1 and ξ_2 are the same whether the self-financing restriction is imposed or not. What about the coefficients β_1 and β_2? The quantities of the bonds are given by

$$\beta_1 = E_0 H_2 - E_0(\xi_2 E_1 \Delta X_2) - \xi_1 E_0 X_1,$$
$$\beta_2 = E_1 H_2 - \xi_2 E_1 X_2.$$

The quantities can be compared to the quantities when the self-financing condition holds, given in (16.44) as

$$\beta_1 = E_0 H_2 - E_0(\xi_2 E_1 \Delta X_2) - \xi_1 E_0 X_1,$$
$$\beta_2 = E_0 H_2 - E_0(\xi_2 E_1 X_2) + E_0(\xi_2 X_1) - \xi_2 X_1 + \xi_1(X_1 - E_0 X_1).$$

We see that β_1 are equal, but β_2 are different.[11]

16.2.3.3 Mean Self-Financing

We have obtained a hedging strategy that is not self-financing, but it is mean self-financing, as defined in (16.5). Indeed,

$$E_1(\beta_2(1 + r)^2 + \xi_2 S_2 - C_2) = 0,$$
$$E_0(\beta_1(1 + r) + \xi_1 S_1 - C_1) = 0,$$

because $E_1(\xi_2 S_2) = \xi_2 E_1 S_2$ and $E_0(\xi_1 S_1) = \xi_1 E_0 S_1$ since ξ_2 is $\mathcal{F}_1$-measurable and ξ_1 is $\mathcal{F}_0$-measurable.

16.3 Implementations of Local Quadratic Hedging

We have derived formulas for the quadratic price and the quadratic hedging coefficient. The formulas are not in a closed form but their application requires numerical methods. In addition, the formulas depend on the knowledge of the unknown data generating mechanism, and we need to use statistical methods to estimate the data generating mechanism.

We implement only the local quadratic hedging, both for the case when the increments are assumed to be independent, and for the case when the increments are assumed to be dependent.

Section 16.3.1 describes the basic setting of historical simulation. Section 16.3.2 describes numerical and statistical methods for the case of independent increments. Section 16.3.3 considers the case of dependent

---

11 We can obtain a modification of the local quadratic hedging without self-financing by changing the optimization problem to be such that the additional cost is minimized. Let us define the expense process (cost process) by

$$E_0 = C_0 = \beta_1 + \xi_1 S_0,$$
$$E_1 = C_1 - W_1 = \beta_2(1 + r) + \xi_2 S_1 - (\beta_1(1 + r) + \xi_1 S_1),$$
$$E_2 = C_2 - W_2 = C_2 - \beta_2(1 + r)^2 + \xi_2 S_2.$$

Let us define the objective function

$$E(C_1 - W_1)^2 + E(C_2 - W_2)^2,$$

which is to be minimized with respect to $\xi_1, \xi_2, \beta_1,$ and β_2. However, this optimization problem is not easier to solve than the global quadratic optimization.

increments. Section 16.3.4 compares the implementations of quadratic pricing and hedging to some benchmarks.

16.3.1 Historical Simulation

To implement quadratic hedging we use historical simulation. Analogously, Monte Carlo simulation could be applied. In Monte Carlo simulation a statistical model is imposed, and sequences of observations are generated from the model. In historical simulation only the previous observations are used.

A similar type of implementation has been described in Potters *et al.* (2001), where price functions $H_t(s)$ and hedging functions $\xi_{t+1}(s)$ are estimated using an expansion with basis functions, whereas we use kernel estimation. Also, we implement a method where the price function and the hedging function depend on volatility, so that they have the form $H_t(s, \sigma)$ and $\xi_{t+1}(s, \sigma)$.

16.3.1.1 Generating Sequences of Observations

We denote the time series of observed historical daily prices by $S_0, \dots, S_N$. The price S_N is the current price. We construct $N - T$ sequences of prices:

$$S_i = (S_{i,i}, \dots, S_{i,i+T}), \quad i = 1, \dots, N - T, \tag{16.59}$$

where

$$S_{i,i+j} = S_N \cdot S_{i+j}/S_i, \quad j = 0, \dots, T.$$

Each sequence consists of $T + 1$ values, and the initial price in each sequence is $S_{i,i} = S_N$.

We may choose to use less than $N - T$ sequences, to make computation faster. Note that $N - T$ sequences are overlapping, so that the use of the all possible $N - T$ sequences may not increase statistical accuracy much, as compared for using a lesser number of sequences. We may construct $M \leq N - T$ sequences of prices, and to get non-overlapping sequences we may choose $M = [(N - T)/T]$, and choose index i to take the values $i = 1 + (l - 1)T$, for $l = 1, \dots, M$.

16.3.1.2 The State Variable

With sequence $S_0, \dots, S_N$ of prices there is an associated sequence $Z_0, \dots, Z_N$ of state variables. Each Z_i can be a vector. We have constructed sequences S_i, which all start at the current stock price S_N. The values of the state variables that correspond to sequence S_i are

$$Z_i, \dots, Z_{i+T}.$$

To utilize the information in the state variables, we use only those sequences S_i that are such that at time i the value of the vector Z_i of state variables is close to the current value Z_N of the state variables. Let $\mathcal{I}$ be the collection of those times:

$$\mathcal{I} = \{i = 1, \dots, N - T : \| Z_N - Z_i \| \le h\},$$

where $h > 0$ is the radius of the window, and $\| \cdot \|$ is the Euclidean distance.

For example, we can choose the state variable to be the logarithm of the current prediction of volatility:

$$Z_i = \log \hat{\sigma}_{i+1},$$

where $\hat{\sigma}_{i+1}$ is estimated using the observed prices $S_0, \dots, S_i$. For instance, we can apply the GARCH(1, 1) volatility estimate.[12] Then $\mathcal{I}$ is defined as

$$\mathcal{I} = \{i = 1, \dots, N - T : |\log(\hat{\sigma}_{N+1}) - \log(\sigma_{i+1})| \le h\},$$

where $h > 0$ is the radius of the window. This is similar to the nonparametric GARCH-pricing in Section 15.3.

16.3.1.3 Heuristic Discussion

We want to solve a series of linear regression problems

$$H_{t+1} = H_t + \xi_{t+1}\Delta X_{t+1} + \text{error}_{t+1}, \tag{16.60}$$

for $t = T - 1, \dots, 0$, where $\Delta X_{t+1} = X_{t+1} - X_t$. These regression problems are conditional on $X_t = x$ and $Z_t = z$. The solutions are functions $\hat{H}_t = \hat{H}_t(x, z)$ and $\hat{\xi}_{t+1} = \hat{\xi}_{t+1}(x, z)$, where x is the discounted value of the stock and z is the value of the state variable. The sample version of the regression problem is

$$H_{t+1}(X_{i,t+1}, Z_{i+t+1}) = H_t(X_{i,t}, Z_{i+t}) + \xi_{t+1}(X_{i,t}, Z_{i+t})\Delta X_{i,t+1} + \text{error}_{i,t+1},$$

where $i = 1, \dots, N - T$, and $t = T - 1, \dots, 0$.

In analogy, consider first the standard linear regression model

$$Y = \alpha + \beta X + \text{error}.$$

---

12 We obtain GARCH(1, 1) volatility estimates recursively by

$$\hat{\sigma}_{i+1}^2 = \hat{\alpha}_{0,i} + \hat{\alpha}_{1,i}R_i^2 + \hat{\beta}_i\hat{\sigma}_i^2,$$

where $R_i = S_i/S_{i-1} - 1$, $i = 1, \dots, N$. Prediction $\hat{\sigma}_{i+1}^2$ is made at time i, and it predicts the volatility at time $i + 1$. In order to obtain initial estimates of parameters α_0, α_1, β, and an initial value σ_0 for the volatility, we assume that there are available observations $S_{-n}, S_{-n+1}, \dots, S_{-1}$. It is reasonable to update the estimates $\hat{\alpha}_{0,i}$, $\hat{\alpha}_{1,i}$, and $\hat{\beta}_i$ sequentially, using data $S_{-n}, S_{-n+1}, \dots, S_i$.

Assume we observe (X_i, Y_i), $i = 1, \ldots, n$, from this model. Then,

$$Y_i = \alpha + \beta X_i + \text{error}_i, \quad i = 1, \ldots n,$$

and we can estimate the constants α and β. Our setting resembles the model

$$Y = \alpha(Z) + \beta(Z)X + \text{error},$$

where Z is an additional random variable, and $\alpha : \mathbf{R} \to \mathbf{R}$ and $\beta : \mathbf{R} \to \mathbf{R}$ are functions. Assume that we observe (X_i, Y_i, Z_i), $i = 1, \ldots, n$, from this model. Then

$$Y_i = \alpha(Z_i) + \beta(Z_i)X_i + \text{error}_i, \quad i = 1, \ldots, n.$$

In order to estimate values $\alpha(z)$ and $\beta(z)$ for a fixed z we cannot use the standard linear regression, because there are no observations from model $Y = \alpha(z) + \beta(z)X + \text{error}$. Instead, we can estimate the functions α and β by localizing into the neighborhood of z. Let $\mathcal{I} = \{i : \| Z_i - z \| \leq h\}$. Now, we can use linear regression for the observations

$$Y_i = \alpha(Z_i) + \beta(Z_i)X_i + \text{error}_i, \quad i \in \mathcal{I}.$$

Note that we need to estimate functions $H_t(s, z)$ and $\xi_{t+1}(s, z)$ only at the points $s = S_{i,t}$ and $z = Z_{i+t}$, $i = 1, \ldots, N - T$.

We need to estimate functions $H_t(x, z)$ and $\xi_{t+1}(x, z)$ of two arguments (where z may be a vector). This can be done in two ways.

1) We can localize with respect to both $x = X_t$ and $z = Z_t$.
2) We can estimate function

$$h_t(x, z) = \frac{H_t(x, z)}{x}. \tag{16.61}$$

Now, it is possible to avoid localization with respect to $x = X_t$, make the localization only with respect to $z = Z_t$, and have available more observations to make the estimation. This is possible for certain $h_t(x, z)$.

Estimation of (16.61) is done by changing model (16.60) to model

$$h_{t+1} = h_t + \xi_{t+1}R_{t+1} + \text{error}_{t+1}, \tag{16.62}$$

where

$$R_{t+1} = \frac{X_{t+1}}{X_t} - 1.$$

An estimate $\hat{h}_t(x, z)$ leads to estimate

$$\hat{H}(x, z) = x\hat{h}_t(x, z).$$

Now the localization with respect to X_t is not necessary when $H_T(X_T, Z_T) = ((1+r)^T X_T - K)_+ = (S_T - K)_+$, since $h_t(X_T, Z_T) = (1+r)^T(1 - K/S_T)_+$. In this case we can ignore the current level of the trajectory of the stock price.

16.3.2 Local Quadratic Hedging Under Independence

We apply the local quadratic hedging and assume independence of the increments. The hedging coefficients are given by the formula (16.49) as

$$\xi_{t+1} = \frac{\text{Cov}_t(H_T, \Delta X_{t+1})}{\text{Var}_t(\Delta X_{t+1})},$$

where $t = 0, \ldots, T-1$. The price is given by the formula (16.50) as

$$H_0 = E_0 H_T - \sum_{t=0}^{T-1} \xi_{t+1} E_0(\Delta X_{t+1}).$$

Here H_T is the discounted payoff of the derivative, $\Delta X_{t-1} = X_{t+1} - X_t$, and X_t are the discounted prices of the risky asset.

Let us assume for notational simplicity that the risk-free rate is zero, so that

$$X_t = S_t.$$

The price H_0 involves only unconditional expectations, whereas hedging coefficients ξ_{t+1} involve conditional expectations. In our implementation of the case of independent increments we make two simplifications, as compared to the previous heuristic discussion. First, the conditioning with respect to the state variable is done only at the time $t = 0$ of writing the option. Second, we do not need to move to the model (16.62), but we can handle the conditioning on the stock price $S_t = s$ by renormalizing the tails of sequences S_i so that they start with value s. This is possible because in the case of independent increments the intermediate values H_t for $t = 1, \ldots, T-1$ do not appear.

16.3.2.1 Unconditional Expectations

To estimate the unconditional expectations $E_0(\cdot)$ we apply sequences S_i in (16.59), where $i \in \mathcal{I}$. These sequences give us the differences and the terminal values

$$\Delta X_{i,t+1} = S_{i,i+t+1} - S_{i,i+t}, \quad H_{i,T} = (S_{i,i+T} - K)_+,$$

where $t = 0, \ldots, T-1$. We estimate the unconditional expectations by

$$E_0(\Delta X_{t+1}) \approx \frac{1}{\#\mathcal{I}} \sum_{i \in \mathcal{I}} \Delta X_{i,t+1}, \quad E_0 H_T \approx \frac{1}{\#\mathcal{I}} \sum_{i \in \mathcal{I}} H_{i,T}.$$

16.3.2.2 Conditional Expectations

To estimate the hedging coefficients ξ_{t+1}, when the stock price is $S_t = s$, we renormalize the tails of the sequences. We define sequences such that the initial price is s, and the number of observations in each sequence is $T - t + 1$, where $t \in \{0, \dots, T-1\}$, that is, the length of the sequences is $T - t$. We define

$$S_i(s, t) = (\tilde{S}_{i,i+t}, \dots, \tilde{S}_{i,i+T}), \quad i = 1, \dots, N - T, \tag{16.63}$$

where

$$\tilde{S}_{i,i+j} = s \cdot S_{i+j}/S_{i+t},$$

for $j = t, \dots, T$. Now the initial price in each sequence is $\tilde{S}_{i,i+t} = s$.

We estimate the conditional expectations $E_t(\cdot) = E(\cdot | S_t)$ by applying sequences $S_i(s, t)$ in (16.63), where $t = 1, \dots, T-1$, $i \in \mathcal{I}$, and $s = S_{i,i+t} \in S_i$. Each such sequence gives the differences and the terminal values

$$\Delta X_{i,t+1}(s, t) = \tilde{S}_{i,i+t+1} - \tilde{S}_{i,i+t},$$
$$H_{i,T}(s, t) = (\tilde{S}_{i,i+T} - K)_+.$$

The conditional expectations $E_t(\cdot) = E(\cdot | S_t)$ are estimated by

$$E(H_T | S_t = s) \approx \frac{1}{\#\mathcal{I}} \sum_{i \in \mathcal{I}} H_{i,T}(s, t),$$

$$E(\Delta X_{t+1} | S_t = s) \approx \frac{1}{\#\mathcal{I}} \sum_{i \in \mathcal{I}} \Delta X_{i,t+1}(s, t),$$

$$E(H_T \Delta X_{t+1} | S_t = s) \approx \frac{1}{\#\mathcal{I}} \sum_{i \in \mathcal{I}} [H_{i,T}(s, t)\, \Delta X_{i,t+1}(s, t)],$$

$$E((\Delta X_{t+1})^2 | S_t = s) \approx \frac{1}{\#\mathcal{I}} \sum_{i \in \mathcal{I}} [\Delta X_{i,t+1}(s, t)]^2.$$

These estimates lead to the estimates of covariances and variances,[13] and we obtain estimates of ξ_{t+1}, which are used to produce an estimate of H_0.

16.3.2.3 Comparison to Black–Scholes

We compare prices and hedging coefficients of local quadratic hedging (with independence assumption) to the Black–Scholes prices and hedging coefficients.

Comparison to Black–Scholes Prices Figure 16.7 shows the ratios of the locally quadratic prices (under independence) to the Black–Scholes prices as a function of the annualized volatility. In panel (a) moneyness is $S/K = 1$ and in panel (b) $S/K = 100/105$. The smoothing parameter is $h = 0.05$ (black), $h = 0.1$ (red), and $h = 0.2$ (blue). The time to expiration is 20 trading days.

---

13 We apply formulas $\text{Cov}(X, Y) = E(XY) - EXEY$ and $\text{Var}(X) = EX^2 - (EX)^2$.

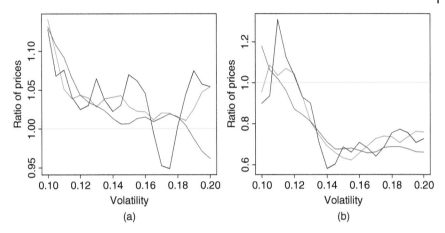

Figure 16.7 *Call price ratios as a function of volatility*. The ratios of the locally quadratic prices under independence to the Black–Scholes prices as a function of the annualized volatility. (a) Moneyness is $S/K = 1$; (b) $S/K = 100/105$. The smoothing parameter is $h = 0.05$ (black), $h = 0.1$ (red), and $h = 0.2$ (blue).

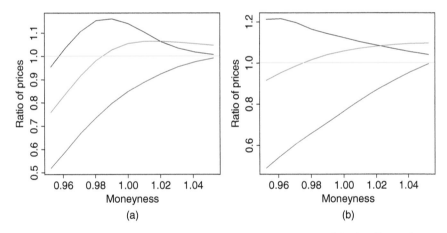

Figure 16.8 *Call price ratios as a function of moneyness*. The ratios of the locally quadratic prices under independence to the Black–Scholes prices as a function of moneyness S/K. (a) $T = 20$; (b) $T = 60$. The annualized volatility is 0.1 (black), 0.2 (red), and 0.3 (blue).

Figure 16.8 shows the ratios of the locally quadratic prices to the Black–Scholes prices as a function of the moneyness S/K. In panel (a) the time to expiration is $T = 20$ trading days, and in panel (b) $T = 60$. The annualized volatility is 0.1 (black), 0.2 (red), and 0.3 (blue). The smoothing parameter is $h = 0.1$.

Figure 16.9 shows the ratios of the locally quadratic prices to the Black–Scholes prices as a function of the smoothing parameter h. In panel (a)

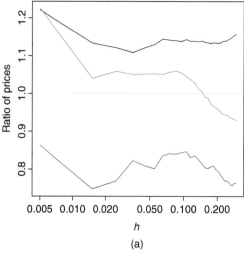

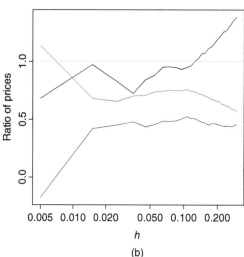

Figure 16.9 *Call price ratios as a function of the smoothing parameter.* The ratios of the locally quadratic prices under independence to the Black–Scholes prices as a function of smoothing parameter h. (a) Moneyness is $S/K = 1$; (b) $S/K = 100/105$. The annualized volatility is 0.1 (black), 0.2 (red), and 0.3 (blue).

moneyness is $S/K = 1$ and in panel (b) $S/K = 100/105$. The annualized volatility is 0.1 (black), 0.2 (red), and 0.3 (blue). Time to expiration is $T = 20$ trading days.

Comparison to Black–Scholes Deltas Figure 16.10 shows the ratios of the locally quadratic hedging coefficients (under independence) to the Black–Scholes deltas as a function of the annualized volatility. In panel (a) moneyness is $S/K = 1$ and in panel (b) $S/K = 100/105$. The smoothing parameter is

Figure 16.10 *Call hedging coefficient ratios as a function of volatility.* The ratios of the locally quadratic hedging coefficients under independence to the Black–Scholes deltas as a function of the annualized volatility. (a) Moneyness is $S/K = 1$; (b) $S/K = 100/105$. The smoothing parameter is $h = 0.05$ (black), $h = 0.1$ (red), and $h = 0.2$ (blue).

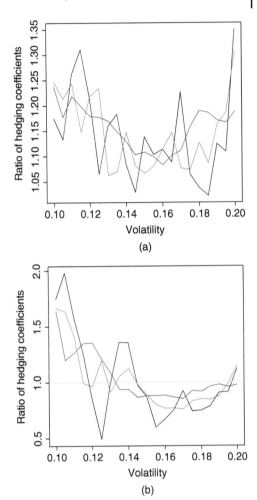

(a)

(b)

$h = 0.05$ (black), $h = 0.1$ (red), and $h = 0.2$ (blue). The time to expiration is 20 trading days.

Figure 16.11 shows the ratios of the locally quadratic prices to the Black–Scholes prices as a function of the moneyness S/K. In panel (a) time to expiration is $T = 20$ trading days, and in panel (b) $T = 60$. The annualized volatility is 0.1 (black), 0.2 (red), and 0.3 (blue). The smoothing parameter is $h = 0.1$.

Figure 16.12 shows the ratios of the locally quadratic hedging coefficients to the Black–Scholes deltas as a function of the smoothing parameter h. In panel (a) moneyness is $S/K = 1$ and in panel (b) $S/K = 100/105$. The

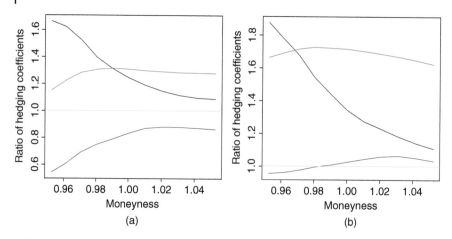

Figure 16.11 *Call hedging coefficient ratios as a function of moneyness.* The ratios of the locally quadratic prices under independence to the Black–Scholes prices as a function of moneyness S/K. (a) $T = 20$; (b) $T = 60$. The annualized volatility is 0.1 (black), 0.2 (red), and 0.3 (blue).

annualized volatility is 0.1 (black), 0.2 (red), and 0.3 (blue). Time to expiration is $T = 20$ trading days.

16.3.3 Local Quadratic Hedging under Dependence

We apply local quadratic hedging without assuming independence of increments. We need to estimate the sequences

$$\xi_{t+1} = \frac{\text{Cov}_t(\Delta X_{t+1}, H_{t+1})}{\text{Var}_t(\Delta X_{t+1})}, \quad H_t = E_t H_{t+1} - \xi_{t+1} E_t \Delta X_{t+1}, \tag{16.64}$$

where $t = 0, \dots, T - 1$. The recursion starts with the known value H_T.

Let us assume for notational simplicity that the risk-free rate is zero, so that

$$X_t = S_t.$$

Let us denote by $s_0, \dots, s_N$ the observed values of the stock S, and let us denote by $z_0, \dots, z_N$ the observed values of the state variable Z. The values in sequence S_i, defined in (16.59), are denoted by

$$S_i = (s_{i,i}, \dots, s_{i,i+T}),$$

where $i = 1, \dots, N - T$. The corresponding sequence of the values of the state variables is

$$z_i, \dots, z_{i+T}.$$

In the case of local quadratic hedging under independence the formulas did not involve the intermediate values H_t for $t = 1, \dots, T - 1$. This simplified the

Figure 16.12 *Call hedging coefficient ratios as a function of smoothing parameter.* The ratios of the locally quadratic hedging coefficients under independence to the Black–Scholes deltas as a function of smoothing parameter h. (a) Moneyness is $S/K = 1$; (b) $S/K = 100/105$. The annualized volatility is 0.1 (black), 0.2 (red), and 0.3 (blue).

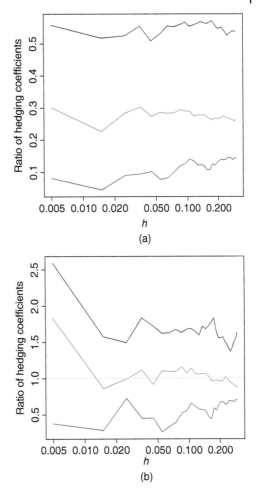

(a)

(b)

computations, and we needed only to renormalize the tails of the price trajectories. Now we use a technique where we move from the increments ΔS_{t+1} to the net returns R_{t+1} of stock prices, and from the values H_t to the values h_t, where

$$R_{t+1} = \frac{S_{t+1}}{S_t} - 1, \quad h_t = \frac{H_t}{S_t}.$$

We can write

$$\xi_{t+1} = \frac{\mathrm{Cov}_t(R_{t+1}, h_{t+1})}{\mathrm{Var}_t(R_{t+1})}, \quad h_t = E_t h_{t+1} - \xi_{t+1} E_t R_{t+1}.$$

We need to estimate the conditional expectations

$$E_t R_{t+1}, \quad E_t h_{t+1}, \quad E_t(R_{t+1} h_{t+1}), \quad E_t R_{t+1}^2.$$

The conditional expectations are interpreted as

$$E_t(\,\cdot\,) = E(\,\cdot\,|X_t, Z_t).$$

16.3.3.1   The Steps of Backward Induction

Consider step $t \in 0, \ldots, T-1$. Assume that we have produced estimates

$$h_{t+1}(s, z), \quad (s, z) \in \mathcal{U}'_{t+1},$$

where

$$\mathcal{U}'_{t+1} = \{(s_{i,i+t+1}, z_{i+t+1}) : i \in \mathcal{I}\}.$$

Step $t = T - 1$ is the first step of the backward induction. When $t = T - 1$, then in the case of a call option

$$h_{t+1}(s, z) = \left(1 - \frac{K}{s}\right)_+,$$

where $s = s_{i,i+T}$, $i \in \mathcal{I}$, are the terminal values of sequences S_i.

We estimate the conditional expectations first with local averaging, and then generalize to kernel estimation.

Local Averaging   Let

$$\mathcal{I}_t(z) = \{j \in \mathcal{I} : \|\, z_{j+t} - z \,\| \leq h\}.$$

Set $\mathcal{I}_t(z)$ contains those indexes $j \in \mathcal{I}$ for which the tth element z_{j+t} is close to z, where $t = 0, \ldots, T-1$. Note that $\mathcal{I}_0(z_N) = \mathcal{I}$. Let

$$\mathcal{V}_t(z) = \{(s_{i,i+t+1}, z_{i+t+1}) : i \in \mathcal{I}_t(z)\}.$$

We estimate for each $(s, z) \in \mathcal{U}_t = \{(s_{i,i+t}, z_{i+t}) : i \in \mathcal{I}\}$ the conditional expectations by[14]

$$E(h_{t+1} \,|\, S_t = s, Z_t = z) \approx \frac{1}{\#\mathcal{V}_t(z)} \sum_{(s_1, z_1) \in \mathcal{V}_t(z)} h_{t+1}(s_1, z_1),$$

$$E(R_{t+1} \,|\, S_t = s, Z_t = z) \approx \frac{1}{\#\mathcal{V}_t(z)} \sum_{(s_1, z_1) \in \mathcal{V}_t(z)} \left(\frac{s_1}{s} - 1\right),$$

$$E\left(R_{t+1}^2 \,|\, S_t = s, Z_t = z\right) \approx \frac{1}{\#\mathcal{V}_t(z)} \sum_{(s_1, z_1) \in \mathcal{V}_t(z)} \left(\frac{s_1}{s} - 1\right)^2,$$

$$E(h_{t+1} R_{t+1} \,|\, S_t = s, Z_t = z)$$
$$\approx \frac{1}{\#\mathcal{V}_t(z)} \sum_{(s_1, z_1) \in \mathcal{V}_t(z)} \left[h_{t+1}(s_1, z_1)\left(\frac{s_1}{s} - 1\right)\right],$$

---

14 Note that the estimate of $E(h_{t+1} \,|\, S_t = s, Z_t = z)$ does not depend on s.

These estimates lead to the estimates of covariances and variances, and we obtain an estimate of $\xi_{t+1}(s,z)$. An estimate for $h_t(s,z)$ is obtained by

$$h_t(s,z) \tag{16.65}$$
$$= \frac{s_1}{s}[E(h_{t+1}\,|\,S_t = s, Z_t = z) - \xi_{t+1}(s,z)E(R_{t+1}\,|\,S_t = s, Z_t = z)]$$

for $(s,z) = (s_{i+t}, z_{i+t})$ and $s_1 = s_{i+t+1}$, where $i \in I_t(z)$. Note that at step $t = 0$ set U_t is a singleton:

$$U_0 = \{(S_N, Z_N)\}.$$

The price is obtained as

$$S_N h_0(S_N, Z_N),$$

and the first hedging coefficient is

$$\xi_1(S_N, Z_N).$$

Kernel Estimation   The estimation of the conditional expectations can be done by using kernel estimation. We estimate for each $(s,z) \in U_t = \{(s_{i,i+t}, z_{i+t}) : i \in I\}$ the conditional expectations using the estimators

$$E(h_{t+1}\,|\,S_t = s, Z_t = z) \approx \sum_{i \in I} p_i(z)\, h_{t+1}(s_{i,i+t+1}, z_{i+t+1}),$$

$$E(R_{t+1}\,|\,S_t = s, Z_t = z) \approx \sum_{i \in I} p_i(z)\left(\frac{s_{i,i+t+1}}{s} - 1\right),$$

$$E\left(R_{t+1}^2\,|\,S_t = s, Z_t = z\right) \approx \sum_{i \in I} p_i(z)\left(\frac{s_{i,i+t+1}}{s} - 1\right)^2,$$

$$E(h_{t+1}R_{t+1}\,|\,S_t = s, Z_t = z)$$
$$\approx \sum_{i \in I} p_i(z)\left[h_{t+1}(s_{i,i+t+1}, z_{i+t+1})\left(\frac{s_{i,i+t+1}}{s} - 1\right)\right],$$

where the weights are defined as

$$p_i(z) = \frac{K_h(z - z_{i+t})}{\sum_{j \in I} K_h(z - z_{t+j})},$$

K_h is the scaled kernel $K_h(x) = K(x/h)$, $K : \mathbf{R}^p \to \mathbf{R}$ is a kernel function, where p is the dimension of vector Z. The previous method of local averaging is obtained as a special case when the kernel function is chosen as

$$K(x) = I_{\{y:\|y\|\leq 1\}}(x).$$

These estimates lead to the estimates of covariances and variances, and we obtain an estimate of $\xi_{t+1}(s,z)$. An estimate of $h_t(s,z)$ is obtained using the formula (16.65).

16.3.3.2 Comparison to Black–Scholes

We compare both prices and hedging coefficients of local quadratic hedging to the Black–Scholes prices and hedging coefficients.

Comparison to Black-Scholes Prices   Figure 16.13 shows the ratios of the locally quadratic prices to the Black–Scholes prices as a function of the annualized volatility. In panel (a) moneyness is $S/K = 1$ and in panel (b) $S/K = 100/105$. The smoothing parameter is $h = 0.05$ (black), $h = 0.1$ (red), and $h = 0.2$ (blue). Time to expiration is 20 trading days.

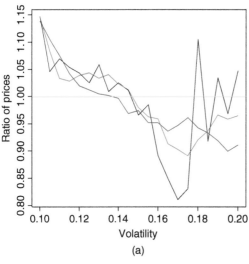

(a)

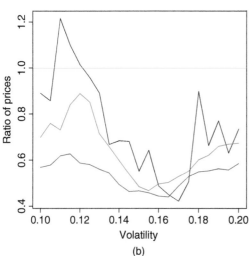

(b)

Figure 16.13 *Call price ratios as a function of volatility under dependence. The ratios of the locally quadratic prices to the Black–Scholes prices as a function of the annualized volatility.* (a) *Moneyness is* $S/K = 1$; (b) $S/K = 100/105$. *The smoothing parameter is* $h = 0.05$ *(black),* $h = 0.1$ *(red), and* $h = 0.2$ *(blue).*

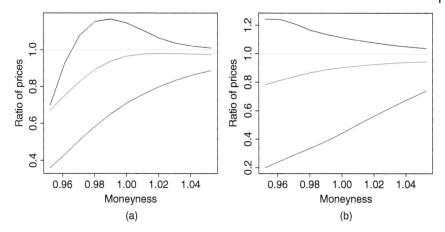

Figure 16.14 *Call price ratios as a function of moneyness: Dependence.* The ratios of the locally quadratic prices under dependence to the Black–Scholes prices as a function of moneyness S/K. (a) $T = 20$; (b) $T = 60$. The annualized volatility is 0.1 (black), 0.2 (red), and 0.3 (blue).

Figure 16.14 shows the ratios of the locally quadratic prices to the Black–Scholes prices as a function of the moneyness S/K. In panel (a) time to expiration is $T = 20$ trading days, and in panel (b) $T = 60$. The annualized volatility is 0.1 (black), 0.2 (red), and 0.3 (blue). The smoothing parameter is $h = 0.1$.

Figure 16.15 shows the ratios of the locally quadratic prices to the Black–Scholes prices as a function of the smoothing parameter h. In panel (a) moneyness is $S/K = 1$ and in panel (b) $S/K = 100/105$. The annualized volatility is 0.1 (black), 0.2 (red), and 0.3 (blue). Time to expiration is $T = 20$ trading days.

Comparison to Black–Scholes Deltas Figure 16.16 shows the ratios of the locally quadratic hedging coefficients (under dependence) to the Black–Scholes deltas as a function of the annualized volatility. In panel (a) moneyness is $S/K = 1$ and in panel (b) $S/K = 100/105$. The smoothing parameter is $h = 0.05$ (black), $h = 0.1$ (red), and $h = 0.2$ (blue). Time to expiration is 20 trading days.

Figure 16.17 shows the ratios of the locally quadratic prices to the Black–Scholes prices as a function of the moneyness S/K. In panel (a) time to expiration is $T = 20$ trading days, and in panel (b) $T = 60$. The annualized volatility is 0.1 (black), 0.2 (red), and 0.3 (blue). The smoothing parameter is $h = 0.1$.

Figure 16.18 shows the ratios of the locally quadratic hedging coefficients to the Black–Scholes deltas as a function of the smoothing parameter h. In panel (a) moneyness is $S/K = 1$ and in panel (b) $S/K = 100/105$. The annualized volatility is 0.1 (black), 0.2 (red), and 0.3 (blue). Time to expiration is $T = 20$ trading days.

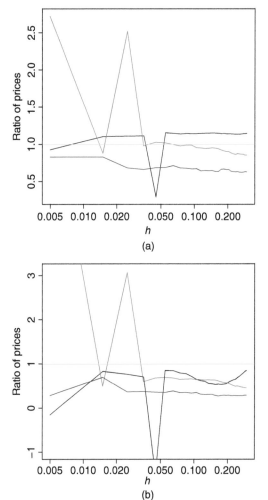

Figure 16.15 *Call price ratios as a function of the smoothing parameter: Dependence.* The ratios of the locally quadratic prices under dependence to the Black–Scholes prices as a function of smoothing parameter h. (a) Moneyness is $S/K = 1$; (b) $S/K = 100/105$. The annualized volatility is 0.1 (black), 0.2 (red), and 0.3 (blue).

16.3.4 Evaluation of Quadratic Hedging

Figure 16.19 shows (a) tail plots and (b) kernel density estimates of hedging errors for call options.[15] The blue curves show the case of Black–Scholes hedging. The local quadratic hedging is done assuming independence and the

---

15 The hedging error e_T of the writer of the option is obtained from (13.10) as $e_T = C_0 + G_T(\xi) - C_T$, where $G_T(\xi) = \sum_{k=1}^{T} \xi_k(S_k - S_{k-1})$, the risk-free rate is $r = 0$, C_0 is the price of the option, C_T is the terminal value of the option, ξ_k are the hedging coefficients, S_k are the stock prices, the current time is denoted by 0, the time to expiration is T days, and hedging is done daily. When hedging is done with a lesser frequency, then we use formula (14.79) for $G_T(\xi)$.

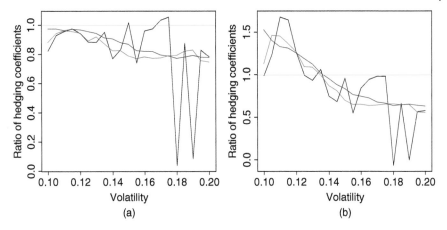

Figure 16.16 *Call hedging coefficient ratios as a function of volatility: Dependence.* The ratios of the locally quadratic hedging coefficients under dependence to the Black–Scholes deltas as a function of the annualized volatility. (a) Moneyness is $S/K = 1$; (b) $S/K = 100/105$. The smoothing parameter is $h = 0.05$ (black), $h = 0.1$ (red), and $h = 0.2$ (blue).

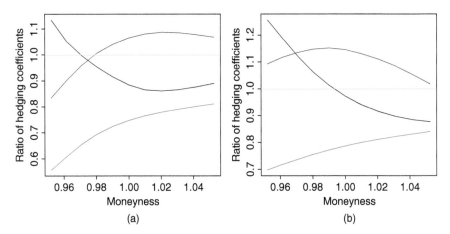

Figure 16.17 *Call hedging coefficient ratios as a function of moneyness: Dependence.* The ratios of the locally quadratic prices under dependence to the Black–Scholes prices as a function of moneyness S/K. (a) $T = 20$; (b) $T = 60$. The annualized volatility is 0.1 (black), 0.2 (red), and 0.3 (blue).

smoothing parameter is $h = 0.1$ (red), $h = 0.5$ (dark green), $h = 1$ (purple), and $h = 10$ (orange). The volatility is in all cases the GARCH$(1, 1)$ volatility. The moneyness of call options is $S/K = 1$. Time to maturity is 20 days and hedging is done every day. Tail plots are defined in Section 3.2.1 and the kernel density estimator is defined in Section 3.2.2. We apply the standard normal kernel function and the smoothing parameter of the density estimator is chosen by

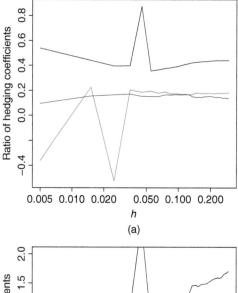

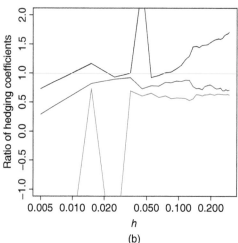

Figure 16.18 *Call hedging coefficient ratios as a function of smoothing parameter: Dependence.* The ratios of the locally quadratic hedging coefficients under dependence to the Black–Scholes deltas as a function of smoothing parameter h. (a) Moneyness is $S/K = 1$; (b) $S/K = 100/105$. The annualized volatility is 0.1 (black), 0.2 (red), and 0.3 (blue).

the normal reference rule. We see from panel (b) that smoothing parameters $h = 0.5 - 1$ lead to similar results, but smoothing parameter $h = 0.1$ leads to a more dispersed distribution. Black–Scholes hedging leads to a more concentrated distribution than the quadratic hedging with independence assumption, but the quadratic hedging leads to a distribution which is skewed to the right in the central area of the distribution, which means that there are more gains than losses for the hedger of the option.

Figure 16.20 considers the case of hedging only once. Panel (a) shows tail plots and panel (b) shows kernel density estimates of hedging errors. The moneyness of call options is $S/K = 1$. Time to maturity is 20 days. The blue curves show the case of Black–Scholes hedging. The red curves show the case of local

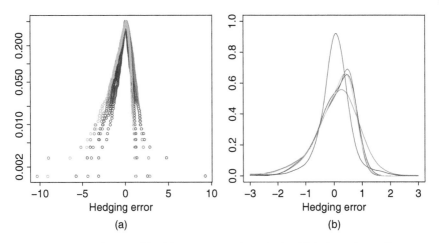

Figure 16.19 *Distribution of hedging errors: Local quadratic with independence.* Shown are (a) tail plots and (b) kernel density estimates of hedging errors. Black–Scholes hedging (blue), local quadratic hedging with independence using the smoothing parameter $h = 0.1$ (red), $h = 0.5$ (dark green), $h = 1$ (purple), and $h = 10$ (orange).

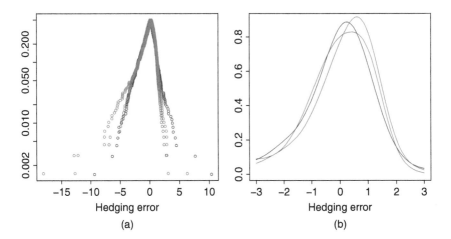

Figure 16.20 *Distribution of hedging errors when hedging is done once: Local quadratic under independence and dependence.* Shown are (a) tail plots and (b) kernel density estimates of hedging errors. Black–Scholes hedging (blue), local quadratic hedging with independence (green), and local quadratic hedging with dependence (red).

quadratic hedging assuming dependence. The green curves show the case of local quadratic hedging assuming independence. The smoothing parameter is in both cases $h = 0.1$. The volatility is in all cases the GARCH(1, 1) volatility.

Figure 16.21 shows (a) tail plots and (b) kernel density estimates of hedging errors. The moneyness of call options is $S/K = 1$. Time to maturity is 20 days

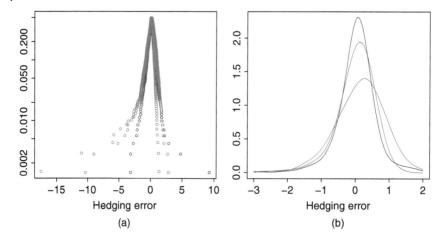

Figure 16.21 *Distribution of hedging errors with daily hedging: Local quadratic under indepen-dence and dependence.* Shown are (a) tail plots and (b) kernel density estimates of hedging errors. Black–Scholes hedging (blue), local quadratic hedging with independence (green), and local quadratic hedging with dependence (red).

and hedging is done every day. The blue curves show the case of Black–Scholes hedging. The red curves show the case of local quadratic hedging assuming dependence. The green curves show the case of local quadratic hedging assuming independence. The smoothing parameter is in both cases $h = 0.1$. The volatility is in all cases the GARCH$(1, 1)$ volatility.

17

Option Strategies

Options can be used to create almost any type of a profit function. Trading with stocks allows the possibility of short selling and leveraging, but options open up a huge number of possibilities for creating a payoff that suits the expectations and the risk profile of an investor. For example, a protective put can be used to protect a portfolio of stocks from negative returns, and a straddle can be used to profit simultaneously from large positive and large negative returns of the stock.

We describe option strategies in three ways: the profit function, the return function, and the return distribution. The profit function shows the profit of the option strategy at the expiration, as a function of the value of the underlying. For example, the profit function of a long call strategy is equal to

$$S_T \mapsto \max\{S_T - K, 0\} - (1 + r)C_0, \tag{17.1}$$

where S_T is the stock price at the expiration, $K > 0$ is the strike price, C_0 is the premium of the call, and $r > -1$ is the interest rate.[1] The return function shows the gross return of the option strategy. For example, the return function of a long call strategy is given by

$$S_T \mapsto \frac{\max\{S_T - K, 0\}}{(1 + r)C_0}. \tag{17.2}$$

The return distribution means the probability distribution of the return of the option strategy. For example, the return distribution of a long call strategy is the probability distribution of the random variable $\max\{S_T - K, 0\}/[(1 + r)C_0]$. The probability distribution can be described by the distribution function, which in this case is

$$x \mapsto P_0\left(\frac{\max\{S_T - K, 0\}}{(1 + r)C_0} \leq x\right), \tag{17.3}$$

---

1 A payout function does not take the premium into account, so that the payout function of a long call strategy is $S_T \mapsto \max\{S_T - K, 0\}$.

Nonparametric Finance, First Edition. Jussi Klemelä.

where $x \in \mathbf{R}$, and P_0 is the conditional probability, conditional on the information available at time 0. The probability distribution of the option return depends on the conditional probability distribution of the underlying S_T, and this probability distribution is unknown. We use both the histogram estimator and the tail plot of the empirical distribution function to estimate the unknown return distribution of the option.

The method of using the return distribution (17.3) is the most intuitive and useful to describe an option strategy, from the three methods (17.1)–(17.3). In fact, the return distribution of the option is directly relevant for the investor who considers including options into the portfolio. On the other hand, the use of the return distribution involves both the problem of estimating the probability distribution and the problem of visualizing the probability distribution.

Option strategies provide an instructive case study for the performance measurement. We get more insight into such concepts as Sharpe ratio, cumulative wealth, and risk aversion by studying the performance measurement of option strategies, instead of just studying the performance measurement of portfolios of stocks.

Section 17.1 shows profit functions of option strategies, which include vertical spreads, strangles, straddles, butterflies, condors, calendar spreads, covered calls, and protective puts. Section 17.2 shows return functions and return distributions of the option strategies, and measures the performance of the option strategies.

17.1 Option Strategies

It is possible to create a large number of profit functions by combining calls and puts with different strike prices and expiration dates. Our examples include vertical spreads, strangles, straddles, butterflies, condors, and calendar spreads. In addition, we discuss how to combine options with the underlying to create protective puts and covered calls.

17.1.1 Calls, Puts, and Vertical Spreads

Calls and puts are the basic building blocks for creating profit functions. Vertical spreads are combinations of calls and puts that limit the downside risk of selling pure calls and puts.

17.1.1.1 Calls and Puts
Figure 17.1(a)–(d) shows profit functions of a long call, long put, short call, and short put. For a call, the profit function is

$$S_T \mapsto \max\{S_T - K, 0\} - (1 + r)C_0,$$

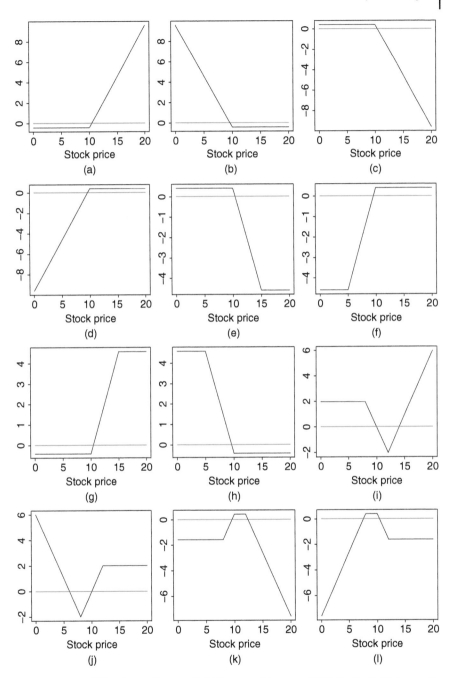

Figure 17.1 *Profit functions.* (a) Long call; (b) long put; (c) short call; (d) short put; (e) short call spread; (f) short put spread; (g) long call spread; (h) long put spread; (i) long 2 × 1 ratio call spread; (j) long 2 × 1 ratio put spread; (k) long call ladder; and (l) long put ladder.

where S_T is the stock price at the expiration, $K > 0$ is the strike price, C_0 is the premium of the call, and $r > -1$ is the interest rate. For the put, the profit function is

$$S_T \mapsto \max\{K - S_T, 0\} - (1 + r)P_0,$$

where P_0 is the premium of the put. When a call is bought, the maximum profit is unlimited. When a put is bought, the maximum profit is equal to the strike price minus the premium. The losses are limited both when a call is bought and when a put is bought.[2]

17.1.1.2 Vertical Spreads

Figure 17.1(e)–(h) shows profit functions of a short call spread, short put spread, long call spread. and long put spread. Let the strike prices satisfy $0 < K_1 < K_2$. Vertical spreads are the following trades:

1) *Short call spread.* Short K_1 call, long K_2 call.
2) *Short put spread.* Long K_1 put, short K_2 put.
3) *Long call spread.* Long K_1 call, short K_2 call.
4) *Long put spread.* Short K_1 put, long K_2 put.

A short call spread has a special importance, because this trade allows us to sell a call option but it makes the maximum possible loss limited, because a call with a higher strike price is bought simultaneously. Selling a put has a limited loss but a short put spread makes the maximum possible loss smaller; see (17.13).

Figure 17.1(i)–(l) shows profit functions of a long 2×1 ratio call spread, long 2×1 ratio put spread, long call ladder, and long put ladder. Ratio spreads are generalizations of simple vertical spreads. Ladders are examples of combinations of three options. The strike prices satisfy $0 < K_1 < K_2 < K_3$. Ratio spreads and ladders are defined as follows:

1) *Long 2×1 ratio call spread.* Short K_1 call, long 2 times K_2 call. (Also called a call backspread.)
2) *Long 2×1 ratio put spread.* Long 2 times K_1 put, short K_2 put. (Also called a put backspread.)
3) *Long call ladder.* Long K_1 call, short K_2 call, short K_3 call.
4) *Long put ladder.* Short K_1 put, short K_2 put, long K_3 put.

---

2 Rising interest rates make calls more attractive than stocks: one has to pay stock immediately whereas one avoids interest rates when buying a call. Rising interest rates make puts less attractive: instead of buying a put one may sell stock short and collect interest. The effect is different with options on futures, because futures contracts do not require cash outlay. High interest rates make options less attractive than the underlying futures contract because one has to pay a premium to buy options. Note that it is possible (and sometimes used) to make a futures types settlement to options, which makes the effect of interest rates on the options price models effectively zero.

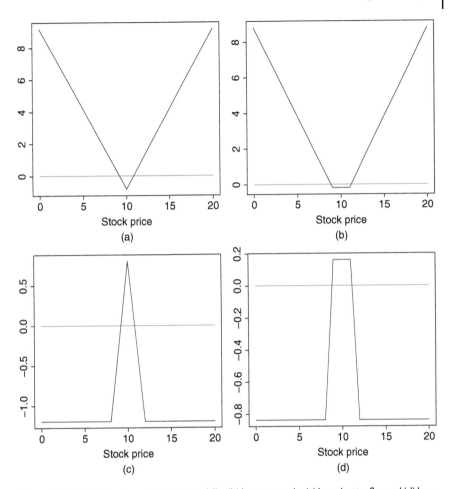

Figure 17.2 *Profit functions.* (a) Long straddle; (b) long strangle; (c) long butterfly; and (d) long condor.

17.1.2 Strangles, Straddles, Butterflies, and Condors

Figure 17.2(a)–(d) shows profit functions of a long straddle, long strangle, long butterfly, and long condor.

Long straddles and strangles are profitable when the underlying makes a large move.

1) *Long straddle.* Long K put and long K call, where K is close to the current stock price.
2) *Long strangle.* The profit function of a long strangle can be constructed in two ways. Let $K_1 < S_0 < K_2$ and let S_0 be the current stock price.
   a) *Long strangle.* Long K_1 put, long K_2 call.
   b) *Long guts.* Long K_1 call, long K_2 put.

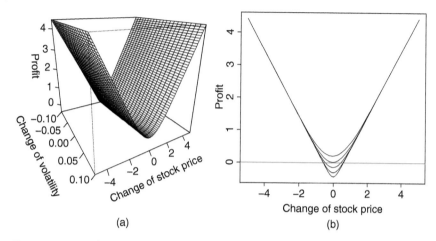

Figure 17.3 *A profit function of a straddle.* (a) A perspective plot of function Profit($\delta S, \delta \sigma$), defined in (17.4); (b) slices $\delta S \mapsto$ Profit($\delta S, \delta \sigma$) for five values of $\delta \sigma$.

Straddles are special cases of strangles and guts: when $K_1 = K_2 = K$, then we obtain a straddle from a strangle or from a guts.

Figure 17.3 shows a two-dimensional profit function of a straddle. Now we consider the profit not as a univariate function of the price of the underlying, but as a two-dimensional function of the change in the price of the underlying and of the change in the volatility. The Black–Scholes prices are used to define the profit function. Panel (a) shows a perspective plot of function

$$\text{Profit}(\Delta S, \Delta \sigma) = C_t(S_0 + \Delta S, K, \sigma_0 + \Delta \sigma, T)$$
$$+ P_t(S_0 + \Delta S, K, \sigma_0 + \Delta \sigma, T)$$
$$- C_0(S_0, K, \sigma_0, T) - P_0(S_0, K, \sigma_0, T), \qquad (17.4)$$

where $C_t(S, K, \sigma, T)$ and $P_t(S, K, \sigma, T)$ are the Black–Scholes prices at time t of a call and a put, when the underlying has value S, strike price is K, σ is the annualized volatility, and T is the time of expiration. Panel (b) shows slices

$$\Delta S \mapsto \text{Profit}(\Delta S, \Delta \sigma)$$

for five values of $\Delta \sigma$. We can see that a long straddle profits also from a rising volatility, and not only from large moves of the underlying.

To profit when the underlying does not move, one can sell a straddle, strangle, or guts. However, these trades have an unlimited downside. Thus, it is useful to apply butterflies and condors, which have a limited downside. Below $0 < K_1 < K_2 < K_3 < K_4$.

1) Long butterflies can be constructed in three ways:
   a) *Call butterfly.* Long K_1 call, short two K_2 calls, long K_3 call.

b) *Put butterfly.* Long K_1 put, short two K_2 puts, long K_3 put.
c) *Long iron butterfly.* Long K_1 put, short K_2 put and call, long K_3 call.[3]
2) Long condors can be constructed in three ways:
a) *Call condor.* Long K_1 call, short K_2 call, short K_3 call, long K_4 call.
b) *Put condor.* Long K_1 put, short K_2 put, short K_3 put, long K_4 put.
c) *Long iron condor.* Long K_1 put, short K_2 call, short K_3 put, long K_4 call.

A long butterfly is obtained from a long condor by taking $K_2 = K_3$. Selling a strangle can be considered as obtainable from a condor by letting $K_1 \to 0$ and $K_4 \to \infty$. Selling a straddle can be considered as obtainable from a butterfly by letting $K_1 \to 0$ and $K_4 \to \infty$.

17.1.3 Calendar Spreads

Calendar spreads allow us to profit from a rising volatility by shorting an option with a shorter time to expiration and going long for an option with a longer time to expiration. Calendar spreads are also called "time spreads" and "horizontal spreads." Diagonal calendar spreads make a simultaneous bet for the direction of the underlying.

1) Long calendar spread:
a) *Call calendar spread.* Short T_1 call, long T_2 call.
b) *Put calendar spread.* Short T_1 put, long T_2 put.
c) *Long straddle calendar spread.* Short T_1 straddle, long T_2 straddle.
2) Long diagonal calendar spread:
a) *Call diagonal calendar spread.* Short T_1, K_1 call, long T_2, K_2 call.
b) *Put diagonal calendar spread.* Short T_1, K_1 put, long T_2, K_2 put.
c) *Long diagonal straddle calendar spread.* Short T_1, K_1 straddle, long T_2, K_2 straddle.

Figure 17.4 shows a profit function of call calendar spread. The profit function is a function of two variables: the change in stock price and the change in volatility. At time 0 we short a call with maturity T_1 and buy a call with maturity T_2. The trade is terminated at T_1. Panel (a) shows a perspective plot and panel (b) shows slices $\Delta S \mapsto \text{Profit}(\Delta S, \Delta\sigma)$ for five values of $\Delta\sigma$. The profit function is

$$\text{Profit}(\Delta S, \Delta\sigma)$$
$$= C_0(S_0, K, \sigma_0, T_1) - C_0(S_0, K, \sigma_0, T_2)$$
$$- (S_0 + \Delta S - K)_+ + C_{T_1}(S_0 + \Delta S, K, \sigma_0 + \Delta\sigma, T_2),$$

---

3 A long iron butterfly is equal to a combination of a short straddle and a long strangle, and it is also equal to a combination of a vertical short call spread and a vertical short put spread.

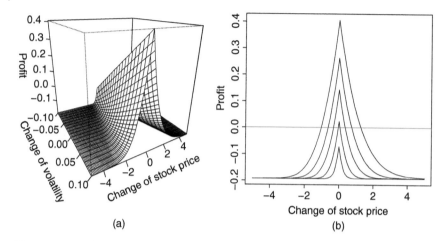

Figure 17.4 *Profit of calendar*. (a) A perspective plot; (b) slices.

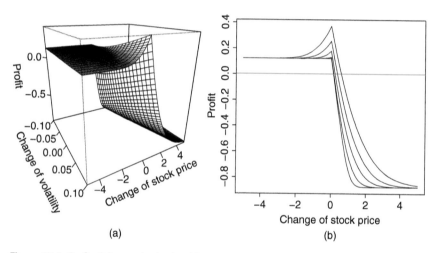

Figure 17.5 *Profit of diagonal calendar*. (a) A perspective plot; (b) slices.

where $C_t(S, K, \sigma, T)$ is the Black–Scholes price at time t of a call option when S is the stock price, K is the strike price, σ is the annualized volatility, and T is the expiration time. Here $S + \Delta S = S_{T_1}$, so that we have $(S_0 + \Delta S - K)_+ = C_{T_1}(S_0 + \Delta S, K, \sigma_0 + \Delta\sigma, T_1)$.

Figure 17.5 shows a profit function of call diagonal calendar spread. At time 0 we short a call with maturity T_1 and strike K_1, and buy a call with maturity T_2 and strike K_2. The trade is terminated at T_1. Panel (a) shows a perspective plot and panel (b) shows slices $\Delta S \mapsto \text{Profit}(\Delta S, \Delta\sigma)$ for five values of $\Delta\sigma$.

The profit function is

$$
\begin{aligned}
\text{Profit}(\Delta S, \Delta\sigma) \\
= C_0(S_0, K_1, \sigma_0, T_1) - C_0(S_0, K_2, \sigma_0, T_2) \\
- (S_0 + \Delta S - K_1)_+ + C_{T_1}(S_0 + \Delta S, K_2, \sigma_0 + \Delta\sigma, T_2),
\end{aligned}
$$

where $K_2 > K_1$.

17.1.4 Combining Options with Stocks and Bonds

A stock can be replicated by a combination of a call and a put. Furthermore, options can be combined with the underlying to make a protective put and a covered call. A bond and a call can be combined to create a position with bounded losses but a stock type upside potential.

17.1.4.1 Replication of the Underlying

The payout of buying a K-call and simultaneously selling a K-put is $S_T - K$:

$$
\max\{S_T - K, 0\} - \max\{K - S_T, 0\} = S_T - K.
$$

The profit of buying a K-call and simultaneously selling a K-put is

$$
S_T - K - (1 + r)(C_0 - P_0),
$$

where C_0 and P_0 are the premiums of the call and the put.

Choosing $K = (1 + r)S_0$ leads to the profit

$$
S_T - (1 + r)S_0.
$$

Indeed, a forward contract to buy stock at time T for the price K is equivalent to buying a K-call and simultaneously selling K-put. This is the so called put–call parity

$$
C_0 - P_0 = S_0 - K(1 + r)^{-1},
$$

studied in Section 14.1.2. Thus, when $K = (1 + r)S_0$, the profit of buying a K-call and simultaneously selling a K-put is $S_T - K$, because the put–call parity gives $C_0 - P_0 = 0$. Thus, the payoff of a stock can be obtained by options.

Conversely, being long K_1-put and short K_2-call, where $K_1 < S_0 < K_2$, is a bet on a falling stock price.[4]

Figure 17.6(a) and (b) shows profit functions of replicating being long and being short of a stock.

---

4 Being long K_1-call, short K_2-put, long K_2-put, and short K_2-call gives a position, whose payoff is $K_2 - K_1$ at the expiration. Thus, there is a possibility for an arbitrage, for suitable premiums. This position is called box.

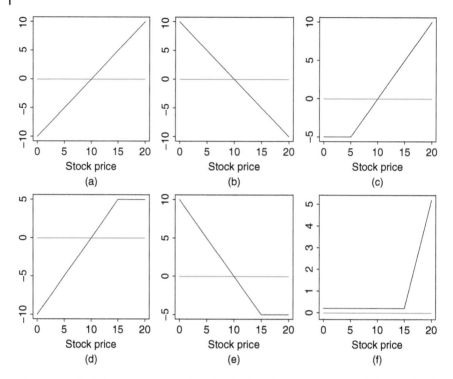

Figure 17.6 *Profit functions.* (a) Long stock; (b) short stock; (c) protective put; (d) covered call; (e) covered shorting; and (f) bond and call.

17.1.4.2 Protective Put and Covered Call

A protective put consists of a simultaneous buying of the underlying and a put on the underlying. The strike price of the put is $K < S_0$, where S_0 is the current price of the underlying. This position gives insurance against a falling price of the underlying, with the cost of paying the premium for the put option.

A covered call consists of a simultaneous buying of the underlying and selling a call with strike price $K > S_0$. A covered call has less risk than the pure position in the underlying, because a premium is obtained from selling the call. On the other hand, selling the call limits the potential upside. Note that a covered call has a similarity with a long call spread.

A covered shorting consists of a simultaneous selling of the underlying and buying a call with strike price $K > S_0$. A covered shorting has less risk than plain shorting, because buying the call makes the loss bounded from below.

Figure 17.6 shows profit functions of a protective put, covered call, and covered shorting in panels (c)–(e).

17.1.4.3 A Bond and A Call

Let a bond and a call be such that the maturity date of the bond is the same as the expiration date of the call. Buying a bond and a call leads to a position where the guaranteed return is smaller than from the pure bond position, because the premium for buying the call has to be subtracted from the profit. On the other hand, there is a considerable upside potential, unlike in the case of the pure bond position. The combination leads to a capital guarantee product.

Figure 17.6(f) shows a profit function of buying a bond and a call. We assume that the profit from the bond is higher than the premium of the call, and thus the profit is always positive. Thus the profit function of Figure 17.6(f) differs from the profit function of the call in Figure 17.1(a), where a loss is possible.

17.2 Profitability of Option Strategies

We discuss how to study the profitability of option strategies that were defined in Section 17.1. First we list the returns of option strategies, then we study the distributions of the returns of the option strategies, and finally compute Sharpe ratios of the option strategies.

What is the gross return of an option strategy? An option strategy is defined by giving its payoff

$$\sum_{i=1}^{N} c_i D_T^i, \tag{17.5}$$

where $c_i \in \mathbf{R}$, and D_T^i are the payoffs of options. For example, the payoff of a long call spread is

$$D_T - D_T',$$

where $D_T = (S_T - K)_+$, $D_T' = (S_T - K')_+$, and $K < K'$. Let us include the possibility of investing in the risk-free rate, and let $(1 + r)$ be the risk-free rate for the period $0 \mapsto T$. Let D_0^i be the premiums of the options. When $\sum_{i=1}^{N} c_i D_0^i \neq 0$, then we can assume without losing generality that

$$\sum_{i=1}^{N} c_i D_0^i > 0.$$

Then the return of the strategy in (17.5) can be written as

$$R_T = (1 - b)(1 + r) + b \frac{\sum_{i=1}^{N} c_i D_T^i}{\sum_{i=1}^{N} c_i D_0^i}, \tag{17.6}$$

where $1 - b$ is the weight of the risk-free rate and b is the weight of the option strategy.

The return can be also written as

$$R_T = \left(1 - a \sum_{i=1}^{N} c_i D_0^i\right)(1 + r) + a \sum_{i=1}^{N} c_i D_T^i, \tag{17.7}$$

where

$$a = \frac{b}{\sum_{i=1}^{N} c_i D_0^i}.$$

A third way to write the return is

$$R_T = \left(1 - \sum_{i=1}^{N} b_i\right)(1+r) + \sum_{i=1}^{N} b_i \frac{D_T^i}{D_0^i}, \tag{17.8}$$

where

$$b_i = a c_i D_0^i.$$

In (17.6)–(17.8), we have assumed that $\sum_{i=1}^{N} c_i D_0^i \neq 0$. In the case when $\sum_{i=1}^{N} c_i D_0^i = 0$, we can combine the option strategy with the risk-free return. We start with the initial wealth $W_0 > 0$ and obtain the wealth

$$W_T = (1+r)W_0 + \xi \sum_{i=1}^{N} c_i D_T^i,$$

where $\xi \in \mathbf{R}$ is the exposure to the option strategy. The return is

$$\frac{W_T}{W_0} = (1+r) + \frac{\xi}{W_0} \sum_{i=1}^{N} c_i D_T^i. \tag{17.9}$$

Note that (17.8) leads to (17.9) with $a = \xi/W_0$.

17.2.1 Return Functions of Option Strategies

We draw return functions

$$S_T \mapsto R_T,$$

where R_T is the gross return of the option strategy. We denote the strike prices

$$K \leq K' \leq K'' \leq \cdots$$

We use the following notation for the payoffs of calls:

$$C_T = \max\{0, S_T - K\}, \quad C_T' = \max\{0, S_T - K'\}, \ldots$$

We use the following notation for the payoffs of puts:

$$P_T = \max\{0, K - S_T\}, \quad P_T' = \max\{0, K' - S_T\}, \ldots$$

The corresponding premiums are $C_0, C_0', C_0'', \ldots$ and $P_0, P_0', P_0'', \ldots$
We draw a blue horizontal line at the level one, because the gross return one means that the wealth does not change. We draw a red horizontal line at the height zero, because in the case of stock trading the gross return zero means

bankruptcy. Note that in option trading we interpret the negative gross return as leading to debt, and the amount deposited in the margin account should be used to pay this debt.

The premiums are chosen to be the Black–Scholes prices, with the annualized volatility 15%. The interest rate is $r = 0$. The initial stock price is $S_0 = 10$. The time to expiration is 6 months (which is $1/2$ in fractions of a year).

17.2.1.1 Calls, Puts, and Vertical Spreads

Figure 17.7(a) shows return functions of buying and selling calls. The gross return is

$$R_T = (1 - b)(1 + r) + b\frac{C_T}{C_0}, \tag{17.10}$$

where $K = S_0$ and $C_T = \max\{0, S_T - K\}$. We show functions for the weights $b = 2$, $b = 1$, $b = 0.5$, and $b = -1$. The profit functions of buying and selling calls are shown in Figure 17.1(a) and (c).

Figure 17.7(b) shows return functions of buying and selling puts. The gross return is

$$R_T = (1 - b)(1 + r) + b\frac{P_T}{P_0}, \tag{17.11}$$

where $K = S_0$ and $P_T = \max\{0, K - S_T\}$. We show functions for the weights $b = 2$, $b = 1$, $b = 0.5$, and $b = -1$. The profit functions of buying and selling puts are shown in Figure 17.1(b) and (d).

Figure 17.7(c) shows return functions of call vertical spreads. The gross return is

$$R_T = (1 - b)(1 + r) + b\frac{C_T - C_T'}{C_0 - C_0'}, \tag{17.12}$$

where $S_0 = K < K'$, $C_T = \max\{0, S_T - K\}$, and $C_T' = \max\{0, S_T - K'\}$. When $b = 1$, we obtain a long call spread. When $b = -1$ we obtain a short call spread. The corresponding profit functions are shown in Figure 17.1(e) and (g). It is of interest to note that a short call vertical spread has a return function which is bounded from below. Indeed, we have that $C_T - C_T' \le K' - K$.[5] Thus, when $b < 0$,

$$R_T \ge (1 - b)(1 + r) + b\frac{K' - K}{C_0 - C_0'}.$$

---

5 It holds that

$$C_T' - C_T = \begin{cases} S_T - K' - (S_T - K), & \text{when } S_T > K', \\ K - S_T, & \text{when } K \le S_T \le K', \\ 0, & \text{when } S_T < K. \end{cases}$$

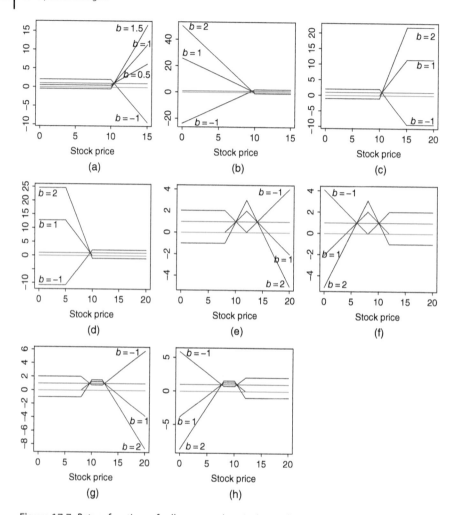

Figure 17.7 *Return functions of calls, puts, and vertical spreads.* (a) Long and short call; (b) long and short put; (c) long and short call spread; (d) long and short put spread; (e) short 2 × 1 ratio call spread; (f) short 2 × 1 ratio put spread; (g) long call ladder; and (h) long put ladder.

Figure 17.7(d) shows return functions of put vertical spreads. The gross return is

$$R_T = (1 - b)(1 + r) + b\frac{P'_T - P_T}{P'_0 - P_0},$$

where $K < K' = S_0$, $P_T = \max\{0, K - S_T\}$, and $P'_T = \max\{0, K' - S_T\}$. When $b = 1$, we obtain a long put spread. When $b = -1$ we obtain a short put spread.

The corresponding profit functions are shown in Figure 17.1(f) and (h). It is of interest to calculate the lower bound for the return of a short put vertical spread. We have that $P'_T - P_T \leq K' - K$. Thus, when $b < 0$,

$$R_T \geq (1 - b)(1 + r) + b\frac{K' - K}{P'_0 - P_0}. \tag{17.13}$$

Figure 17.7(e) shows return functions of 2×1 ratio call spreads. The gross return is

$$R_T = (1 - b)(1 + r) + b\frac{C_T - 2C'_T}{C_0 - 2C'_0}, \tag{17.14}$$

where $K < S_0 < K'$, $C_T = \max\{0, S_T - K\}$, and $C'_T = \max\{0, S_T - K'\}$. When $b = 1$, we obtain a short 2×1 ratio call spread. When $b = -1$ we obtain a long 2×1 ratio call spread. The profit function of a long 2×1 ratio call spread is shown in Figure 17.1(i).

Figure 17.7(f) shows return functions of 2×1 ratio put spreads. The gross return is

$$R_T = (1 - b)(1 + r) + b\frac{P'_T - 2P_T}{P'_0 - 2P_0},$$

where $K < S_0 < K'$, $P_T = \max\{0, K - S_T\}$, and $P'_T = \max\{0, K' - S_T\}$. When $b = 1$, we obtain a short 2×1 ratio put spread. When $b = -1$ we obtain a long 2×1 ratio put spread. The profit function of a long 2×1 ratio put spread is shown in Figure 17.1(j).

Figure 17.7(g) shows return functions of call ladders. The gross return is

$$R_T = (1 - b)(1 + r) + b\frac{C_T - C'_T - C''_T}{C_0 - C'_0 - C''_0}, \tag{17.15}$$

where $K < K' = S_0 < K''$, $C_T = \max\{0, S_T - K\}$, $C'_T = \max\{0, S_T - K'\}$, and $C''_T = \max\{0, S_T - K''\}$. When $b = 1$, we obtain a long call ladder. When $b = -1$ we obtain a short call ladder. The profit function of a long call ladder is shown in Figure 17.1(k).

Figure 17.7(h) shows return functions of put ladders. The gross return is

$$R_T = (1 - b)(1 + r) + b\frac{P''_T - P'_T - P_T}{P''_0 - P'_0 - P_0},$$

where $K < K' = S_0 < K''$, $P_T = \max\{0, K - S_T\}$, $P'_T = \max\{0, K' - S_T\}$, and $P''_T = \max\{0, K'' - S_T\}$. When $b = 1$, we obtain a long put ladder. When $b = -1$ we obtain a short put ladder. The profit function of a long put ladder is shown in Figure 17.1(l).

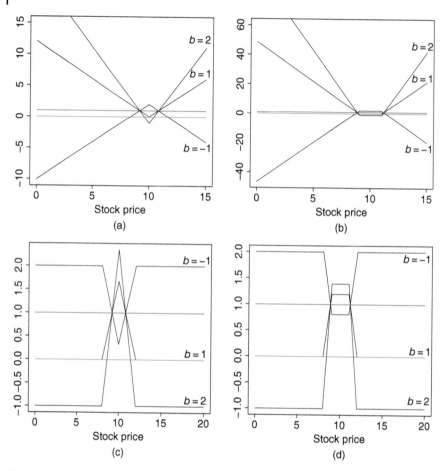

Figure 17.8 *Return functions of straddles, strangles, butterflies, and condors.* (a) Long and short straddles; (b) long and short strangles; (c) long and short butterflies; and (d) long and short condors.

17.2.1.2 Straddles, Strangles, Butterflys, and Condors

Figure 17.8(a) and (b) shows return functions of straddles and strangles. The gross return is

$$R_T = (1 - b)(1 + r) + b\frac{P_T + C_T}{P_0 + C_0}, \tag{17.16}$$

where $P_T = \max\{0, K - S_T\}$ and $C_T = \max\{0, S_T - K'\}$. In panel (a) we have straddles: $K = K' = S_0$. In panel (b) we have strangles: $K < S_0 < K'$. When $b = 1$, we obtain a long straddle and strangle. When $b = -1$ we obtain a short straddle and strangle. The profit functions of a long straddle and a long strangle are shown in Figure 17.2(a) and (b).

Figure 17.8(c) and (d) shows return functions of call butterflies and condors. The gross return is

$$R_T = (1-b)(1+r) + b\frac{C_T - C_T' - C_T'' + C_T'''}{C_0 - C_0' - C_0'' + C_T'''}, \tag{17.17}$$

where $C_T = \max\{0, S_T - K\}$, $C_T' = \max\{0, S_T - K'\}$, $C_T'' = \max\{0, S_T - K''\}$, and $C_T''' = \max\{0, S_T - K'''\}$. In panel (b) we have butterflies: $K < K' = K'' = S_0 < K'''$. In panel (c) we have condors: $K < K' < S_0 < K'' < K'''$. When $b = 1$, we obtain a long butterfly and a long condor. When $b = -1$ we obtain a short butterfly and a short condor. The profit functions of a long butterfly and a long condor are shown in Figure 17.2(c) and (d).

17.2.1.3 Combining Options with Stocks and Bonds

Options can be combined with the underlying to replicate the underlying, to apply a protective put, and to construct a covered call and a covered short. Furthermore, options can be combined with bonds.

Replication of a Stock Let us replicate the stock by a simultaneous buying of a K-call and K-put. The put–call parity implies that the prices C_0 and P_0 of the options satisfy

$$C_0 - P_0 = S_0 - K(1+r)^{-1};$$

see (14.8) for a discussion of the put–call parity. When $K < (1+r)S_0$, then $C_0 - P_0 > 0$, and the return is

$$R_T = (1-b)(1+r) + b\frac{C_T - P_T}{C_0 - P_0}, \tag{17.18}$$

where $b \in \mathbf{R}$.

When $C_0 = P_0$, then we can define the return using (17.9) as

$$R_T = (1+r) + a(C_T - P_T), \tag{17.19}$$

where $a \in \mathbf{R}$.

Figure 17.9(a) shows the return function $S_T \mapsto R_T$ for the return (17.18), when $K = 5$. Note that Figure 17.6(a) and (b) shows profit functions of being long and being short of a stock.

A Protective Put A protective put is a position where the buying of the underlying is combined with buying a put with a strike price $K < S_0$. Let us consider more generally the return

$$R_T = (1-b)(1+r) + b\frac{S_T + P_T}{S_0 + P_0}. \tag{17.20}$$

A protective put is obtained when $b > 0$.

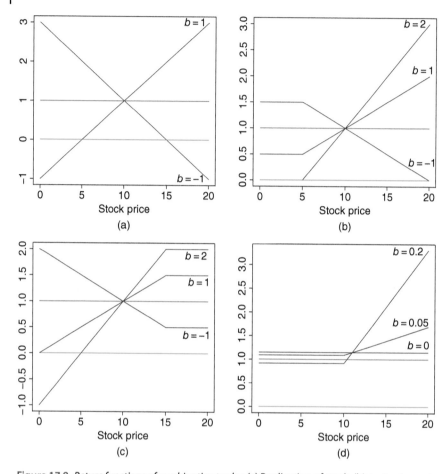

Figure 17.9 *Return functions of combination trades.* (a) Replication of stock; (b) protective put; (c) covered shorting; and (d) bond and call.

Figure 17.9(b) shows a return function of a protective put. In fact, we show the cases $b = 1$, $b = 2$, and $b = -1$.

We can calculate a lower bound to the return of a protective put. In fact,[6]

$$S_T + P_T \geq K.$$

When $b > 0$, the return satisfies

$$R_T \geq (1 - b)(1 + r) + b \, \frac{K}{S_0 + P_0}.$$

---

6 It holds that

$$S_T + P_T = \begin{cases} S_T, & \text{when } S_T > K, \\ S_T + (K - S_T), & \text{when } S_T \leq K. \end{cases}$$

A Covered Call and a Covered Short   A covered call is a position where the buying of the underlying is combined with selling a call with a strike price $K > S_0$. A covered short is a position where the shorting of a stock is combined by the buying of a call with strike price $K > S_0$.

Let us consider returns

$$R_T = (1 - b)(1 + r) + b \frac{S_T - C_T}{S_0 - C_0}. \qquad (17.21)$$

A covered call is obtained when $b > 0$. A covered short is obtained when $b < 0$. Figure 17.9(c) shows return functions for the cases $b = 2$, $b = 1$, and $b = -1$.

Selling a stock has an unbounded maximum loss but in a covered short we buy simultaneously a call option, with strike price $K > S_0$, which makes the maximum possible loss bounded. Selling a call has an unbounded maximum loss but in a covered call we buy simultaneously the stock, which makes the loss bounded when the stock price goes up, although it is possible to lose the total investment, when the stock price goes to zero. The strike price of the covered call satisfies $K > S_0$. The covered call can be used to earn extra return when the stock price does not make a big upside move. We have that[7]

$$S_T - C_T \leq K.$$

When $b < 0$, the return satisfies

$$R_T \geq (1 - b)(1 + r) + b \frac{K}{S_0 - C_0},$$

and thus the return of the covered short is bounded from below. When $b > 0$, the return satisfies

$$R_T \leq (1 - b)(1 + r) + b \frac{K}{S_0 - C_0},$$

and thus the return of the covered call is bounded from above.

A Bond and a Call   A suitable simultaneous buying of a bond and a call creates a capital guarantee product. We assume that the time to maturity of the bond and the time to the expiration of the call are equal. The return is given in (17.10) by

$$R_T = (1 - b)(1 + r) + b \frac{C_T}{C_0}, \qquad (17.22)$$

where $C_T = \max\{0, S_T - K\}$, C_0 is the premium of the call, and r is the net return of the bond. Unlike in (17.10) we take b close to zero, in order to guarantee that the capital is not lost.

Figure 17.9(d) shows return functions for the cases $b = 0$, $b = 0.05$, and $b = 0.2$.

---

7 It holds that

$$S_T - C_T = \begin{cases} S_T - (S_T - K), & \text{when } S_T > K, \\ S_T, & \text{when } S_T \leq K. \end{cases}$$

17.2.2 Return Distributions of Option Strategies

We estimate the return distributions using the S&P 500 daily data, described in Section 2.4.1. The daily data is aggregated to have sampling interval of 20 trading days, and we study options with the time to expiration being 20 days.

The option prices are taken to be the Black–Scholes prices with the volatility being the annualized sample standard deviation of the complete time series of the observations. The risk-free rate is taken to be zero. These simplifications do not prevent us from gaining qualitative insight into the return distributions. The Black–Scholes prices are different from the real market prices, and thus we are not able to obtain precise estimates of the actual return distributions of the past option returns. In particular, the out-of-the-money options tend to have higher market prices than the Black–Scholes prices: this can be seen from the volatility smile, which means that the implied volatilities of the out-of-the-money options have larger implied volatilities than the at-the-money options (see Section 14.3.2).

To estimate the return distributions we use histogram and kernel estimators, as defined in Section 3.2.2. We use the normal reference rule to choose the smoothing parameter of the kernel density estimator. Also, we apply tail plots of the empirical distribution function, as defined in Section 3.2.1.

17.2.2.1 Calls, Puts, and Vertical Spreads

Figure 17.10 shows a return distribution of buying a call option with moneyness $S_0/K = 1$. The return is given in (17.10), where we choose $b = 1$. Panel (a) shows a histogram estimate of the call returns. The red curve shows a kernel estimate of the corresponding S&P 500 returns. Panel (b) shows a tail plot of the empirical distribution function of the option returns with black circles. The red circles show a tail plot of the empirical distribution function of the corresponding S&P 500 returns. The corresponding profit function is shown in Figure 17.1(a) and the return function is shown in Figure 17.7(a). We see that there is a large probability of gross return zero, and small probabilities of high returns.

Figure 17.11 shows the return distribution of buying a put option with moneyness $K/S_0 = 1$. The return is given in (17.11), where we choose $b = 1$. Panel (a) shows a histogram estimate of the put returns. The red curve shows a kernel estimate of the corresponding S&P 500 returns. Panel (b) shows a tail plot of the option returns (black) and a tail plot of the corresponding S&P 500 returns (red). The corresponding profit function is shown in Figure 17.1(b) and the return function is shown in Figure 17.7(b). We see that the return distribution of buying a put option is close to the return distribution of buying a call option.

Figure 17.12 shows the return distribution of selling a call with moneyness $S_0/K = 1$. The return is given in (17.10), where we choose $b = -1$. Panel (a) shows a histogram estimate of the returns. The red curve shows a kernel

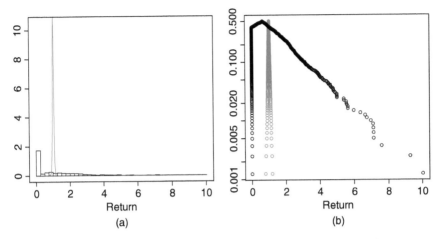

Figure 17.10 *Long call option: Return distribution.* (a) A histogram estimate of call returns (black) and a kernel estimate of S&P 500 returns (red); (b) an empirical distribution function of option returns (black) and S&P 500 returns (red).

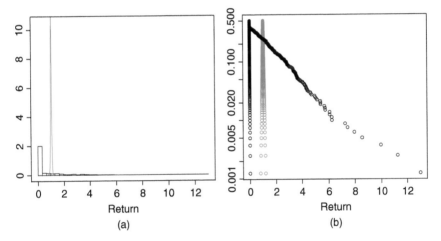

Figure 17.11 *Long put option: Return distribution.* (a) A histogram estimate of call returns (black) and a kernel estimate of S&P 500 returns (red); (b) an empirical distribution function of option returns (black) and S&P 500 returns (red).

estimate of the corresponding S&P 500 returns. Panel (b) shows a tail plot of the option returns (black) and a tail plot of the corresponding S&P 500 returns (red). The corresponding profit function is shown in Figure 17.1(c) and the return function is shown in Figure 17.7(a). We see that the return distribution of selling a call option is a mirror image of the return distribution of buying a call option: there is a large probability of a gross return over one, but small probabilities of quite large negative returns.

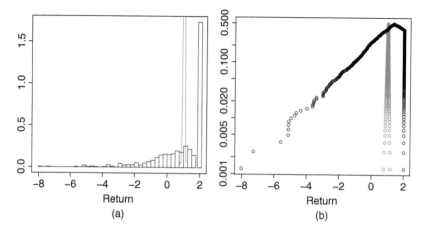

Figure 17.12 *Short call option: Return distribution.* (a) A histogram estimate of option returns (black) and a kernel estimate of S&P 500 returns (red); (b) an empirical distribution function of option returns (black) and S&P 500 returns (red).

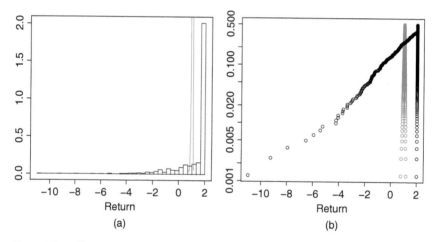

Figure 17.13 *Short put option: Return distribution.* (a) A histogram estimate of option returns (black) and a kernel estimate of S&P 500 returns (red); (b) an empirical distribution function of option returns (black) and S&P 500 returns (red).

Figure 17.13 shows the return distribution of selling a put option with moneyness $K/S_0 = 1$. The return is given in (17.11), where we choose $b = -1$. Panel (a) shows a histogram estimate of the returns. The red curve shows a kernel estimate of the corresponding S&P 500 returns. Panel (b) shows a tail plot of the option returns (black) and a tail plot of the corresponding S&P 500 returns (red). The corresponding profit function is shown in Figure 17.1(d) and the return function is shown in Figure 17.7(b). We see that the return

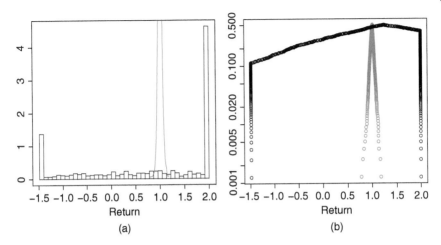

Figure 17.14 *Selling a call spread: Return distribution.* (a) A histogram estimate of option returns (black) and a kernel estimate of S&P 500 returns (red); (b) an empirical distribution function of option returns (black) and S&P 500 returns (red).

distribution of selling a put option is close to the return distribution of selling a call option.

Figure 17.14 shows the return distribution of selling a call spread with $S_0 = 100$, $K = 100$, and $K' = 105$. The return is given in (17.12), where we take $b = -1$. Panel (a) shows a histogram estimate of the returns. The red curve shows a kernel estimate of the corresponding S&P 500 returns. Panel (b) shows a tail plot of the option returns (black) and a tail plot of the corresponding S&P 500 returns (red). The corresponding profit function is shown in Figure 17.1(e) and the return function is shown in Figure 17.7(c). The return distribution is bounded from below, unlike in the case of Figure 17.12, where a call option is sold.

Figure 17.15 shows the return distribution of a 2×1 ratio call spread with $S_0 = 100$, $K = 95$, and $K' = 105$. The return is given in (17.14), where we take $b = -1$. Panel (a) shows a histogram estimate of the return distribution of the option. The red curve shows a kernel estimate of the corresponding S&P 500 returns. Panel (b) shows a tail plot of the option returns (black) and a tail plot of the corresponding S&P 500 returns (red). The corresponding profit function of long position is shown in Figure 17.1(i) and the return function is shown in Figure 17.7(e).

Figure 17.16 shows the return distribution of a short call ladder with $S_0 = 100$, $K = 95$, $K' = 100$, and $K'' = 105$. The return is given in (17.15), where we take $b = -1$. Panel (a) shows a histogram estimate of the return distribution of the option. The red curve shows a kernel estimate of the corresponding S&P 500 returns. Panel (b) shows a tail plot of the option returns (black) and a tail plot

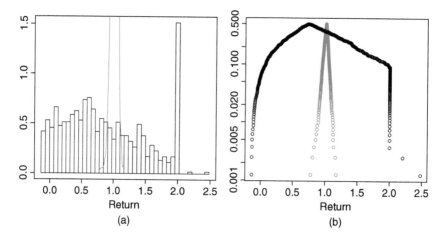

Figure 17.15 *A 2 × 1 ratio call spread: Return distribution.* (a) A histogram estimate of option returns (black) and a kernel estimate of S&P 500 returns (red); (b) an empirical distribution function of option returns (black) and S&P 500 returns (red).

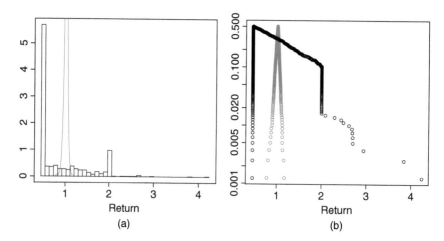

Figure 17.16 *A short call ladder: Return distribution.* (a) A histogram estimate of option returns (black) and a kernel estimate of S&P 500 returns (red); (b) an empirical distribution function of option returns (black) and S&P 500 returns (red).

of the corresponding S&P 500 returns (red). The corresponding profit function of a long position is shown in Figure 17.1(k) and the return function is shown in Figure 17.7(g).

17.2.2.2 Straddles, Strangles, Butterflys, and Condors

Figure 17.17 shows the return distribution of a straddle with $S_0 = 100$ and $K = 100$. The return is given in (17.16), where we take $b = 1$. Panel (a) shows

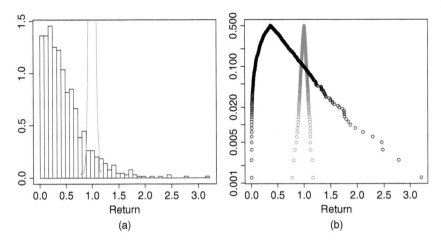

Figure 17.17 *A straddle: Return distribution.* (a) A histogram estimate of option returns (black) and a kernel estimate of S&P 500 returns (red); (b) an empirical distribution function of option returns (black) and S&P 500 returns (red).

a histogram estimate of the return distribution of the option strategy. The red curve shows a kernel estimate of the corresponding S&P 500 returns. Panel (b) shows a tail plot of the option strategy returns (black) and a tail plot of the corresponding S&P 500 returns (red). The corresponding profit function of a long position is shown in Figure 17.2(a) and the return function is shown in Figure 17.8(a).

Figure 17.18 shows a short straddle. The setting is the same as in Figure 17.17.

Figure 17.19 shows the return distribution of a strangle with $S_0 = 100$, $K = 95$, and $K' = 105$. The return is given in (17.16), where we take $b = 1$. Panel (a) shows a histogram estimate of the return distribution of the option strategy. The red curve shows a kernel estimate of the corresponding S&P 500 returns. Panel (b) shows a tail plot of the option strategy returns (black) and a tail plot of the corresponding S&P 500 returns (red). The corresponding profit function of a long position is shown in Figure 17.2(b) and the return function is shown in Figure 17.8(b).

Figure 17.20 shows a short strangle. The setting is the same as in Figure 17.19.

Figure 17.21 shows the return distribution of a butterfly with $S_0 = 100$, $K = 95$, $K' = K'' = 10$, and $K''' = 105$. The return is given in (17.17), where we take $b = 1$. Panel (a) shows a histogram estimate of the return distribution of the option strategy. The red curve shows a kernel estimate of the corresponding S&P 500 returns. Panel (b) shows a tail plot of the option strategy returns (black) and a tail plot of the corresponding S&P 500 returns (red). The corresponding profit function of a long position is shown in Figure 17.2(c) and the return function is shown in Figure 17.8(c).

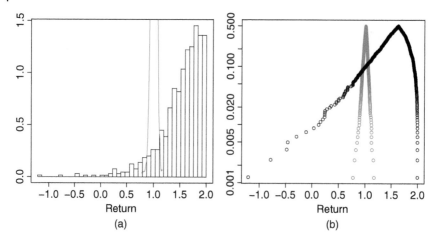

Figure 17.18 *A short straddle: Return distribution.* (a) A histogram estimate of option returns (black) and a kernel estimate of S&P 500 returns (red); (b) an empirical distribution function of option returns (black) and S&P 500 returns (red).

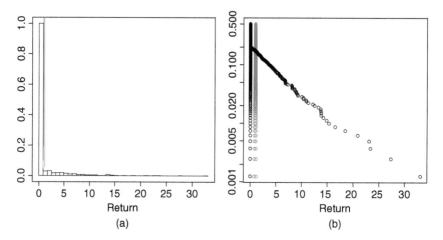

Figure 17.19 *A strangle: Return distribution.* (a) A histogram estimate of option returns (black) and a kernel estimate of S&P 500 returns (red); (b) an empirical distribution function of option returns (black) and S&P 500 returns (red).

Figure 17.22 shows the return distribution of a short butterfly. The setting is the same as in Figure 17.21.

Figure 17.23 shows the return distribution of a condor with $S_0 = 100, K = 90$, $K' = 95$, $K'' = 105$, and $K''' = 110$. The return is given in (17.17), where we take $b = 1$. Panel (a) shows a histogram estimate of the return distribution of

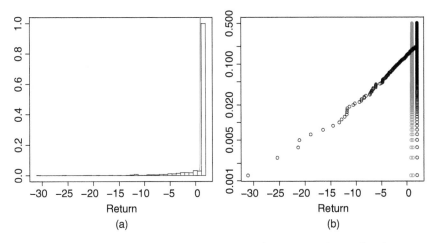

Figure 17.20 *A short strangle: Return distribution.* (a) A histogram estimate of option returns (black) and a kernel estimate of S&P 500 returns (red); (b) an empirical distribution function of option returns (black) and S&P 500 returns (red).

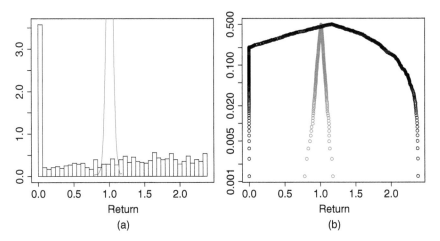

Figure 17.21 *A butterfly: Return distribution.* (a) A histogram estimate of option returns (black) and a kernel estimate of S&P 500 returns (red); (b) an empirical distribution function of option returns (black) and S&P 500 returns (red).

the option strategy. The red curve shows a kernel estimate of the corresponding S&P 500 returns. Panel (b) shows a tail plot of the option strategy returns (black) and a tail plot of the corresponding S&P 500 returns (red). The corresponding profit function of a long position is shown in Figure 17.2(d) and the return function is shown in Figure 17.8(d).

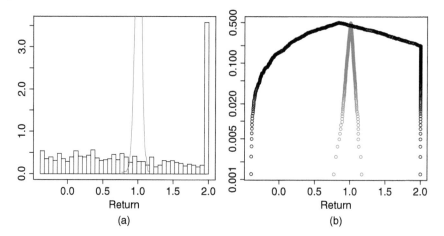

Figure 17.22 *A short butterfly: Return distribution.* (a) A histogram estimate of option returns (black) and a kernel estimate of S&P 500 returns (red); (b) an empirical distribution function of option returns (black) and S&P 500 returns (red).

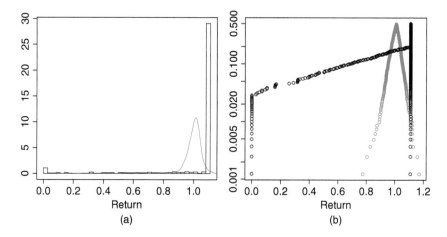

Figure 17.23 *A condor: Return distribution.* (a) A histogram estimate of option returns (black) and a kernel estimate of S&P 500 returns (red); (b) an empirical distribution function of option returns (black) and S&P 500 returns (red).

Figure 17.24 shows the return distribution of a short condor. The setting is the same as in Figure 17.23.

17.2.2.3 A Protective Put and a Covered Call

Figure 17.25 shows the return distribution of a protective put with $S_0 = 100$ and $K = 95$. The return is given in (17.20), where we take $b = 1$. Panel (a) shows a histogram estimate of the return distribution of the option strategy. The red

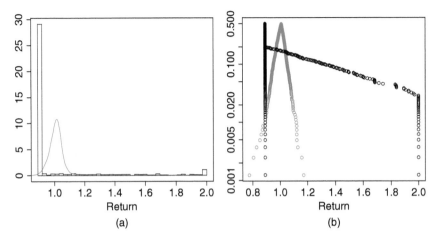

Figure 17.24 *A short condor: Return distribution.* (a) A histogram estimate of option returns (black) and a kernel estimate of S&P 500 returns (red); (b) an empirical distribution function of option returns (black) and S&P 500 returns (red).

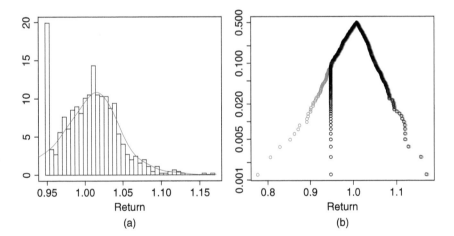

Figure 17.25 *A protective put: Return distribution.* (a) A histogram estimate of option returns (black) and a kernel estimate of S&P 500 returns (red); (b) an empirical distribution function of option returns (black) and S&P 500 returns (red).

curve shows a kernel estimate of the corresponding S&P 500 returns. Panel (b) shows a tail plot of the option strategy returns (black) and a tail plot of the corresponding S&P 500 returns (red). The corresponding profit function is shown in Figure 17.6(c) and the return function is shown in Figure 17.9(c).

Figure 17.26 shows the return distribution of a covered call with $S_0 = 100$ and $K = 105$. The return is given in (17.21), where we take $b = 1$. Panel (a) shows

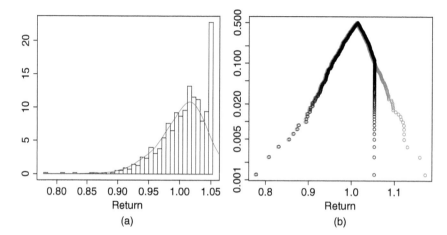

Figure 17.26 *A covered call: Return distribution.* (a) A histogram estimate of option returns (black) and a kernel estimate of S&P 500 returns (red); (b) an empirical distribution function of option returns (black) and S&P 500 returns (red).

a histogram estimate of the return distribution of the option strategy. The red curve shows a kernel estimate of the corresponding S&P 500 returns. Panel (b) shows a tail plot of the option strategy returns (black) and a tail plot of the corresponding S&P 500 returns (red). The corresponding profit function is shown in Figure 17.6(d) and the return function is shown in Figure 17.9(d).

17.2.2.4 A Bond and a Call

Figure 17.27 shows the return distribution of a capital guarantee product with $S_0 = 100$ and $K = 105$. The bond gross return is 1.01. The return is given in (17.22), where we take $b = 0.01$. Panel (a) shows a histogram estimate of the return distribution of the option strategy. The red curve shows a kernel estimate of the corresponding S&P 500 returns. Panel (b) shows a tail plot of the option strategy returns (black) and a tail plot of the corresponding S&P 500 returns (red). The corresponding profit function of is shown in Figure 17.6(f) and the return function is shown in Figure 17.9(d).

17.2.3 Performance Measurement of Option Strategies

We estimate the Sharpe ratios of few option strategies and show the cumulative wealths and wealth ratios.

We use the S&P 500 daily data, described in Section 2.4.1. The option prices are computed using the Black–Scholes formula, with the volatility equal to the sequentially estimated annualized GARCH(1, 1) volatility. The risk-free rate is deduced from the rate of the one-month Treasury bill, using data described in Section 2.4.3.

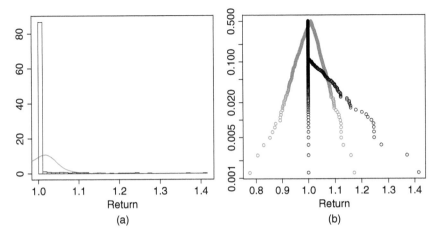

Figure 17.27 *A bond and a call: Return distribution.* (a) A histogram estimate of strategy returns (black) and a kernel estimate of S&P 500 returns (red); (b) an empirical distribution function of strategy returns (black) and S&P 500 returns (red).

The Sharpe ratios are not equal to the Sharpe ratios which are obtained using the market prices of options. Also, we do not take the transaction costs into account. However, studying the performance of option strategies gives insights into the concepts of performance measurement. Also, we can interpret the results as giving information about the properties of the Black–Scholes prices, since the fair option prices should be such that statistical arbitrage is excluded.

17.2.3.1 Covered Call and Protective Put
Figure 17.28 shows the Sharpe ratios of buying (a) covered call and (b) protective put. The x-axis shows the put moneyness, defined as K/S_0, where K is the strike price and S_0 is the stock price at the time of the writing of the call option. The y-axis shows the Sharpe ratio. The time to expiration is 20 days (black), 40 days (red), and 60 days (green). The blue horizontal lines show the Sharpe ratio of S&P 500. We see that the Sharpe ratio of the covered call converges to the Sharpe ratio of the underlying, when the strike price increases, and the Sharpe ratio of the protective put converges to the Sharpe ratio of the underlying, when the strike price decreases.

Figure 17.29 shows the wealth ratios of buying (a) covered call and (b) protective put. The time to expiration is 20 trading days. The strike price is $K = 105$ for the covered call and $K = 95$ for the protective put, when the stock price is $S_0 = 100$.

17.2.3.2 Calls, Puts, and Capital Guarantee Products
Figure 17.30 shows the Sharpe ratios of buying (a) call options and (b) put options. The x-axis shows the put moneyness, defined as K/S_0, where K is the

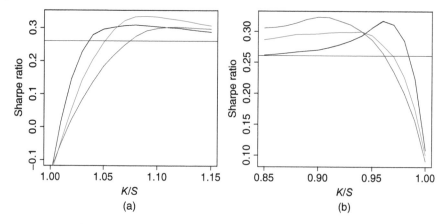

Figure 17.28 *Covered call and protective put: Sharpe ratios.* (a) Covered call and (b) protective put. The Sharpe ratios as a function of the put moneyness K/S, when the time to expiration is 20 (black), 40 (red), and 60 (green) trading days.

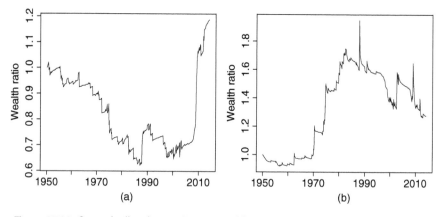

Figure 17.29 *Covered call and protective put: Wealth ratios.* (a) Covered call and (b) protective put. We show the time series of wealth of the option divided by the wealth of S&P 500.

strike price and S_0 is the stock price at the time of the writing of the option. The y-axis shows the Sharpe ratio. The time to expiration is 20 days (black), 40 days (red), and 60 days (green). The blue horizontal lines show the Sharpe ratio of S&P 500. We see that the Sharpe ratios of the call options converge to the Sharpe ratio of the underlying, when the strike price approaches zero. The Sharpe ratios of the call options are larger than the Sharpe ratio of the underlying when the moneyness is around zero. The Sharpe ratios of the put options are lower than the Sharpe ratio of the underlying.

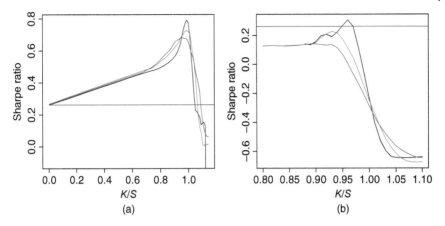

Figure 17.30 *Buying calls and puts: Sharpe ratios.* (a) Long call and (b) long put. The Sharpe ratios as a function of the moneyness, when the time to expiration is 20 (black), 40 (red), and 60 (green) trading days.

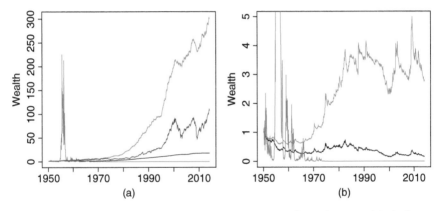

Figure 17.31 *Buying a call and a bond: Wealth.* (a) Time series of cumulative wealths for $b = 0$ (black), $b = 0.01$ (red), and $b = 0.5$ (green). The blue curve is the cumulative wealth of S&P 500. (b) Wealth ratios.

Figure 17.31 considers the capital guarantee product, which is constructed by combining a call and a bond to give return

$$R_T = (1 - b)(1 + r) + b\frac{C_T}{C_0},$$

where $b \in \mathbf{R}$ is the weight. This return was discussed in (17.22). The Sharpe ratio is given in Figure 17.30(a), because the weight b does not change the Sharpe ratio (see Section 10.1.1). We choose b positive but close to zero, in order to guarantee the preservation of the capital. The strike price is $K = 100$ for the call, when the stock price is $S_0 = 100$. The time to expiration is 20 trading days. Panel (a) shows the cumulative wealth for $b = 0$ (black), $b = 0.01$ (red), and $b = 0.5$ (green). The blue curve is the cumulative wealth of S&P 500. Panel (b) shows the wealth ratios where the cumulative wealths of the capital guarantee products are divided by the cumulative wealth of S&P 500. We see that when $b = 0.5$ (green), then the bankruptcy follows quite soon. When $b = 0.01$ (red), then the cumulative wealth is larger than for S&P 500, but with the expense of higher volatility.

18

Interest Rate Derivatives

A zero-coupon bond can be the underlying asset for forwards and options, in the same way as a stock. The price $Z(t, T)$ of a zero-coupon bond is 1 at the expiration, so that $Z(T, T) = 1$; but for $t < T$, the price $Z(t, T)$ is a random variable, in the same way as stock price S_t is a random variable. The forward price of a zero-coupon bond can be found by an arbitrage argument, in the same way as the forward price of a stock. A coupon-bearing bond can be expressed as a portfolio of zero-coupon bonds, so that much of the pricing of coupon-bearing bonds can be done with the help of pricing of zero-coupon bonds.

A forward rate agreement (FRA) allows to change a floating rate L against a fixed rate K, where L is typically a Libor rate. Thus, the payoff of a forward rate agreement is of type

$$L - K.$$

The arbitrage-free price of a forward rate agreement can be found in the same way as the forward price of a zero-coupon bond, or the forward price of a stock. Caplets and floorlets are calls and puts whose underlying is a Libor rate. The payoff of a caplet is of type

$$(L - K)_+ = \max\{L - K, 0\}.$$

A swap is a series of forward rate agreements. Thus, the payoff of a swap is of type

$$\sum_{i=1}^{m}(L_i - K),$$

where $L_i = L(T_{i-1}, T_i)$ are Libor rates for different time periods $[T_{i-1}, T_i]$, whose value is known at time T_{i-1}. A cap is a series of caplets and a floor is a series of floorlets. Thus, the payoff of a cap is of type

$$\sum_{i=1}^{m}(L_i - K)_+.$$

Nonparametric Finance, First Edition. Jussi Klemelä.
© 2018 John Wiley & Sons, Inc. Published 2018 by John Wiley & Sons, Inc.

A swaption is an option on a swap. Thus, the payoff of a payer swaption (call on a swap) is of type

$$\max\left\{\sum_{i=1}^{m}(L_i - K), 0\right\}.$$

In this chapter, we define such concepts as zero-coupon bond, continuously compounded yield, discretely compounded yield, simply compounded yield, yield curve, short rate, discount curve, discount factor, coupon-bearing bond, accrued interest, duration, par yield, forward zero-coupon bond, forward price of a zero-coupon bond, forward rate corresponding to a forward zero-coupon bond, forward rate agreement, forward rate corresponding to a forward rate agreement, instantaneous forward interest rate, swap, forward swap rate, caplet, floorlet, cap, floor, and swaption.

Black–Scholes model has been standard for pricing European derivatives but no such widely applicable model for interest rate derivatives has been found. The pricing and hedging of interest rate derivatives depends on the whole yield curve, and not on any single interest rate, which complicates the pricing and hedging. The main submarkets for derivatives include cap markets and swap markets.

Section 18.1 discusses such basic concepts of fixed income markets as interest rates, zero-coupon bonds, and coupon-bearing bonds. Section 18.2 studies forwards whose underlyings are fixed income instruments. These include forward zero-coupon bonds, forward rate agreements, and swaps. Section 18.3 discusses options whose underlyings are fixed income instruments. These include caps, floors, and swaptions.

18.1 Basic Concepts of Interest Rate Derivatives

First, we discuss various definitions of interest rates. Second, we discuss zero-coupon bonds and coupon-bearing bonds.

Interest rates appear in several contexts.

1) In mathematical finance, the risk-free rate is modeled as a stochastic process, in a similar way as the time series of stock returns. Note that the risk-free rate is a predictable process, which means that the rate is known at the beginning of the period, whereas the stock return is known at the end of the period. Modeling is done in the one-period model, in a multiperiod model, or in a continuous time model.

2) The yield is derived from the price of a zero-coupon bond or from the price of a coupon-bearing bond. The yields are a more convenient way to quote the prices than the dollar value, because we can compare different kinds of bonds with their yields.

3) The coupon payments of a coupon-bearing bond are often quoted as a percentage of the face value, and this percentage is a kind of interest rate.

18.1.1 Interest Rates and a Bank Account

A lender and a borrower can agree for various compounding rules in defining a fixed income investment. Interest rates can be defined as simple, discretely compounded, or continuously compounded. Different definitions of an interest rate lead to different definitions of a bank account. A bank account may also be called a money market account or a cash account.

We denote with

$$T - t \text{ or } \tau(t, T),$$

the time from t to T in fractions of a year. For example, half a year is 0.5. Symbols t and T denote either the real numbers or a calendar date $(d, m, y) = (\text{day, month, year})$. If t and T are real valued, then $\tau(t, T) = T - t$. Otherwise, we have to agree on a day-count convention.[1]

18.1.1.1 Simple Interest Rates and the One-Period Model

In the one-period model, there are two time points $t = t_0$ and $t = t_1$. In the one-period model, the bank account can be defined using the simple interest rate:

$$B_{t_0} = 1, \quad B_{t_1} = 1 + r \cdot \tau(t_0, t_1),$$

where $r > -1$ is the yearly interest rate. In Chapter 13, we had $t_0 = 0$, $t_1 = 1$, and $\tau(t_1, t_2) = \Delta t$. In the following, we denote $t_1 = 0$ and $t_2 = T$.

In the one-period model, it is natural to use the simple interest rate. Rate r is a simple interest rate, if at time t_1 the investment has value

$$(1 + r \cdot \tau(t_0, t_1))P, \tag{18.1}$$

where the principal $P > 0$ is invested at time t_0.

18.1.1.2 Discretely Compounded Interest Rates and the Multiperiod Model

Let us consider time period $[0, T]$. In the multiperiod model, there are $n + 1$ time points

$$t_k = \frac{kT}{n}, \quad k = 0, \dots, n.$$

---

1 Some examples of day-count conventions include the following:

- *Actual/365*. Under this convention, $\tau(t, T)$ is equal to the actual number of days between t and T divided by 365.
- *Actual/360*. Under this convention, $\tau(t, T)$ is equal to the actual number of days between t and T divided by 360.
- *30/360*. Under this convention, we assume that months are 30 days long. Instead of $d_2 - d_1$, we have to use the formula $\max(30 - d_1, 0) + \min(d_2, 30)$. Thus,

$$\tau(t, T) = (y_2 - y_1) + 30(m_2 - m_1 - 1)/360 + [\max(30 - d_1, 0) + \min(d_2, 30)]/360.$$

Here we denote $t = (d_1, m_1, y_1)$, $T = (d_2, m_2, y_2)$.

The bank account can be defined using the discretely compounded interest rate:

$$B_{t_k} = (1 + r\Delta t)^k,$$

where $r > -1$ is the yearly interest rate, and

$$\Delta t = \tau(t_{i-1}, t_i) = \frac{T}{n}.$$

Alternatively, the interest rate can be changing:

$$B_0 = 1, \quad B_{t_k} = \prod_{i=1}^{k}(1 + r_i\Delta t), \quad k = 1, \dots, n,$$

where $r_i > -1$ is interest rate for the period $[t_{i-1}, t_i]$.

In a multiperiod model, it is natural to use a discretely compounded interest rate. Rate $r > 0$ is a discretely compounded interest rate, if the fixed income investment pays

$$\left(1 + \frac{rT}{n}\right)^n P \tag{18.2}$$

at the expiration T, where $n \geq 1$ is an integer, and principal $P > 0$ is invested at time 0. Here r is the yearly interest rate and T is time to maturity in fractions of a year, since we consider period $[0, T]$.[2]

18.1.1.3 Continuously Compounded Interest Rates and the Continuous Time Model

In the continuous time model, the bank account can be defined using the continuously compounded interest rate:

$$B_t = e^{rt}, \quad t \in [0, T],$$

where $r > 0$ is the instantaneous spot rate (spot interest rate, short rate). If the instantaneous spot rate is time varying $r_t > 0$, then

$$B_t = \exp\left(\int_0^t r_s \, ds\right), \quad t \in [0, T].$$

In the Black–Scholes model, r_t is taken to be deterministic and constant. This assumption is justified because usually the variability of interest rates has

---

2 Note that often the notation is used where T is the number of years for the investment, and every year there are m times compounding. Then (18.2) can be written as

$$\left(1 + \frac{r}{m}\right)^{mT} P.$$

We have that $n = mT$.

a contribution of a smaller order to the price of equity derivatives than the variability of the equity. When pricing interest rate products, the variability of r_t has to be taken into account, and it is typically assumed that r_t is stochastic.

In a continuous time model, we use a continuously compounded interest rate. Rate $r > 0$ is the continuously compounded interest rate, if the fixed income investment pays

$$e^{rT} P \qquad (18.3)$$

at the expiration T, when the principal P is invested at time 0. A discretely compounded interest rate approximates the continuously compounded interest rate because

$$\left(1 + \frac{rT}{n}\right)^n \longrightarrow e^{rT},$$

as $n \to \infty$.

If the rate r is continuously compounded, then we get the equivalent rate r_n for the compounding frequency n by solving equation

$$e^{rT} = \left(1 + \frac{r_n T}{n}\right)^n,$$

which gives

$$r_n = \frac{n}{T}(e^{rT/n} - 1).$$

The bank account can be constructed by investing $1 at time 0 into infinitely many zero-coupon bonds with infinitesimal time to maturity, such that one maturity date is the starting date of the next zero-coupon bond, and all zero-coupon bonds together span the time from 0 to t. The prices of the zero-coupon bonds are fixed at time 0.

The bank account with continuous compounding can also be obtained as the solution of the differential equation $\partial B(t)/\partial t = r \cdot B(t)$. The solution is the function $B(t) = B(0)e^{rt}$.

18.1.2 Zero-Coupon Bonds

A zero-coupon bond, or a pure discount bond, is a certificate which gives the owner a nominal amount 1 at the future maturity time T. We denote the price of the bond at time t with

$$Z(t, T), \quad 0 \le t \le T.$$

We have that $Z(T, T) = 1$. When the interest rate is positive, then $Z(t, T) < 1$ for $0 \le t < T$.

18.1.2.1 Yield

The yield of a zero-coupon bond can be defined as continuously compounded, k-times-per-year compounded, or simply compounded.

Continuously Compounded Yield The continuously compounded yield $Y(t, T)$ of the zero-coupon bond $Z(t, T)$ is defined by

$$Y(t, T) = -\frac{\log Z(t, T)}{T - t}.$$

The yield $Y(t, T)$ is also called the continuous zero rate or the zero rate. If $T - t$ is the number of years to maturity, then $Y(t, T)$ is also called the yearly discount rate or the annual zero-coupon rate. The definition of the yield is equivalent with the equation

$$Z(t, T) = \exp\{-Y(t, T)(T - t)\}.$$

Note that if there exists a bank account with deterministic and constant continuously compounded spot rate r from period t to T, then the zero-coupon bond can be written as

$$Z(t, T) = \exp\{-r(T - t)\}, \tag{18.4}$$

because the bank account $e^{-r(T-t)}$ is an investment that has value 1 at time T, just like $Z(t, T)$, and thus to exclude arbitrage, they have to have the same price at the previous times.

k-Times-Per-Year Compounded Yield The k-times-per-year compounded yield is defined by

$$Y^{(k)}(t, T) = \frac{k}{Z(t, T)^{1/[k(T-t)]}} - k.$$

This definition of the yield is equivalent with the equation

$$Z(t, T) = \frac{1}{[1 + Y^{(k)}(t, T)/k]^{k(T-t)}}. \tag{18.5}$$

We have that

$$\lim_{k \to \infty} Y^{(k)}(t, T) = Y(t, T).$$

The annually compounded yield of a zero-coupon bond is obtained by $k = 1$. The simply compounded yield in (18.6) is obtained by $k = 1/(T - t)$, when $T - t \leq 1$.

Simply Compounded Yield The simply compounded spot interest rate is defined by

$$L(t, T) = \frac{1 - Z(t, T)}{(T - t)Z(t, T)}. \tag{18.6}$$

This definition of the yield is equivalent with the equation

$$Z(t, T) = \frac{1}{1 + (T - t)L(t, T)}.$$

The market Libor rates are simply compounded rates.

18.1.2.2 Yield Curve

For a fixed t, the function

$$T \mapsto Y(t, T), \quad T \geq t,$$

is called the yield curve (zero-rate curve, zero-coupon curve, term structure of interest rates), when $T - t$ is larger than a year. When $T - t$ is less than a year, then we consider

$$T \mapsto L(t, T).$$

An increasing yield curve is called normal, and a decreasing yield curve is called inverse. If the market expects the rates to fall, then the yield curve will be inverted.

Statistical techniques are needed to infer a continuous time yield curve from discrete data. Similar methods are needed to estimate the discount curve in (18.8), and forward curve in (18.21). Parametric methods search to model function $f(x) = Y(t, x)$ with a function $f(x, \theta)$, where θ is a vector of parameters. The Nelson–Siegel family models the rate curve by

$$f(x, \theta) = \theta_1 + (\theta_2 + \theta_3 x)e^{-x\theta_4}.$$

The Svensson family models the rate curve by

$$f(x, \theta) = \theta_1 + (\theta_2 + \theta_3 x)e^{-\theta_4 x} + \theta_5 x e^{-\theta_6 x}.$$

The vector parameter θ is found which minimizes

$$\mathcal{L}(\theta) = \sum_j w_j |B_j - B_j(\theta)|^2,$$

where w_j are weights, B_j are the prices for on-the-run bonds and notes available on the day, and $B_j(\theta)$ are the prices which one would get by pricing these bonds and notes using the curve $f(x, \theta)$.

A smoothing spline method estimates the yield curve nonparametrically by finding the curve $\phi(x)$ that approximates $Y(t, x)$. The yield curve ϕ is found which minimizes

$$\mathcal{L}(\phi) = \sum_j w_j |B_j - B_j(\phi)|^2 + \lambda \int |\phi''(x)|^2 \, dx,$$

where $B_j(\phi)$ are the prices which one would get by pricing these bonds and notes using the curve $\phi(x)$, and $\lambda \geq 0$.

A spline method is used by the U.S. Federal Reserve and the Bank of Japan, parametric methods are used by European central banks, and the investment banks use the so called bootstrapping method. More information can be found in Carmona and Tehranchi (2006).

18.1.2.3 Short Rate

The short rate, or instantaneous rate, is defined by

$$r(t) = -\frac{\partial}{\partial T} \log Z(t, T)\bigg|_{T=t}. \tag{18.7}$$

Symbolically, we may write $r(t) = Y(t, t)$, where $Y(t, T)$ is the yield of the bond. We have

$$r(t) = \lim_{T \to t+} Y(t, T),$$

and similarly for $L(t, T)$, and $Y^k(t, T)$.

18.1.2.4 Discount Curve

For a fixed t, the function

$$T \mapsto Z(t, T), \quad T \geq t, \tag{18.8}$$

is called the discount curve (zero-bond curve). The discount curve starts at 1 and is usually decreasing.

18.1.2.5 Discount Factor

The value $Z(t, T)$ can be called the discount ratio, discount factor, or present value, because amount P of cash at time T is worth $Z(t, T)P$ at time t. At time t the value $Z(t, T)$ is known: the random variable $Z(t, T)$ is measurable with respect to $\mathcal{F}_t$.

We can compare this definition of the discount factor to the definition

$$D(t, T) = \exp\left(-\int_t^T r_s \, ds\right),$$

where $r_s > -1$ is instantaneous time varying spot rate. The discount factor $D(t, T)$ is a random variable. If the spot rate r_t is deterministic, then $D(t, T) = Z(t, T)$.

18.1.3 Coupon-Bearing Bonds

Most bonds make regular payments (coupons) before the final payment at the maturity. A coupon bond is a series of payments

$$P_1, \ldots, P_n$$

at times

$$T_1, \ldots, T_n.$$

The terminal payment contains the notional and the final coupon payment.

For example, a 5-year, 4% semiannual coupon bond with $1000 face value makes ten $20 payments every six months and the final payment of $1000. Thus $P_i = \$20$ except the last payment is $1020.

The price $C(t, T)$ of a coupon bond can be written with the help of zero-coupon bonds:

$$C(t, T) = \sum_{i=1}^{n} P_i Z(t, T_i).$$

Let k be the number of coupon payments in one year. The coupon payments are often quoted as a percentage c of the face value P, so that $P_i = cP/k$ and $P_n = (1 + c/k)P$. Now the price of the coupon-bearing bond is

$$C(t, T) = \frac{cP}{k} \sum_{i=1}^{n} Z(t, T_i) + PZ(t, T_n). \tag{18.9}$$

Using the k-times-per-year compounded yield in (18.5), we can write the price as

$$C(t, T) = \sum_{i=1}^{n} \frac{P_i}{(1 + r_i/k)^{k(T_i - t)}}$$

$$= \sum_{i=1}^{n-1} \frac{cP/k}{(1 + r_i/k)^{k(T_i - t)}} + \frac{(1 + c/k)P}{(1 + r_n/k)^{k(T_n - t)}},$$

where we denote the k-times-per-year compounded yield by

$$r_i = Y^{(k)}(t, T_i).$$

It holds often that $k(T_i - t) = i$, so that

$$C(t, T) = \sum_{i=1}^{n-1} \frac{cP/k}{(1 + r_i/k)^i} + \frac{(1 + c/k)P}{(1 + r_n/k)^n}.$$

18.1.3.1 Yield of a Coupon-Bearing Bond

When continuous compounding is used, then the yield of a coupon-bearing bond can be defined as the solution $R(t, T)$ of the equation

$$C(t, T) = \sum_{i=1}^{n} P_i e^{-R(t,T)(T_i - t)}.$$

When discrete compounding is used, then the yield of a coupon-bearing bond can be defined as the solution $R(t, T)$ of the equation

$$C(t, T) = \sum_{i=1}^{n} \frac{P_i}{(1 + R(t, T))^{T_i - t}}.$$

18.1.3.2 Accrued Interest

When a transaction happens between interest payments, then one should take the accrued interest into account. The accrued interest is defined by

$$AC(T_i, t) = \frac{t - T_i}{T_{i+1} - T_i} P_{i+1},$$

when $t \in [T_i, T_{i+1}]$. The clean price is defined by

$$CP(t, T) = C(t, T) - AC(T_i, t), \quad t \in [T_i, T_{i+1}],$$

where $C(t, T) = \sum_{j=i+1}^{n} P_j Z(t, T_j)$.

18.1.3.3 Duration

The duration of a bond is defined by

$$D(t, T) = \frac{1}{C(t, T)} \sum_{i=1}^{n} P_i(T_i - t) e^{-R(t,T)(T_i - t)}.$$

The duration is the effective expiry time: it is the average time of the future cash flows, weighted by the contribution to the bond price.

18.1.3.4 Par Yield

The par yield is defined as the yield (or coupon) of a bond priced at par: it is the value c for which $C(t, T) = P$. Thus, (18.9) shows that the par yield is the value c for which

$$1 = \frac{c}{k} \sum_{i=1}^{n} Z(t, T_i) + Z(t, T_n).$$

A coupon-bearing bond is said to be valued at par if its current market value equals its face value (or par value).

The yield of a coupon paid at par is equal to its coupon rate. If the market value of a bond is less than its face value, we say the bond trades at a discount. Otherwise, the bond trades at a premium.

18.1.3.5 Floating-Rate Notes

A floating-rate note ensures payments at future times $T_1, \ldots, T_n$ of Libor rates that reset at the previous instants $T_0, \ldots, T_{n-1}$. Moreover, the note pays a last cash flow at time T_n consisting of the reimbursement of the notional value. We may write a floating-rate note in terms of zero-coupon bonds as

$$PZ(t, T_n) + P \sum_{i=1}^{n} \tau(T_{i-1}, T_i) L(T_{i-1}, T_i) Z(t, T_i).$$

Replacing the Libor rate $L(T_{i-1}, T_i)$ with the forward interest rate $f(t, T_{i-1}, T_i)$, defined in (18.17), we obtain

$$PZ(t, T_n) + P \sum_{i=1}^{n} \tau(T_{i-1}, T_i) f(t, T_{i-1}, T_i) Z(t, T_i)$$

$$= PZ(t, T_n) + P \sum_{i=1}^{n} [Z(t, T_{i-1}) - Z(t, T_i)]$$

$$= PZ(t, T_0).$$

Using the argument leading to (18.18), which gives the present value of a forward rate agreement, we obtain the present value of the floating-rate note as $PZ(t, T_0)$.

18.2 Interest Rate Forwards

We discuss forward zero-coupon bonds and forward rate agreements. Forward zero-coupon bonds are discussed also in Section 14.1.1.

18.2.1 Forward Zero-Coupon Bonds

A forward zero-coupon bond is an agreement, where at time t two parties want to exchange at a future time T_1 a zero-coupon bond, whose maturity is at T_2, with a cash payment $P = P(t, T_1, T_2)$. At time t, only the agreement is made, the cash payment is made at time T_1, and the zero-coupon bond is received at T_1.

18.2.1.1 The Forward Price
The forward price (the cash payment at T_1) is equal to

$$P(t, T_1, T_2) = \frac{Z(t, T_2)}{Z(t, T_1)}, \tag{18.10}$$

where $Z(t, T_1)$ and $Z(t, T_2)$ are the prices of the zero-coupon bonds with maturities T_1 and T_2. The forward price is the delivery price such that the forward contract has value zero at t. In other words, the forward price is such price that for no cost at t one can agree to buy at T_1 the zero-coupon bond maturing at T_2.

Equation (18.10) is proved in (14.5).

18.2.1.2 Forward Rates Corresponding to Forward Zero-Coupon Bonds
A forward rate $R(t, T_1, T_2)$ is the rate agreed at t, at which borrowing and lending can be made for the period from T_1 to T_2. A forward zero-coupon bond implies a forward rate.

Continuous Compounding Let $P(t, T_1, T_2)$ be the price of a forward zero-coupon bond. We may define the forward rate $R(t, T_1, T_2)$, corresponding to the forward zero-coupon bond, by equation

$$P(t, T_1, T_2) = \exp\{-R(t, T_1, T_2)(T_2 - T_1)\}.$$

By (18.10),

$$P(t, T_1, T_2) = \frac{Z(t, T_2)}{Z(t, T_1)}.$$

Thus,

$$R(t, T_1, T_2) = -\frac{\log Z(t, T_1) - \log Z(t, T_2)}{T_2 - T_1}. \tag{18.11}$$

Letting $T_2 - T_1$ approach zero leads to the instantaneous forward interest rate in (18.19).

Note that if we use (18.4) to write

$$Z(t, T_1) = e^{-r_1(T_1 - t)}, \quad Z(t, T_2) = e^{-r_2(T_2 - t)},$$

then (18.11) leads to

$$R(t, T_1, T_2) = \frac{r_2(T_2 - t) - r_1(T_1 - t)}{T_2 - T_1}. \tag{18.12}$$

Equation (18.12) can also be derived from

$$e^{r_1(T_1 - t)} e^{R(t, T_1, T_2)(T_2 - T_1)} = e^{r_2(T_2 - t)},$$

which should hold to exclude arbitrage, because the rate r_1 for the period from t to T_1 and the forward rate $R(t, T_1, T_2)$, both agreed at t, should as a combination lead to the same result as rate r_2, agreed also at t but for the period from t to T_2.

Discrete Compounding We may define the forward rate $R(t, T_1, T_2)$ by the equation

$$P(t, T_1, T_2) = \frac{1}{(1 + R(t, T_1, T_2))^{T_2 - T_1}}.$$

Then,

$$R(t, T_1, T_2) = P(t, T_1, T_2)^{-1/(T_2 - T_1)} - 1 = \left(\frac{Z(t, T_2)}{Z(t, T_1)}\right)^{-1/(T_2 - T_1)} - 1.$$

If we write

$$Z(t, T_1) = \frac{1}{(1 + r_1)^{T_1 - t}}, \quad Z(t, T_2) = \frac{1}{(1 + r_2)^{T_2 - t}},$$

then

$$R(t, T_1, T_2) = \left(\frac{(1 + r_1)^{T_1 - t}}{(1 + r_2)^{T_2 - t}}\right)^{-1/(T_2 - T_1)} - 1. \tag{18.13}$$

Note that (18.13) can also be derived from

$$(1 + r_1)^{T_1 - t}(1 + R(t, T_1, T_2))^{T_2 - T_1} = (1 + r_2)^{T_2 - t}.$$

Forward Libor Rates Libor rate $L(t, T)$ is defined in (18.6) as the solution of the equation

$$1 = (1 + \tau(t, T)L(t, T))Z(t, T),$$

where $\tau(t, T)$ is the accrual factor (day count fraction).

A forward Libor rate $L(t, T_1, T_2)$ is the interest rate one can contract at time t to put money in money-market account for the time period $[T_1, T_2]$. It is defined by

$$Z(t, T_1) = (1 + \tau(T_1, T_2)L(t, T_1, T_2))Z(t, T_2),$$

which gives

$$L(t, T_1, T_2) = \frac{1}{\tau(T_1, T_2)} \frac{Z(t, T_1) - Z(t, T_2)}{Z(t, T_2)}. \tag{18.14}$$

The time T_1 is called maturity and $T_2 - T_1$ is called tenor of the forward Libor rate. At time T_1, the forward Libor rate $L(T_1, T_1, T_2)$ is fixed (or set), and it is called the spot Libor rate.

18.2.2 Forward Rate Agreements

A forward rate agreement allows to change a fixed rate K against a floating rate $L(T_1, T_2)$.

A forward rate agreement involves three time instants: the current time t (entering time), the expiry time $T_1 > t$ (reset time), and the maturity time $T_2 > T_1$ (payment time). The payoff for the seller at time T_2 is

$$\text{FRA}_{T_2} = P \cdot [K - L(T_1, T_2)] \cdot \tau(T_1, T_2), \tag{18.15}$$

where P is the principal, $L(T_1, T_2)$ is the Libor rate that resets at time T_1 for the payment at T_2, and K is the strike. We assume that rates K and $L(T_1, T_2)$ have the same day-count convention. We call $PL(T_1, T_2)\tau(T_1, T_2)$ the floating leg and $PK\tau(T_1, T_2)$ the fixed leg.

18.2.2.1 The Present Value of a Forward Rate Agreement
The present value of a forward rate agreement is

$$\text{FRA}_t = P \cdot [KZ(t, T_2)\tau(T_1, T_2) - Z(t, T_1) + Z(t, T_2)]. \tag{18.16}$$

This is shown in (14.7).[3]

---

3 Indeed, using (18.6), we can write

$$\text{FRA}_{T_2} = P\left[\tau(T_1, T_2)K - \frac{1}{Z(T_1, T_2)} + 1\right],$$

where the present value of the term $1/Z(T_1, T_2)$ is $Z(t, T_1)$ and the present value of $\tau(T_1, T_2)K + 1$ is $Z(t, T_2)[\tau(T_1, T_2)K + 1]$.

The value of a forward rate agreement at time T_1 is obtained by discounting:

$$\text{FRA}_{T_1} = \frac{\text{FRA}_{T_2}}{1 + L(T_1, T_2)\tau(T_1, T_2)}.$$

18.2.2.2 Forward Rates Corresponding to Forward Rate Agreements

The forward rate is defined to be the strike K that gives zero value to the forward rate agreement: the strike K that gives zero value to (18.16). The forward rate is called also the equilibrium strike or the equilibrium rate. Thus, the simply compounded forward interest rate $f(t, T_1, T_2)$ is defined as

$$f(t, T_1, T_2) = \frac{1}{\tau(T_1, T_2)} \left(\frac{Z(t, T_1)}{Z(t, T_2)} - 1 \right). \tag{18.17}$$

At time T_1, the forward rate is equal to the Libor rate:

$$f(T_1, T_1, T_2) = L(T_1, T_2).$$

Using the definition of forward interest rate, the value of a forward rate agreement can be written as

$$\text{FRA}_t = PZ(t, T_2)\tau(T_1, T_2)[K - f(t, T_1, T_2)]. \tag{18.18}$$

Thus, $f(t, T_1, T_2)$ can be seen as an estimate of the future spot rate $L(T_1, T_2)$.

18.2.2.3 Instantaneous Forward Interest Rate

Instantaneous forward interest rate (the rate for instantaneous borrowing at time T) is defined by

$$f(t, T) = -\frac{\partial}{\partial T} \log Z(t, T)$$

$$= \frac{\partial}{\partial T} [(T - t)Y(t, T)]$$

$$= Y(t, T) + (T - t)\frac{\partial}{\partial T} Y(t, T), \tag{18.19}$$

where the yield $Y(t, T)$ is defined by

$$Z(t, T) = \exp\{-(T - t)Y(t, T)\}.$$

Note that

$$Z(t, T) = \exp \left\{ - \int_t^T f(t, u)\, du \right\}. \tag{18.20}$$

Using the convention $\tau(T_1, T_2) = T_2 - T_1$, we have that the forward rate $f(t, T_1, T_2)$ satisfies

$$
\begin{aligned}
\lim_{T_2 \downarrow T_1} f(t, T_1, T_2) &= -\lim_{T_2 \downarrow T_1} \frac{1}{Z(t, T_2)} \frac{Z(t, T_2) - Z(t, T_2)}{T_2 - T_1} \\
&= -\frac{1}{Z(t, T_2)} \frac{\partial Z(t, T_1)}{\partial T_1} \\
&= -\frac{\partial \log Z(t, T_1)}{\partial T_1} \\
&= f(t, T_1).
\end{aligned}
$$

Note also that $f(t, t) = r(t)$, where short rate $r(t)$ is defined in (18.7).

For a fixed t, the function

$$
T \mapsto f(t, T), \quad T \geq t, \tag{18.21}
$$

is called the forward curve. If the zero-rate curve is normal, then the forward curve lies above the zero-rate curve; if the zero-rate curve is inverted, then the forward curve lies below the zero-rate curve.

18.2.3 Swaps

An interest rate swap (IRS) is a generalization of a forward rate agreement. A swap is a series of FRAs.

A swap is a commitment to exchange the payments originating from a fixed leg and a floating leg. The buyer of a receiver swap obtains payments with a fixed interest rate and pays payments with a variable interest rate. The buyer of a payer swap obtains payments with a variable interest rate and pays payments with a fixed interest rate. Thus, when the fixed leg is paid and the floating leg is received, then an interest rate swap is called a payer (forward-start) interest rate swap (PFS). Otherwise, it is called a receiver (forward-start) interest rate swap (RFS).

A swap spans the period $[T_0, T_m]$, where T_0 is the expiry, and T_m is the maturity. There are time points

$$
t \leq T_0 < T_1 < \cdots < T_m,
$$

where t is the time of the agreement for the swap. At every instant T_{i-1}, $i = 1, \ldots, n$, the fixed leg pays

$$
P \cdot \tau(T_{i-1}, T_i)K,
$$

where K is the fixed interest rate and P is the principal. The floating leg pays

$$P \cdot \tau(T_{i-1}, T_i)L(T_{i-1}, T_i),$$

where $L(T_{i-1}, T_i)$ is the Libor rate.

18.2.3.1 The Value of a Receiver Swap

A receiver swap is a portfolio of forward rate swaps, whose payoff was written in (18.15) as

$$\text{FRA}(T_{i-1}, T_i) = P \cdot [K - L(T_{i-1}, T_i)] \cdot \tau(T_{i-1}, T_i).$$

The value at time t of a forward rate swap was written in (18.18) as

$$\text{FRA}_t(T_{i-1}, T_i) = P \cdot Z(t, T_i)[K - f(t, T_{i-1}, T_i)] \cdot \tau(T_{i-1}, T_i).$$

We can write the value at time t of a receiver swap as

$$\text{RFS}_t = \sum_{i=1}^{m} \text{FRA}_t(T_{i-1}, T_i). \tag{18.22}$$

The forward rate $f(t, T_{i-1}, T_i)$ was written in (18.17) as

$$f(t, T_{i-1}, T_i) = \frac{1}{\tau(T_{i-1}, T_i)} \left(\frac{Z(t, T_{i-1})}{Z(t, T_i)} - 1 \right).$$

Thus,

$$\text{RFS}_t = PK \sum_{i=1}^{m} \tau(T_{i-1}, T_i)Z(t, T_i) - P(Z(t, T_0) - Z(t, T_m)).$$

Note that we have the decomposition of the value of the receiver swap to the value of the fixed leg minus the value of the floating leg:

$$\text{RFS}_t = A_t - \text{FL}_t,$$

where the value of the fixed lag is

$$A_t = P \cdot K \cdot \sum_{i=1}^{m} \tau(T_{i-1}, T_i)Z(t, T_i).$$

The value of the floating leg may be written as

$$\text{FL}_t = P \sum_{i=1}^{m} \tau(T_{i-1}, T_i)f(t, T_{i-1}, T_i)Z(t, T_i)$$

$$= P \sum_{i=1}^{m} [Z(t, T_{i-1}) - Z(t, T_i)]$$

$$= P[Z(t, T_0) - Z(t, T_m)],$$

where we used the formula (18.17) for the simply compounded forward interest rate, $f(t, T_{i-1}, T_i)$. For a spot-starting swap, we have $\text{FL}_t = P[Z(T_0, T_0) - Z(T_0, T_m)]$.

18.2.3.2   Forward Swap Rate

The forward swap rate $SR(t, T_0, T_m)$ (the equilibrium swap rate) is the fixed rate K for which $RFS_t = 0$. This makes an interest rate swap a fair contract at the present time. Thus, we can write the swap rate in terms of forward rates as

$$SR(t, T_0, T_m) = \sum_{i=1}^{m} w_i f(t, T_{i-1}, T_i), \qquad (18.23)$$

where

$$w_i = \frac{Z(t, T_i)\tau(T_{i-1}, T_i)}{\sum_{j=1}^{m} \tau(T_{j-1}, T_j)Z(t, T_j)}.$$

We have also the formula

$$SR(t, T_0, T_m) = \frac{Z(t, T_0) - Z(t, T_m)}{\sum_{i=1}^{m} \tau(T_{i-1}, T_i)Z(t, T_i)}.$$

We can write

$$\frac{Z(t, T_k)}{Z(t, T_0)} = \prod_{j=1}^{k} \frac{Z(t, T_j)}{Z(t, T_{j-1})} = \prod_{j=1}^{k} \frac{1}{1 + \tau_j f_j(t)},$$

where $\tau_j = \tau(T_{j-1}, T_j)$ and $f_j(t) = f(t, T_{j-1}, T_j)$. Thus we arrive at a third formula

$$SR(t, T_0, T_m) = \frac{1 - \prod_{j=1}^{m} \frac{1}{1 + \tau_j f_j(t)}}{\sum_{i=1}^{m} \tau_i \prod_{j=1}^{i} \frac{1}{1 + \tau_j f_j(t)}}.$$

18.2.4   Related Fixed Income Instruments

Bond futures are futures whose underlying is a bond.[4] There are several other fixed income instruments which are constructed from some more basic fixed income instruments.

1) *Convertible bonds.* A convertible bond gives the owner an option to convert the bond into a given number of shares of equity for each unit of the face value.

---

4 The financial instrument underlying the Treasury bill futures contract is a fictitious Treasury bill with exactly 90 days to maturity. The contract has the notional amount of 1 million US dollars. Two years Treasury note futures contract has $200,000 notional value. Ten years Treasury bond futures contract has $100,000 notional value. Euro-Schatz futures are on instruments having remaining term from 1.75 to 2.25 years. Euro-Bobl futures are on instruments having remaining term from 4.5 to 5.5 years. The Euro-Bund futures are on instruments having remaining term from 8.5 to 10.5 years (the delivery window 8.5–11 years). A delivery obligation arising out of a short position may only be fulfilled by the delivery of debt securities with a remaining term on the delivery day within the remaining term of the underlying. A bond is called cheapest to deliver, if it has smallest value among the bonds eligible to be delivered.

2) *Credit default swaps.* A credit default swap is a bilateral financial contract where the seller makes an equalization payment if credit default occurs previously to the stabilized event of the credit. In return, the buyer makes a periodic payment to the seller until either default occurs or maturity of the contract is reached.

3) *Collateralized debt obligations.* Collateralized debt obligations (CDOs) are asset-backed securities whose underlying collateral is a portfolio of corporate or sovereign bonds, or bank loans. CDO securities are divided into tranches, which are treated differently with respect to interest payments and the principal repayments. A subordination scheme is typical: senior CDO notes are paid before mezzanine, and mezzanine is paid before junior. Any residual cash flow is paid to the equity piece.

4) *Basket default swaps.* Basket default swaps (BDSs) are based on several (usually more than five) financial instruments. Typical products are first-to-default (FtD) swaps, and second-to-default (StD) swaps.

5) *Defaultable zero-coupon and coupon-bonds.* The default risk of defaultable zero-coupon and coupon-bonds has to be estimated. The approach of Merton (1974) is to model the price of a defaultable zero-coupon bond as the difference between the value of the firm's assets minus the value of a certain call option. The firm's assets are modeled as an Itô process. The strike price of the call option is equal to the notional of the zero-coupon bond, and the maturity is equal to the maturity of the zero-coupon bond.

In the reduced form or intensity-based models, the default is modeled by the stopping time of some hazard-rate process. Thus, the default process is modeled exogeneously.

18.3 Interest Rate Options

Caplets and floorlets are calls and puts when the underlying is a Libor rate. Caps and floors are portfolios of caplets and floorlets. Swaptions are options when the underlying is a swap. Caps, floors, and swaptions are the basic interest rate derivatives.

18.3.1 Caplets and Floorlets

A caplet is a call option whose underlying is an interest rate. A floorlet is the corresponding put option.

18.3.1.1 Payoffs of Caplets and Floorlets

Let the underlying be the Libor rate $L(T_1, T_2)$. Then the payoff of a caplet at time T_2 is

$$\mathrm{CPL}_{T_2}(K, T_1, T_2) = P \cdot \tau(T_1, T_2) \cdot (L(T_1, T_2) - K)_+, \tag{18.24}$$

where P is the notional value, K is the strike price, and τ is the day count fraction. Time T_1 is called the expiry and T_2 is called the maturity.[5] The payoff of a floorlet is

$$P \cdot \tau(T_1, T_2) \cdot (K - L(T_1, T_2))_+.$$

A company which is Libor indebted can buy an insurance against rising Libor rates by buying a caplet. Indeed,

$$L - (L - K)_+ = \min\{L, K\}$$

and thus when a Libor indebted company enters a cap, it will not pay higher rates than $\min\{L, K\}$.

18.3.1.2 Pricing of Caplets

Caplets can be priced assuming a log-normal distribution for the Libor rate $L(T_1, T_2)$, as in (14.74). This is called Black's model. The corresponding price is given in (14.75).

On the other hand, the price of a caplet can be written in terms of the price of an European put on a zero-coupon bond. The price of a caplet at time $t \le T_1$ is

$$\mathrm{CPL}_t(K, T_1, T_2) = E[D(t, T_2)P \cdot (L(T_1, T_2) - K)_+ \cdot \tau \,|\, \mathcal{F}_t],$$

where the expectation is with respect to the risk-neutral measure, the discount factor is

$$D(t, T_2) = \exp\left\{ -\int_t^{T_2} r_s \, ds \right\},$$

and $\tau = \tau(T_1, T_2)$. Using the iterative conditioning, we can write

$$\mathrm{CPL}_t(K, T_1, T_2) = P \cdot E(D(t, T_1)Z(T_1, T_2) \cdot (L(T_1, T_2) - K)_+ \cdot \tau \,|\, \mathcal{F}_t).$$

Indeed, for $t < s < T$ and for a $\mathcal{F}_s$-measurable random variable H_s,

$$\begin{aligned}
E(D(t, T)H_s \,|\, \mathcal{F}_t) &= E(E(D(t, T)H_s \,|\, \mathcal{F}_s) \,|\, \mathcal{F}_t) \\
&= E(D(t, s)H_s E(D(s, T) \,|\, \mathcal{F}_s) \,|\, \mathcal{F}_t) \\
&= E(D(t, s)H_s Z(s, T) \,|\, \mathcal{F}_t).
\end{aligned}$$

Using the definition (18.6) of Libor rate, we can write the price of the caplet at time t as

$$\begin{aligned}
& \mathrm{CPL}_t(K, T_1, T_2) \\
&= P \cdot E\left(D(t, T_1)Z(T_1, T_2) \cdot \left[\frac{1}{Z(T_1, T_2)} - 1 - K\tau \right]_+ \,\Big|\, \mathcal{F}_t \right) \\
&= P' \cdot E(D(t, T_1) \cdot [K' - Z(T_1, T_2)]_+ \,|\, \mathcal{F}_t),
\end{aligned}$$

---

5 For example, suppose that the 6 month USD Libor rate sets 3% at 1st of March. Then a caplet on the six month USD Libor rate, with the expiry at 1st of March, struck at 2%, with the notional of $1000, pays $1000 × 0.5 × (0.03 − 0.02) at the end of August.

where $P' = P(1 + K\tau)$ and $K' = 1/(1 + K\tau)$. Thus the price of a caplet can be written in terms of the price of a put option on a zero-coupon bond. If we model the risk-neutral distribution of the zero-coupon bond, then we obtain a price for the caplet.

18.3.2 Caps and Floors

A cap is a series of caplets and a floor is a series of floorlets. Thus, an interest rate cap is a derivative in which the buyer receives payments at the end of each such period in which the interest rate exceeds the agreed strike price.

Let us consider time points

$$t \le T_0 < \cdots < T_m,$$

and the corresponding caplet payoffs

$$\text{CPL}_{T_i}(K, T_{i-1}, T_i) = P \cdot \tau(T_{i-1}, T_i) \cdot (L(T_{i-1}, T_i) - K)_+,$$

where $i = 1, \ldots, m$.

A cap payoff is a series of payments $\text{CPL}_{T_i}(K, T_{i-1}, T_i)$ at times T_i. The discounted payoff of a cap can be written as

$$P \sum_{i=1}^{m} D(t, T_i) \tau(T_{i-1}, T_i) [L(T_{i-1}, T_i) - K]_+. \tag{18.25}$$

A cap can be viewed as a payer interest rate swap where each exchange payment is executed only if it has positive value. The discounted payoff of a floor is

$$P \sum_{i=1}^{m} D(t, T_i) \tau(T_{i-1}, T_i) [K - L(T_{i-1}, T_i)]_+.$$

The price of a cap is the sum of the prices of the caplets:

$$\sum_{i=1}^{m} \text{CPL}_t(K, T_{i-1}, T_i).$$

18.3.3 Swaptions

A swaption is an European option on a swap. The buyer of a swaption pays a premium and gets the right to enter an interest rate swap at the given date.

A receiver swaption is a call on a receiver swap. It is an insurance against falling interest rates. A payer swaption is a call on a payer swap. It is an insurance against raising interest rates. The fixed rate of the underlying swap is called the strike of the swaption. In addition, we have to specify the expiration date of the option and the tenor of the swap (time to maturity at the exercise of the option). A swaption is purchased at time t, to enter at the expiry time T_0 a swap spanning the period $[T_0, T_m]$ from the expiry to maturity with the fixed leg paying an annuity with a prespecified coupon K.

A swaption gives the right to enter a swap at a future maturity time. Usually the maturity time is equal to the first reset time. We will denote the maturity time T_0. The underlying interest rate swap length $T_m - T_0$ is called the tenor of the swaption. The payoff of payer swaption at time T_0 can be written by using (18.22) as

$$\mathrm{SWP}_{T_0} = P \cdot \left(\sum_{i=1}^{m} Z(T_0, T_i)\tau(T_{i-1}, T_i)[f(T_0, T_{i-1}, T_i) - K] \right)_+ . \qquad (18.26)$$

Jensen's inequality and the convexity of $x \mapsto (x)_+$ implies

$$\left(\sum a_i(b_i - K) \right)_+ \leq \sum a_i(b_i - K)_+, \quad \text{where } a_i \geq 0.$$

Comparing the discounted payoff of a cap in (18.25) to the payoff of a swaption in (18.26) suggests that the payoff of the swaption is smaller than the value of the cap.

The payer swaption payoff can be expressed in terms of the relevant forward swap rate $\mathrm{SR}(T_0, T_0, T_m)$, defined in (18.23):

$$\mathrm{SWP}_{T_0} = P \cdot (\mathrm{SR}(T_0, T_0, T_m) - K)_+ \sum_{i=1}^{m} \tau(T_{i-1}, T_i)Z(T_0, T_i). \qquad (18.27)$$

A payer swaption is said to be at the money if $K = \mathrm{SR}(T_0, T_0, T_m)$. A payer swaption is said to be in the money if $K < \mathrm{SR}(T_0, T_0, T_m)$ and out of the money if $K > \mathrm{SR}(T_0, T_0, T_m)$. For a receiver swaption, the converse holds.

18.4 Modeling Interest Rate Markets

Interest rate markets can be modeled by modeling the instantaneous forward interest rate, which is done in Heath–Jarrow–Morton (HJM) model, and discussed in Section 18.4.1.

Libor and swap market models are models for the forward interest rate (see Brace *et al.*, 1997, Miltersen *et al.*, 1997, Jamshidian 1997).

Black (1976) introduced the Black model, where a log-normal model is assumed for the short interest rate. We have discussed pricing under this model in (14.74). Other short rate models are listed in Section 18.4.2. Short rate models take the short-term interest rate as the only state variable, but HJM model takes the whole of the current yield curve as summarizing the current information.

It is also possible to directly assume that the prices of zero-coupon bonds are driven by an m-dimensional Wiener process and thus the zero-coupon bonds satisfy

$$dZ(t, T) = \mu(t, T)\, dt + \sum_{j=1}^{d} \sigma_j(t, T)\, dW_j(t),$$

where W_j are Wiener processes. This is called the Itô model.

Interest rate models are studied in Zagst (2002) and in Brigo and Mercurio (2006).

18.4.1 HJM Model

Heath *et al.* (1992) introduced the HJM model. Instantaneous forward interest rate is defined in (18.19). HJM model assumes that the instantaneous forward rates satisfy

$$f(t, T) - f(t_0, T) = \int_{t_0}^t \alpha(s, T)\, ds + \int_{t_0}^t \sigma(s, T)\, dW_s.$$

Using the differential notation, we can write

$$df(t, T) = \alpha(t, T)\, dt + \sigma(t, T)\, dW_t.$$

The model is a special case of a diffusion Markov process, as defined in (5.59). We can assume that the Wiener process is d-dimensional and $\sigma(t, T)$ is a d-dimensional vector, so that $\sigma(t, T)\, dW_t = \sum_{j=1}^d \sigma_j(t, T)\, dW_t^j$.

Let us use $\log Z(t, T) = -\int_t^T f(t, s)\, ds$ as in (18.20). Then it can be shown that the zero-coupon bond follows the model

$$dZ(t, T) = b(t, T)Z(t, T)\, dt + a(t, T)Z(t, T)\, dW_t, \tag{18.28}$$

where

$$a(t, T) = -\int_t^T \sigma(t, s)\, ds$$

and

$$b(t, T) = r(t) - \int_t^T \alpha(t, s)\, ds + \frac{1}{2}\, \|a(t, T)\|^2.$$

Note that $a(t, T)dW_t = \sum_{j=1}^d a_j(t, T)dW_t^j$. The short rate $r(t)$ is defined in (18.7) and it holds that $f(t, t) = r(t)$.[6]

We defined the geometric Brownian motion in (5.63) as a stochastic process $dS_t = \mu S_t\, dt + \sigma S_t\, dW_t$. This is the Black–Scholes model for stock prices, and the pricing of stock options is made under the risk-neutral model with $\mu = r$,

---

6 We obtain first (Leibniz rule) that $Y_t = \log Z(t, T)$ satisfies

$$dY_t = f(t, t)\, dt - \int_t^T df(t, s)\, ds = r(t)\, dt - \int_t^T (\alpha(t, s)dt + \sigma(t, s)dW_t)ds.$$

Second (Fubini's theorem),

$$dY_t = \left(r(t) - \int_t^T \alpha(t, s)\, ds \right) dt - \int_t^T \sigma(t, s)\, ds\, dW_t.$$

Itô's Lemma, given in (5.61), leads to the distribution of $Z(t, T) = \exp\{Y_t\}$.

where $r > 0$ is the risk-free rate (see the discussion after (14.67)). In the similar way, when we assume that the prices of the zero-coupon bonds follow (18.28), then we obtain the risk-neutral model when

$$b(t, T) = r(t) \Leftrightarrow \frac{1}{2} \|a(t, T)\|^2 = \int_t^T \alpha(t, s) \, ds. \qquad (18.29)$$

We have that

$$\|a(t, T)\|^2 = \int_t^T \sigma(t, s)' \, ds \int_t^T \sigma(t, s) \, ds = \sum_{j=1}^d \left(\int_t^T \sigma_j(t, s) \, ds \right)^2.$$

Differentiating (18.29) with respect to T leads to the equation

$$\alpha(t, T) = \sum_{j=1}^d \sigma_j(t, T) \int_t^T \sigma_j(t, s) \, ds = \sigma(t, T)' \int_t^T \sigma(t, s) \, ds.$$

18.4.2   Short-Rate Models

A general short rate model for the spot interest rate $r(t) = r_t$ is given by

$$dr_t = \mu(t, r_t) \, dt + \sigma(t, r_t) \, dW_t.$$

The model is a diffusion Markov process, as defined in (5.59).

- The Vasicek model is

$$dr_t = (\theta - a \cdot r_t) \, dt + \sigma dW_t.$$

- The Hull–White model is

$$dr_t = (\theta(t) - a \cdot r_t) \, dt + \sigma dW_t.$$

  In the Ho–Lee model, $a = 0$.
- The generalized Hull–White model is

$$dr_t = (\theta(t) - a(t) \cdot r_t) \, dt + \sigma(t) r_t^\beta dW_t.$$

  For $\beta = 0$, the model is called generalized Vasicek model and for $\beta = 1/2$, the model is called the generalized Cox–Ingersoll–Ross model.
- The log-normal model or the Black–Derman–Toy model is

$$d \log r_t = (\theta(t) - a(t) \cdot \log r_t) \, dt + \sigma(t) r_t^\beta dW_t.$$

If the bond prices are given by

$$Z(t, T) = \exp\{A(t, T) - r \cdot C(t, T)\},$$

for $t_0 < t < T < T^*$ and for deterministic A and C, then we say that the interest rate market has affine term structure. The Vasicek model and the generalized Hull–White model for $\beta = 0$ and for $\beta = 1/2$ are affine term structure models.

References

Aït-Sahalia, Y. and Duarte, J. (2003) Nonparametric option pricing under shape restrictions. *J. Econom.*, **116**, 9–47.

Aït-Sahalia, Y. and Lo, A. (1998) Nonparametric estimation of state-price densities implicit in financial asset prices. *J. Finance*, **53**, 499–547.

Aït-Sahalia, Y., Wang, Y., and Yared, F. (2001) Do option markets correctly price the probabilities of movement of the underlying asset? *J. Econom.*, **102**, 67–110.

Andersen, T.G., Bollerslev, T., Christoffersen, P.F., and Diebold, F.X. (2006) Volatility and correlation forecasting, in *Handbook of Economic Forecasting* (eds G. Elliott, C.W.J. Granger, and A. Timmerman), North-Holland, Amsterdam, pp. 777–878.

Artzner, P., Delbaen, F., Eber, J.M., and Heath, D. (1999) Coherent measures of risk. *Math. Finance*, **9** (3), 203–228.

Azzalini, A. (1981) A note on the estimation of a distribution function and quantiles by a kernel method. *Biometrika*, **6**, 326–328.

Bachelier, L. (1900) *Théorie de la Spéculation*, Gauthier-Villars, Paris.

Bauwens, L., Laurent, S., and Rombouts, V.K. (2006) Multivariate GARCH models: a survey. *J. Appl. Econ.*, **21**, 79–109.

Bertsimas, D., Kogan, L., and Lo, A.W. (2001) Hedging derivative securities and incomplete markets: an ϵ-arbitrage approach. *Oper. Res.*, **49** (3), 372–397.

Billingsley, P. (2005) *Probability and Measure*, John Wiley & Sons, Inc., New York.

Bingham, N.H. and Kiesel, R. (2004) *Risk-Neutral Valuation: Pricing and Hedging of Financial Derivatives*, Springer-Verlag, Berlin.

Black, F. (1976) The pricing of commodity contracts. *J. Finan. Econ.*, **3**, 167–179.

Black, F. and Scholes, M. (1973) The pricing of options and corporate liabilities. *J. Polit. Econ.*, **81**, 637–659.

Blyth, S. (2014) *An Introduction to Quantitative Finance*, Oxford University Press, New York.

Bollerslev, T. (1986) Generalized autoregressive conditional heteroscedasticity. *J. Econom.*, **31**, 307–327.

Bollerslev, T. (1990) Modeling the coherence in short-run nominal exchange rates: a multivariate generalized ARCH model. *Rev. Econ. Stat.*, **31**, 307–327.

Nonparametric Finance, First Edition. Jussi Klemelä.
© 2018 John Wiley & Sons, Inc. Published 2018 by John Wiley & Sons, Inc.

Bollerslev, T., Engle, R.F., and Wooldridge, J.M. (1988) A capital asset pricing model with time-varying covariances. *J. Polit. Econ.*, **96**, 116–131.

Bouchaud, J.P. (2002) An introduction to statistical finance. *Physica A*, **313**, 238–251.

Bouchaud, J.P. and Potters, M. (2003) *Theory of Financial Risks*, Cambridge University Press, Cambridge.

Bouchaud, J.P. and Sornette, D. (1994) The Black-Scholes option pricing problem in mathematical finance: generalization and extensions for a large class of stochastic processes. *J. Phys. I France*, **4**, 863–881.

Bougerol, P. and Picard, N. (1992) Stationarity of GARCH processes and some nonnegative time series. *J. Econom.*, **52**, 115–127.

Brace, A., Gatarek, D., and Musiela, M. (1997) The market model of interest rate dynamics. *Math. Finance*, **7** (2), 127–155.

Breeden, D.T. and Lizenberger, R.H. (1978) Prices of state-contingent claims implicit in option prices. *J. Bus.*, **51**, 621–651.

Breiman, L. (1993) *Probability*, 2nd edn, SIAM, Philadelphia, PA.

Breymann, W., Dias, A., and Embrechts, P. (2003) Dependence structures for multivariate high-frequency data in finance. *Quant. Finance*, **3**, 1–14.

Brigo, D. and Mercurio, F. (2006) *Interest Rate Models: Theory and Practice*, Springer-Verlag, Berlin.

Brockwell, P.J. and Davis, R.A. (1991) *Time Series: Theory and Methods*, 2nd edn, Springer-Verlag, Berlin.

Brodie, J., Daubechies, I., De Mol, C., Giannone, D., and Loris, I. (2009) Sparse and stable Markowitz portfolios. *Proc. Natl. Acad. Sci. U.S.A.*, **106** (30), 12 267–12 272.

Bühlmann, H., Delbaen, F., Embrechts, P., and Shiryaev, A.N. (1996) No-arbitrage change of measue and conditional Esscher transforms. *CWI Q.*, **9**, 291–317.

Carhart, M.M. (1997) On persistence in mutual fund performance. *J. Finance*, **52**, 57–82.

Carmona, R. and Tehranchi, M. (2006) *Interest Rate Models: An Infinite Dimensional Stochastic Analysis Perspective*, Springer-Verlag, Berlin.

Carr, D.B., Littlefield, R.J., Nicholson, W.L., and Littlefield, J.S. (1987) Scatterplot matrix techniques for large N. *J. Am. Stat. Assoc.*, **82**, 424–436.

Černý, A. (2004a) Dynamic programming and mean-variance hedging in discrete time. *Appl. Math. Finance*, **11** (1), 1–25.

Černý, A. (2004b) *Mathematical Techniques in Finance: Tools for Incomplete Markets*, Princeton University Press, Princeton, NJ.

Chorro, C., Guegan, D., and Ielpo, F. (2012) Option pricing for GARCH-type models with generalized hyperbolic innovations. *Quant. Finance*, **12** (7), 1079–1094.

Christoffersen, P., Heston, S.L., and Jacobs, K. (2006) Option valuation with conditional skewness. *J. Econom.*, **131**, 253–284.

Clayton, D.G. (1978) A model for association in bivariate life tables and its application in epidemiological studies of familial tendency in chronic disease incidence. *Biometrika*, **65**, 141–151.

Cochrane, J.H. (2001) *Asset Pricing*, Princeton University Press, Princeton, NJ.

Coles, S. (2004) *An Introduction to Statistical Modeling of Extreme Values*, Springer-Verlag, Berlin.

Cont, R. (2001) Empirical properties of asset returns: stylized facts and statistical issues. *Quant. Finance*, **1**, 223–236.

Cox, J.C., Ross, S.A., and Rubinstein, M. (1979) Options pricing: a simplified approach. *J. Financ. Econ.*, **7**, 229–263.

Dalang, R.C., Morton, A., and Willinger, W. (1990) Equivalent martingale measures and no-arbitrage in stochastic securities market models. *Stochast. Stochast. Rep.*, **29** (2), 185–201.

Davis, M.H.A. (1997) Option pricing in incomplete markets, in *Mathematics of Derivative Securities* (eds M.A.H. Dempster and S.R. Pliska), Cambridge University Press, Cambridge.

Demeterfi, K., Derman, E., Kamal, M., and Zhou, J. (1999) More Than You Ever Wanted to Know About Volatility Swaps. Tech. Rep., Goldman Sachs.

DeMiguel, V., Garlappi, L., and Uppal, R. (2009) Optimal versus naive diversification: how inefficient is the $1/N$ portfolio strategy? *Rev. Financ. Stud.*, **22**, 1915–1953.

Derman, E. and Taleb, N.N. (2005) The illusions of dynamic replication. *Quant. Finance*, **5** (4), 323–326.

Diebold, F.X. and Mariano, R.S. (1995) Comparing predictive accuracy. *J. Bus. Econ. Stat.*, **13**, 225–263.

Dobric, J. and Schmid, F. (2005) Nonparametric estimation of the lower tail dependence in bivariate copulas. *J. Appl. Stat.*, **32**, 387–407.

Duan, J.C. (1995) The GARCH option pricing model. *Math. Finance*, **5**, 13–32.

Duffie, D. and Richardson, H.R. (1991) Mean-variance hedging in continuous time. *Ann. Appl. Probab.*, **1**, 1–15.

Duffie, D. and Skiadas, C. (1994) Continuous-time security pricing: a utility gradient approach. *J. Math. Econom.*, **23**, 107–131.

El Karoui, N. and Quenez, M.C. (1995) Dynamic programming and pricing of contingent claims in an incomplete market. *SIAM J. Control Optim.*, **33**, 29–66.

Embrechts, P., Klüppelberg, C., and Mikosch, T. (1997) *Modelling Extremal Events*, Springer-Verlag, Berlin.

Engle, R.F. (1982) Autoregressive conditional heteroscedasticity with estimates of the variance of U.K. inflation. *Econometrica*, **50**, 987–1008.

Engle, R.F. (2002) Dynamic conditional correlation: a simple class of multivariate generalized autoregressive conditional heteroskedasticity models. *J. Bus. Econ. Stat.*, **20**, 339–350.

Engle, R.F. and Kroner, K.F. (1995) Multivariate simultaneous generalized ARCH. *Econom. Theory*, **11**, 122–150.

Engle, R.F. and Ng, V. (1993) Measuring and testing the impact of news on volatility. *J. Finance*, **43**, 1749–1778.

Fama, E.F. and French, K.R. (1993) Common risk factors in the returns on stocks and bonds. *J. Financ. Econ.*, **33**, 3–56.

Fama, E.F. and French, K.R. (2010) Luck versus skill in the cross-section of mutual fund returns. *J. Finance*, **65**, 1915–1947.

Fama, E.F. and French, K.R. (2012) Size, value, and momentum in international stock returns. *J. Financ. Econ.*, **105**, 457–472.

Fan, J. and Gu, J. (2003) Semiparametric estimation of value at risk. *Econom. J.*, **6**, 261–290.

Fan, J. and Yao, Q. (2005) *Nonlinear Time Series*, Springer-Verlag, Berlin.

Feller, W. (1957) *An Introduction to Probability Theory and its Applications*, vol. **1**, John Wiley & Sons, Inc., New York.

Feller, W. (1966) *An Introduction to Probability Theory and its Applications*, vol. **2**, John Wiley & Sons, Inc., New York.

Fisher, R.A. and Tippett, L.H.C. (1928) Limiting forms of the frequency distribution of the largest or smallest member of a sample. *Proc. Cambridge Philos. Soc.*, **24**, 180–190.

Föllmer, H. and Leukert, P. (1999) Quantile hedging. *Finance Stochast.*, **3**, 251–273.

Föllmer, H. and Schied, A. (2002) *Stochastic Finance: An Introduction in Discrete Time*, de Gruyter, Berlin.

Föllmer, H. and Schweizer, M. (1989) Hedging by sequential regression: an introduction to the mathematics of option trading. *ASTIN Bull.*, **1**, 147–160.

Föllmer, H. and Sondermann, D. (1986) Hedging of non-redundant contingent claims, in *Contributions to Mathematical Economics* (eds W. Hildenbrand and A. Mas-Colell), North-Holland, Amsterdam, pp. 205–223.

Frahm, G., Junker, M., and Schmidt, R. (2005) Estimating the tail dependence coefficient. *Insurance: Math. Econ.*, **37**, 80–100.

Fung, W. and Hsieh, D.A. (2004) Hedge fund benchmarks: a risk based approach. *Financ. Anal. J.*, **60**, 65–80.

Gerber, H.U. and Shiu, S.W. (1994) Option pricing by Esscher transforms. *Trans. Soc. Actuaries*, **46**, 99–191.

Gijbels, I., Pope, A., and Wand, M.P. (1999) Understanding exponential smoothing via kernel regression. *J. R. Stat. Soc. B*, **61**, 39–50.

Giraitis, L., Kokoszka, P., and Leipus, R. (2000) Stationary ARCH models: dependence structure and central limit theorem. *Econom. Theory*, **16**, 3–22.

Gnedenko, B.V. (1943) Sur la distribution limité du terme d'une série aléatoire. *Ann. Math.*, **44**, 423–453.

Gnedenko, B.V. and Kolmogorov, A.N. (1954) *Limit Distributions for Sums of Independent Random Variables*, Addison Wesley, Reading, MA.

Gourioux, C., Laurent, J.P., and Pham, H. (1998) Mean-variance hedging and numéraire. *Math. Finance*, **8**, 179–200.

Goyal, A. and Welch, I. (2003) Predicting the equity premium with dividend ratios. *Manage. Sci.*, **49**, 639–654.

Goyal, A. and Welch, I. (2008) A comprehensive look at the empirical performance of equity premium prediction. *Rev. Financ. Stud.*, **21**, 1455–1508.

Granger, C.W.J. and Newbold, P. (1977) *Forecasting Economic Time Series*, Academic Press, Orlando, FL.

Györfi, G., Lugosi, G., and Udina, F. (2006) Nonparametric kernel-based sequential investment strategies. *Math. Finance*, **16** (2), 337–357.

Hamilton, J.D. (1994) *Time Series Analysis*, Princeton University Press, Princeton, NJ.

Hammarlid, O. (1998) On minimizing rik in incomplete markets option pricing models. *Int. J. Theor. Appl. Finance*, **1** (2), 227–233.

Harrison, J.M. and Kreps, D.M. (1979) Martingales and arbitrage in multiperiod securities markets. *J. Econ. Theory*, **20**, 381–408.

Harrison, J.M. and Kreps, D.M. (1981) Martingales and stochastic integrals in the theory of continuous trading. *Stochast. Processes Appl.*, **11** (3), 215–260.

Heath, D., Jarrow, R., and Morton, A. (1992) Bond pricing and the term structure of interest rates: a new methodology for contingent claim valuation. *Econometrica*, **60**, 77–105.

Heston, S.L. and Nandi, S. (2000) A closed form GARCH option valuation model. *Rev. Financ. Stud.*, **13**, 585–625.

Hobson, D. (1998) Robust hedging of the lookback option. *Finance Stochast.*, **2**, 329–347.

Hogg, R.V. and Craig, A.T. (1978) *Introduction to Mathematical Statistics*, 4th edn, MacMillan, New York.

Ibragimov, I.A. and Linnik, Y.V. (1971) *Independent and Stationary Sequences of Random Variables*, Walters-Noordhoff, Gröningen.

Ilmanen, A. (2011) *Expected Returns: An Investor's Guide to Harvesting Market Rewards*, John Wiley & Sons, Inc., New York.

Jamshidian, F. (1997) LIBOR and market models and measures. *Finance Stochast.*, **1** (4), 293–330.

Jensen, M.C. (1968) The performance of mutual funds in the period 1945–1964. *J. Finance*, **23**, 389–416.

Jobson, J.D. and Korkie, B.M. (1981) Performance hypothesis testing with the Sharpe and Treynor measures. *J. Finance*, **36** (4), 889–908.

Karatzas, I. (1996) *Lectures on the Mathematics of Finance*, CRM Monograph Series, vol. **8**, AMS, Providence, RI.

Karatzas, I. and Kou, G. (1996) On the pricing of contingent claims under constraints. *Ann. Appl. Probab.*, **6**, 321–369.

Klemelä, J. (2014) *Multivariate Nonparametric Regression and Visualization: With R and Applications to Finance*, John Wiley & Sons, Inc., New York.

Kosowski, R., Timmermann, A., Wermers, R., and White, H. (2006) Can mutual fund "stars" really pick stocks? New evidence from a bootstrap analysis. *J. Finance*, **61**, 2551–2595.

Ledoit, O. and Wolf, M. (2008) Robust performance hypothesis testing with the sharpe ratio. *J. Empir. Finance*, **15**, 850–859.

Lehmann, E.L. (1975) *Nonparametrics: Statistical Methods Based on Ranks*, Holden-Day, San Francisco, CA.

Linton, O.B. (2009) Semiparametric and nonparametric ARCH modeling, in *Handbook of Financial Time Series* (eds T.G. Andersen, R.A. Davis, J.P. Kreiss, and T. Mikosch), Springer, New York, pp. 157–167.

Linton, O.B. and Mammen, E. (2005) Estimating semiparametric ARCH(∞) models by kernel smoothing methods. *Econometrica*, **73**, 771–836.

Magill, M. and Quinzii, M. (1996) *Theory of Incomplete Markets*, MIT Press, Cambridge, MA.

Malevergne, Y. and Sornette, D. (2003) Testing the Gaussian copula hypothesis for financial assets dependences. *Quant. Finance*, **3**, 231–250.

Malevergne, Y. and Sornette, D. (2005) *Extreme Financial Risks: From Dependence to Risk Management*, Springer-Verlag, Berlin.

Mandelbrot, B. (1963) The variation of certain speculative prices. *J. Bus.*, **36** (4), 394–419.

Mantegna, R.N. and Stanley, H.E. (2000) *An Introduction to Econophysics. Correlations and Complexity in Finance*, Cambridge University Press, Cambridge.

Markowitz, H. (1952) Portfolio selection. *J. Finance*, 7, 77–91.

Markowitz, H. (1959) *Portfolio Selection*, John Wiley & Sons, Inc., New York.

Mashal, R. and Zeevi, A.J. (2002) Beyond correlation: extreme co-movements between financial assets. Working paper, Columbia Business School.

McNeil, A.J., Frey, R., and Embrechts, P. (2005) *Quantitative Risk Management: Concepts, Techniques, and Tools*, Princeton University Press, Princeton, NJ.

Meese, R.A. and Rogoff, K. (1988) Was it real? The exchange rate - interest differential relation over the modern floating-rate period. *J. Finance*, **43**, 933–948.

Memmel, C. (2003) Performance hypothesis testing with the Sharpe ratio. *Finance Lett.*, **1**, 21–23.

Menn, C. and Rachev, S.T. (2009) Smoothly truncated stable distributions, GARCH-models, and option pricing. *Math. Methods Oper. Res.*, **69** (3), 411–438.

Merton, R.C. (1973) Theory of rational option pricing. *Bell J. Econ. Manage. Sci.*, **4**, 141–183.

Merton, R.C. (1974) On the pricing of corporate debt: the risk structure of interest rates. *J. Finance*, **2**, 449–470.

Meucci, A. (2009) Managing diversification. *Risk*, **22**, 74–79.

Miltersen, K.R., Sandmann, K., and Sondermann, D. (1997) Closed form solutions for the term structure derivatives with log-normal interest rates. *J. Finance*, **52** (1), 409–430.

Muzy, J.F., Sornette, D., Delour, J., and Arnéodo, A. (2001) Multifractal returns and hierarchical portfolio theory. *Quant. Finance*, **1**, 131–148.

Newey, W.K. and West, K.D. (1987) A simple, positive semi-definite, heteroskedasticity and autocorrelation consistent covariance matrix. *Econometrica*, **55** (3), 703–708.

Partovi, M.H. and Caputo, M. (2004) Principal portfolios: recasting the efficient frontier. *Econ. Bull.*, **7**, 1–10.

Patton, J.A. (2005) Estimation of multivariate models for time series of possibly different lengths. *J. Appl. Econ.*, **21**, 147–173.

Peligrad, M. (1986) Recent advances in the central limit theorems and its weak invariance principle for mixing sequences of random variables (a survey), in *Dependence in Probability and Statistics*, Birkhäuser, Boston, MA, pp. 193–223.

Potters, M., Bouchaud, J.P., and Sestovic, D. (2001) Hedged Monte Carlo: low variance derivative pricing with objective probabilities. *Physica A*, **289**, 517–525.

Priestley, M.B. (1981) *Spectral Analysis and Time Series*, Academic Press, New York.

Rebonato, R. (2007) *Plight of the Fortune Tellers; Why We Need to Manage Financial Risk Differently*, Princeton University Press, Princeton, NJ.

Rogers, L.C.G. (1994) Equivalent martingale measures and no-arbitrage. *Stochast. Stochast. Rep.*, **51**, 41–50.

Schäl, M. (1994) On quadratic cost criteria for option hedging. *Math. Oper. Res.*, **19** (1), 121–131.

Schmidt, R. and Stadtmüller, U. (2006) Nonparametric estimation of tail dependence. *Scand. J. Stat.*, **33**, 307–335.

Schweizer, M. (1994) Approximating random variables by stochastic integrals. *Ann. Probab.*, **22** (3), 1536–1575.

Shadwick, W. and Keating, C. (2002) A universal performance measure. *J. Perform. Meas.*, **6** (3), 59–84.

Sharpe, W.F. (1966) Mutual fund performance. *J. Bus.*, **39** (1), 119–138.

Sheather, S.J. and Marron, J.S. (1990) Kernel quantile estimators. *J. Am. Stat. Assoc.*, **85**, 410–416.

Shiryaev, A.N. (1999) *Essentials of Stochastic Finance: Facts, Models, Theory*, World Scientific, Singapore.

Silvennoinen, A. and Teräsvirta, T. (2009) Multivariate GARCH models, in *Handbook of Financial Time Series* (eds T.G. Andersen, R.A. Davis, J.P. Kreiss, and T. Mikosch), Springer, New York, pp. 201–232.

Silverman, B.W. (1986) *Density Estimation for Statistics and Data Analysis*, Chapman and Hall, London.

Siu, T.K., Tong, H., and Yang, H. (2004) On pricing derivatives under GARCH models: a dynamic Gerber-Shiu approach. *N. Am. Actuar. J.*, **8**, 17–31.

Sklar, A. (1959) Fonctions de répartition à *n* dimensions et leurs marges. *Publ. Inst. Stat. Univ. Paris*, **8**, 229–231.

Sornette, D. (2003) *Why Stock Markets Crash*, Princeton University Press, Princeton, NJ.

Spokoiny, V. (2010) *Local Parametric Methods in Nonparametric Estimation*, Springer-Verlag, Berlin.

Tibshirani, R. (1996) Regression shrinkage and selection via the LASSO. *J. R. Stat. Soc. B*, **58** (1), 267–288.

West, K.D. (1996) Asymptotic inference about predictive ability. *Econometrica*, **64**, 1067–1084.

West, K.D. (2006) Forecast evaluation, in *Handbook of Economic Forecasting* (eds G. Elliott, C.W.J. Granger, and A. Timmerman), North-Holland, Amsterdam, pp. 99–134.

Wolczyńska, G. (1998) An explicit formula for option pricing in discrete incomplete markets. *Int. J. Theor. Appl. Finance*, **1** (2), 283–288.

Zagst, R. (2002) *Interest Rate Management*, Springer-Verlag, Berlin.

Index

Nonparametric Finance, First Edition. Jussi Klemelä.
© 2018 John Wiley & Sons, Inc. Published 2018 by John Wiley & Sons, Inc.